Name	Symbol	Number	Atomic Weight	Notes
Plutonium	Pu	94	*	
Polonium	Po	84	*	
Potassium	K	19	39.0983(1)	
Praseodymium	Pr	59	140.90765(2)	
Promethium	Pm	61	*	
Protactinium	Pa	91	231.03588(2)*	
Radium	Ra	88	*	
Radon	Rn	86	*	
Rhenium	Re	75	186.207(1)	
Rhodium	Rh	45	102.90550(2)	
Rubidium	Rb	37	85.4678(3)	g
Ruthenium	Ru	44	101.07(2)	g
Rutherfordium	Rf	104	*	
Samarium	Sm	62	150.36(3)	g
Scandium	Sc	21	44.955910(8)	
Seaborgium	Sg	106	*	
Selenium	Se	34	78.96(3)	r
Silicon	Si	14	28.0855(3)	r
Silver	Ag	47	107.8682(2)	g
Sodium	Na	11	22.989770(2)	
Strontium	Sr	38	87.62(1)	g,r
Sulfur	S	16	32.065(5)	g,r

Name	Symbol	Number	Atomic Weight	Notes
Tantalum	Ta	73	180.9479(1)	
Technetium	Tc	43	*	
Tellurium	Te	52	127.60(3)	g
Terbium	Tb	65	158.92534(2)	
Thallium	Tl	81	204.3833(2)	
Thorium	Th	90	232.0381(1)*	g
Thulium	Tm	69	168.93421(2)	
Tin	Sn	50	118.710(7)	g
Titanium	Ti	22	47.867(1)	
Tungsten	W	74	183.84(1)	
Ununbium	Uub	112	*	
Ununhexium	Uuh	116	*	
Ununnilium	Uun	110	*	
Ununquadium	Uuq	114	*	
Unununium	Uuu	111	*	
Uranium	U	92	238.02891(3)*	g,m
Vanadium	V	23	50.9415(1)	
Xenon	Xe	54	131.293(6)	g,m
Ytterbium	Yb	70	173.04(3)	g
Yttrium	Y	39	88.90585(2)	
Zinc	Zn	30	65.409(4)	
Zirconium	Zr	40	91.224(2)	g

*Element has no stable nuclides. However, three such elements (Pa, Th, and U) do have a characteristic terrestrial isotopic composition, and for these an atomic weight is tabulated.

†Commercially available Li materials have atomic weights that range between 6.939 and 6.996; if a more accurate value is required, it must be determined for the specific material.

g Geological specimens are known in which the element has an isotopic composition outside the limits for normal material. The difference between the atomic weight of the element in such specimens and that given in the table may exceed the stated uncertainty.

m Modified isotopic compositions may be found in commercially available material because it has been subjected to an undisclosed or inadvertent isotopic fractionation. Substantial deviations in atomic weight of the element from that given in the table can occur.

r Range in isotopic composition of normal terrestrial material prevents a more precise $A_r(E)$ being given; the tabulated $A_r(E)$ value should be applicable to any normal material.

Source: Reprinted with permission from IUPAC, 2003. Copyright 2003 IUPAC.

Reagent Chemicals
Specifications and Procedures

Tenth Edition

ACS Committee on Analytical Reagents
Paul A. Bouis, *Chair*

American Chemical Society Specifications
Official from January 1, 2006

Washington, DC
AMERICAN CHEMICAL SOCIETY
New York   Oxford
OXFORD UNIVERSITY PRESS
2006

Oxford University Press

Oxford   New York

Athens   Auckland   Bangkok   Bogotá   Buenos Aires   Calcutta
Cape Town   Chennai   Dar es Salaam   Delhi   Florence   Hong Kong   Istanbul
Karachi   Kuala Lumpur   Madrid   Melbourne   Mexico City   Mumbai
Nairobi   Paris   São Paulo   Singapore   Taipei   Tokyo   Toronto   Warsaw

and associated companies in

Berlin   Ibadan

Developed and distributed in partnership by the
American Chemical Society and Oxford University Press

Published by Oxford University Press, Inc.
198 Madison Avenue, New York, New York 10016

Library of Congress Cataloging-in-Publication Data

American Chemical Society.
   Reagent chemicals : American Chemical Society specifications, official
from January 1, 2006. —10th ed.
      p.   cm.
   Includes index.
   ISBN 0–8412–3945–2 (alk. paper)
   1. Chemical tests and reagents.   I. Title.
QD77.A54   2005
543'.028'4—dc21   2005045502

1 3 5 7 9 8 6 4 2

Printed in the United States of America
on acid–free paper

Contents

Committee on Analytical Reagents

The following people served on the ACS Committee on Analytical Reagents and contributed to the creation of the Tenth Edition.

Paul A. Bouis, 1987–
Chair, 1992–

Michael Bolgar	1997–	Michael A. Re	1997–
Anne Wilson Coghill	2001–	Nancy S. Simon	1991–
Kishor D. Desai	1981–	Vanaja Sivakumar	2004–
Denise L. Edgren	2004–	Vernon A. Stenger	1962–2001
Christine M. Foster	1998–	*Chair*	1967–1973
Kenneth J. Herwehe	1997–	*Consultant*	1992–2001
Clarence Lowery	1974–	William A. Telliard	1998–
Chair	1985–1991	Robert Thomas	2004–
Loren C. McBride	1983–	Samuel M. Tuthill	1958–2002
Rajendra V. Mehta	1992–	*Chair*	1974–1980
John R. Moody	1986–	*Consultant*	1992–2002
Thyagaraja Parasaran	1996–2004	Thomas Tyner	1998–
Lyle H. Phifer	1997–	Charles M. Wilson	1988–
Paul H. Piscia	1999–		

William E. Schmidt, 1967–
Secretary, 1967–

v

Disclaimer

The reagent chemicals and standards included herein may be hazardous substances, and the use of such reagent chemicals and the application of the various test methods may involve hazardous substances, operations, and equipment. The American Chemical Society (ACS) and the ACS Committee on Analytical Reagents do not purport in this book or in any other publication to specify minimum legal standards or to address all of the risks and safety problems associated with reagent chemicals and standards, their use, or the methods prescribed for testing them.

No warranty, guarantee, or representation is made by ACS or the ACS Committee on Analytical Reagents as to the accuracy or sufficiency of the information contained herein, and ACS and the ACS Committee on Analytical Reagents assume no liability or responsibility in connection therewith. It is the responsibility of whoever uses the reagent chemicals and/or the testing methods set forth in this book to establish appropriate safety and health practices and to determine the applicability of any regulatory standards and/or limitations. Users of this book should consult and comply with pertinent local, state, and federal laws and should consult legal counsel if there are any questions or concerns about the applicable laws, safety issues, and reagent chemicals or the testing methods set forth herein.

Preface

The American Chemical Society (ACS) Committee on Analytical Reagents sets the specifications for most chemicals used in analytical testing. Currently, ACS is the only organization in the world that sets requirements and develops validated methods for determining the purity of reagent chemicals. These specifications have also become the de facto standards for chemicals used in many high-purity applications. Publications and organizations that set specifications or promulgate analytical testing methods—such as the *United States Pharmacopeia* and the U.S. Environmental Protection Agency—specify that ACS Reagent-grade purity be used in their test procedures.

The ACS Committee on Analytical Reagents evolved from the Committee on the Purity of Chemical Reagents, which was established in 1903. Analysts at that time were disturbed by the quality of reagents available and by the discrepancies between labels and the actual purity of the materials. The Committee's role in resolving these issues expanded rapidly after its 1921 publication of specifications for ammonium hydroxide and for hydrochloric, nitric, and sulfuric acids. Specifications appeared initially in *Industrial & Engineering Chemistry* and later in its *Analytical Edition*. In 1941, the existing specifications were reprinted in a single pamphlet. Revisions and new specifications were later gathered into a book, the 1950 edition of *Reagent Chemicals*, and new editions appeared regularly thereafter.

The commonplace introduction of instrumentation into analytical laboratories, beginning in the late 1950s, resulted in dramatic improvements in the sensitivity and accuracy of analytical measurements. As a result, the specifications for reagent chemicals and the tests measuring their purity were improved so that the test methods would be as accurate and cost-effective as possible. The Eighth Edition, which became official in 1993, substantially changed and updated the general procedures and attempted to make the book easier to read. The Ninth Edition, which became official in 2000, continued the trend toward eliminating or simplifying some of the tedious classical procedures for trace analysis and adding instrumental methods where possible.

Tenth Edition

Early in the planning stages for the Tenth Edition of *Reagent Chemicals*, Committee members voiced a desire for one book that met all of their needs for information on analytical reagents. They acknowledged that *Reagent Chemicals* is used in conjunction with other texts for information on the physical properties and the uses of analytical reagents. This resulted in a new direction for *Reagent Chemicals*, the inclusion of general physical properties and analytical uses for each reagent.

The Tenth Edition continues the initiative to simplify the classical chemical methods and substitute instrumental analysis where appropriate. This edition introduces a new instrumental method, inductively coupled plasma mass spectrometry (ICP–MS) for trace metal analysis. It also revises and updates procedures for polarography. Tests have been modified to take into account current laboratory practices and technology, as well as to eliminate the use of environmentally harmful chemicals

Since the Ninth Edition, the Committee on Analytical Reagents has required validation protocols as part of the approval process for adding new reagents. In the Tenth Edition, 32 new reagents and three new classes of standard-grade reference materials are introduced, all of which have validation protocols.

Other improvements in the Tenth Edition are intended to make the book easier to use. Some of these improvements include a CAS number index, a separate index for the standard-grade reference materials, complete assay calculations with titer values, an updated table of atomic weights, frequently used mathematical equations, a quick reference page on how to read a monograph, division of the book into parts, and a detailed table of contents for each part.

Finally, a subtitle has been added to the Tenth Edition. This represents a slight shift in thinking and internal nomenclature. Traditionally, the Committee has referred to the individual entry for each reagent chemical as a specification, and each specification had two components: the requirements and the tests. However, these terms led to some confusion and, beginning with the Tenth Edition, the Committee has adopted the term *monograph* for the complete package of information for each reagent chemical—the general description, specifications, requirements, and tests. The *specification* refers to the purity requirements of the reagent, and the *requirement* is the level of purity required for the reagent to be considered "ACS Reagent Grade".

The Work of the Committee on Analytical Reagents

As new information becomes available, the Committee updates the specifications and tests in this book and occasionally adds new reagents or deletes obsolete ones. Updates are available to the reader on the Internet at http://pubs.acs.org/reagents.

The membership of the Committee was increased substantially during the preparation of the previous two editions in order to provide the specific expertise needed for the new sections, new techniques, and new reagents. Charles M. Wilson leads the subcommittee on new reagents, while William E. Schmidt steers the trace metals subcommittee and Michael A. Re directs the organics subcommittee. Michael Re also leads a new subcommittee on method validation, which is responsible for developing and maintaining the validation policy that takes effect in the Tenth Edition; this subcommittee also reviews and approves all validation protocols and postvalidation documentation.

Special acknowledgment goes to the many volunteers in offices and laboratories for their valuable contributions toward developing and validating some of the analytical procedures that have greatly improved the quality of the test methods. Special thanks also go to Bob Hauserman, Anne Coghill, and Betsy Kulamer for steering the Committee and this edition toward publication, to Paula Bérard and Margaret Brown for their able editorial assistance, and to Susan Kurasz and Bennie Jones for help in preparing the manuscript.

The Committee has formal written operating procedures that describe and govern its operations. Interested parties may obtain a copy of these procedures by addressing their requests to the following:

Secretary, ACS Committee on Analytical Reagents
c/o Books Department
American Chemical Society
1155 16th Street, NW
Washington, DC 20036

Correspondence to the Committee may also be sent by e-mail; a list of current committee members and their e-mail addresses is available at http://pubs.acs.org/reagents.

The Committee urges that any errors observed be reported, invites constructive criticism, and welcomes suggestions, particularly for new reagents and improved test methods. Organizations wishing to adopt the ACS specifications for their own purposes are encouraged to do so by requesting permission. In this way, it is hoped that worldwide harmonization of reagent chemical specifications might occur. Communications on these subjects should be sent to the Secretary at this address. Anyone interested in serving on the Committee should also contact the Secretary.

PAUL A. BOUIS
Chair, ACS Committee on Analytical Reagents

Notice to Readers

As new information becomes available, the Committee on Analytical Reagents updates the specifications in this book and occasionally adds specifications for new reagents. All updates are available to the reader on the World Wide Web at the following URL:

http://pubs.acs.org/reagents

Part 1:
Introduction and Definitions

Overview

The specifications prepared by the Committee on Analytical Reagents of the American Chemical Society are intended to serve for reagents and standard-grade reference materials to be used in precise analytical work of a general nature. Standard-grade reference materials are suitable for preparation of analytical standards used for a variety of applications, including instrument calibration, quality control, analyte identification, method performance, and other applications requiring high-purity materials. (For the sake of brevity, standard-grade reference materials may be referred to as *standards* throughout this book.) It is recognized that there may be special uses for which reagents and standard-grade reference materials conforming to other, or more rigorous, specifications may be needed. Therefore, where known and where feasible, some of the specifications herein include requirements and tests for certain specialized uses. However, it is impossible to include specifications for all such uses, and thus there may be occasions when it will be necessary for the analyst to further purify reagents known to have special purity requirements for certain uses.

The American Chemical Society has adopted the practice of expressing the concentration of chemical solutions in terms of mol/L (moles per liter). However, the concentrations of volumetric solutions used in analytical chemistry usually are expressed as *normality*, N (the number of gram-equivalent weights in each liter of solution), or sometimes as *molarity*, M (the number of gram molecular weights in each liter of solution). In this book, it has been the practice to express such concentrations as either normality or molarity. Two examples showing equivalent concentrations by these two practices are: "...1 N sodium hydroxide (1 mol/L)" and "...1 N sulfuric acid (0.5 mol/L)". However, as can be seen, using mol/L destroys the concept of equivalents, as they are applied in volumetric analysis. Furthermore, it is unwieldy to write a calculation formula where the concentrations of the volumetric solutions are expressed in mol/L. Therefore, the use of normality and molarity are retained in this edition of *Reagent Chemicals*, which makes its practice consistent with that of other chemical testing societies and publications, such as the American Society for Testing and Materials (ASTM), the *United States Pharmacopeia*, and the *Food Chemicals Codex*.

The specifications and the details of tests are based on published work, on the experience of members of the Committee in the examination of reagent chemicals and standards on the market, and on studies of the tests made by members of the Committee. The limits and procedures are designed for application to reagents and standards in freshly opened containers. Reagents and standards in containers of extended age, in containers subject to constant changes in humidity or headspace gas content (as by repetitive opening and closing of the container), or in containers subjected to potential contamination by repeated opening of the container may not conform to the designated requirements. Where the possibility of change due to age, humidity, light, or headspace contamination is recognized, the specification usually contains a warning; nonetheless, the analyst is cautioned to take appropriate steps to ensure the continued purity of the reagents and standards, especially after opening the container.

In determining quality levels to be defined by new or revised specifications, the Committee is guided by the following general principles. When a specification is first prepared, it will usually be based on the highest level of purity (of the reagent or standard to which it applies) that is competitively available. Generally, the term "competitively available" is understood to mean that the material is available from two or more suppliers. If a significantly higher level of purity subsequently becomes available on the same competitive basis, the specification generally will be revised accordingly. There may be cases where a material is available from only one producer. This does not preclude it from becoming an ACS reagent or standard, if a suitable specification can be prepared. If the reagent or standard later becomes available on a competitive basis, it will be appropriate to review the specification for possible revision.

Because the requirements of a specification relating to the content of designated impurities must necessarily be expressed in terms of maximum allowable limits, products conforming to the specification will normally contain less than the maximum allowable proportion of some or all of these impurities. A given preparation of a reagent chemical or standard that has less than the maximum content of one or more impurities permitted by the specification, therefore, is not considered as of higher quality than that defined by the specification.

A lower allowable limit for a given impurity will be adopted only if it is significantly different from the one it is intended to supersede. In general, a new specification for an impurity whose content is not greater than 0.01% will not be considered significantly different unless it decreases the maximum permissible content of the impurity by at least 50%. This principle also will be approximated in the revision of those specifications defined by the term "Passes test".

Tests as written are considered to be applicable only to the accompanying specifications. Modification of a specification, especially if the change is toward a higher level of purity, will necessitate reconsideration, and often revision, of the test to ensure its validity.

The assays and tests described herein constitute the methods upon which the ACS specifications for reagent chemicals and standards are based. The analyst is not prevented, however, from applying alternative methods of analysis. Such methods shall be validated to ensure that they produce results of at least equal reliability. The Committee has developed a policy for validation of analytical methods described in this book, which is presented in the next section, "Classical Methods of Analysis." In the event of doubt or disagreement concerning a substance purported to comply with the ACS specifications, only the methods described herein are applicable.

Classical Methods of Analysis

When the ACS Committee on Analytical Reagents first began work, one of its missions was to bring order and comparability of result to reagent analysis. The Committee set specifications and tests according to the principle that all tests must be performed with procedures that were commonly available to producers and users. Over the years, the Committee has participated in the definite shift toward instrumental methods of analysis that improve test methods and suit current laboratory practices. However, the classical methods of chemical analysis, which have been used to classify chemicals since the days of alchemy, are still very useful to today's chemists.

The classical tests that remain—including gravimetric analysis, volumetric analysis, and physical tests—are generally sufficient to determine the fitness of a reagent for a particular application. The long and robust heritage of classical tests ensures that these tests may be performed under a wide variety of conditions with comparable results. They are usually inexpensive to run compared to instrumental procedures. In addition, the environmental conditions found in many industrial plants are not conducive to instrumental procedures, so the classical tests may be the only viable choice. Moreover, many analysts who use *Reagent Chemicals* do not have access to elaborate instrumentation, even in modern industrial sites. Users in developing nations continue to rely on the classical methods, for reasons of both economics and convenience.

Some of the classical tests are physical in nature, such as the tests for clarity of solution, residue after ignition, insoluble matter, and melting point. These tests are simple and not easily replaced by modern instrumental procedures. The test for melting point is a good example: it is specific and quite indicative of the purity of a chemical. When a melting point *range* is given, the chemical meeting the specification has been shown to be suitable for use in analytical tests where a particular level of purity is required. In this case, the analytical laboratory needs have led to specifications sufficient to meet those needs.

Many of the assay tests in *Reagent Chemicals* are gravimetric or volumetric analytical procedures. With these procedures, the accuracy requirement for the test leads to the use of classical procedures by default, since there are few instrumental procedures with sufficient accuracy and fewer still that are in widespread use. Thus, the classical procedures not only suffice, but also are generally superior to potential instrumental replacements.

The Committee has made an effort to eliminate as many nonspecific classical tests as possible, such as the tests for "substances not precipitated by [...]", which appeared in previous editions. Morever, many very specific and sensitive instrumental tests have been added to the Tenth Edition, and the Committee expects to add many more over time. Each replacement of a classical test entails considerable effort, however, because unexpected complications often arise. Until such replacements become available, the classical tests will remain in this book for good reason: they are quite adequate for their intended purposes and they are well within the capabilities of a competent analyst.

Interpretation of Specifications

The specifications of reagent chemicals can be divided into two main classes: an assay or quantitative determination of the principal or active constituent and the determination of the impurities or minor constituents. The specifications of standard-grade reference materials are divided into identity and assay sections. In some cases, physical properties are specified. A sample reagent monograph appears in Figure 1-1 with notation describing the features used throughout this book.

Physical Properties of Reagents

In addition to the specifications and tests for each reagent, the Tenth Edition of *Reagent Chemicals* contains a new section for each reagent monograph called "General Description." This section provides physical properties for each reagent, including typical appearance, representative analytical use, change in state, aqueous solubility, density, and pK_a.

> *Note:* **The information under "General Description" is *not* to be used as a measurable reagent specification under any circumstances whatsoever.**

Information under "General Description" is provided for the analyst's ease of reference. The following points should be kept in mind. For the sources of information, see the bibliography on physical properties, beginning on page 774.

- Melting points and boiling points have been approximated, even though more exact data may be available.
- Densities for the reagent chemicals have been approximated because exact numbers require exact conditions and the reagents described may not be in the exact form necessary to duplicate literature densities. Small variations in moisture or physical properties have a great effect on the density of both solid and fluid materials.
- Where possible, the solubility of reagents has been approximated based on literature that is available. Some solubility will be described according to a general formula given in Table 1-1.

Each **monograph** begins with the common name for the reagent.

The reagent's formula, formula weight, and CAS number are included.

If there is a common alternate name, it is listed below.

The **General Description** section contains useful (but background) information for each reagent. For more about this section, see page 7.

The **Specifications** section lists the purity requirements for a reagent. These must be met for a reagent to be "ACS Reagent Grade." For more about this section, see pages 9–12.

The **Tests** section outlines the procedure for verifying that a reagent meets the specifications.

Solutions in the tests identified as "reagent," "indicator," "buffer," or "volumetric" are described in Part 3 of this book.

Assays by titration include both the classical statement of milliequivalents and the calculation formula.

Special solutions not included in Part 3 are listed within or following each test.

Some tests are described completely in the monograph.

Ceric Ammonium Nitrate
Ammonium Hexanitratocerate(IV)
$(NH_4)_2Ce(NO_3)_6$ Formula Wt 548.22   CAS No. 16774-21-3

GENERAL DESCRIPTION
Typical appearance: orange-red or orange-yellow solid
Analytical use: oxidimetric standard
Aqueous solubility: 141 g in 100 mL

SPECIFICATIONS
Assay . . . . $\geq$98.5% $(NH_4)_2Ce(NO_3)_6$

Maximum Allowable

Insoluble in dilute sulfuric acid. 0.05%
Chloride (Cl) 0.01%
Phosphate (PO_4) 0.02%
Iron (Fe). 0.005%

TESTS

Assay. (By titration of oxidative capacity of Ce^{IV}). Weigh accurately 2.4–2.5 g, and dissolve in 50 mL of water. From a pipet, add 50 mL of 0.1 N ferrous ammonium sulfate volumetric solution, and swirl until the precipitate that forms is redissolved. Add 10 mL of phosphoric acid and 0.10 mL of diphenylaminesulfonic acid, sodium salt, indicator solution (described below). Titrate at once with 0.1 N potassium dichromate volumetric solution to a change from faint green to violet. Record this titration volume as *A* mL. Pipet 25 mL of 0.1 N ferrous ammonium sulfate volumetric solution into 50 mL of water, and treat it in the same way (with phosphoric acid and indicator but without sample). Titrate to a change from green to gray-blue, and record this titration volume as *B* mL. One milliliter of 0.1 N ferrous ammonium sulfate corresponds to 0.05482 g of $(NH_4)_2Ce(NO_3)_6$.

$$\% \ (NH_4)_2Ce(NO_3)_6 = \frac{[(2B - A\,\text{mL}) \times N\,K_2Cr_2O_7] \times 54.82}{\text{Sample wt (g)}}$$

Diphenylaminesulfonic Acid, Sodium Salt, Indicator Solution. Dissolve 0.10 g of the salt in 100 mL of water.

Insoluble in Dilute Sulfuric Acid. To 5.0 g, add 10 mL of sulfuric acid, stir, and then cautiously add 90 mL of water to dissolve. Heat to boiling, and digest in a preconditioned covered beaker on a hot plate ($\approx$100 °C) for 1 h. Filter through a tared filtering crucible, wash thoroughly, and dry at 105 °C.

Chloride. (Page 35). Use 0.10 g dissolved in 10 mL of water. The comparison is best made by the general method for chloride in colored solutions.

Phosphate. (Page 40, Method 2). Dissolve 0.25 g in 30 mL of dilute sulfuric acid (1 + 9), add hydrogen peroxide until the solution just turns colorless, then boil to destroy excess

Other test methods are described in Part 2 of this book. Page cross references lead to a detailed discussion of the procedure.

Figure 1-1. Features of a typical reagent monograph.

Table 1-1. Classification of Terms Describing Solubility

Descriptive Term	Parts of Solvent Required for 1 Part of Solute
Very soluble	Less than 1
Freely soluble	From 1 to 10
Soluble	From 10 to 30
Sparingly soluble	From 30 to 100
Slightly soluble	From 100 to 1000
Very slightly soluble	From 1000 to 10,000
Practically insoluble or Insoluble	More than 10,000

Source: Reprinted with permission from National Research Council, 1981. Copyright 1981 National Academies Press

Calculation of Results

Because different techniques can be used to obtain results for assays and some secondary tests, the calculation of these results traditionally has been left to the analyst to derive. In order to clarify these calculations, to demonstrate the chemistry employed, and to eliminate ambiguity, the Committee has ensured that all calculations for assay determinations are presented in the Tenth Edition. These calculations follow the general forms, with necessary modifications, that are described in standard works on quantitative analysis (see the bibliography on analytical chemistry, page 772). As much as possible, these calculations will not contain condensed factors or constants that may be confusing to some.

The Committee also has included, thoughout the Tenth Edition, the historic titer method of giving the milliequivalent weight in the statement "one milliliter of 1 N [*titrant*] corresponds to 0.*xxxx* g of [*sample*]." This method allows analysts who already have programmed the calculation to continue doing so and provides a means for new users to do the same.

Assay Specifications

Assay specifications are included for most of the reagent chemicals and all of the standard-grade reference materials in this book. An *assay value*, in the sense used herein, is the content or concentration of a stated major component in the reagent. Unless otherwise specified, assay specifications are on an as-is basis (i.e., without drying, ignition, or other pretreatment of the sample).

Unless described in great detail and carried out with exceptional skill, available assay methods seldom are accurate enough to permit using a weighed quantity of a reagent, which was assayed in an exacting stoichiometric operation. This use of reagent chemicals should be limited to those designated as standards (for example, acidimetric or reductometric standards) because especially exacting assay methods are provided for such reagents.

Except in the case of standards, assays, through their minimum and maximum limits, mainly serve to ensure acceptable consistency of the strength of reagents offered in the marketplace. They are particularly useful, for example, in the requirements for acid–water systems to control strength; for alkalis to limit the content of water and carbonate; for oxidizing and reducing substances that may change strength during storage; and for hydrates to control, within reasonable limits, deviations in the amount of water from that indicated in the formula. If, however, it should be necessary to use such reagents in stoichiometric operations, the user should ascertain that the assay values produced are of sufficient accuracy and precision for such use.

Assayed values for standard-grade reference materials are sufficient for the intended use described in this book. Assay specifications for all standard-grade reference materials include a minimum of two assay methods. Additional assay methods (such as thin-layer chromatography) may include a specification of "Passes test". These secondary methods are intended to support the primary numerical assay methods.

Impurity Specifications

Specifications for impurities are expressed in the following ways: (1) as numerical limits; (2) in terms of the expression "Passes test" with an accompanying approximate numerical limit; or (3) in terms of the expression "Passes test" without an approximate numerical limit. The distinction among these forms of expression is based on the Committee's opinion as to the relative quantitative significance of the prescribed test methods. The methods given for determining conformity to specifications of the first type are considered to yield, in competent hands, what are usually thought of as "quantitative" results, whereas those of the second type can be expected to yield only approximate values. Those in the third category give definitions that cannot be expressed in numbers. It is obvious, however, that these distinctions as to quantitative significance cannot be sharp and that even the numerically expressed specifications are not all defined with equal accuracy. The final and essential definition of any specification, therefore, must reside in the described test method rather than in its numerical expression.

If a test method yields results that are adequately reproducible on repeated trials in different laboratories, it offers a satisfactory definition of the content of an impurity whether or not the result can be expressed by a number. Although the Committee has endeavored to base specifications, so far as possible, on validated test methods that meet this criterion, a considerable number are based on essentially undefined statements such as "no turbidity", "no color", or "the color shall not be completely discharged in x minutes". Although some of the specifications of this kind could be replaced by others based on quantitative comparisons or measurements, to do so would require more costly or time-consuming procedures than appear at this time to be justified. The approach to an ultimate goal of replacing in every instance the word "none" or its equivalent by the expression "maximum allowable...", therefore, is limited both by deficiencies of knowledge and by practical considerations of expediency.

Unlisted Impurities

The primary objective of the Committee in preparing reagent or standard specifications is to assure the user of the strength, quality, and purity of the material. It is, however, manifestly impossible to include in each specification a test for every impurity and contaminant that may be present. The Committee recognizes that for certain uses more stringent or additional requirements may be appropriate, and for such uses, additional testing beyond that described in the specifications should be employed by the user. The Committee's intent in establishing the specifications is to recognize the common uses for which the reagent or standard is employed and to establish requirements that are consistent both with these uses and with the manufacturing processes and quality of the available reagents. The presence of moisture, either as water of crystallization or as an impurity, falls within the purview of contamination unless permitted by an applicable specification.

Although tests for foreign particulate matter are not usually included in the specifications for solid reagents, such matter constitutes contamination. Similarly, although tests for clarity are not usually included in the specifications for liquid reagents or for solutions of solid reagents, the presence of haze, turbidity, or foreign particulate matter also constitutes contamination.

In some instances, residual amounts of substances that have been added as aids in the process of purification may be present. An example is the use of complexing agents to keep certain metal ions in solution during recrystallization. These substances not only are impurities but may interfere with the tests. Certain reagents, such as desiccants and indicators, have requirements that ensure suitability for their intended use. Such reagents may contain impurities that do not interfere with their intended use but that may make these reagents unsuitable for other uses.

When the Committee becomes aware of an unlisted impurity that affects adversely the known or specified uses of reagents or standards, a new specification is added, provided a suitable test method is available. Users of reagents can protect themselves against the effects of unlisted impurities on a specific analytical procedure by applying, ad hoc, an appropriate suitability test.

Identity Specifications

Identity specifications and tests are not included in the monographs for reagent chemicals. If there is any question as to the identity of a chemical, identity can be ascertained by appropriate analytical methods. Identity specifications are included for standard-grade reference materials. For each material, a minimum of one identity method is required. Specifications are designed to unequivocally prove identity. Many of the methods rely on comparisons of data to standardized special libraries, such as the National Institute of Standards and Technology (NIST) infrared and mass spectral libraries. In most instances, a combination of complementary techniques is used.

Accuracy of Measurements

In specifying the weight or volume of sample to be used in the individual test procedures, it is intended, unless otherwise specified in the individual procedure, that the accuracy of measurement be such that the amount of sample used is within 2.0% of the stated amount. Thus, where a 10-g (or -mL) sample is specified, the amount actually taken for analysis must be between 9.8 and 10.2 g (or mL). Similarly, where a test procedure directs that a solution be diluted to a specific volume or that a specified volume of solution be used, it is intended that the volume actually be within 2.0% of the stated amount.

Table 1-2. Rounding Procedures

Requirement	Observed Value	Rounded Value	Pass/Fail
Not less than 98%	97.6	98	pass
	97.5	98	pass
	97.4	97	fail
Not less than 98.0%	97.95	98.0	pass
	97.94	97.9	fail
Not more than 0.01%	0.014	0.01	pass
	0.015	0.02	fail
	0.016	0.02	fail
Not more than 0.02%	0.015	0.02	pass
	0.025	0.03	fail
	0.026	0.03	fail

When the term *weigh accurately* is used, this means weigh to 0.1 mg.

Where the term *pipet* or *buret* is used as a verb, it is intended that the specified volume be taken in a volumetric pipet or buret conforming to the tolerances accepted by NIST, Class A (NIST, 1974). Commercially available, certified, mechanical pipets may be used with proper calibration and maintenance.

The addition of small volumes of liquid reagents is generally stated to the nearest 0.05 mL (0.05 mL, 0.10 mL, 0.15 mL, etc.). Use of the term *drop* to represent 0.05 mL is avoided.

Rounding Procedures

For comparison of analytical results with specifications for assays and impurities, the observed or calculated values are rounded to the number of digits carried in the requirement. The method and the rounding procedure are in accord with those in the *United States Pharmacopeia*. When rounding is required, consider only one digit in the place to the right of the last digit in the limit expression. If this digit is smaller than 5, it is eliminated and the preceding digit is unchanged. If this digit is greater than 5, it is eliminated and the preceding digit is increased by one. If this digit equals 5, the 5 is eliminated and the preceding digit is increased by one. This rounding procedure is illustrated in Table 1-2.

The foregoing procedure conforms to the common electronic calculator and computer procedure of rounding up when the digit to be dropped is 5 or 5 followed by zeros. The rounded value is obtained in a single step by direct rounding of the most precise value available and not in two or more steps of successive rounding. For example, 97.5487 rounds to 97.5 against a requirement of 97.6 and not in two possible steps of 97.55 and then 97.6.

The formula weights and factors for computing results are based on the 2001 International Atomic Weights (shown on the inside of the front cover of this book). Throughout the monographs, the formula weights are rounded to two decimal places.

Validation of Analytical Methods

A method validation process must demonstrate—with evidence to a high degree of assurance—that a specific method will consistently perform according to its intended specified purpose. It is the policy of the ACS Committee on Analytical Reagents to validate any new method described in an individual monograph that is published in or after the Tenth Edition.

Analytical methods published in the Ninth Edition or earlier editions have been tested and proven by both intra- and interlaboratory verifications and will be accepted retrospectively as validated. Major modifications made to existing methods (from the Ninth Edition or earlier) that are incorporated into editions after the Tenth or published in supplements to the Tenth Edition will be validated according to the new policy.

The Committee defines the validation terminology in the following ways:

- Method *specificity* is a measure of the method's ability to accurately measure the analyte in the presence of similar analytes, impurities, or interfering substances.
- Method *linearity* is a measure of the method's ability to provide a linear response across the intended range of the method.
- Method *accuracy* is the nearness of a measurement or result to its true value.
- Method *precision* is a measure of the method's ability to reproduce a measurement or result, and it can include elements of instrument precision, repeatability, and reproducibility.
- Method *verification* (ruggedness) is the ability of a method result to be reproduced between analysts, instruments, days, and laboratories.

Validation of a new analytical method is carried out in the following manner:

1. Validations are initiated by a *method sponsor*, who is responsible for ensuring that validations are performed according to this policy and that all validations are properly documented.
2. A validation protocol is drafted before initiating a validation. The protocol discusses the purpose, scope, procedure, and documentation of results.
   - The *purpose* describes the reason for the protocol.
   - The *scope* defines the boundaries of the protocol and may also define exclusions. If applicable, a reference is provided for the derivation of the method.

■ The *procedure* is a detailed guidance for performing the validation and includes components to illustrate a method's specificity, linearity, accuracy, precision, and ruggedness. For methods where impurity testing is required, the limit of quantitation and limit of detection are also addressed in the protocol, where applicable. (Guidelines for determining detection limits are covered in the next section, "Method Detection Limits".) Each component has acceptance criteria that are defined in the protocol.

■ The *documentation of results* describes the documentation requirements for the protocol, including the requirements for recording raw data and for preparing a validation summary report.

3. The draft of the validation protocol is submitted by the method sponsor to the Validation Subcommittee for approval. The method sponsor then coordinates execution of the validation protocol. The initial validation steps may be performed in the sponsor's laboratory or in a sponsor-designated laboratory.

4. After completion of the intralaboratory portion of the validation, the method sponsor identifies one or more laboratories to perform the interlaboratory verification.

5. Upon completion of all laboratory tests, the method sponsor compiles the validation data, ensures that all data have met the acceptance criteria, and writes the summary report. This report summarizes the results of the validation and contains the following information: an introduction that states the method being validated, the method sponsor, and the interlaboratory participants; a summary of the test results; and observations of any changes or deviations that occurred during protocol performance.

6. The summary report and all supporting data are submitted to the Validation Subcommittee for final review and acceptance of the protocol, the documentation, and the summary report.

7. After final review and approval by the Validation Subcommittee, all reports and documentation are turned over to the Committee Secretary for archiving. A copy of the summary report is made available to requestors.

Method Detection Limits

The purity of reagent chemicals and standard-grade reference materials continues to improve, driven by customer demand and the evolution of manufacturing processes and analytical technology. A clear and practical definition of analytical detection limits is vital to the accurate and precise determination of purity. An excellent review and definition of detection limits has been published. The following discussion is adapted from APHA, 1998. This reference provides other details, including the use of control charts to measure and monitor the accuracy and precision of any analytical procedure.

In case the limit of detection of trace metal impurities in ultratrace reagents needs to be determined, a more relevant method has been published and is recommended. This method is covered in the section on trace and ultratrace elemental analysis, which starts on page 56.

Overview

Detection limits are controversial, principally because of inadequate definitions and confusion of terms. Frequently, the instrumental detection limit is used for the method detection limit, and vice versa. Whatever term is used, most analysts agree that the smallest amount that can be detected above noise in a procedure and within a stated confidence limit is the detection limit. The confidence limits are set so that probabilities of both Type I and Type II errors (described under Determining Detection Limits) are acceptably small.

Current practice identifies several detection limits, each of which has a defined purpose. These are the following:

- instrument detection limit (IDL),
- lower limit of detection (LLD),
- method detection limit (MDL), and
- limit of quantitation (LOQ).

Occasionally the IDL is used as a guide for determining the MDL. The relationship among these limits is approximately IDL:LLD:MDL:LOQ = 1:2:4:10.

Determining Detection Limits

An operating analytical instrument usually produces a signal (noise) even when no sample is present or when a blank is being analyzed. Because any quality assurance program requires frequent analysis of blanks, the mean and standard deviation become well known; the blank signal becomes very precise, i.e., the Gaussian curve of the blank distribution becomes very narrow. The IDL is the constituent concentration that produces a signal noise greater than three standard deviations of the mean noise level or that can be determined by injecting a standard to produce a signal that is five times the signal-to-noise ratio. The IDL is useful for estimating the constituent concentration or amount in an extract needed to produce a signal to permit calculating an estimated method detection limit.

The LLD is the amount of constituent that produces a signal sufficiently large that 99% of the trials with that amount will produce a detectable signal. The LLD can be determined by multiple injections of a standard at near-zero concentration (concentration no greater than five times the IDL). Standard deviation (s) is determined by the usual method. To reduce the probability of a Type I error (false detection) to 5%, multiply s by 1.645 from a cumulative normal probability table. Also, to reduce the probability of a Type II error (false nondetection) to 5%, double this amount to 3.290. As an example, if 20 determinations of a low-level standard yielded a standard deviation of 6 μg/L, the LLD is 3.29 × 6 = 20 μg/L.

The MDL differs from the LLD in that samples containing the constituent of interest are processed through the complete analytical method. The method detection limit is greater than the LLD because of extraction efficiency and extract concentration factors. The MDL can be achieved by experienced analysts operating well-calibrated instruments on a nonroutine basis. For example, to determine the MDL, add a constituent to reagent water, or to the matrix of interest, to make the concentration near the estimated MDL. Analyze seven portions of this solution and calculate s. From a table of the one-sided t distribution, select the value of t for $7 - 1 = 6$ degrees of freedom at the 99% level; this value is 3.14. The product 3.14 × s is the desired MDL.

Although the LOQ is useful within a laboratory, the practical quantitation limit (PQL) has been proposed as the lowest level achievable among laboratories within specified limits during routine laboratory operations. The PQL is significant because different laboratories will produce different MDLs even though using the same analytical procedures, instruments, and sample matrices. The PQL is about five times the MDL and represents a practical and routinely achievable detection limit with a relatively good certainty that any reported value is reliable.

Other Considerations

The descriptions of the individual tests are intended to give all essential details without repetition of considerations that should be obvious to an experienced analyst. A few suggestions are given for precautions and procedures that are particularly applicable to the routine testing of reagent chemicals.

Units of Measure

Throughout this book, the International System of Units (SI, or Système International d'Unités) is used.

Sampling

The main goal in sampling is to obtain and maintain (stabilize) samples that are representative. As a general rule, the analyst should recognize that the degree of error in the sampling portion of the analytical procedure is considerably higher than the degree of error in the methodology (adapted with permission from Schwedt, 1997). To eliminate accidental contamination or possible change in composition, samples for testing must be taken from freshly opened containers.

Containers

The *container* is the device that holds the reagent or standard-grade reference material; it is, or may be, in direct contact with the reagent or standard. The *closure* is part of the container.

The container in which a reagent or standard is sold and/or stored must be suitable for its intended purpose and should not interact physically or chemically with the contained reagent or standard so as to alter its quality (within a reasonable period of time or when mandated by an expiration date) beyond the requirements of the specification.

Containers for solids normally have wide mouths to facilitate both the filling of the container and the removal of the contents. Containers for liquids normally have narrow mouths so the contents may be easily poured into other, frequently smaller, containers.

Before being filled, the container should be free from extraneous particulate matter, otherwise clean, and dry as necessary.

Quality of Reagents

Reagents and standard-grade reference materials used in testing materials included in this book should conform to ACS specifications. Reagents not covered by ACS specifications should be of the best grade obtainable and should be examined carefully for interfering impurities.

Added Substances

Unless otherwise specified for an individual reagent chemical, the reagents described in this volume may contain suitable preservatives or stabilizers, intentionally added to retard or inhibit natural processes of deterioration. Such preservatives or stabilizers may be regarded as suitable only if the following conditions are met:

1. They do not exceed the minimum quantity required to achieve the desired effect.
2. They do not interfere with the tests and assays for the individual reagents unless specifically waived by this book.
3. The presence of any added substance must be declared on the label of the individual package. Unless of a proprietary nature, the name and concentration of any added substance should be stated on the label.

Suitability for Special Purposes

For some reagents, the Sixth and Seventh Editions of *Reagent Chemicals* had separate specifications defining them, such as "suitable for use in ultraviolet spectrophotometry", "suitable for use in determining pesticide residues", or "suitable for use in high-performance liquid chromatography". Beginning with the Eighth Edition, these special-use reagent chemicals have been treated in a single integrated presentation. The seller shall designate in product labeling the suitability for one or more of these special uses on the basis of the relevant specifications and tests.

Blank Tests

Many of the tests are for minute quantities of the impurities sought. Hence, complete blank tests must be made covering the water and other reagents used in each step of the tests—including, for example, filtration and ignition. Frequently, however, the directions stipulate a control, a blank, or other device that corrects for possible impurities in the water and other reagents.

Eliminated Test Procedures

Continuing a trend started with the Eighth Edition, tedious classical procedures are being eliminated and replaced with instrumental methods. The test, "Substances not precipitated by […]" is no longer used in the analysis of reagent chemicals.

Part 2: Analytical Procedures and General Directions

The analytical procedures and general directions for most ACS specifications and tests are described in this section. Where appropriate in the monographs, the tests for particular reagent chemicals or standard-grade reference materials contain page references to individual test procedures described fully in this section. Supplemental information may be obtained from one of the standard references on analytical chemistry listed in the bibliography (see page 772).

Gravimetric Methods

In many of the reagent monographs, it is directed that a precipitate or residue be collected and either dried or ignited so as to provide certain information concerning the purity of the reagent. Except where it is directed otherwise in the specific reagent tests, the following directions and precautions will be used in collecting, drying, and igniting precipitates and residues.

General Considerations

It is imperative that the analyst use the best techniques in performing any of the operations included in this book. Exceptional cleanliness and protection against accidental contamination from dirt, fumes, and analytical containers are rigid requirements for acceptable results. Use of fume or clean-air hoods is recommended whenever possible. The choice of equipment is generally left to the discretion of the analyst, but it must provide satisfactory accuracy and precision. All weighings must be determined with an uncertainty of not more than ±0.0002 g. [Moody (1982) and Zief and Mitchell (1976) wrote the classic works on this subject.]

When the use of a tared container is specified, the container will be carried through a series of operations identical to those used in the procedure, including drying, igniting, cooling in a suitable desiccator, and weighing. In general, desiccators should be charged with indicating-type silica gel. (Table 2-1 compares the different types of desiccators.) Stronger desiccants may be required for certain applications. The length of the drying or ignition and the temperature employed must be the same as specified in the procedure in which the tared equipment is to be used. Where it is directed to dry or ignite to constant weight, two successive weighings may not differ by more than ±0.0002 g, the second weighing following a second drying or ignition period.

In those operations wherein a filtering crucible is specified, a fritted-glass crucible, a porous porcelain crucible, or a crucible having a sponge platinum mat may be used. Certain solutions may attack the filtering vessel; for example, both fritted glass and porcelain are affected by strongly alkaline solutions. Even platinum sponge mats are attacked by hydrochloric acid unless the crucible is first washed with boiling water to remove oxygen.

Table 2-1. Selection Chart for Desiccators

Desiccant	Suitable for Drying	Not Suitable for Drying	Residual Water[a]	Water Content[b]	Regeneration	Reaction Mechanism
Aluminum oxide	Hydrocarbons		0.003	0.2	175 °C	Chemisorption, absorption
Calcium chloride	Ethers, most esters, alkyl halides, aryl halides, saturated hydrocarbons, aromatic hydrocarbons	Alcohols, amines, phenols, aldehydes, amides, amino acids, some esters, ketones	0.4	0.2 ($1H_2O$) 0.3 ($2H_2O$)	None	Hydration
Calcium oxide	Low molecular weight alcohols, amines, ammonia gas	Acidic compounds, esters	0.007	0.3	1000 °C	Chemisorption
Magnesium oxide	Hydrocarbons, aldehydes, alcohols, basic gases, amines	Acidic compounds	0.008	0.5	800 °C	Hydration
Magnesium perchlorate, anhydrous	Inert gas, air	Most organics[c]	0.001	0.2	250 °C with vacuum	Hydration
Magnesium sulfate, anhydrous	Most compounds, including acids, ketones, aldehydes, esters, nitriles		1.0	0.2–0.8	None	Hydration
Molecular sieve, activated, 8–12 Mesh						
Type 3A	Molecules >3 Å in diameter	Molecules >3 Å in diameter		0.18	177–260 °C	Absorption
Indicating, Type 4A	Molecules >4 Å in diameter	Molecules >4 Å in diameter; ethanol, H_2S, CO_2, SO_2, C_2H_4, C_3H_6, C_2H_6, strong acids	0.001	0.18	250 °C	Absorption
Type 5A	Molecules >5 Å in diameter, e.g., branched chain compounds, those with 4 carbon or larger rings	Molecules >5 Å in diameter, e.g., butanol, n-C_4H_8 to n-$C_{22}H_{46}$	0.003	0.18	250 °C	

			Water content[a]	Capacity[b]	Regeneration	Mechanism
Phosphorus pentoxide, granular	Saturated hydrocarbons, aromatic hydrocarbons, ethers, alkyl halides, aryl halides, nitriles, anhydrides	Alcohols, acids, amines, ketones	0.001	0.5	None	Chemisorption
Potassium carbonate, anhydrous	Alcohols, nitriles, ketones, esters, amines	Acids, phenols	0.2	0.2	200 °C	Hydrate formation
Potassium hydroxide pellets	Amines	Acids, phenols, esters, amides, acidic gases	0.3	—	None	Hydration and solution formation
Silica gel, indicating-type	Most organics		0.03	0.2	200–350 °C	Adsorption
Sodium hydroxide pellets	Amines	Acids, phenols, esters, amides	0.16	—	None	Adsorption and solution formation
Sodium sulfate, anhydrous, granular, or powder	Alkyl halides, aryl halides, aldehydes, ketones, acids		12.0	1.2	Usually not	Hydration
Sulfuric acid	Inert gases, air used in desiccators	Too reactive to contact organic materials	0.004	—	None	Hydration
Zinc chloride reagent, broken lump	Hydrocarbons	Ammonia, amines, alcohol	0.9	0.2	110 °C	Hydration

Notes: — indicates indefinite water content.

[a] Milligrams of water per L of dry air.

[b] Grams of water per g of desiccant.

[c] Magnesium perchlorate, anhydrous, may form an explosive compound when exposed to organic vapors.

Source: Reproduced courtesy of Mallinckrodt Baker, Inc.

Collection of Precipitates and Residues

In many of the tests, the amount of residue or precipitate may be so small as to escape easy detection. Therefore, the absence of a weighable residue or precipitate must never be assumed. However small, it must be properly collected, washed, and dried or ignited. The size and type of the filter paper to be used is selected according to the amount of precipitate to be collected, not the volume of solution to be filtered.

Often when barium chloride is added in slight excess to precipitate barium sulfate, the amount of precipitate produced is so small that it is hard to see and collect. If a small amount of a suspension of ashless filter paper pulp is added toward the end of the period of digestion on the steam bath or hot plate, the flocculation and collection of the sulfate are facilitated.

Some precipitates have a strong tendency to "creep", while others, like magnesium ammonium phosphate, stick fast to the walls of the container. The use of the rubber-tip "policeman" is recommended, sometimes supplemented by a small piece of ashless filter paper. The analyst should guard against the possibility of any of the precipitate escaping beyond the upper rim of the filter paper.

Precipitates, especially if recently produced, may be partially dissolved unless proper precautions are taken with both their formation and subsequent washing. In general, precipitates should not be washed with water alone but with water containing a small amount of common ion to decrease the solubility of the precipitate. It is much better to wash with several small portions of washing solution than with fewer and larger portions. Calcium oxalate precipitates should be washed with dilute (about 0.1%) ammonium oxalate solution; therefore, mixed precipitates that may contain calcium oxalate and magnesium ammonium phosphate should be washed with a 1% ammonia solution containing also 0.1% ammonium oxalate.

Weights of Precipitates and Residues

With few exceptions, the weight of the sample used in the test will ensure that at least 0.001 g of the impurity to be determined will be present if the sample fails the test.

Ignition of Precipitates

The proper conditions for obtaining the precipitate are given in the individual test directions.

▶**Procedure for Ignition of Precipitates.** Collect the precipitate on an ashless filter paper of suitable size and porosity (for example, a paper of 2.5-μm retention for a precipitate made up of fine particles). Wash the residue and paper thoroughly with the proper solution, and fold the moist filter paper about the residue. Place the filter paper in a suitable tared crucible that has been preconditioned at the temperature required by the specific test, and dry the paper using a 105 °C oven, a hot plate, an infrared lamp, or (carefully) a gas microburner in a covered crucible. When the paper is dry, char it at the lowest possible temperature; gentle ignition destroys the charred paper. Finally, ignite the contents of the crucible in a vented muffle furnace at 600 ± 25 °C for 15 min. Cool the ignited crucible in a suitable desiccator (using indicating-type silica gel or another dessicator as specified in the individual reagent monograph) and weigh it.

Insoluble Matter

The intent of these tests is to determine the amount of insoluble foreign matter (for example, filter fibers and dust particles) present under test conditions designed to dissolve completely the substance being tested. After dissolution and before filtration, it is a good practice to check visually (using a white background) for insoluble foreign matter.

▶**Procedure for Insoluble Matter.** Prepare a solution of the sample as specified in the individual test directions. Unless otherwise specified, heat to boiling in a covered beaker, and digest on a low-temperature ($\approx$100 °C) hot plate (or a steam bath) for 1 h. Filter the hot solution through a suitable tared, medium-porosity (10–15-μm) filtering crucible. Unless otherwise specified, wash the beaker and filter thoroughly with hot water, dry at 105 °C, cool in a desiccator, and weigh.

Residue after Evaporation

This test is designed to determine the amount of any higher-boiling impurities or nonvolatile dissolved material that may be present in a reagent chemical. It is used chiefly in testing organic solvents and some acids.

The preferred container for the evaporation of almost all reagents is a platinum dish. However, in many cases, other containers (for example, dishes of porcelain, silica, or aluminum) may be found suitable. In a few cases, such as when strong oxidants are evaporated, platinum should not be used.

A low-temperature ($\approx$100 °C) hot plate is usually specified in the tests in this book for evaporations. A steam bath may be used, if the steam is generated from deionized or distilled water in a properly maintained apparatus.

The residue is not dried to constant weight in this test because continued heating may slowly volatilize some of the high-boiling impurities. The drying conditions specified will, however, yield reproducible and reliable results so that drying to constant weight is not necessary.

▶**Procedure for Residue after Evaporation.** Place the quantity of reagent specified in the individual test description in a suitable tared container (see Containers, page 17). Evaporate the liquid gently so that boiling does not occur; do this in a well-ventilated muffle furnace or fume hood, protected from any possibility of contamination. Unless otherwise specified, dry the residue in an oven at 105 °C for 30 min. Cool the container in a suitable desiccator, and weigh. Calculate the percent residue from the weight of the residue and the weight of the sample.

Loss on Ignition

This test is intended to determine the amount of volatile material due to occluded water or process contamination. The procedure follows directives in the Residue after Ignition test, except that the addition of sulfuric acid is eliminated. Specific variances are detailed in the individual monographs. Calculate the percentage of material lost on ignition.

Residue after Ignition

These tests are designed and intended to determine the amount of nonvolatile inorganic material that may be present in a reagent. They are applied to those inorganic reagents that can be sublimed or volatilized without decomposition; among these are reagents such as ammonium salts and some inorganic acids. Various organic reagents are also tested against this specification.

The interpretation of the instruction to "ignite" has varied widely. After considerable study of the various residues involved, a temperature of 600 ± 25 °C for 15 min in a well-ventilated muffle furnace or fume hood has been adopted. This method ensures constant conditions for conversion of the residues to the desired composition without causing appreciable loss of the impurities themselves. It should also be emphasized that certain organic reagents are difficult to volatilize, and slight variations in the test have been suggested where appropriate. Reagents with a high water content should be dried before ignition to prevent loss of sample. It is left to the discretion of the analyst as to the method of heating to be used, such as a hot plate, gas microburner, or infrared lamp. The final ignition at higher temperature should be done under oxidizing conditions, preferably in a muffle furnace.

▶**Procedure for Residue after Ignition.** Ignite the quantity of reagent specified in the individual test description in a tared preconditioned crucible or dish in a well-ventilated hood, protected from air currents. Heating should be gentle and slow at first and should continue at a rate of 1 to 2 h to volatilize inorganic samples completely or to char organic samples thoroughly. If the sample is a liquid, evaporate completely by heating gently without boiling. Cool the crucible, and moisten the residue with 0.5 mL of sulfuric acid, unless specified otherwise. Ignite the crucible until white fumes of sulfur trioxide cease to evolve. Then, finally, ignite at 600 ± 25 °C for 15 min. Cool in a suitable desiccator (using indicating-type silica gel), weigh, and calculate the percentage of residue.

Loss on Drying

This test is designed to determine the amount of volatile material present in a sample due to process contamination, inherent affinity for volatile impurities, or designated moles of hydration. The volatile content must be removable at relatively low temperatures.

▶**Procedure for Loss on Drying.** Accurately weigh the sample in a conditioned and tared vessel compatible with the material being tested (see Containers, page 17). Unless otherwise noted, dry the material to constant weight in an oven at 105 °C, typically for 4 h; care must be taken to avoid sample loss as a result of spattering or crystal fractionation. Cool sample in a desiccator and weigh. Calculate the percentage of material lost on drying.

Titrimetric Methods

In titrimetry, materials or groups of materials are quantified by measuring the volume of a reagent solution with known concentrations of a substance, the *titrimetric solution*. The titrimetric solution is used for a defined, complete chemical conversion with the materials that are to be measured. Adding a reagent until one can recognize the end point of the reaction is known as *titration*. The reagent is called the *titrant*, and the material to be tested is the *sample* (or *analyte*). The chemical conditions that are required as a prerequisite for each titrimetric determination are a defined course of the reaction between the sample and the titrant and the ability to recognize the equivalence point (or *titration end point*). (This paragraph is adapted, with permission, from Schwedt, 1997.)

Potentiometric Titrations

Potentiometric titrations are titrations in which the equivalence point is determined from the rate of change of the potential difference between two electrodes. The potential difference can be measured by any reliable millivoltmeter, including a pH meter that can be read either in pH units or millivolts. (These titrations can also be made by using commercially available automatic titrators, which either plot the complete titration or act to close an electrically operated buret valve exactly at the equivalence point.) The electrode pair, which consists of an indicating electrode and a reference electrode, depends on the type of titration and is indicated in the individual specifications.

Relatively large increments of the titrant may be added until the equivalence point is approached, usually within 1 mL. In the vicinity of the end point, small equal increments (0.1 mL, for example) are added; after allowing sufficient time for the indicator electrode to reach a constant potential, the voltage and volume of titrant are recorded. The titration is continued for several increments of titrant beyond the end point. For acid–base titrations, the pH is recorded instead of the potential difference but is treated the same as potential difference for the determination of the end point.

▶**First Derivative Method for the Determination of the Equivalence Point.**
When the titration is symmetrical about the end point (same rate of change of potential before and after the end point), the position of the end point may be found by a graphical

Table 2-2. Typical Example of the Second Derivative Method for Determination of the Equivalence Point

V (mL)	E (mV)	ΔE	Δ²E
16.00	504		
		6	
16.10	510		+4
		10	
16.20	520		+20
		30	
16.30	550		+60
		90	
16.40	640		−50
		40	
16.50	680		−30
		10	
16.60	690		

$$V(\text{end point}) = 16.30 + 0.10 \times \frac{60}{60 + 50} = 16.35 \text{ mL}$$

plot of $\Delta E/\Delta V$ vs. V, where E is the potential and V is the volume of titrant. The end point corresponds to the maximum value of $\Delta E/\Delta V$. (This is a plot of the first derivative, dE/dV, of $E = f(V)$ vs. the volume of titrant and is referred to as the first derivative method.)

▶ **Second Derivative Method for the Determination of the Equivalence Point.**
The end point can be determined by the use of the second derivative, d^2E/dV^2, of $E = f(V)$ and is the point where d^2E/dV^2 becomes zero. This point can be determined mathematically, assuming there is no significant difference between the average slope $\Delta E/\Delta V$ and the true slope dE/dV. The end point lies in the increment where the ΔE is the greatest. The amount of titrant to be added to the buret reading at the beginning of the interval is found by multiplying the volume of the increment by the factor in which the numerator is the last $+\Delta^2E$ value and the denominator is the sum of the last $+\Delta^2E$ value and the first $-\Delta^2E$ value, disregarding the sign. When the volume increments are identical, the method can be reduced to the use of the first and second differences. This simplification is employed in a typical example shown in Table 2-2.

Ion-Exchange Column Assays

A 100-mL capacity class B buret with Teflon TFE or PTFE stopcock or a commercially available column having 1.5–2-cm bore may be used as a column. For fluoride assays, an acrylic–polyethylene buret having Teflon stopcocks is preferred. All other assays may be carried out in a glass column of the same dimensions.

Note: All water used in the assay must be free of carbon dioxide and ammonia (18 mΩ water from a purification system is suitable).

Column Preparation. Cation-exchange resin, Dowex HCR-W2 H+ form 16–40 mesh or Amberlite IR-120 or equivalent is suitable for assays. About 75 mL of cation-exchange resin is placed in a liter-size container and filled with water, stirred, and let settle for a minute or two. After settling, water is gently decanted, and the washing is repeated at least 4 5 times. In a suitable column, a cotton ball is inserted above the stopcock to prevent resin from plugging the stopcock and to achieve an even flow of the eluate. Following the last washing of the resin and decanting of the wash, the slurry is poured into the column with the stopcock open. The water level is always kept above the resin. Ion-exchange resin in the H+ form should be washed with water until about 50 mL of eluate requires about 0.05 mL of 0.1 N sodium hydroxide to neutralize, using phenolphthalein as an indicator. An ion-exchange resin, which is in Na+ form, should be regenerated by adding approximately 100 mL of 4 N hydrochloric acid through the column and then being washed until 0.05 mL of 0.1 N sodium hydroxide is required to neutralize about 50 mL of eluate, using phenol-phthalein as an indicator solution. About 3–4 assays may be carried out, but drastically lower assays indicate that a new ion-exchange resin is required. Resins may be recharged by repeating the treatment with 4 N hydrochloric acid.

▶**Method for Ion-Exchange Column Assays.** Prepare the sample as directed in the individual reagent tests. Pass the sample solution through a cation-exchange column with water at a rate of about 5 mL/min, and collect about 250 mL of the eluate in a 500-mL titration flask. Wash the resin in the column at a rate of about 10 mL/min into the same titration flask, and titrate with 0.1 N sodium hydroxide. Continue the elution until 50 mL of the eluate requires no further titration.

Water by the Karl Fischer Method

The water content of most of the organic reagents discussed in this book is determined by the Karl Fischer method (for more on Karl Fischer titration, see Schilt, 1991; Scholz, 1984). This method involves the titration of the sample in methanol or any suitable solvent—for example, pyridine, formamide, or petroleum ether—with the Karl Fischer reagent. This reagent consists of iodine, sulfur dioxide, an amine, and a solvent in which the iodine and the sulfur dioxide are rapidly and quantitatively consumed by the water in the sample. The end point is detected either visually from the color change caused by free iodine or electrometrically. The latter method is preferred in most cases and necessary in the case of colored solutions.

Two different techniques can be used to add the iodine needed for the reaction. In the volumetric Karl Fischer titration, the Karl Fischer reagent is added by means of a volumetric buret. With this technique, it is necessary to standardize the reagent often, and the amount of reagent added must be measured accurately. This accuracy is especially important in determining small amounts of water. A small error in delivering this required amount of reagent can lead to a large error in the determined water content.

The development of commercial coulometric instrumentation has enabled the alternative coulometric titration method to be introduced into the analytical laboratory. Coulometry is an electrochemical process involving the generation of various materials in direct proportion to their equivalent weights. As applied to the determination of water by the Karl

Fischer method, the iodine needed in the reaction is generated in situ (inside the actual titration vessel) from an iodide-containing solution. As in the volumetric Karl Fischer method, the excess iodine beyond the end point is usually determined by an electrometric technique. The amount of iodine generated for the reaction is determined accurately and related to the amount of water present in the sample. The coulometric Karl Fischer titration is a more sensitive procedure (i.e., it can determine smaller amounts of water). Because of this, exclusion of extraneous sources of moisture is absolutely necessary. No standardization of reagent is required because this is an absolute method depending only on electrochemical laws and the accurate measurement of electrical current. The method is generally recommended when the samples are liquid and have a low water content.

Volumetric Procedure

The apparatus described here and the Karl Fischer reagent described in Part 3 (see page 102), are suitable for the determination of water by this method. There are, however, a number of commercial titration instruments and Karl Fischer reagents available. These instruments and reagents may be used in conjunction with their manufacturers' instructions.

Apparatus. *Enclosed Titration System.* The Karl Fischer reagent is highly sensitive to water, and exposure to atmospheric moisture must be avoided. The reagent is best stored in a reservoir connected to an automatic buret with all exits to the atmosphere protected with a suitable desiccant—for example, indicating-type silica gel or anhydrous calcium chloride. The closed titration flask, with provision for insertion of platinum–platinum electrodes and insertion of the sample through a septum or removable closure, must be connected to the buret in such a way that atmospheric moisture cannot enter through the joint. Because of the necessity for a closed titration system, stirring is best accomplished magnetically.

Polarizing Unit for Use with the Electrometric End Point. A simple polarizing unit consists of a 1.5-V dry battery connected across a radio-type potentiometer of about 100 ohms. (One platinum electrode is connected to one end of the potentiometer; the other platinum electrode is connected through a microammeter to the movable control of the potentiometer.) By adjustment of the movable contact, the initial polarizing current can be set to the desired value of 100 microamperes through a 2000-ohm precision resistor. This current passes through the platinum–platinum electrodes, which are immersed in the solution to be titrated.

Reagents. *Preparation of the Reagent.* If a commercial reagent is not being used, prepare the reagent according to Mitchell and Smith, 1984.

Commercial Reagents. Stabilized Karl Fischer reagents for volumetric water determinations are available from several suppliers. Pyridine and pyridine-free reagents can be used for water determination. For ketones and aldehydes, specially formulated reagents are available and should be used. Pyridine and pyridine-free reagents have defined water capacities, which should be considered when these reagents are used.

Standardization of Karl Fischer Reagents. The Karl Fischer reagent gradually deteriorates. It should be standardized within an hour before use, or daily if it is in continuous use. The pyridine-free reagents are more stable, and titrant–solvent systems exist that are free from deterioration; however, a daily check may be necessary. A Karl Fischer solvent system that uses indicating-type silica gel is recommended.

Method A. Place 25–50 mL of methanol in the titration flask, and titrate with Karl Fischer reagent to the electrometric end point. The deflection should be maintained for at least 30 s. This is a blank titration on the water contained in the methanol and the titration flask. Record the volume. Add 0.05 mL of water, accurately weighed, and again titrate with Karl Fischer reagent to the end point. Calculate the strength of the reagent in milligrams of water per milliliter of Karl Fischer reagent from the weight of water added and the net volume of Karl Fischer reagent used.

Method B. Use the same general procedure but add a weighed amount of reagent sodium tartrate dihydrate instead of the water, and calculate the Karl Fischer titration factor (expressed in milligrams of water per milliliter) by the following:

$$\text{Karl Fischer titration factor (mg/mL)} = \frac{\text{Sodium tartrate wt (mg)} \times 0.1566}{\text{Karl Fischer reagent (mL)}}$$

Sodium tartrate does not dissolve completely in methanol and thus can cause premature, transient end points near the true end point. Hence the titration must be performed slowly, and the suspension must be well stirred when near the final end point.

►**Volumetric Procedure for Samples Using Karl Fischer Reagent, Method 1.** Titrate 25–50 mL of methanol or other suitable solvent, as specified, to an end point with the Karl Fischer reagent, as in the standardization. Unstopper the titration flask and rapidly add an accurately weighed portion of the solid or liquid sample to be tested. (For liquids, a weight buret or syringe is convenient or, when the density is known, a measured volume of sample may be introduced with a syringe or pipet.) Quickly restopper the flask, and titrate again with Karl Fischer reagent to an end point. The percentage of water in the sample is calculated from the known strength of the Karl Fischer reagent, the net volume used, and the sample weight.

Coulometric Procedure

Apparatus. Follow the instrument manufacturer's instructions for specific operation procedures.

Reagents. Pyridine and pyridine-free coulometric reagents have defined water capacities. Several suppliers offer solutions that are used for ensuring the performance of coulometric instruments. These usually are nonhygroscopic mixtures of solvents with a known water content. Reagent water may be used if the small amount required is weighed accurately.

►**Coulometric Procedure for Samples Using Karl Fischer Reagent, Method 2.** Titrate the sample cell to dryness. Activate the sample titration procedure, and add the specified amount of sample with a syringe (if liquid) or as a weighed solid. Determine the weight of the liquid sample by weighing the syringe before and after injection or by calculating from the volume injected and the known sample density. The instrument will indicate the end of the titration, and the amount of water in the sample is automatically calculated.

Colorimetry and Turbidimetry

Conditions and quantities for colorimetric or turbidimetric tests have been chosen to produce colors or turbidities that can be observed easily. These conditions and quantities approach but do not reach the minimum that can be observed. Five minutes, unless some other time is specified, must be allowed for the development of the colors or turbidities before the comparisons are made. If solutions of samples contain any turbidity or insoluble matter that might interfere with the later observation of colors or turbidities, they must be filtered before the addition of the reagent used to produce the color or turbidity. Conditions of the tests will vary from one reagent to another, but most of the tests fall close to the limits indicated in Table 2-3.

Ammonium (Test for Ammonia and Amines)

▶**Procedure for Ammonium.** Dissolve the specified quantity of sample in 60 mL of ammonia-free water in a Kjeldahl flask connected through a spray trap to a condenser, the end of which dips below the surface of 10 mL of 0.1 N hydrochloric acid. Add 20 mL of freshly boiled sodium hydroxide solution (10%), distill 35 mL, and dilute the distillate with water as directed. Add 2 mL of freshly boiled sodium hydroxide solution (10%), mix, add 2 mL of Nessler reagent, and again mix. Any color should not be darker than that produced in a standard containing the amount of ammonium ion (NH_4) specified in the individual test description in an equal volume of solution containing 2 mL of the sodium hydroxide solution and 2 mL of Nessler reagent.

Arsenic

This colorimetric comparative procedure is based upon the reaction between silver diethyldithiocarbamate and arsine.

Interferences. Metals or salts of metals—such as chromium, cobalt, copper, mercury, molybdenum, nickel, palladium, and silver—may interfere with the evolution of arsine.

Antimony, which forms stibine, may produce a positive interference in color development with silver diethyldithiocarbamate solution. Although potassium iodide and stan-

Table 2-3. Usual Limits for Colorimetric or Turbidimetric Standards

Sample	Quantity Used for Comparison (mg)	Approximate Volume (mL)
Ammonium (NH_4)	0.01	50
Arsenic (As)	0.002–0.003	3
Barium (Ba) chromate test	0.05–0.10	50
Chloride (Cl)	0.01	20
Copper(Cu)		
Dithizone extraction	0.006	10–20
Hydrogen sulfide	0.02	50
Pyridine thiocyanate	0.02	10–20
Sodium sulfide	0.02	20–50
Heavy metals (as Pb)	0.02	50
Iron (Fe)		
1,10-Phenanthroline	0.01	25
Thiocyanate	0.01	50
Lead (Pb)		
Chromate	0.05	25
Dithizone	0.002	10–20
Sodium sulfide	0.01–0.02	50
Manganese (Mn)	0.01	50
Nitrate (NO_3)		
Brucine sulfate	0.005–0.02	50
Diphenylamine	0.005	25
Phosphate (PO_4)		
Direct molybdenum blue	0.02	25
Molybdenum blue ether extraction	0.01	25
Sulfate (SO_4)	0.04 0.06	12
Zinc (Zn)		
Dithizone extraction	0.006–0.01	10–20

nous chloride in the generator tend to repress the evolution of stibine at low levels of antimony, at higher levels repression is incomplete, a situation that can lead to erroneously high results for arsenic. Interference from antimony, however, can be essentially eliminated by adding ferric ion to the generator. An arsenic test employing this addition is included in the monograph for phosphoric acid.

Antimony can be determined, if desired, by atomic absorption or by differential pulse polarography. An atomic absorption procedure for antimony is also included in the monograph for phosphoric acid.

Apparatus. The apparatus (see Figure 2-1) consists of an arsine generator (a) fitted with a scrubber unit (c) and an absorber tube (e), with standard-taper or ground-glass ball-and-socket joints (b and d) between the units. Alternatively, any apparatus embodying the principle of the assembly described and illustrated may be used.

Sample Solution. Dissolve the specified weight of sample in water and dilute with water to 35 mL, or use the solution prepared as directed in the individual reagent standard.

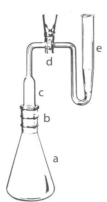

Figure 2-1. Arsenic test apparatus.

▶**Procedure for Arsenic.** Unless otherwise specified in the individual test description, proceed as follows: To the sample, add 20 mL of dilute sulfuric acid (1 + 4), 2 mL of potassium iodide reagent solution (16.5 g of KI per 100 mL), and 0.5 mL of stannous chloride reagent solution (4 g of $SnCl_2 \cdot 2H_2O$ in 10 mL of hydrochloric acid), and mix. Allow the mixture to stand for 30 min at room temperature. Pack the scrubber tube (c in Figure 2-1) with two swabs of cotton previously wetted with a saturated solution of lead acetate, freed from excess lead acetate solution by squeezing, and dried in a vacuum. Allow a small space between the two swabs. Place 3.0 mL of silver diethyldithiocarbamate solution (1 g in 200 mL of clear colorless pyridine) in the absorber tube (e in Figure 2-1) and 3.0 g of granulated zinc (No. 20 mesh) in the generator flask, and immediately connect the scrubber–absorber assembly to the flask. Place the generator flask in a water bath maintained at 25 ± 3 °C, and swirl the flask gently at 10-min intervals. (The addition of a small amount of isopropyl alcohol to the generator flask may promote uniformity of the rate of gas evolution.) After 45 min, disconnect the tubing from the generator flask, and transfer the silver diethyldithiocarbamate solution to a suitable color-comparison tube. Any red color in the silver diethyldithiocarbamate solution of the sample should not exceed the color of an arsenic standard solution equivalent to the limit specified in the individual monograph.

> *Note:* The color comparison may be made in a photometer at the wavelength of maximum absorbance occurring between 535 and 540 nm. If a photometer is used, the silver diethyldithiocarbamate solution should be free from any trace of impurity (before the test is started). The use of capped cells is recommended.

Chloride

The general test procedure cited here applies to the majority of reagent chemicals having a chloride specification.

Perform the test for the sample and the standard in glass tubes or cylinders of the same diameter, matched as closely as practicable. Use identical quantities of the same reagents for the sample and the standard solutions. If after dissolution the solution is not clear, filter it through a chloride-free filter paper. Prepare the filter paper by washing with water until the filtrate gives a negative test for chloride. The filter may also be prewashed with dilute nitric acid (1 + 3).

Experience has shown that visual turbidimetric comparisons are best made between solutions containing 0.01 mg of chloride ion in a volume of 20 mL.

▶**Procedure for Chloride.** Unless otherwise stated, dissolve the specified quantity of sample in water and dilute with water to 20 mL. If the substance is already in solution, dilute with water to 20 mL. Filter if necessary through chlorine-free filter paper. For the standard, use 1 mL of the chloride ion (Cl) standard solution, and dilute with water to 20 mL. To each tube, add 1 mL of nitric acid and 1 mL of silver nitrate reagent solution. Mix, allow to stand for 5 min protected from sunlight, and compare. The solutions can best be viewed visually against a black background. Any turbidity in the solution of the sample should not exceed that in the standard.

For samples yielding colored solutions, dissolve the sample, prepare the standard, and treat the two solutions as directed in the specification. The comparison is best made by viewing both turbidities through the same depth and color of solution. Do this as follows: Prepare a test tube with the specified quantity of sample in a volume of water equal to the total volume of the test, and superimpose it over the test tube that contains the standard turbidity. Place a third tube containing the same volume of water below the tube containing the sample and added reagents. The comparison tubes may be machine-made vials, long style, of about 20-mL capacity.

Dithizone

Because dithizone is extremely sensitive to many metal ions, all glassware used must be specially cleaned. After the usual cleaning, it should be thoroughly rinsed with warm dilute nitric acid (1 + 1) and finally rinsed with reagent water. All glassware used in the preparation and storage of reagents and in performing the tests must be made of lead- and zinc-free glass. Alternatively, for solution storage, polyethylene containers may be used.

Heavy Metals (as Lead)

The heavy metals test is designed to limit the common metallic impurities (Ag, As, Bi, Cd, Cu, Hg, Mo, Pb, Sb, and Sn) that produce colors when sulfide ion is added to slightly acid solutions that contain them. In this test, such heavy metals are expressed "as lead" by comparing the color developed in a test solution with that developed in a lead standard/control solution. The optimum conditions for proper performance of the test are: pH of the solution between 3 and 4, total volume of 45 mL, use of pH 3.5 ammonium acetate dilution buffer solution, use of freshly prepared hydrogen sulfide water, and use of a standard/control solution containing 0.02 mg of lead ion (Pb). These conditions are rarely modified in the individual procedures, and it would only be after a method validation that the modifications would be suitable for the individual reagent. The color comparisons should be made, using matched 50-mL color-comparison tubes, by viewing vertically over a white background. In some instances, an appropriate amount of the sample is added to the standard, thereafter called the control (see page 107), because the color developed in the test may be affected by the sample being tested. In other instances, no such effort occurs and the standard may be prepared without any of the sample. In general, the individual standard procedures provide only for the preparation of a sample solution. The remainder of the test directions, being the same for all reagents, are provided in this chapter.

Two general test procedures are provided. Method 1 is to be used unless otherwise specified in the individual standard. This method is used, generally, for those substances that yield clear, colorless solutions under the specified test conditions. Method 2 is used, generally, only for certain organic compounds. An alternate procedure (page 37) is provided for cases where hydrogen sulfide is not suitable.

Unless otherwise directed in the individual specification, the test shall be carried out as follows.

▶**Procedure for Heavy Metals, Method 1.** Test in a well-ventilated fume hood.

    Ammonium Acetate, pH 3.5, Dilution Buffer Solution. Dissolve 5.0 g of ammonium acetate in about 450 mL of water in a 500-mL beaker. Using a pH meter, while stirring, adjust the pH to 3.5 with a slow addition of acetic acid, glacial. If needed, 10% ammonium hydroxide reagent solution may be used to readjust the pH. Dilute to 500 mL with water, and store in a plastic bottle. This dilution buffer is used to prepare the standard/control and sample solutions; dilute to volume after pH is adjusted.

    Test Solution. Using the solution prepared as directed in the individual reagent monograph, adjust the pH to between 3 and 4 (by using a pH meter) with 1 N acetic acid reagent solution or 10% ammonium hydroxide reagent solution. Dilute with pH 3.5 ammonium acetate dilution buffer solution to 40 mL, if necessary, and mix.

    Standard/Control Solution. To 20 mL of pH 3.5 ammonium acetate dilution buffer solution, or to the portion of that solution specified in the individual standard, add 0.02 mg of lead ion (Pb). [When a control solution is specified in the individual reagent monograph and a direction is given therein to add a weight of lead ion (Pb), no further lead shall be added here.] Dilute with water to 25 mL, and mix. Adjust the pH (within 0.1 unit) to the value established for the test solution (using a pH meter) with 1 N acetic acid reagent solution or 10% ammonium hydroxide reagent solution, dilute with pH 3.5 ammonium acetate dilution buffer solution to 40 mL, and mix.

To each of the tubes containing the test solution and the standard/control solution, add 5 mL of freshly prepared hydrogen sulfide water, and mix. Any brown color produced within 5 min in the test solution should not be darker than that produced in the standard/control solution.

▶**Procedure for Heavy Metals, Method 2.** Test in a well-ventilated fume hood.

    Test Solution. Transfer the quantity of substance specified in the individual reagent monograph to a suitable crucible, add sufficient sulfuric acid to wet the sample, and carefully heat at a low temperature until thoroughly charred. (The crucible may be loosely covered with a suitable lid during the charring.) Add to the carbonized mass 2 mL of nitric acid and 0.25 mL of sulfuric acid, and heat cautiously until white fumes of sulfur trioxide are no longer evolved. Ignite, preferably in a well-ventilated muffle furnace, at 500–600 °C until the carbon is burned off completely. Cool, add 4 mL of 6 N hydrochloric acid, and cover. Digest on a hot plate (≈100 °C) for 15 min, uncover, and slowly evaporate on the hot plate to dryness. Moisten the residue with 0.05 mL of hydrochloric acid, add 10 mL of hot water, and digest for 2 min. Add 10% ammonium hydroxide reagent solution dropwise, until the solution is just alkaline, and dilute with water to 25 mL. Adjust the pH of the solution to between 3 and 4 (by using a pH meter) with 1 N acetic acid reagent solution, dilute with pH 3.5 ammonium acetate dilution buffer solution (described in Method 1) to 40 mL, and mix.

Standard/Control Solution. To 20 mL of water, add 0.02 mg of lead ion (Pb), and dilute with water to 25 mL. Adjust the pH (within 0.1 unit) to the value established for the test solution (by using a pH meter) with 1 N acetic acid reagent solution or 10% ammonium hydroxide reagent solution, dilute with pH 3.5 ammonium acetate dilution buffer solution (described in Method 1) to 40 mL, and mix.

To each of the tubes containing the test solution and the standard/control solution, add 5 mL of freshly prepared hydrogen sulfide water, and mix. Any brown color produced in the test solution within 5 min should not be darker than that produced in the standard/control solution.

Alternate Sulfide Reagent Solution and Use.

Sodium Sulfide Solution. (Note: Before sodium sulfide is weighed, any excessive liquid on the reagent should be removed on filter paper.) Dissolve 2.5 ± 0.05 g of sodium sulfide, nonahydrate, in a mixture of 5 mL of water and 15 mL of glycerol. Store the solution in a small, tightly closed, light-resistant bottle. Use within 3 months. Make fresh when in doubt.

Add 0.05 mL or 1 drop of sodium sulfide solution to the test solution and the standard/control solution, mix well, and allow to stand for 5 min. Compare both solutions against a white background. Any brown color produced in the test solution within 5 min should not be darker than that produced in the standard/control solution.

Note: The final volume of the sample and control solutions must be the same in both Method 1 and Method 2 when the pH 3.5 ammonium acetate dilution buffer solution is used.

Iron

Two procedures are provided for the determination of iron. Method 1, which uses thiocyanate, is usually performed in 50 mL of solution containing 2 mL of hydrochloric acid, to which 30–50 mg of ammonium peroxydisulfate crystals and 3 mL of 30% ammonium thiocyanate reagent solution are added. These quantities of acid and thiocyanate were selected so that the production of the red color would not be affected significantly by anything except the iron. The use of peroxydisulfate to oxidize the iron prevents fading of the color and eliminates the need for immediate comparison with the standard. For certain alkaline earth salts, permanganate is used to oxidize the iron.

Method 2, which uses 1,10-phenanthroline, is used generally for phosphate reagents. It is usually performed by adjusting the pH (when necessary) to between 4 and 6, adding 6 mL of hydroxylamine hydrochloride reagent solution and 4 mL of 1,10-phenanthroline reagent solution, and diluting with water to 25 mL. Because development of the color may be slow, the comparison of the color produced is not made until 60 min after addition of the 1,10-phenanthroline reagent solution. The hydroxylamine hydrochloride reagent solution is the unacidified reagent solution and not the acidified solution used in the dithizone test for lead and other metals.

The two general methods, which are referred to in the individual reagent monographs, are described here. Where a difference appears between the general test method and the directions set forth in the individual reagent monograph, the directions in the monograph are to be followed.

▶**Procedure for Iron, Method 1 (Ammonium Thiocyanate).** Dissolve the specified amount of sample in 40 mL of water or use the sample solution prepared as directed in the individual test description. Unless otherwise directed in the individual specification, add 2 mL of hydrochloric acid and, if necessary, dilute with water to 50 mL. Add 30–50 mg of ammonium peroxydisulfate crystals and 3 mL of 30% ammonium thiocyanate reagent solution and mix. Any red color produced should not exceed that produced by 0.01 mg of iron in an equal volume of solution containing the quantities of reagents used in the test.

▶**Procedure for Iron, Method 2 (Hydroxylamine and 1,10-Phenanthroline).** Dissolve the specified amount of sample in 10 mL of water, or use the sample solution prepared as directed in the individual test description. Add 6 mL of hydroxylamine hydrochloride reagent solution and 4 mL of 1,10-phenanthroline reagent solution, and dilute with water to 25 mL. Any red color produced within 1 h should not exceed that produced by 0.01 mg of iron in an equal volume of solution containing the quantities of reagents used in the test.

Nitrate

Four different tests are applied to the determination of nitrate: brucine sulfate, diphenylamine, indigo carmine, and phenoldisulfonic acid. General procedures are provided below for brucine sulfate and diphenylamine.

▶**Procedure for Nitrate, Method 1 (Brucine Sulfate).** It is recognized that the sensitivity of brucine sulfate for determining small amounts of nitrate varies with the conditions of each particular test. Therefore, the tests are designed so that the sensitivity of the brucine sulfate is determined under the particular conditions of the test. The test must be followed carefully to ensure the development of a reproducible color.

Sample Solution A. Dissolve the specified quantity or volume of the sample in 3 mL of water by heating in a boiling water bath. Dilute with brucine sulfate reagent solution to 50 mL.

Control Solution B. To the specified volume of nitrate ion (NO_3) standard solution, add the same quantity of sample as in sample solution A and sufficient water to make a volume of 3 mL. Dissolve the mixture by heating in a boiling water bath. Dilute with brucine sulfate reagent solution to 50 mL.

Blank Solution C. Use 50 mL of brucine sulfate reagent solution.

Heat the three solutions in a preheated (boiling) water bath for 15 min with periodic gentle swirling. Cool rapidly in an ice water bath to room temperature. Set a spectrophotometer at 410 nm and, using 1-cm quartz cells, adjust the instrument to read zero absorbance with blank solution C in the light path, then determine the absorbance of sample solution A and of control solution B. Calculate the nitrate content as:

$$\% \, NO_3 = \frac{\text{Absorbance of } A}{\text{Absorbance of } B - \text{Absorbance of } A} = \% \text{ maximum allowable}$$

▶**Procedure for Nitrate, Method 2 (Diphenylamine).** The prescribed tests must be followed carefully to ensure the development of a reproducible and stable color. Under slightly different conditions, the oxidation reaction may produce a somewhat different color, which is not stable.

Place the specified quantity of the sample in a dry beaker. Cool the beaker thoroughly in an ice bath, and add 22 mL of sulfuric acid that has been cooled to ice-bath temperature. Allow the mixture to warm to room temperature, and swirl the beaker at intervals to effect gentle dissolution with slow evolution of the acid vapor. Prepare a standard by evaporating to dryness a solution containing the specified quantity of nitrate ion (NO_3) standard solution and 0.01 g of anhydrous sodium carbonate. Treat the residue exactly as the sample. Add 3 mL of diphenylamine reagent solution to each solution, and digest on a hot plate ($\approx$100 °C) for 90 min. Compare the color of the solutions visually. Any blue color produced in the solution of the sample should not exceed that in the standard.

Nitrogen Compounds: Test for Ammonia, Amines, and Nitrogen Compounds Reduced by Aluminum

▶**Procedure for Nitrogen Compounds.** Dissolve the specified quantity of sample in 60 mL of ammonia-free water in a Kjeldahl flask connected through a spray trap to a condenser, the end of which dips below the surface of 10 mL of 0.1 N hydrochloric acid. Add 10 mL of freshly prepared 10% sodium hydroxide reagent solution and 0.5 g of aluminum wire (previously cleaned with 1 N sodium hydroxide, followed by rinsing with deionized water), in small pieces, to the Kjeldahl flask. Allow to stand for 1 h protected from loss of and exposure to ammonia. Distill 35 mL, and dilute the distillate with water to 50 mL. Add 2 mL of freshly prepared 10% sodium hydroxide reagent solution, mix, add 2 mL of Nessler reagent solution, and again mix. Any color should not be darker than that produced in a standard containing the amount of nitrogen specified in the individual test description and treated exactly like the sample. Commercially available apparatus may be used.

Phosphate

Three general procedures are provided to test for phosphate: direct molybdenum blue (Method 1), extracted molybdenum blue (Method 2), and precipitation (Method 3). In Method 2, conditions are established that allow only the phosphomolybdate to be extracted into ether, thus eliminating any interference from arsenate and silicate.

▶**Procedure for Phosphate, Method 1 (Direct Molybdenum Blue).** Prepare   the sample solution as directed in the individual test description. Add 1 mL of ammonium molybdate reagent solution and 1 mL of 4-(methylamino)phenol sulfate reagent solution, and allow to stand at room temperature for 2 h. Any blue color should not exceed that produced by 0.02 mg of phosphate ion (PO_4) in an equal volume of solution containing the quantities of reagents used in the test.

▶**Procedure for Phosphate, Method 2 (Extracted Molybdenum Blue).** Prepare the sample solution as directed in the individual test description. Transfer to a separatory funnel, add 35 mL of ether, shake vigorously, and allow the layers to separate. Draw off and discard the aqueous layer unless it is indicated in the individual specification that the aqueous layer should be saved for the silica determination. Wash the ether layer twice with 10 mL of dilute hydrochloric acid (1 + 9), drawing off and discarding the aqueous layer each time. Add 0.2 mL of a freshly prepared 2% stannous chloride reagent solution. If the solution is cloudy, shake with a small amount of dilute hydrochloric acid (1 + 9) to clear. Any blue color should not exceed that produced by a standard containing the concentration of phosphate ion (PO_4) cited in the individual test description and treated exactly as is the sample.

▶**Procedure for Phosphate, Method 3 (Precipitation).** Dissolve 20.0 g of sample in 100 mL of dilute ammonium hydroxide (1 + 4). Prepare a standard containing 0.10 mg of phosphate ion (PO_4) in 100 mL of dilute ammonium hydroxide (1 + 4). To each solution, add dropwise, with rapid stirring, 3.5 mL of ferric nitrate reagent solution, and allow to stand about 15 min. If necessary, warm gently to coagulate the precipitate, but do not boil. Filter and wash several times with dilute ammonium hydroxide (1 + 9). Dissolve the precipitate by pouring 60 mL of warm dilute nitric acid (1 + 3) through the filter and catching the solution in a 250-mL glass-stoppered conical flask. Add 13 mL of ammonium hydroxide, warm the solution to 40 °C, and add 50 mL of ammonium molybdate–nitric acid reagent solution. Shake vigorously for 5 min, and allow to stand for 2 h at 40 ± 5 °C. Any yellow precipitate obtained from the sample should not exceed that obtained from the standard.

Silicate

Small amounts of silica or silicate are determined by a molybdenum blue method. Conditions have been established that allow the extraction of the silicomolybdate and thus eliminate any interference from arsenate or phosphate.

Sulfate

Note: Use 12% barium chloride reagent solution unless stated otherwise; for reagent monographs for phosphates, use 40% barium chloride reagent solution.

▶**Procedure for Sulfate, Method 1.** Unless otherwise stated in the reagent monograph, dissolve (0.005 ÷ % maximum allowable) g of sample in a minimum volume of water. This gives a sample size that provides 0.05 mg of sulfate ion (SO_4) at limit. The standard also contains 0.05 mg of sulfate ion (SO_4). To determine the sample weight in grams, divide 0.005 by the maximum allowable (in actual percent). For example, if the maximum allowable is 0.01%, 0.005 is divided by 0.01 to arrive at 0.50 g of sample (use a four-place balance for a sample size less than 1.0 g). Add 1 mL of dilute hydrochloric acid (1 + 19). Filter through a prewashed small filter paper, and wash with two 2-mL portions of water.

Dilute to 10 mL, and add 1 mL of barium chloride reagent solution. Any turbidity should not exceed that produced by 0.05 mg of sulfate ion (SO_4) in an equal volume of solution containing the quantities of reagents used in the test. Unless otherwise specified, compare turbidity 10 min after adding the barium chloride to the sample and standard solutions.

▶**Procedure for Sulfate, Method 2.** Unless otherwise stated in the specification, dissolve (0.005 ÷ % maximum allowable) g of sample in water, and evaporate to dryness on a hot plate (≈100 °C). Proceed as in Method 1.

▶**Procedure for Sulfate, Method 3.** Unless otherwise stated in the specification, add 10 mg of sodium carbonate to (0.005 ÷ % maximum allowable) g of sample, and evaporate to dryness on a hot plate (≈100 °C). Proceed as in Method 1.

▶**Procedure for Sulfate, Method 4.** Unless otherwise stated in the specification, dissolve (0.005 ÷ % maximum allowable) g of sample in water, evaporate to 5 mL, add 1 mL of bromine water, and evaporate to dryness on a hot plate (≈100 °C). Proceed as in Method 1.

Measurement of Physical Properties

Boiling Range

The boiling-range procedure is essentially an empirical method. Hence adherence to the specified procedural details is required to obtain consistent and reproducible results. The recommended method is that developed by ASTM, and the analyst is referred to ASTM D1078 (ASTM, 2003) for information concerning the description of the apparatus and the proper performance of the test. If the liquid to be tested is extremely volatile or produces noxious fumes during distillation, suitable cooling and venting facilities must be provided.

Color (APHA)

The color intensity of liquids may be estimated rapidly by using platinum–cobalt standards, as described in ASTM D1209 (ASTM, 2000) (see also APHA, 1998). This method is particularly applicable to those materials in which the color-producing substances have light-absorption characteristics nearly identical with those in the standards. Colors having hues other than light yellow or reddish yellow cannot be determined with these standards.

The platinum–cobalt color standards contain carefully controlled amounts of potassium chloroplatinate and cobaltous chloride. Each platinum–cobalt color unit is equivalent to 1 mg of platinum per liter of solution (1 ppm), and the standards are named accordingly. For example, the No. 20 platinum–cobalt standard contains 20 ppm of platinum. These platinum–cobalt standards are also called APHA (for American Public Health Association) and Hazen standards.

Apparatus. The apparatus for this measurement consists of a series of Nessler tubes, a color comparator, and a light source. These tubes must match each other with respect to the color of the glass and height of the graduation mark. They must be fitted with suitable closures to prevent loss of liquid by evaporation and contamination of the standards by dust or dirt. The most commonly used type of Nessler tube is the 100-mL tall-form tube. However, for some samples, particularly those having darker colors, a better color match may be obtained by using 100-mL short-form Nessler tubes.

For the most accurate estimation of color, it is desirable to use a comparator constructed so that white light is reflected from a white glass plate with equal intensity through the longitudinal axes of the tubes being compared. The tubes are shielded so that no light enters the tubes from the side. In most cases, satisfactory estimates of the color can be obtained by holding the sample and standards close to each other over a white plate.

The best source of light is generally considered to be diffuse daylight. However, for general routine analyses, the use of a titrating lamp equipped with a daylight-type fluorescent tube is satisfactory.

APHA No. 500 Platinum–Cobalt Standard. (This standard is also commercially available.) Add approximately 500 mL of reagent water to a 1000-mL volumetric flask, add 100 mL of hydrochloric acid, and mix well. Weigh 1.245 g of potassium chloroplatinate and 1.000 g of cobalt chloride hexahydrate to the nearest milligram, and transfer to the flask. Swirl until complete dissolution is effected, dilute to the mark with reagent water, and mix thoroughly. The spectral absorbance of this APHA No. 500 standard must fall within the limits given below when measured in a suitable spectrophotometer, with a 1-cm light path and reagent water as the reference liquid in a matched cell.

Wavelength (nm)	Absorbance
430	0.110 to 0.120
455	0.130 to 0.145
480	0.105 to 0.120
510	0.055 to 0.065

APHA Platinum–Cobalt Standards. Prepare the required color standards by diluting the volume of APHA No. 500 platinum–cobalt standard listed in Table 2-4 with reagent water to a total volume of 100 mL. The use of a buret is recommended in measuring the No. 500 standard. This series of standards is usually sufficient to permit an experienced analyst to make color comparisons with the necessary precision and accuracy. If a more exact estimate of color is desired, additional standards may be prepared to supplement those given by using proportional amounts of the No. 500 platinum–cobalt standard.

▶**Procedure for Color (APHA).** Transfer 100 mL of the sample to a matched 100-mL tall-form Nessler tube. (If the sample is turbid, filter or centrifuge before filling the tube to remove visible turbidity.) Compare the color of the sample with the colors of the series of platinum–cobalt standards in matching Nessler tubes. View vertically down through the tubes against a white background. Report as the color the number of the APHA standard that most nearly matches the sample. In the event that the color lies midway between two standards, report the darker of the two.

Conductivity (Specific Conductance)

Conductivity is a measure of the ability of a material to conduct electricity. For liquids, this ability depends on the presence of ionic species (electrolytes) in a solution to carry an electric current. *Conductance* is defined as the reciprocal of resistance. The unit of conductance is the mho, which is the conductance of a body through which 1 A of current flows when

Table 2-4. APHA Platinum–Cobalt Color Standard Preparation

APHA Pt–Co Color Standard Number	APHA No. 500 Pt–Co Standard (mL to dilute to 100 mL)	APHA Pt–Co Color Standard Number	APHA No. 500 Pt–Co Standard (mL to dilute to 100 mL)
0	0.00	80	16.00
1	0.20	90	18.00
3	0.60	100	20.00
5	1.00	120	24.00
10	2.00	140	28.00
15	3.00	160	32.00
18	3.60	180	36.00
20	4.00	200	40.00
25	5.00	250	50.00
30	6.00	300	60.00
35	7.00	350	70.00
40	8.00	400	80.00
50	10.00	450	90.00
60	12.00	500	100.00
70	14.00		

the potential difference is 1 V. This unit is also referred to as the reciprocal ohm (ohm^{-1}) or the siemens (S). *Conductivity*, also called *specific conductance*, is defined as the reciprocal of the resistance of a one meter cube of liquid at a specified temperature ($\Omega^{-1} \cdot m^{-1}$).

Laboratory conductivity measurements are used to assess the degree of mineralization of distilled or deionized water. The most convenient unit for reporting water conductivity is the μmho/cm, or the numerically equal $\mu S \cdot cm^{-1}$. Numerous conductivity instruments are available with various means of reporting units of conductance that can be converted to the previously mentioned units of measure. For the purposes described in this book, the cell should be standardized by using known low-concentration solutions of potassium chloride in ion-free water. Conductivity reference data can be obtained in Lind et al., 1959.

Density

Density is a fundamental physical property of all substances; it is defined as the mass, not the weight, of a unit volume of a substance. The volume of a substance changes with temperature, so density is always given at a specified temperature. *Specific gravity* is defined as the density of a liquid substance relative to the density of water (Kenkel, 2003). In the General Description section of individual monographs in this book, the density of many reagents is given as an approximate value, and the reader is referred to other available sources for exact values.

Density (g/mL) may be determined by any reliable method in use in the testing laboratory, such as by a pycnometer or a density meter. A satisfactory method uses a 5-mL U-shaped bicapillary pycnometer, as described by Smith and others (1950) and later adopted by ASTM. Density meters that measure the natural vibrational frequency of a substance and relate the amplitude to changes in the density now are routinely used and have replaced the pycnometer in many laboratories.

▶**Procedure for Density.** Tare a pycnometer, the volume of which has been accurately determined at 25 °C. Fill the pycnometer with the liquid to be tested and immerse in a constant-temperature (25 ± 0.1 °C) water bath for at least 30 min. Observe the volume of liquid at this temperature. Remove the pycnometer from the bath, wipe it dry, and weigh.

$$D \text{ at } 25\,°C = \frac{W}{V} + 0.0010$$

where D is density in g/mL; W is weight, in grams, of liquid in the pycnometer at 25 °C; V is volume, in mL, of liquid in the pycnometer at 25 °C; and 0.0010 is the correction factor to compensate for buoyancy.

The correction of 0.0010 for air buoyancy is accurate to ±0.0001 g/mL if the density varies from 0.7079 g/mL (anhydrous ethyl ether) to 1.585 g/mL (carbon tetrachloride) and the pycnometer is weighed at a temperature of from 20 to 30 °C at an atmospheric pressure of from 720 to 775 mm of mercury.

Freezing Point

Place 15 mL of sample in a test tube (20 × 150 mm) in which is centered a thermometer traceable to NIST. The sample tube is centered by corks in an outer tube about 38 × 200 mm. Cool the whole apparatus, without stirring, in a bath of shaved or crushed ice with water enough to wet the outer tube. When the temperature is about 3 °C below the normal freezing point of the reagent, stir to induce freezing, and read the thermometer every 30 s. The temperature that remains constant for 1 to 2 min is the freezing point.

Melting Point

Where required, the melting point may be determined by the capillary tube method. Calibration of the method using any commercially available apparatus with the use of reference standards is required. Thermometers traceable to NIST are also required.

▶**Procedure for Melting Point.** Grind the material to be tested to a very fine powder. If it contains water of hydration, render it anhydrous by drying as directed. If the substance contains no water of hydration, dry it over a suitable desiccant. Charge a capillary tube that is about 10 cm long with an internal diameter of 1.0 ± 0.2 mm, walls about 0.2–0.3 mm thick, and one end that has been heat-sealed. The height of the sample in the tube should be about 2.5–3.5 mm. Moderate tapping on a hard surface may be required to pack the sample.

Raise the temperature to about 10 °C below the expected melting point, and continue to increase the heat at a rate of 1 ± 0.5 °C/min until the solid coalesces and is completely melted. Record the temperature.

Particle Size

When a mesh or a mesh range is stated on the label for a reagent chemical, the label shall include reference to a coarse sieve and to a finer sieve. The sieve number (mesh) is related to the sieve opening as indicated in Table 2-5.

When tested according to the following procedures, at least 95% of the material shall pass through the coarse sieve and at least 70% shall be retained on the finer sieve. This requirement applies only to materials that are 60-mesh or coarser.

Table 2-5.　Sieve Numbers and Sieve Openings

Sieve No. (mesh)	Sieve Opening (mm)
2	9.52
4	4.76
8	2.38
10	2.00
20	0.84
30	0.59
40	0.42
50	0.297
60	0.250
80	0.177

▶**Procedure for Particle Size.** The sieves used in this procedure shall be those known as the U.S. Standard Sieve Series. Details of the standardization of such sieves can be found in ASTM E11 (ASTM, 2004).

Place 25–100 g of the material to be tested on the appropriate coarser standard sieve, which is mounted above the finer sieve, to which a close-fitting pan is attached. Place a cover on the coarser sieve, and shake the stack in a rotary horizontal direction and vertically by tapping on a hard surface for not less than 20 min or until sifting is practically complete. Weigh accurately the amount of material remaining on each sieve.

An alternate procedure may be used in which the screening through the standard sieves is carried out in a mechanical sieve shaker. This shaker reproduces the circular and tapping motion given to the testing sieves in hand sifting, but with a mechanical action. Follow the directions provided by the manufacturer of the shaker.

Specific Rotation

▶**Procedure for Specific Rotation.** To measure optical rotation, any suitable commercially available instrument may be used. Manufacturer's supplied standards or certified standards must be used to verify accuracy of the instrument at the time of optical rotation determination. Specific rotation is to be calculated from the observed optical rotations in the prepared test solution, as directed in the test. Unless otherwise specified, measurements of optical rotation are made using the sodium D-line (a doublet at 589.0 nm and 589.6 nm) at 25 °C. For optical rotation measurements using a polarimeter, a minimum of five readings is required. Blank correction consists of measuring the optical rotation of the solvent used to dissolve the sample. For a photoelectric instrument, a single, stable reading of the sample solution is taken and corrected for the solvent blank. Specified temperature of the test solution is to be maintained within 0.2 °C during the determination. Calculation:

$$[\alpha]_D^{25°} = \frac{A \times 100}{L \times \text{Sample wt (g)}}$$

where $[\alpha]$ is specific rotation at 25 °C, A is corrected specific rotation, and L is the length of the polarimeter cell in decimeters.

Direct Electrometric Methods

pH Potentiometry

pH Range

This pH requirement is intended to limit the amount of free acid or alkali that is allowed in a reagent. Sodium chloride, for example, has a requirement that the pH of a 5% solution should be from 5.0 to 9.0 at 25.0 °C. This requirement limits the free hydrochloric acid or sodium hydroxide to about 0.001%. The effect of excess acid or base on the pH of a 5% solutions of salts of strong acids and bases is given in Table 2-6.

The pH of the solution is determined by means of any suitable pH meter equipped with a glass electrode and a reference half cell, usually a saturated calomel electrode (SCE), while the solution is protected from absorption of atmospheric gases. The meter and electrodes are standardized at pH 4.00 at 25.0 °C with 0.05 M NIST potassium hydrogen phthalate and at pH 9.18 at 25.0 °C with 0.01 M NIST sodium borate decahydrate. Because the pH values are reported to only 0.1 pH unit, the meter and electrodes are considered to be in satisfactory working order if they show an error of less than 0.05 pH unit in the pH range of 4.00 to 9.18. (Use pH from NIST-certified value.) For more accurate measurement of a test solution, the meter and electrodes should be standardized against a buffer standard solution whose pH is close to the pH of the test solution.

For reference purposes, Table 2-7 presents a list of acid–base indicators with their visual transition intervals.

Table 2-6. Effect of Excess Acid or Base on the pH of 5% Solution of Salts of Strong Acids and Bases

Excess Acid/Base	0.01%, pH	0.001%, pH	0.0001%, pH
HCl	3.9	4.9	5.9
H_2SO_4	4.0	5.0	6.0
HNO_3	4.1	5.1	6.1
HBr	4.2	5.2	6.2
$HClO_4$	4.3	5.3	6.3
KOH	9.9	8.9	7.9
NaOH	10.1	9.1	8.1

Table 2-7. Acid–Base Indicators, by Visual Transition Interval

Visual Transition Interval (pH)	Indicator	Color Change
0.1–2.0	Malachite green hydrochloride	Yellow to blue-green
0.1–2.3	Methyl green	Yellow to blue-green
0.2–1.0	Picric acid	Colorless to yellow
0.2–1.8	Cresol red	Red to yellow
0.8–2.6	Crystal violet	Green to violet
1.0–3.1	Basic fuchsin	Purple to red
1.2–2.8	m-Cresol purple	Red to yellow
1.2–2.8	Thymol blue	Red to yellow
1.2–2.8	Thymol blue, sodium salt	Red to yellow
1.4–2.6	Orange IV	Red to yellow
1.5–3.2	Methyl violet 2B	Blue to violet
3.0–4.6	Bromphenol blue	Yellow to blue
3.0–4.6	Bromphenol blue, sodium salt	Yellow to blue
3.0–5.2	Congo red	Purple to red
3.2–4.4	Methyl orange, sodium salt	Red to yellow
4.0–5.4	Bromocresol green	Yellow to blue
4.0–5.4	Bromocresol green, sodium salt	Yellow to blue
4.0–6.0	Alizarin red S	Yellow to blue
4.2–6.2	Methyl red hydrochloride	Pink to yellow
4.2–6.2	Methyl red, sodium salt	Pink to yellow
4.8–6.4	Chlorophenol red	Yellow to red
5.2–6.8	Bromocresol purple, sodium salt	Yellow to purple
5.8–7.2	Alizarin	Yellow to red
6.0–7.0	Nitrazine yellow	Yellow to blue
6.0–7.5	Bromthymol blue	Yellow to blue
6.0–7.6	Bromthymol blue, sodium salt	Yellow to blue
6.0–7.6	Bromthymol blue, solution	Yellow to blue
6.0–8.0	Rosolic acid	Yellow to red
6.6–8.6	m-Nitrophenol	Colorless to yellow
6.8–8.0	Neutral red	Red to yellow
6.8–8.2	Phenol red	Yellow to red
6.8–8.2	Phenol red, sodium salt	Yellow to red
7.0–8.8	Cresol red	Yellow to red
7.4–9.0	m-Cresol purple	Yellow to purple
8.0–9.2	Thymol blue	Yellow to blue
8.0–9.2	Thymol blue, sodium salt	Yellow to blue
8.0–10.0	Phenolphthalein	Colorless to red
8.6–10.0	Thymolphthalein	Colorless to blue
10.2–13.0	Nile blue A (sulfate)	Blue to red
10.5–12.5	Malachite green hydrochloride	Blue to colorless
11.0–13.0	Alizarin	Red to purple
11.0–13.0	Azo violet	Yellow to violet
11.4–13.0	5,5'-Indigodisulfonic acid, disodium salt	Blue to yellow
12.0–14.0	Acid fuchsin	Red to colorless
12.0–14.0	Clayton yellow	Yellow to red

Source: Reproduced courtesy of Mallinckrodt Baker, Inc.

Table 2-8. pH of a 5% Solution of Reagents at 25.0 °C

Salt	pH	Salt	pH
Ammonium acetate	7.0	Potassium nitrate	7.0
Ammonium bromide	4.9	Potassium phosphate, monobasic	4.2
Ammonium chloride	4.7	Potassium sodium tartrate,	
Ammonium nitrate	4.8	tetrahydrate	8.4
Ammonium phosphate,		Potassium sulfate	7.3
monobasic	4.1	Potassium thiocyanate	7.0
Ammonium sulfate	5.2	Sodium acetate, anhydrous	8.9
Ammonium thiocyanate	4.9	Sodium acetate trihydrate	8.9
Barium chloride, dihydrate	6.9	Sodium bromide	7.0
Barium nitrate	6.9	Sodium chloride	7.0
Calcium chloride, dihydrate	6.6	Sodium molybdate dihydrate	8.1
Calcium nitrate, tetrahydrate	6.6	Sodium nitrate	7.0
Lithium perchlorate	7.0	Sodium phosphate, dibasic,	
Magnesium nitrate, hexahydrate	6.6	heptahydrate	9.0
Magnesium sulfate, heptahydrate	6.8	Sodium phosphate, monobasic,	
Manganese chloride, tetrahydrate	5.4	monohydrate	4.2
Potassium acetate	9.0	Sodium pyrophosphate,	
Potassium bromate	7.0	decahydrate	10.4
Potassium bromide	7.0	Sodium sulfate	7.2
Potassium chloride	7.0	Sodium tartrate dihydrate	8.4
Potassium chromate	9.3	Sodium thiosulfate, pentahydrate	7.9
Potassium iodate	7.0	Zinc sulfate heptahydrate	5.4
Potassium iodide	7.0		

▶**Procedure for pH of a 5% Solution at 25.0 °C.** Dissolve 5.0 g of the sample in 100 mL of carbon dioxide-free water while protecting the solution from absorption of carbon dioxide from the atmosphere. The pH measurement is made as previously described.

For reference purposes, Table 2-8 reports the pH of a 5% solution of the pure reagent chemical. This value was determined experimentally for each salt by dissolving 10.0 g of the reagent in approximately 200 mL of water, then making the solution slightly acid and titrating with standard alkali. Another similar solution was prepared, made slightly alkaline, and titrated with standard acid. Graphs were constructed for each titration by plotting pH vs. milliliters of titrating solution. The average of the two end points so determined is reported as the pH of a 5% solution containing no free acid or alkali.

pH of Other Concentrations

Directions for the preparation of solutions at concentrations of other than 5% are given in individual specifications. The pH measurements are made as previously described.

pH of Buffer Standard Solutions

For reagents that are suitable for use as pH standards, concentrations are stated in the individual specifications. They are expressed in the same units (molal) as the NIST standard reference materials to which they are compared. The meter and electrodes must be capable of a precision of 0.005 pH unit at the specified pH for satisfactory results.

Polarographic Analysis

Polarography offers a rapid and sensitive method of determining a number of ions or compounds that are electrolytically reducible, usually at a dropping mercury cathode, in a solution with appropriate electrical conductivity. Commercial instruments are designed to perform both original direct current (dc) polarographic methods and modern voltammetric extensions such as differential pulse polarography, square wave polarography, and mercury drop stripping analysis, which permit higher sensitivity and greater resolution. Instrumentation incorporates computer-controlled mercury drop electrodes designed to maintain the quality of ultrapure mercury.

dc Polarography

The measurements depend on obtaining the curve of current transported vs. applied potential as the latter is scanned in a negative direction. Typical apparatus consists of a reference half cell (usually an SCE), an electrolyte container or cell that can be de-aerated by bubbling with nitrogen or argon, the dropping mercury electrode, a source of uniformly increasing potential, means of converting the resulting current to a measurable signal, and a recorder with suitable ranges of sensitivity down to 1 μA full scale.

As the voltage applied to the cell rises, the residual current of the supporting electrolyte increases very gradually until the potential at the dropping electrode reaches the reduction potential of the most easily reduced species, ordinarily the substance under determination. At this point the current rises rapidly above the residual value to a limiting value dependent on the concentration of the electroactive species. The difference between the residual current and the limiting current comprises the diffusion current of the electroactive species. The resultant current–voltage curve, which is overall S-shaped, is called a *polarographic curve* or *polarogram*. The midpoint in the rise of the curve, one-half the distance between the residual current and the limiting current, is called the *half-wave potential* and is characteristic of the electroactive substance. Ultimately, as the applied voltage continues to rise, the current increases again as a result of the reduction of the supporting electrolyte.

Because the observed current is also a function of the size of the growing mercury drop at the cathode, the curve will show pronounced oscillations as the drops grow and fall. Most instruments incorporate means of damping out such oscillations. Reading of the current is simplified with the damped curve, but on some early-model instruments the potential readings may be shifted by up to 0.1 V from those observed without damping. A typical damped polarographic curve is shown in Figure 2-2. Calibration is performed with known solutions of the material to be determined.

Directions for individual reagent chemicals normally include appropriate means of treating the sample to convert it to a suitable electrolyte, free from oxygen and containing any needed pH buffer and maximum suppressor. *Maxima* are abnormally high current readings that may distort a polarogram—generally at the top of the step in the wave. Most maxima can be suppressed by gelatin, some indicators, or a surface-active agent. Use of too

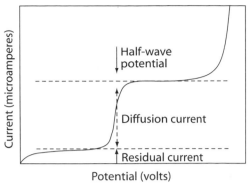

Figure 2-2. Typical dc polarographic curve.

much suppressor, however, may lower the diffusion current of the material being determined. Maxima can be avoided if the method is sensitive and the concentration of the substance being determined is low.

Typical substances that may be determined polarographically in this book include many metals (especially cadmium, copper, lead, and zinc), some anions (such as bromate, iodate, and nitrate), and several types of organic materials (such as aldehydes, ketones, and peroxides).

Differential Pulse Polarography

The detection limits in conventional dc polarography, typically 10^{-5} M, are set by the charging current resulting from the continuous growth of the mercury-drop electrode. One approach to improving the sensitivity of polarographic methods has been the development of pulse methods that reduce the contribution of charging current to the overall measured current. In pulse polarography, instead of applying a continuously increasing potential (as in dc polarography), a voltage pulse is applied to the mercury drop near the end of its drop time. The reason for this can be seen in Figure 2-3. The faradaic current, i_f (or that due to the reduction of the electroactive species), increases continuously during the time of the drop and is greatest at that moment just before the drop falls. The charging current (i_c), however, decreases steadily over the time of the drop. The ratio of faradaic to charging current (i.e., the sensitivity) can be optimized by sampling the current at the end of the drop time.

In differential pulse polarography, a voltage pulse, typically 50 ms, is applied to the electrode during the last portion of the drop's time (which is controlled by a mechanical drop dislodger and is commonly 0.5, 1.0, or 2.0 s). When the voltage is first applied, the charging current is very large, but it decays exponentially. The pulses have a constant amplitude of between 5 and 100 mV, and they are superimposed on a slowly increasing voltage ramp, as shown in Figure 2-4. The current is measured twice: immediately preceding the pulse and near the end of the drop time. The overall response output to the recorder is the difference in the two currents (Δi) sampled. The plot of Δi as a function of voltage is peak-shaped, as shown in Figure 2-5. By measuring the difference in current

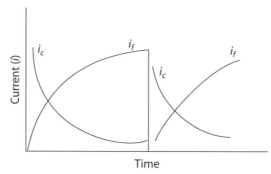

Figure 2-3. Comparison of faradaic current (i_f) and charging current (i_c) vs. time.

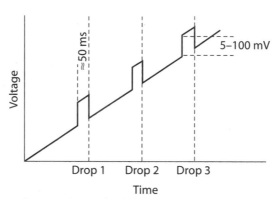

Figure 2-4. Voltage pulse excitation for differential pulse polarography.

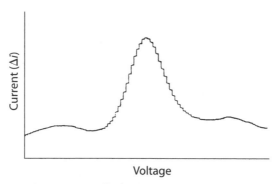

Figure 2-5. Differential pulse polarogram.

before the pulse and toward the end of the pulse, the contribution of the charging current to the overall measured current is significantly reduced. In addition, the backgrounds are flat, as opposed to the sloping backgrounds found in dc and some other polarographic techniques. Because the polarograms are peak-shaped, the limiting currents are easy to determine in differential pulse polarography. The reduction in charging current results in detection limits as low as 10^{-8} M.

Instruments suitable for differential pulse polarography and other voltammetric techniques are commercially available from several manufacturers. The instrument settings and conditions should be set in accordance with the manufacturer's recommendations, and these may not necessarily agree with those described in the next section.

Square Wave Polarography

In this method, a square wave with a typical pulse period of 5 ms (compared to 50 ms in differential pulse polarography) is superimposed on a voltage staircase having a step height of 10 mV. The polarographic signal produced by the pulse incorporates both forward and reverse currents; sensitivity is thus enhanced. In addition, the entire polarogram can be measured within the life of one mercury drop, as little as 0.5 s. This feature favors the use of square wave polarography for detection of electroactive species in columns for liquid chromatography.

Mercury Drop Stripping Analysis

An analyte from a dilute solution is concentrated with reproducible stirring for a fixed period into one drop of mercury by reduction at controlled potential. Stirring is stopped for a short equilibration interval, and the direction of the voltage sweep is reversed to measure the polarographic signal during oxidation. The metals such as cadmium, copper, lead, and zinc and other amalgamating metals attain sensitivities as high as 10^{-11} M.

Specific Polarographic Procedures

▶**Polarographic Procedure for Ammonium.** Transfer the specified amount of sample to each of two 25-mL volumetric flasks containing 5 mL of water. Neutralize, if necessary, with the appropriate quantity of ammonia-free 6 N sodium hydroxide (prepared by boiling and cooling 6 N sodium hydroxide solution). To one of the flasks, add the indicated quantity of ammonium ion. To each, add 5 mL of acetate buffer (to establish pH 4) and 5 mL of ammonia-free formaldehyde (both described below). Heat on a hot plate ($\approx 100\ ^\circ$C) for 5 min with occasional shaking, cool, and dilute to volume with water. Transfer a suitable portion to the cell, deoxygenate for 5 min with inert gases such as nitrogen or argon, and record the polarogram from –0.60 to –1.10 V vs. SCE. The following instrumental settings have proved to be satisfactory: drop time, 1 s; scan rate, 5 mV/s; sensitivity, 0.5 μA full scale; modulation amplitude, 50 mV. The peak for the ammonium–formaldehyde derivative occurs at about –0.85 V, but its position is dependent on pH.

The peak for the sample should not exceed one-half of the peak for the sample plus standard.

Sodium Acetate Buffer. Dissolve 10.4 g of sodium acetate trihydrate in 200 mL of water, and add 32 mL of glacial acetic acid.

Ammonia-Free Formaldehyde. Add 50 g of a cation-exchange resin in hydrogen form (e.g., Dowex HCR-W2, 50–100 mesh) to 250-mL bottle of 37% formaldehyde solution. Place the bottle in a shaker for 1 h, or let it stand for several hours and mix occasionally.

▶**Polarographic Procedure for Bromate and Iodate.** Add to the polarographic cell an appropriate volume of solution prepared as described in the specification. Deoxygenate for 5 min with inert gases such as nitrogen or argon, and record the polarogram from −0.6 V to −1.8 V vs. SCE. The following conditions are satisfactory: drop time, 1 s; scan rate, 5 mV/s; sensitivity, 0.5 μA full scale; modulation amplitude, 25 mV. The peak occurs at about −1.25 V for iodate and −1.6 V for bromate. If the ionic strength is not sufficiently high, the bromate wave becomes more negative and the cathodic wave may interfere. In this case, calcium chloride may be added, and a reagent blank correction may be applied.

The peak for the sample should not exceed one-half of the peak for the sample plus standard.

▶**Polarographic Procedure for Carbonyl Compounds.** To the specified quantity of sample in each of two 25-mL volumetric flasks, add 10 mL of phosphate buffer (described below) and 2 mL of 2% hydrazine sulfate reagent solution. To one of the flasks, add the indicated quantities of standard solutions. Dilute each to volume with water and mix. Promptly transfer a suitable portion to the cell, deoxygenate for 5 min with nitrogen, and record the polarogram from −0.6 V to −1.6 V vs. SCE. The following conditions are satisfactory: drop time, 1 s; scan rate, 5 mV/s; sensitivity, 0.5 μA full scale; modulation amplitude, 50 mV.

The peak heights for the sample should not be greater than one-half of the peak heights for the sample plus standards. Approximate peak potentials for the hydrazones of known carbonyl compounds are as follows: acetaldehyde, −1.1 V; acetone, −1.3 V; 2-butanone, −1.30 V; butyraldehyde, −1.20 V; formaldehyde, −1.0 V; propionaldehyde, −1.1 V. When formaldehyde alone is specified as the standard, the sum of the carbonyl peak heights in the sample should not be greater than the height due to added formaldeyde in the sample plus standard.

Phosphate Buffer. Dissolve 10.0 g each of monobasic sodium phosphate monohydrate and anhydrous dibasic sodium phosphate in water, and dilute with water to 500 mL.

▶**Polarographic Procedure for Lead (and Cadmium) in Zinc Compounds.** Prepare a solution with the prescribed quantities of sample and hydrochloric acid in water to make 25.0 mL. Prepare a standard with the same volume of hydrochloric acid and the specified quantities of lead (and cadmium) in 25.0 mL. Place 10 mL of either solution in the polarographic cell, deoxygenate with nitrogen for 5 min, and record the polarogram from −0.25 to −0.80 V vs. SCE. The following conditions are satisfactory: drop time, 1 s; scan rate, 5 mV/s; sensitivity, 0.5 μA full scale; modulation amplitude, 50 mV. Read the peak heights for lead at approximately −0.44 V and those for cadmium at approximately −0.63 V.

The peak(s) for the sample should not be greater than the peak(s) for the standard.

▶**Polarographic Procedure for Nitrite.** Dissolve the prescribed weight of sample in 20 mL of mixed reagent solution (described below), and dilute with water to 25 mL. Prepare a standard with the specified quantity of nitrite and 20 mL of reagent solution, also diluted to 25 mL. Place a suitable volume of either solution in the polarographic cell, deoxygenate for 5 min with nitrogen, and record the polarogram from −0.30 to −0.80 V vs. Ag/AgCl electrode. The following conditions are satisfactory: drop time, 0.5 s; scan rate, 10 mV/s; sensitivity, 0.2 µA full scale; modulation amplitude, 50 mV. Read the peak height at approximately −0.52 V. The peak for the sample should not be greater than that for the standard.

> *Mixed Reagent Solution.*   Prepare by adding 5 mL of solution A, 10 mL of solution B, and 10 mL of solution C to 100 mL of water in an amber bottle. A fresh mixture should be prepared each week.
>
> *Solution A.*   Dissolve 0.044 g of diphenylamine in 40 mL of methanol, and dilute to 100 mL with water.
>
> *Solution B.*   Use 0.1 N potassium thiocyanate.
>
> *Solution C.*   Dilute 7.0 mL of 60% perchloric acid to 250 mL with water.

Trace and Ultratrace Elemental Analysis

The determination of individual elements can be performed by a number of atomic spectroscopic (AS) techniques, which differ mainly in their sensitivity, selectivity, sample throughput capability, and cost. The main trace element techniques used in this book for measuring the purity of reagent chemicals include atomic absorption spectroscopy (AAS) and inductively coupled plasma optical emission spectroscopy (ICP–OES). Atomic absorption is recommended for most of the reagent chemicals, whereas plasma emission spectroscopy is used mainly for the multielement trace analysis of reagents. It should also be emphasized that many laboratories are now using inductively coupled plasma mass spectrometry (ICP–MS) for ultratrace metal determinations. Not only does this rapid multielement technique have much lower detection limits than the other AS techniques, but also it offers the exciting capability of isotopic measurements.

General Background

The majority of routine trace metal analysis today is carried out by either atomic absorption, atomic emission, or mass spectroscopic techniques. Flame atomic absorption (flame AAS), electrothermal atomic absorption (ETAA), cold vapor atomic absorption spectroscopy (CVAAS), hydride-generation atomic absorption spectroscopy (HGAAS), ICP–OES, and ICP–MS are the principal instrumental techniques used for trace metal analysis. In this book, AAS is recommended when the determination of a small number of elements in solution is required. Flame AAS is the most cost-effective analytical technique for the determination of common metallic impurities in most reagent chemicals, while ETAA is used when the detection capability of flame AAS is not acceptable. CVAAS is used for the determination of mercury, whereas HGAAS is used mainly for the volatile elements such as selenium. ICP–OES is usually the technique of choice when rapid, multielement analysis of solutions is required. The ultratrace reagents in this book are analyzed by ICP–OES for approximately 25 of the most laboratory-significant metallic impurities. Sample preparation typically involves dissolving and preconcentrating the chemicals by some kind of sample evaporation and/or matrix removal procedure before analysis. Interferences will be discussed later, but typically flame AAS has fewer interferences than ICP–OES.

The fundamental principles of atomic absorption, plasma optical emission, and plasma mass spectroscopy are well documented in the literature, so only the information pertaining to the specifications for the analyte elements of interest will be covered in this book. (See the bibliography of useful works on trace analysis beginning on page 774.) For that reason, specific instrumental parameters for each technique will not be given, because operating procedures will vary depending on the design of the instrument.

Reagents and Standards

Precautions should be taken to avoid contamination. Sample containers and glassware should be carefully cleaned and stored. The highest purity water should be used, obtained preferably from a mixed-bed strong acid, strong-base ion-exchange cartridge capable of producing water with an electrical resistivity of 18 MΩ-cm. It is strongly recommended that Teflon or ultraclean plasticware be used to avoid leaching of metals from laboratory glassware.

Accurate trace element standard solutions should be used for the preparation of working calibration standards. These are commercially available from the manufacturers of high-purity standards and should be traceable to NIST standards. Concentrations of 1–10 g/L are usually recommended for storage. However, if lower concentration standards are being used, these solutions should be discarded after 24 h because of the strong likelihood of the metals' being adsorbed by the walls of glassware—unless experience has shown that they are stable for a longer period of time. Alternatively, the appropriate plasticware should be used for long-term storage of these solutions.

The Ultratrace Environment

The analysis of reagent chemicals intended for use in ultratrace applications must be done under "clean air" conditions, since the limit of detection is very often dependent on the procedural "blank" obtained. The specifications used to determine the cleanliness of airborne particulates in a chemical laboratory is covered by ISO standard 14644, which recently replaced the older U.S. federal standard 209E. Conditions that conform to the air quality standards, as outlined in Class 1000 or better, where the airborne particle count is fewer than 1000 particles of 0.6 μm diameter per cubic centimeter of air, are generally considered sufficient for analysis of ultratrace reagents by flame AAS or ICP–OES. However, it is widely accepted that the more sensitive techniques, such as ETAA or ICP–MS, require Class 100 air quality or better (that is, an airborne particle count of fewer than 100 particles of 0.6 μm diameter per cubic centimeter of air). These ultraclean air conditions are usually achieved by the use of high-efficiency particulate filters in conjunction with proper laboratory design. Clean air conditions can be used in either an entire laboratory or in a laminar flow hood. Very often, in the absence of such facilities, clean air is supplied only to the specific area where contamination of a sample might occur. Extensive texts on these subjects are available, and the reader is strongly encouraged to examine these references before undertaking any ultratrace analysis.

Limit of Detection

The determination of the limit of detection (LOD) for ultratrace analysis should be done using recognized methodology, such as those used by the U.S. Environmental Protection Agency or the SEMI Standards organization. The SEMI method is a regression-based approach (SEMI, 2005), which uses the calibration data to determine the LOD; estimator software is available as a download at http://www.airproducts.com/products/specialtygases/northamerica/download.htm (accessed February 23, 2005). This guide, although applicable to many analytical techniques, such as gas and liquid chromatography, has been specifically adapted for trace metal analysis of semiconductor-grade chemicals of similar quality to the ultratrace reagents found in this book. Typical detection limits for the major AS techniques are shown in Table 2-9.

Table 2-9. Typical Atomic Spectroscopic Detection Limits (parts per billion or μg/L)

Elements	Flame AAS	CVAA/HGAA	ETAA	ICP–OES	ICP–MS
Aluminum (Al)	45	–	0.1	1	0.005
Antimony (Sb)	45	0.15	0.05	2	0.0009
Arsenic (As)	150	0.03	0.05	2	0.0006
Barium (Ba)	15	–	0.35	0.03	0.00002
Beryllium (Be)	1.5	–	0.008	0.09	0.003
Bismuth (Bi)	30	0.03	0.05	1	0.0006
Boron (B)	1,000	–	20	1	0.003
Bromine (Br)	–	–	–	–	0.2
Cadmium (Cd)	0.8	–	0.002	0.1	0.00009
Calcium (Ca)	1.5	–	0.01	0.05	0.0002
Carbon (C)	–	–	–	–	0.8
Cerium (Ce)	–	–	–	1.5	0.0002
Cesium (Cs)	15	–	–	–	0.0003
Chlorine (Cl)	–	–	–	–	12
Chromium (Cr)	3	–	0.004	0.2	0.0002
Cobalt (Co)	9	–	0.15	0.2	0.0009
Copper (Cu)	1.5	–	0.014	0.4	0.0002
Dysprosium (Dy)	50	–	–	0.5	0.0001
Erbium (Er)	60	–	–	0.5	0.0001
Europium (Eu)	30	–	–	0.2	0.00009
Fluorine (F)	–	–	–	–	372
Gallium (Ga)	75	–	–	1.5	0.0002
Gadolinium (Gd)	1,800	–	–	0.9	0.0008
Germanium (Ge)	300	–	–	1	0.001
Gold (Au)	9	–	0.15	1	0.0009
Hafnium (Hf)	300	–	–	0.5	0.0008
Holmium (Ho)	60	–	–	0.4	0.00006
Indium (In)	30	–	–	1	0.0007
Iodine (I)	–	–	–	–	0.002
Iridium (Ir)	900	–	3.0	1	0.001
Iron (Fe)	5	–	0.06	0.1	0.0003
Lanthanum (La)	3,000	–	–	0.4	0.0009

Continued on next page

Table 2-9. *(continued)*

Elements	Flame AAS	CVAA/HGAA	ETAA	ICP–OES	ICP–MS
Lead (Pb)	15	–	0.05	1	0.00004
Lithium (Li)	0.8	–	0.06	0.3	0.001
Lutetium (Lu)	1,000	–	–	0.1	0.00005
Magnesium (Mg)	0.15	–	0.004	0.04	0.0003
Manganese (Mn)	1.5	–	0.005	0.1	0.00007
Mercury (Hg)	300	0.009	0.6	1	0.016
Molybdenum (Mo)	45	–	0.03	0.5	0.001
Neodymium (Nd)	1,500	–	–	2	0.0004
Nickel (Ni)	6	–	0.07	0.5	0.0004
Niobium (Nb)	1,500	–	–	1	0.0006
Osmium (Os)	–	–	–	6	–
Palladium (Pd)	30	–	0.09	2	0.0005
Phosphorus (P)	75,000	–	130	4	0.1
Platinum (Pt)	60	–	2.0	1	0.002
Potassium (K)	3	–	0.005	1	0.0002
Praseodymium (Pr)	7,500	–	–	2	0.00009
Rhenium (Re)	750	–	–	0.5	0.0003
Rhodium (Rh)	6	–	–	5	0.0002
Rubidium (Rb)	3	–	0.03	5	0.0004
Ruthenium (Ru)	100	–	1.0	1	0.0002
Samarium (Sm)	3,000	–	–	2	0.0002
Scandium (Sc)	30	–	–	0.1	0.004
Selenium (Se)	100	0.03	0.05	4	0.0007
Silicon (Si)	90	–	1.0	10	0.03
Silver (Ag)	1.5	–	0.005	0.6	0.002
Sodium (Na)	0.3	–	0.005	0.5	0.0003
Strontium (Sr)	3	–	0.025	0.05	0.00002
Sulfur (S)	–	–	–	10	28
Tantalum (Ta)	1,500	–	–	1	0.0005
Terbium (Tb)	900	–	–	2	0.00004
Tellurium (Te)	30	0.03	0.1	2	0.0008
Thorium (Th)	–	–	–	2	0.0004
Tin (Sn)	150	–	0.1	2	0.0005
Titanium (Ti)	75	–	0.35	0.4	0.003
Thallium (Tl)	15	–	0.1	2	0.0002
Thulium (Tm)	15	–	–	0.6	0.00006
Tungsten (W)	1,500	–	–	1	0.005
Uranium (U)	15,000	–	–	10	0.0001
Vanadium (V)	60	–	0.1	0.5	0.0005
Ytterbium (Yb)	8	–	–	0.1	0.0002
Yttrium (Y)	75	–	–	0.2	0.0002
Zinc (Zn)	1.5	–	0.02	0.2	0.0003
Zirconium (Zr)	450	–	–	0.5	0.0003

Note: – indicates that the technique is not optimal for that particular element.

Source: Reproduced with permission from PerkinElmer, 2004. Copyright 2004 PerkinElmer LAS.

Atomic Absorption Spectroscopy

AAS is used when accurate determinations of only a few elements in solution are required. AAS has been chosen as the most cost-effective analytical technique for trace metal analysis of many reagents contained in this book, because it is a very sensitive technique capable of detection in the parts-per-million range with flame atomization. However, if lower levels are encountered, then electrothermal atomization should be employed, which is capable of measurement at the parts-per-billion level.

Flame AAS is predominantly a single-element technique that uses a flame, such as from air–acetylene or nitrous oxide–acetylene, to generate ground-state atoms of the element of interest. The liquid sample is aspirated via a nebulizer and a spray chamber into the flame, where desolvation takes place and ground-state atoms are formed. The ground-state atoms in the sample absorb light of a specific wavelength from an element-specific, hollow cathode lamp source. The amount of light absorbed, which is directly related to the number of ground-state atoms, is measured by a monochromator (optical system) and detected by a photomultiplier tube or solid-state detector, which converts the photons into an electrical pulse. The sample absorbance signal for a particular element is compared to known calibration standards in order to determine the concentration of that element in the sample. Flame AAS typically uses about 2–5 mL/min of liquid sample.

ETAA—also known as graphite furnace atomic absorption, or GFAA—is also a single-element technique, although multielement instrumentation is now available. It works on the same principle as flame AAS, except that the flame is replaced by a small heated tungsten filament or graphite tube. The other major difference is that a very small sample (typically 50 µL) is injected automatically onto the filament or into the tube, rather than aspirated via a nebulizer and a spray chamber as in flame AAS. Because the ground-state atoms are concentrated in a smaller area than a flame, a higher degree of absorption takes place. The result is that ETAA offers about 100 to 1000 times lower detection limits than flame AAS. Although ETAA is widely used by the ultratrace element community, no methodology is presented in this book because flame AAS is sensitive enough to meet most of the reagent chemical specifications. However, if the ETAA technique is available, there is no reason why it cannot be used, assuming that the correct sample preparation and instrument methodology is applied.

Conventional flame AAS spectrometers are equipped with a turret to accommodate multiple hollow cathode lamp sources. The burner is slotted and will vary in design and path length depending on the gases used. The monochromator will have a resolution of at least 0.1 nm. The signal from the photomultiplier tube detector is processed by a microprocessor and displayed on a digital readout or a terminal screen. To compensate for nonspecific absorption in the flame, a deuterium lamp background corrector or equivalent is used.

In this book, typical instrument parameters, such as flame type, gas flows, absorption wavelengths, slit widths, and background correction methods are given as guidelines in the procedures for individual reagents. However, the user should always refer to the manufacturers' operator manual or application "cookbook" in order to achieve optimized conditions for a particular determination.

Sensitivity and Detection Limits

Sensitivity in AAS is defined as the concentration of a test element in an aqueous solution that will produce absorption of 1% or 0.0044 absorbance units (AU). It is normally expressed in μg/mL or μg/g per 1% absorption. Note that this expression is related to the slope of the calibration curve. The sensitivity of the instrument is normally optimized during the method setup. A tabulation of expected absorbance vs. standard concentration is shown in Table 2-10. There are many different definitions and procedures for determining detection limits, depending on the data quality objectives of the analysis. The detection limit is often referred to as signal-to-background noise, and for a 99% confidence level, it is typically defined as three times the standard deviation of 10 replicates of the calibration blank (expressed in concentration units). To achieve the best detection limits, it is generally recognized that instrumental parameters will need to be optimized, based on the application requirements.

Instrumentation

The manufacturer's instruction manual should be followed for detailed operating procedures and for matters of routine maintenance, cleaning, and calibration. The analyst should be familiar with the literature available in the public domain and be aware of the various interferences and potential sources of error. The procedures for the analysis of individual reagents in this book attempt to take into account the major interferences as much as possible. However, the analyst should always be aware of these interferences and ensure that the instrument being used is compensating for them.

Interferences

The major interferences in atomic absorption spectrometry include

- *Sample transport:* Surface tension, viscosity, acid content, and amount of dissolved solids may affect the nebulization rate and influence the sensitivity of an instrument by altering the concentration of the absorbing analyte in the flame. This is not such a serious problem in flame AAS, but if present, it can be compensated for by diluting the sample, matching matrix/acid content, using the method of standard additions, or monitoring an internal standard element.
- *Chemical:* With some samples, the analyte forms a strong chemical bond with other matrix components in the sample. This might necessitate the use of a much hotter nitrous oxide–acetylene flame or the addition of a complexing or releasing agent to change the chemistry of the analyte and/or the matrix.
- *Ionization:* Some of the easily ionizable elements, such as sodium and potassium, readily ionize in a flame, which affects the number of ground-state atoms formed. This effect results in anomalous curvature of the calibration curve and can produce either positive or negative errors in the results. The addition of ionization buffers can minimize this effect.

Table 2-10. Expected Absorbance Values vs. Standard Concentrations for a Group of Analytes

Element/Analyte	Wavelength (nm)	Sensitivity Check[a] (mg/L)	Linear Range[b] (mg/L)	Minimum[c] (mg/L)	Minimum (mg/25 mL)	Maximum[d] (mg/25 mL)	Recommended Standard (mg/25 mL)	Expected AU
Aluminum	396.2	15.0	100.0	1.5	0.038	2.50	0.25–0.50	0.13–0.26
Antimony	217.6	15.0	100.0	1.5	0.038	2.50	0.10–0.20	0.07–0.14
Barium	553.6	10.0	50.0	1.0	0.025	1.25	0.10–0.20	0.06–0.12
Bismuth	223.1	10.0	50.0	1.0	0.025	1.25	0.10–0.20	0.12–0.24
Cadmium	228.8	0.5	3.0	0.05	0.0013	0.075	0.01–0.02	0.18–0.36
Calcium	422.7	0.5	3.0	0.05	0.0013	0.075	0.02–0.04	0.25–0.50
Cobalt	240.7	2.5	15.0	0.25	0.0063	0.375	0.02–0.04	0.08–0.16
Copper	324.8	1.5	10.0	0.15	0.0038	0.250	0.02–0.04	0.11–0.22
Iron	248.3	2.5	15.0	0.25	0.0063	0.375	0.05–0.10	0.20–0.40
Lead	217.0	5.0	30.0	0.50	0.013	0.750	0.10–0.20	0.16–0.32
Lithium	670.8	1.0	5.0	0.10	0.0025	0.125	0.01–0.02	0.11–0.22
Magnesium	285.2	0.15	1.0	0.015	0.0004	0.025	0.005–0.01	0.25–0.50
Manganese	279.5	1.0	5.0	0.10	0.0025	0.125	0.02–0.04	0.22–0.44
Molybdenum	313.3	15.0	100.0	1.5	0.038	2.50	0.25–0.50	0.13–0.26
Nickel	232.0	4.0	20.0	0.40	0.01	0.50	0.05–0.10	0.18–0.36
Potassium	766.5	0.4	2.0	0.04	0.001	0.050	0.01–0.02	0.20–0.40
Silver	328.1	1.5	10.0	0.15	0.0038	0.250	0.02–0.04	0.11–0.22
Sodium	589.0	0.15	1.0	0.015	0.0004	0.025	0.005–0.01	0.25–0.50
Strontium	460.7	2.0	10.0	0.20	0.005	0.250	0.05–0.10	0.35–0.70
Zinc	213.9	0.4	2.0	0.04	0.001	0.050	0.01–0.02	0.20–0.40

Notes: The data are based on instrument-specific conditions. In some instances, the recommended standard addition is lower than cited in this table. Dilution may be modified to meet maximum allowable limits.

[a] Sensitivity check is the concentration giving approximately 0.2 AU.

[b] Linear range is the upper concentration of linear range.

[c] Minimum is the concentration giving 0.02 AU (sensitivity check divided by 10).

[d] Maximum is the upper limit in mg per 25 mL (linear range divided by 40).

■ *Molecular absorption:* Molecular absorption at the resonance wavelength or by light scattering is caused by very small, unvolatilized particles in the flame. This type of interference becomes appreciable at 250–300 nm and increasingly severe at lower wavelengths. Scattering of radiant energy from particles in the flame is usually negligible above 300 nm or at sample matrix concentrations on the order of 10 g/L or less. Molecular absorption is usually compensated for by some kind of background correction technique. For an element with a resonance line in the ultraviolet region (<350 nm), the correction can be performed with continuum source background correction, using a deuterium arc lamp. For an element with a resonance line in the visible region (>350 nm), an alternative approach known as the "adjacent-line" technique needs to be used. In this method, the background absorption is determined at a wavelength adjacent to or close to the analyte resonance line, so the background absorption is measured at a wavelength where there is no significant atomic absorption by that element. Continuum source background correction is usually sufficient for flame AAS work, while an adjacent-line technique, such as Zeeman background correction, is preferred for ETAA work, where high levels of structured background are sometimes encountered.

Analysis

In this book, the predominant approach used for trace metal analysis of chemical reagents in solution is the method of standard additions. This approach is used to ensure that the matrix has a minimal effect on the result. The method of standard additions uses the principle of making up the calibration standards in the same matrix as the sample by spiking the sample with fixed concentrations of analytes, corresponding to the number of points on the calibration curve. For standard additions to be applicable at low concentrations, it is essential that the sample solution, along with the spiked additions, generate absorbance signals in the linear part of the calibration curve, which is usually on the order of 0.01–0.50 AU. The method of standard additions does not eliminate the interference, but it does ensure that the elements of interest behave in a similar manner in both the sample and the calibration standards. However, if it is known that the chemical (matrix) effects are negligible, a calibration line using aqueous standards can be used when the slope and the slope of the standard-addition calibration are identical.

As indicated above, reagent-specific conditions for AAS analysis are listed under the individual reagents. This information includes, but is not limited to, wavelength, sample preparation, standard-addition methodology, flame type, and background correction.

▶**Procedure for Flame AAS.** Prepare the sample for analysis by dissolving in water or by following a reagent-specific procedure. Prepare four solutions for standard-addition calibration: the reagent blank (which may be only water), the sample solution, and two standard-addition solutions. The recommended additions usually correspond to the specification limit and half the limit, although in some cases the limit and twice the limit are used. The signal for the reagent blank preparation can be subtracted from the signals of the three other solutions, or the auto-zero feature of the instrument can be employed. The three results can be treated graphically or mathematically to determine the analyte concentration in the sample. Most modern instruments have standard-addition software that will calculate this automatically. When several elements are to be determined, they can be added simultaneously to the standard-addition solutions. A separate set of standard additions may be required if different sample weights are used in order to stay in the linear absorbance range.

Example. Figure 2-6 shows a typical procedure, as it appears in this book, that could be used for trace or ultratrace elemental analysis of a reagent chemical, using the method of standard additions, which is illustrated here.

To a first set of three 25-mL volumetric flasks, add 2.0 mL (0.20 g) of sample stock solution. To two of the flasks, add the specified amounts of calcium and sodium shown in the table in Figure 2-6. Add 2 mL of 5% potassium chloride solution to all three flasks, and dilute to the mark with deionized water.

To a second set of three 25-mL volumetric flasks, add 10.0 mL (1.0 g) of sample stock solution. To two of the flasks, add the specified amounts of potassium and strontium shown in the table in Figure 2-6. Dilute the contents of the three flasks to the mark with deionized water.

To a third set of three 25-mL volumetric flasks, add 20.0 mL (2.0 g) of sample stock solution. To two of the flasks, add the specified amounts of manganese shown in the table in Figure 2-6. Dilute the contents of the three flasks to the mark with deionized water.

Analyze the solutions by means of a suitable atomic absorption spectrophotometer, using the conditions outlined in the table in Figure 2-6. Calculate the trace metal content of the sample by the method of standard additions.

SPECIFICATIONS

Maximum Allowable

Potassium (K) 0.005%
Sodium (Na) 0.005%
Strontium (Sr). 0.005%

TESTS

Potassium, Sodium, and Strontium. (By flame AAS, page 63).

> **Sample Stock Solution.** Dissolve 10.0 g of sample with water in a 100-mL volumetric flask and dilute to the mark with deionized water (1 mL = 0.10 g).

Element	Wavelength (nm)	Sample Wt (g)	Standard Added (mg)	Flame Type*	Background Correction
K	766.5	1.0	0.025; 0.05	A/A	No
Na	589.0	0.20	0.005; 0.01	A/A	No
Sr	460.7	1.0	0.05; 0.10	N/A	No

*A/A is air/acetylene; N/A is nitrous oxide/acetylene.

Figure 2-6. Sample specifications and tests for procedure using flame AAS.

Mercury Determination by CVAAS

The ultratrace determination of mercury in aqueous solution by atomic absorption is based on its reduction by a suitable reducing agent to the free element and transportation of the vaporized form of the element via a stream of inert gas into a quartz cell at room temperature (instead of the burner head used in flame AAS). A hollow cathode lamp can be used, but to maximize sensitivity a more intense mercury electrodeless discharge lamp is used to generate wavelength-specific photons of light. The mercury ground-state atoms in the quartz cell then absorb the wavelength-specific light emitted by the electrodeless discharge lamp. Calibration is carried out in the normal way by comparing the absorbance of the sample with known mercury calibration standards. (The method of standard additions can also be used.) Commercially available mercury analyzers based on atomic fluorescence or amalgamation with a gold film may be substituted if the test method is validated. Care must be taken to guard against contamination of samples and solutions by mercury. Glassware should be washed with nitric acid $(1 + 3)$, followed by thorough rinsing with water before analysis. However, dedicated plasticware or Teflon is strongly recommended for this type of analysis.

▶**Procedure for Mercury Determination by CVAAS.** Some details of sample preparation, including the use of reagent blanks, are described in the specific test for each reagent. Carry out all preparations in a fume hood. Samples are normally prepared in a set of three 100-mL volumetric flasks containing sample sizes described in the individual reagent monograph. Add mercury ion (Hg) to the second and third flasks. Use ultrapure reagents, including ultrapure water. If necessary, adjust sample weight and final volume according to the sensitivity of the mercury analyzer system. The manufacturer's manual may suggest sensitivity or general guidelines.

Transfer the sample solution, or an aliquot thereof, to the aeration vessel before adding any reducing agent. If the sample has been treated with permanganate, rinse the container with the specified quantity of hydroxylamine hydrochloride reagent solution, and add the rinsings to the aeration vessel with enough water to bring the level to the calibration line. Adjust the instrument to give a smooth baseline by flowing carrier gas through the bypass and cell and focusing the radiation of the 253.7-nm mercury resonance line through the cell onto the detector. Introduce stannous chloride solution as specified into the aeration vessel (generally 2.5% stannous chloride reagent solution, but follow the procedure recommended by the manufacturer's manual), and set the valve to direct the carrier gas through the vessel and the cell. Allow the bubbling to continue until the instrument returns to its zero point. Before running the next sample, drain and rinse the aeration vessel. Periodically remove any deposit of stannic oxide.

Selenium Determination by HGAAS

HGAAS is commonly used for trace-level analysis of hydride-forming elements such as antimony, arsenic, and selenium because it gives better sensitivity for these elements than flame AAS. Selenium analysis, for example, using conventional flame AAS is not sensitive because its analytical wavelength (196.0 nm) is too close to the vacuum ultraviolet region where absorption by air and flame gases occurs. In this book, HGAAS is used for the determination of selenium in ultratrace reagents and is based on selenium's conversion to a volatile hydride (SeH_2) by the sodium borohydride reagent and aspiration into an atomic absorption spectrometer fitted with a heated quartz atomization cell.

Equipment

AAS systems equipped with a hydride reaction cell and an externally mounted, heated quartz cell or a quartz cell with an internal fuel-rich oxygen–hydrogen or air–hydrogen flame are recommended for this analysis.

Sensitivity and Interferences

Quartz atomization cells are required for ultratrace analysis because they minimize background noise caused by the flame. The formation or presence of chlorine during the hydride generation will result in low values due to the reoxidation of Se^{IV} to Se^{VI}. Only Se^{IV} readily forms a hydride using sodium borohydride.

Analysis

Selenium determination is carried out using the method of standard additions. The reagent-specific conditions can be found under the individual reagent. Selenium is determined at the 196.0-nm wavelength and depends on the instrument setup; 1.0 µg of selenium in 100-mL volume gives approximately 0.12 AU.

▶ **Procedure for Selenium Determination by HGAAS.** Prepare the sample for analysis by following the reagent-specific procedure. The method of standard additions is the preferred analytical approach for selenium in order to minimize chemical or matrix-induced interferences. However, for this method to be applicable at low concentrations, it is essential that the sample solution, along with the spiked additions, generate selenium absorbance signals in the linear part of the calibration curve. It should also be emphasized that the method of standard additions does not eliminate the interference, but it does ensure that the selenium behaves in a similar manner in both the sample and the calibration standards. However, if it is known that the chemical (matrix) effects are negligible, a calibration line using aqueous standards can be used instead of the standard addition calibration curve. The specific standard additions for selenium can be found under the individual reagents.

Plasma Emission Spectroscopy

Optical emission spectroscopy using inductively coupled plasma as the excitation source is an extremely rapid technique to carry out multielement analysis in various sample matrices. Whereas AA is used to determine small numbers of elements in solution, ICP–OES is useful for the determination of many elements or when high sample throughput is required. In order to achieve the specifications for ultratrace reagents in this book, samples need to be preconcentrated by evaporation under clean-air conditions and made up to volume using 1% nitric acid before analysis by ICP–OES.

ICP–OES is a multielement technique that uses an inductively coupled plasma to ionize and excite ground-state atoms to the point where they emit wavelength-specific photons of light, characteristic of a particular element. The emitted photons are focused and measured by a high-resolving optical system and a photosensitive detector. This emission signal is directly related to the concentration of that element in the sample. The analytical temperature of an ICP is about 6000–7000 K, compared to a flame, which is typically 2500–4000 K. Detection limits for ICP–OES are generally similar to flame AAS for the majority of elements, but are much better for refractory and rare earth elements because of the higher excitation temperature. Sample volume requirements are on the order of 1 mL/min.

The inductively coupled plasma is created in a quartz torch, which consists of three concentric tubes: the outer tube, the middle tube, and the sample injector. The torch can be either one piece, in which all three tubes are connected, or a demountable design, in which the tubes and the sample injector are separate. Argon gas (known as the *plasma gas*) is passed between the outer and middle tubes at a flow rate of ~12–17 L/min. A second argon gas flow (*auxiliary gas*) passes between the middle tube and the sample injector at ~1 L/min and is used to change the position of the base of the plasma relative to the tube and the injector. A third gas flow (*nebulizer gas*), also at ~1 L/min, brings the sample, in the form of a fine-droplet aerosol, from the sample introduction system (nebulizer and spray chamber) and physically punches a channel through the center of the plasma, where the excitation takes place. The sample injector is often made from inert materials, such as alumina, if highly corrosive materials are being analyzed.

The principles of plasma formation are well documented in the literature. First, a tangential (spiral) flow of argon gas is directed between the outer and middle tube of a quartz torch. A load coil (usually copper) surrounds the top end of the torch and is connected to a radio frequency (RF) generator. When RF power (typically 750–1500 watts, depending on the sample) is applied to the load coil, an alternating current oscillates within the coil at a rate corresponding to the frequency of the generator, which in most commercial instruments is either 27 or 40 MHz. This RF oscillation of the current in the coil causes an intense electromagnetic field to be created in the area at the top of the torch. With argon gas flowing through the torch, a high-voltage spark is applied to the gas, causing some electrons to be stripped from their argon atoms. These electrons, which are caught up and accelerated in the magnetic field, then collide with other argon atoms, stripping off still more electrons. This collision-induced ionization of the argon continues in a chain reaction, breaking down the gas into argon atoms, argon ions, and electrons, forming the ICP discharge. The amount of energy required to generate argon ions in this process is on the order of 15.8 eV, which is enough energy to ionize and excite more than 70 of the elements in the periodic table that produce emission spectra mainly in the visible region (200–900 nm). However, spectrometers are now available that extend the wavelength coverage down into the low ultraviolet range (120–190 nm), where elements such as sulfur, phosphorus, aluminum, and the halogens can be determined.

The emission produced from the plasma excitation process is then focused onto the entrance slit of a high-resolution optical system. Traditionally, these optics have either been based on a scanning monochromator (also known as a sequential system) or a dispersing polychromator (also known as a simultaneous system). The sequential system typically provides greater wavelength selection and flexibility, while the simultaneous system allows higher sample throughput. In both cases, very sensitive photomultiplier tubes are used as the detector to convert the photons into an electrical signal. More recently, spectrometers that use charge-coupled device or charge-injected device solid-state detectors have become commercially available. These instruments are usually equipped with high-resolution Echelle gratings that disperse the spectrum into two dimensions. The benefit of coupling solid-state detection to high-resolution, two-dimensional spectrometers is that virtually any wavelength can be accessed very quickly, which makes them very flexible and extremely fast.

In addition to the different optical designs, it should also be noted that there are two main types of ICP viewing configuration available: a conventional radial design, where the torch is positioned vertically (sometimes known as *side-on viewing*), and an axial configuration, where the torch is positioned horizontally (sometimes known as *end-on viewing*). Radial-view ICP can typically achieve similar detection limits to flame AAS, for the majority of elements (with the exception of the refractory and rare earth elements). The benefit of the axial design is that more photons are seen by the detector, and as a result, detection limits can be as much as 5 to 10 times lower, depending on the design of the instrument. However, the disadvantage of the axial approach is that more severe matrix interferences are observed, which means that matrix matching can be critical and, with some complex matrices, internal standardization is often required.

Interferences, Detection Limits, and Sensitivity

Interference effects, limits of detection, sensitivity, and linear dynamic range must be investigated and established for each individual analyte line on the particular instrument being used in analysis of the individual reagent. In general, ICP–OES is mostly prone to spectral-type overlaps and matrix suppression effects produced by other matrix components in the sample. These types of interferences must be fully investigated in the method-development stage, before accurate multielement determinations can be carried out. Refer to the manufacturer's application-specific methodology along with reference literature for sensitivity, detection limits, and linear dynamic range of the emission lines being used for the analysis. Special attention must be paid to the possibility of external contamination during sample preparation and analysis, especially if ultratrace determinations are being carried out.

Instrumentation

A grating instrument with sufficient resolution to separate the relevant spectral emission lines of the elements of interest for individual reagents is required. No specific procedure is given because operating details will vary with instrument design.

Analysis

Multielement standard solutions should be prepared on the day of analysis by dilution of a stock standard solution. They should be matrix-matched to the sample solution, which is typically in the order of 1% nitric acid. The elemental components should be selected based on the chemistry of the individual elements to provide stable solutions for the required concentration ranges. Analyze the samples versus matrix-matched calibration standards. If a new or unusual matrix is encountered, the method of standard additions should be used. However, the standard addition will not detect coincident spectral overlap. If this overlap occurs, an alternate wavelength or method is recommended for verification. The suggested analytical lines for determination of the various metals in ultratrace reagents

Table 2-11. Suggested Plasma Emission Lines

Element	Wavelength (nm)	Element	Wavelength (nm)
Aluminum (Al)	396.15	Manganese (Mn)	257.60
Barium (Ba)	455.40	Molybdenum (Mo)	202.02
Boron (B)	249.77	Potassium (K)	766.49
Cadmium (Cd)	214.44	Silicon (Si)	212.41
Calcium (Ca)	393.27	Sodium (Na)	589.59
Chromium (Cr)	205.55	Strontium (Sr)	407.77
Cobalt (Co)	228.63	Tin (Sn)	189.99
Copper (Cu)	324.75	Titanium (Ti)	334.94
Iron (Fe)	238.20	Vanadium (V)	292.40
Lead (Pb)	220.35	Zinc (Zn)	213.86
Lithium (Li)	670.78	Zirconium (Zr)	339.20
Magnesium (Mg)	279.55		

are shown in Table 2-11. Other analytical lines may be used, depending on the sensitivity requirements.

▶**Procedure for ICP–OES.** The achievable detection limits can be lowered by either preconcentrating the sample using clean-air evaporation techniques and/or by matrix removal using some kind of matrix-specific ion-exchange chemical procedure. Reliability and accuracy can be improved by the use of a blank and the analysis of a control sample. The sample size will depend on the detection limits required and the capability of the instrument used. A spectral background correction technique, such as interelement correction or multicomponent spectral subtraction, might be required, especially with complex samples. Refer to the analytical procedure for the individual reagent for any additions to the sample and blank solutions. Check the performance of the instrument on a regular basis to ensure that adequate sensitivity is being obtained. When in doubt, consult the operator's manual.

The following sample preparation has given satisfactory results for the appropriate ultratrace reagents in this book. Special sampling conditions can be found under the trace metal test for individual reagents.

Sample Preparation and Analysis. In a clean-air environment, place 100 g of sample in one 100-mL PTFE (or equivalent) evaporating dish and a reagent blank in another. Slowly evaporate on a hot plate ($\approx$100 °C), avoiding loss of sample by effervescence or spattering, until approximately one drop of liquid remains. Do not allow the sample to evaporate to complete dryness. Cool, transfer quantitatively to individual volumetric flasks using a 1% solution of ultratrace-grade nitric acid for rinsing and dilution to volume. Analyze the reagent sample by ICP–OES, using suitable calibration standards and a reagent blank.

ICP–MS

Since it was first developed more than 20 years ago, ICP mass spectrometry (ICP–MS) has enabled trace metal detection capability at the parts per trillion (ppt) level. Even though ICP–MS can broadly determine the same suite of elements as the other atomic spectroscopic techniques, it has clear advantages in its multielement characteristics, speed of analysis, detection limits, and isotopic capability. Although this book does not specify the use of ICP–MS, it is acceptable to use this technique for the ultratrace reagents as long as method validation has been done.

Several different ICP–MS designs are available today. They share many components—such as the nebulizer, spray chamber, plasma torch, and detector—but can differ quite significantly in the design of the interface, ion focusing system, mass separation device, and vacuum chamber. As in ICP–OES, the sample is delivered into a nebulizer, where it is converted into a fine aerosol with argon gas. The fine droplets of the aerosol are separated from larger droplets by means of a spray chamber. The fine aerosol then emerges from the spray chamber and is transported into the plasma torch via a sample injector.

The plasma torch in ICP–MS functions differently from in ICP–OES. The plasma is formed in a similar way, but this is where the similarity ends. In ICP–OES, the plasma is used to generate photons of light, whereas in ICP–MS, the plasma torch is used to generate positively charged ions. The production and detection of large quantities of positively charged ions gives ICP–MS its characteristic low-parts-per-trillion detection capability—about three or four orders of magnitude better than ICP–OES.

Once the ions are produced in the plasma, they are directed into the mass spectrometer via the interface region, which is maintained at a vacuum of 1–2 torr with a mechanical pump. This interface region consists of two metallic cones called the *sampler* and the *skimmer cone*, each with a small orifice (0.6–1.2 mm) to allow the ions to pass through to the ion optics, where they are guided into the mass separation device.

After the ions have been successfully extracted from the interface region, they are directed into the main vacuum chamber by a series of electrostatic lenses, called *ion optics*. The operating vacuum in this region is maintained at about 10^{-3} torr with a turbomolecular pump. There are many different designs of the ion optic region, but they serve the same function: to electrostatically focus the ion beam toward the mass separation device while stopping photons, particulates, and neutral species from reaching the detector.

The ion beam containing all the analyte and matrix ions exits the ion optics and passes into the mass separation device, which is maintained at a vacuum of 10^{-6} torr with a second turbomolecular pump. The many mass separation devices include quadrupole, magnetic sector, time-of-flight, and collision/reaction cell technology. Their principles of operation differ, but they all allow analyte ions of a particular mass-to-charge ratio to pass through the detector and reject all the nonanalyte ions, interfering species, and matrix components.

The last step is converting the ions into an electrical signal with an ion detector. The most common design today is called a *discrete dynode detector*, which contains a series of metal dynodes along the length of the detector. In this design, when the ions emerge from the mass filter, they impinge on the first dynode and are converted into electrons. As the electrons are attracted to the next dynode, electron multiplication takes place, which results in a very high stream of electrons emerging from the final dynode. This electronic signal is then processed by the data-handling system and converted into analyte concentration using ICP–MS calibration standards. Most detection systems used can handle up to eight orders of dynamic range, which means they can be used to analyze samples from the parts-per-trillion level up to a few hundred parts per million.

Interferences

Although the detection capability of ICP–MS is significantly better than that of other AS techniques, it is prone to certain interferences. Although these interferences are reasonably well understood, it can often be difficult and time-consuming to compensate for them, particularly in complex sample matrices. Having prior knowledge of the interferences associated with a particular set of samples will often dictate the sample preparation steps and the instrumental methodology used to analyze them. Interferences are generally classified into two major groups, spectral-based and matrix-based. Each has the potential to be problematic, but modern instrumentation and good software, combined with optimized analytical methodologies, have minimized their negative impact on trace element determinations by ICP–MS.

- *Spectral interferences.* These are probably the most serious type of interferences in ICP–MS. The most common ones are known as polyatomic or molecular spectral interferences, which are produced by the combination of two or more atomic ions. They are caused by a variety of factors but are usually associated with the argon gas, matrix components in the solvent or sample, other elements in the sample, or entrained oxygen or nitrogen from the surrounding air. Simple polyatomic interferences are handled using mathematical correction equations, whereas severe ones have to be compensated for by using either cool plasma conditions, collision/reaction cells, or in extreme cases, by using a high-resolution spectrometer.

- *Matrix Interferences.* These are classified into three different categories. The first and simplest to overcome is a sample transport effect and is a physical suppression of the analyte signal, brought on by the level of dissolved solids or acid concentration in the sample. The second type of matrix suppression is caused when the sample affects the ionization conditions in the plasma discharge. This results in the signal being suppressed by varying amounts, depending on the concentration of the matrix components. The third type of matrix interference, called a space-charge effect, is caused by suppression of a high-mass matrix element on the transmission of a lower mass analyte ion through the ion optics. The susceptibility of ICP–MS to matrix-induced interferences means that internal standardization is nearly always used for the quantitation of unknown samples.

Instrumentation and Methodology

An instrument with sufficient resolving power to separate the mass spectrum of the elements of interest is required. No specific methodology is provided here since operating procedure will vary depending on whether the instrument design is based on quadrupole, collision/reaction cell, double focusing magnetic sector, or time-of-flight technology. The manufacturer's instruction manual should be followed for detailed operating procedures and for matters of routine maintenance, cleaning, and calibration. The analyst should be familiar with current literature and aware of the various interferences and potential sources of error for the analytes of interest.

Chromatography

Chromatography is an analytical technique used in quantitative determination of purity of most organic and an increasing number of inorganic reagent chemicals and standard-grade reference materials. The broad scope of chromatography allows it to be used in the separation, identification, and assay of diverse chemical species, ranging from simple metal ions to compounds of complex molecular structure, such as proteins.

In chromatography, the separation of individual components in a mixture is achieved when a mobile phase is passed over a stationary phase. Differences in affinities of various substances for these phases result in their separation.

Chromatography can be divided into two main branches, depending on whether the mobile phase is a gas or a liquid. Gas chromatography is principally used for analysis of volatile, thermally stable materials. Liquid chromatography is particularly useful for analysis of nonvolatile or thermally unstable organic substances. Ion chromatography, a technique in which anions and cations can be determined by using the principles of ion exchange, is a form of liquid chromatography. Thin-layer chromatography, often called planar chromatography, is also a form of liquid chromatography.

Gas Chromatography

Gas chromatography (GC) is used in this monograph to determine the assay and/or the trace impurities in both organic reagents and standards. Gas chromatography may be subdivided into gas–liquid and gas–solid chromatography. Gas–liquid chromatography is by far the most widely used form of gas chromatography.

The heart of a GC system is the column, which is contained in an oven operated in either the isothermal or the temperature-programmed mode. Columns packed with a nonvolatile liquid phase coated on a porous solid support were used extensively in early editions of this book. In this edition, only capillary columns, constructed of fused silica onto which is bonded the liquid phase, are used because of their greater resolving power and chemical inertness.

Conventional capillary gas chromatography uses long, narrow-bore columns with an inside diameter (i.d.) from 0.22 to 0.32 mm, coated or bonded with a thin film of the liquid phase. This arrangement results in high resolution but low sample capacity. Special injection techniques and hardware are used. The introduction of wide-bore capillary columns, which have an i.d. of typically 0.53 mm and a relatively thick film of the liquid phase (1–5 μm), can allow a laboratory to use a standard packed-column instrument and conditions while gaining the advantages of capillary technology.

Direct flash vaporization or on-column injection of the sample with standard-gauge needles can be used with capillary columns. Two types of flash vaporization injection techniques can be used: a split mode of operation, in which part of the sample is vented from the injector, or a splitless mode, in which a smaller volume is injected and no portion of the sample is vented from the injector. The splitless mode, using a 0.1-μL sample size, is recommended for the assay of most reagent solvents; the split mode often is recommended for analysis of most of the standard-grade reference materials.

The conventional split injector is a flash vaporization device. The liquid plug, introduced with a syringe, is immediately volatilized, and a small fraction of the resultant vapor enters the column while the major portion is vented to waste. To perform a GC analysis using a split injection technique, the GC instrument should be configured such that a preheated carrier gas, controlled by a pressure regulator or a combination of a flow controller and a back pressure regulator, enters the injector. The flow is divided into two streams. One stream of carrier gas flows upward and purges the septum. The septum purge flow is controlled by a needle valve. Septum purge flow rates are usually between 3 and 5 mL/min. A high flow of carrier gas enters the vaporization chamber, which is a glass or quartz liner, where the vaporized sample is mixed with the carrier gas. The mixed stream is split at the column inlet, and only a small fraction enters the column. A needle valve or flow controller regulates the split ratio.

Split ratios (measured column flow/measured inlet flow) typically range from 1:50 to 1:500 for conventional capillary columns (0.22–0.32 mm i.d.). For high sample capacity columns, such as wide-bore columns and/or thick-film columns, low split ratios (1:5–1:50) are commonly used.

The method most commonly recommended for analysis of pure chemical components (for example, matrices with a narrow constant boiling range) is the hot needle, fast sample introduction. In this method, the sample is taken into the syringe barrel (typically 2–5 μL in a 10-μL syringe) without leaving an air plug between the sample and the plunger. After insertion into the injection zone, the needle is allowed to heat up for 3 to 5 s. This period of time is sufficient for the needle to be heated to the injector temperature. Then, the sample is injected by rapidly pushing the plunger down (fast injection), after which the needle is withdrawn from the injector within one second. Either manual or automatic sample introductions can be used. The measurement reproducibility will be enhanced by not varying the injected volume, which typically should be 0.5–2.0 μL. The use of an automatic injection system can significantly enhance measurement precision. Also, loosely packing the injection liner with deactivated glass wool or glass beads can provide thorough mixing between sample and carrier gas, yielding less sample discrimination and better measurement precision. However, analysts should be aware of adsorption and decomposition.

In conjunction with wide-bore columns, on-column injection can minimize sample degradation while increasing the reproducibility of results. On-column injection allows the injection of a liquid sample directly into the inlet of the column. Excellent quantitative precision and accuracy for thermally labile compounds and wide volatility range samples have been reported. Sample sizes range from 0.5 μL to 2 μL. The injection should be performed as fast as possible, with the column oven temperature below or equal to the boiling point of the solvent.

After injection, the liquid is allowed to form a stable film (flooded zone), which takes several seconds. If the solutes to be analyzed differ much in boiling points, in comparison to the solvent, ballistic heating to high temperature is allowed. On the other hand, if the solute boiling points do not differ too much compared to the solvent, temperature programming is applied to fully exploit the solvent effect. When peak splitting and/or peak distortion is observed, the connection of a retention gap can provide the solution.

If the composition of the sample is not that complex, the use of wide-bore columns is recommended, particularly since automated injection becomes easier. If high resolution with automated injection is needed, a deactivated but uncoated wide-bore precolumn (20–50 cm in length) should be connected to a conventional analytical narrow-bore column. Hydrogen is the carrier gas of choice. If H_2 cannot be used for safety reasons, helium may be substituted. High carrier-gas velocities (50–80 cm/s H_2, 30–50 cm/s He) ensure negligible band broadening. High-purity grades of carrier gases should always be used.

In comparison to vaporizing injectors (that is, split and splitless), there are some disadvantages to on-column injections. The two primary disadvantages pertain to sample pretreatment. First, because the sample is introduced directly onto the column, relatively "clean" samples must be prepared. Nonvolatile and less volatile materials collect at the head of the column, causing a loss of separation efficiency; therefore, sample cleanup is a prerequisite. Second, many samples may be too concentrated for on-column injection and will need to be diluted.

Wide-bore columns can be used either in a high-resolution mode (carrier-gas flow rates <10 mL/min) to obtain optimum resolution of sample components or at higher flow rates (10–30 mL/min), which will generate packed-column–quality separations in a shorter time. The increased length of capillary columns provides better separation of components than traditional packed columns. Improved resolution reduces the types of columns needed for most analytical requirements to as few as three: high, moderate, and low polarity.

A wide variety of detectors are used to quantify and/or identify the components in the eluent from the column. Thermal conductivity detectors (TCDs) and flame ionization detectors (FIDs) are examples of general detectors that provide a linear response to most organic compounds. Electron capture detectors (ECDs) and photoionization are examples of class-specific detectors often used in trace environmental analysis. Mass spectrometric-type detectors are used to provide positive component identification. The detector output after amplification is used to produce the *chromatogram*, a plot of component response vs. elution time. Modern systems convert the analog output to digital form, which allows for further manipulation and interpretation of the data.

Results from a chromatogram, when used to determine the assay of a reagent or standard, are often expressed as area percent. A response factor correction for each component is required for the most accurate results, especially when the sample components differ markedly in their detector response. An internal standard reduces error due to variations in injection quantities, column conditions, and detector conditions. Use of a TCD or FID minimizes the need to correct for response and is usually sufficient for determining reagent or standard assay.

The use of control charts to aid the analyst in visualizing chromatographic variability is suggested. To certify that assay results are valid, analysts should use a system suitability test as described below.

Recommended gas chromatography procedures begin on page 80.

Gas Chromatography–Mass Spectrometry

Gas chromatography–mass spectrometry (GC–MS) is one of the techniques used in the identity requirement for standard-grade reference materials. Mass spectrometers are used for many kinds of chemical analysis, especially those where identity or proof of chemical structure is critical. Examples range from environmental analysis to the analysis of petroleum products and biological materials, including the products of genetic engineering. Mass spectrometers use the difference in mass-to-charge ratio (m/e) of ionized atoms or molecules to separate them from each other. Mass spectrometry is therefore useful for quantitation of atoms or molecules and also for determining chemical and structural information about molecules. Molecules have distinctive fragmentation patterns that provide structural information to identify structural components. The largest peak in a mass spectrum is called the *base peak* and is assigned an arbitrary height of 100. The remaining peaks are then normalized to the base peak.

The general operation of a mass spectrometer is

1. creating gas-phase ions;
2. separating the ions in space or time based on their mass-to-charge ratio; and
3. measuring the quantity of ions of each mass-to-charge ratio.

The ion separation power of a mass spectrometer is described by the resolution, which is defined as $R = m/\Delta m$, where m is the ion mass and Δm is the difference in mass between two resolvable peaks in a mass spectrum with similar mass values. For example, a mass specrometer with a resolution of 1000 can resolve an ion with an m/e of 100.0 from an ion with an m/e of 100.1.

In general, a mass spectrometer consists of an ion source, a mass-selective analyzer, and an ion detector. Since mass spectrometers create and manipulate gas-phase ions, they operate in a high-vacuum system. The magnetic-sector, quadrupole, and time-of-flight designs also require extraction and acceleration ion optics to transfer ions from the source region into the mass analyzer.

GC–MS is used in this book to verify the identity of various standard-grade reference materials. The specific procedure is given under the individual standards.

Liquid Chromatography

Many reagent chemicals and standard-grade reference materials described in this book can be assayed by and/or tested for suitability for use in liquid chromatography (LC). Liquid chromatography is a technique in which the sample interacts with both the liquid mobile phase and the stationary phase to effect a separation. An LC system consists of a pump that delivers a liquid, usually a solvent for the sample, at a constant flow rate through a sample injector, a column, and a detector. The solvent is called the mobile phase. Typically, the flow rate is 1–10 mL/min for conventional liquid chromatography. In isocratic operation, the composition of the mobile phase is kept constant, while in gradient elution, it is varied during the analysis. Gradient elution is required when the sample mixture contains components with a wide range of affinity for the stationary phase. In the isocratic mode, the purity of the solvents is less critical, as the impurities are adsorbed–desorbed at a constant rate, whereas in gradient elution impurities may result in extraneous peaks and/or shifts in the baseline. This characteristic necessitates the incorporation of a gradient elution test in this book to verify the quality of a solvent when used in this most stringent LC system mode of operation.

In liquid chromatography, a dilute solution of the sample to be analyzed is introduced into the mobile phase via the sample injector and enters the column, where it is separated into its individual components. The columns are usually steel or glass, densely packed with semirigid organic gels or rigid inorganic silica microspheres, to which a variety of substrates can be chemically bonded and whose typical particle size is 3–10 μm.

Several mechanisms of separation are possible in liquid chromatography. In *adsorption chromatography*, separation is based on adsorption–desorption kinetics, whereas in *partition chromatography*, the separation is based on partitioning of the components between the mobile and stationary phases. Ion exchange is the dominant mechanism in ion chromatography. In practice, a successful separation may involve a combination of separation mechanisms.

A commonly used term, coined by early chromatographers for describing separations dominated by adsorption–desorption, is *normal phase*. In normal phase chromatography, the stationary phase is strongly polar (for instance, silica or aminopropyl), and the mobile phase is less polar (for instance, hexane). Polar components are thus retained on the column longer than less-polar materials.

A second term of historical origin is *reverse phase*. Reverse phase generally applies to separations dominated by partition chromatography, and the elution of components in a reverse-phase separation are more or less reversed from the order that would be obtained in a normal phase separation. Reverse phase, the more widely used mode of liquid chromatography, uses a nonpolar (hydrophobic) stationary phase, such as C-18 (octadecyl) chemically bonded to silica, while the mobile phase is a polar liquid such as acetonitrile–water. Hydrophobic (nonpolar) components are retained longer than hydrophilic (polar) components. The elution properties of the mobile phase can be adjusted to modify a separation by addition of appropriate ionic modifiers. These modifiers can be chosen for their ability to either suppress or enhance ion formation in the sample. When enhancement is chosen, the separation mechanism may involve both ion exchange and partition.

The wide range of available stationary phases in combination with changes in mobile-phase composition makes liquid chromatography a very flexible separation-assay technique. The back pressure on the pump is several hundred to several thousand psi, depending upon the particle size of the packing material, mobile-phase flow rate, and viscosity.

The column eluent is continuously monitored by a sensitive detector chosen to respond either to the sample component alone (for instance, an ultraviolet photometer) or to a change in some physical property of the mobile phase due to the presence of the solute (for example, a differential refractometer). Other detectors such as electrochemical, fluorescence, etc., are also used for more specialized applications. When photometric detectors are used, solvents are often specified in absorbance units at a specific wavelength. The response of the detector is related to the concentration or weight of the solute and is displayed on a recorder. A computer is usually interfaced to the LC system to control the method and to collect and analyze data.

Recommended liquid chromatography procedures begin on page 84.

Ion Chromatography

Ion chromatography (IC) is a subset of liquid chromatography. Whereas conventional liquid chromatography is mainly used for the analysis of nonionic organic compounds, ion chromatography separates and determines ionic species or ionizable compounds, both organic and inorganic. Under ideal conditions, quantitation to the sub-part-per-billion level is attainable. For trace analysis, water and reagents of the highest quality need to be used. Deionized water of >18 M$\Omega \cdot$ cm resistivity is recommended.

For the testing of reagent chemicals, ion chromatography is ordinarily applied to the determination of anions present as minor impurities. Generally, the major component should be eliminated or greatly reduced in concentration to avoid overloading the anion-exchange column used for separation.

IC columns are usually 5–25 cm long and 2–10 mm in diameter. Columns may be either metallic or plastic. Packing materials can be either anionic or cationic resins, depending on the separation desired, and there are several variations within each type. Typical particle sizes range from 5 to 20 μm.

In many anion applications of ion chromatography, buffered, aqueous, mobile phases (eluents) of approximately millimole strength are used. In most cases, IC analyses are carried out isocratically, and gradient elutions are used only for the more complex separations. Typical eluent flow rates for ion chromatography are 1–2 mL/min, with system operating pressures at 1000 psi or less. Microbore columns are operated at lower flow rates of 0.2–0.5 mL/min.

Standard and sample injections are normally made by using a loop injection system. The most widely used detection method for ion chromatography is conductivity. However, other detection methods, including ultraviolet-visible spectroscopy, electrochemistry, and fluorescence can be used. Conductivity is a universal detection mode for ions, whereas the other detectors provide selective, analyte-specific detection.

In conductimetric IC analysis, the impact of the significant background conductivity of the eluent on analyte determinations is minimized by chemical or electronic suppression. Chemical suppression involves notably reducing the eluent conductivity, as well as enhancing analyte response by means of a chemical reaction. Several innovative techniques have been developed to achieve these changes, including packed-column and membrane-based devices. Electronic suppression minimizes eluent background noise via the design of the electronic circuitry in the conductivity detector, although the actual background eluent conductivity is not reduced, as it is with chemical suppression.

Because of the two different modes of suppression, the columns and eluents used also fall into two categories. Columns for chemical suppression typically have higher ion-exchange capacities than those used for electronic suppression. Eluents used in chemical suppression also differ from those used in electronic suppression in that they are defined by the chemical reactions that occur in the suppression of the eluent.

A discussion of ion chromatography procedures begins on page 87.

Thin-Layer Chromatography

Thin-layer chromatography (TLC) is a very simple form of solid–liquid adsorption chromatography. It is probably the quickest, easiest, and most frequently applied technique for determining purity of organic compounds. As in other forms of liquid chromatography, the sample interacts with a liquid mobile phase and a stationary phase to effect partitioning of components based on their affinity to the solid and liquid phase. Thin-layer chromatography serves many purposes in the laboratory because of its simplicity. It is commonly used in organic synthesis to monitor chemical reactions. Starting materials, intermediates, and products often elute differently, and product formation or starting-material disappearance may be observed.

Another very common use is in purity determinations of organic compounds. Typically, it is not used as a quantitative technique, but it is useful for looking for impurities. In many cases, thin-layer chromatography is very sensitive and can be used to detect impurities of less than 1% in the sample. A single spot on a TLC plate is a good indication of purity. Thin-layer chromatography is also applied in selected requirements for standard-grade reference materials as a technique for purity confirmation. The technique is used with other complementary purity assays to screen standard-grade reference materials for impurities. Standard-grade reference materials have passed the purity test when, having used the specific conditions in the method, an analyst can observe a single spot.

In thin-layer chromatography, the stationary phase is spread as a thin layer over glass or plastic. Calcium sulfate or an organic polymer (such as starch) is added to the solid phase to bind the solid to the glass or plastic. Often, fluorescent material is added to the solid phase to aid in detection of analytes. TLC plates are commercially available or can be prepared in the laboratory. Glass or liquid plates are available commercially as sheets that are cut into strips for use. Plates can be prepared in the laboratory by preparing a slurry of the solid phase with a solvent, such as chloroform or methanol, and applying a thin coat over a glass plate. Plates are most effective when dried in an oven before use. The most common stationary phases used in thin-layer chromatography are silica gel and alumina. Reverse-phase TLC plates are also available for elution of polar compounds. It is common to perform TLC analysis with both silica and alumina to maximize the effectiveness of the technique.

Performing TLC analysis requires minimal equipment. The plates are cut into strips approximately 10 cm in height. The sample is dissolved in a volatile solvent and applied to the bottom of the plate. This is accomplished by applying or "spotting" the solution about 0.5 cm from the bottom of the plate using a thin capillary tube. The plate is placed into a development chamber that contains enough elution solvent to come just below the sample spot. The solvent then travels up the plate and moves the components of the sample at different rates according to their affinity. When the solvent is about 1 cm from the upper end of the plate, the plate is removed, the solvent line is marked, and the plate is allowed to dry.

Detection of spots on the TLC plate is accomplished in a number of ways. The most common method is to view the plate under ultraviolet (UV) light. Compounds that fluoresce will be detected. Alternatively, the plates can contain a fluorescent material within the solid phase. In this case, the plate will fluoresce, and compounds that do not fluoresce will appear as black spots on the plate. Another common method is to use iodine as a complexing agent. Many organic compounds form charge-transfer complexes with iodine, resulting in a dark-colored species. The TLC plate is placed in a chamber containing iodine and "developed" for several minutes. Upon removal from the chamber, the plates will contain dark spots; these spots should be immediately circled with a pencil, since the complex is reversible and the spots will fade away. In other detection methods, the plate is sprayed with a solution to "stain" the analytes. Common reagents are sulfuric acid solution, which will char many organic compounds, and potassium permanganate solution, which will oxidize many organic compounds. Combinations of two or more methods may be used to detect a broader range of analytes.

It is often useful to determine the distance that a particular compound has moved up the plate in relation to the solvent front. This value is called the retention factor and designated R_f. The R_f value is calculated by measuring the distance that the analyte moved from the origin and dividing that by the distance the solvent moved from the origin. For purity assays, it is desirable to have an R_f value in the range of 0.3–0.5.

Recommended thin-layer chromatography procedures begin on page 86.

Recommended Procedures

Assay Methods

Procedures published in this book list the parameters and values established to give satisfactory results. These procedures should be considered adequate only for determining the constituents specified in the individual reagent monographs, and they may not necessarily separate other components in a given sample. An attempt is made to keep each procedure for reagents as simple as possible, while still retaining the desired sensitivity, because simplicity may lead to wider usage. The procedures used for standard-grade reference materials are, in general, more specific, since it is necessary to separate similar eluting impurities.

The list of parameters for gas chromatography includes such variables as column type, diameter and length of column, carrier-gas flow rate, detector type, and operating parameters. The list for liquid chromatography includes such variables as diameter and length of column, packing type and particle size, mobile-phase composition and flow rate, detector type, and operating parameters. For reagents, the parameters represent one set of conditions that result in reproducible analyses. Because of differences among instruments, one or more of the stated parameters may require adjustment for any given instrument. Therefore these parameters, except for the column and type of detector, should be regarded as guides to aid the analyst in establishing the optimum conditions for a particular instrument. Relative retention times are given as an aid in peak identification. Exact reproduction of those times is not essential. Parameters given for the standard-grade reference materials section, however, must be adhered to exactly as stated in the test.

Gas Chromatography

The volatile organic reagents in this book can be assayed by gas chromatography. A single set of instrument conditions shown here has been chosen and found to give satisfactory results for the assay of these reagents. Three columns of varying polarity can be used to achieve the desired separation: type I, low polarity, methyl silicone, 5 μm; type II, moderate polarity, mixed cyano, phenylmethyl silicone, 1.5 μm; and type III, high polarity, polyethylene glycol, 1 μm. The recommended column type and reagent-specific conditions can be found listed alphabetically under the individual reagents. The retention times and relative retention times vs. methanol of the organic reagents are listed alphabetically (Table 2-12) and in increasing order (Table 2-13).

▶ **Procedure for Gas Chromatography.**

Column: 30-m × 0.53-mm i.d. fused silica capillary coated with a film that has been surface-bonded and cross-linked

Column Temperature: 40 °C isothermal for 5 min, then programmed to 220 °C at 10 °C/min, hold 2 min

Injector Temperature: 150–220 °C

Detector Temperature: 250 °C

Sample Size: 0.2 μL splitless

Carrier gas: Helium at 3 mL/min

Detector: Thermal conductivity or flame ionization

Standard-grade reference materials must be assayed by multiple methods. Specific GC methods are given under most classes of standard-grade reference materials. Methods differ mainly in the type of stationary phase employed and the physical dimensions of the capillary column.

Identity by Gas Chromatography–Mass Spectrometry

Standard-grade reference materials must be identified by multiple methods. GC–MS is one of the primary identity methods used. Specific conditions are found under each class of standard-grade reference materials.

Liquid Chromatography

Reagents and standard-grade reference materials can be analyzed by various modes of liquid chromatography. In this book, reagents not suitable for assay by gas chromatography are analyzed by liquid chromatography. Some standard-grade reference materials are used for LC applications. Specific tests for LC suitability for these materials are described in the

Table 2-12. GC Retention Time of Reagents in Alphabetical Order

Compounds	Column of Choice	Type I, Nonpolar		Type II, Med. Polar		Type III, High Polar	
		RT	RR	RT	RR	RT	RR
Acetaldehyde	I	2.9	0.97	2.2	0.89	2.8	0.4
Acetic acid	I	8.8	2.9	–	–	–	–
Acetic anhydride	I	11.4	3.8	–	–	–	–
Acetone	I	4.9	1.6	3.4	1.4	4.1	0.6
Acetonitrile	I	4.7	1.6	4.4	1.8	8.7	1.4
Acetyl chloride	I	5.2	1.7	–	–	–	–
2-Aminoethanol	I	10.6	3.5	9.9	4.0	–	–
Aniline	I	19.5	6.5	17.3	7.0	21.9	3.4
Benzene	I	11.2	3.7	6.7	2.7	6.9	1.1
Benzyl alcohol	I	20.6	6.8	18.6	7.5	23.5	3.7
2-Butanone	I	8.5	2.8	6.0	2.4	6.0	1.0
Butyl acetate	I	15.3	5.1	11.7	4.7	10.6	1.7
Butyl alcohol	I	11.1	3.7	9.0	3.6	12.3	1.9
t-Butyl alcohol	I	6.3	2.1	4.1	1.7	6.5	1.0
Carbon disulfide	I	7.4	2.5	3.0	1.2	3.0	0.5
Carbon tetrachloride	I	11.2	3.7	6.0	2.4	5.3	0.8
Chlorobenzene	I	16.6	5.5	12.5	5.0	16.5	2.6
Chloroform	I	9.2	3.1	6.0	2.4	8.9	1.4
Cyclohexane	I	11.4	3.8	5.4	2.2	3.0	0.5
Cyclohexanone	I	17.5	5.8	14.9	6.0	15.4	2.4
1,2-Dichloroethane	I	10.3	3.4	7.4	3.0	10.0	1.6
Dichloromethane	I	6.2	2.0	3.5	1.4	6.4	1.0
Diethanolamine	I	20.6	6.9	20.0	8.0	–	–
Diethylamine	I	11.2	3.0	5.0	2.5	–	–
N,N-Dimethylformamide	I	14.2	4.7	13.6	5.5	16.2	2.5
Dimethylsulfoxide	III	15.7	5.2	16.0	6.4	20.5	3.2
Dioxane	I	12.2	4.1	9.1	3.6	10.3	1.6
Ethyl acetate	I	9.2	3.1	5.8	2.3	5.6	0.9
Ethyl alcohol	I	4.3	1.4	3.2	1.3	7.4	1.2
Ethyl ether	I	5.6	1.9	2.7	1.1	2.4	0.4
Ethylbenzene	I	17.0	5.6	12.5	5.0	11.8	1.8
Formamide	III	–	–	–	–	23.0	3.7
2-Furancarboxyaldehyde	I	15.8	5.2	14.2	5.7	17.9	2.8
Glycerol	II	19.8	6.6	19.7	7.9	–	–
2-Hexane	I	8.6	2.9	3.2	1.3	2.3	0.4
n-Hexane	I	9.2	3.1	3.5	1.4	2.3	0.4
Isobutyl alcohol	I	10.0	3.3	7.9	3.2	11.2	1.7
Isopentyl alcohol	I	13.2	4.4	11.0	4.4	13.5	2.1
Isopropyl alcohol	I	5.4	1.8	3.8	1.5	7.3	1.1
Isopropyl ether	I	9.1	3.0	2.6	1.0	2.6	0.4
Methanol	I	3.0	1.0	2.5	1.0	6.4	1.0
2-Methoxyethanol	III	10.2	3.4	8.4	3.4	13.2	2.1
Methyl-tert-butyl ether	I	11.2	3.0	5.9	2.0	2.8	0.4
4-Methyl-2-pentanone	I	13.3	4.4	10.2	4.1	8.9	1.4
1-Methyl-2-pyrrolidone	I	20.6	6.8	19.3	7.7	21.8	3.4
Nitrobenzene	I	22.0	7.3	19.4	7.8	22.1	3.5
Nitromethane	III	7.2	2.4	7.0	2.8	12.2	1.9

Continued on next page

Table 2-12. *(continued)*

Compounds	Column of Choice	Type I, Nonpolar		Type II, Med. Polar		Type III, High Polar	
		RT	RR	RT	RR	RT	RR
1-Octanol	I	21.2	7.0	17.8	7.2	19.2	3.0
2-Octanol	I	20.2	6.7	16.0	6.4	18.2	2.9
1-Pentanol	I	14.2	4.7	11.8	4.7	14.3	2.2
Perchloroethylene	I	15.8	5.2	10.7	4.3	9.4	1.5
1,2-Propanediol	III	13.4	4.5	13.4	5.4	19.8	3.1
Perchloroethylene	I	15.8	5.2	10.7	4.3	9.4	1.5
Propionic acid	III	14.0	4.6	12.5	5.0	18.5	2.9
η-Propyl alcohol	III	7.6	2.5	5.9	2.4	10.1	1.6
Pyridine	I	13.4	4.5	10.6	4.2	13.2	2.1
Quinoline	II	–	–	121.6	8.7	25.0	3.9
Tetrahydrofuran	I	9.8	3.3	5.8	2.3	5.1	0.8
Toluene	I	14.4	4.8	10.0	4.0	9.7	1.5
1,1,1-Trichloroethane	I	10.6	3.5	6.0	2.4	5.4	0.9
Trichloroethylene	I	12.3	4.1	7.8	3.1	8.4	1.3
1,1,2-Trichlorotrifluoroethane	I	6.7	2.2	2.7	1.1	2.4	0.4
2,2,4-Trimethylpentane	I	12.4	4.1	5.9	2.4	2.6	0.4
1,2-Xylene	I	17.8	5.9	13.3	5.4	12.8	2.0
1,3-Xylene	I	17.2	5.7	12.6	5.1	12.0	1.9
1,4-Xylene	I	16.9	5.6	12.4	5.0	11.6	1.8

Note: RT is retention time; RR is relative retention (methanol 1.0); and – indicates data not available.

Table 2-13. GC Retention Time of Reagents by Chronological Order on Type I Column

Compounds	Column of Choice	Type I, Nonpolar		Type II, Med. Polar		Type III, High Polar	
		RT	RR	RT	RR	RT	RR
Acetaldehyde	I	2.9	0.97	2.2	0.89	2.8	0.4
Methanol	I	3.0	1.0	2.5	1.0	6.4	1.0
Ethyl alcohol	I	4.3	1.4	3.2	1.3	7.4	1.2
Acetonitrile	I	4.7	1.6	4.4	1.8	8.7	1.4
Acetone	I	4.9	1.6	3.4	1.4	4.1	0.6
Acetyl chloride	I	5.2	1.7	–	–	–	–
Isopropyl alcohol	I	5.4	1.8	3.8	1.5	7.3	1.1
Ethyl ether	I	5.6	1.9	2.7	1.1	2.4	0.4
Dichloromethane	I	6.2	2.0	3.5	1.4	6.4	1.0
t-Butyl alcohol	I	6.3	2.1	4.1	1.7	6.5	1.0
1,1,2-Trichlorotrifluoroethane	I	6.7	2.2	2.7	1.1	2.4	0.4
Nitromethane	III	7.2	2.4	7.0	2.8	12.2	1.9
Carbon disulfide	I	7.4	2.5	3.0	1.2	3.0	0.5
η-Propyl alcohol	III	7.6	2.5	5.9	2.4	10.1	1.6
2-Butanone	I	8.5	2.8	6.0	2.4	6.0	1.0
2-Hexane	I	8.6	2.9	3.2	1.3	2.3	0.4
Acetic acid	I	8.8	2.9	–	–	–	–
Isopropyl ether	I	9.1	3.0	2.6	1.0	2.6	0.4
Chloroform	I	9.2	3.1	6.0	2.4	8.9	1.4
Ethyl acetate	I	9.2	3.1	5.8	2.3	5.6	0.9
n-Hexane	I	9.2	3.1	3.5	1.4	2.3	0.4
Tetrahydrofuran	I	9.8	3.3	5.8	2.3	5.1	0.8

Continued on next page

Table 2-13. *(continued)*

Compounds	Column of Choice	Type I, Nonpolar RT	RR	Type II, Med. Polar RT	RR	Type III, High Polar RT	RR
Isobutyl alcohol	I	10.0	3.3	7.9	3.2	11.2	1.7
2-Methoxyethanol	III	10.2	3.4	8.4	3.4	13.2	2.1
1,2-Dichloroethane	I	10.3	3.4	7.4	3.0	10.0	1.6
2-Aminoethanol	I	10.6	3.5	9.9	4.0	–	–
1,1,1-Trichloroethane	I	10.6	3.5	6.0	2.4	5.4	0.9
Butyl alcohol	I	11.1	3.7	9.0	3.6	12.3	1.9
Benzene	I	11.2	3.7	6.7	2.7	6.9	1.1
Carbon tetrachloride	I	11.2	3.7	6.0	2.4	5.3	0.8
Diethylamine	I	11.2	3.0	5.0	2.5	–	–
Methyl-*tert*-butyl ether	I	11.2	3.0	5.9	2.0	2.8	0.4
Acetic anhydride	I	11.4	3.8	–	–	–	–
Cyclohexane	I	11.4	3.8	5.4	2.2	3.0	0.5
Dioxane	I	12.2	4.1	9.1	3.6	10.3	1.6
Trichloroethylene	I	12.3	4.1	7.8	3.1	8.4	1.3
2,2,4-Trimethylpentane	I	12.4	4.1	5.9	2.4	2.6	0.4
Isopentyl alcohol	I	13.2	4.4	11.0	4.4	13.5	2.1
4-Methyl-2-pentanone	I	13.3	4.4	10.2	4.1	8.9	1.4
1,2-Propanediol	III	13.4	4.5	13.4	5.4	19.8	3.1
Pyridine	I	13.4	4.5	10.6	4.2	13.2	2.1
Propionic acid	III	14.0	4.6	12.5	5.0	18.5	2.9
1-Pentanol	I	14.2	4.7	11.8	4.7	14.3	2.2
N,N-Dimethylformamide	I	14.2	4.7	13.6	5.5	16.2	2.5
Toluene	I	14.4	4.8	10.0	4.0	9.7	1.5
Butyl acetate	I	15.3	5.1	11.7	4.7	10.6	1.7
Dimethylsulfoxide	III	15.7	5.2	16.0	6.4	20.5	3.2
2-Furancarboxyaldehyde	I	15.8	5.2	14.2	5.7	17.9	2.8
Perchloroethylene	I	15.8	5.2	10.7	4.3	9.4	1.5
Chlorobenzene	I	16.6	6.6	12.5	5.0	16.5	2.6
1,4-Xylene	I	16.9	5.6	12.4	5.0	11.6	1.8
Ethylbenzene	I	17.0	5.6	12.5	5.0	11.8	1.8
1,3-Xylene	I	17.2	5.7	12.6	5.1	12.0	1.9
Cyclohexanone	I	17.5	5.8	14.9	6.0	15.4	2.4
1,2-Xylene	I	17.8	5.9	13.3	5.4	12.8	2.0
Aniline	I	19.5	6.5	17.3	7.0	21.9	3.4
Glycerol	II	19.8	6.6	19.7	7.9	–	–
2-Octanol	I	20.2	6.7	16.0	6.4	18.2	2.9
Benzyl alcohol	I	20.6	6.8	18.6	7.5	23.5	3.7
Diethanolamine	I	20.6	6.9	20.0	8.0	–	–
1-Methyl-2-pyrrolidone	I	20.6	6.8	19.3	7.7	21.8	3.4
1-Octanol	I	21.2	7.0	17.8	7.2	19.2	3.0
Nitrobenzene	I	22.0	7.3	19.4	7.8	22.1	3.5
Quinoline	II	–	–	21.6	8.7	25.0	3.9
Formamide	III	–	–	–	–	23.0	3.7

Note: RT is retention time; RR is relative retention (methanol 1.0); and – indicates data not available.

class specifications. Thin-layer chromatography is used to confirm the purity of standard-grade reference materials. Ion chromatography is used to determine the anion level of some reagents.

Solvent Suitability for Liquid Chromatography. Solvents used in liquid chromatography are tested for suitability in gradient elution analysis. The objective is to assess the performance suitability of the solvents under routine LC operating conditions. The operating parameters found under the specific reagent were selected to simulate a typical gradient run. The reverse gradient portion of the chromatogram is of little practical interest in actual sample analysis. Thus, only the peaks eluted in the "up-gradient" portion of the program are recorded (e.g., water to acetonitrile or water to methanol). Commercially available reverse-phase columns come in a variety of supports and stationary phases. Thus, parameters such as monomeric vs. polymeric phase, endcapped vs. nonendcapped, particle size, and %C loading are specified.

▶ Procedure for Liquid Chromatography.

Column: Octadecyl (C-18 polymeric phase), 250×4.6 mm i.d., 5 μm, 12% C loading, endcapped

Mobile Phase: A. Solvent to be tested

                              B. LC reagent-grade water

Conditions: Detector: Ultraviolet at 254 nm

                              Sensitivity: 0.02 AUFS (absorbance units full scale)

Gradient Elution:

1. Program the LC system to develop a solvent–water gradient as follows:

Time (min)	Flow (mL/min)	%A (Solvent)	%B (Water)
0	2.0	20	80
30	2.0	20	80
50	2.0	100	0
60	2.0	100	0

2. Start the program after achieving a stable baseline.

3. Disregard the results of the first program. Repeat the program and use these results in the Step 4 evaluation.

4. No peak should be greater than 25% of full scale (0.005 absorbance units).

Reagent-specific conditions can be found under the individual reagents.

Extraction–Concentration

Solvent Suitability for Extraction–Concentration. Trace-level analysis, such as for pollutants in the environment and toxic chemicals in the workplace, frequently requires the use of solvent-extraction procedures to isolate the analyte. In some cases, the solvent is then removed to concentrate the analyte for detection and quantification. It is critical that the solvent be of the highest purity, so that potential interferences during subsequent chromatographic analysis are minimized. For GC analysis, a pure solvent is usually characterized by a small or narrow solvent "front" and the absence of extraneous peaks under the temperature-programmed conditions commonly used for these analyses.

In this book, the purity of the solvent is specified as suitable for use with FID (universal) and ECD (class-specific) detectors. For analysis of solvents that are not compatible with a given detector, as in the case of halogenated solvents and ECD, a solvent-exchange step is used before GC analysis. In LC analysis using gradient elution and UV detection, a pure solvent would not show extraneous peaks or cause unusual baseline drift when analyzed as a sample. In this book, this suitability is specified by measuring the absorbance through a particular spectral region for each solvent. The solvent should give a smooth curve and contain no extraneous absorptions throughout the specified region.

▶ Procedure for ECD and FID Suitability.

Preconcentration. Preconcentration of the solvent is required to reach the specification levels set for the ECD and FID. Preconcentration can be accomplished with a conventional Kuderna–Danish evaporator-concentrator or a vacuum rotary evaporator. The solvent should never be evaporated to dryness.

> **Procedure.** Accurately transfer 500 mL of test solvent to a 1-L evaporator. Concentrate (100×) to approximately 5 mL. If solvent exchange is necessary, add two separate 50-mL portions, and evaporate to 5 mL each time.

> **Analysis.** Analyze the concentrated solvent and an external standard by using a gas chromatograph equipped with wide-bore capillary columns and ECD and FID detectors. The parameters cited here have given satisfactory results.

Instrument for GC–ECD.

> *Column:* 30-m × 530-μm i.d. fused silica capillary, coated with 1.5-μm film of 5% diphenyl, 94% dimethyl, 1% vinyl polysiloxane that has been surface-bonded and cross-linked
>
> *Column Temperature:* 40–250 °C at 15 °C/min; hold at 250 °C for 15 min
>
> *Injector Temperature:* 250 °C
>
> *Detector Temperature:* 320 °C
>
> *Carrier Gas:* Helium at 2–5 mL/min
>
> *Sample Size:* 5 μL
>
> *Standard:* 1 μg/L of heptachlor epoxide in hexane
>
> *Standard Size:* 5 μL

Instrument for GC–FID.

> *Column:* 30-m × 530-μm i.d. fused silica capillary, coated with 1.5-μm film of 5% diphenyl, 94% dimethyl, 1% vinyl that has been surface-bonded and cross-linked
>
> *Column Temperature:* 40–250 °C at 15 °C/min; hold at 250 °C for 15 min
>
> *Injector Temperature:* 250 °C
>
> *Detector Temperature:* 250 °C

Carrier Gas: Helium at 2–5 mL/min

Sample Size: 5 μL

Standard: 1 mg/L of 2-octanol in hexane

Standard Size: 5 μL

Results. The purity of the solvents is expressed in micrograms per liter for FID and nanograms per liter for ECD, measured against an external standard appropriate for the analysis being performed. In this book, 2-octanol is used for FID and heptachlor epoxide is used for ECD. Measure the peak area-height for all peaks inside a window of 0.5–2 times the retention time of 2-octanol for GC–FID suitability. The sum of the peaks must be no greater than 10 μg/L, with no single peak greater than 5.0 μg/L. Measure the peaks for all peaks inside a window of 0.2–2 times the retention time of heptachlor epoxide for GC–ECD suitability. The sum of the peaks must be no greater than 10 ng/L, with no single peak greater than 5.0 ng/L. For pesticide residue analysis, the standards might also be a series of pesticides, or, for drinking water, a series of halomethanes. Detailed procedures for such analysis can be found in many excellent references.

LC Suitability

▶**Procedure for Absorbance.** Measure the absorbance of the sample in a 1-cm cell against water (as the reference liquid) in a matched cell from 400 nm to the cutoff of the reagent solvent. The absorbance should be less than 1.00 at the cutoff wavelength and not exceed the absorbances at each wavelength specified in the requirements of the reagent.

Thin-Layer Chromatography

▶**Procedure for Thin-Layer Chromatography.** Prepare a solvent system as specified in the individual reagent monograph. Prepare also a developing chamber that can be sealed, and fill it to a depth of approximately 3 mm with the solvent system. Cut a piece of filter paper to the appropriate size and place it inside the chamber, then close the chamber. Shake the chamber gently to ensure saturation of the filter paper and chamber. Prepare a sample solution in a volatile solvent, such as methanol or dichloromethane, at a concentration of about 1–10%. Dry a TLC plate approximately 10 cm in height in an oven at 110 °C, and store it in a desiccator before use.

Place a spot of the sample solution onto the plate approximately 0.5 cm from the bottom using a capillary tube. The diameter of the spot should be 1–2 mm. Allow the solvent to evaporate, and repeat the spotting procedure one or two more times, each time allowing the solvent to dry (the amount of sample applied will vary depending on sample concentration). Excess sample will result in loss of resolution because the sample will "streak" up the plate or to a broad and undefined spot. Carefully place the plate into the developing chamber on the opposite side from the filter paper. Make sure that the sample spot is above the solvent level. Allow the solvent to climb to about 1 cm from the top of the plate. Remove the plate, and mark the solvent front. Detect the analytes as specified in the method. The R_f value should be in the range of 0.3–0.5.

Ion Chromatography

▶**Procedure for Ion Chromatography.** Ion chromatography is used in this book to determine the anion content of various reagents. The specific procedure is given under the individual reagent monograph. The methods are based on the initial removal of any anionic matrix, when one is present. The use of a chemical suppression method is recommended. The water used in the preparation of the standards and as a diluent should be on the order of 0.06 micromho cm^{-1}. The use of a preconcentration column can dramatically lower the concentration at which anions can be determined.

Calibration for GC and LC Systems

Because the response of a detector is not always the same for equal concentrations of all components, an instrument must be calibrated to obtain concentrations in terms of weight percentages. Correction of area values to weight values is accomplished by the use of calibration factors (also referred to as response factors) that may be obtained in one of the following ways.

Standard-Addition Technique. The sample is analyzed by the prescribed procedure before and after the addition of a measured amount of the component of interest. The ratio of the area obtained for that component, before and after the addition, combined with the weight of added component, can be used to calculate the calibration factor.

Calibration Mixture. A mixture is prepared to correspond as closely as possible in composition to the sample to be analyzed. For the preparation of a calibration mixture, it is desirable, but not always possible, to start with components of negligible impurity. When the impurity content cannot be ignored, it may be established by an independent means. The mixture is analyzed by the prescribed procedure, and the area of each peak is measured. The calibration factor for each component is the ratio of its weight or concentration to its area. For an unidentified component, the calibration factor cannot be determined and is usually assumed to be the same as that of a known component in the calibration mixture. The frequency of calibration is left up to the analyst and should follow normal good laboratory practices.

System Suitability Tests for GC and LC Systems

To determine whether a chromatographic system can provide acceptable, repeatable results, it can be subjected to a suitability test at the time of use. Underlying such a test is the concept that electronics, equipment, column, and standard or samples, coupled with sample handling, constitute a single system. Specific data are collected from repeated injections of the assay preparation or standard preparation. The results are compared with limiting values for key parameters, such as peak symmetry, efficiency, precision, and resolution.

The most useful test parameter is the precision of replicate injections of the analytical reference solution, prepared as directed under the individual reagent. The precision of replicate injections is expressed as the relative standard deviation as follows:

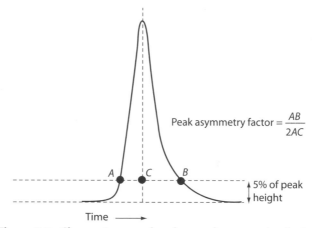

Figure 2-7. Chromatogram showing peak asymmetry factor.

$$S_R(\%) = \left(\frac{100}{\overline{X}}\right)\left[\sum_{i=1}^{N}\frac{(X_i - \overline{X})^2}{(N-1)}\right]^{1/2}$$

where S_R is the relative standard deviation in percent, $\overline{X}$ is the mean of the set of N measurements, and X_i is an individual measurement. When an internal standard is used, the measurement X_i usually refers to the measurement of the relative area, A_S:

$$X_I = A_S = a_R/a_i$$

where a_R is the peak area of the substance being analyzed and a_i is the peak area of the internal standard.

The peak asymmetry value is useful and can be limited by specifying an asymmetry factor. The asymmetry factor, T, is defined as the ratio of the distance from the leading edge to the trailing edge divided by twice the distance from the leading edge to the peak perpendicular measured at 5% of the peak height (see Figure 2-7). Therefore,

$$T = AB/2AC$$

This asymmetry factor is unity for a symmetrical peak and increases in value as the asymmetry becomes more pronounced.

Resolution, R, can be specified to ensure separation of closely eluting components or to establish the general separation efficiency of the system. R can be calculated as follows:

$$R = \frac{1.178(t_2 - t_1)}{W_{(h/2)_1} + W_{(h/2)_2}}$$

where t_2 and t_1 are the retention times of the two component peaks, and $W_{(h/2)_1}$ and $W_{(h/2)_2}$ are the half-height widths of the two peaks in units of time. An R value of 1.0 means that the resolution is 98% complete. This condition is sufficient, in most cases, for peak area calculations. An R value of 1.5 or greater represents baseline, or complete, separation of the peaks.

Infrared Spectroscopy

Infrared (IR) spectroscopy is an absorption method widely used in both qualitative and quantitative analysis. The infrared region of the spectrum includes wave numbers ranging from about 12,800 to 10 cm^{-1}. The IR range can be separated into three regions: the near-infrared (12,800–4,000 cm^{-1}), the mid-infrared (4,000–200 cm^{-1}), and the far-infrared (200–10 cm^{-1}). Most analytical applications fall in the mid-infrared region of the spectrum. IR spectroscopy has primarily been used to assist in identification of organic compounds. The IR spectrum of an organic compound is a unique physical property and can be used to identify unknowns by interpretation of characteristic absorbances and comparison to spectral libraries. IR spectroscopy is also used in quantitative techniques. Because of its sensitivity and selectivity, IR spectroscopy can be used to quantitate analytes in complex mixtures. Quantitative analysis is used extensively in detection of industrial pollutants in the environment. The near-infrared region has been used for quantitative applications, such as determination of water in glycerol and determination of aromatic amines in complex mixtures. The IR technique is discussed here primarily for application in identification of organic compounds and will focus on the mid-infrared region. Instrumental operating procedures are not given because they will vary depending on instrument design. A brief discussion of the theory will be followed by a discussion of instrumentation, sample handling techniques, and qualitative analysis.

General Background

Unlike UV and visible spectroscopy, which use larger energy absorbances from electronic transitions, IR spectroscopy relies on the much smaller energy absorbances that occur between various vibrational and rotational states. When molecular vibrational or rotational events occur that cause a net dipole moment, IR absorption can occur. Molecular vibrations can be classified as either *stretching* or *bending*. Stretching is a result of continuous changing distances in a bond between two atoms. Bending refers to a change in the angle between two bonds. Bending motions include scissoring, rocking, wagging, and twisting. The various types of vibrations and rotations absorb at different frequencies within the infrared region, thus resulting in unique spectral properties for different molecular species.

Instrumentation

Basic instrumentation for IR spectroscopy includes a radiation source, wavelength selector, sample container, detector, and signal processor. Continuous sources are used in the mid-infrared region and include the incandescent wire source, the Nernst glower, and the Globar. The Nernst glower is hotter and brighter than the incandescent wire. The Globar is a rod of silicon carbide that is electrically heated and that typically requires a water cooling system. The Globar provides greater output than the Nernst glower in the region below 5 µm. Otherwise, their spectral energies are similar. Newer technologies have also used various lasers as sources for infrared applications.

The three general types of infrared detectors are thermal, pyroelectric, and photoconducting detectors. Thermal detectors measure very minute temperature changes as a method of detection. They operate over a wide range of wavelengths and can be operated at room temperature. The main disadvantages of thermal detectors are slow response and less sensitivity when compared to other detectors. Pyroelectric detectors are specialized thermal detectors that provide faster response and more sensitivity. Photoconducting detectors rely on interactions between incident photons and a semiconductor material. These detectors also provide increased sensitivity and faster response.

The two main types of instruments used for qualitative analysis are dispersive-grating spectrophotometers and multiplex instruments that use Fourier transform. Wavelength selection in dispersive-grating instruments is accomplished using filters, prisms, or reflection gratings. Traditional instruments have consisted of a filter-grating or prism-grating system that covers the range between 4000 and 650 cm^{-1}. The most common is a double-beam instrument that uses reflection grating for dispersing radiation.

Fourier transform has been applied to infrared spectroscopy and is now a very common technique. Fourier transform infrared (FTIR) spectroscopy offers enhanced sensitivity compared to dispersive IR spectroscopy. Signal-to-noise ratios are often improved by an order of magnitude. Most commercial FTIR instruments are based on the Michelson interferometer and require a computer interface to perform the Fourier transform and process data. Using an interferometer provides an increase in resolution and results in accurate and reproducible frequency determinations. This method offers an advantage when using background subtraction techniques. Most Fourier transform instruments are now benchtop size and relatively easy to maintain. These instruments are largely replacing traditional dispersive instruments.

Sample Handling

Sample containers and handling can present a challenge in the infrared region. Materials used to produce cuvettes are not transparent and cannot be used. Cells prepared from alkali halides such as sodium chloride are widely used due to their transparent properties in the infrared region. A common problem with sodium chloride cells is that they absorb moisture and become fogged. Polishing is required to restore the cells to a more transparent state.

Liquids may be analyzed in their neat form by placing a small amount of sample on a sodium chloride plate and then placing a second plate on top to form a sample film. The plates are then placed in an appropriate holder in the sample compartment of the instrument. This technique provides adequate spectra for qualitative use. Alternatively, the use of silver chloride or silver bromide "disposable" windows has gained widespread acceptance for use with liquids and Nujol (a heavy hydrocarbon oil) mulls.

Solutions of liquid or solid materials can also be analyzed by IR spectroscopy. Solvents should be chosen that do not have absorbances in the region of interest. Unfortunately, no solvent is completely transparent in the mid-infrared region. With double-beam instruments, a reference cell containing blank solvent can be used. Common moderate absorbances will not be observed. Solvent transmission should always be above 10% when using a solvent reference cell. Influences of solvent on the absorbance of the solute should be considered. For example, hydrogen bonding of alcohols or amines with the solvent may affect characteristic vibrational frequency of the functional group. When practical, it is desirable to analyze neat materials for qualitative analysis.

A method commonly used for analysis of neat solid samples is the *mull* technique. The technique consists of grinding the material into a fine powder and then dispersing it into a liquid or solid matrix to form a mull. Liquid mulls have been formed by combining the powdered analyte with Nujol. The liquid mull is analyzed between salt plates as described above. The disadvantage of Nujol is that hydrocarbon bands may interfere with analyte absorbances. A second method of forming a mull involves grinding the powdered analyte with dry potassium bromide and forming a disk. The ratio of analyte to potassium bromide is usually about 1:100. The materials are ground together using a mortar and pestle or a small ball mill. The mixture is then pressed in a die at 10,000–15,000 psi to form a small transparent disk and analyzed. Care must be taken when preparing the disk to protect it from moisture. It is common to see absorbances for moisture when using potassium bromide disks.

Another technique for handling samples is the use of disposable sample cards. These commercially available cards contain an IR-transparent material on which a neat liquid or solution can be placed for analysis. An analyte can be dissolved in a volatile solvent, placed on the card, and the solvent evaporated to form a thin coating of sample, which can then be analyzed. This technique may be used when a limited amount of analyte material is available.

Qualitative Analysis

The most widely used application of IR spectroscopy is for qualitative analysis of organic compounds. Compounds have unique spectra that depend on molecular attributes. A common method of interpreting IR spectra is to consider two regions: the functional group frequency region (3600–1200 cm^{-1}) and the "fingerprint" region (1200–600 cm^{-1}). A combination of interpreting the functional group region and comparing the fingerprint region with those in spectral libraries provides, in many cases, sufficient evidence to positively identify a compound.

The functional group region provides evidence of functional groups in a molecule based on the absorbance frequency. Tables are available in standard spectra manuals that provide ranges of absorbances for specific functional groups. Common groups with characteristic absorbances include aldehydes, ketones, esters, alkenes, alkynes, alcohols, amines, amides, carboxylic acids, nitro groups, and nitriles. While most functional groups fall in the 3600–1200 cm^{-1} range, some can also fall in the fingerprint region. For example, C–O bonds can be around 1000 cm^{-1}, and C–Cl absorbances are typically found in the range of 600–800 cm^{-1}.

The fingerprint region is often unique to the analyte. In this region, small differences in structure can lead to differences in absorbances. Most single bonds absorb in this area, and differences in skeletal structure of molecules will result in frequency and intensity differences. The fingerprint region is most effectively used by comparison to existing spectra. Many commercial suppliers of IR instrumentation also offer searchable electronic spectral libraries. Computer-based search systems offer a rapid method of comparing unknown samples to known spectra. The systems usually yield a list of possible compounds ranked in order of best fit of the unknown spectrum to the library spectrum.

▶**Procedure for IR Spectroscopy.** IR spectroscopy is used in selected standard-grade reference materials specifications as a technique for identity confirmation. The criteria under which standard-grade reference materials pass identity tests are based on observation of characteristic absorbances and by comparison to a spectral library, if one is available. Each entry to be analyzed by IR spectroscopy will contain a minimum of three absorbances that must be present to pass the test. In addition, NIST should be used as a source to obtain comparison spectra. A standard-grade reference material will pass the comparison test if all absorbances found in the NIST spectrum are present in the test spectrum. The NIST spectral library contains many of the standard-grade reference materials listed in this book, and spectral comparisons should be made if NIST spectra are available. NIST spectra are available by contacting NIST or by accessing NIST's Web site, http://webbook. nist.gov/chemistry/. Data represented in the NIST Infrared Spectral Library were acquired by analysis of samples in the vapor phase. Methods are written for analysis in either the solid or liquid phase. In some instances, vapor-phase analysis provides differences in absorbance frequency and intensity when compared to solid- or liquid-phase analysis. Comparison of test spectra to NIST spectra should be limited to the relative presence of absorbances and should not be an exact comparison of absorbance frequency or intensity. Specific frequency absorbances should match those stated in the compound tables.

Part 3:
Solutions and Mixtures
Used in Tests

Reagents, Buffers, and Indicators

Throughout the monographs in this book, the term *reagent solution* is used to designate solutions described in this section. The tests and limits in the monographs are based on the use of the solutions in the strength indicated below.

Wherever the use of ammonium hydroxide or an acid is prescribed with no indication of strength or dilution, the reagent is to be used at full strength as described in its monograph. Dilutions are indicated either by the percentage of some constituent or by the volumes of reagents and water mixed to prepare a dilute reagent. Dilute acid or ammonium hydroxide $(1 + x)$ means a dilute solution prepared by mixing 1 volume of the strong acid or ammonium hydroxide with x volumes of reagent water.

Unless otherwise indicated, the reagent solutions are prepared and diluted with reagent water by using standard class A volumetric pipets and flasks. Weights are measured on a four-place calibrated balance.

For reference purposes, Table 3-1 presents reagents useful in the preparation of buffers for a given pH.

Reagent Water

Throughout the monographs for reagent chemicals, the term *water* means distilled water or deionized water that meets the requirements of Water, Reagent, page 716. However, for specific applications—such as UV determinations or liquid and ion chromatography—ASTM Type I reagent water (ASTM, 1999) or water for which the suitability has been determined should be used. In tests for nitrogen compounds, water should be "ammonia-free" or "nitrogen-free". Water for use in analysis of ultratrace metals must meet the requirements on page 718. For some tests, freshly boiled water must be used to ensure freedom from material absorbed from the air, such as ammonia, carbon dioxide, or oxygen.

Carbon Dioxide-Free Water. Carbon dioxide-free water can be prepared by purging reagent water with carbon dioxide-free air (using a gas dispersion tube) or nitrogen for at least 15 min or by boiling reagent water vigorously for at least 5 min and allowing it to cool while protected from absorption of carbon dioxide from the atmosphere; alternatively, fresh 18-MΩ deionized water can be used.

Table 3-1. Reagents Useful in Preparing Buffers of a Specific pH Range

pH Range	pK_a	Reagent
0–2		Hydrochloric acid; nitric acid; perchloric acid
0.3–5.3	1.3, 4.4	Oxalic acid, dihydrate; sodium oxalate; potassium tetroxalate, dihydrate
0.9–2.9, 5.2–7.2	1.9, 6.2	Maleic acid
1.1–1.8		Potassium chloride
1.1–3.1, 6.2–10.1	2.1, 7.2, 12.4	Phosphoric acid; potassium phosphate, monobasic; potassium phosphate, dibasic; potassium phosphate, tribasic, *n*-hydrate; sodium phosphate, monobasic, monohydrate; sodium phosphate, dibasic, 7-hydrate; sodium phosphate, tribasic, 12-hydrate
1.8–3.8	2.8	Monochloroacetic acid; chloroacetic acid, sodium salt
1.9–6.4	2.9, 5.4	Phthalic acid; potassium biphthalate
2.0–5.4	3.0, 4.4	*d*-Tartaric acid; potassium tartrate, ½-hydrate
2.1–7.4	3.1, 4.8, 6.4	Citric acid, anhydrous; citric acid, monohydrate; potassium citrate, monohydrate
2.8–4.8	3.8	Formic acid; sodium formate
3.2–6.6	4.2, 5.6	Succinic acid
3.6–5.6	4.6	Acetic acid; sodium acetate, trihydrate
4.1–6.1	5.1	Hexamethylenetetramine
5.3–7.3, 9.3–11.3	6.3, 10.3	Carbonic acid; potassium bicarbonate; potassium carbonate, anhydrous; sodium bicarbonate; sodium carbonate, anhydrous; sodium carbonate, monohydrate
6.5–8.5	7.5	Imidazole
6.8	4.6 (acetic acid); 9.3 (ammonium hydroxide)	Ammonium acetate
6.8–8.8	7.8	Triethanolamine; TRIS hydrochloride
7.1–9.1	8.1	*N,N*-bis(2-hydroxyethyl)glycine (bicine); 4-(2-hydroxyethyl)-1-piperazineethanesulfonic acid (HEPES)
7.8–9.8	8.8	2-Amino-2-methyl-1,2-propanediol (AMP)
8.2–10.2	9.2	Boric acid; sodium borate, 10-hydrate
8.3–10.3	9.3	Ammonium hydroxide; ammonium chloride
12–14		Potassium hydroxide, 45% solution; sodium hydroxide, 50% solution
12.4 (saturated solution)		Calcium hydroxide

Source: Reproduced courtesy of Mallinckrodt Baker, Inc.

Water (Suitable for LC Gradient Elution Analysis)

Analyze the sample by using the following gradient elution conditions. Use ACS reagent-grade or better acetonitrile.

Column: Octadecyl (C-18 polymeric phase), 250 × 4.6 mm i.d., 5 μm, 12% C loading, endcapped

Mobile Phase: A. Water to be tested

B. ACS reagent-grade acetonitrile

Conditions: Detector: Ultraviolet at 254 nm

Sensitivity: 0.02 AUFS

Gradient Elution:

1. Program the liquid chromatograph to develop a water–acetonitrile gradient as follows:

Time (min)	Flow (mL/min)	% A (water)	% B (acetonitrile)
0	2.0	100	0
20	2.0	100	0
40	2.0	0	100
50	2.0	0	100

2. Equilibrate the system at 100% water. Start the program after achieving a stable baseline. Repeat the gradient program, and use these results to evaluate the water. No peak should be greater than 10% of full scale (0.002 absorbance units).

Solutions and Mixtures

Note: Required quantities may be made having the same concentrations as below.

Acetic Acid, 1 N. Dilute 29 mL of glacial acetic acid with water to 500 mL.

Alcohol. *Alcohol* refers to the monograph for ethyl alcohol, page 308. This is distinguished from the monograph for reagent alcohol, page 567, which is a denatured form of ethyl alcohol.

Alizarin Red S, 1%. Dissolve 0.250 g of alizarin red S in water, dilute with water to 250 mL, filter, and store in glass.

Ammonia–Cyanide, Lead-Free, 2%. In a well-ventilated fume hood, dissolve 2.0 g of potassium cyanide in 15 mL of ammonium hydroxide, and dilute with water to 100 mL. Remove lead by shaking the solution with small portions of dithizone extraction solution until the dithizone retains its original green color; discard the extraction solution. Store the ammonia–cyanide solution in a polyethylene bottle.

Ammoniacal Buffer, pH 10. Dissolve 90 g of ammonium chloride in 375 mL of 28–30% ammonium hydroxide, and dilute to 500 mL with water. (The pH of a 1 + 10 dilution with water should be about 10.)

Ammonium Acetate–Acetic Acid Buffer. Dissolve 40 g of ammonium acetate in water, add 29 mL of glacial acetic acid, and dilute with water to 500 mL. (The pH of a 1 + 10 dilution with water should be about 4.5.)

Ammonium Acetate Buffer. Dissolve 30 g of ammonium acetate in water, and dilute to 100 mL with water.

Ammonium Citrate, Lead-Free, 40%. Dissolve 40 g of citric acid in 100 mL of water, and make alkaline to phenol red with ammonium hydroxide. Remove lead by shaking the solution with small portions of dithizone extraction solution until the dithizone solution retains its original green color; discard the extraction solution. Store the ammonium citrate solution in a polyethylene bottle.

Ammonium Hydroxide, 2.5% NH_3. Dilute 50 mL of ammonium hydroxide with water to 500 mL.

Ammonium Hydroxide, 10% $\tilde{N}H_3$. Dilute 200 mL of ammonium hydroxide with water to 500 mL.

Ammonium Metavanadate, 0.25%. Dissolve 1.25 g of ammonium metavanadate in 250 mL of boiling water, cool, and add 10 mL of nitric acid. Dilute with water to 500 mL, and store in a polyethylene or Teflon bottle.

Ammonium Molybdate, 5%. Dissolve 5 g of ammonium molybdate tetrahydrate in water, and dilute with water to 100 mL. Store in a polyethylene bottle. Discard if a precipitate forms.

Ammonium Molybdate, 10%. Dissolve 5 g of ammonium molybdate in water, and dilute to 50 mL.

Ammonium Molybdate–Nitric Acid Solution. Mix thoroughly 50 g of molybdic acid, 85%, in 120 mL of water, and add 70 mL of ammonium hydroxide. Filter, and add 30 mL of nitric acid. Cool, and pour, with constant stirring, into a cool mixture of 200 mL of nitric acid and 480 mL of water. Add 0.1 g of ammonium phosphate dissolved in 5 mL of water, allow to stand for 24 h, and filter through glass wool.

Ammonium Molybdate–Sulfuric Acid Solution, 5%. (For determination of phosphate.) Dissolve 5.0 g of ammonium molybdate tetrahydrate in 50 mL of 10% sulfuric acid, and dilute with water to 100 mL.

Ammonium Nitrate, 10%. Dissolve 10.0 g of ammonium nitrate in water, and dilute with water to 100 mL.

Ammonium Oxalate, 4%. Dissolve 20 g of ammonium oxalate monohydrate in water, and dilute with water to 500 mL.

Ammonium Phosphate, 13%. Dissolve 65 g of dibasic ammonium phosphate in water, and dilute with water to 500 mL.

Ammonium Thiocyanate, 30%. Dissolve 150 g of ammonium thiocyanate in water, and dilute with water to 500 mL.

Aqua Regia. In a fume hood, cautiously mix 5 mL of nitric acid in 15 mL of hydrochloric acid. Store in a Teflon or glass bottle.

Ascorbic Acid, 5%. Dissolve 5.0 g of ascorbic acid in water, and dilute with water to 100 mL. Prepare fresh solution daily.

Barium Chloride, 12%. (For determination of sulfate.) Dissolve 60 g of barium chloride dihydrate in water, filter, and dilute with water to 500 mL.

Barium Chloride, 40%. (For determination of sulfate in phosphate salts.) Dissolve 40 g of barium chloride dihydrate in 80 mL of water, filter, and dilute with water to 100 mL.

Barium Nitrate, 0.1 M. Dissolve 0.654 g of barium nitrate in water, and dilute with water to 25.0 mL.

Benedict's Solution. Dissolve 87 g of sodium citrate dihydrate and 50 g of anhydrous sodium carbonate in 400 mL of water. Heat to aid dissolution, filter if necessary, and dilute with water to 425 mL. Dissolve 8.65 g of copper sulfate pentahydrate in 50 mL of water. Add this solution, with constant stirring, to the alkaline citrate solution, and dilute with water to 500 mL.

Bromine Water. (A saturated aqueous solution of bromine.) In a well-ventilated fume hood, add about 10 mL of liquid bromine to 75 mL of water in a 100-mL bottle so that when the mixture is shaken, undissolved bromine remains in a separate phase. Store in an amber glass bottle.

Bromphenol Blue Indicator, 0.10%. Dissolve 0.10 g of the sodium salt form of bromphenol blue in water, and dilute with water to 100 mL (pH 3.0–4.6).

Bromthymol Blue Indicator, 0.10%. Dissolve 0.10 g of bromthymol blue in 100 mL of dilute alcohol (1 + 1), and filter if necessary (pH 6.0–7.6).

Brucine Sulfate. Dissolve 0.30 g of brucine sulfate in dilute ACS Reagent-grade sulfuric acid (2 + 1), previously cooled to room temperature, and dilute to 500 mL with the dilute acid. If necessary, nitrate-free acid should be prepared as follows: In a well-ventilated fume hood, dilute the concentrated sulfuric acid (about 96% H_2SO_4) to about 80% H_2SO_4 by adding it to water, heat to dense fumes of sulfur trioxide, and cool. Repeat the dilution and fuming three or four times.

Cadmium Nitrate, 3%. Dissolve 3.0 g of cadmium nitrate tetrahydrate in water, and dilute with water to 100 mL.

Chloramine-T. Dissolve 10.0 mg of chloramine-T (sodium *p*-toluenesulfonchloramide) trihydrate in 100.0 mL of water. Prepare fresh.

Chromotropic Acid, 0.01%. Dissolve 0.02 g of recrystallized chromotropic acid (see below) in concentrated, nitrate-free sulfuric acid, and dilute to 200 mL with the acid. Store in an amber bottle.

To recrystallize chromotropic acid. Add solid sodium sulfate to a saturated aqueous solution of chromotropic acid. Filter the resultant crystals using suction with a Buchner funnel, wash with ethyl alcohol, and air dry. Transfer the crystals to a beaker, add sufficient water to redissolve, and then add just enough sodium sulfate to reprecipitate the salt. Filter as before, wash with ethyl alcohol, and dry at a temperature no higher than 80 °C. Store in an amber bottle.

Note: If an orange or red precipitate forms in the solution of the sample before the addition of the reagent, discard the solution and start again. Immediate addition of the reagent should prevent formation of such a precipitate.

Chromotropic Acid, 1%. Dissolve 1.0 g of chromotropic acid in water, and dilute with water to 100 mL.

Cobalt Chloride, 6%. Dissolve 5.95 g of cobalt chloride hexahydrate and 2.5 mL of hydrochloric acid in 20 mL of water, and dilute to 100 mL with water.

Crystal Violet Indicator, 1%. Dissolve 100 mg of crystal violet in 10 mL of glacial acetic acid.

Cupric Sulfate, 6%. Dissolve 6.24 g of cupric sulfate pentahydrate and 2.5 mL of hydrochloric acid in 20 mL of water, and dilute to 100 mL with water.

Dichlorofluorescein Indicator, 0.1%. Dissolve 100 mg of dichlorofluorescein in 50 mL of alcohol, add 2.5 mL of 0.1 N sodium hydroxide, mix, and dilute with water to 100 mL.

5,5-Dimethyl-1,3-cyclohexanedione, 5% in Alcohol. Dissolve 5.0 g of 5,5-dimethyl-1,3-cyclohexanedione in alcohol, and dilute with alcohol to 100 mL.

Dimethylglyoxime, 1%. Dissolve 1.0 g of dimethylglyoxime in alcohol, and dilute with alcohol to 100 mL.

***N,N*-Dimethyl-*p*-phenylenediamine.** Add 50 mL of sulfuric acid to approximately 175 mL of water. Cool to room temperature, add 0.25 g of *N,N*-dimethyl-*p*-phenylenediamine (*p*-aminodimethylaniline) monohydrochloride, and dilute with water to 250 mL.

Diphenylamine. Dissolve 10 mg of colorless diphenylamine in 100 mL of sulfuric acid. In a separate beaker, dissolve 2 g of ammonium chloride in 200 mL of water. Cool both solutions in an ice bath, and cautiously add the sulfuric acid solution to the water solution, taking care to keep the resulting solution cold. The solution should be nearly colorless.

Dithizone Extraction Solution. Dissolve 15 mg of dithizone in 500 mL of chloroform and 5 mL of alcohol. Store the solution in a refrigerator.

Dithizone Indicator Solution. Dissolve 26 mg of dithizone in 100 mL of alcohol. Store in a refrigerator, and use within 2 months.

Dithizone Test Solution. Dissolve 5 mg of dithizone in 500 mL of chloroform. Keep the solution in a polyethylene bottle, protected from light and stored in a refrigerator.

Eosin Y, 0.50%. Dissolve 50 mg of eosin Y in 10 mL of water.

Eriochrome Black T Indicator. Grind 0.20 g of Eriochrome Black T to a fine powder with 20 g of potassium chloride.

(Ethylenedinitrilo)tetraacetic Acid, Disodium Salt, 0.2 M. Dissolve 37.2 g of (ethylenedinitrilo)tetraacetic acid, disodium salt, dihydrate, in 475 mL of water, and dilute with water to 500 mL.

Ferric Ammonium Sulfate Indicator, 8%. Dissolve 8.0 g of crystals of ferric ammonium sulfate dodecahydrate in water, and dilute with water to 100 mL. A few drops of sulfuric acid may be added, if necessary, to clear the solution.

Ferric Chloride, 4.5%. Dissolve 4.50 g of ferric chloride hexahydrate and 2.5 mL of hydrochloric acid in water, and dilute to 100 mL with water.

Ferric Chloride, 5%. Dissolve 50 g of ferric chloride hexahydrate in dilute hydrochloric acid (1 + 99), and dilute with the dilute acid to 500 mL.

Ferric Nitrate, 17%. Dissolve 17 g of ferric nitrate nonahydrate in 100 mL of water.

Ferric Sulfate, 12%. Add 13 mL of sulfuric acid to 60 g of ferric sulfate n-hydrate. Add about 350 mL of water, digest on a hot plate ($\approx$ 100 °C) to dissolve, cool to room temperature, and dilute with water to 500 mL.

Ferroin Indicator, 0.025 M. Dissolve 0.70 g of ferrous sulfate heptahydrate and 1.5 g of 1,10-phenanthroline in 100 mL of water.

Hexamethylenetetramine Solution, Saturated. Dissolve about 20 g of hexamethylenetetramine in 100 mL of water.

Hydrazine Sulfate, 2%. Dissolve 2.0 g of hydrazine sulfate in water, and dilute with water to 100 mL.

Hydrochloric Acid, 10%. Dilute 118 mL of hydrochloric acid with water to 500 mL.

Hydrochloric Acid, 20%. Dilute 235 mL of hydrochloric acid with water to 500 mL.

Hydrochloric Acid, 6 M. Dilute 250 mL of hydrochloric acid with water to 500 mL.

Hydrogen Peroxide, 3%. Dilute 10 mL of hydrogen peroxide to 100 mL with water. Prepare fresh at time of use.

Hydrogen Sulfide Water. In a well-ventilated fume hood, saturate 100 mL of water by bubbling hydrogen sulfide gas for 1 min. This solution must be freshly prepared. Store in a fume hood. Securely close container immediately after use.

Hydroxy Naphthol Blue Indicator. Use the mixture with sodium chloride that is commercially supplied, or grind 0.20 g of hydroxy napthol blue with 30 g of sodium chloride.

Hydroxylamine Hydrochloride, 10%. Dissolve 10.0 g of hydroxylamine hydrochloride in water, and dilute with water to 100 mL.

Hydroxylamine Hydrochloride for Dithizone Test. Dissolve 20 g of hydroxylamine hydrochloride in about 65 mL of water, and add 0.15 mL of thymol blue indicator solution. Add ammonium hydroxide until a yellow color appears. Add 5 mL of a 4% solution of sodium diethyldithiocarbamate. Mix thoroughly, and allow to stand for 5 min. Extract with successive portions of chloroform until no yellow color is developed in the chloroform layer when the extract is shaken with a dilute solution of a copper salt. Add hydrochloric acid until the indicator turns pink, and dilute with water to 100 mL.

Indigo Carmine, 0.10%. Dissolve 0.10 g of sample indigo carmine, dried at 105 °C, in a mixture of 80 mL of water and 10 mL of sulfuric acid, and dilute with water to 100 mL.

Jones Reductor. In a well-ventilated fume hood, cover a 250-g portion of 20-mesh zinc with reagent water in a 1-L suction flask (see apparatus description below). Pour a solution containing 11 g of mercuric chloride in 100 mL of hydrochloric acid into the flask; slowly mix and shake the system for about 2 min. Pour off the solution, and wash the amalgam thoroughly with hot tap water, and then with reagent water. The column is charged with six 250-g portions.

> *Apparatus.* Use a dispensing buret, about 22 in. long and 2 in. in diameter, equipped with a glass stopcock and a delivery tube, 6 mm wide and 3.5 in. long. The reductor is charged with an 8-in. column of 20-mesh amalgamated zinc (1500 g) and, on top of this, a 6-in. column of larger (1–2 cm) amalgamated zinc (about 750 g). The delivery tube is connected to a 1-L flask through a two-hole rubber stopper. One hole is used as an inlet; the other functions as an outlet for carbon dioxide gas.

Karl Fischer Volumetric Reagent. (Before making Karl Fischer reagent, see the discussion of commercial reagents starting on page 30.) In a well-ventilated fume hood, dissolve 254 g of iodine in 807 mL of pyridine in a 3-L glass-stoppered bottle, and add 2 L of methanol. To prepare the active reagent, add 1 L of foregoing stock to a 2-L bottle, and cool by placing the bottle in a slurry of ice pieces. Add carefully about 45 mL of liquid sulfur dioxide, collected in a calibrated cold trap, and stopper the bottle. Shake the mixture until it is homogeneous, and set aside for 24 h before use.

Lead Acetate, 10%. Cautiously dissolve 25 g of lead acetate trihydrate in water. If necessary, add a few drops of acetic acid to clear the solution, and dilute with water to 250 mL.

Lithium Chloride, 30%. Dissolve 6.1 g of lithium chloride in water, and dilute with water to 20 mL. Prepare fresh.

Metalphthalein-Screened Indicator, 0.2%. Dissolve 0.18 g of metalphthalein and 0.02 g of naphthol green B in water containing 0.5 mL of ammonium hydroxide, and dilute to 100 mL with water.

Methyl Orange Indicator, 0.10%. Dissolve 0.10 g of methyl orange in 100 mL of water (pH 3.2–4.4).

Methyl Red Indicator, 0.10%. Dissolve 0.10 g of methyl red in 100 mL of alcohol. See page 443 for a description of the three forms of methyl red (pH 4.2–6.2).

4-(Methylamino)phenol Sulfate, 2%. (For determination of phosphates.) Dissolve 2.0 g of 4-(methylamino)phenol sulfate in 100 mL of water. To 10 mL of this solution, add

90 mL of water and 20 g of sodium bisulfite. Confirm the suitability of the reagent solution by the following test: Add 1 mL of this reagent solution to each of four solutions containing 25 mL of 0.5 N sulfuric acid and 1 mL of ammonium molybdate–sulfuric acid reagent solution. Add 0.005 mg of phosphate ion (PO_4) to one of the solutions, 0.01 mg to a second, and 0.02 mg to a third. Allow to stand at room temperature for 2 h. The solutions in the three tubes should show readily perceptible differences in blue color corresponding to the relative amounts of phosphate added, and the one to which 0.005 mg of phosphate was added should be perceptibly bluer than the blank.

Methylthymol Blue Indicator. Grind 0.20 g of methylthymol blue to a fine powder with 20 g of potassium nitrate.

Murexide Indicator. Grind 0.20 g of murexide to a fine powder with 20 g of potassium nitrate.

Nessler Reagent. In a well-ventilated fume hood, dissolve 72 g of sodium hydroxide in 350 mL of water. Dissolve 25 g of red mercuric iodide and 20 g of potassium iodide in 100 mL of water. Pour the iodide solution into the hydroxide solution, and dilute with water to 500 mL. Allow to settle, and use the clear supernatant liquid. Nessler reagent prepared by any of the recognized methods may be used, provided that it has equal sensitivity. *Note:* After testing is done, the solutions should be kept separate as mercury waste, following local and state laws.

Nitric Acid, 1%. Dilute 5.3 mL of nitric acid with water to 500 mL.

Nitric Acid, 10%. Dilute 53 mL of nitric acid with water to 500 mL.

Oxalic Acid, 4%. Dissolve 20 g of oxalic acid dihydrate in water, and dilute with water to 500 mL.

PAN Indicator, 0.10%. Dissolve 0.05 g of 1-(2-pyridylazo)-2-naphthol in 50 mL of reagent alcohol.

PAR Indicator, 0.10%. Dissolve 0.10 g of 4-(2-pyridylazo)resorcinol in 100 mL of alcohol.

pH 6.5 Buffer. Dissolve 10.1 g of anhydrous sodium phosphate, dibasic, and 3.05 g of citric acid monohydrate in water, and dilute with water to 500 mL.

1,10-Phenanthroline, 0.10%. Dissolve 0.10 g of 1,10-phenanthroline monohydrate in 100 mL of water containing 0.1 mL of 10% hydrochloric acid reagent solution.

Phenol Red Indicator I, 0.10%. Dissolve 0.10 g of phenol red in 100 mL of alcohol, and filter if necessary (pH 6.8–8.2)

Phenol Red Indicator II, pH 4.7. (For determination of bromide.) Dissolve 33 mg of phenol red in 1.5 mL of 2 N sodium hydroxide solution, dilute with water to 100 mL (this is solution A). Dissolve 25 mg of ammonium sulfate in 235 mL of water; add 105 mL of 2 N sodium hydroxide solution and 135 mL of 2 N acetic acid (this is solution B). Add 25 mL of solution A to solution B, and mix. Using a pH meter, adjust the pH of this solution to 4.7 if necessary.

Phenoldisulfonic Acid. In a well-ventilated fume hood, dissolve 5 g of phenol in 30 mL of sulfuric acid, add 15 mL of fuming sulfuric acid (15% SO_3), and heat at 100 °C for 2 h. Cool, and store in a glass bottle.

Phenolphthalein Indicator, 1%. Dissolve 1.0 g of phenolphthalein in 100 mL of alcohol (pH 8.0–10.0).

Potassium Chloride, 2.5%. Dissolve 2.5 g of potassium chloride in water, and dilute with water to 100 mL.

Potassium Chromate, 10%. Dissolve 10.0 g of potassium chromate in water, and dilute with water to 100 mL.

Potassium Cyanide, Lead-Free, 2.5%. In a well-ventilated fume hood, dissolve 5.0 g of potassium cyanide in sufficient water to make 10 mL. Remove lead by shaking with portions of dithizone extraction solution. Part of the dithizone remains in the aqueous phase but can be removed, if desired, by shaking with chloroform. Dilute the potassium cyanide solution with water to 50 mL.

Potassium Dichromate, 10%. Dissolve 10.0 g of potassium dichromate in water, and dilute with water to 100 mL.

Potassium Ferricyanide, 5%. Dissolve 2.0 g of potassium ferricyanide in 40 mL of water. Prepare the solution at the time of use.

Potassium Ferrocyanide, 10%. Dissolve 4.0 g of potassium ferrocyanide trihydrate in 40 mL of water. Prepare the solution at the time of use.

Potassium Hydroxide, 0.5 N in Methanol. Dissolve 18 g of potassium hydroxide in 10 mL of water, and dilute with methanol to 500 mL. Allow the solution to stand in a stoppered bottle for 24 h. Decant the clean supernatant solution into a bottle provided with a tight-fitting stopper. Store in a polyethylene or Teflon bottle.

Potassium Iodide, 10%. (For determination of free chlorine.) Dissolve 1.0 g of potassium iodide in water, and dilute with water to 10 mL. Prepare the solution at the time of use.

Potassium Iodide, 16.5%. (For determination of arsenic.) Dissolve 16.5 g of potassium iodide in water, and dilute with water to 100 mL.

Potassium Permanganate, 5%. Dissolve 2.5 g of potassium permanganate in 50 mL of water.

Silver Diethyldithiocarbamate, 0.5%. Dissolve 1.0 g of silver diethyldithiocarbamate in 200 mL of freshly distilled pyridine.

Silver Nitrate, 1.7%. Dissolve 8.5 g of silver nitrate in 500 mL of water. Store in an amber bottle.

Sodium Acetate, 10%. Dissolve 10 g of sodium acetate trihydrate in water, and dilute with water to 100 mL.

Sodium Borohydride, 0.6%. Dissolve 0.6 g of sodium borohydride and 1.0 g of 50% sodium hydroxide, and dilute with stirring to 100 mL with water.

Sodium Carbonate, 1%. Dissolve 5.0 g of sodium carbonate, anhydrous, in 450 mL of water, and dilute with water to 500 mL. Store in a plastic bottle.

Sodium Citrate, 1 M. Dissolve 147 g of sodium citrate dihydrate in 450 mL of water, and dilute with water to 500 mL.

Sodium Cyanide, 10%. In a well-ventilated hood, dissolve 10 g of sodium cyanide in water, and dilute with water to 100 mL.

Sodium Diethyldithiocarbamate. Dissolve 0.25 g of sodium diethyldithiocarbamate in water, and dilute with water to 250 mL.

Sodium Hydroxide, 10%. Dissolve 50 g of sodium hydroxide in water, and dilute with water to 500 mL. Store in a polyethylene or Teflon bottle.

Sodium Hydroxide, Ammonia-Free, 6 N. Dissolve 24 g of sodium hydroxide in 100 mL of water.

Sodium Sulfite, 10%. Dissolve 2.5 g of sodium sulfite in water, and dilute with water to 20 mL. Keep in a tightly closed bottle. This solution should be freshly prepared.

Stannous Chloride, 2%. Dissolve 0.50 g of stannous chloride dihydrate in hydrochloric acid, and dilute with hydrochloric acid to 25 mL. Store in a polyethylene or Teflon bottle.

Stannous Chloride, 40%. Dissolve 20.0 g of stannous chloride dihydrate in 50 mL of hydrochloric acid. Store this solution in a polyethylene or Teflon container, and use within 3 months.

Starch Indicator, 0.5%. Mix 1.0 g of soluble starch with 10 mg of red mercuric iodide and enough cold water to make a thin paste, add 200 mL of boiling water, and boil for 1 min while stirring. Cool before use.

Sulfuric Acid, 0.5 N. In a well-ventilated fume hood, slowly add 7.5 mL of sulfuric acid to 375 mL of water, and dilute to 500 mL.

Sulfuric Acid, 4 N. In a well-ventilated fume hood, slowly add 60 mL of sulfuric acid to 375 mL of water, and dilute to 500 mL.

Sulfuric Acid, 10%. In a well-ventilated fume hood, slowly add 30 mL of sulfuric acid to 375 mL of water, cool, and dilute with water to 500 mL.

Sulfuric Acid, 25%. In a well-ventilated fume hood, slowly add 70 mL of sulfuric acid to 375 mL of water, cool, and dilute with water to 500 mL.

Sulfuric Acid, Chloride- and Nitrate-Free. In a well-ventilated fume hood, gently fume sulfuric acid in a crucible or dish for at least 30 min. Allow to cool, and transfer to a tightly capped glass bottle for storage.

Tartaric Acid, 10%. Dissolve 50 g of tartaric acid in 400 mL of water, dilute with water to 500 mL, and filter into a plastic bottle.

Thymol Blue Indicator, 0.10%. Dissolve 0.10 g of thymol blue in 100 mL of alcohol (acid range, pH 1.2–2.8; alkaline range, pH 8.0–9.2).

Thymolphthalein Indicator, 0.10%. Dissolve 0.10 g of thymolphthalein in 100 mL of alcohol (pH 8.8–10.5).

Titanium Tetrachloride. In a well-ventilated fume hood, cool separately, in small beakers surrounded by crushed ice, 5 mL of 20% hydrochloric acid and 5 mL of clear, colorless titanium tetrachloride. Add the titanium tetrachloride dropwise to the chilled hydrochloric acid. Allow the mixture to stand at ice temperature until all of the solid dissolves, and then dilute the solution with 20% hydrochloric acid to 500 mL.

Triton X-100, 0.20%. Dissolve 0.20 g of Triton X-100 (polyethylene glycol ether of isooctylphenol) in water, and dilute with water to 100 mL.

Variamine Blue B Indicator, 1%. Dissolve 0.20 g of variamine blue B in 20 mL of water, and stir for 5 min.

Xylenol Orange Indicator Mixture. Grind 0.20 g of xylenol orange (either free acid or sodium salt form) to a fine powder with 20 g of potassium nitrate. Alternatively, use a solution of 0.1 g of xylenol orange in 100 mL of alcohol (for the acid form) or water (for the salt form), depending on the solution matrix.

Control, Standard, and Stock Solutions

Control Solutions

Many of the tests for impurities require the comparison of the color or the turbidity produced under specified conditions by an impurity ion with the color or the turbidity produced under similar conditions by a known amount of the impurity. Three types of solutions are used to determine amounts of impurities in reagent chemicals: (1) the blank, (2) the standard, and (3) the control.

(1) Blank. A solution containing the quantities of solvents and reagents used in the test.

(2) Standard. A solution containing the quantities of solvents and reagents used in the test plus an added known amount of the impurity to be determined.

(3) Control. A solution containing the quantities of solvents and reagents used in the test plus an added known amount of the impurity to be determined and a known amount of the reagent chemical being tested. Some of the chemical being tested must be added to the solution because it is known that the chemical interferes with the test.

Throughout this book, the expressions "for the control, …" or "for the standard, add X mg of Y ion" mean to add the proper aliquot of one of the standard solutions listed below. These solutions must be prepared at the time of use or checked often enough to make certain that the tests will not be affected by changes in strength of the standard solution during storage. Calibrated volumetric glassware must be used in preparing these solutions.

Standard and Stock Solutions

Note: In place of standard solutions prepared as described below, commercially available standards may be used after proper dilution. Required quantities may be made per usage or in smaller quantities, as long as the concentration remains the same.

Acetaldehyde (0.1 mg of CH_3CHO in 1 mL). Prepare in water by appropriate dilution of an assayed acetaldehyde solution.

Acetone (0.1 mg of $(CH_3)_2CO$ in 1 mL). Dilute 6.4 mL of acetone with water to 1 L. Dilute 2.0 mL of this solution with water to 100 mL. Prepare the solution at the time of use.

Aluminum (0.01 mg of Al in 1 mL). Dissolve 0.100 g of metallic aluminum in 10 mL of dilute hydrochloric acid (1 + 1), and dilute with water to 100 mL. To 10 mL of this solution, add 25 mL of dilute hydrochloric acid (1 + 1), and dilute with water to 1 L.

Ammonium (0.01 mg of NH_4 in 1 mL). Dissolve 0.296 g of ammonium chloride in water, and dilute with water to 100 mL. Dilute 10 mL of this solution with water to 1 L.

Antimony (0.1 mg of Sb in 1 mL). Dissolve 0.2743 g of antimony potassium tartrate hemihydrate in 100 mL of water. Add 100 mL of hydrochloric acid, and dilute with water to 1 L.

Arsenic (0.001 mg of As in 1 mL). Dissolve 0.132 g of arsenic trioxide in 10 mL of 10% sodium hydroxide reagent solution, neutralize with 10% sulfuric acid reagent solution, add an excess of 10 mL, and dilute with water to 1 L. To 10 mL of this solution, add 10 mL of 10% sulfuric acid reagent solution, and dilute with water to 1 L.

Barium (0.1 mg of Ba in 1 mL). Dissolve 0.178 g of barium chloride dihydrate in water, and dilute with water to 1 L.

Bismuth (0.01 mg of Bi in 1 mL). Dissolve 0.232 g of bismuth trinitrate pentahydrate in 10 mL of dilute nitric acid (1 + 9), and dilute with water to 100 mL. Dilute 10 mL of this solution plus 10 mL of nitric acid with water to 1 L.

Bromate (0.1 mg of BrO_3 in 1 mL). Dissolve 0.131 g of potassium bromate in water, and dilute with water to 1 L.

Bromide (0.1 mg of Br in 1 mL). Dissolve 1.49 g of potassium bromide in water, and dilute with water to 100 mL. Dilute 10 mL of this solution with water to 1 L.

Cadmium (0.025 mg of Cd in 1 mL). Dissolve 0.10 g of cadmium chloride 2.5-hydrate in water, add 1 mL of hydrochloric acid, and dilute with water to 500 mL. To 25 mL of this solution, add 1 mL of hydrochloric acid, and dilute with water to 100 mL.

Calcium I (0.1 mg of Ca in 1 mL). Dissolve 0.250 g of calcium carbonate in 20 mL of water and 5 mL of 10% hydrochloric acid reagent solution, and dilute with water to 1 L.

Calcium II (0.01 mg of Ca in 1 mL). Dilute 100 mL of calcium I standard solution with water to 1 L.

Carbon (4 mg of C in 1 mL). Dissolve 8.5 g of potassium hydrogen phthalate in water, and dilute to 1 L.

Carbonate (0.3 mg of CO_3, 0.2 mg of CO_2, or 0.06 mg of C in 1 mL). Dissolve 0.53 g of sodium carbonate in carbon dioxide-free water, and dilute with carbon dioxide-free water to 1 L. Prepare the solution at the time of use.

Chloride (0.01 mg of Cl in 1 mL). Dissolve 0.165 g of sodium chloride in water, and dilute with water to 100 mL. Dilute 10 mL of this solution with water to 1 L.

Chlorine (0.01 mg of Cl$_2$ in 1 mL). Dilute 1.0 mL of a 2.5% sodium hypochlorite solution with water to 1 L. Correct for any assay not equal to 2.5%. Standardize the sodium hypochlorite solution before use by the following method: Pipet 3 mL into a tared, glass-stoppered flask. Weigh accurately, and add 50 mL of water, 2 g of potassium iodide, and 10 mL of acetic acid. Titrate the liberated iodine with 0.1 N sodium thiosulfate volumetric solution, adding 3 mL of starch indicator solution near the end point. Prepare this solution at the time of use. One milliliter of 0.1 N sodium thiosulfate corresponds to 3.723 mg of NaOCl.

Chlorobenzene (0.1 mg of C$_6$H$_5$Cl in 1 mL). Dilute 1.0 mL of chlorobenzene to 100 mL with 2,2,4-trimethylpentane. Dilute 9.0 mL of this solution to 100 mL with 2,2,4-trimethylpentane.

Chromate (0.01 mg of CrO$_4$ in 1 mL). Dissolve 1.268 g of potassium dichromate in water, and dilute with water to 1 L. Dilute 1.0 mL of this solution to 100 mL.

Copper Standard for Dithizone Test (0.001 mg of Cu in 1 mL). Dilute 1 mL of the copper stock solution with water to 100 mL. This dilute solution must be prepared immediately before use.

Copper Stock Solution (0.1 mg of Cu in 1 mL). Dissolve 0.393 g of cupric sulfate pentahydrate in water, and dilute with water to 1 L.

Ferric Iron (0.1 mg of Fe^{3+} in 1 mL). Dissolve 0.863 g of ferric ammonium sulfate dodecahydrate in water, and dilute with water to 1 L.

Ferrous Iron (0.01 mg of Fe^{2+} in 1 mL). Dissolve 0.702 g of ferrous ammonium sulfate hexahydrate in 10 mL of 10% sulfuric acid reagent solution, and dilute with water to 100 mL. Before use, make a further dilution: To 1.0 mL of the first solution, dilute with water to 100.0 mL containing 0.10 mL of sulfuric acid.

Fluoride Standard Solution (0.005 mg in 1 mL). Transfer 5.0 mL of the fluoride stock solution to a 1-L plastic volumetric flask, dilute with water to the mark, and store in a plastic bottle. Prepare the solution at the time of use.

Fluoride Stock Solution (1 mg of F in 1 mL). Dissolve 2.21 g of sodium fluoride previously dried at 110 °C for 2 h, and dissolve with 200 mL of water in a 400-mL plastic beaker. Transfer to a 1-L volumetric flask, dilute with water to the mark, and store the solution in a plastic bottle.

Formaldehyde I (0.1 mg of HCHO in 1 mL). Dilute 2.7 g (2.7 mL) of 37% formaldehyde solution to 1 L with water (prepare freshly as needed). Dilute 10 mL of this solution to 100 mL.

Formaldehyde II (0.01 mg of HCHO in 1 mL). Dilute 10 mL of formaldehyde I standard solution with water to 100 mL. Prepare the solution at the time of use.

Hydrogen Peroxide (0.1 mg of H$_2$O$_2$ in 1 mL). Dilute 3.0 mL of 30% hydrogen peroxide with water to 100 mL. Dilute 1.0 mL of this solution with water to 100 mL. Prepare the solution at the time of use.

Iodate I (0.10 mg of IO$_3$ in 1 mL). Dissolve 0.123 g of potassium iodate in water, and dilute to 1 L.

Iodate II (0.01 mg of IO$_3$ in 1 mL). Dilute 10 mL of iodate I standard solution to 100 mL.

Iodide (0.01 mg of I in 1 mL). Dissolve 0.131 g of potassum iodide in water, and dilute with water to 100 mL. Dilute 10 mL of this solution with water to 1 L.

Iron (0.01 mg of Fe in 1 mL). Dissolve 0.702 g of ferrous ammonium sulfate hexahydrate in 10 mL of 10% sulfuric acid reagent solution, and dilute with water to 100 mL. To 10 mL of this solution, add 10 mL of 10% sulfuric acid reagent solution, and dilute with water to 1 L.

Lead Standard for Dithizone Test (0.001 mg of Pb in 1 mL). Dilute 1 mL of lead stock solution to 100 mL with dilute nitric acid (1 + 99). This solution must be prepared immediately before use.

Lead Standard for Heavy Metals Test (0.01 mg of Pb in 1 mL). Dilute 10 mL of lead stock solution to 100 mL with water. This dilute solution must be prepared at the time of use.

Lead Stock Solution (0.1 mg of Pb in 1 mL). Dissolve 0.160 g of lead nitrate in 100 mL of dilute nitric acid (1 + 99), and dilute with water to 1 L. The solution should be prepared and stored in containers free from lead. Its strength should be checked every few months to determine whether the lead content has changed by reaction with the container.

Magnesium I (0.01 mg of Mg in 1 mL). Dissolve 1.014 g of clear crystals of magnesium sulfate heptahydrate in water, and dilute with water to 100 mL. Dilute 10 mL of this solution with water to 1 L.

Magnesium II (0.002 mg of Mg in 1 mL). Dilute 100 mL of magnesium I standard solution with water to 500 mL. Prepare the solution at the time of use.

Manganese (0.01 mg of Mn in 1 mL). Dissolve 3.08 g of manganese sulfate monohydrate in water, and dilute with water to 1 L. Dilute 10 mL of this solution with water to 1 L.

Mercury (0.05 mg of Hg in 1 mL). Dissolve 1.35 g of mercuric chloride in water, add 8 mL of hydrochloric acid, and dilute with water to 1 L. To 50 mL of this solution, add 8 mL of hydrochloric acid, and dilute with water to 1 L.

Methanol (0.1 mg of CH$_3$OH in 1 mL). Dilute 5.1 mL of methanol with water to 200 mL. Dilute 5.0 mL of this solution with water to 1 L. Prepare the solution at the time of use.

Molybdenum (0.01 mg of Mo in 1 mL). Dissolve 0.150 g of molybdenum trioxide in 10 mL of dilute ammonium hydroxide (1 + 9), and dilute with water to 100 mL. Dilute 10 mL of this solution with water to 1 L.

Nickel (0.01 mg of Ni in 1 mL). Dissolve 0.448 g of nickel sulfate hexahydrate in water, and dilute with water to 100 mL. Dilute 10 mL of this solution with water to 1 L.

Nitrate (0.01 mg of NO$_3$ in 1 mL). Dissolve 0.163 g of potassium nitrate in water, and dilute with water to 100 mL. Dilute 10 mL of this solution with water to 1 L.

Nitrilotriacetic Acid (10 mg of (HOCOCH$_2$)$_3$N in 1 mL). Transfer 1.0 g of nitrilotriacetic acid to a 100-mL volumetric flask. Dissolve in 10 mL of 10% potassium hydroxide solution, and dilute to the mark with water.

Nitrite I (0.01 mg of NO$_2$ in 1 mL). Dissolve 0.150 g of sodium nitrite in water, and dilute with water to 100 mL. Dilute 1.0 mL of this solution with water to 100 mL. This solution should be freshly prepared.

Nitrite II (0.001 mg of NO$_2$ in 1 mL). Dilute 0.1 mL of nitrite I standard solution with water to 100 mL. This solution must be prepared at the time of use.

Nitrogen (0.01 mg of N in 1 mL). Dissolve 0.382 g of ammonium chloride in water, and dilute with water to 100 mL. Dilute 10 mL of this solution with water to 1 L.

Phosphate (0.01 mg of PO$_4$ in 1 mL). Dissolve 0.143 g of monobasic potassium phosphate in water, and dilute with water to 100 mL. Dilute 10 mL of this solution with water to 1 L.

Platinum–Cobalt Standard (APHA No. 500). See page 43.

Potassium (0.01 mg of K in 1 mL). Dissolve 0.191 g of potassium chloride in water, and dilute with water to 1 L. Dilute 100 mL of this solution with water to 1 L.

Potassium Bromide (0.003 mg of KBr in 1 mL). Dissolve 3.0 mg of potassium bromide in 1 L of water. Prepare fresh before use.

Selenium (1 μg of Se in 1 mL). Transfer 0.10 mL of selenium standard (1000 μg/g Se, commercially available standard) to a 100-mL volumetric flask, and dilute to the mark with water. Prepare fresh before use.

Silica (0.01 mg of SiO$_2$ in 1 mL). Dissolve 0.473 g of sodium silicate in 100 mL of water in a platinum or polyethylene dish. Dilute 10 mL of this solution (polyethylene graduate) with water to 1 L in a polyethylene bottle.

Silver (0.1 mg of Ag in 1 mL). Dissolve 0.157 g of silver nitrate in water, and dilute with water to 1 L. Store in an opaque plastic bottle.

Sodium (0.01 mg of Na in 1 mL). Dissolve 0.254 g of sodium chloride in water, and dilute with water to 1 L. Dilute 100 mL of this solution with water to 1 L.

Strontium I (0.1 mg of Sr in 1 mL). Dilute 10 mL of strontium stock solution with water to 100 mL.

Strontium II (0.01 mg of Sr in 1 mL). Dilute 10 mL of strontium stock solution with water to 1 L.

Strontium Stock Solution (1 mg of Sr in 1 mL). Dissolve 0.242 g of strontium nitrate in water, and dilute with water to 100 mL.

Sulfate (0.01 mg of SO$_4$ in 1 mL). Dissolve 0.148 g of anhydrous sodium sulfate in water, and dilute with water to 100 mL. Dilute 10 mL of this solution with water to 1 L.

Sulfide (0.01 mg of S in 1 mL). (*Note:* If liquid is observable on crystals, blot the crystals with filter paper, then weigh.) Dissolve 0.75 g of sodium sulfide nonahydrate in water, and dilute with water to 100 mL. Dilute 10 mL of this solution with water to 1 L. This solution must be prepared immediately before use.

Tin (0.05 mg of Sn in 1 mL). Dissolve 0.100 g of metallic tin (Sn) in 10 mL of dilute hydrochloric acid (1 + 1), and dilute with water to 100 mL. Dilute 5 mL of this solution with dilute hydrochloric acid (1 + 9) to 100 mL. This solution must be prepared immediately before use.

Titanium I (0.1 mg of Ti in 1 mL). Cool separately, in small beakers surrounded by crushed ice, 3.96 g (2.3 mL) of titanium tetrachloride and 25 mL of 20% hycrochloric acid. Add dropwise the titanium tetrachloride to the chilled hydrochloric acid. Allow the mixture to stand at ice temperature until all the solid dissolves. Transfer to a 1-L volumetric flask, and dilute to the mark with water. Dilute 10 mL of this solution to 100 mL in a volumetric flask.

Titanium II (0.01 mg of Ti in 1 mL). Dilute 10 mL of titanium I standard solution to 100 mL in a volumetric flask. Prepare fresh at time of use.

Zinc Standard for Dithizone Tests (0.001 mg of Zn in 1 mL). Dilute 1 mL of zinc stock solution with water to 100 mL. This dilute solution must be prepared immediately before use.

Zinc Stock Solution (0.1 mg of Zn in 1 mL). Dissolve 0.124 g of zinc oxide in 10 mL of dilute sulfuric acid (1 + 9), and dilute with water to 1 L.

Volumetric Solutions

Throughout the monographs in this book, the term *volumetric solution* is used to designate a solution prepared and standardized as described in this section.

Two background references that the analyst may find helpful are ASTM E200-97(2001)e1 and the section on Volumetric Solutions in *United States Pharmacopeia* (see the bibliographies on analytical chemistry, page 772, and on physical properties, page 774). Various textbooks on chemical analysis also deal with volumetric solutions and volumetric analysis, on both a theoretical and a practical basis.

General Information

The concentrations of volumetric solutions usually are expressed as N, normality (the number of gram equivalent weights in each liter of solution), although concentration sometimes is expressed as M, molarity (the number of gram molecular weights in each liter of solution).

The accuracy of many analytical procedures is dependent on the manner in which such solutions are prepared, standardized, and stored. The directions that follow describe the preparation of these solutions and their standardization against primary standards or against already-standardized solutions. Alternatively, in cases where suitable primary standard chemicals are available, volumetric solutions may be prepared by dissolving accurately weighed portions of these substances to make an accurately known volume of solution. All volumetric solutions, of course, must be thoroughly mixed as part of their preparation. Stronger or weaker solutions than those described are prepared and standardized in the same general manner, using proportional amounts of reagents and standards. Lower strengths frequently may be prepared by accurately diluting a stronger solution. However, when necessary because of the accuracy requirements of the analysis being performed, volumetric solutions prepared in this manner should be standardized before use against a primary standard or by comparison with an appropriate volumetric solution of known strength.

Certain reagents, such as arsenic trioxide, reductometric standard, are designated as *standards* in this book. Certified standards of a number of chemical compounds are available as standard reference materials (SRMs) from NIST.

It is desirable to standardize a solution in such a way that the end point is observed at the same pH value (or oxidation–reduction potential) anticipated during subsequent use.

For example, a sodium hydroxide solution found to be 0.1000 N when standardized against potassium hydrogen phthalate to phenolphthalein (pH 8.5) may behave like 0.1002 N when used for titrating a strong acid to methyl red (pH 5). Only a small part of this difference is due to the NaOH required to bring pure water from pH 5 to 8.5. In this case, the error can be avoided by titrating the strong acid to phenolphthalein (pH 8.5), the same indicator that was used in standardizing the sodium hydroxide.

Storage

Glass containers are suitable for the storage of most volumetric solutions. However, poly-olefin containers are recommended for alkaline solutions. Volumetric solutions are stable for varying lengths of time, depending on their chemical nature and their concentration. Dilute solutions are likely to be less stable than those that are 0.1 N or stronger. If there is any doubt about the reliability of a solution, it should be restandardized at the time of use. Volumetric solutions should be stored appropriately to avoid environmental interaction that may adversely affect the stability of the assigned normality or molarity. Opaque containers and carbon dioxide or oxygen traps may be appropriate under certain conditions.

Temperature Considerations

If possible, volumetric solutions should be prepared, standardized, and used at 25 °C. If a titration is carried out at a temperature different from that at which the solution was standardized, a temperature correction may be needed, depending on the accuracy requirement of the analysis being performed. If the temperature of use is higher than the temperature of standardization, the correction is to be subtracted from the standardization value; if the temperature of use is lower than the temperature of standardization, the correction is to be added. For guidance with respect to temperature corrections, consider that a 5 °C temperature change will alter the normalities of solutions approximately as follows: For aqueous solutions 0.1 N or less, by 1 part in 1000; 0.5 N, by 1.2 parts in 1000; and 1 N, by 1.4 parts in 1000. For nonaqueous solutions, based on the expansion coefficients of the solvents, by 5.4 parts in 1000 for glacial acetic acid; 5.5 parts in 1000 for dioxane; and 6.2 parts in 1000 for methanol.

Errors caused by temperature differences, as well as by improper drainage of burets or pipets, can be avoided if measurements are made on a weight basis. With a sensitive direct-reading balance and the use of squeeze bottles, titrations by weight have become very practical and precise.

Correction Factors

It is not necessary that volumetric solutions have the identical normalities (or molarities) indicated in this section, or as specified in many of the standardization and assay calculations in this book. It is necessary, however, that the exact value be known, because if it is different from the nominal value specified, a correction factor will have to be applied. Thus, if a calculation is based on an exactly 0.1 N volumetric solution, and the solution actually used is 0.1008 N, its volume must be multiplied by a factor of 1.008 to obtain the corrected volume of 0.1 N solution to use in the calculation. Similarly, if the actual normality is 0.0984 N, the factor is 0.984.

Replication

Because the results obtained by titrations depend on the reliability of the volumetric solutions used, it is imperative that the latter be standardized at least in triplicate, with particular care, and preferably by experienced analysts. Triplicate standardizations for solutions from 0.1 N to 1 N should agree at least to within 2 parts in 1000. Triplicates for 0.05 N solutions may differ by as much as 10 parts in 1000. If deviations are greater than indicated, the determinations should be repeated until the criteria are met.

Solutions

Note: When "measure accurately" is specified in this section, the volumetric ware used shall conform to the tolerance accepted by NIST (see the bibliography on measurement techniques, page 773).

Acetic Acid, 1 N. Add 61 g (58.1 mL) of glacial acetic acid to a 1-L volumetric flask, dilute to volume with water, and mix thoroughly. Standardize as follows: Measure accurately 40 mL of the solution into a 250-mL conical flask, add 0.10 mL of phenolphthalein indicator solution, and titrate with freshly standardized 1 N sodium hydroxide volumetric solution to a permanent pink color.

$$N = \frac{mL \times N\ NaOH}{mL\ CH_3COOH}$$

Ammonium Thiocyanate, 0.1 N. Dissolve 7.61 g of ammonium thiocyanate in 100 mL of water. Transfer to a 1-L volumetric flask, dilute to volume with water, and mix thoroughly. Standardize as follows: Measure accurately 40 mL of freshly standardized 0.1 N silver nitrate volumetric solution into a 250-mL conical flask containing 50 mL of water. Add 2 mL of nitric acid and 2 mL of ferric ammonium sulfate indicator solution. While stirring, titrate with ammonium thiocyanate solution to the first appearance of a red-brown color.

$$N = \frac{mL \times N\ AgNO_3}{mL\ NH_4SCN}$$

Bromine, 0.1 N, Solution I. (Prepared from bromine.) In a well-ventilated hood, pipet 2.6 mL of bromine into a 1-L glass-stoppered volumetric flask containing 25 g of potassium bromide and 5 mL of hydrochloric acid dissolved in 500 mL of water. Stopper and swirl until the bromine is dissolved, dilute to volume with water, and mix thoroughly. Standardize as follows: Measure accurately 40 mL of the solution into a 250-mL iodine flask containing 100 mL of water. Add 3 g of potassium iodide, and titrate the liberated iodine with freshly standardized 0.1 N sodium thiosulfate volumetric solution. Add 3 mL of starch indicator solution near the end of the titration, and continue to the absence of the blue starch–iodine complex.

$$N = \frac{mL \times N\ Na_2S_2O_3}{mL\ Br_2}$$

Bromine, 0.1 N, Solution II. (Prepared as a potassium bromate–potassium bromide mixture.) Transfer 2.8 g of potassium bromate, plus 15 g of potassium bromide, to a 1-L volumetric flask, dilute to volume with water, and mix thoroughly. Standardize as follows: Measure accurately 40 mL of the solution into a 250-mL iodine flask, and dilute with 100 mL of water. Add 3 g of potassium iodide, stopper the flask, and swirl the contents carefully; then add 3 mL of hydrochloric acid, stopper the flask again, swirl vigorously, and allow to stand in the dark for 5 min. Titrate the liberated iodine with freshly standardized 0.1 N sodium thiosulfate volumetric solution. Add 3 mL of starch indicator solution near the end of the titration, and continue to the absence of the blue starch–iodine complex.

$$N = \frac{mL \times N \ Na_2S_2O_3}{mL \ Br_2}$$

Ceric Ammonium Sulfate, 0.1 N. Place 63.26 g of ceric ammonium sulfate dihydrate in a 1500-mL beaker, and slowly add 30 mL of sulfuric acid. Stir to a smooth paste, and cautiously add water in portions of 20 mL or less, with stirring, until the salt is dissolved. Dilute, if necessary, to approximately 500 mL, and mix thoroughly by stirring. Allow to stand for at least 8 h. If a residue has formed, or the solution is turbid, filter through a fine-porosity sintered-glass filter (do not filter through paper or similar material). Transfer the filtrate to a 1-L volumetric flask, dilute to volume, and mix thoroughly. Standardize as follows: Weigh accurately into a 250-mL conical flask 0.21 g of arsenic trioxide, reductometric standard (NIST SRM 83), previously dried for 12 h at 110 °C. Add 15 mL of 1 N sodium hydroxide, warm gently, and swirl to hasten dissolution, being certain that no particles of arsenic trioxide remain on the sides of the flask. Add 100 mL of water and 10 mL of 25% sulfuric acid solution, followed by 0.1 mL of 1,10-phenanthroline reagent solution and 0.2 mL of 0.1 N ferrous ammonium sulfate solution. Add 0.1 mL of osmium tetroxide solution [prepared by dissolving, in a well-ventilated hood, 0.1 g of osmium tetroxide in 40 mL of 0.1 N sulfuric acid (*caution*: osmium tetroxide vapor is extremely hazardous)], and titrate slowly with the ceric solution to a change from pink to pale blue. Perform a blank titration on all of the reagents except the arsenic trioxide, and subtract the volume of ceric sulfate solution consumed in the blank titration from that consumed in the first titration.

$$N = \frac{g \ As_2O_3 \times (Assay / 100)}{0.04946 \times [mL(sample) - mL(blank)] \ Ce^{IV} \ solution}$$

Cupric Sulfate, 0.1 M. Weigh accurately 24.97 g of cupric sulfate pentahydrate. Dissolve in 500 mL of water, and quantitatively transfer to a 1-L volumetric flask. Dilute to volume with water, and mix thoroughly. Standardize as follows: Using a 50-mL buret, measure 40.0 mL of the cupric sulfate solution into a 250-mL beaker. From a buret, add 45.0 mL of freshly standardized 0.1 M EDTA and 50 mL of methanol. Stir magnetically and, using a pH meter, adjust the pH of the solution to 5 with saturated aqueous ammonium acetate. Maintain the pH between 5.1 and 5.3 throughout the titration. Add 0.1 mL of PAN indicator solution, and continue to titrate with the 0.1 M copper sulfate volumetric solution to a permanent blue end point.

$$N = \frac{mL \times M \ EDTA}{mL \ CuSO_4 \cdot 5H_2O \ (total)}$$

EDTA, 0.1 M. Dissolve 37.22 g of $Na_2EDTA \cdot 2H_2O$ in 500 mL of water. Transfer quantitatively to a 1-L volumetric flask, dilute to volume with water, and mix thoroughly. Standardize as follows: Weigh accurately 0.40 g of calcium carbonate, chelometric standard (NIST SRM 915) previously dried at 210 °C for 4 h. Transfer to a 400-mL beaker, and add water. Cover the beaker with a watch glass, and introduce 4 mL of hydrochloric acid (1 + 1) from a pipet inserted between the lip of the beaker and the edge of the watch glass. Swirl the contents of the beaker to dissolve the calcium carbonate. Wash down the sides of the beaker and the watch glass with water, and dilute to 200 mL. While stirring, add from a 50-mL buret about 30 mL of EDTA standard solution. Adjust the solution to pH 12–13 with 10% sodium hydroxide and add 300 mg of hydroxy naphthol blue indicator mixture. Continue the titration with the EDTA volumetric standard solution to a blue end point.

$$N = \frac{(g \, CaCO_3) \times (Assay \, / \, 100)}{0.10009 \times mL \, EDTA \, solution}$$

Ferrous Ammonium Sulfate, 0.1 N. Dissolve 39.21 g of ferrous ammonium sulfate hexahydrate in 200 mL of 25% sulfuric acid. Transfer quantitatively to a 1-L volumetric flask, dilute to volume with water, and mix thoroughly. On each day of use, standardize as follows: Measure accurately 40 mL of the solution into a 250-mL conical flask, add 0.1 mL of 1,10-phenanthroline reagent solution, and titrate with freshly standardized 0.1 N ceric ammonium sulfate volumetric solution to the change from pink to pale blue.

$$N = \frac{mL \times N \, Ce^{IV} \, solution}{mL \, Fe^{II} \, solution}$$

Ferrous Sulfate, 0.1 N. Dissolve 27.80 g of clear ferrous sulfate heptahydrate crystals in about 500 mL of water containing 50 mL of sulfuric acid. Transfer quantitatively to a 1-L volumetric flask, dilute to volume with water, and mix thoroughly. On each day of use, standardize as follows: Measure accurately 40 mL of the solution into a 250-mL conical flask, add 60 mL of water, and titrate with freshly standardized 0.1 N potassium permanganate to a pink end point. Perform a blank titration using only water.

$$N = \frac{[mL(sample) - mL(blank)] \times N \, KMnO_4}{mL \, FeSO_4}$$

Hydrochloric Acid, 1 N. Add 85 mL of hydrochloric acid to 50 mL of water in a 1-L volumetric flask. Cool to room temperature, dilute to volume with water, and mix thoroughly. Standardize as follows: Weigh accurately about 2.2 g of sodium carbonate, alkalimetric standard (NIST SRM 351), that previously has been heated at 285 °C for 2 h. Dissolve in 100 mL of water, and add 0.1 mL of methyl red indicator solution. Add the acid slowly from a buret, with stirring, until the solution becomes faintly pink. Heat the solution to boiling, cool, and again titrate until the solution becomes faintly pink. Repeat this procedure until the faint color is no longer discharged on further boiling.

$$N = \frac{g \, Na_2CO_3 \times (Assay \, / \, 100)}{0.05300 \times mL \, HCl}$$

Iodine, 0.1 N. Place approximately 40 g of potassium iodide and 10 mL of water in a large, glass-stoppered weighing bottle. Mix the contents by swirling; allow to come to room temperature, and then weigh accurately. Add approximately 12.7 g of assayed iodine (page 370), previously weighed on a rough balance, and reweigh the stoppered bottle and contents to obtain the exact weight of the iodine added. Mix, and transfer quantitatively to a 1-L amber volumetric flask. Dilute to volume with water, and mix thoroughly. Standardize as follows: Weigh accurately 0.21 g of arsenic trioxide reductometric standard (NIST SRM 83) previously dried for 12 h at 110 °C into a 250-mL iodine flask. Add 10 mL of 1.0 N sodium hydroxide, warm gently, and swirl to hasten dissolution, being certain that no particles of arsenic trioxide remain on the sides of the flask, and add 0.15 mL of phenolphthalein indicator solution. Neutralize with 6 M hydrochloric acid, and add 1 mL in excess of the acid. Add 75 mL of water and 2 g of sodium bicarbonate dissolved in 20 mL of water. Titrate with the iodine solution until within about 2 mL of the anticipated equivalence point. Then add 3 mL of starch indicator solution, and continue the titration to the first permanent blue color of the starch–iodine complex. Store in an opaque or amber glass bottle.

$$N = \frac{g\,As_2O_3 \times (Assay\,/\,100)}{0.04946 \times mL\,I_2}$$

Lead Nitrate, 0.1 M. Dissolve 33.12 g of lead nitrate in 300 mL of water. Transfer quantitatively to a 1-L volumetric flask, dilute to the mark with water, and mix thoroughly. Standardize as follows: Measure accurately 40 mL of the solution into a 400-mL beaker. Add 0.15 mL of acetic acid and 15 mL of a saturated aqueous solution of hexamethylenetetramine. Add a few milligrams of xylenol orange indicator mixture, and titrate with freshly standardized 0.1 M EDTA volumetric solution to a yellow color.

$$N = \frac{mL \times M\,EDTA}{mL\,Pb(NO_3)_2}$$

Oxalic Acid, 0.1 N. Dissolve 6.30 g of oxalic acid dihydrate in 200 mL of water in a 1-L volumetric flask, dilute to volume with water, and mix thoroughly. Standardize as follows: Measure accurately 40 mL of the solution in a 250-mL conical flask, add 50 mL of water and 7 mL of sulfuric acid. Add slowly, with stirring, 30 mL of freshly standardized 0.1 N potassium permanganate volumetric solution. Heat the solution to 70 °C and continue the titration until the pink color persists for 30 s.

$$N = \frac{mL \times N\,KMnO_4}{mL\,oxalic\,acid}$$

Perchloric Acid in Glacial Acetic Acid, 0.1 N. Add slowly, with stirring, 8.5 mL of 70% perchloric acid to a mixture of 500 mL of glacial acetic acid and 21 mL of acetic anhydride. (*Caution*: Perchloric acid in contact with some organic materials can form explosive mixtures. Use safety goggles and rubber gloves when preparing the solution. Rinse any glassware that has been in contact with perchloric acid before setting it aside.) Cool, dilute to volume in a 1-L volumetric flask with glacial acetic acid, and mix thoroughly. Wait 30 min, and determine the water content of the solution by titrating 30 g with Karl Fischer reagent (page 102). If the water content is greater than 0.05%, add a small amount of acetic anhydride, mix thoroughly, and again determine the water. Repeat until the water content is between 0.02%

and 0.05%. In a similar manner, if the original water content is below 0.02%, add water to bring the water content into the 0.02–0.05% range. Allow the solution to stand for 24 h, and check the water content to be certain that it remains in this range. Standardize as follows: Weigh accurately about 0.7 g of potassium hydrogen phthalate, acidimetric standard (NIST SRM 84), previously lightly crushed and dried for 2 h at 120 °C. Dissolve in 50 mL of glacial acetic acid, add 0.1 mL of crystal violet indicator solution, and titrate with the perchloric acid solution until the violet color changes to blue-green. Perform a blank titration on 50 mL of glacial acetic acid in the same manner, and subtract the volume of perchloric acid solution consumed in the blank titration from that consumed in the first titration.

$$N = \frac{g\ KHC_8H_4O_4\ (Assay\ /\ 100)}{0.20423 \times [mL(sample) - mL(blank)]\ HClO_4}$$

Potassium Bromate, 0.1 N. Weigh accurately 2.78 g of potassium bromate, transfer to a 1-L volumetric flask, dilute to volume with water, and mix thoroughly. Standardize as follows: Measure accurately 40 mL of the solution into a 250-mL iodine flask, add 3 g of potassium iodide and 3 mL of hydrochloric acid, stopper the flask, and allow to stand in the dark for 5 min. Titrate the liberated iodine with freshly standardized 0.1 N sodium thiosulfate volumetric solution. Add 3 mL of starch indicator solution near the end of the titration, and continue to the absence of the blue starch–iodine complex.

$$N = \frac{mL \times N\ Na_2S_2O_3}{mL\ KBrO_3}$$

Potassium Dichromate, 0.1 N. Weigh accurately 4.90 g of potassium dichromate, previously dried for 2 h at 110 °C. Transfer to a 1-L volumetric flask, dilute to volume with water, and mix thoroughly. Standardize as follows: Measure accurately 40 mL of the solution into a 250-mL iodine flask, and add 100 mL of water, 3 g of potassium iodide, and 3 mL of hydrochloric acid. Stopper immediately, and allow to stand in the dark for 5 min. Titrate the liberated iodine with freshly standardized 0.1 N sodium thiosulfate solution. Add 3 mL of starch indicator solution near the end of the titration, and continue to the absence of the blue starch–iodine complex.

$$N = \frac{mL \times N\ Na_2S_2O_3}{mL\ K_2Cr_2O_7}$$

Potassium Hydroxide, Methanolic, 0.1 N. Dissolve 6.8 g of potassium hydroxide in a minimum amount of water (about 5 mL) in a 1-L volumetric flask. Dilute to volume with methanol, and mix thoroughly. Allow to stand in a container protected from carbon dioxide for 24 h. Carefully decant the clear solution, leaving any residue behind, into a suitable container, mix well, and store protected from carbon dioxide. Standardize as follows: Weigh accurately about 0.85 g of potassium hydrogen phthalate, acidimetric standard (NIST SRM 84), previously lightly crushed and dried for 2 h at 120 °C. Dissolve in 100 mL of carbon dioxide-free water, add 0.15 mL of phenolphthalein indicator solution, and titrate with the methanolic potassium hydroxide solution to a permanent faint pink color.

$$N = \frac{g\ KHC_8H_4O_4 \times (Assay\ /\ 100)}{0.20423 \times mL\ of\ methanolic\ KOH\ solution}$$

Potassium Iodate, 0.05 M (0.3 N). Weigh exactly 10.70 g of potassium iodate, previously dried at 110 °C to constant weight, transfer to a 1-L volumetric flask, dilute to volume with water, and mix thoroughly. Standardize as follows: To 15.0 mL of solution in a 250-mL iodine flask, add 3 g of potassium iodide and 3 mL hydrochloric acid previously diluted with 10 mL of water. Stopper immediately, and allow to stand in the dark for 5 min. Then add 50 mL of cold water, and titrate the liberated iodine with freshly standardized 0.1 N sodium thiosulfate. Add 3 mL of starch indicator solution near the end of the titration, and continue to the absence of the blue starch–iodine complex.

$$M = \frac{\text{mL} \times \text{N Na}_2\text{S}_2\text{O}_3}{\text{mL KIO}_3 \times 6}$$

$$N = \frac{\text{mL} \times \text{N Na}_2\text{S}_2\text{O}_3}{\text{mL KIO}_3}$$

Potassium Permanganate, 0.1 N. Dissolve 3.3 g of potassium permanganate in 1 L of water in a 1500-mL conical flask, heat to boiling, and boil gently for 15 min. Allow to cool, stopper, and keep in the dark for 48 h. Then filter through a fine-porosity sintered-glass filter, and place in a glass-stoppered amber bottle. Standardize as follows: Weigh accurately about 0.26 g of sodium oxalate (NIST SRM 40) previously dried for 2 h at 105 °C), transfer to a 500-mL conical flask containing 250 mL of water and 7 mL of sulfuric acid, and heat to 70 °C. Titrate immediately with the permanganate solution to a pink color that persists for 15 s. The temperature at the end of the titration should not be less than 60 °C. Perform a blank titration containing 250 mL of water and 7 mL of sulfuric acid in the same manner, and subtract the volume of the potassium permanganate consumed in the blank from the first titration. Store in an opaque or amber glass container.

$$N = \frac{\text{g Na}_2\text{C}_2\text{O}_4 \times (\text{Assay} / 100)}{0.06700 \times [\text{mL(sample)} - \text{mL(blank)}] \text{ KMnO}_4}$$

Potassium Thiocyanate, 0.1 N. Weigh exactly 9.72 g of potassium thiocyanate, previously dried for 2 h at 110 °C, transfer to a 1-L volumetric flask, dilute to volume, and mix thoroughly. Standardize as follows: Measure accurately 40 mL of freshly standardized 0.1 N silver nitrate volumetric solution into a 250-mL conical flask and add 100 mL of water, 1 mL of nitric acid, and 2 mL of ferric ammonium sulfate indicator solution. Titrate with the thiocyanate solution, with agitation, to a permanent light pinkish-brown color of the supernatant solution.

$$N = \frac{\text{mL} \times \text{N AgNO}_3}{\text{mL KSCN}}$$

Silver Nitrate, 0.1 N. Weigh exactly 16.99 g of silver nitrate, previously dried for 1 h at 100 °C, transfer to a 1-L amber volumetric flask, dilute to volume, and mix thoroughly. Standardize as follows: Weigh accurately about 0.33 g of potassium chloride (NIST SRM 999) ignited at 500 °C. Place in a 250-mL beaker, and dissolve in 100 mL of water. Add 2 mL of dichlorofluorescein indicator solution. With stirring, titrate with silver nitrate solution until the silver chloride flocculates and the mixture acquires a faint pink color. Store in an opaque container.

$$N = \frac{g \, KCl \times (Assay / 100)}{0.07455 \times mL \, AgNO_3}$$

Sodium Acetate in Glacial Acetic Acid, 0.1 N. Place 8.20 g of sodium acetate, anhydrous, previously dried to constant weight at 120 °C, in a 1-L volumetric flask. Add 500 mL of glacial acetic acid and 10 mL of acetic anhydride, and dissolve. Dilute to volume with glacial acetic acid, and mix thoroughly. Standardize as follows: Measure accurately 40 mL of freshly standardized perchloric acid in glacial acetic acid volumetric solution, and add 0.1 mL of crystal violet indicator solution. While stirring, titrate with the sodium acetate until the solution changes from an emerald green to a violet color.

$$N = \frac{mL \times N \, HClO_4}{mL \, C_2H_3NaO_2}$$

Sodium Hydroxide, 1 N. Dissolve 162 g of sodium hydroxide in 150 mL of carbon dioxide-free water, cool the solution to room temperature. Prepare a 1 N solution by diluting 54.5 mL of the concentrated solution to 1 L with carbon dioxide-free water. Mix thoroughly, and store in a tight polyolefin container. Standardize as follows: Weigh accurately about 8.5 g of potassium hydrogen phthalate, acidimetric standard (NIST SRM 84), previously lightly crushed and dried for 2 h at 120 °C. Dissolve in 100 mL of carbon dioxide-free water, add 0.15 mL of phenolphthalein indicator solution, and titrate with the sodium hydroxide solution to a permanent faint pink color.

$$N = \frac{g \, KHC_8H_4O_4 \times (Assay / 100)}{0.20423 \times mL \, NaOH}$$

Sodium Methoxide in Methanol, 0.01 N. Dilute 20.0 mL of sodium methoxide in methanol, 0.5 M methanolic solution, with methanol to 1 L, and mix thoroughly. Standardize as follows: Measure accurately 40 mL of 0.01 N sulfuric acid volumetric solution into a suitable container, add 0.10 mL of thymol blue indicator solution, and titrate with the sodium methoxide in methanol solution to a blue end point.

$$N = \frac{mL \times N \, H_2SO_4}{mL \, CH_3ONa}$$

Sodium Nitrite, 0.1 N. Weigh accurately 6.90 g of sodium nitrite. Dissolve in 100 mL of water in a 1-L volumetric flask, dilute to volume, and mix thoroughly. Standardize as follows: Add 5 mL of sulfuric acid to 200 ml of water, and while the solution is still warm, add 0.1 N potassium permanganate until a faint pink color persists for 2 min. Measure accurately 50 mL of freshly standardized 0.1 N potassium permanganate, transfer to the solution, and mix gently. Pipet, slowly and with constant agitation, 40 mL of the sodium nitrite solution, holding the tip of the pipet well below the surface of the liquid. Add 3 g of potassium iodide, and titrate the liberated iodine from the excess of potassium permanganate with 0.1 N sodium thiosulfate. Add 3 mL of starch indicator near the end point, and titrate to the absence of the blue starch–iodine complex. Perform a complete blank using all reagents except the 0.1 N sodium nitrite solution.

$$N = \frac{[mL(blank) - mL(sample)\ Na_2S_2O_3] \times N\ KMnO_4}{mL\ NaNO_2}$$

Sodium Thiosulfate, 0.1 N. Dissolve 24.82 g of sodium thiosulfate pentahydrate and 200 mg of sodium carbonate, anhydrous, in 1 L of recently boiled and cooled water, and mix thoroughly. Allow to stand for 24 h and standardize as follows: Weigh accurately about 0.21 g of NIST potassium dichromate, oxidimetric standard (NIST SRM 136), previously crushed and dried at 110 °C for 2 h. Dissolve in 100 mL of water in a glass-stoppered iodine flask. Swirl to dissolve the sample, remove the stopper, and quickly add 3 g of potassium iodide, 2 g of sodium bicarbonate, and 5 mL of hydrochloric acid. Insert the stopper, swirl slightly to release excess carbon dioxide, cover the stopper with water, and allow to stand in the dark for 10 min. Rinse the stopper and the inner walls of the flask with water and titrate the liberated iodine with the sodium thiosulfate until the solution becomes faintly yellow. Add 3 mL of starch indicator solution, and continue the titration to the disappearance of the blue starch–iodine complex. Restandardize the solution frequently.

$$N = \frac{g\ K_2Cr_2O_7 \times (Assay\ /\ 100)}{0.04904 \times mL\ Na_2S_2O_3}$$

Sulfuric Acid, 1 N. Add slowly (use caution), with stirring, 30 mL of sulfuric acid to about 500 mL of water. Allow the mixture to cool to room temperature, dilute to 1 L, and mix thoroughly. Standardize against sodium carbonate, alkalimetric standard, as directed for hydrochloric acid, 1 N.

Zinc Chloride, 0.1 M. Weigh accurately 6.54 g of zinc, and dissolve in 80 mL of 10% hydrochloric acid. Warm if necessary to complete dissolution, cool, dilute with water to volume in a 1-L volumetric flask, and mix thoroughly. Standardize as follows: To 40.0 mL of freshly standardized 0.1 M EDTA in a suitable beaker, add 5 mL of pH 10 ammoniacal buffer and 0.15 mL of xylenol orange indicator. Titrate with the zinc chloride solution to a magenta purple color that remains for 30 s.

$$M = \frac{mL \times M\ EDTA}{mL\ ZnCl_2}$$

Zinc Sulfate, 0.1 M. Weigh accurately 28.76 g of zinc sulfate heptahydrate, and dissolve in 500 mL of water in a 1-L volumetric flask. Dilute to the mark with water, and mix thoroughly. Standardize as follows: To 40.0 mL of freshly standardized 0.1 M EDTA volumetric solution, in a suitable beaker, add 10 mL of ammonium acetate–acetic acid buffer solution, 50 mL of ethyl alcohol, and 2 mL of dithizone indicator solution. Titrate with the zinc sulfate solution to a pink end point.

$$M = \frac{mL \times M\ EDTA}{mL\ ZnSO_4}$$

Part 4:
Monographs for
Reagent Chemicals
General Descriptions, Specifications, and Tests

Acetaldehyde

Ethanal

CH₃CHO   **Formula Wt 44.05**   **CAS No.75-07-0**

GENERAL DESCRIPTION

Typical appearance: liquid with pungent odor
Analytical use: standard in chromatography and polarography
Change in state (approximate): boiling point, 21 °C
Aqueous solubility: miscible with water
Density: 0.79

SPECIFICATIONS

Assay . . . . ≥99.5% CH₃CHO

Maximum Allowable

Residue after evaporation 0.005%
Titrable acid . . . . 0.008 meq/g

TESTS

Assay. Analyze the sample by gas chromatography using the general parameters cited on page 80. The following specific conditions are also required.

> *Column:* Type I, methyl silicone

> *Detector:* Flame ionization

Measure the area under all peaks, and calculate the area percent for acetaldehyde.

Residue after Evaporation. (Page 25). Evaporate 40 g (51 mL) to dryness in a tared, preconditioned platinum dish on a hot plate (≈100 °C) in a well-ventilated hood, and dry the residue at 105 °C for 30 min.

Titrable Acid. Work in a fume hood. Chill 25 g (32 mL) of sample in a graduated cylinder in an ice bath. To a 250-mL Erlenmeyer flask, add about 25 mL of deionized water and 75 g of deionized ice. Add 0.5 mL of phenolphthalein indicator solution, and titrate with 0.1 N sodium hydroxide to the first perceptible pink color that persists for 15 s. Add the chilled sample and titrate immediately to the same faint pink color. Not more than 2.0 mL of 0.1 N sodium hydroxide should be required.

Acetic Acid, Glacial

CH₃COOH   Formula Wt 60.05   CAS No. 64-19-7

GENERAL DESCRIPTION

Typical appearance: liquid with a sharp odor

Analytical use: aqueous and nonaqueous acid–base titrations

Change in state (approximate): boiling point, 118 °C; freezing point, 16 °C

Aqueous solubility: miscible with water

pK_a: 4.8

SPECIFICATIONS

Assay . . . . ≥99.7% CH₃COOH

	Maximum Allowable
Color (APHA)	10
Dilution test.	Passes test
Residue after evaporation	0.001%
Acetic anhydride [(CH₃CO)₂O].	0.01%
Chloride (Cl)	1 ppm
Sulfate (SO₄)	1 ppm
Heavy metals (as Pb).	0.5 ppm
Iron (Fe).	0.2 ppm
Substances reducing dichromate	Passes test
Substances reducing permanganate	Passes test
Titrable base	0.0004 meq/g

TESTS

Assay and Acetic Anhydride. Analyze the sample by gas chromatography using the general parameters cited on page 80. The following specific conditions are also required.

    Column: Type I, methyl silicone

Measure the area under all peaks, and calculate the area percent for acetic acid and acetic anhydride. Correct for water content.

Color (APHA). (Page 43).

Dilution Test. Dilute 1 volume of the acid with 3 volumes of water and allow to stand for 1 h. The solution should be as clear as an equal volume of water.

Residue after Evaporation. (Page 25). Evaporate 100 g (95 mL) to dryness in a tared, preconditioned dish on a hot plate (≈100 °C), and dry the residue at 105 °C for 30 min.

Chloride. (Page 35). Dilute 10 g (9.5 mL) with 10 mL of water, and add 1 mL of silver nitrate reagent solution. Prepare a standard containing 0.01 mg of chloride ion (Cl) in 20 mL of water, and add 1 mL of silver nitrate reagent solution. Evaporate the solutions to dryness on a hot plate (≈100 °C). Dissolve the residues with 0.5 mL of ammonium hydroxide, dilute with 20 mL of water, and add 1.5 mL of nitric acid. Any turbidity in the solution of the sample should not exceed that of the standard.

Sulfate. To 48 mL (50 g) of sample, add 2 mL of 1% sodium carbonate solution, and evaporate to dryness on a hot plate ($\approx$100 °C). Dissolve the residue in 1.0 mL of 10% hydrochloric acid and 15 mL of water. Dilute to 25 mL with water. For the control, take 0.10 mg of sulfate ion (SO_4) in 20 mL of water, add 1.0 mL of 10% hydrochloric acid and dilute to 25 mL. To sample and control solutions, add 1 mL of 12% barium chloride solution. Compare after 10 min. Sample turbidity should not exceed that of the control solution.

Heavy Metals. (Page 36, Method 1). To 40 g (38 mL), add about 10 mg of sodium carbonate, evaporate to dryness on a hot plate ($\approx$100 °C), dissolve the residue in about 20 mL of water, and dilute with water to 25 mL.

Iron. (Page 38, Method 1). To 50 g (48 mL), add 10 mg of sodium carbonate, and evaporate to dryness. Dissolve the residue in 2 mL of hydrochloric acid, dilute with water to 50 mL, and use the solution without further acidification.

Substances Reducing Dichromate. To 10 mL, add 1.0 mL of 0.1 N potassium dichromate, and cautiously add 10 mL of sulfuric acid. Cool the solution to room temperature, and allow to stand for 30 min. While the solution is swirled, dilute slowly and cautiously with 50 mL of water, cool, and add 1 mL of freshly prepared 10% potassium iodide reagent solution. Titrate the liberated iodine with 0.1 N sodium thiosulfate, using starch as the indicator. Not more than 0.40 mL of the 0.1 N potassium dichromate should be consumed (not less than 0.60 mL of the 0.1 N sodium thiosulfate should be required). Correct for a complete blank.

Substances Reducing Permanganate. Add 40 g (38 mL) of the sample to 10 mL of water. Cool to 15 °C, add 0.30 mL of 0.1 N potassium permanganate, and allow to stand at 15 °C for 10 min. The pink color should not be entirely discharged.

Titrable Base. To 25 g (24 mL) add 0.10 mL of a solution of 1 g of crystal violet (or methyl violet) in 100 mL of glacial acetic acid. The color should be violet. Titrate the solution with 0.1 N perchloric acid in glacial acetic acid to a green color. Not more than 0.10 mL should be required.

Acetic Acid, Ultratrace
CH_3COOH   **Formula Wt 60.05**   **CAS No. 64-19-7**

Suitable for use in ultratrace elemental analysis.

Note: Reagent must be packaged in a preleached Teflon bottle and used in a clean laboratory environment to maintain purity.

GENERAL DESCRIPTION

Typical appearance: clear liquid with a sharp odor

Analytical use: trace metal analysis

Change in state (approximate): boiling point, 118 °C; freezing point, 16 °C

Aqueous solubility: miscible with water

pK_a: 4.8

SPECIFICATIONS

Assay . . . . ≥90% CH_3COOH

Maximum Allowable

Chloride (Cl)	1 ppm
Sulfate (SO_4)	1 ppm
Mercury (Hg)	1 ppb
Aluminum (Al)	1 ppb
Barium (Ba)	1 ppb
Boron (B)	5 ppb
Cadmium (Cd)	1 ppb
Calcium (Ca)	1 ppb
Chromium (Cr)	1 ppb
Cobalt (Co)	1 ppb
Copper (Cu)	1 ppb
Iron (Fe)	5 ppb
Lead (Pb)	1 ppb
Lithium (Li)	1 ppb
Magnesium (Mg)	1 ppb
Manganese (Mn)	1 ppb
Molybdenum (Mo)	1 ppb
Potassium (K)	1 ppb
Silicon (Si)	5 ppb
Sodium (Na)	5 ppb
Strontium (Sr)	1 ppb
Tin (Sn)	1 ppb
Titanium (Ti)	1 ppb
Vanadium (V)	1 ppb
Zinc (Zn)	1 ppb
Zirconium (Zr)	1 ppb

TESTS

Assay. (By acid–base titrimetry). The assay of acetic acid used for ultratrace metal analysis may vary, depending on the method of purification. Tare a glass-stoppered flask containing about 30 mL of water. Quickly add about 3 mL of the sample, stopper, and weigh accurately. Dilute to about 50 mL, add 0.15 mL of methyl orange indicator solution, and titrate with 1 N sodium hydroxide. One milliliter of 1 N sodium hydroxide corresponds to 0.06005 g of acetic acid.

$$\% \text{ Acetic acid} = \frac{(\text{mL} \times \text{N NaOH}) \times 6.005}{\text{Sample wt (g)}}$$

Chloride. (Page 35). Dissolve 10.0 g in 10 mL of water.

Sulfate. To 48 mL (50 g) of sample, add 2 mL of 1% sodium carbonate solution, and evaporate to dryness on a hot plate ($\approx$100 °C). Dissolve the residue in 1.0 mL of 10% hydrochloric acid and 15 mL of water. Dilute to 25 mL with water. For the control, take 0.05 mg of sulfate ion (SO_4) in 20 mL of water, add 1.0 mL of 10% hydrochloric acid, and dilute to 25 mL. To sample and control solutions, add 1 mL of 12% barium chloride solution. Compare after 10 min. Sample turbidity should not exceed that of the control solution.

Mercury. (By CVAAS, page 65). To a set of three 100-mL volumetric flasks containing about 35 mL of water, add 20.0 g of sample. To the second and third flasks, add mercury ion (Hg) standards of 10 ng (0.5 ppb) and 20 ng (1.0 ppb), respectively. Add 5 mL of nitric acid to all three flasks, and dilute to the mark with water. Mix. Zero the instrument with the blank, and determine the mercury content using a suitable mercury analyzer system (1.0 mL of 0.01 μg/mL Hg = 10 ng).

Trace Metals. Determine the aluminum, barium, boron, cadmium, calcium, chromium, cobalt, copper, iron, lead, lithium, magnesium, manganese, molybdenum, potassium, silicon, sodium, strontium, tin, titanium, vanadium, zinc, and zirconium by the ICP–OES method described on page 69.

Acetic Anhydride

$$CH_3-\overset{\displaystyle O}{\overset{\|}{C}}-O-\overset{\displaystyle O}{\overset{\|}{C}}-CH_3$$

(CH₃CO)₂O   **Formula Wt 102.09**   **CAS No. 108-24-7**

GENERAL DESCRIPTION

Typical appearance: liquid with a pungent odor
Analytical use: preparation of anhydrous acetic acid in nonaqueous titrimetry
Change in state (approximate): boiling point, 139 °C
Aqueous solubility: gradually dissolves with formation of acetic acid
Density: 1.08

SPECIFICATIONS

Assay	≥97.0% $(CH_3CO)_2O$

Maximum Allowable

Residue after evaporation	0.003%
Chloride (Cl)	5 ppm
Phosphate (PO_4)	0.001%
Sulfate (SO_4)	5 ppm
Heavy metals (as Pb)	2 ppm
Iron (Fe)	5 ppm
Substances reducing permanganate	Passes test

TESTS

Assay. Analyze the sample by gas chromatography using the general parameters cited on page 80. The following specific conditions are also required.

   Column: Type I, methyl silicone

Measure the area under all peaks and calculate the area percent for acetic anhydride.

Residue after Evaporation. (Page 25). Evaporate 50 g (46 mL) in a tared, preconditioned dish on a hot plate ($\approx$100 °C), and dry the residue at 105 °C for 30 min.

For the Determination of Chloride, Phosphate, Sulfate, Heavy Metals, Iron, and Substances Reducing Permanganate

Sample Solution A. Dilute 40 g (37 mL) of the sample with water to 200 mL (1 mL = 0.2 g).

Chloride. (Page 35). Use 10 mL of sample solution A (2-g sample).

Phosphate. (Page 40, Method 1). Evaporate 10 mL of sample solution A (2-g sample) to dryness on a hot plate ($\approx$100 °C), and dissolve the residue in 25 mL of 0.5 N sulfuric acid.

Sulfate. (Page 41, Method 3). Use 50 mL of sample solution A (10-g sample).

Heavy Metals. (Page 36, Method 1). To 50 mL of sample solution A (10-g sample), add about 1.0 mL of 1% sodium carbonate reagent solution, evaporate to dryness on a hot plate ($\approx$100 °C), dissolve the residue in about 20 mL of water, and dilute with water to 25 mL.

Iron. (Page 38, Method 1). To 10 mL of sample solution A (2-g sample), add 10 mg of sodium carbonate, and evaporate to dryness. Dissolve the residue in 2 mL of hydrochloric acid, dilute with water to 50 mL, and use the solution without further acidification.

Substances Reducing Permanganate. To 10 mL of sample solution A (2-g sample), add 0.4 mL of 0.1 N potassium permanganate, and allow to stand for 5 min. The pink color should not be entirely discharged.

Acetone
2-Propanone

$$CH_3-\overset{\displaystyle O}{\overset{\displaystyle \|}{C}}-CH_3$$

(CH$_3$)$_2$CO   **Formula Wt 58.08**   **CAS No. 67-64-1**

Suitable for general use or in ultraviolet spectrophotometry. Product labeling shall designate the uses for which suitability is represented on the basis of meeting the relevant specifications and tests. The ultraviolet spectrophotometry specifications include all of the specifications for general use.

GENERAL DESCRIPTION

Typical appearance: liquid with a characteristic odor
Analytical use: liquid chromatography; cleaning glassware; extraction of solid waste
Change in state (approximate): boiling point, 56 °C
Aqueous solubility: miscible with water
Density: 0.79

SPECIFICATIONS

General Use

Assay . . . . $\geq$99.5% $(CH_3)_2CO$

Maximum Allowable

Color (APHA)	10
Residue after evaporation	0.001%
Solubility in water	Passes test
Titrable acid	0.0003 meq/g
Titrable base	0.0006 meq/g
Aldehyde (as HCHO)	0.002%
Isopropyl alcohol	0.05%
Methanol	0.05%
Substances reducing permanganate	Passes test
Water (H_2O)	0.5%

Specific Use

Ultraviolet Spectrophotometry

Wavelength (nm)	Absorbance (AU)
400	0.01
350	0.02
340	0.10
330	1.00

TESTS

Assay, Isopropyl Alcohol, and Methanol. Analyze the sample by gas chromatography using the general parameters cited on page 80. The following specific conditions are also required.

Column: Type I, methyl silicone

Measure the area under all peaks, and calculate the area percent for acetone, isopropyl alcohol, and methanol. Correct for water content.

Color (APHA). (Page 43).

Residue after Evaporation. (Page 25). Evaporate 100 g (127 mL) to dryness in a tared, preconditioned dish on a hot plate ($\approx$100 °C), and dry the residue at 105 °C for 30 min.

Solubility in Water. Mix 30 g (38 mL) with 38 mL of carbon dioxide-free water. The solution should remain clear for 30 min. Reserve this solution for the test for titrable acid.

Titrable Acid. Add 0.10 mL of phenolphthalein indicator solution to the solution prepared in the preceding test. Not more than 1.0 mL of 0.01 N sodium hydroxide should be required to produce a pink color.

Titrable Base. Mix 18 g (23 mL) with 25 mL of water, and add 0.05 mL of methyl red indicator solution. Not more than 1.0 mL of 0.01 N hydrochloric acid should be required to produce a red color.

Aldehyde. Dilute 2 g (2.5 mL) with water to 10 mL. Prepare a standard containing 0.04 mg of formaldehyde in 10 mL of water and a complete reagent blank in 10 mL of water alone. To each, add 0.15 mL of a freshly prepared 5% solution of 5,5-dimethyl-1,3-cyclohexanedione in alcohol. Evaporate each on a hot plate ($\approx$100 °C) until the acetone is volatilized. Dilute each with water to 10 mL, and cool quickly in an ice bath while stirring vigorously. Any turbidity in the solution of the sample should not exceed that in the standard. If any turbidity develops in the reagent blank, the 5,5-dimethyl-1,3-cyclohexanedione should be discarded, and the entire test rerun with a fresh solution.

Substances Reducing Permanganate. Add 0.05 mL of 0.1 N potassium permanganate to 10 mL of the sample, and allow to stand for 15 min at 25 °C. The pink color should not be entirely discharged, when compared with an equal volume of water.

Water. (Page 31, Method 1). Use 25 mL (20 g) of the sample and a methanol-free system to prevent ketal formation with liberation of water.

Ultraviolet Spectrophotometry. Use the procedure on page 86 to determine the absorbance.

Acetonitrile
CH_3CN Formula Wt 41.05   CAS No. 75-05-8

Suitable for general use or in high-performance liquid chromatography, extraction–concentration analysis, or ultraviolet spectrophotometry. Product labeling shall designate the uses for which suitability is represented on the basis of meeting the relevant specifications and tests. The ultraviolet spectrophotometry and liquid chromatography suitability specifications include all of the specifications for general use. The extraction–concentration suitability specifications include only the general use specifications for color.

GENERAL DESCRIPTION

Typical appearance: clear, colorless liquid; slight but characteristic odor
Analytical use: solvent
Change in state (approximate): boiling point, 81 °C
Aqueous solubility: miscible
pK_a: 25

SPECIFICATIONS

General Use

Assay . . . . ≥99.5% CH_3CN

Maximum Allowable

Color (APHA) 10
Residue after evaporation 0.005%
Titrable acid . . . . 8 µeq/g
Titrable base . . . . 0.6 µeq/g
Water (H_2O) 0.3%

Specific Use

Ultraviolet Spectrophotometry

Wavelength (nm)	Absorbance (AU)
254	0.01
220	0.05
190	1.00

Liquid Chromatography Suitability

Absorbance . . . . Passes test
Gradient elution . . . . Passes test

Extraction–Concentration Suitability

Absorbance . . . . Passes test
GC–FID . . . . Passes test
GC–ECD . . . . Passes test

TESTS

Assay. Analyze the sample by gas chromatography using the general parameters cited on page 80. The following specific conditions are also required.

Column: Type I, methyl silicone

Measure the area under all peaks, and calculate the acetonitrile content in area percent. Correct for water content.

Color (APHA). (Page 43).

Residue after Evaporation. (Page 25). Evaporate 39 g (50 mL) to dryness in a tared, preconditioned platinum dish on a hot plate (≈100 °C) in a well-ventilated hood, and dry the residue at 105 °C for 30 min.

Titrable Acid. Mix 10 g (13 mL) with 13 mL of carbon dioxide-free water. Add 0.10 mL of phenolphthalein indicator solution. Not more than 8.0 mL of 0.01 N sodium hydroxide solution should be required in the titration.

Titrable Base. To 78 g (100 mL) in a 250-mL conical flask, add 0.30 mL of 0.05% solution of bromocresol green in alcohol and 0.10 mL of methyl red indicator solution. Titrate with 0.01 N hydrochloric acid to a light orange-pink end point. Not more than 4.5 mL of 0.01 N hydrochloric acid should be required.

Water. (Page 31, Method 1). Use 10 mL (7.8 g) of the sample.

Ultraviolet Spectrophotometry. Use the procedure on page 86 to determine the absorbance.

Liquid Chromatography Suitability. Analyze the sample by using the gradient elution procedure cited on page 84. Use the procedure on page 86 to determine the absorbance.

Extraction–Concentration Suitability. Analyze the sample by using the general procedure cited on page 85. In the sample preparation step, exchange the sample with ACS grade 2,2,4-trimethylpentane suitable for extraction–concentration. Use the procedure on page 86 to determine the absorbance.

Acetyl Chloride
Ethanoic Acid Chloride
CH_3COCl Formula Wt 78.50   CAS No. 75-36-5

GENERAL DESCRIPTION
Typical appearance: clear, colorless liquid with a pungent odor
Analytical use: acetylation agent
Change in state (approximate): boiling point, 51 °C
Aqueous solubility: decomposes with formation of acetic acid and hydrochloric acid
Density: 1.11

SPECIFICATIONS
Assay . . . . $\geq$98.5% CH_3COCl

Maximum Allowable

Color (APHA) 20
Residue after evaporation 0.005%
Insoluble matter 0.0025%
Phosphate (PO_4) 0.002%
Heavy metals (as Pb) . . . . 5 ppm
Iron (Fe) . . . . 5 ppm

TESTS

Assay. Analyze the sample by gas chromatography using the general parameters cited on page 80. The following specific conditions are also required.

Column: Type I, methyl silicone

Detector: Flame ionization

Column Temperature: Initial 75 °C for 5 min, then increased at a rate of 10 °C per minute to a final temperature of 200 °C

Color (APHA). (Page 43).

Residue after Evaporation. (Page 25). Evaporate 20 g (18 mL) to dryness in a tared, preconditioned dish on a hot plate (≈ 100 °C), and dry the residue at 105 °C for 30 min.

Insoluble Matter. (Page 25). Add 40 g (36 mL) cautiously, with stirring, to 150 mL of water, and dilute to 200 mL with water. The solution is clear and colorless. Continue with the procedure as described on page 25. Retain the filtrate separate from the washings for the tests for phosphate, heavy metals, and iron.

Phosphate. (Page 40, Method 1). To 5 mL of the solution from the test for insoluble matter (1.0-g sample), add 5 mg of anhydrous sodium carbonate, and evaporate to dryness. Take up the residue in 20 mL of 0.5 N sulfuric acid. For the standard, use 0.02 mg of phosphate and 5 mg of sodium carbonate, and treat exactly as the sample.

Heavy Metals. (Page 36, Method 1). Use 20 mL (4-g sample) of the solution retained from the test for insoluble matter.

Iron. (Page 38, Method 1). Use 10 mL (2.0-g sample) of the solution retained from the test for insoluble matter.

Aluminum

Al	Atomic Wt 26.98	CAS No. 7429-90-5

GENERAL DESCRIPTION

Typical appearance: tin-white, malleable, ductile metal with somewhat bluish tint
Analytical use: reduction of nitrate in nitrogen compounds
Change in state (approximate): melting point, 660 °C

SPECIFICATIONS

	Maximum Allowable
Insoluble in dilute hydrochloric acid	0.05%
Copper (Cu)	0.02%
Iron (Fe)	0.1%
Manganese (Mn)	0.002%
Titanium (Ti)	0.03%
Nitrogen compounds (as N)	0.001%
Silicon (Si)	0.1%

TESTS

Insoluble in Dilute Hydrochloric Acid. Place a 1-L beaker containing 75 mL of water in a fume hood, and add 10.0 g of sample. Cautiously, and with stirring, add 30 mL of hydrochloric acid (exothermic reaction), and then add an additional 70 mL in 10-mL increments. If necessary, one or more additional 5-mL portions of hydrochloric acid may be added to complete the dissolution. Wash the sides of the sample beaker with about 10 mL of water. Boil the solution for 20 min, then cool, and dilute to 190 mL. Filter through a tared filtering crucible. Retain the filtrate, without washings, for sample solution A. Thoroughly wash the contents of the crucible, and dry the residue at 105 °C.

For the Determination of Copper, Iron, Manganese, and Titanium

Sample Solution A. Dilute the filtrate, without washings, obtained in the test for insoluble in dilute hydrochloric acid with water to 200 mL in a volumetric flask (1 mL = 0.05 g).

Copper, Iron, and Manganese. (By flame AAS, page 63). Use sample solution A.

Element	Wavelength (nm)	Sample Wt (g)	Standard Added (mg)	Flame Type*	Background Correction
Cu	324.8	0.10	0.02; 0.04	A/A	Yes
Fe	248.3	0.10	0.10; 0.20	A/A	Yes
Mn	279.5	1.00	0.02; 0.04	A/A	Yes

*A/A is air/acetylene.

Titanium. To 20 mL of sample solution A (1-g sample) add 3 mL of phosphoric acid and 0.5 mL of 30% hydrogen peroxide. Any color should not exceed that produced by 0.3 mg of titanium in an equal volume of solution containing the quantities of reagents used in the test.

Nitrogen Compounds. (Page 39). Use 2.5 g dissolved in 60 mL of freshly boiled 10% sodium hydroxide reagent solution. For the standard, use 0.5 g of aluminum and 0.02 mg of nitrogen.

Silicon.

> *Note:* The molybdenum blue test for silicon is extremely sensitive; therefore, all solutions that are more alkaline than pH 4 should be handled in containers other than glass or porcelain. Platinum dishes and plastic graduated cylinders are recommended. The 10% solution of sodium hydroxide is prepared and measured in plastic.

Dissolve 0.25 g in 10 mL of 10% sodium hydroxide reagent solution. Add the aluminum to the sodium hydroxide solution in small portions, particularly if the sample is finely divided. When the vigorous reaction has subsided, digest the covered container on a hot plate ($\approx$100 °C) until dissolution is complete. Uncover, and evaporate to a moist residue. Dissolve the residue in 10 mL of water, and add 0.05–0.10 mL of 30% hydrogen peroxide. Boil the solution gently to decompose the excess peroxide, cool, and dilute with water to 100 mL. Adjust the acidity of the solution to pH 2 (external indicator) by adding dilute sulfuric acid (1 + 1). Digest on a hot plate ($\approx$100 °C) until any precipitate is dissolved and the solution is clear. If necessary, add more of the dilute sulfuric acid (1 + 1). When dissolution is complete, cool and dilute with water to 250 mL. Prepare a blank solution by treating 10 mL of the 10% sodium hydroxide reagent solution exactly as described above, but omit the 0.25-g sample. Dilute 5.0 mL of the sample solution with water to 80 mL. For the standard, add 0.01 mg of silica (0.005 mg of silicon) to 5 mL of the blank solution and dilute with water to 80 mL. To each solution, add 5.0 mL of a freshly prepared 10% solution of ammonium molybdate. Adjust the pH of each solution to 1.7–1.9 (using a pH meter) with dilute hydrochloric acid (1 + 9) or silica-free ammonium hydroxide. Heat the solutions just to boiling, cool, add 20 mL of hydrochloric acid, and dilute with water to 110 mL. Transfer the solutions to separatory funnels, add 50 mL of butyl alcohol, and shake vigorously. Allow the layers to separate and draw off and discard the aqueous phase. Wash the butyl alcohol three times with 20-mL portions of dilute hydrochloric acid (1 + 99), discarding each aqueous phase. Add 0.5 mL of a freshly prepared 2% stannous chloride reagent solution, and shake. If the butyl alcohol is turbid, wash it with 10 mL of dilute hydrochloric acid (1 + 99). Any blue color in the butyl alcohol from the solution of the sample should not exceed that in the butyl alcohol from the standard.

Aluminum Ammonium Sulfate Dodecahydrate
Ammonium Alum
$AlNH_4(SO_4)_2 \cdot 12H_2O$ Formula Wt 453.33   CAS No. 7784-26-1

GENERAL DESCRIPTION
Typical appearance: colorless solid
Analytical use: buffer
Change in state (approximate): melting point, 93 °C
Aqueous solubility: 15 g in 100 mL at 20 °C

SPECIFICATIONS
Assay . . . . .98.0–102.0% $AlNH_4(SO_4)_2 \cdot 12H_2O$

	Maximum Allowable
Insoluble matter	0.005%
Chloride (Cl)	0.001%
Heavy metals (as Pb)	0.001%
Calcium (Ca)	0.05%
Iron (Fe)	0.001%
Potassium (K)	0.05%
Sodium (Na)	0.01%

TESTS

Assay. (By complexometry). Weigh accurately 0.8 g, and transfer to a 400-mL beaker. Moisten with 1 mL of glacial acetic acid, add 50 mL of water, 40.0 mL of 0.1 M EDTA volumetric solution, and 20 mL of ammonium acetate–acetic acid buffer solution. Warm on a steam bath until solution is complete, and boil gently for 5 min. Cool, add 50 mL of ethyl alcohol and 2 mL of dithizone indicator solution, and titrate with 0.1 M zinc chloride volumetric solution to a bright rose-pink color. Perform a blank titration of 40.0 mL of 0.1 M EDTA volumetric solution, using the same quantities of reagents as for the assay. One milliliter of 0.1 M EDTA consumed by the sample corresponds to 0.04533 g of $AlNH_4(SO_4)_2 \cdot 12H_2O$.

$$\% \ AlNH_4(SO_4)_2 \cdot 12H_2O = \frac{[(40.0 \times M \ EDTA) - (mL \times M \ ZnCl_2)] \times 45.33}{Sample \ wt \ (g)}$$

Insoluble Matter. (Page 25). Use 10 g dissolved in 150 mL of water containing 5 mL of 25% sulfuric acid.

Chloride. (Page 35). Use 1.0 g.

Heavy Metals. (Page 36, Method 1). Dissolve 2.0 g in 20 mL of water.

Calcium, Iron, Potassium, and Sodium. (By flame AAS, page 63).

Sample Stock Solution. Dissolve 10.0 g in 100 mL of water (1 mL = 0.1 g).

Element	Wavelength (nm)	Sample Wt (g)	Standard Added (mg)	Flame Type*	Background Correction
Ca	422.7	0.20	0.05; 0.10	N/A	No
Fe	248.3	5.0	0.025; 0.05	A/A	Yes
K	766.5	0.10	0.025; 0.05	A/A	No
Na	589.0	0.20	0.01; 0.02	A/A	No

*A/A is air/acetylene; N/A is nitrous oxide/acetylene.

Aluminum Nitrate Nonahydrate

$Al(NO_3)_3 \cdot 9H_2O$ **Formula Wt 375.13**   **CAS No. 7784-27-2**

GENERAL DESCRIPTION

Typical appearance: colorless, deliquescent solid
Analytical use: metal standard solution; extraction of uranium
Change in state (approximate): melting point, 73 °C
Aqueous solubility: 64 g in 100 mL at 25 °C

SPECIFICATIONS

Assay . . . . .98.0–102.0% $Al(NO_3)_3 \cdot 9H_2O$

Maximum Allowable

Insoluble matter 0.005%
Chloride (Cl). 0.001%
Sulfate (SO_4) 0.005%
Calcium (Ca) 0.005%
Magnesium (Mg) 0.001%
Potassium (K) 0.002%
Sodium (Na). 0.005%
Heavy metals (as Pb). 0.001%
Iron (Fe). 0.002%

TESTS

Assay. (By indirect complexometric titration of Al^{III}). Weigh accurately 5 g, and transfer to a 250-mL volumetric flask. Dissolve in about 75 mL of water, add 5 mL of hydrochloric acid, dilute to volume with water, and mix. Place a 25.0-mL aliquot of this solution in a 250-mL beaker, add 50 mL of water, and mix. While stirring, add 40.0 mL of 0.1 M EDTA volumetric solution and 20 mL of ammonium acetate–acetic acid buffer solution. Heat the solution to near boiling for 5 min, and cool. (The pH should be 4.0–5.0 and should be maintained in that range by adding more buffer during the titration if necessary.) Add 50 mL of alcohol and 2 mL of dithizone indicator solution, and titrate the excess EDTA with 0.1 M zinc chloride volumetric solution until the color changes from a green-violet to a rose-pink. One milliliter of 0.1 M zinc chloride corresponds to 0.03751 g of Al(NO$_3$)$_3 \cdot$ 9H$_2$O.

$$\% \ Al(NO_3)_3 \cdot 9H_2O = \frac{[(40.0 \times M \ EDTA) - (mL \times M \ ZnCl_2)] \times 37.51}{Sample \ wt \ (g) \ / \ 10}$$

Insoluble Matter. (Page 25). Dissolve 20 g in 150 mL of water and filter immediately without digesting.

Chloride. (Page 35). Use 1.0 g.

Sulfate. Dissolve 4.0 g in 10 mL of hydrochloric acid, and evaporate to dryness on a hot plate ($\approx$100 °C). Take up the residue in 10 mL of hydrochloric acid, and again evaporate to dryness. Dissolve the residue in 100 mL of water, add a slight excess of ammonium hydroxide, and boil the solution until the odor of ammonia is entirely gone. Dilute with water to 200 mL, filter, and evaporate 50 mL of the filtrate to 10 mL. To the evaporated filtrate, add 1 mL of dilute hydrochloric acid (1 + 19) and 1 mL of 12% barium chloride reagent solution. Any turbidity should not exceed that produced by 0.05 mg of sulfate ion (SO$_4$) in an equal volume of solution containing 1 mL of dilute hydrochloric acid (1 + 19) and 1 mL of 12% barium chloride reagent solution. Compare 10 min after adding the barium chloride to the sample and standard solutions.

Calcium, Magnesium, Potassium, and Sodium.   (By flame AAS, page 63).

Sample Stock Solution. Dissolve 20.0 g of sample in 80 mL of water, add 2 mL of nitric acid (1 + 1), transfer to a 100-mL volumetric flask, dilute to the mark with water, and mix (1 mL = 0.20 g).

Element	Wavelength (nm)	Sample Wt (g)	Standard Added (mg)	Flame Type*	Background Correction
Ca	422.7	2.0	0.05; 0.10	N/A	No
Mg	285.2	2.0	0.01; 0.02	A/A	Yes
K	766.5	2.0	0.02; 0.04	A/A	No
Na	589.0	0.40	0.01; 0.02	A/A	No

*A/A is air/acetylene; N/A is nitrous oxide/acetylene.

Heavy Metals. (Page 36, Method 1). Dissolve 4.0 g in 40 mL of water, and dilute with water to 48 mL. Use 36 mL to prepare the sample solution, and use the remaining 12 mL to prepare the control solution.

Iron. (Page 38, Method 1). Dissolve 1.0 g in 50 mL of water, and use 25 mL of this solution.

Aluminum Potassium Sulfate Dodecahydrate

$AlK(SO_4)_2 \cdot 12H_2O$ Formula Wt 474.39   CAS No. 7784-24-9

GENERAL DESCRIPTION

Typical appearance: colorless solid

Analytical use: buffer; flocculating reagent

Change in state (approximate): melting point, 92 °C

Aqueous solubility: 5.7 g in 100 mL at 0 °C; extremely soluble in hot water

SPECIFICATIONS

Assay . . . . .98.0–102.0% $AlK(SO_4)_2 \cdot 12H_2O$

Maximum Allowable

Insoluble matter 0.005%

Chloride (Cl). . . . . .5 ppm

Ammonium (NH_4) 0.005%

Heavy metals (as Pb). 0.001%

Iron (Fe). 0.001%

Sodium (Na). 0.02%

TESTS

Assay. (By indirect complexometric titration of Al^{III}). Weigh accurately 10 g, and transfer to a 250-mL volumetric flask. Dissolve in about 75 mL of water, add 5 mL of hydrochloric acid, dilute to volume with water, and mix. Place a 25.0-mL aliquot of this solution in a 250-mL beaker, add 50 mL of water, and mix. While stirring, add 40.0 mL of 0.1 M EDTA volumetric solution and 20 mL of ammonium acetate–acetic acid buffer solution. Heat the solution to near boiling for 5 min, and cool. (The pH should be 4.0–5.0 and should be maintained in that range by adding more buffer during the titration if necessary.) Add 50 mL of alcohol and 2 mL of dithizone indicator solution, and titrate the excess of EDTA with 0.1 M zinc chloride volumetric solution until the color changes from a green-violet to a rose-pink. One milliliter of 0.1 M zinc chloride corresponds to 0.04744 g of $AlK(SO_4)_2 \cdot 12H_2O$.

$$\% \, AlK(SO_4)_2 \cdot 12H_2O = \frac{[(40.0 \times M \text{ EDTA}) - (mL \times M \text{ ZnCl}_2)] \times 47.44}{\text{Sample wt (g)} / 10}$$

Insoluble Matter. (Page 25). Use 20 g dissolved in 150 mL of hot water. Filter immediately.

Chloride. (Page 35). Use 2.0 g dissolved in warm water.

Ammonium. Dissolve 1.0 g in 100 mL of ammonia-free water. To 20 mL of this solution, add 10% sodium hydroxide reagent solution until the precipitate first formed is redissolved. Dilute with water to 50 mL, and add 2 mL of Nessler reagent. Any color should not exceed that produced by 0.01 mg of ammonium ion (NH_4) in an equal volume of solution containing the quantities of reagents used in the test.

Heavy Metals. (Page 36, Method 1). Dissolve 5.0 g in 40 mL of water, and dilute with water to 50 mL. Use 30 mL to prepare the sample solution, and use 10 mL to prepare the control solution.

Iron. (Page 38, Method 1). Use 1.0 g.

Sodium. (By flame AAS, page 63).

> *Sample Stock Solution.* Dissolve 1.0 g of sample in a 100-mL volumetric flask, and dilute to the mark with water (1 mL = 0.01 g).

Element	Wavelength (nm)	Sample Wt (g)	Standard Added (mg)	Flame Type*	Background Correction
Na	589.0	0.05	0.01; 0.02	A/A	No

*A/A is air/acetylene.

Aluminum Sulfate, Hydrated

Al$_2$(SO$_4$)$_3$ · (14–18)H$_2$O

for 18-hydrate:	Formula Wt 666.41	CAS No. 7784-31-8

Note: This reagent is available in degrees of hydration ranging from 14 to 18 molecules of water.

GENERAL DESCRIPTION

Typical appearance: white solid

Analytical use: decolorizing, deodorizing, or precipitating agent; clarifying agent for fats and oils

Change in state (approximate): melts when gradually heated

Aqueous solubility: miscible

SPECIFICATIONS

Assay (as the labeled hydrate) 98.0–102.0%

	Maximum Allowable

Insoluble matter 0.01%
Chloride (Cl) 0.005%
Calcium (Ca) 0.01%
Magnesium (Mg). 0.002%
Potassium (K) 0.005%
Sodium (Na) 0.02%
Heavy metals (as Pb) 0.001%
Iron (Fe). 0.002%

TESTS

Assay. (By indirect complexometric titration of Al^{III}). Weigh accurately 8 g and transfer to a 250-mL volumetric flask. Dissolve in about 75 mL of water, add 5 mL of hydrochloric acid, dilute to volume with water, and mix. Place a 25.0-mL aliquot of this solution in a 250-mL beaker, add 50 mL of water, mix, and then add 40.0 mL of standard 0.1 M EDTA and 20 mL of ammonium acetate–acetic acid buffer solution. Heat the solution to near boiling for 5 min and cool. (The pH should be 4.0–5.0 and should be maintained in that range by adding more buffer during the titration if necessary.) Add 50 mL of alcohol and 2 mL of dithizone indicator solution, and titrate the excess of EDTA with 0.1 M zinc chloride volumetric solution until the color changes from a green-violet to a rose-pink. One millili-ter of 0.1 M zinc chloride corresponds to 0.02972 g of $Al_2(SO_4)_3 \cdot 14H_2O$, 0.03152 g of $Al_2(SO_4)_3 \cdot 16H_2O$, and 0.03332 g of $Al_2(SO_4)_3 \cdot 18H_2O$.

$$\% \ Al_2(SO_4)_3 \cdot (14\text{--}18)H_2O = \frac{[(40.0 \times M \ EDTA) - (mL \times M \ ZnCl_2)] \times F}{\text{Sample wt (g)} / 10}$$

where $F = 29.72$ for $14H_2O$, 31.52 for $16H_2O$, and 33.32 for $18H_2O$.

Insoluble Matter. (Page 25). Use 20 g dissolved in 150 mL of water containing 5 mL of 25% sulfuric acid.

Chloride. (Page 35). Use 0.20 g.

Calcium, Magnesium, Potassium, and Sodium. (By flame AAS, page 63).

> **Sample Stock Solution.** Dissolve 10.0 g of sample in 80 mL of water, add 2 mL of nitric acid (1 + 1), transfer to a 100-mL volumetric flask, dilute to the mark with water, and mix (1 mL = 0.10 g).

Element	Wavelength (nm)	Sample Wt (g)	Standard Added (mg)	Flame Type*	Background Correction
Ca	422.7	1.0	0.05; 0.10	N/A	No
Mg	285.2	1.0	0.01; 0.02	A/A	Yes
K	766.5	1.0	0.025; 0.05	A/A	No
Na	589.0	0.10	0.01; 0.02	A/A	No

*A/A is air/acetylene; N/A is nitrous oxide/acetylene.

Heavy Metals. (Page 36, Method 1). Dissolve 4.0 g in 40 mL of water, and dilute with water to 48 mL. Use 36 mL to prepare the sample solution, and use the remaining 12 mL to prepare the control solution.

Iron. (Page 38, Method 1). Dissolve 1.0 g in 40 mL of water; use 20 mL of this solution, and 1 mL of hydrochloric acid.

4-Aminoantipyrine
Ampyrone

C$_{11}$H$_{13}$N$_3$O   **Formula Wt 203.25**   **CAS No. 83-07-8**

GENERAL DESCRIPTION

Typical appearance: yellow solid

Analytical use: indicator for trace phenol determinations in water

Change in state (approximate): melting point, 107–109 °C

Aqueous solubility: soluble

SPECIFICATIONS

Assay . . . . ≥98.0% C$_{11}$H$_{13}$N$_3$O

Sensitivity to phenol . . . . Passes test

Maximum allowable

Residue after ignition 0.1%

Loss on drying 0.5%

TESTS

Assay. (Total alkalinity by indirect nonaqueous titration). Weigh accurately about 0.8 g of sample, and dissolve in 100 mL of glacial acetic acid. Add 50.0 mL of 0.1 N perchloric acid in glacial acetic acid volumetric solution and 0.05 mL of crystal violet indicator solution. Titrate the excess perchloric acid with 0.1 N sodium acetate in glacial acetic acid volumetric solution to a violet end point. One milliliter of 0.1 N HClO$_4$ corresponds to 0.02033 g of 4-aminoantipyrine.

$$\% \ C_{11}H_{13}N_3O = \frac{[(mL \times N \ HClO_4) - (mL \times N \ CH_3COONa)] \times 20.33}{Sample \ wt \ (g)}$$

Sensitivity to Phenol. Prepare a 500-mL water blank and 500-mL phenol standards containing 0.005 mg (described below), 0.010 mg, 0.020 mg, 0.030 mg, 0.040 mg, and 0.050 mg of phenol. To the blank and standards, add 12.0 mL of 0.5 N ammonium hydroxide (described below), and adjust the pH of each to 7.9 ± 0.1 with phosphate buffer solution (described below) using a suitable pH meter. About 10 mL of the phosphate buffer is required. Transfer to a 1-L separatory funnel, add 3.0 mL of 4-aminoantipyrine sample solution (described below), mix well, add 3.0 mL of potassium ferricyanide solution (described below), mix well, and let the color develop for 3 min. The solutions should be clear and light yellow. Extract each solution immediately with 25.0 mL of chloroform; shake the separatory funnel at least 10 times. Allow the layers to separate, and shake again at least 10 times. Allow to fully separate. Draw off and filter the chloroform extracts through filter paper, and collect the dry extracts. Determine the absorbance of the extracts using a suitable spectrophotometer at 460 nm. Run the standards in 1-cm cells, using the blank as the reference. Plot the absorbance against mg phenol concentration. The graph should be linear with a correlation greater than 0.99.

> **Phenol Standard (0.005 mg of C₆H₅OH in 1 mL).** Dissolve 0.50 g of solid phenol in water, and dilute to 1L. Dilute 10.0 mL of this solution to 1L.

> **Ammonium Hydroxide Solution, 0.5 N.** Dilute 3.5 mL of fresh concentrated ammonium hydroxide to 100 mL with water.

> **4-Aminoantipyrine Sample Solution.** Dissolve 2.0 g of sample in water, and dilute to 100 mL. Prepare at time of use.

> **Potassium Ferricyanide Solution.** Dissolve 8.0 g of $K_3Fe(CN)_6$ in water, and dilute to 100 mL. Filter and store in an amber glass bottle. Prepare fresh weekly.

> **Phosphate Buffer Solution.** Dissolve 10.45 g of potassium phosphate, dibasic, and 7.23 g of potassium phosphate monobasic in water, and dilute to 100 mL.

Residue after Ignition. (Page 26). Ignite 1.0 g.

Loss on Drying. (Page 26). Use 1 g. Dry at 90 °C.

2-Aminoethanol
Monoethanolamine
$HOCH_2CH_2NH_2$ Formula Wt 61.08   CAS No. 141-43-5

GENERAL DESCRIPTION
Typical appearance: viscous, hygroscopic liquid with ammoniacal odor
Analytical use: buffer; removal of carbon dioxide and hydrogen sulfide from gas mixtures
Change in state (approximate): boiling point, 170 °C
Aqueous solubility: miscible
Density: 1.01
pK_a: 9.4

SPECIFICATIONS
Assay . . . . ≥99.0% $HOCH_2CH_2NH_2$

Maximum Allowable

Color (APHA) 15
Water (H₂O) 0.30%
Iron (Fe) . . . . 5 ppm
Heavy metals (as Pb) . . . . 5 ppm

TESTS

Assay. Analyze the sample by gas chromatography using the general parameters cited on page 80. The following specific conditions are also required.

Column: Type I, methyl silicone

Measure the area under all peaks, and calculate the 2-aminoethanol content in area percent. Correct for water content.

Color (APHA). (Page 43).

Water. (Page 31, Method 2). Use 0.10 mL (0.1 g) of the sample.

Iron. (Page 38, Method 1). Use 2.0 g (2.0 mL) in 40 mL of water and 5 mL of hydrochloric acid.

Heavy Metals. (Page 36, Method 1). Use 4.0 g (4.0 mL) in 30 mL of water and 6 mL of hydrochloric acid.

4-Amino-3-hydroxy-1-naphthalene Sulfonic Acid
1-Amino-2-naphthol-4-sulfonic Acid
$H_2N(HO)C_{10}H_5SO_3H$ **Formula Wt 239.25**   **CAS No. 116-63-2**

GENERAL DESCRIPTION

Typical appearance: white to slightly brownish-pink powder
Analytical use: determination of phosphate
Change in state (approximate): melting point, 285 °C
Aqueous solubility: insoluble

SPECIFICATIONS

Assay . . . . ≥98.0% $H_2N(HO)C_{10}H_5SO_3H$

Maximum Allowable

Solubility in sodium carbonate . . . . Passes test
Residue after ignition 0.1%
Sulfate (SO₄) 0.2%
Sensitivity to phosphate. . . . . Passes test

TESTS

Assay. (By acid–base titrimetry). Weigh accurately about 0.5 g and transfer to a beaker. Add 100 mL of dimethyl sulfoxide, and while stirring, add 25 mL of water in small portions. Insert the electrodes of a suitable pH meter, and titrate with 0.1 N sodium hydroxide, establishing the end point potentiometrically. Correct for a blank determination. One milliliter of 0.1 N sodium hydroxide corresponds to 0.02392 g of $H_2N(HO)C_{10}H_5SO_3H$.

$$\% \ H_2N(HO)C_{10}H_5SO_3H = \frac{\{[mL\ (sample) - mL\ (blank)] \times N\ NaOH\} \times 23.92}{Sample\ wt\ (g)}$$

Solubility in Sodium Carbonate. Dissolve 0.1 g in 3 mL of 1% sodium carbonate reagent solution, and dilute with water to 10 mL. The solution should be clear and dissolution complete, or nearly so.

Residue after Ignition. (Page 26). Ignite 1.0 g.

Sulfate. Transfer 0.5 g to a beaker, add 25 mL of water and 0.10 mL of hydrochloric acid, and heat on a hot plate ($\approx$100 °C) for 10 min. Cool, dilute with water to 50 mL, and filter. To 10 mL of the clear filtrate, add 1 mL of 1 N hydrochloric acid and 2 mL of 12% barium chloride reagent solution. Any turbidity should not exceed that produced by 0.2 mg of sulfate ion (SO_4) in an equal volume of solution containing the quantities of reagents used in the test. Compare 10 min after adding the barium chloride to the sample and standard solutions.

Sensitivity to Phosphate. Dissolve 0.10 g in a mixture of 50 mL of water and 10 g of sodium bisulfite, warm if necessary to complete dissolution, and filter. Add 1.0 mL of this solution to a test solution prepared as follows: To 20 mL of water, add 0.02 mg of phosphate ion (PO_4) and 2 mL of 25% sulfuric acid reagent solution, mix, and then add 1.0 mL of ammonium molybdate reagent solution. A distinct blue color should be produced in the test solution in 5 min.

Ammonium Acetate

CH₃COONH₄　　　　　　**Formula Wt 77.08**　　　　　**CAS No. 631-61-8**

GENERAL DESCRIPTION

Typical appearance: hygroscopic solid
Analytical use: buffer solution; determination of lead and iron; separating lead sulfate from other sulfates
Change in state (approximate): melting point, 114 °C
Aqueous solubility: very soluble

SPECIFICATIONS

Assay . . . . $\geq$97% CH₃COONH₄
pH of a 5% solution at 25.0 °C 6.7–7.3

	Maximum Allowable
Insoluble matter	0.005%
Residue after ignition	0.01%
Chloride (Cl)	5 ppm
Nitrate (NO₃)	0.001%
Sulfate (SO₄)	0.001%
Heavy metals (as Pb)	5 ppm
Iron (Fe)	5 ppm

TESTS

Assay. (By alkalimetry for ammonium). Weigh accurately about 2.5 g of sample, and dissolve in 50 mL of water in a 500-mL Erlenmeyer flask. Add exactly 50.0 mL of 1 N sodium hydroxide, place a filter funnel loosely in the neck of the flask, and boil until all the ammonia is expelled (about 10–15 min) as determined with litmus paper. Cool, and add 0.15 mL of thymol blue indicator solution. Titrate the excess sodium hydroxide with 1 N sulfuric acid. One milliliter of 1 N sodium hydroxide corresponds to 0.07708 g of CH_3COONH_4.

$$\% \ CH_3COONH_4 = \frac{[(50.0 \times N \ NaOH) - (mL \times N \ H_2SO_4)] \times 7.708}{\text{Sample wt (g)}}$$

pH of a 5% Solution at 25.0 °C. (Page 49).

Insoluble Matter. (Page 25). Use 20 g dissolved in 200 mL of water.

Residue after Ignition. (Page 26). Ignite 10 g.

Chloride. (Page 35). Use 2.0 g.

Nitrate. (Page 38, Method 1). For sample solution A, use 1.0 g. For control solution B, use 1.0 g and 1 mL of nitrate ion (NO_3) standard solution.

For the Determination of Sulfate, Heavy Metals, and Iron

Sample Solution A. Dissolve 15 g in 15 mL of water, and add about 10 mg of sodium carbonate, 2 mL of hydrogen peroxide, 15 mL of hydrochloric acid, and 25 mL of nitric acid.

Blank Solution B. Prepare a solution containing the quantities of reagents used in sample solution A, omitting only the sample.

Digest each in covered beakers on a hot plate (≈ 100 °C) until reaction ceases, uncover, and evaporate to dryness. Dissolve each in 3 mL of 1 N acetic acid plus 15 mL of water, filter through a small filter, and dilute with water to 30 mL (1 mL = 0.5 g).

Sulfate. (Page 40, Method 1). Use 10 mL of sample solution A (5-g sample).

Heavy Metals. (Page 36, Method 1). Dilute 8.0 mL of sample solution A (4-g sample) with water to 25 mL. Use 8.0 mL of blank solution B to prepare the standard solution.

Iron. (Page 38, Method 1). Use 4.0 mL of sample solution A (2-g sample). Prepare the standard from 4.0 mL of blank solution B.

Ammonium Benzoate
Benzoic Acid, Ammonium Salt

$C_7H_9NO_2$ **Formula Wt 139.16**   **CAS No. 1863-63-4**

GENERAL DESCRIPTION

Typical appearance: solid
Analytical use: determination of benzoate stabilizers in feeds
Change in state (approximate): melting point, 198 °C
Aqueous solubility: very soluble

SPECIFICATIONS

Assay . . . . $\geq$98.0% $C_7H_9NO_2$
pH of a 5% solution at 25.0 °C 5.0–8.0

	Maximum Allowable
Insoluble matter	0.005%
Residue after ignition	0.005%
Substances reducing permanganate	Passes test
Chlorine compounds (as Cl)	0.01%
Heavy metals (as Pb)	0.001%
Iron (Fe)	0.001%

TESTS

Assay. (By alkalimetry for ammonium). Weigh accurately about 5.0 g of sample, and dissolve in 50 mL of water in a 500-mL Erlenmeyer flask. Add exactly 50.0 mL of 1 N sodium hydroxide, place a filter funnel loosely in the neck of the flask, and boil until all the ammonia is expelled (about 10–15 min) as determined with litmus paper. Cool, and add 0.15 mL of thymol blue indicator solution. Titrate the excess sodium hydroxide with 1 N sulfuric acid. One milliliter of 1 N sodium hydroxide corresponds to 0.13916 g of $C_7H_9NO_2$.

$$\% \; C_7H_9NO_2 = \frac{[(50.0 \times \text{N NaOH}) - (\text{mL} \times \text{N H}_2\text{SO}_4)] \times 13.916}{\text{Sample wt (g)}}$$

pH of a 5% Solution at 25.0 °C. (Page 49).

Insoluble Matter. (Page 25). Use 20.0 g in 200 mL of water.

Residue after Ignition. (Page 26). Heat 20.0 g in a preconditioned platinum dish at a temperature sufficient to volatilize. When volatilized, cool and continue as described.

Substances Reducing Permanganate. Add 1.0 g of the sample to 10 mL of water. Add 0.2 mL of 0.1 N potassium permanganate, and let stand for 5 min. The pink color should not be entirely discharged.

Chlorine Compounds. Mix 1.0 g with 0.5 g of sodium carbonate, and add 10–15 mL of water. Evaporate and ignite until charred. Extract with 20 mL of water and 5.5 mL of nitric acid. Filter through a chloride-free filter, and wash with two 10-mL portions of water. Dilute with water to 100 mL. Use 10 mL for the test. Continue as in the test for chloride, page 35.

Heavy Metals. (Page 36, Method 1). Volatilize 2.0 g over a low flame, cool, and add 5 mL of nitric acid and 10 mg of sodium carbonate. Evaporate to dryness on a hot plate ($\approx$100 °C), dissolve the residue in water, and dilute to 25 mL. For the control, add 0.02 mg of lead to 5 mL of nitric acid and 10 mg of sodium carbonate, and evaporate on a hot plate. Dissolve the residue, and dilute to 25 mL.

Iron. (Page 38, Method 2). Use 1.0 g.

Ammonium Bromide

NH_4Br **Formula Wt 97.94**   **CAS No. 12124-97-9**

GENERAL DESCRIPTION

Typical appearance: white solid

Analytical use: photochemical reactions; precipitation of silver salts

Change in state (approximate): sublimation point, 542 °C

Aqueous solubility: 68 g in 100 mL at 10 °C

SPECIFICATIONS

Assay . . . . ≥99.0% NH_4Br

pH of a 5% solution at 25.0 °C 4.5–6.0

Maximum Allowable

Insoluble matter 0.005%

Residue after ignition 0.01%

Bromate (BrO_3) 0.002%

Chloride (Cl) 0.2%

Iodide (I) . . . . Passes test

Sulfate (SO_4) 0.005%

Barium (Ba) 0.002%

Heavy metals (as Pb) . . . . 5 ppm

Iron (Fe) . . . . 5 ppm

TESTS

Assay. (By indirect argentimetric titration of bromide). Weigh accurately about 0.4 g of sample in 50 mL of water. Add 10 mL of 10% nitric acid reagent solution and 50.0 mL of 0.1 N silver nitrate volumetric solution, and titrate the excess silver nitrate with 0.1 N ammonium thiocyanate volumetric solution, using 0.5 mL of ferric sulfate reagent solution as an indicator. One milliliter of 0.1 N silver nitrate consumed corresponds to 0.009794 g of NH_4Br.

$$\% \, NH_4Br = \frac{[(50.0 \times N \, AgNO_3) - (mL \times N \, NH_4SCN)] \times 9.794}{Sample \, wt \, (g)}$$

pH of a 5% Solution at 25.0 °C. (Page 49).

Insoluble Matter. (Page 25). Use 20 g dissolved in 150 mL of water.

Residue after Ignition. (Page 26). Ignite 20 g.

Bromate. (Page 54). Use 5.0 g of sample in 25 mL of solution. For the standard, add 0.10 mg of bromate ion (BrO_3).

Chloride. (Page 35). Dissolve 0.50 g in 15 mL of dilute nitric acid (1 + 2) in a small flask. Add 3 mL of 30% hydrogen peroxide and digest on a hot plate ($\approx$100 °C) until the solution is colorless. Wash down the sides of the flask with a little water, digest for an additional 15 min, cool, and dilute with water to 200 mL. Dilute 2.0 mL of the solution with water to 20 mL.

Iodide. Dissolve 5.0 g in 20 mL of water. Add 1 mL of chloroform, 0.15 mL of 10% ferric chloride reagent solution, and 0.25 mL of 10% sulfuric acid, and shake for 1 min. No violet tint should be produced in the chloroform. (Limit about 0.005%)

Sulfate. (Page 40, Method 1). Allow 30 min for turbidity to form.

Barium. For the sample, dissolve 6.0 g in 15 mL of water. For the control, dissolve 1.0 g in 15 mL of water, and add 0.1 mg of barium. To each solution, add 5 mL of glacial acetic acid, 5 mL of 30% hydrogen peroxide, and 1 mL of hydrochloric acid. Digest in a covered beaker on a hot plate ($\approx$100 °C) until reaction ceases, uncover, and evaporate to dryness. Dissolve each residue in 15 mL of water, filter if necessary, and dilute with water to 23 mL. Add 2 mL of 10% potassium dichromate reagent solution, then add ammonium hydroxide until the orange color is just dissipated and the yellow color persists. Add 25 mL of methanol, stir vigorously, and allow to stand for 10 min. Any turbidity in the solution of the sample should not exceed that in the control.

Heavy Metals. (Page 36, Method 1). Dissolve 6.0 g in about 20 mL of water, and dilute with water to 30 mL. Use 25 mL to prepare the sample solution, and use the remaining 5.0 mL to prepare the control solution.

Iron. (Page 38, Method 1). Use 2.0 g.

Ammonium Carbonate

CAS No. 8000-73-5

Note: This product is a mixture of variable proportions of ammonium bicarbonate and ammonium carbamate.

GENERAL DESCRIPTION

Typical appearance: white powder with odor of ammonia
Analytical use: pH adjustment
Change in state (approximate): volatilizes at 60 °C
Aqueous solubility: 25 g in 100 mL at 15 °C

SPECIFICATIONS

Assay . . . . $\geq$30.0% NH_3

Maximum Allowable

Insoluble matter 0.005%
Nonvolatile matter 0.01%
Chloride (Cl) . . . . 5 ppm
Sulfur compounds (as SO_4) 0.002%
Heavy metals (as Pb) . . . . 5 ppm
Iron (Fe). . . . . 5 ppm

TESTS

Assay. (By indirect acid–base titrimetry). Tare a glass-stoppered flask containing about 25 mL of water. Add 2.0–2.5 g of the sample, and weigh accurately. Add slowly 50.0 mL of 1 N hydrochloric acid, and titrate the excess with 1 N sodium hydroxide, using methyl orange indicator. One milliliter of 1 N hydrochloric acid corresponds to 0.01703 g of NH_3.

$$\% \, NH_3 = \frac{[(50.0 \times N \, HCl) - (mL \times N \, NaOH)] \times 1.703}{Sample \, wt \, (g)}$$

Insoluble Matter. (Page 25). Use 20 g dissolved in 100 mL of water.

Nonvolatile Matter. To 20 g in a tared, preconditioned dish add 10 mL of water, volatilize on a hot plate ($\approx$100 °C), and dry for 1 hour at 105 °C. Retain the residue for the test for heavy metals.

Chloride. (Page 35). Dissolve 2.0 g in 25 mL of hot water, add 1.0 mL of sodium carbonate reagent solution, and evaporate to dryness on a hot plate ($\approx$100 °C). Dissolve the residue in 20 mL of water.

Sulfur Compounds. Dissolve 2.0 g in 20 mL of water, add about 10 mg of sodium carbonate, and evaporate to dryness. Dissolve the residue in a slight excess of hydrochloric acid, add 2 mL of bromine water, and again evaporate to dryness. Dissolve the residue in 4 mL of water plus 1 mL of dilute hydrochloric acid (1 + 19). Filter through a small filter, wash with two 2-mL portions of water, dilute with water to 10 mL, and add 1 mL of 12% barium chloride reagent solution. Any turbidity should not exceed that produced by 0.04 mg of sulfate ion (SO_4) in an equal volume of solution containing the quantities of reagents used in the test. Compare 10 min after adding the barium chloride to the sample and standard solutions.

Heavy Metals. (Page 36, Method 1). Dissolve the residue from the test for nonvolatile matter in 5 mL of 1 N acetic acid, and dilute with water to 50 mL. Dilute 10 mL of this solution with water to 25 mL.

Iron. (Page 38, Method 1). To 2.0 g, add 5 mL of water and evaporate on a hot plate ($\approx$100 °C). Dissolve the residue in 2 mL of hydrochloric acid, add 20 mL of water, and filter if necessary. Use this solution without further acidification.

Ammonium Chloride

NH₄Cl   **Formula Wt 53.49**   **CAS No. 12125-02-9**

GENERAL DESCRIPTION

Typical appearance: white solid
Analytical use: electrolyte
Change in state (approximate): sublimation point, 335 °C
Aqueous solubility: 39.5 g in 100 mL at 25 °C

SPECIFICATIONS

Assay . . . . ≥99.5% NH₄Cl
pH of a 5% solution at 25.0 °C 4.5–5.5

	Maximum Allowable
Insoluble matter	0.005%
Residue after ignition.	0.01%
Calcium (Ca)	0.001%
Magnesium (Mg)	5 ppm
Heavy metals (as Pb).	5 ppm
Iron (Fe).	2 ppm
Phosphate (PO₄).	2 ppm
Sulfate (SO₄)	0.002%

TESTS

Assay. (By argentimetric titration of chloride content). Weigh accurately about 0.25 g, and dissolve in 50 mL of water in a 250-mL glass-stoppered flask. Add 1 mL of dichlorofluorescein indicator solution. While stirring, titrate with 0.1 N silver nitrate volumetric solution. The silver chloride flocculates, and the mixture changes to a pink color. One milliliter of 0.1 N silver nitrate corresponds to 0.005349 g of NH₄Cl.

$$\% \text{ NH}_4\text{Cl} = \frac{(\text{mL} \times \text{N AgNO}_3) \times 5.349}{\text{Sample wt (g)}}$$

pH of a 5% Solution at 25.0 °C. (Page 49).

Insoluble Matter. (Page 25). Use 20.0 g dissolved in 200 mL of water.

Residue after Ignition. (Page 26). Ignite 10.0 g.

Calcium and Magnesium. (By flame AAS, page 63).

> **Sample Stock Solution.** Dissolve 10.0 g of sample in 80 mL of water, and warm to room temperature. Transfer to a 100-mL volumetric flask, and dilute to the mark with water (1 mL = 0.10 g).

Element	Wavelength (nm)	Sample Wt (g)	Standard Added (mg)	Flame Type*	Background Correction
Ca	422.7	2.0	0.02; 0.04	N/A	No
Mg	285.2	2.0	0.01; 0.02	A/A	Yes

*A/A is air/acetylene; N/A is nitrous oxide/acetylene.

For the Determination of Heavy Metals, Iron, Phosphate, and Sulfate

Sample Solution A. Dissolve 25.0 g in 75 mL of water, and add about 10 mg of sodium carbonate, 10 mL of hydrogen peroxide, and 50 mL of nitric acid.

Blank Solution B. To 75 mL of water, add about 10 mg of sodium carbonate, 10 mL of hydrogen peroxide, and 50 mL of nitric acid.

Digest sample solution A and blank solution B in covered beakers on a hot plate ($\approx$100 °C) until reaction ceases, uncover, and evaporate to dryness. Dissolve the two residues in separate 5-mL portions of 1 N acetic acid and 50 mL of water, filter if necessary, and dilute with water to 100 mL (1 mL of sample solution A = 0.25 g).

Heavy Metals. (Page 36, Method 1). Dilute 16 mL of sample solution A (4-g sample) with water to 25 mL. Use 16 mL of blank solution B to prepare the standard solution.

Iron. (Page 38, Method 1). Use 20 mL of sample solution A (5-g sample). Prepare the standard with 20 mL of blank solution B.

Phosphate. (Page 40, Method 1). Evaporate 40 mL of sample solution A (10-g sample) to dryness on a hot plate ($\approx$100 °C). Dissolve the residue in 25 mL of approximately 0.5 N sulfuric acid. Use 40 mL of blank solution B to prepare the standard solution.

Sulfate. (Page 40, Method 1). Use 10 mL of sample solution A (2.5-g sample). Use 10 mL of blank solution B to prepare the standard solution.

Ammonium Citrate, Dibasic

Citric Acid, Diammonium Salt; 2-Hydroxy-1,2,3-propanetricarboxylic Acid, Diammonium Salt

$(NH_4)_2HC_6H_5O_7$ **Formula Wt 226.19**   **CAS No. 3012-65-5**

GENERAL DESCRIPTION

Typical appearance: colorless solid
Analytical use: determination of phosphates
Aqueous solubility: very soluble

SPECIFICATIONS

Assay . . . . 98.0–103.0% $(NH_4)_2HC_6H_5O_7$

	Maximum Allowable
Insoluble matter	0.005%
Residue after ignition	0.01%
Chloride (Cl)	0.001%
Heavy metals (as Pb)	5 ppm
Iron (Fe)	0.001%
Oxalate (C_2O_4)	Passes test
Phosphate (PO_4)	5 ppm
Sulfur compounds (as SO_4)	0.005%

TESTS

Assay. (By alkalimetry for ammonium). Weigh accurately about 3.0 g of sample and dissolve in 50 mL of water in a 500-mL Erlenmeyer flask. Add exactly 50.0 mL of 1 N sodium hydroxide volumetric solution, place a filter funnel loosely in the neck of the flask, and boil until all the ammonia is expelled (about 10–15 min) as determined with litmus paper. Cool, and add 0.15 mL of thymol blue indicator solution. Titrate the excess sodium hydroxide with 1 N sulfuric acid. One milliliter of 1 N sodium hydroxide corresponds to 0.07540 g of $(NH_4)_2HC_6H_5O_7$.

$$\% \ (NH_4)_2HC_6H_5O_7 = \frac{[(50.0 \times N \ NaOH) - (mL \times N \ H_2SO_4)] \times 7.540}{Sample \ wt \ (g)}$$

Insoluble Matter. (Page 25). Use 20 g dissolved in 200 mL of water.

Residue after Ignition. (Page 26). Ignite 10 g.

Chloride. (Page 35). Use 1.0 g.

Heavy Metals. (Page 36, Method 1). Dissolve 6.0 g in about 20 mL of water, add 5 mL of dilute hydrochloric acid (1 + 1), and dilute with water to 30 mL. Use 25 mL to prepare the sample solution, and use the remaining 5.0 mL to prepare the control solution.

Iron. (Page 38, Method 1). Use 1.0 g.

Oxalate. Dissolve 5.0 g in 25 mL of water, and add 3 mL of glacial acetic acid and 2 mL of 10% calcium acetate solution. No turbidity or precipitate should appear after standing 4 h. (Limit about 0.05%)

Phosphate. (Page 40, Method 1). Mix 4.0 g with 0.5 g of magnesium nitrate hexahydrate in a platinum dish and ignite. Dissolve the residue in 5 mL of water, add 5 mL of nitric acid, and evaporate to dryness. Dissolve the residue in 25 mL of approximately 0.5 N sulfuric acid, and continue as described.

Sulfur Compounds. To 1.0 g, add 1 mL of hydrochloric acid and 3 mL of nitric acid. Prepare a standard containing 0.05 mg of sulfate ion (SO_4), 1 mL of hydrochloric acid, and 3 mL of nitric acid. Digest each in a covered beaker on a hot plate ($\approx$100 °C) until the reaction ceases, remove the covers, and evaporate to dryness. Add 0.2–0.5 mg of ammonium vanadate and 10 mL of nitric acid, and digest in covered beakers on the hot plate until reactions cease. Remove the covers, and evaporate to dryness. Add 10 mL more of nitric acid, and repeat the digestions and evaporations. Add 5 mL of dilute hydrochloric acid $(1 + 1)$, and evaporate to dryness. Dissolve the residues in 4 mL of water plus 1 mL of dilute hydrochloric acid $(1 + 19)$, and filter through a small filter. Wash with two 2-mL portions of water, dilute with water to 10 mL, and add 1 mL of 12% barium chloride reagent solution to each. Any turbidity in the solution of the sample should not exceed that in the standard. Compare 10 min after adding the barium chloride to the sample and standard solutions.

Ammonium Dichromate

$(NH_4)_2Cr_2O_7$	Formula Wt 252.07	CAS No. 7789-09-5

Note: This reagent may be stabilized by the addition of 0.5–3.0% water.

GENERAL DESCRIPTION

Typical appearance: orange-red solid
Analytical use: oxidimetric standard
Change in state (approximate): decomposes at 185 °C
Aqueous solubility: 35.7 g in 100 mL at 20 °C

SPECIFICATIONS

Assay (dried basis). . . . . ≥99.5% $(NH_4)_2Cr_2O_7$

Maximum Allowable

Insoluble matter	0.005%
Loss on drying	3.0%
Chloride (Cl)	0.005%
Sulfate (SO$_4$)	0.01%
Calcium (Ca)	0.002%
Iron (Fe).	0.002%
Sodium (Na)	0.005%

TESTS

Assay. (By iodometry). Weigh accurately about 1.0 g, previously dried to constant weight at 105 °C, transfer to a 100-mL volumetric flask, dissolve in water, dilute to volume, and mix well. Pipet a 10.0-mL aliquot to a glass-stoppered flask, add 50 mL of water, 10 mL of 10% sulfuric acid reagent solution, and 3.0 g of potassium iodide. Mix, stopper, and allow to stand in the dark for 10 min. Titrate the liberated iodine with 0.1 N sodium thiosulfate, adding 3 mL of starch indicator solution near the end point, which is a greenish-blue color. One milliliter of 0.1 N sodium thiosulfate is equivalent to 0.004201 g of $(NH_4)_2Cr_2O_7$.

$$\% \ (NH_4)_2Cr_2O_7 = \frac{(mL \times N \ Na_2S_2O_3) \times 4.201}{Sample \ wt \ (g) \ / \ 10}$$

Insoluble Matter. (Page 25). Dissolve 20 g in 200 mL of water.

Loss on Drying. Crush lightly and weigh accurately 2.0 g, and dry in a preconditioned dish for 2 h at 105 °C.

Chloride. (Page 35). Dissolve 0.20 g in 10 mL of water, filter if necessary through a small chloride-free filter, and add 1 mL of ammonium hydroxide and 1 mL of silver nitrate reagent solution. Prepare a standard containing 0.01 mg of chloride ion (Cl) in 10 mL of water, and add 1 mL of ammonium hydroxide and 1 mL of silver nitrate reagent solution. Add 2 mL of nitric acid to each. The comparison is best made by the general method for chloride in colored solutions, page 35.

Sulfate. Dissolve 10.0 g in 250 mL of water, filter if necessary, and heat to boiling. Add 25 mL of a solution containing 1.0 g of barium chloride and 2 mL of hydrochloric acid per 100 mL of solution. Digest in a covered beaker on a hot plate ($\approx$100 °C) for 2 h and allow to stand for at least 8 h. If a precipitate is formed, filter, wash thoroughly, and ignite. Fuse the residue with 1.0 g of sodium carbonate. Extract the fused mass with water, and filter off the insoluble residue. Add 5 mL of hydrochloric acid to the filtrate, dilute with water to about 200 mL, heat to boiling, and add 10 mL of alcohol. Digest in a covered beaker on the hot plate until reduction of chromate is complete, as indicated by the change to a clear green or colorless solution. Neutralize the solution with ammonium hydroxide, and add 2 mL of hydrochloric acid. Heat to boiling, and add 10 mL of 12% barium chloride reagent solution. Digest in a covered beaker on a hot plate ($\approx$100 °C) for 2 h, and allow to stand for at least 8 h. Filter, wash thoroughly, and ignite. The weight of the precipitate should not exceed the weight obtained in a complete blank test by more than 0.0024 g. If the original precipitate of barium sulfate weighs less than the requirement permits, the fusion with sodium carbonate is not necessary.

Calcium, Iron, and Sodium. (By flame AAS, page 63).

> **Sample Stock Solution.** Dissolve 20.0 g in water and dilute with water to 100 mL (1 mL = 0.2 g).

Element	Wavelength (nm)	Sample Wt (g)	Standard Added (mg)	Flame Type*	Background Correction
Ca	422.7	2.0	0.02; 0.04	N/A	No
Fe	248.3	4.0	0.04; 0.08	A/A	Yes
Na	589.0	0.20	0.005; 0.01	A/A	No

*A/A is air/acetylene; N/A is nitrous oxide/acetylene.

Ammonium Fluoride

NH₄F   **Formula Wt 37.04**   **CAS No. 12125-01-8**

GENERAL DESCRIPTION

Typical appearance: colorless or white, deliquescent solid
Analytical use: etchant; extracting agent

Change in state (approximate): sublimes on heating
Aqueous solubility: 100 g in 100 mL at 0 °C

SPECIFICATIONS

Assay . . . . ≥98.0% NH$_4$F

Maximum Allowable

Insoluble matter 0.005%
Residue after ignition 0.01%
Chloride (Cl) 0.001%
Sulfate (SO$_4$) 0.005%
Heavy metals (as Pb) . . . . 5 ppm
Iron (Fe). . . . . 5 ppm

TESTS

Assay. (By alkalimetry for ammonium). Weigh accurately about 1.0 g of sample, and dissolve in 50 mL of water in a 500-mL Erlenmeyer flask. Add exactly 50.0 mL of 1 N sodium hydroxide volumetric solution, place a filter funnel loosely in the neck of the flask, and boil until all the ammonia is expelled (about 10–15 min) as determined with litmus paper. Cool, and add 0.15 mL of thymol blue indicator solution. Titrate the excess sodium hydroxide with 1 N sulfuric acid. One milliliter of 1 N sodium hydroxide corresponds to 0.03704 g of NH$_4$F.

$$\% \, \mathrm{NH_4F} = \frac{[(50.0 \times \mathrm{N\,NaOH}) - (\mathrm{mL} \times \mathrm{N\,H_2SO_4})] \times 3.704}{\text{Sample wt (g)}}$$

Insoluble Matter. (Page 25). Use 20 g dissolved in 200 mL of hot water.

Residue after Ignition. (Page 26). Ignite 10 g in a tared, preconditioned platinum crucible or dish.

Chloride. Dissolve 1.0 g in a mixture of 10 mL of water and 1 mL of nitric acid in a platinum dish. In a test tube or small beaker, mix 20 mL of water and 1 mL of silver nitrate reagent solution, add this solution to the sample solution, and mix. Any turbidity should not exceed that produced by 0.01 mg of chloride ion (Cl) in an equal volume of solution containing the quantities of reagents used in the test.

Sulfate. (Page 41, Method 2). Use 10 mL of hydrochloric acid in a platinum dish, and perform four evaporations.

Heavy Metals. (Page 36, Method 1). Dissolve 5.0 g in about 25 mL of water, add 2 g of sodium acetate, and dilute with water to 30 mL. Adjust the pH with short-range pH paper. Use 1.0 g of sample and 2 g of sodium acetate to prepare the control solution.

Iron. (Page 38, Method 1). Treat 2.0 g in a platinum dish with 10 mL of dilute hydrochloric acid (1 + 1), and evaporate on a hot plate (≈100 °C) to dryness. Repeat the evaporation with a second portion of the dilute acid. Warm the residue with 2 mL of hydrochloric acid, dilute with water to 50 mL, and use this solution without further acidification. Prepare the standard from the residue remaining from the evaporation of 10 mL of hydrochloric acid.

Ammonium Hydroxide
Aqueous Ammonia
NH₄OH   **Formula Wt 35.05**   **CAS No. 1336-21-6**

GENERAL DESCRIPTION

Typical appearance: clear liquid with a suffocating odor
Analytical use: buffers; pH adjustment
Change in state (approximate): boiling point, 100 °C
Aqueous solubility: soluble

SPECIFICATIONS

Appearance . . . . Passes test
Assay (as NH_3) 28.0–30.0%

Maximum Allowable

Residue after ignition 0.002%
Carbon dioxide (CO_2) 0.002%
Chloride (Cl) . . . . 0.5 ppm
Nitrate (NO_3) . . . . 2 ppm
Phosphate (PO_4) . . . . 2 ppm
Sulfate (SO_4) . . . . 2 ppm
Heavy metals (as Pb) . . . . 0.5 ppm
Iron (Fe) . . . . 0.2 ppm
Substances reducing permanganate . . . . Passes test

TESTS

Appearance. Mix the material in the original container, pour 10 mL into a test tube (20 × 150 mm), and compare with distilled water in a similar tube. The liquids should be equally clear and free from suspended matter; looking across the columns by means of transmitted light should reveal no apparent difference in color between the two liquids.

Assay. (By acidimetry). Tare a small glass-stoppered Erlenmeyer flask containing 35 mL of water. Using a measuring pipet, without suction, introduce barely under the surface about 2 mL of the ammonium hydroxide, stopper, and weigh accurately. Titrate with 1 N hydrochloric acid, using methyl red indicator solution. One milliliter of 1 N hydrochloric acid is equivalent to 0.01703 g of NH_3.

$$\% \, NH_3 = \frac{(mL \times N \, HCl) \times 1.703}{Sample \, wt \, (g)}$$

Residue after Ignition. (Page 26). Evaporate 50 g (56 mL) and ignite.

Carbon Dioxide. Dilute 10.0 g (11 mL) of the sample with 10 mL of water free from carbon dioxide, and add 5 mL of clear saturated barium hydroxide solution. Any turbidity should not be greater than is produced when the same quantity of barium hydroxide solution is added to 21 mL of carbon dioxide-free water containing 0.3 mg of carbonate standard (0.2 mg of carbon dioxide).

Chloride, Nitrate, Phosphate, and Sulfate. (By ion chromatography, page 87). Add 2 mL of 1% sodium carbonate reagent solution to 10 g (11 mL) of sample, and evaporate to dryness. Prepare also a blank and a standard with 0.005 mg chloride and 0.02 mg each of nitrate, phosphate, and sulfate. Take up in water, and dilute each to 10 mL in a volumetric flask. Analyze 50-μL aliquots by ion chromatography, and measure the various peaks. The differences between the sample and the blank should not be greater than the corresponding differences between the standard and the sample.

Heavy Metals. (Page 36, Method 1). To 40.0 g (44 mL), add 10 mL of 1% sodium carbonate reagent solution, evaporate to dryness on a hot plate (≈ 100 °C), dissolve the residue in about 20 mL of water, and dilute with water to 25 mL.

Iron. (Page 38, Method 1). To 50 g (56 mL), add about 10 mg of sodium carbonate, and evaporate to dryness on a hot plate (≈ 100 °C). Dissolve the residue in 3 mL of hydrochloric acid, dilute with water to 50 mL, and use this solution without further acidification.

Substances Reducing Permanganate. Dilute 3 mL of the sample with 5 mL of water, and add 50 mL of 10% sulfuric acid reagent solution. Add 0.05 mL of 0.1 N potassium permanganate, heat to boiling, and keep at this temperature for 5 min. The pink color should not be entirely discharged.

Ammonium Hydroxide, Ultratrace
NH₄OH   **Formula Wt 35.05**   **CAS No. 1336-21-6**

Suitable for use in ultratrace elemental analysis.

Note: Reagent must be packaged in a preleached Teflon bottle and used in a clean laboratory environment to maintain purity.

GENERAL DESCRIPTION
Typical appearance: clear, colorless liquid with a suffocating odor
Analytical use: trace metal analysis
Change in state (approximate): boiling point, 100 °C
Aqueous solubility: soluble

SPECIFICATIONS
Assay (as NH₃). 20–30%

                                              Maximum Allowable

Chloride (Cl) . . . . 0.5 ppm
Phosphate (PO₄) . . . . 2 ppm
Sulfate (SO₄) . . . . 2 ppm
Mercury (Hg). . . . . 1 ppb
Aluminum (Al) . . . . 1 ppb

Barium (Ba) . . . . 1 ppb
Boron (B). . . . . 5 ppb
Cadmium (Cd) . . . . 1 ppb
Calcium (Ca) . . . . 1 ppb
Chromium (Cr) . . . . 1 ppb
Cobalt (Co). . . . . 1 ppb
Copper (Cu) . . . . 1 ppb
Iron (Fe). . . . . 5 ppb
Lead (Pb). . . . . 1 ppb
Lithium (Li). . . . . 1 ppb
Magnesium (Mg) . . . . 1 ppb
Manganese (Mn) . . . . 1 ppb
Molybdenum (Mo) . . . . 1 ppb
Potassium (K) . . . . 1 ppb
Silicon (Si). . . . . 5 ppb
Sodium (Na). . . . . 5 ppb
Strontium (Sr) . . . . 1 ppb
Tin (Sn) . . . . 1 ppb
Titanium (Ti) . . . . 1 ppb
Vanadium (V) . . . . 1 ppb
Zinc (Zn) . . . . 1 ppb
Zirconium (Zr). . . . . 1 ppb

TESTS

Assay. (By acidimetry). Tare a small glass-stoppered Erlenmeyer flask containing 35 mL of water. Using a measuring pipet, without suction, introduce barely under the surface about 2 mL of the ammonium hydroxide, stopper, and weigh accurately. Titrate with 1 N hydrochloric acid, using methyl red indicator solution. One milliliter of 1 N hydrochloric acid is equivalent to 0.01703 g of NH_3.

$$\% \ NH_3 = \frac{(mL \times N \ HCl) \times 1.703}{Sample \ wt \ (g)}$$

Chloride, Phosphate, and Sulfate. See test for ammonium hydroxide on page 159.

Mercury. (By CVAAS, page 65). To a set of three 100-mL volumetric flasks containing about 35 mL of water, add 10.0 g of sample. Cool the samples in an ice water bath for about 15 min, and with caution slowly add to all three flasks 10.0 mL of nitric acid. Cool to room temperature. To the second and third flasks, add mercury ion (Hg) standards of 10 ng (1.0 ppb) and 20 ng (2.0 ppb), respectively. Dilute to the mark with water. Mix. Zero the instrument with the blank, and determine the mercury content using a suitable mercury analyzer system (1.0 mL of 0.01 µg/mL Hg = 10 ng).

Trace Metals. Determine the aluminum, barium, boron, cadmium, calcium, chromium, cobalt, copper, iron, lead, lithium, magnesium, manganese, molybdenum, potassium, silicon, sodium, strontium, tin, titanium, vanadium, zinc, and zirconium by the ICP–OES method described on page 69.

Ammonium Iodide

NH₄I   **Formula Wt 144.94**   **CAS No. 12027-06-4**

GENERAL DESCRIPTION

Typical appearance: colorless or white solid; may become pale yellow with storage
Analytical use: photochemical reactions
Change in state (approximate): when heated, it partly decomposes and partly sublimes
Aqueous solubility: 170 g in 100 mL at 20 °C.

SPECIFICATIONS

Assay . . . . ≥99.0% NH₄I

Maximum Allowable

Insoluble matter 0.005%
Residue after ignition 0.05%
Chloride and bromide (as Cl) 0.005%
Phosphate (PO₄) 0.001%
Sulfate (SO₄) 0.05%
Barium (Ba) 0.002%
Heavy metals (as Pb) 0.001%
Iron (Fe) . . . . 5 ppm

TESTS

Assay. (By oxidation–reduction titration of iodide). Weigh to the nearest 0.1 mg about 0.3 g of sample, and dissolve with about 20 mL of water in a 250-mL glass-stoppered titration flask. Add 30 mL of hydrochloric acid and 5 mL of chloroform. Cool, if necessary, and titrate with 0.05 M potassium iodate solution until the iodine color disappears from the aqueous layer. Stopper, shake vigorously for 30 s, and continue the titration, shaking vigorously after each addition of the iodate until the iodine color in the chloroform is discharged. One milliliter of 0.05 M potassium iodate corresponds to 0.007247 g of NH₄I.

$$\% \, NH_4I = \frac{(mL \times M \, KIO_3) \times 14.494}{Sample \, wt \, (g)}$$

Insoluble Matter. (Page 25). Use 20 g dissolved in 200 mL of water.

Residue after Ignition. (Page 26). Ignite 2.0 g.

Chloride and Bromide. Dissolve 1.0 g in 100 mL of water in a distilling flask. Add 1 mL of hydrogen peroxide and 1 mL of phosphoric acid, heat to boiling, and boil gently until all the iodine is expelled and the solution is colorless. Cool, wash down the sides of the flask, and add 0.5 mL of hydrogen peroxide. If an iodine color develops, boil until the solution is colorless and for 10 min longer. If no color develops, boil for 10 min, filter if necessary through a chloride-free filter, and dilute with water to 100 mL. Dilute 20 mL with water to 23 mL, and add 1 mL of nitric acid and 1 mL of silver nitrate reagent solution. Any turbidity should not exceed that produced by 0.01 mg of chloride ion (Cl) in an equal volume of solution containing the quantities of nitric acid and silver nitrate used in the test.

For the Determination of Phosphate, Sulfate, Barium, Heavy Metals, and Iron

Sample Solution A. Dissolve 20.0 g in 40 mL of water in a 600-mL beaker, and add about 10 mg of sodium carbonate. Add 16 mL of hydrochloric acid and 32 mL of nitric acid, cover with a watch glass, and warm on a hot plate ($\approx$100 °C). When the rapid evolution of iodine ceases, add an additional 16 mL of nitric acid and 24 mL of hydrochloric acid. Digest in the covered beaker until the bubbling ceases, remove the watch glass, and evaporate to dryness.

Blank Solution B. Evaporate to dryness the quantities of acids and sodium carbonate used to prepare sample solution A.

Dissolve the residues of both solutions in separate 20-mL portions of water, filter if necessary, and dilute each with water to 100 mL (1 mL of sample solution A = 0.2 g).

Phosphate. (Page 40, Method 1). Evaporate 10 mL of sample solution A (2-g sample) to dryness on a hot plate ($\approx$100 °C). Dissolve the residue in 25 mL of approximately 0.5 N sulfuric acid. Use 10 mL of blank solution B to prepare the standard solution.

Sulfate. (Page 40, Method 1). Use 0.5 mL of sample solution A. Use 0.5 mL of blank solution B to prepare the standard solution.

Barium. To 20 mL of sample solution A (4-g sample) add 5 mL of a 1% solution of potassium sulfate. For the standard, add 0.08 mg of barium and 5 mL of a 1% solution of potassium sulfate to 20 mL of blank solution B. Any turbidity in the solution of the sample should not exceed that in the standard. Compare 10 min after adding the potassium sulfate to the sample and standard solutions.

Heavy Metals. (Page 36, Method 1). Dilute 10 mL of sample solution A (2-g sample) with water to 25 mL. Use 10 mL of blank solution B to prepare the standard solution.

Iron. (Page 38, Method 1). Use 10 mL of sample solution A (2-g sample). Prepare the standard with 10 mL of blank solution B.

Ammonium Metavanadate
NH₄VO₃ $\qquad$ **Formula Wt 116.98** $\qquad$ **CAS No. 7803-55-6**

GENERAL DESCRIPTION
Typical appearance: white to pale yellow solid
Analytical use: combustion analysis of carbon, hydrogen, and nitrogen
Change in state (approximate): melting point, 200 °C
Aqueous solubility: 0.44 g in 100 mL at 18 °C

SPECIFICATIONS

Assay . . . . >99.0% NH_4VO_3

Maximum Allowable

Solubility in ammonium hydroxide . . . . Passes test
Carbonate (CO_3). . . . . Passes test
Chloride (Cl) 0.2%
Sulfate (SO_4) 0.05%

TESTS

Assay. (By oxidation–reduction titration of vanadium). Weigh accurately about 0.4 g, dissolve in 50 mL of warm water, and add 1 mL of sulfuric acid and 30 mL of sulfurous acid. Boil gently until the sulfur dioxide is expelled, then boil for 15 min longer. Cool, dilute with water to 100 mL, and titrate with 0.1 N potassium permanganate. One milliliter of 0.1 N potassium permanganate corresponds to 0.01170 g of NH_4VO_3.

$$\% \ NH_4VO_3 = \frac{(mL \times N \ KMnO_4) \times 11.70}{Sample \ wt \ (g)}$$

Solubility in Ammonium Hydroxide. Dissolve 5.0 g in a mixture of ammonium hydroxide and 250 mL of hot water, and add ammonium hydroxide until the reagent dissolves. Heat the solution to boiling. The solution should be clear.

Carbonate. To 0.5 g, add 1 mL of water and 2 mL of 10% hydrochloric acid reagent solution. No effervescence should be produced. (Limit about 0.3%)

Chloride. (Page 35). Dissolve 0.5 g in 50 mL of hot water, add 2 mL of nitric acid, and let stand for 1 h. Filter through a chloride-free filter, wash with a few milliliters of water, and dilute with water to 100 mL; use 1.0 mL.

Sulfate. (Page 40, Method 1). Dissolve 0.5 g in 40 mL of hot water, add 2 mL of 10% hydrochloric acid reagent solution and 1.5 g of hydroxylamine hydrochloride, and heat at 60 °C for 5 min. Filter through a suitable filter paper, cool, and dilute with water to 100 mL. To 20 mL, add 1 mL of dilute hydrochloric acid (1 + 19), and continue as directed.

Ammonium Molybdate Tetrahydrate

Ammonium Heptamolybdate Tetrahydrate

$(NH_4)_6Mo_7O_{24} \cdot 4H_2O$ **Formula Wt 1235.86**   **CAS No. 12054-85-2**

GENERAL DESCRIPTION

Typical appearance: colorless or white solid; sometimes with a slight green or yellow tint
Analytical use: determination of phosphates, arsenates, and lead
Change in state (approximate): loses one water at 90 °C; decomposes at 190 °C
Aqueous solubility: soluble

SPECIFICATIONS

Assay (as MoO$_3$) 81.0–83.0%

Maximum Allowable

Insoluble matter 0.005%

Chloride (Cl). 0.002%

Nitrate (NO$_3$) . . . . Passes test

Arsenate, phosphate, and silicate (as SiO$_2$). 0.001%

Phosphate (PO$_4$). . . . . 5 ppm

Sulfate (SO$_4$) 0.02%

Heavy metals (as Pb). 0.001%

Magnesium (Mg) 0.005%

Potassium (K) 0.01%

Sodium (Na). 0.01%

TESTS

Assay. (By complexometric titration). Weigh accurately about 0.7 g, and dissolve in 100 mL of water. Adjust the pH of the solution to 4.0 with 1% nitric acid reagent solution. Add saturated hexamethylenetetramine reagent solution to a pH of 5–6, using a pH meter. Heat the solution to 60 °C, and titrate with 0.1 M lead nitrate volumetric solution. Add 0.2 mL of 0.1% PAR indicator solution, and titrate from yellow color to the first permanent pink end point. One milliliter of 0.1 M lead nitrate corresponds to 0.01439 g of MoO$_3$.

$$\% \, MoO_3 = \frac{[mL \times N \, Pb(NO_3)_2] \times 14.394}{Sample \, wt \, (g)}$$

Insoluble Matter. (Page 25). Use 20 g dissolved in 200 mL of water.

Chloride. (Page 35). Use 0.5 g.

Nitrate. Dissolve 1.0 g in 10 mL of water containing 5 mg of sodium chloride. Add 0.10 mL of indigo carmine reagent solution and 10 mL of sulfuric acid. The blue color should not be completely discharged in 5 min. (Limit about 0.003%)

Arsenate, Phosphate, and Silicate. Dissolve 2.5 g in 70 mL of water. For the control, dissolve 0.5 g in 70 mL of water and add 0.02 mg of silica (SiO$_2$). These solutions are prepared in containers other than glass to avoid excessive silica contamination. Adjust the pH to between 3 and 4 (pH paper) with dilute hydrochloric acid (1 + 9), then transfer to glass containers and treat each solution as follows: add 1–2 mL of bromine water and adjust the pH to between 1.7 and 1.9 (using a pH meter) with dilute hydrochloric acid (1 + 9). Heat just to boiling, but do not boil, and cool to room temperature. Dilute with water to 90 mL, add 10 mL of hydrochloric acid, and transfer to a separatory funnel. Add 1 mL of butyl alcohol and 30 mL of 4-methyl-2-pentanone, shake vigorously, and allow the phases to separate. Draw off and discard the aqueous phase and wash the ketone phase three times with 10-mL portions of dilute hydrochloric acid (1 + 99), discarding each aqueous phase. To the washed ketone phase, add 10 mL of dilute hydrochloric acid (1 + 99) to which has just been added 0.2 mL of a freshly prepared 2% solution of stannous chloride in hydrochloric acid. Any blue color in the solution of the sample should not exceed that in the control.

Phosphate. (Page 40, Method 3).

Sulfate. (Page 41, Method 2). Use 5 mL of nitric acid.

Heavy Metals. Dissolve 2.0 g in about 20 mL of water, add 10 mL of 10% sodium hydroxide reagent solution and 2 mL of ammonium hydroxide, and dilute with water to 40 mL. For the control, add 0.01 mg of lead ion (Pb) to 10 mL of the solution. Add 10 mL of freshly prepared hydrogen sulfide water to each. Any color in the solution of the sample should not exceed that in the control.

Magnesium, Potassium, and Sodium. (By flame AAS, page 63).

> **Sample Stock Solution.** Dissolve 5.0 g of sample in 60 mL of water, and add 20 mL of 10% sulfuric acid. Transfer to a 100-mL volumetric flask, and dilute to the mark with water (1 mL = 0.05 g).

Element	Wavelength (nm)	Sample Wt (g)	Standard Added (mg)	Flame Type*	Background Correction
Mg	285.2	0.20	0.01; 0.02	A/A	Yes
K**	766.5	0.20	0.01; 0.02	A/A	No
Na	589.0	0.20	0.01; 0.02	A/A	No

*A/A is air/acetylene.
**The means of absorbing radiation below 650 nm, such as a red filter and the use of an alkali salt to suppress ionization, is critical for this analysis.

Ammonium Nitrate

NH_4NO_3 **Formula Wt 80.04**   **CAS No. 6484-52-2**

GENERAL DESCRIPTION

Typical appearance: hygroscopic solid
Analytical use: freezing mixtures
Change in state (approximate): melting point, 169 °C
Aqueous solubility: 190 g in 100 mL at 20 °C

SPECIFICATIONS

Assay . . . . 95.0% NH_4NO_3
pH of a 5% solution at 25.0 °C 4.5–6.0

Maximum Allowable

Insoluble matter 0.005%
Residue after ignition 0.01%
Chloride (Cl) . . . . 5 ppm
Nitrite (NO_2) . . . . Passes test
Phosphate (PO_4) . . . . 5 ppm
Sulfate (SO_4) 0.002%

Heavy metals (as Pb). . . . . .5 ppm
Iron (Fe). . . . . .2 ppm

TESTS

Assay. (By alkalimetry for ammonium). Weigh, to the nearest 0.1 mg, 3.0 g of sample, and dissolve in 50 mL of water in a 500-mL Erlenmeyer flask. Add exactly 50.0 mL of 1 N sodium hydroxide volumetric solution, place a filter funnel loosely in the neck of the flask, and boil until all the ammonia is expelled (about 10–15 min) as determined with litmus paper. Cool, and add 0.15 mL of thymol blue indicator solution. Titrate the excess sodium hydroxide with 1 N sulfuric acid. One milliliter of 1 N sodium hydroxide corresponds to 0.08004 g of NH_4NO_3.

$$\% NH_4NO_3 = \frac{[(mL \times N\ NaOH) - (mL \times N\ H_2SO_4)] \times 8.004}{Sample\ wt\ (g)}$$

pH of a 5% Solution at 25.0 °C. (Page 49).

Insoluble Matter. (Page 25). Use 20 g dissolved in 200 mL of water.

Residue after Ignition. (Page 26). Ignite 10 g.

Chloride. (Page 35). Use 2.0 g.

Nitrite. (Page 55). Dissolve 0.150 g of sodium nitrite in water, and dilute with water to 100 mL. Dilute 1.0 mL of this solution with water to 100 mL. This solution should be freshly prepared. Use 1.0 g of sample and 0.005 mg of nitrite. (Limit about 5 ppm)

Phosphate. (Page 40, Method 1). Dissolve 4.0 g in 25 mL of approximately 0.5 N sulfuric acid.

For the Determination of Sulfate, Heavy Metals, and Iron

Sample Solution A. Dissolve 20 g in 20 mL of water, add 10 mg of sodium carbonate, 5 mL of 30% hydrogen peroxide, 20 mL of hydrochloric acid, and 20 mL of nitric acid.

Blank Solution B. Prepare a solution containing the quantities of reagents used in sample solution A.

Digest each in covered beakers on a hot plate ($\approx$100 °C) until reaction ceases, uncover, and evaporate to dryness. Dissolve the residues in 5 mL of 1 N acetic acid and 50 mL of water, filter if necessary, and dilute to 100 mL with water (1 mL = 0.2 g).

Sulfate. (Page 40, Method 1). Use 12.5 mL of sample solution A (2.5-g sample). Use 12.5 mL of blank solution B to prepare the standard solution.

Heavy Metals. (Page 36, Method 1). Dilute 20 mL of sample solution A (4-g sample) with water to 25 mL. Use 20 mL of blank solution B to prepare the standard solution.

Iron. (Page 38, Method 1). Use 25 mL of sample solution A (5-g sample). Prepare the standard solution with 25 mL of blank solution B.

Ammonium Oxalate Monohydrate
Ethanedioic Acid, Diammonium Salt Monohydrate
$(COONH_4)_2 \cdot H_2O$ **Formula Wt 142.11**   **CAS No. 6009-70-7**

GENERAL DESCRIPTION
Typical appearance: colorless solid
Analytical use: determination of calcium, lead, and rare earth metals
Change in state (approximate): decomposes at about 128 °C
Aqueous solubility: 5 g in 100 mL at 20 °C

SPECIFICATIONS
Assay . . . . .99.0–101.0% $(COONH_4)_2 \cdot H_2O$

Maximum Allowable

Insoluble matter 0.005%
Residue after ignition 0.02%
Chloride (Cl) 0.002%
Sulfate (SO_4) 0.002%
Heavy metals (as Pb) . . . . 5 ppm
Iron (Fe). . . . . 2 ppm

TESTS

Assay. (By oxidation–reduction titration of oxalate). Transfer 5.0 g of previously crushed sample, accurately weighed, to a 500-mL volumetric flask. Dissolve it in water, dilute with water to volume, and mix. Pipet 25.0 mL of the solution into 70 mL of water and 5 mL of sulfuric acid in a beaker. Titrate slowly with 0.1 N potassium permanganate until about 25 mL has been added, then heat the mixture to about 70 °C, and complete the titration. One milliliter of 0.1 N potassium permanganate corresponds to 0.007106 g of $(COONH_4)_2 \cdot H_2O$.

$$\% \ (COONH_4)_2 \cdot H_2O = \frac{(mL \times N \ KMnO_4) \times 7.106}{\text{Sample wt (g)} \ / \ 20}$$

Insoluble Matter. (Page 25). Use 20.0 g dissolved in 400 mL of hot water.

Residue after Ignition. (Page 26). Ignite 5.0 g.

Chloride. Dissolve 2.0 g in water plus 10 mL of nitric acid, filter if necessary through a chloride-free filter, and dilute with water to 100 mL. To 25 mL of the solution, add 1 mL of silver nitrate reagent solution. Any turbidity should not exceed that produced by 0.01 mg of chloride ion (Cl), in an equal volume of solution containing the quantities of reagents used in the test.

For the Determination of Sulfate, Heavy Metals, and Iron

Sample Solution A. Digest 20.0 g of sample plus 10 mL of 1% sodium carbonate reagent solution with 40 mL of nitric acid plus 35 mL of hydrochloric acid in a covered beaker on a hot plate ($\approx$100 °C) until no more bubbles of gas are evolved. Remove the cover, and evaporate until a small amount of crystals forms in the beaker. Add 10 mL of 30% hydrogen peroxide, cover the beaker, and digest on the hot plate until reaction ceases. Add an additional 10 mL of 30% hydrogen peroxide, re-cover the beaker, digest on a hot plate ($\approx$100 °C) until reaction ceases, remove the cover, and evaporate to dryness. Add 5 mL of dilute hydrochloric acid (1 + 1), cover, digest on the hot plate for 15 min, remove the cover, and evaporate to dryness. Dissolve the residue in about 50 mL of water, filter if necessary, and dilute with water to 100 mL (1 mL = 0.2 g).

Blank Solution B. Prepare a similar solution containing the quantities of reagents used in preparing sample solution A and treated in the identical manner.

Sulfate. (Page 40, Method 1). Use 12.5 mL of sample solution A (2.5-g sample). Use 12.5 mL of blank solution B to prepare the standard solution.

Heavy Metals. (Page 36, Method 1). Dilute 20 mL of sample solution A (4-g sample) with water to 25 mL. Use 20 mL of blank solution B to prepare the standard solution.

Iron. (Page 38, Method 1). Use 25 mL of sample solution A (5-g sample). Prepare the standard solution from 25 mL of blank solution B.

Ammonium Peroxydisulfate
Ammonium Persulfate
$(NH_4)_2S_2O_8$ Formula Wt 228.19   CAS No. 7727-54-0

Note: Because of its inherent instability, this reagent may be expected to decrease in strength and to increase in acidity during storage. After storage for some time, the reagent may fail to meet the specified requirements for assay and acidity. Store in a dry, cool place.

GENERAL DESCRIPTION
Typical appearance: white or colorless solid
Analytical use: detection and determination of manganese and iron
Change in state (approximate): melting point, 120 °C with decomposition
Aqueous solubility: 58 g in 100 mL at 20 °C

SPECIFICATIONS
Assay . . . . $\geq$98.0% $(NH_4)_2S_2O_8$
Maximum Allowable
Insoluble matter 0.005%
Residue after ignition. 0.05%
Titrable free acid . . . . .0.04 meq/g

Chloride and chlorate (as Cl) 0.001%
Heavy metals (as Pb) 0.005%
Iron (Fe). 0.001%
Manganese (Mn) . . . . 0.5 ppm

TESTS

Assay. (By oxidation–reduction titration of persulfate). Weigh accurately about 0.4 g and add it to 50.0 mL of 0.1 N ferrous ammonium sulfate volumetric solution in a glass-stoppered flask. Stopper the flask, allow it to stand for 1 h with frequent shaking, and titrate the excess ferrous ammonium sulfate with 0.1 N potassium permanganate volumetric solution. Run a complete blank. One milliliter of 0.1 N ferrous ammonium sulfate corresponds to 0.01141 g of $(NH_4)_2S_2O_8$.

$$\% \ (NH_4)_2S_2O_8 = \frac{\{[mL \ (blank) - mL \ (sample)] \times N \ KMnO_4\} \times 11.41}{Sample \ wt \ (g)}$$

Insoluble Matter. (Page 25). Use 20 g dissolved in 200 mL of water.

Residue after Ignition. (Page 26). Ignite 5.0 g.

Titrable Free Acid. Dissolve 10.0 g in 100 mL of water, and add 0.01 mL of methyl red indicator solution. If a red color is produced, not more than 4.0 mL of 0.1 N sodium hydroxide should be required to discharge it.

Chloride and Chlorate. (Page 35). Mix 1.0 g with 1 g of sodium carbonate, and heat until no more gas is evolved. Dissolve the residue in 20 mL of water, and neutralize with nitric acid.

Heavy Metals. (Page 36, Method 1). Gently ignite 2.0 g in a porcelain or silica crucible or dish (not in platinum), and to the residue add 1 mL of hydrochloric acid, 1 mL of nitric acid, and about 10 mg of sodium carbonate. Evaporate to dryness on a hot plate ($\approx$100 °C), dissolve the residue in about 20 mL of water, and dilute with water to 50 mL. Dilute 10 mL of the solution with water to 25 mL.

Iron. (Page 38, Method 1). To 1.0 g, add 5 mL of water and 10 mL of hydrochloric acid, and evaporate to dryness. Dissolve the residue in 2 mL of hydrochloric acid, dilute with water to 50 mL, and use this solution without further acidification. Prepare the standard solution from the residue remaining from evaporation of 10 mL of hydrochloric acid.

Manganese. Gently ignite 20.0 g in a dish (but not in platinum) until the sample is decomposed, and finally ignite at 600 °C until the sample is volatilized. Dissolve the residue by boiling for 5 min with 35 mL of water plus 10 mL of nitric acid, 5 mL of sulfuric acid, and 5 mL of phosphoric acid, and cool. Prepare a standard containing 0.01 mg of manganese ion (Mn) in an equal volume of solution containing the quantities of reagents used to dissolve the residue. To each solution add 0.25 g of potassium periodate, boil gently for 5 min, and cool. Any pink color in the solution of the sample should not exceed that in the standard.

Ammonium Phosphate, Dibasic
Diammonium Hydrogen Phosphate

$(NH_4)_2HPO_4$ Formula Wt 132.06 CAS No. 7783-28-0

GENERAL DESCRIPTION

Typical appearance: colorless or white solid

Analytical use: buffer solutions

Change in state (approximate): melting point, 155 °C, decomposes

Aqueous solubility: 1 g in 1.7 mL at about 25 °C

SPECIFICATIONS

Assay . . . . $\geq$98.0% $(NH_4)_2HPO_4$

pH of a 5% solution at 25.0 °C 7.7–8.1

	Maximum Allowable
Insoluble matter	0.005%
Chloride (Cl)	0.001%
Nitrate (NO_3)	0.003%
Sulfate (SO_4)	0.01%
Heavy metals (as Pb)	0.001%
Iron (Fe)	0.001%
Calcium (Ca)	0.001%
Magnesium (Mg)	0.0005%
Potassium (K)	0.005%
Sodium (Na)	0.005%

TESTS

Assay. (By acid–base titration). Weigh accurately about 0.5 g, and dissolve in 50 mL of water. Titrate with 0.1 N hydrochloric acid to the potentiometric end point. One milliliter of 0.1 N hydrochloric acid corresponds to 0.01321 g of $(NH_4)_2HPO_4$.

$$\% \ (NH_4)_2HPO_4 = \frac{(mL \times N \ HCl) \times 13.21}{Sample \ wt \ (g)}$$

pH of a 5% Solution at 25.0 °C. (Page 49).

Insoluble Matter. (Page 25). Use 20.0 g dissolved in 200 mL of water. Save the filtrate separate from the washings for the test for ammonium hydroxide precipitate.

Chloride. (Page 35). Use 1.0 g of sample and 3 mL of nitric acid.

Nitrate. (Page 38, Method 1). For sample solution A, use 0.50 g. For control solution B, use 0.50 g and 1.5 mL of nitrate ion (NO_3) standard solution.

Sulfate. Dissolve 1.0 g in 15 mL of water, and add 4.0 mL of 10% hydrochloric acid (solution approximately pH 2). Filter through a washed filter paper, and add 10 mL of water through the same filter paper. For the control, add 0.10 mg of sulfate ion (SO_4) in 20 mL of water, and add 1.0 mL of (1 + 19) hydrochloric acid. Dilute both to 35 mL with water, add 5.0 mL of 40% barium chloride reagent solution, and mix. Observe turbidity after 10 min. Sample solution turbidity should not exceed that of the control solution.

Heavy Metals. (Page 36, Method 1). Dissolve 4.0 g in 10 mL of water, add 15 mL of 2 N hydrochloric acid, and dilute with water to 32 mL. Use 24 mL to prepare the sample solution, and use the remaining 8.0 mL to prepare the control solution.

Iron. (Page 38, Method 2). Use 1.0 g.

Calcium, Magnesium, Potassium, and Sodium. (By flame AAS, page 63).

Sample Stock Solution. Dissolve 20.0 g of sample in 60 mL of water, add 20 mL of nitric acid (1 + 1), transfer to a 100-mL volumetric flask, and mix (1 mL = 0.20 g).

Element	Wavelength (nm)	Sample Wt (g)	Standard Added (mg)	Flame Type*	Background Correction
Ca	422.7	4.0	0.02; 0.04	N/A	No
Mg	285.2	4.0	0.01; 0.02	A/A	Yes
K	766.5	1.0	0.05; 0.10	A/A	No
Na	589.0	0.10	0.005; 0.01	A/A	No

*A/A is air/acetylene; N/A is nitrous oxide/acetylene.

Ammonium Phosphate, Monobasic
Ammonium Dihydrogen Phosphate
$NH_4H_2PO_4$ **Formula Wt 115.03**   **CAS No. 7722-76-1**

GENERAL DESCRIPTION

Typical appearance: colorless solid
Analytical use: buffer solutions
Change in state (approximate): melting point, 190 °C
Aqueous solubility: 1 g in 2.5 mL at 25 °C

SPECIFICATIONS

Assay . . . .≥98.0% $NH_4H_2PO_4$
pH of a 5% solution at 25.0 °C 3.8–4.4

Maximum Allowable

Insoluble matter 0.005%
Chloride (Cl) . . . . 5 ppm
Nitrate (NO_3) 0.001%
Sulfate (SO_4) 0.01%
Heavy metals (as Pb) . . . . 5 ppm

Iron (Fe)..0.001%
Calcium (Ca) 0.001%
Magnesium (Mg)..0.0005%
Potassium (K) 0.005%
Sodium (Na)..0.005%

TESTS

Assay. (By acid–base titration). Weigh accurately about 2 g, and dissolve in 50 mL of water. Add exactly 5.0 mL of 1 N hydrochloric acid, and stir until the sample is completely dissolved. Titrate the excess acid, stirring constantly, with 1 N sodium hydroxide to the inflection point occurring at about pH 4, as measured with a pH meter.

Calculate A, the volume of 1 N hydrochloric acid consumed by the sample. Continue the titration with 1 N sodium hydroxide to the inflection point occurring at about pH 8. Calculate B, the volume of 1 N sodium hydroxide required in the titration between the two inflection points.

Calculation:

$$A = (5.0 \times N \text{ of } HCl) - (V_1 \times N \text{ of } NaOH)$$

where V_1 is mL of sodium hydroxide needed to reach the first inflection point.

$$B = (V_2 - V_1)(N \text{ of } NaOH)$$

where V_2 is mL of sodium hydroxide needed to reach the second inflection point from the beginning of the titration.

$$\% \ NH_4H_2PO_4 = \frac{B \times 11.503}{\text{Sample wt (g)}}$$

pH of a 5% Solution at 25.0 °C. (Page 49).

Insoluble Matter. (Page 25). Use 20.0 g dissolved in 200 mL of water. Save the filtrate separate from the washings for the test for ammonium hydroxide precipitate.

Chloride. (Page 35). Use 2.0 g of sample and 3 mL of nitric acid.

Nitrate. (Page 38, Method 1). For sample solution A, use 1.5 g. For control solution B, use 1.5 g and 1.5 mL of nitrate ion (NO_3) standard solution.

Sulfate. Dissolve 1.0 g in 15 mL of water, and add 2.5 mL of 10% hydrochloric acid (solution approximately pH 2). Filter through a washed filter paper, and add 10 mL of water through the same filter paper. For the control, add 0.10 mg of sulfate ion (SO_4) in 20 mL of water, and add 1.0 mL of (1 + 19) hydrochloric acid. Dilute both to 35 mL with water, add 5.0 mL of 40% barium chloride reagent solution, and mix. Observe turbidity after 10 min. Sample solution turbidity should not exceed that of the control solution.

Heavy Metals. (Page 36, Method 1). Dissolve 6.0 g in 30 mL of water. Use 25 mL to prepare the sample solution, and use the remaining 5 mL to prepare the control solution.

Iron. (Page 38, Method 2). Use 1.0 g. Add 3 mL of dilute ammonium hydroxide (1 + 4) to both the sample and the standard solutions.

Calcium, Magnesium, Potassium, and Sodium.   (By flame AAS, page 63).

Sample Stock Solution. Dissolve 20.0 g of sample in 75 mL of water, add 2 mL of nitric acid (1 + 1), transfer to a 100-mL volumetric flask, and mix (1 mL = 0.20 g).

Element	Wavelength (nm)	Sample Wt (g)	Standard Added (mg)	Flame Type*	Background Correction
Ca	422.7	4.0	0.02; 0.04	N/A	No
Mg	285.2	4.0	0.01; 0.02	A/A	Yes
K	766.5	1.0	0.05; 0.10	A/A	No
Na	589.0	0.10	0.005; 0.01	A/A	No

*A/A is air/acetylene; N/A is nitrous oxide/acetylene.

Ammonium Sulfamate

$NH_4OSO_2NH_2$ **Formula Wt 114.13**   **CAS No. 7773-06-0**

GENERAL DESCRIPTION

Typical appearance: colorless hygroscopic solid
Analytical use: source for nitrous oxide
Change in state (approximate): melting point, 130 °C
Aqueous solubility: 200 g in 100 mL at 20 °C

SPECIFICATIONS

Assay . . . . $\geq$98.0% $NH_4OSO_2NH_2$
Melting point . . . . Within a range of 2.0 °C, including 133.0 °C

Maximum Allowable

Insoluble matter 0.02%
Residue after ignition 0.10%
Heavy metals (as Pb) . . . . 5 ppm

TESTS

Assay.   (By oxidation–reduction titration). Weigh accurately about 0.35 g, and transfer to a 250-mL flask. Add 75 mL of water and 5 mL of sulfuric acid, swirl to dissolve the sample, and titrate slowly with 0.1 N sodium nitrite volumetric solution, shaking the flask vigorously from time to time. Titrate dropwise near the end point, shaking after each addition, until a blue color is produced immediately when a glass rod dipped into the titrated solution is streaked on starch–iodide test paper. One milliliter of 0.1 N sodium nitrite corresponds to 0.005705 g of $NH_4OSO_2NH_2$.

$$\% \ NH_4OSO_2NH_2 = \frac{(mL \times N \ NaNO_2) \times 5.705}{Sample \ wt \ (g)}$$

Melting Point.   (Page 45).

Insoluble Matter.   (Page 25). Use 5.0 g dissolved in 100 mL of water.

Residue after Ignition. (Page 26). Ignite 1.0 g.

Heavy Metals. (Page 36, Method 1). Dissolve 6.0 g in 20 mL of water, and dilute with water to 30 mL. Use 25 mL to prepare the sample solution, and use the remaining 5 mL to prepare the control solution.

Ammonium Sulfate

$(NH_4)_2SO_4$	Formula Wt 132.14	CAS No. 7783-20-2

GENERAL DESCRIPTION

Typical appearance: colorless powder or solid
Analytical use: freezing mixtures
Change in state (approximate): melting point, 280 °C with decomposition
Aqueous solubility: 75 g in 100 mL at 20 °C

SPECIFICATIONS

Assay . . . . ≥99.0% $(NH_4)_2SO_4$
pH of a 5% solution at 25.0 °C 5.0–6.0

Maximum Allowable

Insoluble matter 0.005%
Residue after ignition. 0.005%
Chloride (Cl). . . . . .5 ppm
Nitrate (NO_3) 0.001%
Phosphate (PO_4). . . . . .5 ppm
Heavy metals (as Pb). . . . . .5 ppm
Iron (Fe). . . . . .5 ppm

TESTS

Assay. (By alkalimetry for ammonium). Weigh, to the nearest 0.1 mg, 2.5 g of sample, and dissolve in 50 mL of water in a 500-mL Erlenmeyer flask. Add exactly 50.0 mL of 1 N sodium hydroxide, place a filter funnel loosely in the neck of the flask, and boil until all the ammonia is expelled (about 10–15 min) as determined with litmus paper. Cool, and add 0.15 mL of thymol blue indicator solution. Titrate the excess sodium hydroxide with 1 N sulfuric acid volumetric solution. One milliliter of 1 N sodium hydroxide corresponds to 0.06607 g of $(NH_4)_2SO_4$.

$$\% \ (NH_4)_2SO_4 = \frac{[(50.0 \times N \ NaOH) - (mL \times N \ H_2SO_4)] \times 6.607}{\text{Sample wt (g)}}$$

pH of a 5% Solution at 25.0 °C. (Page 49).

Insoluble Matter. (Page 25). Use 20 g dissolved in 200 mL of water.

Residue after Ignition. (Page 26). Ignite 20 g.

Chloride. (Page 35). Use 2.0 g.

Nitrate. (Page 38, Method 1). For sample solution A, use 1.0 g. For control solution B, use 1.0 g and 1 mL of nitrate ion (NO_3) standard solution.

Phosphate. (Page 40, Method 1). Dissolve 4.0 g in 25 mL of 0.5 N sulfuric acid.

Heavy Metals. (Page 36, Method 1). Dissolve 6.0 g in 20 mL of water, and dilute with water to 30 mL. Use 25 mL to prepare the sample solution, and use the remaining 5.0 mL to prepare the control solution.

Iron. (Page 38, Method 1). Use 2.0 g.

Ammonium Sulfide Solution
$(NH_4)_2S$ **Formula Wt 68.14**   **CAS No. 12135-76-1**

GENERAL DESCRIPTION
Typical appearance: yellow liquid
Analytical use: reagent for trace heavy metal analysis
Aqueous solubility: soluble

SPECIFICATIONS
Assay . . . . .20.0–24.0% $(NH_4)_2S$

Maximum Allowable
Residue after ignition 0.04%
Carbonate (CO_2) 0.005%
Chloride (Cl) 0.005%

TESTS

Assay. (By titration of reductive capacity). Weigh accurately about 3.0 g, and in a 250-mL volumetric flask, dilute with water through which nitrogen has been freshly bubbled to expel oxygen. Dilute to volume with similar water, and displace air from the headspace with nitrogen. Stopper, and mix thoroughly. Place a 50.0-mL aliquot of this solution into a mixture of 50.0 mL of 0.1 N iodine and 25 mL of 0.1 N hydrochloric acid in 400 mL of water. Be sure that the tip of the pipet is below the surface of the iodine when adding the ammonium sulfide. Titrate the excess of iodine with 0.1 N sodium thiosulfate volumetric solution, adding 3 mL of starch indicator solution near the end point of the titration. One milliliter of 0.1 N iodine corresponds to 0.00341 g of $(NH_4)_2S$.

$$\% \, (NH_4)_2S = \frac{[(50 \times N \, I_2) - (mL \times N \, Na_2S_2O_3)] \times 3.41}{\text{Sample wt (g)} / 5}$$

Residue after Ignition. (Page 26). Evaporate 5.0 mL (5.0 g) to dryness, and proceed as directed.

Carbonate. Dilute 4.0 mL (4.0 g) of sample to 20 mL in a suitable Nessler tube. For the standard, dilute 1.0 mL of the carbonate (0.2 mg/mL of CO_2) standard solution to 20 mL in a matched Nessler tube. Add 3 mL of 12% barium chloride reagent solution, and gently heat both solutions for 1 min in a boiling water bath. The turbidity in the sample should not exceed that in the standard.

Chloride. (Page 35). Dilute 0.5 mL (0.5 g) to 20 mL with water, add 2 mL of hydrogen peroxide, and evaporate to dryness. Dissolve the residue in water to make 50 mL. Use 20 mL for the sample solution.

Ammonium Thiocyanate
NH_4SCN Formula Wt 76.12   CAS No. 1762-95-4

GENERAL DESCRIPTION
Typical appearance: colorless, deliquescent solid
Analytical use: determination of iron, mercury, and silver
Change in state (approximate): melting point, 149 °C
Aqueous solubility: 165 g in 100 mL at 20 °C

SPECIFICATIONS
Assay . . . . ≥97.5% NH_4SCN
pH of a 5% solution at 25.0 °C 4.5–6.0

	Maximum Allowable
Insoluble matter	0.005%
Residue after ignition	0.025%
Chloride (Cl)	0.005%
Sulfate (SO_4)	0.005%
Heavy metals (as Pb)	5 ppm
Iron (Fe)	3 ppm
Iodine-consuming substances	0.004 meq/g

TESTS

Assay. (By argentimetric titration of thiocyanate content). Weigh, to the nearest 0.1 mg, 7 g of sample. Dissolve with 100 mL of water in a 1-L volumetric flask, dilute with water to the mark, and mix well. Transfer a 50.0-mL aliquot to a 250-mL glass-stoppered iodine-type flask, add 15 mL of 10% nitric acid reagent solution, then add with agitation exactly 50.0 mL of 0.1 N silver nitrate volumetric solution, and stir vigorously. Add 2 mL of ferric ammonium sulfate indicator solution, and titrate the excess silver nitrate with 0.1 N ammonium thiocyanate volumetric solution. Near the end point, stir after the addition of each drop. One milliliter of 0.1 N silver nitrate corresponds to 0.007612 g of NH_4SCN.

$$\% \, NH_4SCN = \frac{[(50.0 \times N \, AgNO_3) - (mL \times N \, NH_4SCN)] \times 7.612}{Sample \, wt \, (g) \, / \, 20}$$

pH of a 5% Solution at 25.0 °C. (Page 49).

Insoluble Matter. (Page 25). Use 20 g dissolved in 150 mL of water.

Residue after Ignition. (Page 26). Ignite 3.3 g. Retain the residue for the test for iron.

Chloride. (Page 35). Dissolve 1.0 g in 20 mL of water in a small flask. Add 10 mL of 25% sulfuric acid reagent solution and 7 mL of 30% hydrogen peroxide. Evaporate to 20 mL by boiling in a well-ventilated hood, add 17 mL of water, and evaporate again. Repeat until all the cyanide has been volatilized. Cool, filter if necessary through a chloride-free filter, and dilute with water to 100 mL. Use 20 mL of this solution.

Sulfate. (Page 40, Method 1). Use 1.0-g sample, and allow 30 min for turbidity to form.

Heavy Metals. (Page 36, Method 1). Dissolve 6.0 g in 20 mL of water, and dilute with water to 30 mL. Use 25 mL to prepare the sample solution, and use the remaining 5.0 mL to prepare the control solution.

Iron. (Page 38, Method 1). To the residue after ignition add 3 mL of dilute hydrochloric acid (1 + 1), cover with a watch glass, and digest on a hot plate ($\approx$100 °C) for 15–20 min. Remove the watch glass and evaporate to dryness. Dissolve the residue in 2 mL of hydrochloric acid, filter if necessary, and dilute with water to 50 mL. Use the solution without further acidification.

Iodine-Consuming Substances. Dissolve 5.0 g in 50 mL of water plus 1.7 mL of 10% sulfuric acid reagent solution. Add 1 g of potassium iodide and 2 mL of starch indicator solution, and titrate with 0.01 N iodine solution. Not more than 2.0 mL of the iodine solution should be required.

Aniline
Benzenamine

$C_6H_5NH_2$ **Formula Wt 93.13**   **CAS No. 62-53-3**

Note: This reagent darkens to a reddish-brown color on storage.

GENERAL DESCRIPTION
Typical appearance: oily liquid
Analytical use: solvent
Change in state (approximate): boiling point, 184 °C
Aqueous solubility: 3.6 g in 100 mL at 20 °C
Density: 1.02
pK_a: 4.6

SPECIFICATIONS
Assay . . . . $\geq$99.0% $C_6H_5NH_2$

Maximum Allowable

Color (APHA) 250
Residue after ignition. 0.005%
Chlorobenzene (C_6H_5Cl) 0.01%
Hydrocarbons. . . . . .Passes test
Nitrobenzene ($C_6H_5NO_2$). . . . . .Passes test

TESTS

Assay and Chlorobenzene. Analyze the sample by gas chromatography using the general parameters cited on page 80. The following specific conditions are also required.

Column: Type I, methyl silicone

Measure the area under all peaks and calculate the area percent for aniline and chlorobenzene.

Color (APHA). (Page 43).

Residue after Ignition. (Page 26). In a preconditioned, tared crucible, evaporate 20 g (20 mL) to dryness in the hood, and ignite.

Hydrocarbons. Mix 5 mL with 10 mL of hydrochloric acid. The solution should be clear after dilution with 15 mL of cold water.

Nitrobenzene.

Sample Solution A. Place 10 mL of methanol in a 25-mL glass-stoppered graduated cylinder.

Control Solution B. Place 10 mL of methanol in a 25-mL glass-stoppered graduated cylinder, and add 1 mL of a methanol solution containing 0.10 mg of nitrobenzene per mL.

Immerse the cylinders in a beaker of water at or below room temperature, and add 10 mL of the aniline and 2.5 mL of hydrochloric acid to each. Mix well, and bring to room temperature.

Transfer a portion of sample solution A to a polarographic cell, and deaerate with nitrogen or hydrogen. Record the polarogram from –0.2 to –0.7 V vs. SCE with a current sensitivity of 0.02 μA/mm. Repeat this procedure with control solution B. The diffusion current in sample solution A should not exceed the difference in diffusion current between control solution B and sample solution A. (Limit about 0.001%)

Anthrone
9(10*H*)-Anthracenone

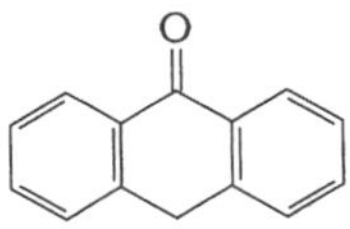

$C_{14}H_{10}O$ **Formula Wt 194.23**   **CAS No. 90-44-8**

GENERAL DESCRIPTION

Typical appearance: pale yellow solid
Analytical use: colorimetric determination of sugar and animal starches
Change in state (approximate): melting point, 153–159 °C
Aqueous solubility: insoluble

SPECIFICATIONS

Melting point . . . . No more than a 5° range, including 156 °C
Sensitivity to carbohydrates . . . . Passes test

Maximum Allowable

Absorbance of reagent solution . . . . Passes test
Solubility in ethyl acetate. . . . . Passes test

TESTS

Melting Point. (Page 45).

Sensitivity to Carbohydrates. Mark four 25-mL volumetric flasks as 0 µg (blank), 250 µg, 500 µg, and 750 µg. Add the glucose standard (described below) and water as specified in the following table, and dilute to volume with the reagent solution (described below). Place all flasks in a vigorously boiling water bath for 7.0 min exactly. Chill the flasks quickly to room temperature, and determine the absorbance of each at 620 nm against the blank in 1.00-cm cells. A plot of the absorbance against concentration should be linear, and the absorbance of the 750-µg flask should not be less than 0.90.

> **Glucose Standard.** (1000 µg in 1 mL). Dissolve 1.00 g of glucose in water, and dilute with water to 100 mL. Pipet 10.0 mL of this solution (1 mL = 10 mg) into a 100-mL volumetric flask, and dilute to volume with water.

> **Reagent Solution.** Slowly add 132 mL of sulfuric acid to 68 mL of ice-cold water in an ice bath. Allow the mixture to cool to room temperature, add 0.100 g of the sample, and stir if necessary to dissolve.

Flask (µg)	Glucose Standard (mL)	Water (mL)
0	0.00	0.75
250	0.25	0.50
500	0.50	0.25
750	0.75	0.00

Absorbance of Reagent Solution. Measure the absorbance of the reagent solution (see the test for sensitivity to carbohydrates) in a spectrophotometer, using 1.00-cm cells, with sulfuric acid as a reference. Record the absorbance at 425 and 620 nm, respectively. The absorbance should not exceed 0.75 at 425 nm and 0.045 at 620 nm.

Solubility in Ethyl Acetate. Dissolve 1 g in 50 mL of freshly distilled ethyl acetate in a dry 50-mL glass-stoppered graduated cylinder. The mixture should not be heated, dissolution should be complete, and the resulting solution should be clear.

Antimony Trichloride
Antimony(III) Chloride
SbCl₃   Formula Wt 228.12   CAS No. 10025-91-9

GENERAL DESCRIPTION
Typical appearance: colorless, deliquescent solid
Analytical use: chemical microscopy of drugs
Change in state (approximate): melting point, 73 °C

SPECIFICATIONS
Assay . . . . ≥99.0% SbCl₃

Maximum Allowable
Insoluble in chloroform 0.05%
Sulfate (SO₄) 0.005%
Arsenic (As) 0.02%
Calcium (Ca) 0.005%
Copper (Cu) 0.001%
Iron (Fe) 0.002%
Lead (Pb) 0.005%
Potassium (K) 0.01%
Sodium (Na) 0.02%

TESTS

Assay. (By oxidation–reduction titration). Weigh accurately 0.5 g, and in a glass-stoppered Erlenmeyer flask, dissolve in 5 mL of 10% hydrochloric acid reagent solution. When dissolution is complete, add a solution of 4.0 g of potassium sodium tartrate tetrahydrate in 30 mL of water. Stopper and swirl. Add 50 mL of a cold saturated solution of sodium bicarbonate, and titrate immediately with 0.1 N iodine volumetric solution, adding 3 mL of starch indicator solution near the end of the titration. One milliliter of 0.1 N iodine corresponds to 0.01140 g of SbCl₃.

$$\% \ SbCl_3 = \frac{(mL \times N \ I_2) \times 11.40}{Sample \ wt \ (g)}$$

Insoluble in Chloroform. Dissolve 5.0 g in 25 mL of chloroform, filter through a tared, preconditioned filtering crucible, wash the crucible with several portions of chloroform, and dry at 105 °C.

> *Note:* Weigh quickly to avoid moisture absorption and the formation of insoluble oxychlorides. The use of anhydrous, ethanol-free chloroform is recommended.

Sulfate. Dissolve 10.0 g in the minimum volume of hydrochloric acid required to achieve complete dissolution, dilute with water to 75 mL, neutralize with ammonium hydroxide, and filter. Add 2 mL of hydrochloric acid to the filtrate, dilute with water to 100 mL, and heat to boiling. Add 10 mL of 12% barium chloride reagent solution, digest in a covered beaker on a hot plate ($\approx$100 °C) for 2 h, and allow to stand for at least 8 h. If a precipitate is formed, filter, wash thoroughly, and ignite. Correct for the weight obtained in a complete blank test. Calculate the sulfate content of the sample from the weight of barium sulfate thus obtained.

Arsenic. (Page 34). Dissolve 1.0 g in 5 mL of hydrochloric acid, pour this solution into a solution of 2 g of stannous chloride dihydrate in 2 mL of hydrochloric acid, and allow to stand for at least 8 h. If no precipitate is visible, the arsenic content is less than the limit and the test need not be completed. If a precipitate has formed, filter it on a glass-fiber filter. Wash the beaker and the precipitate with two 5-mL portions of hydrochloric acid and then with water, passing the washings through the filter. Discard the filtrate. Dissolve the precipitate by passing a mixture of 5 mL of hydrochloric acid and 0.15 mL of bromine water through the filter, and wash the filter with 5 mL of hydrochloric acid and then with water. Collect the filtrate and washings in the beaker in which the precipitation was made, warm the solution on a hot plate ($\approx$100 °C) to remove the excess bromine, and dilute with water to 1 L. Dilute 15 mL of this solution with water to 35 mL. For the standard, use 0.003 mg of arsenic (As).

Calcium, Copper, Iron, Lead, Potassium, and Sodium. (By flame AAS, page 63).

Sample Stock Solution. Dissolve 20.0 g of sample in 80 mL of hydrochloric acid (1 + 3). Transfer to a 100-mL volumetric flask, and dilute to the mark with hydrochloric acid (1 + 3) (1 mL = 0.20 g).

Blank Solution. Use an appropriate amount of hydrochloric acid (1 + 3) as a blank solution to zero the instrument.

Element	Wavelength (nm)	Sample Wt (g)	Standard Added (mg)	Flame Type*	Background Correction
Ca	422.7	0.80	0.02; 0.04	N/A	No
Cu	324.8	4.0	0.02; 0.04	A/A	Yes
Fe	248.3	4.0	0.08; 0.16	A/A	Yes
Pb	217.0	4.0	0.10; 0.20	A/A	Yes
K	766.5	0.10	0.01; 0.02	A/A	No
Na	589.0	0.10	0.01; 0.02	A/A	No

*A/A is air/acetylene; N/A is nitrous oxide/acetylene.

Arsenic Trioxide

Arsenic(III) Oxide, Reductometric Standard

As$_2$O$_3$ **Formula Wt 197.84**   **CAS No. 1327-53-3**

GENERAL DESCRIPTION

Typical appearance: white solid
Analytical use: reductometric standard

Change in state (approximate): melting point, 655 °C

Aqueous solubility: crystalline forms, 2 g in 100 mL at 20 °C; amorphous forms, 3.3 g in 100 mL at 20 °C

SPECIFICATIONS

Assay . . . . 99.95–100.05% As_2O_3

Maximum Allowable

Residue after ignition. 0.02%

Insoluble in dilute hydrochloric acid. 0.01%

Chloride (Cl). 0.005%

Sulfide (S) . . . . Passes test

Antimony (Sb). 0.05%

Lead (Pb). 0.002%

Iron (Fe). . . . . .5 ppm

TESTS

Assay. (By oxidation–reduction titration of arsenic). This assay is based on a direct titration with NIST Standard Reference Material Potassium Dichromate. Accurately weighed portions of the arsenic trioxide are reacted with accurately weighed portions of the NIST potassium dichromate of such size as to provide a small excess of the latter. The excess potassium dichromate is determined by titration with standardized ferrous ammonium sulfate solution, using a potentiometrically determined end point.

> *Caution:* Because of the small tolerance in assay limits for this reductometric standard, extreme care must be observed in the weighing, transferring, and titrating operations in the following procedure. Strict adherence to the specified sample weights and final titration volumes is absolutely necessary. It is recommended that the titrations be run at least in duplicate. Duplicate values for the assay of this material should agree within 2 parts in 5000 to be acceptable for averaging.

Procedure. Place 2.00 g of sample in a weighing bottle, dry for 1 h at 105 °C, and cool for 2 h in a desiccator. Weigh accurately to within 0.1 mg, and transfer to a 400-mL beaker. Weigh the bottle again, and determine by difference to the nearest 0.1 mg the weight of arsenic trioxide.

Weigh accurately to within 0.1 mg an amount of NIST potassium dichromate equivalent to a slight excess of the weight of arsenic trioxide taken, and transfer it to a 50-mL beaker. Dissolve the arsenic trioxide in 20 mL of 20% sodium hydroxide solution, add 100 mL of water and 10 mL of dilute sulfuric acid (1 + 1), and stir. Transfer the potassium dichromate to the solution, using water to loosen any crystals adhering to the beaker. Stir until all of the potassium dichromate is dissolved, and let stand for 10 min.

Add 20 mL of dilute sulfuric acid (1 + 1), dilute with water to 300 mL, and stir. Titrate the excess potassium dichromate with standardized 0.02 N ferrous ammonium sulfate solution (described below), using a platinum indicator electrode and a calomel reference electrode for the measurement of the potential difference in millivolts. The end point of the titration is determined potentiometrically, using the second derivative method (page 28). A correction for "blank" oxidants or reductants must be applied by titrating an accurately weighed 40-mg portion of the NIST potassium dichromate with the standardized ferrous ammonium sulfate in an equal volume of solution containing the quantities of reagents used in this assay. The "blank" is equal to the calculated milliequivalents of potassium dichromate minus the calculated milliequivalents of ferrous ammonium sulfate. A positive blank correction results from other oxidants and a negative correction from other reductants.

$$M_1 = \left[\frac{W_1}{49.0307} - VN + B \right] \times 1.00032F$$

where M_1 = number of milliequivalents of potassium dichromate equivalent to arsenic trioxide; W_1 = weight of potassium dichromate in milligrams; V = volume of ferrous ammonium sulfate solution in milliliters; N = normality of ferrous ammonium sulfate solution in equivalents per liter; B = blank correction in milliequivalents (may be either positive or negative); and F = assay value of the NIST standard in percent divided by 100. (The equivalent formula weight of potassium dichromate is 49.0307, and the conversion factor for air to vacuum weight of potassium dichromate is 1.00032.)

Calculate the assay value of the arsenic trioxide from

$$A = \% \, As_2O_3 = \frac{(49.4603)(M_1)}{(1.00017)W_2} 100$$

where W_2 = weight of arsenic trioxide in milligrams and 1.00017 is the conversion factor for air to vacuum weight of arsenic trioxide.

Transfer a 40-mg portion of the potassium dichromate, accurately weighed to within 0.1 mg, to a 400-mL beaker. Add 300 mL of dilute sulfuric acid (1 + 19), and stir to dissolve the salt. Titrate the solution with the 0.02 N ferrous ammonium sulfate solution, using a potentiometrically determined end point. The concentration of the ferrous ammonium sulfate, in equivalents per liter (N), is calculated from the following equation:

$$N = \left[\frac{W}{49.03} \right] \left[\frac{1000}{V} \right]$$

where N = normality of ferrous ammonium sulfate; W = weight of potassium dichromate in grams; and V = volume of ferrous ammonium sulfate solution in milliliters.

Ferrous Ammonium Sulfate Solution, 0.02 N. Weigh 8.0 g of ferrous ammonium sulfate hexahydrate, and transfer to a 1-L flask. Add 200 mL of dilute sulfuric acid (1 + 19), and after the salt is dissolved, dilute to the mark with the dilute sulfuric acid. Mix thoroughly.

Residue after Ignition. (Page 26). Ignite 5.0 g in a tared, preconditioned platinum dish. Retain the residue for the test for iron.

Insoluble in Dilute Hydrochloric Acid. To 10 g, add 90 mL of dilute hydrochloric acid (3 + 7) and 10 mL of hydrogen peroxide. Cover with a watch glass and allow to stand until the sample goes into solution, heating if necessary. Heat to boiling, digest in a covered beaker on a hot plate ($\approx$100 °C) for 1 h, and filter through a tared, preconditioned filtering crucible. Retain the filtrate for the antimony and lead test, wash the insoluble matter thoroughly, and dry at 105 °C.

Chloride. (Page 35). Dissolve 1.0 g in 10 mL of dilute ammonium hydroxide $(1 + 2)$ with the aid of gentle heating, and dilute with water to 100 mL. Neutralize 20 mL with nitric acid.

Sulfide. Dissolve 1.0 g in 10 mL of 10% sodium hydroxide reagent solution, and add 0.05 mL of 10% lead acetate reagent solution. The color should be the same as that of an equal volume of sodium hydroxide solution to which only the lead acetate is added. (Limit about 0.001%)

Antimony and Lead. (By flame AAS, page 63).

> *Sample Stock Solution.* Dilute the filtrate retained from the insoluble in dilute hydrochloric acid test with water to the mark in a 200-mL volumetric flask (1 mL = 0.05 g).

Element	Wavelength (nm)	Sample Wt (g)	Standard Added (mg)	Flame Type*	Background Correction
Sb	217.6	1.0	0.5; 1.0	A/A	Yes
Pb	217.0	1.0	0.05; 0.10	A/A	Yes

*A/A is air/acetylene.

Iron. (Page 38, Method 1). To the residue from the test for residue after ignition, add 3 mL of dilute hydrochloric acid $(1 + 1)$, and warm. Add 5 mL of hydrochloric acid, dilute with water to 125 mL, and use 50 mL of this solution without further acidification.

Ascorbic Acid

$C_6H_8O_6$ **Formula Wt 176.13**   **CAS No. 50-81-7**

GENERAL DESCRIPTION

Typical appearance: white solid
Analytical use: antioxidant
Change in state (approximate): melting point, 191 °C with decomposition
Aqueous solubility: 30 g in 100 mL at 25 °C

SPECIFICATIONS

Assay . . . . $\geq$99.0% $C_6H_8O_6$
Specific rotation $[\alpha]_D^{25°}$ . . . . +21.0° ± 0.5°

Maximum Allowable

Residue after ignition. 0.1%
Heavy metals (as Pb). 0.002%
Iron (Fe). 0.001%

TESTS

Assay. (By oxidation–reduction titration). Weigh accurately 0.4 g, and in a conical flask, dissolve in a mixture of 100 mL of water (made oxygen-free by bubbling nitrogen gas through it) and 25 mL of 10% sulfuric acid. Nitrogen gas used in this procedure must be of a high purity grade. Swirl and titrate immediately with 0.1 N iodine, adding 3 mL of starch indicator near the end of the titration. One milliliter of 0.1 N iodine corresponds to 0.008806 g of $C_6H_8O_6$.

$$\% \ C_6H_8O_6 = \frac{(mL \times N\,I_2) \times 8.806}{Sample\ wt\ (g)}$$

Specific Rotation. (Page 46). Weigh accurately about 10 g, dissolve in 90 mL of oxygen-free water in a 100-mL volumetric flask, and dilute with water to volume. Adjust the temperature of the solution to 25 °C. Observe the optical rotation in a polarimeter at 25 °C using the sodium line, and calculate the specific rotation.

Residue after Ignition. (Page 26). Ignite 1.0 g in a tared, preconditioned crucible. Moisten the char with 1 mL of sulfuric acid. Retain the residue for the test for iron.

Heavy Metals. (Page 36, Method 2). Use 1.0 g.

Iron. (Page 38, Method 1). To the residue from the test for residue after ignition, add 3 mL of dilute hydrochloric acid (1 + 1) and 0.10 mL of nitric acid, cover with a watch glass, and digest on a hot plate (≈100 °C) for 15–20 min. Remove the watch glass, and evaporate to dryness. Dissolve the residue in a mixture of 2 mL of hydrochloric acid and 10 mL of water, dilute with water to 50 mL, and use without further acidification.

Aurin Tricarboxylic Acid, [tri]Ammonium Salt
5-[(3-Carboxy-4-hydroxyphenyl)(3-carboxy-4-oxo-2,5-cyclohexadien-1-ylidene)methyl]-2-hydroxybenzoic Acid, Triammonium Salt; Aluminon

$(HOC_6H_3COONH_4)_2C{:}C_6H_3(COONH_4){:}O$ **Formula Wt 473.43**   **CAS No. 569-58-4**

GENERAL DESCRIPTION

Typical appearance: yellowish-brown or brownish-red solid
Analytical use: detection and colorimetric estimation of aluminum in water, tissues, and foods
Aqueous solubility: freely soluble

SPECIFICATIONS

Sensitivity to aluminum .... Passes test

Maximum Allowable

Insoluble matter 0.1%
Residue after ignition.. 0.2%

TESTS

Sensitivity to Aluminum. Dissolve 25 mg in 20 mL of water and 0.1 mL of dilute ammonium hydroxide (10% NH_3). Add 0.1 mL of this solution and 0.1 mL of acetic acid to 10 mL of a solution containing 0.001 mg of aluminum ion (Al). A distinct pink color should appear within 15 min.

Insoluble Matter. (Page 25). Use 1.0 g dissolved in 100 mL of water plus 0.5 mL of dilute ammonium hydroxide (10% NH_3).

Residue after Ignition. (Page 26). Ignite 0.50 g.

Barium Acetate

$(CH_3COO)_2Ba$	Formula Wt 255.42	CAS No. 543-80-6

GENERAL DESCRIPTION

Typical appearance: colorless or white solid
Analytical use: catalyst for organic reactions
Change in state (approximate): melting point, 450 °C
Aqueous solubility: 72 g in 100 mL at 20 °C

SPECIFICATIONS

Assay .... 99.0–102.0% $(CH_3COO)_2Ba$

Maximum Allowable

Insoluble matter 0.01%
Chloride (Cl)... 0.001%
Oxidizing substances (as NO_3) 0.005%
Calcium (Ca) 0.05%
Potassium (K) 0.003%
Sodium (Na)... 0.005%
Strontium (Sr) 0.2%
Heavy metals (as Pb).. 5 ppm
Iron (Fe).. 0.001%

TESTS

Assay. (By complexometry for barium). Weigh accurately about 1.0 g, transfer to a 400-mL beaker, and dissolve in 100 mL of carbon dioxide-free water. Add 100 mL of ethanol, 10 mL of ammonium hydroxide, and 3.0 mL of metalphthalein-screened indicator solution. Titrate immediately with standard 0.1 M EDTA to a color change from magenta to gray-green. One milliliter of 0.1 M EDTA corresponds to 0.02554 g of $(CH_3COO)_2Ba$.

$$\% \ (CH_3COO)_2Ba = \frac{(mL \times M \ EDTA) \times 25.54}{Sample \ wt \ (g)}$$

Insoluble Matter. (Page 25). Use 10 g dissolved in 100 mL of water.

Chloride. (Page 35). Use 1.0 g.

Oxidizing Substances. Place 0.10 g in a dry beaker. Cool the beaker thoroughly in an ice bath, and add 22 mL of sulfuric acid that has been cooled to ice-bath temperature. Allow the mixture to warm to room temperature, and swirl the beaker at intervals to effect gentle dissolution with slow evolution of the acetic acid vapor. When dissolution is complete, add 3 mL of diphenylamine reagent solution and digest on a hot plate ($\approx$100 °C) for 90 min. Prepare a standard by evaporating to dryness a solution containing 0.005 mg of nitrate (0.5 mL of the nitrate standard solution) and 0.01 g of sodium carbonate. Treat the residue exactly like the sample. Any color produced in the solution of the sample should not exceed that in the standard.

Calcium, Potassium, Sodium, and Strontium. (By flame AAS, page 63).

Sample Stock Solution A. Dissolve 10.0 g of sample in sufficient water in a beaker, add 5 mL of nitric acid, and transfer to a 100-mL volumetric flask. Dilute to the mark with water (1 mL = 0.10 g).

Sample Stock Solution B. Pipet 10.0 mL of sample stock solution A into a 100-mL volumetric flask, and dilute to the mark with water (1 mL = 0.01 g). Add 2.0 mL of a 2.5% potassium chloride reagent solution to the flask for calcium and strontium.

Element	Wavelength (nm)	Sample Wt (g)	Standard Added (mg)	Flame Type*	Background Correction
Ca	422.7	0.04	0.02; 0.04	N/A	No
K	766.5	1.0	0.02; 0.04	A/A	No
Na	589.0	0.10	0.005; 0.01	A/A	No
Sr	460.7	0.05	0.05; 0.10	N/A	No

*A/A is air/acetylene; N/A is nitrous oxide/acetylene.

Heavy Metals. (Page 36, Method 1). Dissolve 6.0 g in about 15 mL of water, add 8 mL of dilute hydrochloric acid (1 + 1), and dilute with water to 30 mL. Use 25 mL to prepare the sample solution, and use the remaining 5.0 mL to prepare the control solution.

Iron. Dissolve 1.0 g in 40 mL of water. For the standard, add 0.01 mg of iron (Fe) to 40 mL of water. Add 2 mL of hydrochloric acid to each, dilute with water to 50 mL, and add 0.10 mL of 0.1 N potassium permanganate volumetric solution. Allow to stand for 5 min, and add 3 mL of ammonium thiocyanate reagent solution. Any red color in the solution of the sample should not exceed that in the standard.

Barium Carbonate

BaCO₃   **Formula Wt 197.34**   **CAS No. 513-77-9**

GENERAL DESCRIPTION

Typical appearance: white solid

Analytical use: preparation of barium standard solutions

Change in state (approximate): decomposes at about 1300 °C into barium oxide and carbon dioxide

Aqueous solubility: insoluble

SPECIFICATIONS

Assay . . . . 99.0–101.0% BaCO₃

Maximum Allowable

Insoluble in dilute hydrochloric acid.	0.015%
Chloride (Cl).	0.002%
Water-soluble titrable base	0.002 meq/g
Oxidizing substances (as NO₃)	0.005%
Sulfide (S).	0.001%
Calcium (Ca)	0.05%
Potassium (K)	0.005%
Sodium (Na).	0.02%
Strontium (Sr)	0.7%
Heavy metals (as Pb).	0.001%
Iron (Fe).	0.002%

TESTS

Assay. (By acid–base titrimetry). Weigh to the nearest 0.1 mg, 4 g of sample. Transfer to a 250-mL beaker, add 50 mL of water, and mix. Cautiously add exactly 50.0 mL of 1 N hydrochloric acid, boil, and cool. Add a few drops of bromphenol blue indicator solution, and titrate the excess acid with 1 N sodium hydroxide volumetric solution to the blue end point. One milliliter of 1 N hydrochloric acid corresponds to 0.0987 g of BaCO₃.

$$\% \text{ BaCO}_3 = \frac{[(50.0 \times \text{N HCl}) - (\text{mL} \times \text{N NaOH})] \times 9.87}{\text{Sample wt (g)}}$$

Insoluble in Dilute Hydrochloric Acid. Cautiously dissolve 10 g in 100 mL of dilute hydrochloric acid (1 + 9), and dilute with water to 200 mL. Ignore any slight haze that is produced. Filter through a tared, preconditioned filtering crucible, wash thoroughly with dilute hydrochloric acid (1 + 9), and dry at 105 °C.

Chloride. To 1.0 g, add 20 mL of water, and add dropwise, with stirring, about 2–3 mL of nitric acid. Filter if necessary through a chloride-free filter, wash with a little hot water, and dilute with water to 50 mL. To 25 mL of the solution, add 1 mL of silver nitrate reagent solution. Any turbidity should not exceed that produced by 0.01 mg of chloride ion (Cl) in an equal volume of solution containing the quantities of reagents used in the test.

Water-Soluble Titrable Base. Shake 5.0 g for 5 min with 50 mL of carbon dioxide-free water, cool, and filter. To 25 mL of the filtrate, add 0.10 mL of phenolphthalein indicator solution. If a pink color is produced, it should be discharged by 0.50 mL of 0.01 N hydrochloric acid.

Oxidizing Substances. Place 0.10 g in a dry beaker. Cool the beaker thoroughly in an ice bath, and add 22 mL of sulfuric acid that has been cooled to ice-bath temperature. Allow the mixture to warm to room temperature, and swirl the beaker at intervals to effect gentle dissolution with slow evolution of carbon dioxide. When dissolution is complete, add 3 mL of diphenylamine reagent solution, and digest on a hot plate ($\approx$100 °C) for 90 min. Prepare a standard by evaporating to dryness a solution containing 0.005 mg of nitrate (0.5 mL of the nitrate standard solution) and 0.01 g of sodium carbonate. Treat the residue exactly like the sample. Any color produced in the solution of the sample should not exceed that in the standard.

Sulfide. Dissolve 1.0 g in 10 mL of dilute acetic acid (1 + 4). As soon as dissolution is complete, add 1 mL of silver nitrate reagent solution. Any color after 5 min should not exceed that produced by 0.01 mg of sulfide ion (S) in an equal volume of solution containing the quantities of reagents used in the test.

Calcium, Potassium, Sodium, and Strontium. (By flame AAS, page 63).

    Sample Stock Solution A. Cautiously dissolve 10.0 g of sample in 50 mL of dilute hydrochloric acid (1 + 3) in a beaker. Transfer to a 100-mL volumetric flask, and dilute with water to the mark (1 mL = 0.10 g).

    Sample Stock Solution B. Pipet 10.0 mL of sample stock solution A into a 100-mL volumetric flask, and dilute to the mark with water (1 mL = 0.01 g). Add 2.0 mL of a 2.5% potassium chloride reagent solution to the flask for calcium and strontium.

Element	Wavelength (nm)	Sample Wt (g)	Standard Added (mg)	Flame Type*	Background Correction
Ca	422.7	0.05	0.025; 0.05	N/A	Yes
K	766.5	0.40	0.02; 0.04	A/A	No
Na	589.0	0.05	0.01; 0.02	A/A	No
Sr	460.7	0.01	0.035; 0.07	N/A	Yes

*A/A is air/acetylene; N/A is nitrous oxide/acetylene.

Heavy Metals. (Page 36, Method 1). Cautiously dissolve 5.0 g in 30 mL of dilute hydrochloric acid (1 + 4), and evaporate to dryness on a hot plate ($\approx$100 °C). Dissolve the residue in about 20 mL of water, and dilute with water to 25 mL. Dilute 15 mL with water to 25 mL, and use to prepare the sample solution. Use 5.0 mL of the remaining solution to prepare the control solution.

Iron. Cautiously dissolve 1.0 g in 15 mL of dilute hydrochloric acid (1 + 2). For the standard, add 0.02 mg of iron (Fe) to 15 mL of dilute hydrochloric acid (1 + 2). Evaporate both solutions to dryness on a hot plate ($\approx$100 °C). Dissolve each residue in about 20 mL of water, add 4 mL of hydrochloric acid, and dilute with water to 100 mL. To 50 mL of each solution, add 0.10 mL of 0.1 N potassium permanganate, allow to stand for 5 min, and add 3 mL of ammonium thiocyanate reagent solution. Any red color in the solution of the sample should not exceed that in the standard.

Barium Chloride Dihydrate

$BaCl_2 \cdot 2H_2O$ **Formula Wt 244.26**   **CAS No. 10326-27-9**

GENERAL DESCRIPTION

Typical appearance: white solid

Analytical use: sulfate determination

Change in state (approximate): melting point of the anhydrous salt, 960 °C

Aqueous solubility: 36 g in 100 mL at 20 °C

SPECIFICATIONS

Assay . . . . $\geq$99.0% $BaCl_2 \cdot 2H_2O$
Loss on drying 14.0–16.0%
pH of a 5% solution at 25.0 °C 5.2–8.2

                                                   Maximum Allowable

Insoluble matter 0.005%
Oxidizing substances (as NO_3) 0.005%
Calcium (Ca) 0.05%
Potassium (K) 0.0025%
Sodium (Na). 0.005%
Strontium (Sr) 0.1%
Heavy metals (as Pb). . . . . .5 ppm
Iron (Fe). . . . . .2 ppm

TESTS

Assay. (By complexometric titration of barium). Weigh accurately 0.8 g of sample, and dissolve in 200 mL of water. Add, while stirring, 2.0 mL of 6 M hydrochloric acid reagent solution, and from a 50-mL buret, add 25.0 mL of 0.1 M EDTA volumetric solution. Add 100 mL of methanol, 25 mL of ammonium hydroxide, and about 80–100 mg of metalphthalein-screened indicator solution. Immediately continue titration until color changes from violet to clear colorless. The color change is gradual. One milliliter of 0.1 M EDTA corresponds to 0.02443 g of $BaCl_2 \cdot 2H_2O$.

$$\% \ BaCl_2 \cdot 2H_2O = \frac{(mL \times M \ EDTA) \times 24.43}{Sample \ wt \ (g)}$$

Loss on Drying. Weigh accurately about 1.0 g, and dry in a preconditioned weighing bottle at 150 °C to constant weight. (The theoretical loss for $BaCl_2 \cdot 2H_2O$ is 14.75%.)

pH of a 5% Solution at 25.0 °C. (Page 49).

Insoluble Matter. (Page 25). Use 20 g dissolved in 200 mL of water. The solution should be clear and colorless.

Oxidizing Substances. Place 0.10 g in a dry beaker. Cool the beaker thoroughly in an ice bath, and add 22 mL of sulfuric acid that has been cooled to ice-bath temperature. Allow the mixture to warm to room temperature, and swirl the beaker at intervals to effect slow dissolution with gentle evolution of hydrogen chloride. When dissolution is complete, add 3 mL of diphenylamine reagent solution, and digest on a hot plate ($\approx$100 °C) for 90 min. Prepare a standard by evaporating to dryness a solution containing 0.005 mg of nitrate (0.5 mL of the standard nitrate solution) and 0.01 g of sodium carbonate. Treat the residue exactly like the sample. Any color produced in the solution of the sample should not exceed that in the standard.

Calcium, Potassium, Sodium, and Strontium. (By flame AAS, page 63).

Sample Stock Solution A. Dissolve 10.0 g of sample in sufficient water in a beaker, add 5 mL of nitric acid, and transfer to a 100-mL volumetric flask. Dilute to the mark with water (1 mL = 0.10 g).

Sample Stock Solution B. Pipet 10.0 mL of sample stock solution A into a 100-mL volumetric flask and dilute to the mark with water (1 mL = 0.01 g). Add 2.0 mL of a 2.5% potassium chloride reagent solution to the flask for calcium and strontium.

Element	Wavelength (nm)	Sample Wt (g)	Standard Added (mg)	Flame Type*	Background Correction
Ca	422.7	0.04	0.02; 0.04	N/A	No
K	766.5	1.0	0.02; 0.04	A/A	No
Na	589.0	0.10	0.005; 0.01	A/A	No
Sr	460.7	0.05	0.05; 0.10	N/A	No

*A/A is air/acetylene; N/A is nitrous oxide/acetylene.

Heavy Metals. (Page 36, Method 1). Dissolve 6.0 g in about 20 mL of water, and dilute with water to 30 mL. Use 25 mL to prepare the sample solution, and use the remaining 5.0 mL to prepare the control solution.

Iron. Dissolve 5.0 g in 40 mL of water plus 2 mL of hydrochloric acid. Prepare a standard containing 0.01 mg of iron (Fe) in 40 mL of water plus 2 mL of hydrochloric acid. Add 0.10 mL of 0.1 N potassium permanganate to each, dilute with water to 50 mL, and allow to stand for 5 min. Add 3 mL of ammonium thiocyanate reagent solution to each. Any red color in the solution of the sample should not exceed that in the standard.

Barium Diphenylamine Sulfonate

Diphenylamine-4-sulfonic Acid, Barium Salt

$C_{24}H_{20}BaN_2O_6S_2$ Formula Wt 633.89   CAS No. 6211-24-1

GENERAL DESCRIPTION

Typical appearance: white to gray solid
Analytical use: indicator for redox titrations
Aqueous solubility: slightly soluble

SPECIFICATIONS

Residue after ignition (as sulfate)......................................35.0–38.0%
Clarity of solution ....Passes test
Sensitivity as redox indicator...Passes test

TESTS

Residue after Ignition. Accurately weigh 1.0 g in a preconditioned, tared crucible. Add 2 mL of sulfuric acid in 5 mL of water, and swirl gently. Using a high temperature setting, heat on a hot plate until fuming ceases. Ignite at 600 ± 25 °C until a white residue is obtained. Cool and weigh.

Clarity of Solution. Weigh 0.15 g, and dissolve in 100 mL of water by heating. The solution should be complete and colorless. Reserve this solution for test for sensitivity as redox indicator.

Sensitivity as Redox Indicator. To 25 mL of water in a 50-mL test tube, add 10 mL of 4 N sulfuric acid, 5 mL of phosphoric acid, 0.05 mL of 0.01 N ferrous ammonium sulfate, and 0.05 mL of the solution reserved from the clarity of solution test. The addition of 0.10 mL of 0.01 N potassium dichromate should produce a violet color. The color should be completely discharged by the addition of 0.10 mL of 0.01 N ferrous ammonium sulfate.

Barium Hydroxide Octahydrate

$Ba(OH)_2 \cdot 8H_2O$ Formula Wt 315.46   CAS No. 12230-71-6

GENERAL DESCRIPTION

Typical appearance: colorless or white solid that absorbs carbon dioxide from air

Analytical use: carbonate-free base in alkalimetry
Change in state (approximate): melting point, 78 °C
Aqueous solubility: 7 g per 100 mL at 20 °C

SPECIFICATIONS

Assay . . . . ≥98.0% $Ba(OH)_2 \cdot 8H_2O$

Maximum Allowable

Carbonate (as $BaCO_3$). 2.0%
Insoluble in dilute hydrochloric acid 0.01%
Chloride (Cl) 0.001%
Sulfide (S). . . . . Passes test
Calcium (Ca) 0.05%
Potassium (K) 0.01%
Sodium (Na) 0.01%
Strontium (Sr). 0.8%
Heavy metals (as Pb) . . . . 5 ppm
Iron (Fe). 0.001%

TESTS

Assay. (By acid–base titrimetry). Weigh accurately 4–5 g-sample, and dissolve in about 200 mL of carbon dioxide-free water. Add 0.10 mL of phenolphthalein indicator solution, and titrate with 1 N hydrochloric acid volumetric solution. Reserve this solution for the determination of carbonate. Each milliliter of 1 N hydrochloric acid corresponds to 0.1577 g of $Ba(OH)_2 \cdot 8H_2O$.

$$\% \ Ba(OH)_2 \cdot 8H_2O = \frac{(mL \times N \ HCl) \times 15.77}{Sample \ wt \ (g)}$$

Carbonate. To the solution reserved from the test for assay, add 5.00 mL of 1 N hydrochloric acid, heat to boiling, boil gently to expel all of the carbon dioxide, and cool. Add 0.10 mL of methyl orange indicator solution, and titrate the excess acid with 1 N sodium hydroxide volumetric solution. Each milliliter of 1 N hydrochloric acid consumed corresponds to 0.09867 g of $BaCO_3$.

$$\% \ BaCO_3 = \frac{[(5.0 \times N \ HCl) - (mL \times N \ NaOH)] \times 9.867}{Sample \ wt \ (g)}$$

Insoluble in Dilute Hydrochloric Acid. Dissolve 10 g in 100 mL of dilute hydrochloric acid (1 + 9). Heat the solution to boiling, then digest in a covered beaker on a hot plate (≈100 °C) for 1 h. Filter through a tared, preconditioned filtering crucible, wash thoroughly, and dry at 105 °C.

Chloride. Dissolve 2.0 g in 25 mL of water, add 0.05 mL of phenolphthalein indicator solution, and neutralize with nitric acid. If the solution is not perfectly clear, filter it through a chloride-free filter, and dilute the filtrate with water to 50 mL. To 25 mL of the solution, add 1 mL of nitric acid and 1 mL of silver nitrate reagent solution. Any turbidity should not exceed that produced by 0.01 mg of chloride ion (Cl) in an equal volume of solution containing the quantities of reagents used in the test.

Sulfide. Dissolve 1.0 g in 8 mL of warm water and add 0.25 mL of alkaline lead solution (prepared by adding 10% sodium hydroxide solution to a 10% lead acetate solution until the precipitate is redissolved) and 2 mL of glacial acetic acid. No darkening should occur. (Limit about 0.001%)

Calcium, Potassium, Sodium, and Strontium.   (By flame AAS, page 63).

Sample Stock Solution A. Dissolve 10.0 g of sample in water in a beaker, add 5 mL of nitric acid, and transfer to a 100-mL volumetric flask. Dilute to the mark with water (1 mL = 0.10 g).

Sample Stock Solution B. Pipet 10.0 mL of sample stock solution A into a 100-mL volumetric flask, and dilute to the mark with water (1 mL = 0.01 g). Add 2.0 mL of a 2.5% potassium chloride reagent solution to the flask for calcium and strontium.

Element	Wavelength (nm)	Sample Wt (g)	Standard Added (mg)	Flame Type*	Background Correction
Ca	422.7	0.04	0.02; 0.04	N/A	No
K	766.5	0.10	0.01; 0.02	A/A	No
Na	589.0	0.05	0.005; 0.01	A/A	No
Sr	460.7	0.01	0.04; 0.08	N/A	No

*A/A is air/acetylene; N/A is nitrous oxide/acetylene.

Heavy Metals.   (Page 36, Method 1). To 5.0 g in a 150-mL beaker, add 25 mL of water, mix, and cautiously add 10 mL of hydrochloric acid. Evaporate to dryness on a hot plate ($\approx$100 °C), dissolve the residue in about 2 mL of water, and dilute with water to 25 mL. For the control, treat 1.0 g of the sample and 0.02 mg of lead ion (Pb) exactly as the 5.0 g of sample.

Iron.   Dissolve 1.0 g in 40 mL of water. For the standard, add 0.01 mg of iron ion (Fe) to 40 mL of water. Add 2.5 mL of hydrochloric acid to each, dilute with water to 50 mL, and add 0.10 mL of 0.1 N potassium permanganate solution. Allow to stand for 5 min, and add 3 mL of ammonium thiocyanate reagent solution. Any red color in the solution of the sample should not exceed that in the standard.

Barium Nitrate

$Ba(NO_3)_2$ Formula Wt 261.35   CAS No. 10022-31-8

GENERAL DESCRIPTION

Typical appearance: colorless solid
Analytical use: sulfate determination
Change in state (approximate): melting point, 592 °C
Aqueous solubility: 9.0 g in 100 mL at 20 °C

SPECIFICATIONS

Assay . . . . ≥99.0% Ba(NO$_3$)$_2$
pH of a 5% solution at 25.0 °C 5.0–8.0

Maximum Allowable

Insoluble matter 0.01%
Chloride (Cl) . . . . 5 ppm
Calcium (Ca) 0.05%
Potassium (K) 0.005%
Sodium (Na) 0.005%
Strontium (Sr) 0.1%
Heavy metals (as Pb) . . . . 5 ppm
Iron (Fe) . . . . 2 ppm

TESTS

Assay. (By complexometry for barium). Weigh accurately about 0.20 g, transfer to a 400-mL beaker, and dissolve in 100 mL of freshly boiled (carbon dioxide-free) water. Add 100 mL of ethanol, 10 mL of ammonium hydroxide, and 0.3 mL of metalphthalein-screened indicator solution. Titrate immediately with 0.1 M EDTA volumetric solution to a color change from magenta to gray-green. One milliliter of 0.1 M EDTA corresponds to 0.02614 g of Ba(NO$_3$)$_2$.

$$\% \, Ba(NO_3)_2 = \frac{(mL \times M \, EDTA) \times 26.14}{Sample \, wt \, (g)}$$

pH of a 5% Solution at 25.0 °C. (Page 49).

Insoluble Matter. (Page 25). Use 10 g dissolved in 150 mL of hot water.

Chloride. Dissolve 2.0 g in 30 mL of warm water, filter if necessary through a chloride-free filter, and add 0.10 mL of nitric acid and 1 mL of silver nitrate reagent solution. Any turbidity should not exceed that produced by 0.01 mg of chloride ion (Cl) in an equal volume of solution containing the quantities of reagents used in the test.

Calcium, Potassium, Sodium, and Strontium. (By flame AAS, page 63).

Sample Stock Solution. Dissolve 10.0 g of sample in sufficient water in a beaker, add 5 mL of nitric acid, and transfer to a 100-mL volumetric flask. Dilute to the mark with water (1 mL = 0.10 g). Add 2.0 mL of a 2.5% potassium chloride reagent solution to the flask for calcium and strontium.

Element	Wavelength (nm)	Sample Wt (g)	Standard Added (mg)	Flame Type*	Background Correction
Ca	422.7	0.10	0.05; 0.10	N/A	No
K	766.5	0.10	0.02; 0.04	A/A	No
Na	589.0	0.10	0.005; 0.01	A/A	No
Sr	460.7	0.10	0.10; 0.20	N/A	No

*A/A is air/acetylene; N/A is nitrous oxide/acetylene.

Heavy Metals. Dissolve 4.0 g in about 50 mL of water, and dilute with water to 60 mL. For the control, add 0.01 mg of lead to 15 mL of the solution, and dilute with water to 45 mL. For the sample, use the remaining 45-mL portion. Adjust the pH of the control and sample solutions to between 3 and 4 (using a pH meter) with 1 N acetic acid or 10% ammonium hydroxide reagent solution, dilute with water to 48 mL, and mix. Add 5 mL of freshly prepared hydrogen sulfide water to each, and mix. Any color in the solution of the sample should not exceed that in the control.

Iron. Dissolve 5.0 g in 45 mL of hot water plus 2 mL of hydrochloric acid, and allow to cool. Prepare a standard containing 0.01 mg of iron ion (Fe) in 45 mL of water plus 2 mL of hydrochloric acid. Add 0.10 mL of 0.1 N potassium permanganate volumetric solution to each, allow to stand for 5 min, and add 3 mL of 30% ammonium thiocyanate reagent solution. Any red color in the solution of the sample should not exceed that in the standard.

Bathocuproine
2,9-Dimethyl-4,7-diphenyl-1,10-phenanthroline

$C_{26}H_{20}N_2$ **Formula Wt 360.46**   **CAS No. 4733-39-5**

GENERAL DESCRIPTION

Typical appearance: off-white to buff solid
Analytical use: reagent for colorimetric determination of copper
Change in state (approximate): melting point, 281–282 °C
Aqueous solubility: insoluble

SPECIFICATIONS

Melting point . . . . .2° range, within 279–283 °C
Clarity of solution . . . . .Passes test
Sensitivity to copper . . . . .Passes test

TESTS

Melting Point. (Page 45).

Clarity of Solution. Weigh 0.10 g, and add to 50 mL of methanol. The solution should be complete. Reserve this solution for test for sensitivity to copper.

Sensitivity to Copper. Dilute the sample from the clarity of solution test to 100 mL with alcohol. In a 25-mL flask, prepare a standard aliquot containing 0.005 mg of copper, 1.0 mL of hydroxylamine hydrochloride reagent solution, and 2.0 mL of 10% sodium acetate reagent solution. Add 0.10 mL of the prepared sample solution. A yellow-orange color should develop (maximum absorbance about 479 nm).

Benzene

C₆H₆   **Formula Wt 78.11**   **CAS No. 71-43-2**

Suitable for general use or in ultraviolet spectrophotometry. Product labeling shall designate the uses for which suitability is represented on the basis of meeting the relevant specifications and tests. The ultraviolet spectrophotometry specifications include all of the specifications for general use.

GENERAL DESCRIPTION

Typical appearance: clear liquid
Analytical use: reference for quantitating compounds
Change in state (approximate): boiling point, 80 °C
Aqueous solubility: 0.1 g in 100 mL at 20 °C
Density: 0.88

SPECIFICATIONS

General Use

Assay . . . . ≥99.0% C₆H₆

Maximum Allowable

Color (APHA) 10
Residue after evaporation 0.001%
Substances darkened by sulfuric acid . . . . Passes test
Thiophene (limit about 1 ppm) . . . . Passes test
Sulfur compounds (as S) 0.005%
Water (H₂O) 0.05%

Specific Use

Ultraviolet Spectrophotometry

Wavelength (nm)	Absorbance (AU)
380 to 400	0.01
350	0.02
330	0.04
300	0.10
290	0.30
280	1.00

TESTS

Assay. Analyze the sample by gas chromatography using the general parameters cited on
page 80. The following specific conditions are also required.

    Column: Type I, methyl silicone

Measure the area under all peaks, and calculate the benzene content in area percent. Correct for water content.

Color (APHA). (Page 43).

Residue after Evaporation. (Page 25). Evaporate 100 g (115 mL) to dryness in a tared, preconditioned dish on a hot plate ($\approx$100 °C), and dry the residue at 105 °C for 30 min.

Substances Darkened by Sulfuric Acid. Shake 25 mL with 15 mL of sulfuric acid for 15–20 s, and allow to separate. Neither the benzene nor the acid should be darkened.

Thiophene. Add 5.0 mL of freshly prepared isatin solution (described below) to a dry, clean 50-mL porcelain crucible. Carefully overlay with 5.0 mL of the sample, allow to stand undisturbed for 1 h, and compare with a blank containing 5.0 mL of isatin solution in a 50-mL porcelain crucible. No bluish-green color should be present.

> **Isatin Solution.** Dissolve 0.25 g of isatin in 25 mL of sulfuric acid (solution A). Dissolve 0.25 g of ferric chloride in 1 mL of water; dilute to 50 mL with sulfuric acid, and allow the evolution of gas to cease (solution B). To 2.5 mL of solution A, add 5.0 mL of solution B, and dilute to 100 mL with sulfuric acid.

Sulfur Compounds. In a well-ventilated fume hood, place 30 mL of 0.5 N potassium hydroxide in methanol in a conical flask, add 5.2 g (6.0 mL) of the sample, and boil the mixture gently for 30 min under a reflux condenser, avoiding the use of a rubber stopper or connection. Detach the condenser, dilute with 50 mL of water, and heat on a hot plate ($\approx$100 °C) until the benzene and methanol are evaporated. Add 50 mL of bromine water, and heat for 15 min longer. Transfer the solution to a beaker, neutralize with dilute hydrochloric acid (1 + 3), add an excess of 1 mL of the acid, and concentrate to about 50 mL. Filter if necessary, heat the filtrate to boiling, add 5 mL of 12% barium chloride reagent solution, digest in a covered beaker on a hot plate ($\approx$100 °C) for 2 h, and allow to stand for at least 8 h. If a precipitate is formed, filter, wash thoroughly, ignite, and weigh. Correct for the weight obtained in a complete blank test.

Water. (Page 31, Method 2). Use 100 μL (88 mg) of the sample.

Ultraviolet Spectrophotometry. Use the procedure on page 86 to determine the absorbance.

1,3-Benzenediol
Resorcinol

$C_6H_4(OH)_2$ **Formula Wt 110.11**   **CAS No. 108-46-3**

GENERAL DESCRIPTION

Typical appearance: colorless or white solid; may become pink on exposure to air and light
Analytical use: reagent for zinc

Change in state (approximate): melting point, 109–111 °C
Aqueous solubility: 125 g in 100 mL at 20 °C
Density: 1.27

SPECIFICATIONS

Assay . . . . 99.0–100.5% $C_6H_4(OH)_2$
Melting point . . . . 110.0–112.0 °C

Maximum Allowable

Insoluble matter 0.005%
Residue after ignition 0.01%
Titrable acid . . . . 0.004 meq/g

TESTS

Assay. (By bromination). Weigh accurately about 1.5 g, transfer to a 500-mL volumetric flask, dissolve in water, and dilute with water to volume. Stopper and mix thoroughly. Place a 25-mL aliquot in a 500-mL iodine flask. Add 50.0 mL of 0.1 N bromine, 50 mL of water, and 5 mL of hydrochloric acid. Stopper, shake, and allow to stand in the dark for 15 min, shaking occasionally. Add 5 mL of 16.5% potassium iodide reagent solution, taking care to prevent loss of bromine. Stopper and shake. Rinse the stopper, neck, and inner walls of the flask with about 25 mL of water. Titrate the liberated iodine with 0.1 N sodium thiosulfate, adding 3 mL of starch indicator solution near the end of the titration. One milliliter of 0.1 N bromine corresponds to 0.001835 g of $C_6H_4(OH)_2$.

$$\% \ C_6H_4(OH)_2 = \frac{[(50.0 \times N \ Br_2) - (mL \times N \ Na_2S_2O_3)] \times 1.835}{\text{Sample wt (g)} / 20}$$

Melting Point. (Page 45).

Insoluble Matter. (Page 25). Use 20 g dissolved in 150 mL of water.

Residue after Ignition. (Page 26). Ignite 10 g in a tared, preconditioned crucible or dish.

Titrable Acid. Dissolve 1 g in 50 mL of water, add 0.15 mL of methyl orange indicator solution, and neutralize with 0.02 N sodium hydroxide. Add 5.0 g of sample to the neutralized solution, and titrate with 0.02 N sodium hydroxide. Not more than 1.0 mL of 0.02 N sodium hydroxide should be consumed in the titration.

Benzoic Acid

C_6H_5COOH   **Formula Wt 122.12**   **CAS No. 65-85-0**

GENERAL DESCRIPTION

Typical appearance: white or colorless solid
Analytical use: standard in volumetric and colorimetric analysis
Change in state (approximate): melting point, 122 °C
Aqueous solubility: 0.3 g in 100 mL at 20 °C
Density: 1.32
pK_a: 4.2

SPECIFICATIONS

Assay . . . .≥99.5% C_6H_5COOH
Freezing point . . . . .122–123 °C

Maximum Allowable

Residue after ignition. 0.005%
Insoluble in methanol 0.005%
Chlorine compounds (as Cl). 0.005%
Sulfur compounds (as S) 0.002%
Heavy metals (as Pb). . . . . .5 ppm
Substances reducing permanganate . . . . .Passes test

TESTS

Assay. (By acid–base titrimetry). Dissolve about 0.5 g of the sample, accurately weighed, in 25 mL of 50% alcohol previously neutralized with 0.1 N sodium hydroxide, using phenolphthalein indicator solution. Titrate with 0.1 N sodium hydroxide. One milliliter of 0.1 N sodium hydroxide corresponds to 0.01221 g of C_6H_5COOH.

$$\% \ C_6H_5COOH = \frac{(mL \times N \ NaOH) \times 12.21}{Sample \ wt \ (g)}$$

Freezing Point. Place 12–15 g in a test tube (20 × 150 mm), and heat gently to melt the benzoic acid. Insert an accurate thermometer in the test tube so that the bulb of the thermometer is centrally located, and heat gently to about 130 °C. Place the tube containing the melted sample in a larger test tube (38 × 200 mm), and keep it centered by means of corks. Allow the sample to cool slowly. If crystallization does not start when the temperature is about 120 °C, stir gently with the thermometer to start freezing, and read the thermometer every half minute. The temperature that remains constant for 2–3 min is the freezing point.

Residue after Ignition. (Page 26). Heat 20 g in a preconditioned platinum dish at a temperature sufficient to volatilize the acid slowly, but do not char or ignite it. When nearly all the acid has been volatilized, cool, and continue as described.

Insoluble in Methanol. Dissolve 20 g in 200 mL of methanol, and digest under complete reflux for 30 min. Filter through a tared, preconditioned filtering crucible, wash thoroughly with methanol, and dry at 105 °C.

Chlorine Compounds. (Page 35). Mix 1.0 g with 0.5 g of sodium carbonate, and add 10–15 mL of water. Evaporate on a hot plate (≈100 °C), and ignite until the mass is thoroughly charred, avoiding an unduly high temperature. Extract the fusion with 20 mL of water and 5.5 mL of nitric acid. Filter through a chloride-free filter, wash with two 10-mL portions of water, and dilute with water to 50 mL. Dilute 10 mL with water to 20 mL.

Sulfur Compounds. Mix 2.0 g with 1.0 g of sodium carbonate, and add in small portions 15 mL of water. Evaporate and thoroughly ignite, protected from sulfur in the flame. Treat the residue with 20 mL of water and 2 mL of 30% hydrogen peroxide, and heat on a hot plate ($\approx$100 °C) for 15 min. Add 5 mL of hydrochloric acid, and evaporate to dryness on the hot plate. Dissolve the residue in 10 mL of water, filter, wash with two 5-mL portions of water, and dilute with water to 25 mL. Add to this solution 0.5 mL of 1 N hydrochloric acid and 2 mL of 12% barium chloride reagent solution. Any turbidity should not exceed that in a standard prepared as follows: treat 1 g of sodium carbonate with 2 mL of 30% hydrogen peroxide and 5 mL of hydrochloric acid, and evaporate to dryness on a hot plate ($\approx$100 °C). Dissolve the residue and 0.12 mg of sulfate ion (SO_4) in sufficient water to make 25 mL, and add 0.5 mL of 1 N hydrochloric acid and 2 mL of 12% barium chloride reagent solution. Compare 10 min after adding the barium chloride to the sample and standard solutions.

Heavy Metals. (Page 36, Method 1). Volatilize 4.0 g over a low flame, cool, and add 5 mL of nitric acid and 1.0 mL of 1% sodium carbonate reagent solution. Evaporate to dryness on a hot plate ($\approx$100 °C), dissolve the residue in 20 mL of water, and dilute with water to 25 mL. For the control, add 0.02 mg of lead to 5 mL of nitric acid and about 10 mg of sodium carbonate, evaporate to dryness on a hot plate ($\approx$100 °C), dissolve the residue in 20 mL of water, and dilute with water to 25 mL.

Substances Reducing Permanganate. Dissolve 1.0 g in 100 mL of water containing 1 mL of sulfuric acid. Heat to 85 °C (steam bath temperature), and add 0.50 mL of 0.1 N potassium permanganate solution. The pink color should not be entirely discharged in 5 min.

Benzoyl Chloride

| C_6H_5COCl | Formula Wt 140.57 | CAS No. 98-88-4 |

GENERAL DESCRIPTION

Typical appearance: clear, fuming liquid with a pungent odor

Analytical use: derivatizing agent

Change in state (approximate): boiling point, 198 °C

Aqueous solubility: decomposes in water
Density: 1.21

SPECIFICATIONS

Assay . . . . 98.0–100.5% C_6H_5COCl
Freezing point . . . . –2.0 to 0.0 °C

Maximum Allowable

Residue after ignition. 0.005%
Phosphorus compounds (as P) 0.002%
Heavy metals (as Pb). 0.001%
Iron (Fe). 0.001%

TESTS

Assay. (By acid–base titrimetry). Weigh accurately about 2 mL in a glass-stoppered flask, add 50.0 mL of 1 N sodium hydroxide volumetric solution, and stopper the flask. Allow to stand, with frequent agitation, until the sample has dissolved. Add 0.15 mL of phenolphthalein indicator solution, and titrate the excess sodium hydroxide with 1 N hydrochloric acid. One milliliter of 1 N sodium hydroxide corresponds to 0.07028 g of C_6H_5COCl.

$$\% \; C_6H_5COCl = \frac{[(50.0 \times N \; NaOH) - (mL \times N \; HCl) \times 7.028}{\text{Sample wt (g)}}$$

Freezing Point. Place 15 mL in a test tube (20 × 150 mm) in which is centered an accurate thermometer. Center the sample tube by corks in an outer tube about 38 × 200 mm. Cool the whole apparatus, without stirring, in a bath of shaved or crushed ice mixed with sufficient salt and water to provide a temperature of –5 °C. When the temperature is about –2 °C, stir to start the freezing and read the thermometer every 30 s. The temperature that remains constant for 1–2 min is the freezing point.

Residue after Ignition. (Page 26). Evaporate 40 g (33 mL) to dryness in a tared, preconditioned dish, adding sufficient sulfuric acid to wet the sample. Add to the carbonized mass 2 mL of nitric acid and 0.25 mL of sulfuric acid, and heat cautiously until white fumes of sulfur trioxide are no longer evolved. Ignite at 600 °C for 1 h. Retain the residue to prepare sample solution A for the determination of heavy metals and iron.

Phosphorus Compounds. To 1 g (0.85 mL), add 7 mL of water and 3 mL of nitric acid. Heat the mixture to boiling, boil for 2 min, cool, dilute with water to 20 mL, and filter. To 10 mL of the filtrate, add 5 mL of 10% sulfuric acid reagent solution, and evaporate to fumes of sulfur trioxide. Cool, dilute with water to 25 mL, add 1 mL of ammonium molybdate reagent solution and 1 mL of 4-(methylamino)phenol sulfate reagent solution, and allow to stand for 2 h at room temperature. Any blue color should not exceed that produced by 0.03 mg of phosphate ion (PO_4) in an equal volume of solution containing the quantities of reagents used in the test.

For the Determination of Heavy Metals and Iron

Sample Solution A. Add 10 mL of 10% hydrochloric acid reagent solution to the residue obtained in the test for residue after ignition, and evaporate to dryness on a hot plate ($\approx$100 °C). Dissolve the residue in 0.5 mL of hydrochloric acid and 25 mL of hot water, filter if necessary, and dilute with water to 200 mL. (1 mL = 0.2 g).

Heavy Metals. (Page 36, Method 1). Use 10 mL of sample solution A (2-g sample).

Iron. (Page 38, Method 1). Use 5.0 mL of sample solution A (1-g sample).

Benzyl Alcohol
Benzenemethanol

$$CH_2OH$$

$C_6H_5CH_2OH$	Formula Wt 108.14	CAS No. 100-51-6

GENERAL DESCRIPTION
Typical appearance: clear, colorless liquid
Analytical use: preparation of alkoxides
Change in state (approximate): flash point, 100 °C; boiling point, 205 °C
Aqueous solubility: 4 g in 10 mL at 20 °C
Density: 1.05

SPECIFICATIONS
Assay . . . . $\geq$99.0% $C_6H_5CH_2OH$

Maximum Allowable

Color (APHA). 20
Residue after ignition 0.005%
Acetophenone ($C_6H_5COCH_3$). 0.02%
Benzaldehyde (C_6H_5CHO) 0.01%

TESTS

Assay, Acetophenone, and Benzaldehyde. Analyze the sample by gas chromatography using the parameters cited on page 80. The following specific conditions are also required.

   Column: Type III, polyethylene glycol

Measure the area under all peaks, and calculate the area percent for benzyl alcohol, acetophenone, and benzaldehyde.

Color (APHA). (Page 43).

Residue after Ignition. (Page 26). Use 25 g (24 mL).

2,2'-Bipyridine

α,α'-Dipyridyl

$C_{10}H_8N_2$ Formula Wt 156.19   CAS No. 366-18-7

GENERAL DESCRIPTION

Typical appearance: crystal from dilute alcohol
Analytical use: determination of ferrous iron
Change in state (approximate): melting point, 69 °C; boiling point, 272–273 °C
Aqueous solubility: soluble in about 200 parts water

SPECIFICATIONS

Melting point . . . . ≥70 °C
Clarity and absorbance of solution . . . . Passes test
Sensitivity to ferrous iron . . . . Passes test

TESTS

Melting Point. (Page 45).

Clarity and Absorbance of Solution. Dissolve 5.0 g in 10.0 mL of methanol. The solution should be clear. Measure the absorbance in a 1-cm cell at 400 nm. The absorbance should not exceed 0.30 AU. Save this solution for the sensitivity to ferrous iron test.

Sensitivity to Ferrous Iron. To 0.10 mL of iron ion (Fe) standard solution and 0.5 mL of hydroxylamine hydrochloride reagent solution, add 0.1 mL of the solution from the clarity and absorbance of solution test. A distinct red color should develop.

Bismuth Chloride

Bismuth (III)-Chloride
$BiCl_3$ Formula Wt 315.34   CAS No. 7787-60-2

GENERAL DESCRIPTION

Typical appearance: white solid
Analytical use: catalyst for organic reagents
Change in state (approximate): melting point, 230 °C
Aqueous solubility: decomposes into BiOCl

SPECIFICATIONS

Assay . . . . 98.0–102.0% $BiCl_3$

Maximum Allowable

Insoluble matter 0.01%
Nitrate (NO_3) 0.01%
Sulfate (SO_4) 0.01%
Arsenic (As) 0.001%
Calcium (Ca) 0.005%
Copper (Cu) 0.001%
Iron (Fe) 0.005%
Lead (Pb) 0.01%
Potassium (K) 0.01%
Sodium (Na) 0.02%
Silver (Ag) 0.002%

TESTS

Assay. (By indirect complexometric titration of bismuth). Weigh accurately about 1.2 g of sample, and transfer to a 250-mL beaker. Dissolve in 5 mL of hydrochloric acid. Add 50.0 mL of 0.1 M EDTA volumetric solution, and add ammonium acetate buffer solution to a pH of 4.5 as measured using a pH meter. Add 75 mL of alcohol and 2 mL of dithizone indicator solution, and titrate with standard 0.1 M zinc chloride volumetric solution to a pink end point. One milliliter of 0.1 M EDTA corresponds to 0.03153 g of $BiCl_3$.

$$\% \ BiCl_3 = \frac{[(50 \times M \ EDTA) - (mL \times M \ ZnCl_2)] \times 31.53}{Sample \ wt \ (g)}$$

Insoluble Matter. (Page 25). Use 20 g dissolved in 100 mL of dilute hydrochloric acid (1 + 4).

Nitrate. (Page 38, Method 1).

> *Sample Solution A.* Dissolve 0.20 g in 1 mL of 6 M hydrochloric acid and 2 mL of water by heating on a hot plate ($\approx$100 °C). Cautiously add 20 mL of nitrate-free sulfuric acid (2 + 1), and cool in an ice bath. Dilute to 50 mL with brucine sulfate reagent solution.

> *Control Solution B.* Dissolve 0.20 g in 1 mL of 6 M hydrochloric acid and 2 mL of the nitrate ion (NO_3) standard solution by heating on a hot plate ($\approx$100 °C). Cautiously add 20 mL of nitrate-free sulfuric acid (2 + 1), and cool in an ice bath. Dilute to 50 mL with brucine sulfate reagent solution.

> *Blank Solution C.* To 1 mL of 7 M hydrochloric acid, cautiously add 20 mL of nitrate-free sulfuric acid (2 + 1), and cool in an ice bath. Dilute to 50 mL with brucine sulfate reagent solution.

Continue with the procedure, starting with "Heat the three solutions…".

Sulfate. Dissolve 3.0 g in 5 mL of warm hydrochloric acid, dilute with warm water to 100 mL, and nearly neutralize with 10% ammonium hydroxide. Filter and wash the residue with hot water. Dilute the filtrate to 150 mL. Transfer 100 mL to an evaporating dish, evaporate to about 15 mL, add 0.5 g of sodium carbonate, evaporate to dryness, and ignite gently. Cool, add 15 mL of hot water, neutralize with 10% hydrochloric acid, filter, and wash with water to 20 mL. To the combined filtrate and washings, add 1 mL of 1 N hydrochloric acid and 2 mL of 12% barium chloride reagent solution, and allow to stand for 10 min. Any turbidity should not exceed that of a standard prepared as follows: To 0.5 g of sodium carbonate, add 0.2 mg of sulfate ion (SO_4) and 2 mL of hydrochloric acid, and evaporate to dryness on a hot plate ($\approx$100 °C). Dissolve the residue in 20 mL of water, 1 mL of 1 N hydrochloric acid, and 2 mL of 12% barium chloride reagent solution.

Arsenic. (Page 34). Mix 0.5 g with 5 mL of water, and cautiously add 2 mL of sulfuric acid. Heat the mixture until white fumes of sulfur trioxide are evolved copiously. Cool, cautiously add 10 mL of water, and evaporate to strong fumes of sulfur trioxide. Cool, cautiously add another 10 mL of water, and again evaporate to strong fumes of sulfur trioxide. Cool, and cautiously wash the solution into a generator flask with sufficient water to make the volume 35 mL. For the standard, use 0.005 mg of arsenic.

Calcium, Copper, Iron, Lead, Potassium, and Sodium. (By flame AAS, page 63).

Sample Stock Solution. Dissolve 10.0 g of sample in 20 mL of hydrochloric acid and 20 mL of water. Transfer to a 100-mL volumetric flask, and dilute to the mark with water (1 mL = 0.10 g).

Element	Wavelength (nm)	Sample Wt (g)	Standard Added (mg)	Flame Type*	Background Correction
Ca	422.7	2.0	0.05; 0.10	N/A	No
Cu	324.7	2.0	0.02; 0.04	A/A	Yes
Fe	248.3	2.0	0.10; 0.20	A/A	Yes
Pb	217.0	2.0	0.20; 0.40	A/A	Yes
K	766.5	0.10	0.01; 0.02	A/A	No
Na	589.0	0.10	0.01; 0.02	A/A	No

*A/A is air/acetylene; N/A is nitrous oxide/acetylene.

Silver. Add 1.0 g to a mixture of 5 mL of nitric acid and 15 mL of water, and add 0.05 mL of hydrochloric acid. Any turbidity should not exceed that of a standard containing 0.02 mg of silver ion treated as the sample.

Bismuth Nitrate Pentahydrate
Bismuth(III) Nitrate Pentahydrate
$Bi(NO_3)_3 \cdot 5H_2O$ **Formula Wt 485.07**   **CAS No. 10035-06-0**

GENERAL DESCRIPTION
Typical appearance: colorless solid
Analytical use: trace metal analysis
Aqueous solubility: decomposes

SPECIFICATIONS

Assay . . . . ,≥98.0% $Bi(NO_3)_3 \cdot 5H_2O$

Maximum Allowable

Insoluble matter 0.005%
Arsenic (As) 0.001%
Chloride (Cl) 0.001%
Sulfate (SO_4) 0.005%
Calcium (Ca) 0.005%
Copper (Cu). 0.002%
Iron (Fe). 0.001%
Lead (Pb) 0.002%
Potassium (K) 0.01%
Sodium (Na) 0.02%
Silver (Ag). 0.001%

TESTS

Assay. (By complexometric titration for bismuth). Weigh accurately about 1.8 g, and transfer to a 400-mL beaker. Dissolve in 5 mL of nitric acid, and add about 250 mL of water. Add about 50 mg of ascorbic acid and 50 mg of xylenol orange indicator mixture. Titrate with standard 0.1 M EDTA volumetric solution to canary yellow. (The solution color changes from purple to salmon pink just before the canary yellow end point). One milliliter of 0.1 M EDTA corresponds to 0.04851 g of $Bi(NO_3)_3 \cdot 5H_2O$.

$$\% \ Bi(NO_3)_3 \cdot 5H_2O = \frac{(mL \times M \ EDTA) \times 48.51}{Sample \ wt \ (g)}$$

Insoluble Matter. (Page 25). Use 20 g dissolved in 100 mL of dilute nitric acid (1 + 4).

Arsenic. (Page 34). Mix 0.5 g with 5 mL of water, and cautiously add 2 mL of sulfuric acid. Heat the mixture until white fumes of sulfur trioxide are evolved copiously. Cool, cautiously add 10 mL of water, and evaporate to strong fumes of sulfur trioxide. Cool, cautiously add another 10 mL of water, and again evaporate to strong fumes of sulfur trioxide. Cool, and cautiously wash the solution into a generator flask with sufficient water to make the volume 35 mL. The solution should show not more than 0.005 mg of arsenic.

Chloride. (Page 35). Use 1.0 g.

Sulfate. Dissolve 3.0 g in 5 mL of warm hydrochloric acid, dilute with water to 100 mL, and nearly neutralize with 10% ammonium hydroxide. Filter, and wash the residue and the filter with hot water to 150 mL. Transfer 100.0 mL to an evaporating dish, evaporate to about 15 mL, add 0.5 g of sodium carbonate, evaporate to dryness, and ignite gently. Cool, add 15 mL of hot water, neutralize with 10% hydrochloric acid, filter, and wash with water to 20 mL. Add to the combined filtrate and washings 1 mL of 1 N hydrochloric acid and 2 mL of 12% barium chloride reagent solution, and allow to stand for 10 min. Any turbidity should not exceed that of a standard prepared as follows. To 0.5 g of sodium carbonate, add 0.1 mg of sulfate ion (SO_4) and 2 mL of hydrochloric acid, and evaporate to dryness on a hot plate (≈100 °C). Dissolve the residue in 20 mL of water, and add 1 mL of 1 N hydrochloric acid and 2 mL of 12% barium chloride reagent solution.

Calcium, Copper, Iron, Lead, Potassium, and Sodium. (By flame AAS, page 63).

Sample Stock Solution. Dissolve 10.0 g of sample in 20 mL of nitric acid and 20 mL of water. Transfer to a 100-mL volumetric flask, and dilute to the mark with water (1 mL = 0.1 g).

Element	Wavelength (nm)	Sample Wt (g)	Standard Added (mg)	Flame Type*	Background Correction
Ca	422.7	2.0	0.05; 0.10	N/A	No
Cu	324.7	2.0	0.02; 0.04	A/A	Yes
Fe	248.3	2.0	0.02; 0.04	A/A	Yes
Pb	217.0	2.0	0.04; 0.08	A/A	Yes
K	766.5	0.10	0.01; 0.02	A/A	No
Na	589.0	0.10	0.01; 0.02	A/A	No

*A/A is air/acetylene; N/A is nitrous oxide/acetylene.

Silver. Dissolve 1.0 g in a mixture of 5 mL of nitric acid and 15 mL of water, and add 0.05 mL of hydrochloric acid. Any turbidity should not exceed that of a standard containing 0.01 mg of silver.

Boric Acid

H_3BO_3 **Formula Wt 61.83**   **CAS No. 10043-35-3**

GENERAL DESCRIPTION

Typical appearance: colorless or white solid
Analytical use: primary standard; buffer
Change in state (approximate): melting point, 171 °C
Aqueous solubility: 5 g in 100 mL at 21 °C
pK_a: 9.2

SPECIFICATIONS

Assay . . . . $\geq$99.5% H_3BO_3

Maximum Allowable

Insoluble in methanol 0.005%
Nonvolatile with methanol 0.05%
Chloride (Cl). 0.001%
Phosphate (PO_4). 0.001%
Sulfate (SO_4) 0.01%
Heavy metals (as Pb). 0.001%
Iron (Fe). 0.001%
Calcium (Ca) 0.005%

TESTS

Assay. (By acid–base titrimetry). Weigh, to the nearest 0.1 mg, 2.5 g of sample, and transfer to a 250-mL glass-stoppered flask. Dissolve in a mixture of 50 mL of water and 75 mL of glycerol, previously neutralized to a pH of 9.0 with 0.1 N sodium hydroxide, using a pH meter. Titrate with 1 N sodium hydroxide volumetric solution to a pH of 9.0. One milliliter of 1 N sodium hydroxide corresponds to 0.06183 g of H_3BO_3.

$$\% \ H_3BO_3 = \frac{(mL \times N \ NaOH) \times 6.183}{Sample \ wt \ (g)}$$

Insoluble in Methanol. Heat 20 g with 200 mL of methanol under complete reflux until the acid is dissolved, then reflux for 30 min. Filter through a tared, preconditioned filtering crucible, wash thoroughly with hot methanol, and dry at 105 °C.

Nonvolatile with Methanol. To 2.0 g of the sample in a platinum dish, add 25 mL of methanol and 0.5 mL of hydrochloric acid, and evaporate to dryness. Repeat the evaporation twice, each with 15 mL of methanol and 0.3 mL of hydrochloric acid. Add to the residue 0.10 mL of sulfuric acid, and ignite at 600 ± 25 °C for 15 min.

Chloride. (Page 35). Use 1.0 g dissolved in 20 mL of warm water.

For the Determination of Phosphate, Sulfate, Heavy Metals, and Iron

Sample Solution A. To 10 g, add 10 mg of sodium carbonate, 100 mL of methanol, and 5 mL of hydrochloric acid. Digest in a covered beaker on a hot plate ($\approx$100 °C) to effect dissolution, then uncover, and evaporate to dryness. Add 75 mL of methanol and 5 mL of hydrochloric acid, and repeat the digestion and evaporation. Dissolve the residue in 5 mL of 1 N acetic acid, and digest in a covered beaker on a hot plate ($\approx$100 °C) for 15 min. Filter, and dilute with water to 100 mL (1 mL = 0.1 g).

Phosphate. (Page 40, Method 1). Add 0.5 mL of nitric acid to 20 mL of sample solution A (2-g sample), and evaporate to dryness. Dissolve the residue in 25 mL of approximately 0.5 N sulfuric acid, and continue as described.

Sulfate. (Page 40, Method 1). Use 5.0 mL of sample solution A (0.5-g sample).

Heavy Metals. (Page 36, Method 1). Dilute 20 mL of sample solution A (2-g sample) with water to 25 mL.

Iron. (Page 38, Method 1). Use 10 mL of sample solution A (1-g sample).

Calcium. (By flame AAS, page 63).

Sample Stock Solution. To 20 g, add 20 mg of sodium carbonate, 200 mL of methanol, and 10 mL of hydrochloric acid. Digest in a covered beaker on a hot plate ($\approx$100 °C) to effect solution, then uncover, and evaporate to dryness. Add 75 mL of methanol and 5 mL of hydrochloric acid, and repeat the digestion and evaporation. Dissolve the residue in 5 mL of 1 N acetic acid, and digest in a covered beaker on a hot plate ($\approx$100 °C) for 15 min. Filter, and dilute with water to volume in a 200-mL volumetric flask (1 mL = 0.1 g).

Element	Wavelength (nm)	Sample Wt (g)	Standard Added (mg)	Flame Type*	Background Correction
Ca	422.7	1.0	0.05; 0.10	N/A	No

*N/A is nitrous oxide/acetylene.

Bromine

Br_2 Formula Wt 159.808   CAS No. 7726-95-6

GENERAL DESCRIPTION

Typical appearance: dark reddish-brown fuming liquid
Analytical use: brominating agent
Change in state (approximate): boiling point, 60 °C
Aqueous solubility: 3.13 g in 100 mL at 30 °C

SPECIFICATIONS

Assay . . . . $\geq$99.5% Br_2

Maximum Allowable

Residue after evaporation 0.005%
Chlorine (Cl). 0.05%
Iodine (I) 0.001%
Sulfur compounds (as S) 0.001%
Heavy metals (as Pb). . . . . 2 ppm
Nickel (Ni). . . . . 5 ppm

TESTS

Assay. (By iodometric titration). Weigh a 250-mL volumetric flask containing 15 g of potassium iodide in 45 mL of water. Add 1 mL of sample. Stopper promptly, mix by shaking, and reweigh. Dilute to volume with water, and mix thoroughly. Place a 25.0-mL aliquot of this solution in a conical flask, and titrate the liberated iodine with 0.1 N sodium thiosulfate volumetric solution, adding 3 mL of starch indicator solution near the end of the titration. One milliliter of 0.1 N sodium thiosulfate corresponds to 0.007990 g of Br_2.

$$\% \ Br_2 = \frac{(mL \times N \ Na_2S_2O_3) \times 7.99}{Sample \ wt \ (g) \ / \ 10}$$

Residue after Evaporation. (Page 25). In a fume hood, evaporate 31 g (10 mL) to dryness from a tared, preconditioned container on a hot plate ($\approx$100 °C), and dry the residue at 105 °C for 30 min. Reserve the residue for the test for heavy metals.

Chlorine. To each of two 250-mL wide-mouth conical flasks, add 0.5 mL of dilute sulfuric acid (1 + 4), 5 mL of potassium bromide solution (1.50 g of KBr per liter), and 35 mL of water. For the sample, add 3 g (1.0 mL) of the bromine to one of the flasks. For the standard, add 1.5 mg of chloride ion (3.2 mg of KCl) to the other. Digest each on a hot plate ($\approx$100 °C) until the sample solution is colorless, to each add 4 mL of potassium peroxydisulfate solution (1.0 g per 100 mL), and wash down the sides of the flasks with a little water. Digest again on a hot plate ($\approx$100 °C) for 30–40 min. Cool and dilute each with water to 100 mL. Transfer 1.0 mL of each solution into separate 50-mL beakers, and add 22 mL of water, 1 mL of nitric acid, and 1 mL of silver nitrate reagent solution to each. Any turbidity in the solution of the sample should not exceed that in the standard.

Iodine. To each of two 250-mL wide-mouth conical flasks, add 1.0 mL of 10% potassium chloride reagent solution, 50 mL of water, and a few silicon carbide boiling chips. For the sample, add 10.5 g (3.5 mL) of the bromine to one of the flasks. For the control, add 0.3 mL of the bromine and 0.1 mg of iodide ion (0.13 mg of KI) to the other flask. Boil cautiously in a hood to remove the excess bromine, adding water as required to maintain a volume of not less than 50 mL. When no trace of yellow color remains in either flask, cool, and add 5 mL of 10% potassium iodide reagent solution and 5 mL of dilute sulfuric acid (1 + 4) to each. Any color in the solution of the sample should not exceed that in the control.

Sulfur Compounds. To 8 g (2.6 mL), add 5 mL of water, and evaporate to dryness on a hot plate ($\approx$100 °C). To the residue, add 5 mL of dilute hydrochloric acid (1 + 19), filter if necessary, and dilute to 50 mL with water. To 10 mL of this residue solution, add 1 mL of 12% barium chloride reagent solution. Any turbidity should not exceed that produced by 0.05 mg of sulfate ion (SO_4) in an equal volume of solution containing the quantities of reagents used in the test. Compare 10 min after adding the barium chloride to the sample and standard solutions. Reserve the remaining residue solution for the test for nickel.

Heavy Metals. For the sample, use the residue reserved from the test for residue after evaporation. For the standard, use a solution containing 0.06 mg of lead. To each, add 3 mL of nitric acid, 1 mL of water, and 1.0 mL of 1% sodium carbonate reagent solution. Evaporate to dryness on a hot plate ($\approx$100 °C), dissolve the residues in about 20 mL of water, and dilute with water to 25 mL. Adjust the pH of the standard and sample solutions to between 3 and 4 (using a pH meter) with 1 N acetic acid reagent solution or 10% ammonium hydroxide reagent solution, dilute with water to 40 mL, and mix. Add 10 mL of freshly prepared hydrogen sulfide water to each, and mix. Any color in the solution of the sample should not exceed that in the standard.

Nickel. To 25 mL of the residue solution reserved from the test for sulfur compounds, add 5 mL of bromine water. Stir, and add ammonium hydroxide (1 + 1) until the bromine color is discharged. Add 5 mL of 1% dimethylglyoxime reagent solution in ethyl alcohol and 5 mL of 10% sodium hydroxide reagent solution. Any red color should not exceed that produced by 0.02 mg of nickel ion (Ni) in an equal volume of solution containing the quantities of reagents used in the test. Compare 10 min after adding the dimethylglyoxime to the sample and standard solutions.

Bromocresol Green

3′,3″,5′,5″-Tetrabromo-*m*-cresolsulfonphthalein; 4,4′-(3*H*-2,1-Benzoxathiol-3-ylidene)bis-[(2,6-dibromo-3-methyl)phenol] *S,S*-Dioxide

$C_{21}H_{14}Br_4O_5S$ (sultone)	Formula Wt 698.02	CAS No. 76-60-8
$C_{21}H_{13}Br_4O_5SNa$ (sodium salt)	Formula Wt 720.00	CAS No. 62625-32-5

Note: This monograph applies to both the sultone and the salt forms of this indicator.

GENERAL DESCRIPTION

Typical appearance: white to yellow solid
Analytical use: indicator
Change in state (approximate): melting point, 218–219 °C
Aqueous solubility: slightly soluble

SPECIFICATIONS

Clarity of solution . . . . Passes test
Visual transition interval . . . . pH 3.8 (yellow) to pH 5.4 (blue)

TESTS

Clarity of Solution. If the indicator is the sultone form, dissolve 0.1 g in 100 mL of alcohol. If the indicator is a salt form, dissolve 0.1 g in 100 mL of water. Not more than a faint trace of turbidity or insoluble matter should remain. Reserve the solution for the test for visual transition interval.

Visual Transition Interval. Dissolve 1 g of potassium chloride in 100 mL of water. Adjust the pH of the solution to 3.8 (using a pH meter) with 0.01 N hydrochloric acid. Add 0.10–0.30 mL of the 0.1% solution reserved from the test for clarity of solution. The color of the solution should be yellow. Titrate the solution with 0.01 N sodium hydroxide to a pH of 4.5 (using the pH meter). The color of the solution should be green. Continue the titration to a pH of 5.4. The color of the solution should be blue.

Bromocresol Purple

5′,5″-Dibromo-*o*-cresolsulfonphthalein

$C_{21}H_{16}Br_2O_5S$ (sultone)	**Formula Wt 540.24**	**CAS No. 115-40-2**
$C_{21}H_{15}Br_2O_5SNa$ (sodium salt)	**Formula Wt 562.20**	**CAS No. 62625-30-3**

Note: This monograph applies to both the sultone and the salt forms of this indicator.

GENERAL DESCRIPTION

Typical appearance: slightly yellow solid
Analytical use: pH indicator

Change in state (approximate): melting point, 241–242 °C (sultone)

Aqueous solubility: sultone, practically insoluble; salt, soluble

SPECIFICATIONS

Clarity of solution . . . . Passes test

Visual transition interval . . . . pH 5.2 (yellow) to pH 6.8 (purple)

TESTS

Clarity of Solution. If the indicator is the sultone form, dissolve 0.10 g in 100 mL of alcohol. If the indicator is the salt form, dissolve 0.10 g in 100 mL of water. The resulting solution is clear, and no appreciable amount of insoluble residue remains. Reserve the solution for the visual transition interval test.

Visual Transition Interval. Dissolve 1 g of potassium chloride in 100 mL of water. Adjust the pH of the solution to 4.0 with 0.01 N hydrochloric acid using a pH meter. Add 0.15 mL of the 0.1% solution reserved from the test for clarity of solution. Titrate the solution with 0.01 N sodium hydroxide to a pH of 5.2 (using a pH meter). The color of the solution should be yellow. Continue the titration to a pH of 6.8. The color of the solution should be purple.

Bromphenol Blue

3′,3″,5′,5″ -Tetrabromophenolsulfonphthalein; 4,4′-(3H-2,1-Benzoxathiol-3-ylidene)bis-[2,6-dibromophenol] S,S-Dioxide

$C_{19}H_{10}Br_4O_5S$ (sultone)   **Formula Wt 669.96**   **CAS No. 115-39-9**

$C_{19}H_9Br_4O_5SNa$ (sodium salt)   **Formula Wt 691.94**   **CAS No. 34725-61-6**

Note: This monograph applies to both the sultone and the salt forms of this indicator.

GENERAL DESCRIPTION

Typical appearance: pale orange solid

Analytical use: indicator

Change in state (approximate): decomposes at 279 °C

Aqueous solubility: 0.4 g in 100 mL at 25 °C

SPECIFICATIONS

Clarity of solution . . . . .Passes test
Visual transition interval . . . . .pH 3.0 (yellow) to pH 4.6 (blue)

TESTS

Clarity of Solution. If the indicator is the sultone form, dissolve 0.1 g in 100 mL of alcohol. If the indicator is the salt form, dissolve 0.1 g in 100 mL of water. Not more than a faint trace of turbidity or insoluble matter should remain. Reserve the solution for the test for visual transition interval.

Visual Transition Interval. Dissolve 1 g of potassium chloride in 100 mL of water. Adjust the pH of the solution to 3.0 (using a pH meter) with 0.01 N hydrochloric acid. Add 0.10–0.30 mL of the 0.1% solution of the indicator reserved from the test for clarity of solution. The color of the solution should be yellow with a slight greenish hue. Titrate the solution with 0.01 N sodium hydroxide to a pH of 3.4 (using the pH meter). The color of the solution should be green. Continue the titration to pH 4.6. The color of the solution should be blue.

Bromthymol Blue

3′,3″-Dibromothymolsulfonphthalein; 4,4′-(3H-2,1-Benzoxathiol-3-ylidene)bis-[2-bromo-3-methyl-6-(1-methylethyl)phenol] S,S-Dioxide

$C_{27}H_{28}Br_2O_5S$ (sultone)   **Formula Wt 624.38**   **CAS No. 76-59-5**
$C_{27}H_{27}Br_2O_5SNa$ (sodium salt)   **Formula Wt 646.36**   **CAS No. 34722-90-2**

Note: This monograph applies to both the sultone and salt forms of this indicator.

GENERAL DESCRIPTION

Typical appearance: cream-colored solid
Analytical use: indicator
Change in state (approximate): melting point, 200–202 °C
Aqueous solubility: sparingly soluble

SPECIFICATIONS

Clarity of solution . . . . .Passes test
Visual transition interval . . . . .pH 6.0 (yellow) to pH 7.6 (blue)

TESTS

Clarity of Solution. If the indicator is the sultone form, dissolve 0.1 g in 100 mL of alcohol. If the indicator is a salt form, dissolve 0.1 g in 100 mL of water. Not more than a faint trace of turbidity or insoluble matter should remain. Reserve the solution for the test for visual transition interval.

Visual Transition Interval. Dissolve 1 g of potassium chloride in 100 mL of water. Adjust the pH of the solution to 6.0 (using a pH meter as described) with 0.01 N hydrochloric acid or sodium hydroxide. Add 0.10–0.30 mL of the 0.1% solution reserved from the test for clarity of solution. The color of the solution should be yellow, with not more than a faint trace of green color. Titrate the solution with 0.01 N sodium hydroxide to a pH of 6.7 (using a pH meter). The color of the solution should be green. Continue the titration to pH 7.6. The color of the solution should be blue.

Brucine Sulfate Heptahydrate

2,3-Dimethoxystrychnidin-10-one Sulfate Heptahydrate

$\cdot 7H_2O$
$\cdot H_2SO_4$

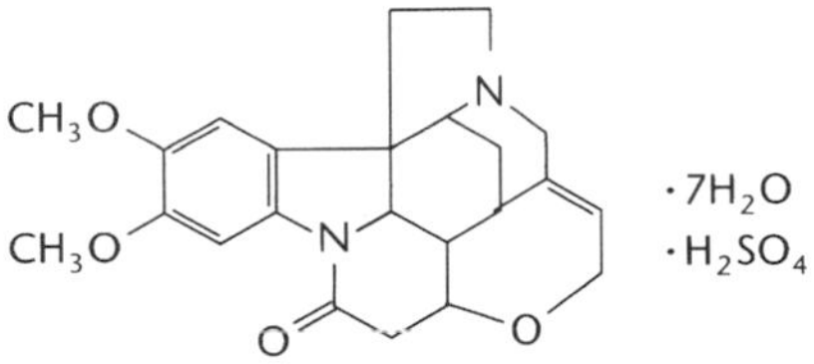

$(C_{23}H_{26}N_2O_4)_2 \cdot H_2SO_4 \cdot 7H_2O$ **Formula Wt 1013.11**   **CAS No. 5787-00-8**

GENERAL DESCRIPTION

Typical appearance: colorless or white solid
Analytical use: determination of nitrate
Change in state (approximate): melting point, 180 °C, decomposes
Aqueous solubility: 1 g in 75 g of cold water, 10 g of hot water

SPECIFICATIONS

Sensitivity to nitrate . . . . Passes test

Maximum Allowable

Clarity of solution . . . . Passes test
Loss on drying 13.0%
Residue after ignition 0.1%

TESTS

Sensitivity to Nitrate.

Test Solutions 1, 2, 3, and 4. Prepare a sample solution by dissolving 0.6 g in dilute sulfuric acid (2 + 1), as described in Method 1 on page 38. Place 50 mL of this sample solution in each of four test tubes, and add nitrate ion (NO_3) to each as follows: No. 1, none; No. 2, 0.01 mg; No. 3, 0.02 mg; and No. 4, 0.03 mg.

Heat the four test tubes in a preheated (boiling) water bath for 10 min. Cool rapidly in an ice bath to room temperature. Set a spectrophotometer at 410 nm and, using 1-cm cells, adjust the instrument to read 0 absorbance with solution 1 in the light path. Determine the absorbance for solutions 2, 3, and 4 at this adjustment, using similar cells. The absorbances for solutions 2, 3, and 4 should not be less than 0.025, 0.050, and 0.075, respectively, and the plot of absorbances versus nitrate concentrations should be linear.

Clarity of Solution. Dissolve 1 g in 100 mL of water, and digest in a covered beaker on a hot plate ($\approx$100 °C) for 1 h. The solution should be as colorless and clear as an equal volume of water.

Loss on Drying. Weigh accurately about 1 g, and dry on a tared, preconditioned dish at 105 °C for 6 h.

Residue after Ignition. (Page 26). Ignite 1.0 g. Omit the addition of sulfuric acid.

2-Butanone
Methyl Ethyl Ketone

$$CH_3CH_2-\overset{\overset{\displaystyle O}{\|}}{C}-CH_3$$

$CH_3COCH_2CH_3$ **Formula Wt 72.11**   **CAS No. 78-93-3**

GENERAL DESCRIPTION
Typical appearance: clear liquid
Analytical use: solvent
Change in state (approximate): boiling point, 80 °C
Aqueous solubility: 37 g in 100 mL
Density: 0.81

SPECIFICATIONS
Assay . . . . $\geq$99.0% $CH_3COCH_2CH_3$
Maximum Allowable
Color (APHA) 15
Residue after evaporation 0.0025%
Titrable acid. . . . . .0.0005 meq/g
Water (H_2O) 0.20%

TESTS

Assay. Analyze the sample by gas chromatography using the general parameters cited on page 80. The following specific conditions are also required.

Column: Type I, methyl silicone

Measure the area under all peaks, and calculate the 2-butanone content in area percent. Correct for water content.

Color (APHA). (Page 43).

Residue after Evaporation. (Page 25). Use 80 g (100 mL) in a tared, preconditioned beaker on a hot plate ($\approx$100 °C). When the liquid has evaporated, complete the drying at 105 °C to constant weight after cooling in a desiccator.

Titrable Acid. Mix 40 g (50 mL) with 50 mL of carbon dioxide-free water, and add 0.15 mL of phenolphthalein indicator solution. Not more than 2.0 mL of 0.01 N sodium hydroxide should be required to produce a pink color.

Water. (Page 31, Method 1). Use 25 mL (20 g) of the sample and a methanol-free system to prevent ketal formation with liberation of water.

n-Butyl Acetate
$CH_3COO(CH_2)_3CH_3$ **Formula Wt 116.16**   **CAS No. 123-86-4**

GENERAL DESCRIPTION
Typical appearance: colorless, clear liquid
Analytical use: solvent
Change in state (approximate): boiling point, 126 °F
Aqueous solubility: 0.7 g in 100 mL
Density: 0.88

SPECIFICATIONS
Assay . . . . $\geq$99.5% $CH_3COO(CH_2)_3CH_3$

Maximum Allowable

Color (APHA) 10
Residue after evaporation 0.001%
Titrable acid . . . . 0.0016 meq/g
Substances darkened by sulfuric acid . . . . Passes test
Water (H_2O) 0.1%
n-Butyl alcohol (C_4H_9OH) 0.2%

TESTS

Assay and *n*-Butyl Alcohol. Analyze the sample by gas chromatography using the general parameters cited on page 80. The following specific conditions are also required.

Column: Type I, methyl silicone

Measure the area under all peaks, and calculate the area percent for *n*-butyl acetate and *n*-butyl alcohol. Correct for water content.

Color (APHA). (Page 43).

Residue after Evaporation. (Page 25). Evaporate 100 g (114 mL) to dryness in a tared, preconditioned dish on a hot plate (≈100 °C), and dry the residue at 105 °C for 30 min.

Titrable Acid. Neutralize 25 mL of reagent alcohol, to which 0.05 mL of bromthymol blue indicator solution has been added, with 0.01 N sodium hydroxide to a blue-green color. Add 25 g (28 mL) of the sample, mix, and titrate with 0.01 N sodium hydroxide to the same color. Not more than 4.0 mL of 0.01 N sodium hydroxide should be required.

Substances Darkened by Sulfuric Acid. Cool 5 mL of the sample to below 10 °C. Cautiously add, dropwise, 5 mL of 95.0 ± 0.5% sulfuric acid. The color of the solution should not exceed that of a color standard composed of 9 volumes of water plus one volume of a color standard containing 3.25 g of cobalt chloride hexahydrate, 2.25 g of ferric chloride hexahydrate, and 3 mL of hydrochloric acid per 100 mL of solution.

Water. (Page 31, Method 1). Use 25 g (28 mL) of the sample.

Butyl Alcohol
1-Butanol
$CH_3(CH_2)_2CH_2OH$ **Formula Wt 74.12**   **CAS No. 71-36-3**

GENERAL DESCRIPTION
Typical appearance: clear, colorless liquid
Analytical use: solvent
Change in state (approximate): boiling point, 118 °C
Aqueous solubility: 7.4 g in 100 mL
Density: 0.81

SPECIFICATIONS
Assay . . . . ≥99.4% $CH_3(CH_2)_2CH_2OH$
Maximum Allowable
Color (APHA) 10
Residue after evaporation 0.005%
Titrable acid. . . . . .0.0008 meq/g
Carbonyl compounds (as butyraldehyde) 0.01%
Water (H_2O) 0.1%
Butyl ether ($C_8H_{18}O$) 0.2%

TESTS

Assay and Butyl Ether. Analyze the sample by gas chromatography using the general parameters cited on page 80. The following specific conditions are also required.

Column: Type I, methyl silicone

Measure the areas under all peaks, and calculate the butyl alcohol and butyl ether content in area percent. Correct for water content.

Color (APHA). (Page 43).

Residue after Evaporation. (Page 25). Evaporate 20 g (25 mL) to dryness in a tared, preconditioned platinum dish on a hot plate ($\approx$100 °C), and dry the residue at 105 °C for 30 min.

Titrable Acid. To 60 g (74 mL) in a glass-stoppered flask, add 0.10 mL of phenolphthalein indicator solution, and titrate with 0.01 N methanolic potassium hydroxide to a pink end point that persists for at least 15 s. Not more than 5.0 mL of 0.01 N methanolic potassium hydroxide should be consumed.

Carbonyl Compounds. (Page 54). Use 1.0 mL (0.80 g) sample and 0.08 mg of butaraldehyde.

Water. (Page 31, Method 1). Use 30 mL (24 g) of the sample.

tert-Butyl Alcohol
2-Methyl-2-propanol

$$CH_3 - \underset{\underset{CH_3}{|}}{\overset{\overset{CH_3}{|}}{C}} - OH$$

(CH$_3$)$_3$COH   **Formula Wt 74.12**   **CAS No. 75-65-0**

Note: Since the melting point of this reagent is close to 25 °C, the sample may be warmed to about 30 °C for testing.

GENERAL DESCRIPTION

Typical appearance: colorless solid
Analytical use: solvent
Change in state (approximate): melting point, 25 °C; boiling point, 82 °C
Aqueous solubility: soluble
Density: 0.79

SPECIFICATIONS

Assay . . . . $\geq$99.0% (CH$_3$)$_3$COH

Maximum Allowable

Color (APHA). 20
Residue after evaporation 0.003%
Titrable acid . . . . 0.001 meq/g

Carbonyl compounds (as formaldehyde) 0.01%
Water (H$_2$O) 0.1%

TESTS

Assay. Analyze the sample by gas chromatography using the parameters cited on page 80. The following specific conditions are also required.

   Column: Type I, methyl silicone

Measure the areas under all peaks, and calculate the *tert*-butyl alcohol content in area percent. Correct for water content.

Color (APHA). (Page 43).

Residue after Evaporation. (Page 25). Evaporate 60 g (77 mL) of sample to dryness in a tared, preconditioned evaporating dish on a hot plate ($\approx$100 °C), and dry the residue at 105 °C for 30 min.

Titrable Acid. To 35 g (45 mL) in a glass-stoppered Erlenmeyer flask, add 0.15 mL of phenolphthalein indicator solution, and shake for 1 min. No pink color should appear. Titrate the solution with 0.01 N methanolic potassium hydroxide to a faint pink color that persists for at least 15 s. Not more than 3.5 mL of 0.01 N methanolic potassium hydroxide should be required.

Carbonyl Compounds. (Page 54). Use 1.3 mL (1.0 g) sample. For the standard, use 0.10 mg of formaldehyde.

Water. (Page 31, Method 1). Use 45 mL (35 g) of the sample.

Cadmium Chloride, Anhydrous

CdCl$_2$ **Formula Wt 183.35**   **CAS No. 10108-64-2**

GENERAL DESCRIPTION

Typical appearance: hygroscopic solid
Analytical use: preparation of cadmium standard; analysis of sulfides; testing of pyridine
   bases
Change in state (approximate): melting point, 568 °C
Aqueous solubility: freely soluble

SPECIFICATIONS

Assay . . . . $\geq$99.0% CdCl$_2$
                                                                    Maximum Allowable
Insoluble matter 0.01%
Nitrate and nitrite (as NO$_3$) 0.003%

Sulfate (SO_4) 0.01%
Ammonium (NH_4) 0.01%
Calcium (Ca) 0.01%
Copper (Cu) 0.001%
Lead (Pb) 0.005%
Potassium (K) 0.02%
Sodium (Na) 0.05%
Zinc (Zn) 0.05%
Iron (Fe) 0.001%

TESTS

Assay. (By complexometric titration of cadmium). Weigh accurately about 0.4 g of sample, and dissolve in about 200 mL of water. Add 15 mL of a saturated hexamethylenetetramine reagent solution and 50 mg of xylenol orange indicator mixture. Titrate with standard 0.1 M EDTA volumetric solution to a color change of red-purple to lemon yellow. One milliliter of 0.1 M EDTA corresponds to 0.01833 g of $CdCl_2$.

$$\% \; CdCl_2 = \frac{(mL \times M \; EDTA) \times 18.33}{Sample \; wt \; (g)}$$

Insoluble Matter. (Page 25). Use 10 g dissolved in 150 mL of water. Retain the filtrate and washings for the test for sulfate.

Nitrate and Nitrite. (Page 38, Method 1).

Sample Solution A. In a 50-mL centrifuge tube, dissolve 0.30 g in 3 mL of water by heating on a hot plate ($\approx$100 °C). Add 20 mL of nitrate-free sulfuric acid (2 + 1), and cool in an ice bath. Centrifuge for 10 min, decant the solution into a 50-mL volumetric flask, and dilute to 50 mL with brucine sulfate reagent solution.

Control Solution B. In a 50-mL centrifuge tube, dissolve 0.30 g in 2 mL of water and 1 mL of nitrate ion (NO_3) standard solution by heating on a hot plate ($\approx$100 °C). Add 20 mL of nitrate-free sulfuric acid (2 + 1), and cool in an ice bath. Centrifuge for 10 min, decant the solution into a 50-mL volumetric flask, and dilute to 50 mL with brucine sulfate reagent solution.

Continue with the procedure, starting with the preparation of blank solution C.

Sulfate. (Page 40, Method 1). Use 1.0 g of sample retained from the test for insoluble matter. Allow 30 min for turbidity to form.

Ammonium. Dissolve 1.0 g in 80 mL of water, and add with stirring 20 mL of freshly boiled 10% sodium hydroxide reagent solution. After allowing the precipitate to settle, dilute 10 mL of the clear supernatant liquid with water to 50 mL, and add 2 mL of Nessler reagent. Any color should not exceed that produced by 0.01 mg of ammonium ion (NH_4) in an equal volume of solution containing 4 mL of 10% sodium hydroxide reagent solution and 2 mL of the Nessler reagent.

Calcium, Copper, Lead, Potassium, Sodium, and Zinc. (By flame AAS, page 63).

Sample Stock Solution. Dissolve 10.0 g of sample in a 100-mL volumetric flask, add 5 mL of nitric acid, and dilute to the mark with water (1 mL = 0.1 g).

Element	Wavelength (nm)	Sample Wt (g)	Standard Added (mg)	Flame Type*	Background Correction
Ca	422.7	0.20	0.02; 0.04	N/A	No
Cu	324.7	2.0	0.02; 0.04	A/A	Yes
Pb	217.0	2.0	0.10; 0.20	A/A	Yes
K	766.5	0.20	0.02; 0.04	A/A	No
Na	589.0	0.04	0.01; 0.02	A/A	No
Zn	213.9	0.04	0.01; 0.02	A/A	Yes

*A/A is air/acetylene; N/A is nitrous oxide/acetylene.

Iron. (Page 38, Method 1). Use 1.0 g.

Cadmium Chloride, Crystals
$CdCl_2 \cdot 2\frac{1}{2} H_2O$ **Formula Wt 228.35**   **CAS No. 7790-78-5**

GENERAL DESCRIPTION
Typical appearance: colorless solid
Analytical use: standard solutions
Change in state (approximate): melting point, 568 °C
Aqueous solubility: freely soluble

SPECIFICATIONS
Assay . . . . 79.5–81.0% $CdCl_2$

Maximum Allowable

Insoluble matter 0.005%
Nitrate and nitrite (as NO_3) 0.003%
Sulfate (SO_4) 0.005%
Ammonium (NH_4) 0.005%
Calcium (Ca) 0.005%
Copper (Cu) . . . . 5 ppm
Lead (Pb). 0.005%
Potassium (K) 0.02%
Sodium (Na). 0.05%
Zinc (Zn) 0.05%
Iron (Fe). . . . . 5 ppm

TESTS

Assay. (By complexometric titration of cadmium). Weigh accurately about 0.8 g of sample, and dissolve in about 200 mL of water. Add 15 mL of a saturated hexamethylenetetramine reagent solution and 50 mg of xylenol orange indicator mixture. Titrate with 0.1 M EDTA volumetric solution to a color change of red-purple to lemon yellow. One milliliter of 0.1 M EDTA corresponds to 0.01833 g of $CdCl_2$.

$$\% \; CdCl_2 = \frac{(mL \times M \; EDTA) \times 18.33}{Sample \; wt \; (g)}$$

Insoluble Matter. (Page 25). Use 20 g dissolved in 150 mL of water. Reserve the filtrate for the test for sulfate.

Nitrate and Nitrite. (Page 38, Method 1).

Sample Solution A. In a 50-mL centrifuge tube, dissolve 0.30 g in 3 mL of water by heating on a hot plate ($\approx$100 °C). Add 20 mL of nitrate-free sulfuric acid (2 + 1), and cool in an ice bath. Centrifuge for 10 min, decant the solution into a 100-mL flask, and dilute with 50 mL of brucine sulfate reagent solution.

Control Solution B. In a 50-mL centrifuge tube, dissolve 0.30 g in 2 mL of water and 1 mL of nitrate ion (NO_3) standard solution by heating on a hot plate ($\approx$100 °C). Add 20 mL of nitrate-free sulfuric acid (2 + 1), and cool in an ice bath. Centrifuge for 10 min, decant the solution into a 100-mL flask, and dilute with 50 mL of brucine sulfate reagent solution.

Continue with the procedure, starting with the preparation of blank solution C.

Sulfate. (Page 40, Method 1). Use 2.0 g of sample. Allow 30 min for turbidity to form.

Ammonium. Dissolve 1.0 g in 80 mL of water, and add with stirring, 20 mL of freshly boiled 10% sodium hydroxide reagent solution. After allowing the precipitate to settle, dilute 20 mL of the clear supernatant liquid with water to 50 mL, and add 2 mL of Nessler reagent. Any color should not exceed that produced by 0.01 mg of ammonium ion (NH_4) in an equal volume of solution containing 4 mL of the 10% sodium hydroxide reagent solution and 2 mL of the Nessler reagent.

Calcium, Copper, Lead, Potassium, Sodium, and Zinc. (By flame AAS, page 63).

Sample Stock Solution. Dissolve 10.0 g of sample in a 100-mL volumetric flask, add 5 mL of nitric acid, and dilute to the mark with water (1 mL = 0.10 g).

Element	Wavelength (nm)	Sample Wt (g)	Standard Added (mg)	Flame Type*	Background Correction
Ca	422.7	2.0	0.05; 0.10	N/A	No
Cu	324.7	2.0	0.01; 0.02	A/A	Yes
Pb	217.0	2.0	0.10; 0.20	A/A	Yes
K	766.5	0.20	0.02; 0.04	A/A	No
Na	589.0	0.04	0.01; 0.02	A/A	No
Zn	213.9	0.04	0.01; 0.02	A/A	Yes

*A/A is air/acetylene; N/A is nitrous oxide/acetylene.

Iron. (Page 38, Method 1). Use 2.0 g.

Cadmium Nitrate Tetrahydrate

$Cd(NO_3)_2 \cdot 4H_2O$ Formula Wt 308.47   CAS No. 10022-68-1

GENERAL DESCRIPTION

Typical appearance: solid
Analytical use: production of other cadmium compounds
Change in state (approximate): melting point, 59 °C
Aqueous solubility: very soluble

SPECIFICATIONS

Assay . . . . $\geq$ 99.0% $Cd(NO_3)_2 \cdot 4H_2O$

Maximum Allowable

Insoluble matter	0.005%
Chloride (Cl).	0.005%
Sulfate (SO_4)	0.002%
Calcium (Ca)	0.02%
Copper (Cu)	0.002%
Lead (Pb).	0.005%
Magnesium (Mg)	0.02%
Zinc (Zn)	0.05%
Iron (Fe).	0.001%

TESTS

Assay. (By complexometric titration of cadmium). Weigh accurately about 1.2 g of sample, and dissolve in about 100 mL of water. Add 15 mL of a saturated hexamethylene-tetramine reagent solution and 50 mg of xylenol orange indicator mixture. Titrate with 0.1 M EDTA volumetric solution to a color change of red-purple to yellow. One milliliter of 0.1 M EDTA corresponds to 0.03085 g of $Cd(NO_3)_2 \cdot 4H_2O$.

$$\% \ Cd(NO_3)_2 \cdot 4H_2O = \frac{(mL \times M \ EDTA) \times 30.85}{Sample \ wt \ (g)}$$

Insoluble Matter. (Page 25). Use 20 g in 200 mL of water.

Chloride. (Page 35). Use 0.20 g.

Sulfate. (Page 40, Method 1). Evaporate 2.5 g of sample with 5 mL of hydrochloric acid. Dissolve the residue in a minimum of water. Continue the procedure beginning with "Add 1 mL of dilute hydrochloric acid (1 + 19)…".

Calcium, Copper, Lead, Magnesium, and Zinc. (By flame AAS, page 63).

   Sample Stock Solution A. Dissolve 10.0 g of sample with water in a 100-mL volumetric flask, and dilute to the mark with water (1 mL = 0.10 g).

   Sample Stock Solution B. Transfer 10.0 mL of sample stock solution A to a 100-mL volumetric flask, and dilute to volume with water (1 mL = 0.01 g).

Element	Wavelength (nm)	Sample Wt (g)	Standard Added (mg)	Flame Type*	Background Correction
Ca	422.7	0.2	0.02; 0.04	N/A	No
Cu	324.8	2.0	0.02; 0.04	A/A	No
Pb	217.0	2.0	0.05; 0.10	A/A	Yes
Mg	285.2	0.04	0.004; 0.008	A/A	Yes
Zn	213.9	0.04	0.01; 0.02	A/A	Yes

*A/A is air/acetylene; N/A is nitrous oxide/acetylene.

Iron. (Page 38, Method 1). Dissolve 1.0 g of sample in 30 mL of water. Add 5 mL of hydrochloric acid, boil for 2 min, cool, and dilute with water to 40 mL.

Cadmium Sulfate

Cadmium Sulfate, Anhydrous

CdSO$_4$ **Formula Wt 208.47**   **CAS No. 10124-36-4**

GENERAL DESCRIPTION

Typical appearance: colorless solid
Analytical use: catalyst in Marsh test for arsenic; determination of hydrogen sulfide
Change in state (approximate): melting point, 1000 °C
Aqueous solubility: freely soluble

SPECIFICATIONS

Assay . . . . ≥99.0% CdSO$_4$

Maximum Allowable

Insoluble matter 0.005%
Loss on drying 1.0%
Chloride (Cl) 0.001%
Nitrate and nitrite (as NO$_3$). 0.003%
Calcium (Ca) 0.01%
Copper (Cu). 0.002%
Lead (Pb) 0.003%
Potassium (K) 0.02%
Sodium (Na) 0.05%
Zinc (Zn). 0.05%
Iron (Fe). 0.001%

TESTS

Assay. (By complexometric titration of cadmium). Weigh accurately about 0.9 g, transfer to a 400-mL beaker, and dissolve in 200 mL of water. Add 15 mL of a saturated hexamethylenetetramine reagent solution and 50 mg of xylenol orange indicator mixture. Titrate with

0.1 M EDTA volumetric solution to a color change of red-purple to lemon yellow. One milliliter of 0.1 M EDTA corresponds to 0.02085 g of $CdSO_4$.

$$\% \ CdSO_4 = \frac{(mL \times M \ EDTA) \times 20.85}{Sample \ wt \ (g)}$$

Insoluble Matter. (Page 25). Use 20 g dissolved in 150 mL of water.

Loss on Drying. Weigh accurately about 1 g, and dry at 150 °C for 4 h.

Chloride. (Page 35). Use 1.0 g.

Nitrate and Nitrite. (Page 38, Method 1). For sample solution A, use 0.50 g. For control solution B, use 0.50 g and 1.5 mL of nitrate ion (NO_3) standard solution.

Calcium, Copper, Lead, Potassium, Sodium, and Zinc. (By flame AAS, page 63).

> *Sample Stock Solution A.* Dissolve 10.0 g of sample with water in a 100-mL volumetric flask, and dilute to the mark with water (1 mL = 0.10 g).

> *Sample Stock Solution B.* Transfer 10.0 mL of sample stock solution A to a 100-mL volumetric flask, and dilute to the mark with water (1 mL = 0.01 g).

Element	Wavelength (nm)	Sample Wt (g)	Standard Added (mg)	Flame Type*	Background Correction
Ca	422.7	0.20	0.02; 0.04	N/A	No
Cu	324.7	2.0	0.04; 0.08	A/A	Yes
Pb	217.0	2.0	0.04; 0.08	A/A	Yes
K	766.5	0.20	0.02; 0.04	A/A	No
Na	589.0	0.04	0.01; 0.02	A/A	No
Zn	213.9	0.04	0.01; 0.02	A/A	Yes

*A/A is air/acetylene; N/A is nitrous oxide/acetylene.

Iron. (Page 38, Method 1). Dissolve 1.0 g in 30 mL of water, add 5 mL of hydrochloric acid, and dilute with water to 50 mL. Use this solution without further acidification.

Cadmium Sulfate, $\frac{8}{3}$-Hydrate
Cadmium Sulfate, Crystals
$CdSO_4 \cdot \frac{8}{3}H_2O$ **Formula Wt 256.52**   **CAS No. 7790-84-3**

GENERAL DESCRIPTION
Typical appearance: odorless solid
Analytical use: preparation of cadmium standards
Change in state (approximate): on heating, loses water above 40 °C, forming monohydrate by 80 °C; does not become anhydrous on further heating
Aqueous solubility: freely soluble

SPECIFICATIONS

Assay . . . . 98.0–102.0% $CdSO_4 \cdot \frac{8}{3}H_2O$

Maximum Allowable

Insoluble matter 0.005%
Chloride (Cl) 0.001%
Nitrate and nitrite (as NO_3). 0.003%
Calcium (Ca) 0.005%
Copper (Cu). 0.002%
Lead (Pb) 0.003%
Potassium (K) 0.01%
Sodium (Na) 0.02%
Zinc (Zn). 0.05%
Iron (Fe). 0.001%

TESTS

Assay. (By complexometric titration of cadmium). Weigh accurately 1.0 g, transfer to a 400-mL beaker, and dissolve in 175 mL of water. Add 15 mL of saturated hexamethylene-tetramine reagent solution and 50 mg of xylenol orange indicator mixture. Titrate with 0.1 M EDTA volumetric solution to a color change of red-purple to lemon yellow. One milliliter of 0.1 M EDTA corresponds to 0.02565 g of $CdSO_4 \cdot \frac{8}{3}H_2O$.

$$\% \ CdSO_4 \cdot \tfrac{8}{3}H_2O = \frac{(mL \times M \ EDTA) \times 25.65}{Sample \ wt \ (g)}$$

Insoluble Matter. (Page 25). Use 20 g dissolved in 150 mL of water.

Chloride. (Page 35). Use 1.0 g.

Nitrate and Nitrite. (Page 38, Method 1). For sample solution A, use 0.50 g. For control solution B, use 0.50 g and 1.5 mL of nitrate ion (NO_3) standard solution.

Calcium, Copper, Lead, Potassium, Sodium, and Zinc. (By flame AAS, page 63).

> **Sample Stock Solution A.** Dissolve 20.0 g of sample with water in a 100-mL volumetric flask, and dilute to the mark with water (1 mL = 0.20 g).

> **Sample Stock Solution B.** Transfer 5.0 mL of sample stock solution A to a 100-mL volumetric flask, and dilute to the mark with water (1 mL = 0.01 g).

Element	Wavelength (nm)	Sample Wt (g)	Standard Added (mg)	Flame Type*	Background Correction
Ca	422.7	0.40	0.02; 0.04	N/A	No
Cu	324.7	4.0	0.04; 0.08	A/A	Yes
Pb	217.0	4.0	0.12; 0.24	A/A	Yes
K	766.5	0.40	0.02; 0.04	A/A	No
Na	589.0	0.05	0.005; 0.01	A/A	No
Zn	213.9	0.04	0.01; 0.02	A/A	Yes

*A/A is air/acetylene; N/A is nitrous oxide/acetylene.

Iron. (Page 38, Method 1). Dissolve 1.0 g in 30 mL of water, add 5 mL of hydrochloric acid, and dilute with water to 50 mL. Use this solution without further acidification.

Calcium Acetate Monohydrate

Ca(CH₃COO)₂ · H₂O   **Formula Wt 176.18**   **CAS No. 5743-26-0**

GENERAL DESCRIPTION

Typical appearance: odorless, colorless solid

Analytical use: precipitation of oxalates

Change in state (approximate): melting point, above 160 °C, decomposes to form acetone and calcium carbonate

Aqueous solubility: 34.7 g in 100 mL at 20 °C

SPECIFICATIONS

Assay . . . . ≥99.0% Ca(CH₃COO)₂ · H₂O

Maximum Allowable

Insoluble matter	0.005%
Alkalinity	Passes test
Titrable acid.	0.035 meq/g
Chloride (Cl).	0.001%
Sulfate (SO₄)	0.01%
Barium (Ba)	0.01%
Heavy metals (as Pb).	0.005%
Iron (Fe).	0.001%
Magnesium (Mg)	0.05%
Potassium (K)	0.01%
Sodium (Na).	0.02%
Strontium (Sr)	0.05%

TESTS

Assay. (By complexometric titration for calcium). Weigh accurately 0.70 g, transfer to a 250-mL beaker, and dissolve in 150 mL of water. While stirring, add from a 50-mL buret about 30 mL of 0.1 M EDTA volumetric solution. Add 10% sodium hydroxide reagent solution until the pH is above 12 (using a pH meter). Add about 300 mg of hydroxy naphthol blue indicator mixture, and continue the titration with the 0.1 M EDTA to a blue color. One milliliter of 0.1 M EDTA corresponds to 0.01762 g of Ca(CH₃COO)₂ · H₂O.

$$\% \ \text{Ca(CH}_3\text{COO)}_2 \cdot \text{H}_2\text{O} = \frac{(\text{mL} \times \text{M EDTA}) \times 17.62}{\text{Sample wt (g)}}$$

Insoluble Matter. (Page 25). Use 20 g dissolved in 200 mL of water.

Alkalinity and Titrable Acid. Dissolve 2.0 g in 25 mL of carbon dioxide-free water, and add 0.10 mL of phenolphthalein indicator solution. The sample solution should show no pink color due to excess alkalinity. Titrate the solution with 0.10 N sodium hydroxide until a pink color is produced after shaking. Not more than 0.70 mL of 0.10 N sodium hydroxide should be required in the titration.

Chloride. (Page 35). Dissolve 1.0 g in 20 mL of water.

Sulfate. (Page 40, Method 1). Use 0.50 g.

Barium. Dissolve 2.0 g in 15 mL of water, add 2 drops of glacial acetic acid, filter, and add to the filtrate 0.3 mL of 10% potassium dichromate reagent solution. Compare the turbidity to a standard containing 0.2 mg of barium treated exactly as the sample.

Heavy Metals. (Page 36, Method 1). Dissolve 0.40 g in 20 mL of water.

Iron. (Page 38, Method 1). Dissolve 1.0 g in 20 mL of water.

Magnesium, Potassium, Sodium, and Strontium. (By flame AAS, page 63).

> *Sample Stock Solution.* Dissolve 2.0 g in 80 mL of water containing 2 mL of nitric acid. Transfer to a 100-mL volumetric flask, and dilute to the mark with water (1 mL = 0.02 g).

Element	Wavelength (nm)	Sample Wt (g)	Standard Added (mg)	Flame Type*	Background Correction
Mg	285.2	0.02	0.01; 0.02	A/A	Yes
Na	589.0	0.05	0.01; 0.02	A/A	No
K	766.5	0.20	0.02; 0.04	A/A	No
Sr	460.7	0.20	0.10; 0.20	N/A	No

*A/A is air/acetylene; N/A is nitrous oxide/acetylene.

Calcium Carbonate

$CaCO_3$ **Formula Wt 100.09**   **CAS No. 471-34-1**

GENERAL DESCRIPTION

Typical appearance: white solid
Analytical use: determining halogens in organic compounds
Change in state (approximate): melting point, 825 °C
Aqueous solubility: insoluble

SPECIFICATIONS

Assay (dried basis). . . . . ≥99.0% $CaCO_3$

Maximum Allowable

Insoluble in dilute hydrochloric acid 0.01%
Chloride (Cl) 0.001%
Fluoride (F) 0.0015%
Sulfate (SO_4) 0.01%
Ammonium (NH_4). 0.003%
Heavy metals (as Pb) 0.001%

Iron (Fe). 0.003%
Barium (Ba) 0.01%
Magnesium (Mg) 0.02%
Potassium (K) 0.01%
Sodium (Na). 0.1%
Strontium (Sr) 0.1%

TESTS

Assay. (By complexometric titration of calcium). Weigh accurately about 0.4 g, previ-ously dried at 300 °C for 4 h, and transfer to a 400-mL beaker. Cover the beaker with a watch glass. Add 2 mL of 20% hydrochloric acid from a pipet placed between the lip of the beaker and the watch glass. Swirl the beaker to aid dissolution. With water, wash down the inner wall of the beaker, the outer surface of the pipet, and the watch glass. Dilute to about 100 mL with water. While stirring, add from a 50-mL buret about 30 mL of 0.1 M EDTA volumetric solution. Adjust the solution to above pH 12 (using a pH meter with a suitable high-pH glass electrode) with 10% sodium hydroxide reagent solution. Add 300 mg of hydroxy naphthol blue indicator mixture, and continue the titration immediately with 0.1 M EDTA to a blue color. One milliliter of 0.1 M EDTA corresponds to 0.01001 g of $CaCO_3$.

$$\% \, CaCO_3 = \frac{(mL \times M \, EDTA) \times 10.01}{Sample \, wt \, (g)}$$

Insoluble in Dilute Hydrochloric Acid. Add 10 g to about 100 mL of water, swirl, slowly and carefully add 20 mL of hydrochloric acid, and dilute with water to 150 mL. Heat the solution to boiling, boil gently to expel the carbon dioxide, and digest in a covered beaker on a hot plate ($\approx$100 °C) for 1 h. Filter through a tared filtering crucible, wash thor-oughly, and dry at 105 °C.

Chloride. (Page 35). Dissolve 1.0 g in 10 mL of water plus 2 mL of nitric acid, filter if necessary through a small chloride-free filter, wash with hot water, and dilute with water to 20 mL.

Fluoride. Dissolve 1.0 g in 50 mL of water and 20 mL of 1 N hydrochloric acid in a 250-mL beaker. For the standards, add 1.0, 2.0, 3.0, 5.0, 10.0, and 15.0 mL of fluoride ion (F) standard solution to 250-mL glass beakers containing 50 mL of water and 5 mL of 1 N hydrochloric acid. Boil the sample and six standard solutions for a few seconds, cool rap-idly, and transfer to plastic beakers. To each solution, add 10 mL of 1 M sodium citrate reagent solution and 10 mL of 0.2 M (ethylenedinitrilo)tetraacetate, disodium salt, reagent solution, and mix. If necessary, adjust the pH to 5.5 ± 0.1 with dilute hydrochloric acid (1 + 9) or dilute sodium hydroxide (1 + 9). Transfer to 100-mL volumetric flasks, dilute to the mark with water, mix, and pour into plastic beakers. For each solution, measure imme-diately the potential of a fluoride electrode versus a reference electrode using a pH meter with an expanded scale or an appropriate potentiometer. Plot a 2-cycle semilogarithmic calibration curve with micrograms of fluoride per 100 mL of solution on the logarithmic scale. Determine the concentration of fluoride from the calibration curve.

Sulfate. Cautiously dissolve 1.0 g of sample with 10 mL of dilute hydrochloric acid (1 + 1) in a 100-mL beaker. Add 0.05 mL of bromine water, and evaporate to dryness. Dissolve the residue in 5 mL of water, and evaporate to dryness. Dissolve the residue in 15 mL of water, add 1 mL of dilute hydrochloric acid (1 + 19), and filter through a small, washed filter paper. Add two 3-mL portions of water to the same filter. For the control, take 0.10 mg of sulfate ion (SO_4) in 15 mL of water, and add 1 mL of dilute hydrochloric acid (1 + 19). Dilute both solutions to 25 mL, add 1 mL of 12% barium chloride reagent solution, and mix. Compare after 10 min. Sample turbidity should not exceed that of the control solution.

For the Determination of Ammonium, Heavy Metals, and Iron

Sample Solution A. In a 400-mL beaker, cautiously dissolve 20 g in 100 mL of dilute hydrochloric acid (1 + 1), and evaporate on a hot plate ($\approx 100\ ^{\circ}C$) to dryness or a moist residue. Dissolve the residue in about 100 mL of water, filter, and dilute with water to 200 mL (1 mL = 0.1 g).

Ammonium. Dilute 10 mL (1-g sample) of sample solution A with water to 80 mL, add 20 mL of 10% sodium hydroxide reagent solution, stopper, mix well, and allow to stand for 1 h. Decant 50 mL through a filtering crucible that has been washed with 10% sodium hydroxide reagent solution, and add to the filtrate 2 mL of Nessler reagent. Any color should not exceed that produced by 0.015 mg of ammonium ion (NH_4) in an equal volume of solution containing the quantities of reagents used in the test.

Heavy Metals. (Page 36, Method 1). Use 30 mL (3-g sample) of sample solution A for the sample solution, and use 10 mL (1-g sample) of sample solution A to prepare the control solution.

Iron. (Page 38, Method 1). Use 3.3 mL of sample solution A (0.33-g sample).

Barium, Magnesium, Potassium, Sodium, and Strontium. (By flame AAS, page 63).

Sample Stock Solution B. Cautiously dissolve 10.0 g of sample in 50 mL of nitric acid (1 + 3), transfer to a 100-mL volumetric flask, and dilute to the mark with water (1 mL = 0.10 g).

Sample Stock Solution C. Transfer 5.0 mL (0.50 g) of sample stock solution B to a 100-mL volumetric flask, and dilute to the mark with water (1 mL = 0.005 g).

Element	Wavelength (nm)	Sample Wt (g)	Standard Added (mg)	Flame Type*	Background Correction
Ba	553.6	2.0	0.10; 0.20	N/A	No
Mg	285.2	0.10	0.01; 0.02	A/A	Yes
K	766.5	0.50	0.05; 0.10	A/A	No
Na	589.0	0.01	0.01; 0.02	A/A	No
Sr	460.7	0.10	0.10; 0.20	N/A	No

*A/A is air/acetylene; N/A is nitrous oxide/acetylene.

Calcium Carbonate, Low in Alkalis

$CaCO_3$ **Formula Wt 100.09**   **CAS No. 471-34-1**

GENERAL DESCRIPTION

Typical appearance: white solid
Analytical use: determination of halogen in organic compounds
Change in state (approximate): melting point, 825 °C
Aqueous solubility: insoluble

SPECIFICATIONS

Assay (dried basis). . . . . ≥99.0% $CaCO_3$

Maximum Allowable

Insoluble in dilute hydrochloric acid. 0.01%
Chloride (Cl). 0.001%
Fluoride (F). 0.0015%
Sulfate (SO_4) 0.005%
Ammonium (NH_4) 0.003%
Heavy metals (as Pb). 0.001%
Iron (Fe). 0.002%
Barium (Ba) 0.01%
Magnesium (Mg) 0.01%
Potassium (K) 0.01%
Sodium (Na). 0.01%
Strontium (Sr) 0.1%

TESTS

Assay. (By complexometric titration of calcium). Weigh accurately about 0.4 g, previously dried at 300 °C for 4 h, and transfer to a 400-mL beaker. Cover the beaker with a watch glass. Add 2 mL of 20% hydrochloric acid from a pipet placed between the lip of the beaker and the watch glass. Swirl the beaker to aid dissolution. With water, wash down the inner wall of the beaker, the outer surface of the pipet, and the watch glass. Dilute to about 100 mL with water. While stirring, add, from a 50-mL buret, about 30 mL of 0.1 M EDTA volumetric solution. Adjust the solution to above pH 12 (using a pH meter) with 10% sodium hydroxide reagent solution. Add 300 mg of hydroxy naphthol blue indicator mixture and continue the titration immediately with 0.1 M EDTA to a blue color. One milliliter of 0.1 M EDTA corresponds to 0.01001 g of $CaCO_3$.

$$\% \ CaCO_3 = \frac{(mL \times M \ EDTA) \times 10.01}{Sample \ wt \ (g)}$$

Insoluble in Dilute Hydrochloric Acid. In a 400-mL beaker, add 20 g to about 200 mL of water, swirl, slowly and carefully add 40 mL of hydrochloric acid, and dilute with water to 300 mL. Heat the solution to boiling, boil gently to expel the carbon dioxide, and digest in a covered beaker on a hot plate (≈ 100 °C) for 1 h. Filter through a tared filtering crucible, wash thoroughly, and dry at 105 °C.

Chloride. (Page 35). Dissolve 1.0 g in 10 mL of water plus 2 mL of nitric acid, filter if necessary through a small chloride-free filter, wash with hot water, and dilute with water to 20 mL.

Fluoride. Dissolve 1.0 g in 50 mL of water and 25 mL of 1 N hydrochloric acid in a 250-mL beaker. For the standards, add 1.0, 2.0, 3.0, 5.0, 10.0, and 15.0 mL of standard fluoride solution to 250-mL glass beakers containing 50 mL of water and 5 mL of 1 N hydrochloric acid. Boil the sample and six standard solutions for a few seconds, cool rapidly, and transfer to plastic beakers. To each solution, add 20 mL of 1 M sodium citrate reagent solution and 10 mL of 0.2 M (ethylenedinitrilo)tetraacetate, disodium salt dihydrate, reagent solution, and mix. If necessary, adjust the pH to 5.5 ± 0.1 with dilute hydrochloric acid (1 + 9) or dilute sodium hydroxide (1 + 9). Transfer to 100-mL volumetric flasks, dilute to the mark with water, mix, and pour into plastic beakers. For each solution, measure immediately the potential of a fluoride electrode versus a reference electrode using a pH meter with an expanded scale or an appropriate potentiometer. Plot a 2-cycle semilogarithmic calibration curve with micrograms of fluoride per 100 mL of solution on the logarithmic scale. Determine the concentration of fluoride from the calibration curve.

Sulfate. Cautiously dissolve 2.0 g of sample with 15 mL of (1 + 1) hydrochloric acid in a 100-mL beaker. Add 0.05 mL of bromine water, and evaporate to dryness. Dissolve the residue in 5 mL of water, and evaporate to dryness. Dissolve the residue in 15 mL of water, add 1 mL of (1 + 19) hydrochloric acid, and filter through a small, washed filter paper. Add two 3-mL portions of water to the same filter. For the control, take 0.10 mg of sulfate ion (SO_4) in 15 mL of water, and add 1 mL of (1 + 19) hydrochloric acid. Dilute both solutions to 25 mL, add 1 mL of 12% barium chloride reagent solution, and mix. Compare after 10 min. Sample turbidity should not exceed that of the control solution.

For the Determination of Ammonium, Heavy Metals, and Iron

Sample Solution A. In a 400-mL beaker, cautiously dissolve 20 g in 100 mL of dilute hydrochloric acid (1 + 1), and evaporate on a hot plate (≈ 100 °C) to dryness or a moist residue. Dissolve the residue in about 100 mL of water, filter, and dilute with water to 200 mL in a volumetric flask (1 mL = 0.10 g).

Ammonium. Dilute 10 mL (1-g sample) of sample solution A with water to 80 mL, add 20 mL of 10% sodium hydroxide reagent solution, stopper, mix well, and allow to stand for 1 h. Decant 50 mL through a filtering crucible that has been washed with 10% sodium hydroxide reagent solution, and add to the filtrate 2 mL of Nessler reagent. Any color should not exceed that produced by 0.015 mg of ammonium ion (NH_4) in an equal volume of solution containing the quantities of reagents used in the test.

Heavy Metals. (Page 36, Method 1). Use 30 mL (3-g sample) of sample solution A to prepare the sample solution, and use 10 mL of sample solution A (1-g sample) to prepare the control solution.

Iron. (Page 38, Method 1). Use 5.0 mL of sample solution A (0.5-g sample).

Barium, Magnesium, Potassium, Sodium, and Strontium. (By flame AAS, page 63).

> **Sample Stock Solution B.** Cautiously dissolve 10.0 g of sample in 50 mL of nitric acid (1 + 3), transfer to a 100-mL volumetric flask, and dilute to the mark with water (1 mL = 0.10 g).
>
> **Sample Stock Solution C.** Transfer 5.0 mL (0.50 g) of sample stock solution B to a 100-mL volumetric flask, and dilute to the mark with water (1 mL = 0.005 g).

Element	Wavelength (nm)	Sample Wt (g)	Standard Added (mg)	Flame Type*	Background Correction
Ba	553.6	2.0	0.10; 0.20	N/A	No
Mg	285.2	0.10	0.01; 0.02	A/A	Yes
K	766.5	0.50	0.05; 0.10	A/A	No
Na	589.0	0.10	0.01; 0.02	A/A	No
Sr	460.7	0.10	0.10; 0.20	N/A	No

*A/A is air/acetylene; N/A is nitrous oxide/acetylene.

Calcium Carbonate, Chelometric Standard

$CaCO_3$ Formula Wt 100.09   CAS No. 471-34-1

GENERAL DESCRIPTION

Typical appearance: white solid
Analytical use: chelometric standard
Change in state (approximate): melting point, 825 °C
Aqueous solubility: insoluble

SPECIFICATIONS

Assay (dried basis). . . . . 99.95–100.05% $CaCO_3$

Maximum Allowable

Insoluble in dilute hydrochloric acid. 0.01%
Chloride (Cl). 0.001%
Fluoride (F). 0.0015%
Sulfate (SO_4) 0.005%
Ammonium (NH_4) 0.003%
Barium (Ba) 0.01%
Heavy metals (as Pb). 0.001%
Iron (Fe). 0.002%
Magnesium (Mg) 0.01%

Potassium (K) 0.01%
Sodium (Na) 0.01%
Strontium (Sr) 0.1%

TESTS

Assay. (By complexometric titration for calcium). This method is a comparative proce-dure in which the calcium carbonate to be assayed is compared with NIST Standard Refer-ence Material Calcium Carbonate (SRM 915). A minimum of two samples and two stand-ards should be run according to this procedure.

Procedure. Dry the sample and NIST standard at 110 °C for 2 h. Weigh (to the nearest 0.1 mg) 0.4000 ± 0.0040 g of calcium carbonate, and place in a 400-mL beaker with a mag-netic stirring bar and 50 mL of water. Cover with a watch glass, and add 12 mL of 10% hydrochloric acid with the pipet tip under the watch glass. Wash down the watch glass and the sides of the beaker. All of the sample must be dissolved.

Weigh (to the nearest 0.1 mg) 1.4400 ± 0.0100 g of 0.02 M EDTA, disodium salt dihy-drate (described below) (see Note 1). Add the EDTA directly to the calcium solution, dilute to about 125 mL, and add 15 mL of 5 M potassium hydroxide (described below). This addition should dissolve all the EDTA and leave a clear solution. Add 0.05–0.10 mL of 0.02% calcein indicator solution (described below), and titrate the excess calcium with 0.0200 M EDTA under UV illumination in a dark room or box (see Note 2). The end point is extremely sharp from a bright yellow-green fluorescence to black.

$$
\% \, CaCO_3 = \frac{\left[\dfrac{\left[Wt \, EDTA \, (g) \right] + (0.37226 \times mL \times M)}{Wt \, sample \, CaCO_3 \, (g)} \right]}{\left[\dfrac{\left[Wt \, EDTA \, (g) \right] + (0.37226 \times mL \times M)}{\left[Wt \, NIST \, CaCO_3 \, (g) \right] \times NIST \, assay \, / 100} \right]}
$$

Note 1. This method is independent of the quality of the disodium EDTA used, but it is recommended that reagent-quality material be selected, then crushed, and mixed thoroughly before use. Do not dry the EDTA, because water loss will change the proportions needed for the test. A disodium EDTA dihy-drate having an assay of 99.9–100.1% is commercially available. ACS Reagent is 99–101% and suitable for this test.

Note 2. The UV lamp should have dark blue glass (little visible light) and should be placed at a right angle to the viewer, shielded from the viewer's eyes.

Calcein Indicator Solution, 0.02%. Dissolve 20 mg of calcein in 10 mL of water by addition of a few drops of 5 M potassium hydroxide (described below), and dilute with water to 100 mL. This solution is not stable at room temperature for more than 1 day. Frozen solutions are stable, so the indicator solution may be split into portions in plastic bottles and frozen, and a fresh bottle may be thawed as needed.

EDTA, Disodium Salt Dihydrate, 0.02 M. Dissolve 7.446 g of reagent-quality (ethylenedinitrilo)tetraacetic acid disodium salt dihydrate in water, and dilute with water to 1 L. Standardize against a calcium solution prepared by dissolving 2.002 g (weighed to the nearest 0.1 mg) of NIST calcium carbonate in water by addition of 6 mL of concentrated hydrochloric acid, followed by dilution to exactly 1 L. The molarity of this solution should be calculated from the weight taken. The standardization of a 25-mL aliquot of the calcium solution will require 10 mL of 5 M potassium hydroxide (described below) and 0.05–0.10 mL of calcein indicator solution, as in the procedure cited.

Potassium Hydroxide, 5 M. Dissolve 56 g of potassium hydroxide in water, and dilute with water to 200 mL.

Insoluble in dilute hydrochloric acid and other tests except assay are the same as for calcium carbonate, low in alkalis, page 232.

Calcium Chloride Desiccant
CaCl$_2$ Formula Wt 110.98   CAS No. 10043-52-4

GENERAL DESCRIPTION
Typical appearance: hygroscopic solid
Analytical use: desiccant
Change in state (approximate): melting point, 772 °C
Aqueous solubility: freely soluble

SPECIFICATIONS
Assay . . . . ≥96.0% CaCl$_2$
Maximum Allowable
Titrable base . . . . 0.006 meq/g

TESTS

Assay. (By complexometric titration of calcium). Weigh accurately about 2.5 g in a 250-mL beaker. Dissolve in 100 mL of water, and add 5 mL of 10% hydrochloric acid. Transfer to a 250-mL volumetric flask, dilute to the mark with water, and mix. Pipet 50.0 mL of the sample solution to a 400-mL beaker, add 75 mL of water, and add while stirring 20 mL of 0.1 M EDTA from a 50-mL buret. Adjust the pH using a pH meter to above 12 with 10% sodium hydroxide reagent solution. Add 300 mg of hydroxy naphthol blue indicator mixture and continue the titration immediately with 0.1 M EDTA volumetric solution to a blue color. One milliliter of 0.1 M EDTA corresponds to 0.0111 g of CaCl$_2$.

$$\% \ CaCl_2 = \frac{(mL \times M \ EDTA) \times 11.1}{Sample \ wt \ (g) \ / \ 5}$$

Titrable Base. Dissolve 5.0 g in 50 mL of water, and add 0.10 mL of phenolphthalein indicator solution. If any pink color is produced, it should be discharged by not more than 3.0 mL of 0.01 N hydrochloric acid.

Calcium Chloride Dihydrate

$CaCl_2 \cdot 2H_2O$ **Formula Wt 147.01**   **CAS No. 10035-04-8**

GENERAL DESCRIPTION

Typical appearance: colorless or white hygroscopic solid

Analytical use: preparation of calcium standard solutions; electrolyte

Aqueous solubility: 326 g in 100 mL at 60 °C

SPECIFICATIONS

Assay . . . . 99.0–105.0% $CaCl_2 \cdot 2H_2O$

pH of a 5% solution at 25.0 °C 4.5–8.5

	Maximum Allowable
Insoluble matter	0.01%
Oxidizing substances (as NO_3)	0.003%
Sulfate (SO_4)	0.01%
Ammonium (NH_4)	0.005%
Barium (Ba)	0.005%
Heavy metals (as Pb)	5 ppm
Iron (Fe).	0.001%
Magnesium (Mg).	0.005%
Potassium (K)	0.01%
Sodium (Na)	0.02%
Strontium (Sr).	0.1%

TESTS

Assay. (By complexometric titration of calcium). Weigh accurately about 2.5 g in a 250-mL beaker. Dissolve in 100 mL of water, and add 5 mL of 10% hydrochloric acid. Transfer to a 250-mL volumetric flask, dilute to the mark with water, and mix. Pipet 50.0 mL of the sample solution to a 400-mL beaker, add 75 mL of water, and while stirring add 20 mL of 0.1 M EDTA from a 50-mL buret. Adjust the pH using a pH meter to above 12 with 10% sodium hydroxide reagent solution. Add 300 mg of hydroxy naphthol blue indicator mixture, and continue the titration immediately with 0.1 M EDTA volumetric solution to a blue color. One milliliter of 0.1 M EDTA corresponds to 0.014701 g of $CaCl_2 \cdot 2H_2O$.

$$\% \; CaCl_2 \cdot 2H_2O = \frac{(mL \times M \; EDTA) \times 14.701}{Sample \; wt \; (g) \, / \, 5}$$

pH of a 5% Solution at 25.0 °C. (Page 49).

Insoluble Matter. (Page 25). Dissolve 20 g in 200 mL of water.

Oxidizing Substances. Place 0.20 g in a beaker. Cool the beaker thoroughly in an ice bath, and add 22 mL of sulfuric acid that has been cooled to ice-bath temperature. Allow the mixture to warm to room temperature, and swirl the beaker at intervals to effect gentle dissolution with slow evolution of hydrogen chloride. When dissolution is complete, add 3 mL of diphenylamine reagent solution and digest on a hot plate ($\approx$100 °C) for 90 min. Prepare a standard by evaporating to dryness a solution containing 0.006 mg of nitrate (0.6 mL of the nitrate standard solution) and 0.01 g of sodium carbonate. Treat the residue exactly like the sample. Any color produced in the solution of the sample should not exceed that in the standard.

Sulfate. Dissolve 1.0 g of sample in 20 mL of water, and add 1.0 mL of 10% hydrochloric acid. Filter through a small, washed filter paper. Add two 3-mL portions of water through the filter paper. For the control, take 0.10 mg of sulfate ion (SO_4) in 25 mL of water, and add 1.0 mL of 10% hydrochloric acid. Dilute both solutions to 35 mL with water, and add 1 mL of 12% barium chloride solution. Compare turbidity after 10 min. Sample turbidity should not exceed that of the control solution.

For the Determination of Ammonium, Barium, Heavy Metals, and Iron

Sample Solution A. Dissolve 50 g in about 200 mL of water, filter if necessary, and dilute with water to 250 mL in a volumetric flask (1 mL = 0.2 g).

Ammonium. Dilute 10 mL (1-g sample) of sample solution A with water to 80 mL, add 20 mL of 10% sodium hydroxide reagent solution, stopper, mix well, and allow to stand for 1 h. Decant 50 mL through a filtering crucible that has been washed with 10% sodium hydroxide reagent solution and add to the filtrate 2 mL of Nessler reagent. Any color should not exceed that produced by 0.015 mg of ammonium ion (NH_4) in an equal volume of solution containing the quantities of reagents used in the test.

Barium. For the sample, add 2 g of sodium acetate and 0.05 mL of glacial acetic acid to 15 mL (3-g sample) of sample solution A. For the control, add 2 g of sodium acetate, 0.05 mL of glacial acetic acid, and 0.1 mg of barium ion to 5 mL (1-g sample) of sample solution A, and dilute with water to 15 mL. To each solution, add 2 mL of 10% potassium dichromate reagent solution, and allow to stand for 15 min. Any turbidity in the solution of the sample should not exceed that in the control.

Heavy Metals. (Page 36, Method 1). Use 30 mL (6-g sample) of sample solution A to prepare the sample solution, and use 10 mL (2-g sample) of sample solution A to prepare the control solution.

Iron. (Page 38, Method 1). Use 5.0 mL (1-g sample) of sample solution A.

Magnesium, Potassium, Sodium, and Strontium. (By flame AAS, page 63).

Sample Stock Solution. Dissolve 2.0 g of sample in 50 mL of water and 5 mL of nitric acid in a 200-mL volumetric flask, and dilute to the mark with water (1 mL = 0.01 g).

Element	Wavelength (nm)	Sample Wt (g)	Standard Added (mg)	Flame Type*	Background Correction
Mg	285.2	0.20	0.01; 0.02	A/A	Yes
K	766.5	0.20	0.02; 0.04	A/A	No
Na	589.0	0.05	0.01; 0.02	A/A	No
Sr	460.7	0.05	0.05; 0.10	N/A	No

*A/A is air/acetylene; N/A is nitrous oxide/acetylene.

Calcium Hydroxide

Ca(OH)$_2$ **Formula Wt 74.09**   **CAS No. 1305-62-0**

GENERAL DESCRIPTION

Typical appearance: white solid
Analytical use: absorbant for carbon dioxide
Aqueous solubility: slightly soluble

SPECIFICATIONS

Assay . . . . ≥95.0% Ca(OH)$_2$ and ≤3.0% CaCO$_3$

Maximum Allowable

Insoluble in hydrochloric acid. 0.03%
Chloride (Cl) 0.03%
Sulfur compounds (as SO$_4$) 0.1%
Heavy metals (as Pb) 0.003%
Iron (Fe). 0.05%
Magnesium (Mg). 0.5%
Potassium (K) 0.05%
Sodium (Na) 0.05%
Strontium (Sr). 0.05%

TESTS

Assay. (By acid–base titrimetry). Weigh accurately about 1.4 g of sample, mix with 100 mL of carbon dioxide-free water, and add 0.15 mL of phenolphthalein indicator solution. Titrate with 1 N hydrochloric acid volumetric solution to the disappearance of the pink color (end point = *A* mL). (The sample is initially suspended in the solution, but dissolves as the acid is added.) Add 0.2 mL of methyl orange indicator solution, and continue the titration to a pinkish-orange color (end point = *B* mL). One milliliter of 1 N hydrochloric acid corresponds to 0.03705 g of Ca(OH)$_2$ and 0.05005 g of CaCO$_3$.

$$\% \text{ Ca(OH)}_2 = \frac{(A \text{ mL} \times \text{N HCl}) \times 3.705}{\text{Sample wt (g)}}$$

$$\% \text{ CaCO}_3 = \frac{[(B - A) \text{ mL} \times \text{N HCl}] \times 5.005}{\text{Sample wt (g)}}$$

Insoluble in Hydrochloric Acid. Add 5.0 g to 100 mL of water, add 25 mL of hydrochloric acid, heat the solution to boiling, and digest in a covered beaker on a hot plate ($\approx$100 °C) for 1 h. Filter through a tared filtering crucible, wash thoroughly, and dry at 105 °C.

Chloride. (Page 35). Dissolve 0.17 g in 10 mL of water plus 1 mL of nitric acid, dilute with water to 50 mL, and dilute 10 mL of this solution with water to 20 mL.

Sulfur Compounds. Mix 0.34 g with 20 mL of water, add 2 mL of bromine water, and digest in a dish on a hot plate ($\approx$100 °C) for 10 min. Add 3 mL of hydrochloric acid, boil to expel the excess bromine, and evaporate to dryness. Dissolve the residue in 10 mL of 10%

hydrochloric acid reagent solution, and dilute to 100 mL with water. Filter if necessary. Dilute 5 mL (0.05-g sample) of this solution to 10 mL, and add 1 mL of 12% barium chloride reagent solution. Any turbidity should not exceed that in a standard prepared as follows: Add 2 mL of bromine water to 20 mL of water, and digest on a hot plate ($\approx$100 °C) for 10 min. Add 3 mL of hydrochloric acid, boil to expel the excess bromine, and evaporate to dryness. Dissolve the residue in 10 mL of 10% hydrochloric acid, and dilute to 100 mL with water. Filter if necessary. To 5 mL of 10% hydrochloric acid, add 0.05 mg of sulfate ion (SO_4), dilute to 10 mL, and add 1 mL of 12% barium chloride reagent solution.

For the Determination of Heavy Metals and Iron

Sample Solution A. To 3.0 g, add 40 mL of water, mix, cautiously add 10 mL of hydrochloric acid and 3 mL of nitric acid, and evaporate on a hot plate ($\approx$100 °C) to dryness. Take up the residue with 1 mL of 10% hydrochloric acid reagent solution and 30 mL of hot water, filter, and wash with a few mL of water. Add 0.10 mL of phenolphthalein indicator solution and sufficient 1 N sodium hydroxide volumetric solution to produce a pink color, then add sufficient 1 N hydrochloric acid to discharge the pink color, and dilute with water to 60 mL (10 mL = 0.5 g).

Heavy Metals. (Page 36, Method 1). Use 20 mL (1-g sample) of sample solution A to prepare the sample solution, and use 6.7 mL (0.33-g sample) of sample solution A to prepare the control solution.

Iron. (Page 38, Method 1). Dilute 4.0 mL of sample solution A (0.2-g sample) with water to 100 mL, and use 10 mL of this dilution.

Magnesium, Potassium, Sodium, and Strontium. (By flame AAS, page 63).

Stock Solution B. Dissolve carefully with stirring 2.0 g of sample in 30 mL of water containing 5 mL of nitric acid. Cool to room temperature. Transfer to a 100-mL volumetric flask, and dilute to the mark with water (1 mL = 0.02 g).

Sample Stock Solution C. Transfer 5.0 mL of stock solution B to a 100-mL volumetric flask, and dilute to the mark with water (1 mL = 0.001 g).

Element	Wavelength (nm)	Sample Wt (g)	Standard Added (mg)	Flame Type*	Background Correction
Mg	285.2	0.002	0.01; 0.02	A/A	Yes
K	766.5	0.02	0.01; 0.02	A/A	No
Na	589.0	0.02	0.01; 0.02	A/A	No
Sr	460.7	0.20	0.10; 0.20	N/A	No

*A/A is air/acetylene; N/A is nitrous oxide/acetylene.

Calcium Nitrate Tetrahydrate

$Ca(NO_3)_2 \cdot 4H_2O$ Formula Wt 236.15   CAS No. 13477-34-4

GENERAL DESCRIPTION

Typical appearance: colorless, hygroscopic solid
Analytical use: preparation of calcium standard solutions
Change in state (approximate): melting point, 42 °C
Aqueous solubility: 66 g in 100 mL at 30 °C

SPECIFICATIONS

Assay . . . . 99.0–103.0% $Ca(NO_3)_2 \cdot 4H_2O$
pH of a 5% solution at 25.0 °C 5.0–7.0

Maximum Allowable

Insoluble matter 0.005%
Chloride (Cl) 0.005%
Nitrite (NO_2) 0.001%
Sulfate (SO_4) 0.002%
Barium (Ba) 0.005%
Heavy metals (as Pb) . . . . 5 ppm
Iron (Fe). . . . . 5 ppm
Magnesium (Mg). 0.05%
Potassium (K) 0.005%
Sodium (Na) 0.01%
Strontium (Sr). 0.05%

TESTS

Assay. (By complexometric titration of calcium). Weigh accurately about 0.90 g, transfer to a 250-mL beaker, and dissolve in 150 mL of water. While stirring, add from a 50-mL buret about 30 mL of 0.1 M EDTA volumetric solution. Then adjust the solution to above pH 12 with 10% sodium hydroxide reagent solution. Add about 300 mg of hydroxy naphthol blue indicator mixture, and continue the titration immediately with 0.1 M EDTA to a blue color. One milliliter of 0.1 M EDTA corresponds to 0.02362 g of $Ca(NO_3)_2 \cdot 4H_2O$.

$$\% \ Ca(NO_3)_2 \cdot 4H_2O = \frac{(mL \times M \ EDTA) \times 23.62}{Sample \ wt \ (g)}$$

pH of a 5% Solution at 25.0 °C. (Page 49).

Insoluble Matter. (Page 25). Dissolve 20 g in 200 mL of water.

Chloride. (Page 35). Dissolve 1.0 g in 50 mL of water, and dilute 10 mL of the solution with water to 20 mL.

Nitrite. (Page 55). Use 1.0 g of sample and 0.01 mg of nitrite.

Sulfate. (Page 41, Method 2). Use 6 mL of hydrochloric acid.

Barium. For the sample, dissolve 2.0 g in 15 mL of water, add 2 g of sodium acetate and 0.05 mL of glacial acetic acid, and filter if necessary. For the standard, add 0.1 mg of barium to 15 mL of water, then add 2 g of sodium acetate and 0.05 mL of glacial acetic acid. Add to each solution 2 mL of 10% potassium dichromate reagent solution, mix, and let stand for 15 min. Any turbidity in the solution of the sample should not exceed that in the standard.

Heavy Metals. (Page 36, Method 1). Dissolve 6.0 g in about 20 mL of water, and dilute with water to 30 mL. Use 25 mL to prepare the sample solution, and use the remaining 5.0 mL to prepare the control solution.

Iron. (Page 38, Method 1). Use 2.0 g.

Magnesium, Potassium, Sodium, and Strontium. (By flame AAS, page 63).

> **Sample Stock Solution.** Dissolve 4.0 g of sample in 80 mL of water. Transfer to a 100-mL volumetric flask, and dilute to the mark with water (1 mL = 0.04 g).

Element	Wavelength (nm)	Sample Wt (g)	Standard Added (mg)	Flame Type*	Background Correction
Mg	285.2	0.02	0.01; 0.02	A/A	Yes
K	766.5	0.20	0.01; 0.02	A/A	No
Na	589.0	0.10	0.01; 0.02	A/A	No
Sr	460.7	0.20	0.10; 0.20	N/A	No

*A/A is air/acetylene; N/A is nitrous oxide/acetylene.

Calcium Sulfate Dihydrate

$CaSO_4 \cdot 2H_2O$ Formula Wt 172.17   CAS No. 10101-41-4

GENERAL DESCRIPTION

Typical appearance: white solid
Analytical use: determination of oxalates
Change in state (approximate): dehydrates partially at 70 °C and completely at 160 °C
Aqueous solubility: 0.24 g in 100 mL at 20 °C

SPECIFICATIONS

Assay . . . . 98.0–102.0% $CaSO_4 \cdot 2H_2O$

Maximum Allowable

Insoluble in dilute hydrochloric acid. 0.02%
Chloride (Cl). 0.005%
Nitrate (NO_3) . . . . Passes test

Carbonate (CO_3). . . . . Passes test
Heavy metals (as Pb) 0.002%
Iron (Fe). 0.001%
Magnesium (Mg). 0.02%
Potassium (K) 0.005%
Sodium (Na) 0.02%
Strontium (Sr). 0.05%

TESTS

Assay. (By complexometric titration for calcium). Dissolve about 0.60 g, accurately weighed, in 100 mL of water and 5 mL of hydrochloric acid; boil if necessary to effect dissolution. Cool and stir, preferably with a magnetic stirrer, while adding the following reagents in the cited order: 0.5 mL of 2,2′,2″-nitrilotriethanol (i.e., triethanolamine), 300 mg of hydroxy naphthol blue indicator mixture, and (from a 50-mL buret) about 30 mL of 0.1 M EDTA volumetric solution. Now add 10% sodium hydroxide reagent solution until the initial red color changes to clear blue; then continue the addition dropwise until the pH is above 12. Continue the titration immediately with 0.1 M EDTA to a blue color. One milliliter of 0.1 M EDTA corresponds to 0.017217 g of $CaSO_4 \cdot 2H_2O$.

$$\% \ CaSO_4 \cdot 2H_2O = \frac{(mL \times M \ EDTA) \times 17.217}{Sample \ wt \ (g)}$$

Insoluble in Dilute Hydrochloric Acid. In a 400-mL beaker, dissolve 4.0 g in a mixture of 100 mL of water and 20 mL of hydrochloric acid, heating on a hot plate ($\approx$100 °C) if necessary until dissolution is complete. Filter through a tared filtering crucible, wash thoroughly with hot 10% hydrochloric acid, followed by water, and dry at 105 °C.

Chloride. (Page 35). Dissolve 0.20 g in a mixture of 10 mL of water and 7 mL of nitric acid, warming if necessary to complete dissolution. Filter, if necessary, through a chloride-free filter, and dilute with water to 20 mL.

Nitrate. Mix 1.0 g with 9 mL of water and 1 mL of 0.5% sodium chloride solution. Add 0.10 mL of indigo carmine reagent solution, mix, and add 10 mL of sulfuric acid. The blue color should not be completely discharged in 10 min. (Limit about 0.003%)

Carbonate. Mix 1.0 g with 5 mL of water, and slowly add 2 mL of hydrochloric acid. The evolution of a gas should not be observed.

Heavy Metals. (Page 36, Method 1). Add 2.0 g to a mixture of 25 mL of water and 5 mL of hydrochloric acid, heat the solution to boiling, and nearly neutralize with ammonium hydroxide. Filter, wash with water to about 40 mL, complete the neutralization with 10% ammonium hydroxide reagent solution, and dilute with water to 50 mL. Use 25 mL for the sample solution. To prepare the standard, use 25 mL of the following solution: add 0.04 mg of lead to 5 mL of hydrochloric acid, neutralize with ammonium hydroxide, and dilute with water to 50 mL.

Iron. (Page 38, Method 1). Boil 1.0 g with 6 mL of hydrochloric acid and 40 mL of water, cool, filter, and wash with water to 50 mL. Test the solution without further acidification.

Magnesium, Potassium, Sodium, and Strontium. (By flame AAS, page 63).

Sample Stock Solution. Dissolve 2.0 g in a mixture of 50 mL of water and 10 mL of hydrochloric acid by heating on a hot plate until dissolution is complete. Cool to room temperature. Transfer to a 100-mL volumetric flask, and dilute to the mark with water (1 mL = 0.02 g).

Element	Wavelength (nm)	Sample Wt (g)	Standard Added (mg)	Flame Type*	Background Correction
Mg	285.2	0.05	0.01; 0.02	A/A	Yes
K	766.5	0.20	0.01; 0.02	A/A	No
Na	589.0	0.05	0.01; 0.02	A/A	No
Sr	460.7	0.20	0.10; 0.20	N/A	No

*A/A is air/acetylene; N/A is nitrous oxide/acetylene.

Calmagite
3-Hydroxy-4-[(2-hydroxy-5-methyl-phenyl)azo]-1-naphthalenesulfonic Acid

$C_{17}H_{14}N_2O_5S$	**Formula Wt 358.37**	**CAS No. 3147-14-6**

Note: This reagent contains varying amounts of salt. It is often sold as a solid indicator adsorbed on NaCl.

GENERAL DESCRIPTION

Typical appearance: red crystals from acetone
Analytical use: indicator in titration of calcium or magnesium with EDTA
Aqueous solubility: soluble

SPECIFICATIONS

Clarity of solution . . . . Passes test
Suitability for complexometric titration . . . . Passes test

TESTS

Clarity of Solution. Dissolve 0.10 g in 100 mL of water. The solution should be complete and free of turbidity.

Suitability for Complexometric Titration. To 25 mL of water, add 2 mL of pH 10 ammoniacal buffer solution, and add 0.50 mL of the solution from the clarity of solution test.

A blue color should form. Add 0.10 mg of magnesium. The color should change to red. Add 0.1 mL of 0.1 M EDTA volumetric solution. The color should change back to blue.

Carbon Disulfide

CS_2 **Formula Wt 76.13**   **CAS No. 75-15-0**

Note: Carbon disulfide should be supplied and stored in amber glass containers and protected from direct sunlight.

GENERAL DESCRIPTION

Typical appearance: clear, colorless liquid

Analytical use: solvent

Change in state (approximate): boiling point, 46 °C

Aqueous solubility: 0.3 g in 100 mL at 20 °C

Density: 1.26

SPECIFICATIONS

Assay . . . . ≥99.9% CS_2

Maximum Allowable

Color (APHA) 10

Residue after evaporation 0.002%

Hydrogen sulfide (H_2S) . . . . Passes test

Sulfur dioxide (SO_2) . . . . Passes test

Water (H_2O) 0.05%

TESTS

Assay. Analyze the sample by gas chromatography using the general parameters cited on page 80. The following specific conditions are also required.

Column: Type I, methyl silicone

Measure the area under all peaks, and calculate the carbon disulfide content in area percent. Correct for water content.

Color (APHA). (Page 43).

Residue after Evaporation. (Page 25). In a well-ventilated fume hood, evaporate 50 g (40 mL) to dryness in a tared, preconditioned dish at 50–60 °C, and heat at 60 °C for 1 h.

Hydrogen Sulfide and Sulfur Dioxide. In a well-ventilated fume hood, shake vigorously 25 g (20 mL) with 0.2 mL of 0.01 N iodine for 15 s in a glass-stoppered cylinder. The pink color should not disappear. (Limit for hydrogen sulfide is about 1.5 ppm; limit for sulfur dioxide is about 2.5 ppm.)

Water. (Page 31, Method 2). Use 20 µL (25 mg) of the sample.

Ceric Ammonium Nitrate

Ammonium Hexanitratocerate(IV)

$(NH_4)_2Ce(NO_3)_6$ Formula Wt 548.22   CAS No. 16774-21-3

GENERAL DESCRIPTION

Typical appearance: orange-red or orange-yellow solid

Analytical use: oxidimetric standard

Aqueous solubility: 141 g in 100 mL

SPECIFICATIONS

Assay . . . . ≥98.5% $(NH_4)_2Ce(NO_3)_6$

Maximum Allowable

Insoluble in dilute sulfuric acid. 0.05%

Chloride (Cl). 0.01%

Phosphate (PO_4). 0.02%

Iron (Fe). 0.005%

TESTS

Assay. (By titration of oxidative capacity of Ce^{IV}). Weigh accurately 2.4–2.5 g, and dissolve in 50 mL of water. From a pipet, add 50 mL of 0.1 N ferrous ammonium sulfate volumetric solution, and swirl until the precipitate that forms is redissolved. Add 10 mL of phosphoric acid and 0.10 mL of diphenylaminesulfonic acid, sodium salt, indicator solution (described below). Titrate at once with 0.1 N potassium dichromate volumetric solution to a change from faint green to violet. Record this titration volume as *A* mL. Pipet 25 mL of 0.1 N ferrous ammonium sulfate volumetric solution into 50 mL of water, and treat it in the same way (with phosphoric acid and indicator but without sample). Titrate to a change from green to gray-blue, and record this titration volume as *B* mL. One milliliter of 0.1 N ferrous ammonium sulfate corresponds to 0.05482 g of $(NH_4)_2Ce(NO_3)_6$.

$$\% \ (NH_4)_2Ce(NO_3)_6 = \frac{[(2B - A \text{ mL}) \times N \ K_2Cr_2O_7] \times 54.82}{\text{Sample wt (g)}}$$

> *Diphenylaminesulfonic Acid, Sodium Salt, Indicator Solution.* Dissolve 0.10 g of the salt in 100 mL of water.

Insoluble in Dilute Sulfuric Acid. To 5.0 g, add 10 mL of sulfuric acid, stir, and then cautiously add 90 mL of water to dissolve. Heat to boiling, and digest in a preconditioned covered beaker on a hot plate (≈100 °C) for 1 h. Filter through a tared filtering crucible, wash thoroughly, and dry at 105 °C.

Chloride. (Page 35). Use 0.10 g dissolved in 10 mL of water. The comparison is best made by the general method for chloride in colored solutions.

Phosphate. (Page 40, Method 2). Dissolve 0.25 g in 30 mL of dilute sulfuric acid (1 + 9), add hydrogen peroxide until the solution just turns colorless, then boil to destroy excess peroxide. Cool, and dilute with water to 50 mL. Dilute 10 mL with water to 60 mL, and adjust the pH between 2 and 3 (using pH paper) with ammonium hydroxide.

> *Note:* This neutralization must be made with care to avoid formation of a permanent precipitate that would void the test. If this happens, discard the solution and start with another 10 mL.

Add 0.5 g of ammonium molybdate, and adjust the pH to 1.8 (using a pH meter) with dilute hydrochloric acid (1 + 9). Heat the solution to boiling and cool to room temperature. Dilute with water to 90 mL, add 10 mL of hydrochloric acid, and continue as described. Carry along a standard containing 0.01 mg of phosphate ion (PO_4) in 10 mL of dilute sulfuric acid (3 + 47) treated exactly as the 10 mL of sample.

Iron. (Page 38, Method 2). Dissolve 1.0 g in 30 mL of dilute sulfuric acid (1 + 9), and add, dropwise, 3% hydrogen peroxide solution until the yellow color disappears. For the standard, add 0.05 mg of iron ion (Fe) to 30 mL of dilute sulfuric acid (1 + 9) and the same volume of 3% hydrogen peroxide solution used for the sample. To each solution, add ammonium hydroxide until the pH is between 1 and 3, cool to room temperature, and adjust the pH to 3.5 (using a pH meter). Dilute each solution with water to 50 mL, and use 10 mL of each.

Ceric Ammonium Sulfate Dihydrate
Ammonium Tetrasulfatocerate Dihydrate
$(NH_4)_4Ce(SO_4)_4 \cdot 2H_2O$ **Formula Wt 632.58**   **CAS No. 10378-47-9**

GENERAL DESCRIPTION

Typical appearance: orange-yellow solid
Analytical use: preparation of cerium IV oxidimetric solutions

SPECIFICATIONS

Assay . . . . ≥94% $(NH_4)_4Ce(SO_4)_4 \cdot 2H_2O$

Maximum Allowable

Insoluble in dilute sulfuric acid 0.05%
Phosphate (PO_4) 0.03%
Iron (Fe) 0.01%

TESTS

Assay. (By titration of the oxidative capacity of cerium IV). Weigh accurately 2.4–2.5 g, and place in a beaker. Cover with 10 mL of water, and add cautiously 5 mL of sulfuric acid. Stir thoroughly, and add water to complete dissolution. Dilute to 75–100 mL with water, add 0.05 mL of 0.025 M ferroin indicator solution, and titrate with 0.1 N ferrous ammonium sulfate to the red end point. One milliliter of 0.1 N ferrous ammonium sulfate corresponds to 0.06326 g of $(NH_4)_4Ce(SO_4)_4 \cdot 2H_2O$.

$$\% \ (NH_4)_4Ce(SO_4)_4 \cdot 2H_2O = \frac{[mL \times N \ Fe(NH_4)_2(SO_4)_2] \times 63.26}{\text{Sample wt (g)}}$$

Insoluble in Dilute Sulfuric Acid. To 2.5 g, add 5 mL of sulfuric acid, stir, and then cautiously add 90 mL of water to dissolve. Heat to boiling, and digest on a hot plate ($\approx$100 °C) for 1 h. Filter through a tared filtering crucible, and wash thoroughly, first with 2% sulfuric acid, then with water. Dry at 105 °C.

Phosphate. (Page 40, Method 2). Dissolve 0.75 g by adding 1 mL of sulfuric acid and stirring in 30 mL of water. Add hydrogen peroxide dropwise until just colorless, boil to destroy the excess peroxide, cool, and dilute to 50 mL. Dilute a 2-mL aliquot to 60 mL, and adjust the pH to between 2 and 3 (using pH paper) with ammonium hydroxide.

> *Note:* This neutralization must avoid the formation of a precipitate, which would void the test.

Add 0.5 g of ammonium molybdate, and adjust the pH to 1.8 (using a pH meter) with 10% hydrochloric acid. Heat to boiling and cool to room temperature. Dilute with water to 90 mL, add 10 mL of hydrochloric acid, and continue as described. Carry along a standard containing 0.01 mg of phosphate ion (PO_4) in 2.0 mL of sulfuric acid (1 + 29) treated as the 2.0-mL aliquot of the sample.

Iron. (Page 38, Method 2). Dissolve 1 g by adding 2 mL of sulfuric acid and stirring in 30 mL of water. Add dropwise 3% hydrogen peroxide until the color disappears. For the standard, add 0.1 mg of iron to 30 mL of water containing 2 mL of sulfuric acid and the same volume of hydrogen peroxide used in the test. Dilute the standard and sample each to 100 mL, and use 10-mL aliquots.

Chloramine-T Trihydrate
N-Chloro-4-toluenesulfonamide, Sodium Salt, Trihydrate

$H_3CC_6H_4SO_2NClNa \cdot 3H_2O$ **Formula Wt 281.69**   **CAS No. 127-65-1**

GENERAL DESCRIPTION
Typical appearance: white solid
Analytical use: detection of bromate and halogens
Change in state (approximate): loses water on drying
Aqueous solubility: 14 g in 100 mL

SPECIFICATIONS
Assay . . . . 98.0–103.0% $C_7H_7ClNNaO_2S \cdot 3H_2O$
pH of a 5% solution at 25.0 °C 8.0–10.0
Suitability for determination of bromide . . . . Passes test
Clarity of aqueous solution . . . . Passes test

Maximum Allowable

Insoluble in alcohol. 1.5%

TESTS

Assay. (By iodometric titration). Weigh accurately about 0.5 g, transfer to a glass stoppered Erlenmeyer flask, and dissolve in 100 mL of water. Add 5 mL of acetic acid and 2 g of potassium iodide, and allow to stand in the dark for 10 min. Titrate the liberated iodine with 0.1 N sodium thiosulfate volumetric solution, using starch indicator solution. One milliliter of 0.1 N sodium thiosulfate corresponds to 0.01408 g of $C_7H_7ClNNaO_2S \cdot 3H_2O$.

$$\% \ C_7H_7ClNNaO_2S \cdot 3H_2O = \frac{(mL \times N \ Na_2S_2O_3) \times 14.08}{Sample \ wt \ (g)}$$

pH of a 5% Solution at 25.0 °C. (Page 49). Reserve the solution for the test for clarity of aqueous solution.

Suitability for Determination of Bromide. Transfer 0.01 mg and 0.02 mg of bromide ion (Br) (2.0 mL and 4.0 mL of diluted bromide standard solution, described below) to two color-comparison tubes, and dilute each to 50 mL with water. For the blank, use 50 mL of water. To each, add 2.0 mL of acetate buffer solution, 2.0 mL of phenol red solution, and 0.5 mL of chloramine-T solution (all described below), mixing thoroughly immediately after each addition. Twenty minutes after adding the chloramine-T solution, add, with mixing, 0.5 mL of 2 M sodium thiosulfate solution (described below). The color of the solution containing 0.01 mg of bromide should be distinctly more reddish purple than the blank, and the purple color should increase with the higher bromide concentration.

> *Acetate Buffer Solution.* Dissolve 9.0 g of sodium chloride and 6.8 g of sodium acetate trihydrate in about 50 mL of water. Add 3.0 mL of glacial acetic acid, and dilute to 100 mL. The pH should be 4.6–4.7.
>
> *Bromide Standard Solution, diluted, 0.005 mg of Br per mL.* Dilute 5 mL of bromide ion (Br) standard solution to 100 mL with water.
>
> *Chloramine-T Solution.* Dissolve 0.50 g of the sample in water, and dilute to 100 mL with water.
>
> *Phenol Red Solution.* Dissolve 21 mg of the sodium salt of phenol red in water, and dilute to 100 mL.
>
> *Sodium Thiosulfate Solution, 2 M.* Dissolve 49.6 g of sodium thiosulfate pentahydrate in water, and dilute to 100 mL.

Clarity of Aqueous Solution. The solution prepared for the test for pH of a 5% solution should be free from turbidity and insolubles, and not more than slightly yellow in color.

Insoluble in Alcohol. Dissolve 5.0 g in 100 mL of reagent alcohol, and stir for 30 min. Filter the solution through a tared medium-porosity sintered-glass filter, and wash with 5 mL of reagent alcohol. Dry the filter at 105 °C to constant weight.

Chloroacetic Acid

Monochloroacetic Acid; Chloroethanoic Acid

$ClCH_2COOH$ **Formula Wt 94.50**   **CAS No. 79-11-8**

GENERAL DESCRIPTION

Typical appearance: colorless or white deliquescent solid

Analytical use: preparation of metal derivatives

Change in state (approximate): exists in α, β, and γ forms, with melting points of 63 °C, 55–56 °C, and 50 °C, respectively

Aqueous solubility: very soluble

Density: 1.58

pK_a: 2.9

SPECIFICATIONS

Assay . . . . $\geq$99.0% $ClCH_2COOH$

Maximum Allowable

Insoluble matter	0.01%
Residue after ignition.	0.02%
Carbonyl compounds (as acetone)	0.02%
Other carbonyl compounds	0.01%
Chloride (Cl).	0.01%
Sulfate (SO_4)	0.02%
Heavy metals (as Pb).	0.001%
Iron (Fe).	0.002%
Substances darkened by sulfuric acid	Passes test

TESTS

Assay. (By acid–base titrimetry). Weigh accurately about 3 g, transfer to a conical flask, and dissolve in about 50 mL of water. Add 0.15 mL of phenolphthalein indicator solution, and titrate with 1 N sodium hydroxide volumetric solution. One milliliter of 1 N sodium hydroxide corresponds to 0.0945 g of $ClCH_2COOH$.

$$\% \ ClCH_2COOH = \frac{(mL \times N \ NaOH) \times 9.45}{Sample \ wt \ (g)}$$

Insoluble Matter. (Page 25). Use 10 g in 100 mL of water.

Residue after Ignition. (Page 26). Use 5.0 g. Retain the residue for the test for iron.

Carbonyl Compounds. (Page 54). Use 1.0 g of sample diluted with 4 mL of water, and neutralize to pH between 6.5 and 6.8 with 10% sodium hydroxide reagent solution. For the standard, use 0.20 mg of acetone and 0.10 mg of acetylaldehyde.

Chloride. (Page 35). Use 0.1 g.

Sulfate. Dissolve 5.0 g in 30 mL of water, add 25 mg of anhydrous sodium carbonate, evaporate to dryness, and heat over a low flame until the chloroacetic acid is volatilized. Dissolve the residue in 50 mL of water, and neutralize with hydrochloric acid. Filter if necessary, wash, and dilute with water to 100 mL. To 5 mL (0.25-g sample), add 1 mL of dilute hydrochloric acid (1 + 19). Dilute to 10 mL, and add 1 mL of 12% barium chloride reagent solution. Any turbidity produced in the sample solution after 10 min should not exceed that produced by 0.05 mg of sulfate ion (SO_4) in an equal volume of solution containing the quantities of reagents used in the test.

Heavy Metals. (Page 36, Method 1). Dissolve 5.0 g in 30 mL of water, neutralize with (1 + 1) ammonium hydroxide, and dilute with water to 50 mL. For the test solution, use 30 mL. For the standard–control solution, add 0.02 mg of lead to 10 mL.

Iron. (Page 38, Method 1). To the residue after ignition, add 3 mL of hydrochloric acid, cover with a watch glass, and digest on a hot plate ($\approx$100 °C) for 15 min. Remove the cover, and evaporate to dryness. Dissolve the residue with 1 mL of hydrochloric acid and dilute with water to 100 mL. Use 10 mL (0.5-g sample) as the sample solution.

Substances Darkened by Sulfuric Acid. Heat 1.0 g with 10 mL of 95.0 ± 0.5% sulfuric acid at 50 °C for 5 min. The solution should not have more than a faintly brown color.

Chlorobenzene
Monochlorobenzene

C_6H_5Cl	Formula Wt 112.56	CAS No. 108-90-7

GENERAL DESCRIPTION

Typical appearance: clear liquid

Analytical use: solvent

Change in state (approximate): boiling point, 132 °C

Aqueous solubility: 0.05 g in 100 mL at 20 °C

Density: 1.11

SPECIFICATIONS

Assay . . . . ≥99.5% C_6H_5Cl

Maximum Allowable

Color (APHA). 30

Residue after evaporation 0.02%

Titrable acid . . . . 0.004 meq/g

TESTS

Assay. Analyze the sample by gas chromatography using the general parameters cited on page 80. The following specific conditions are also required.

Column: Type I, methyl silicone

Measure the area under all peaks, and calculate the chlorobenzene content in area percent.

Color (APHA). (Page 43).

Residue after Evaporation. (Page 25). Evaporate 20 g (18 mL) in a tared, preconditioned dish on a hot plate (≈100 °C), and dry the residue at 105 °C for 30 min.

Titrable Acid. Neutralize 200 mL of methanol to a methyl red end point with 0.1 N sodium hydroxide. Add 25 g (23 mL) of sample, and titrate with 0.1 N sodium hydroxide to the same methyl red end point. Not more than 1.0 mL should be required.

Chloroform
Trichloromethane
$CHCl_3$ **Formula Wt 119.38**   **CAS No. 67-66-3**

Suitable for general use or in high-performance liquid chromatography, extraction–concentration analysis, or ultraviolet spectrophotometry. Product labeling shall designate the uses for which suitability is represented on the basis of meeting the relevant specifications and tests. The ultraviolet spectrophotometry and liquid chromatography suitability specifications include all of the specifications for general use. The extraction–concentration suitability specifications include only the general use specification for color.

Note: Chloroform should be supplied and stored in amber glass containers and protected from direct sunlight. This solvent usually contains additives to retard decomposition. Typical additives include alcohol or mixed amylenes. The substances darkened by sulfuric acid test is not required for chloroform containing mixed amylenes.

GENERAL DESCRIPTION
Typical appearance: clear, colorless liquid
Analytical use: solvent; extraction solvent
Change in state (approximate): boiling point, 61.0 °C
Aqueous solubility: 0.5 g in 100 mL at 25 °C
Density: 1.48

SPECIFICATIONS
General Use
Assay . . . . ≥99.8% $CHCl_3$

Maximum Allowable
Color (APHA) 10
Residue after evaporation 0.001%
Acetone and aldehyde . . . . .Passes test

Acid and chloride . . . . Passes test
Free chlorine (Cl) . . . . Passes test
Lead (Pb) . . . . 0.05 ppm
Substances darkened by sulfuric acid . . . . Passes test

Specific Use

Ultraviolet Spectrophotometry

Wavelength (nm)	Absorbance (AU)
290 to 400	0.01
270	0.05
260	0.15
255	0.25
245	1.00

Liquid Chromatography Suitability

Absorbance. . . . . Passes test

Extraction–Concentration Suitability

Absorbance. . . . . Passes test
GC–FID . . . . Passes test
GC–ECD . . . . Passes test

TESTS

Assay. Analyze the sample by gas chromatography using the general parameters cited on page 80. The following specific conditions are also required.

Column: Type I, methyl silicone

Measure the area under all peaks, and calculate the chloroform and active additive content in area percent. The assay is the sum of the chloroform content plus all known active additives.

Color (APHA). (Page 43).

Residue after Evaporation. (Page 25). Evaporate 100 g (67 mL) to dryness in a tared, preconditioned dish on a hot plate ($\approx$100 °C), and dry the residue at 105 °C for 30 min.

Acetone and Aldehyde. Shake 15 mL with 20 mL of ammonia-free water for 5 min in a separatory funnel. Allow the layers to separate, and transfer 10 mL of the aqueous layer to a 125-mL glass-stoppered flask containing 40 mL of ammonia-free water. Adjust and maintain the temperature of this solution at 25 $\pm$ 1 °C. Add 5 mL of Nessler reagent solution, and allow to stand for 5 min. No turbidity or precipitate should develop. [Limit about 0.005% as $(CH_3)_2CO$]

Acid and Chloride. Shake 25 g (17 mL) with 25 mL of water for 5 min, allow the liquids to separate, and draw off the aqueous phase. Add a small piece of blue litmus paper to 10 mL of the aqueous phase. The blue litmus paper should not change color. Add 0.25 mL of silver nitrate reagent solution to another 10 mL of the aqueous phase. No turbidity should be produced in this solution.

Free Chlorine. Shake 10 mL for 2 min with 10 mL of water to which 0.10 mL of freshly made 10% potassium iodide reagent solution has been added, and allow to separate. The lower layer should not show a violet tint.

Lead.

> *Note:* When doing this test, all glassware must be carefully cleaned and thoroughly rinsed with warm dilute nitric acid (1 + 1) to remove any adsorbed lead, and finally rinsed with distilled water. All glassware used in the preparation and storage of reagents and in performing the test must be made of lead-free glass.

For the sample, transfer 40 g (27 mL) to a separatory funnel, add 20 mL of dilute nitric acid (1 + 99), and shake vigorously for 1 min. Allow the phases to separate, draw off and discard the chloroform phase, and use the aqueous phase for the solution of the sample. For the standard, prepare a solution containing 0.002 mg of lead ion in 20 mL of dilute nitric acid (1 + 99) in another separatory funnel. To each solution, in a well-ventilated fume hood, add 4 mL of ammonium cyanide, lead-free, reagent solution and 5 mL of dithizone extraction solution in chloroform, and shake vigorously for 30 s. Allow the phases to separate, draw off the dithizone–chloroform phase into a clean, dry comparison tube, and compare the color to a tube containing only the chloroform using a white background. The purplish hue in the solution of the sample resulting from any red lead dithizonate present should not exceed that in the standard.

Substances Darkened by Sulfuric Acid. To 40 mL in a separatory funnel, add 5 mL of sulfuric acid, shake the mixture vigorously for 5 min, and allow the liquids to separate completely. The chloroform layer should be colorless. The acid layer should have no more color than 5 mL of a color standard of the following composition: 0.4 mL of cobalt chloride reagent solution, 1.6 mL of 4.5% ferric chloride reagent solution, 0.4 mL of cupric sulfate reagent solution, and 17.6 mL of water.

Ultraviolet Spectrophotometry. Use the procedure on page 86 to determine the absorbance.

Liquid Chromatography Suitability. Use the procedure on page 86 to determine the absorbance.

Extraction–Concentration Suitability. Analyze the sample, using the general procedure cited on page 85. In the sample preparation step, exchange the sample with ACS-grade isooctane suitable for extraction–concentration. Use the procedure on page 86 to determine the absorbance.

Chloroplatinic Acid Hexahydrate
Hexachloroplatinic Acid Hexahydrate; Platinic Chloride Hexahydrate

$H_2PtCl_6 \cdot 6H_2O$　　　　　　　Formula Wt 517.91　　　　　　CAS No. 16941-12-1

GENERAL DESCRIPTION

Typical appearance: brownish-yellow solid
Analytical use: determination of potassium

Change in state (approximate): melting point, 60 °C
Aqueous solubility: soluble

SPECIFICATIONS

Assay . . . . ≥37.50% Pt

Maximum Allowable

Solubility in alcohol . . . . Passes test
Alkali and other salts (as sulfates) 0.05%
Suitability for potassium determinations . . . . Passes test

TESTS

Assay. (By electrolytic determination of platinum content). Using precautions to prevent absorption of moisture, weigh accurately 1 g, and transfer to a tared, preconditioned platinum dish of about 100-mL capacity. Dissolve in 80 mL of water, and add 2 mL of sulfuric acid. Cover the dish with a split watch glass, and electrolyze the solution for 4 h at a current of 0.5 A and a temperature 55–60 °C and then for 1½ h at 1 A, using a rotating platinum anode and the platinum dish as the cathode. Wash the cover several times during the electrolysis. When the electrolysis is complete, transfer the solution to a second tared, preconditioned platinum dish, and reserve for the alkali and other salts test. Rinse the platinum dish and deposited platinum with water, dry, and ignite at 600 °C for 5 min. The weight of the deposit should not be less than 37.50% of the sample weight. All weighings must be determined to ±0.00002 g.

$$\% \text{ Pt} = \frac{\text{Residue wt (g)} \times 100}{\text{Sample wt (g)}}$$

Solubility in Alcohol. Dissolve 1.0 g in 10 mL of alcohol, and allow to stand, with occasional stirring, for 15 min. The solution should contain no more than traces of insoluble matter.

Alkali and Other Salts. Evaporate the solution retained from the assay in a tared, preconditioned dish, and ignite at 600 ± 25 °C for 30 min.

Suitability for Potassium Determinations. Dissolve 7.456 g of potassium chloride, previously dried at 105 °C, in water, and dilute with water to 1 L in a volumetric flask. To a 10-mL aliquot of this solution, add 0.100–0.110 g of sodium chloride, and dilute with water to 100 mL. Heat on a hot plate ($\approx$100 °C), and add between 0.90 and 1.00 g of the chloroplatinic acid crystals dissolved in about 1 mL of water. Evaporate the solution on a hot plate ($\approx$100 °C) to a moist residue. (Do not evaporate to dryness, as this will render the sodium salt insoluble.) Cool, and add 10 mL of absolute ethyl alcohol, crush the crystals with a glass stirring rod flattened on the end, and allow the mixture to stand for 30 min. Filter by decantation through a previously dried and weighed fritted-glass crucible. Add 10 mL of absolute ethyl alcohol to the residue, grind the residue in the beaker, and filter again by decantation. When the filtrate runs clear, transfer the precipitate to the crucible, and wash with three 10-mL portions of alcohol. Discard the filtrate. Dry the filtering crucible in an oven at 105 °C for 1 h, cool, and weigh. The weight of the residue must not be less than 0.2410 g nor more than 0.2450 g.

Chromium Potassium Sulfate Dodecahydrate

Chromium(III) Potassium Sulfate Dodecahydrate

$CrK(SO_4)_2 \cdot 12H_2O$ Formula Wt 499.40   CAS No. 7788-99-0

GENERAL DESCRIPTION

Typical appearance: violet solid

Analytical use: preparation of chromium standard solutions

Change in state (approximate): melting point, 89 °C

Aqueous solubility: 20 g in 100 mL cold water

SPECIFICATIONS

Assay . . . . 98.0–102.0% as $CrK(SO_4)_2 \cdot 12H_2O$

Maximum Allowable

Insoluble matter 0.01%

Chloride (Cl). 0.002%

Aluminum (Al) 0.02%

Iron (Fe). 0.01%

Ammonium (NH_4) 0.01%

Heavy metals (as Pb). 0.01%

TESTS

Assay. (By iodometric titration for oxidative capacity of Cr^{VI}). Weigh accurately about 0.65 g, and transfer to a 500-mL iodine flask with 50 mL of water. Add 5 mL of 30% hydrogen peroxide and 15 mL of 10% sodium hydroxide reagent solution. Dilute with 100 mL of water, and boil until the solution color is yellow. Add dropwise 5 mL of 5% aqueous nickel sulfate hexahydrate solution. When the vigorous evolution of oxygen has ceased, boil for 3 min and cool. Add 1.0 g of potassium iodide and 5 mL each of phosphoric and hydrochloric acids. Mix well, and titrate the liberated iodine with 0.1 N sodium thiosulfate, adding 3 mL of starch indicator solution near the end of the titration. One milliliter of 0.1 N sodium thiosulfate corresponds to 0.01665 g of $CrK(SO_4)_2 \cdot 12H_2O$.

$$\% \ CrK(SO_4)_2 \cdot 12H_2O = \frac{(mL \times N \ Na_2S_2O_3) \times 16.65}{Sample \ wt \ (g)}$$

Insoluble Matter. (Page 25). Use 10 g dissolved in 100 mL of water.

Chloride. Dissolve 1.0 g in 10 mL of water, heat the solution to boiling, and add 2 mL of ammonium hydroxide to the hot, constantly stirred solution. Boil gently to expel excess ammonia, filter through a small chloride-free filter, wash with hot water until the volume of filtrate and washings is about 45 mL, dilute with water to 50 mL, and mix. To 25 mL of this solution, add 1.5 mL of nitric acid and 1 mL of silver nitrate reagent solution. Any turbidity should not exceed that produced by 0.01 mg of chloride ion (Cl) in an equal volume of solution containing the quantities of reagents used in the test.

Aluminum and Iron. (By flame AAS, page 63).

> **Sample Stock Solution.** Dissolve 20.0 g of sample in 60 mL of water, transfer to a 100-mL volumetric flask, and dilute to the mark with water (1 mL = 0.2 g).

Element	Wavelength (nm)	Sample Wt (g)	Standard Added (mg)	Flame Type*	Background Correction
Al	309.3	4.0	0.40; 0.80	N/A	Yes
Fe	248.3	4.0	0.20; 0.40	A/A	Yes

*A/A is air/acetylene; N/A is nitrous oxide/acetylene.

Ammonium. (By colorimetry, page 32). Dilute the distillate from a 1.0-g sample with water to 50 mL, and use 5 mL of the dilution. For the standard, use 0.01 mg of ammonium ion (NH_4).

Heavy Metals. (Page 36, Method 1). Dissolve 1.0 g with 30 mg of mercuric chloride in 40 mL of water. Add 15 mL of hydrogen sulfide water, stir, and allow to stand for 30 min. Filter and wash thoroughly with hydrogen sulfide water containing 1% potassium sulfate. Ignite the precipitate at low temperature in a small porcelain dish in a well-ventilated hood to char the paper, then at 525 ± 25 °C for 30 min. To the residue, add 1 mL of hydrochloric acid, 1 mL of nitric acid, and about 30 mg of sodium carbonate, and evaporate to dryness on a hot plate ($\approx$100 °C). Dissolve the residue in about 20 mL of water, and dilute with water to 100 mL. Use 20 mL to prepare the sample solution.

Chromium Trioxide
Chromium(VI) Oxide
CrO_3 **Formula Wt 99.99**   **CAS No. 1333-82-0**

GENERAL DESCRIPTION

Typical appearance: dark red or almost-black solid
Analytical use: oxidizing agent in organic chemistry
Change in state (approximate): melting point, 197 °C
Aqueous solubility: 167 g in 100 mL at 20 °C

SPECIFICATIONS

Assay . . . . $\geq$98.0% CrO_3

Maximum Allowable

Insoluble matter 0.01%
Chloride (Cl) 0.005%
Nitrate (NO_3) 0.05%
Sulfate (SO_4) 0.005%
Aluminum (Al) 0.02%

Barium (Ba) 0.01%
Iron (Fe). 0.02%
Sodium (Na). 0.02%

TESTS

Assay. (By iodometric titration for oxidative capacity). Weigh accurately about 5 g, transfer to a 1-L volumetric flask, dissolve in water, dilute with water to volume, and mix thoroughly. Place a 25-mL aliquot of this solution in a glass-stoppered conical flask, and dilute with 100 mL of water. Add 5 mL of dilute sulfuric acid (1 + 1) and 3 g of potassium iodide, and allow to stand in the dark for 15 min. Dilute with 100 mL of water, and titrate the liberated iodine with 0.1 N sodium thiosulfate volumetric solution, adding 3 mL of starch indicator solution near the end of the titration. Correct for a complete blank. One milliliter of 0.1 N sodium thiosulfate corresponds to 0.003333 g of CrO_3.

$$\% \, CrO_3 = \frac{(mL \times N \, Na_2S_2O_3) \times 3.333}{Sample \, wt \, (g) \, / \, 40}$$

Insoluble Matter. (Page 25). Use 10.0 g dissolved in 100 mL of water.

Chloride. Dissolve 1.0 g in water, filter if necessary through a chloride-free filter, and dilute with water to 50 mL. To 10 mL of this solution, add 1.5 mL of ammonium hydroxide and 1 mL of silver nitrate reagent solution, mix, and add 2 mL of nitric acid. Any turbidity should not exceed that produced in a standard containing 1 mL of ammonium hydroxide, 1 mL of silver nitrate reagent solution, 2 mL of nitric acid, and 0.01 mg of added chloride ion (Cl). The comparison is best made by the general method for chloride in colored solutions, page 35.

Nitrate.

Sample Stock Solution. Dissolve 1.0 g in about 20 mL of water. Add 10% sodium hydroxide reagent solution until the color of the solution just turns light yellow, then add 0.2 mL of glacial acetic acid to change the color to orange (pH about 6.0–6.5). Slowly add a solution of 5 g of lead acetate in about 10 mL of water, dilute to 50 mL with water, mix, and allow to stand for 15 min. Transfer to a 50-mL centrifuge tube, centrifuge, and decant through a filter. *This solution must be clear and colorless.*

Note: Two filter papers may be required to obtain a clear, colorless filtrate. Use a Nessler tube to check.

Sample Solution A. Dissolve 0.1 g of mercuric acetate in 2.0 mL of sample stock solution plus 1.0 mL of water in a dry test tube. Add 0.05 g of urea, and shake to dissolve. Put the tube in an ice bath, and immediately, but slowly, add 7 mL of cold 0.01% chromotropic acid reagent solution while swirling. Keep the tube in the ice bath an additional 2–3 min, remove, and let stand for 30 min, swirling occasionally.

Control Solution B. Dissolve 0.1 g of mercuric acetate in 2.0 mL of sample stock solution in a dry test tube. Add 1.0 mL of a solution containing 0.02 mg of nitrate ion (NO_3) per mL, then add 0.05 g of urea, and shake to dissolve. Put the tube in an ice bath, and immediately, but slowly, add 7 mL of cold 0.01% chromotropic acid reagent solution while swirling. Keep the tube in the ice bath an additional 2–3 min, remove, and let stand for 30 min, swirling occasionally.

Blank Solution C. Dissolve 0.1 g of mercuric acetate in 3.0 mL of water in a dry test tube. Add 0.05 g of urea, shake to dissolve, and put the tube in an ice bath. Add 7 mL of cold 0.01% chromotropic acid reagent solution while swirling, remove from the bath, and let stand for 30 min.

Transfer sample solution A, control solution B, and blank solution C to dry 15-mL centrifuge tubes, and centrifuge until the supernatant liquid is clear. Set a spectrophotometer at 405 nm, and using 1-cm cells, adjust the instrument to read zero (absorbance units) with blank solution C in the light path, then determine the absorbance of sample solution A. Adjust the instrument to read zero (absorbance units) with sample solution A in the light path and determine the absorbance of control solution B. The absorbance of sample solution A should not exceed that of control solution B.

Sulfate. Dissolve 10.0 g in 350 mL of water, and add 5.0 g of sodium carbonate. Heat to boiling, and add 35 mL of a solution containing 1 g of barium chloride and 2 mL of hydrochloric acid in 100 mL of water. Digest in a covered beaker on a hot plate ($\approx$100 °C) for 2 h, and allow to stand for at least 8 h. If a precipitate is formed, filter, wash thoroughly, and ignite. Fuse the ignited precipitate with 1 g of sodium carbonate. Extract the fused mass with water, and filter out the insoluble residue. Add 5 mL of hydrochloric acid to the filtrate, dilute with water to about 200 mL, heat to boiling, and add 10 mL of alcohol. Digest on a hot plate ($\approx$100 °C) until the reduction of chromate is complete, as indicated by the change to a clear green or colorless solution. Neutralize the solution with ammonium hydroxide, and add 2 mL of hydrochloric acid. Heat to boiling, and add 10 mL of 12% barium chloride reagent solution. Digest in a covered beaker on a hot plate ($\approx$100 °C) for 2 h, and allow to stand for at least 8 h. Filter, wash thoroughly, and ignite. Correct for the weight obtained in a complete blank test.

Aluminum, Barium, Iron, and Sodium. (By flame AAS, page 63).

Sample Stock Solution A. Dissolve 10.0 g of sample in about 75 mL of water, transfer to a 100-mL volumetric flask, dilute to the mark with water, and mix (1 mL = 0.10 g).

Sample Stock Solution B. Transfer exactly 10.0 mL of sample stock solution A to a 100-mL volumetric flask, dilute to the mark with water, and mix (1 mL = 0.01 g).

Element	Wavelength (nm)	Sample Wt (g)	Standard Added (mg)	Flame Type*	Background Correction
Al	309.3	2.0	0.20; 0.40	N/A	No
Ba	553.6	2.0	0.10; 0.20	N/A	No
Fe	248.3	2.0	0.20; 0.40	A/A	No
Na	589.0	0.01	0.01; 0.02	A/A	No

*A/A is air/acetylene; N/A is nitrous oxide/air.

Chromotropic Acid, Disodium Salt

4,5-Dihydroxy-2,7-naphthalenedisulfonic Acid, Disodium Salt

(HO)$_2$C$_{10}$H$_4$(SO$_3$Na)$_2$ · 2H$_2$O (disodium salt)	Formula Wt 400.29	CAS No. 5808-22-0
4,5-(OH)$_2$C$_{10}$H$_4$-2,7-(SO$_3$H)$_2$ (free acid)	Formula Wt 320.29	CAS No. 148-25-4

Note: Two forms of this reagent are available: the disodium salt dihydrate and the free acid. This specification applies to both forms, but the sample weight of the acid should be changed to 80% of that for the salt.

GENERAL DESCRIPTION

Typical appearance: white to off-white solid

Analytical use: determination of nitrate and formaldehyde

Change in state (approximate): melting point, 300 °C

Aqueous solubility: very soluble

SPECIFICATIONS

Clarity of solution . . . . Passes test
Sensitivity to nitrate . . . . Passes test
Sensitivity to formaldehyde . . . . Passes test

TESTS

Clarity of Solution. Use 5.0 g of the salt or 4.0 g of the acid dissolved in 100 mL of hot water. Add a few drops of 10% sodium hydroxide reagent solution if necessary. The solution should be free of turbidity.

For the Determination of Sensitivities to Nitrate and Formaldehyde

Sample Solution A. Dissolve 0.20 g of the salt (0.16 g of the acid) in 100 mL of sulfuric acid.

Sensitivity to Nitrate. Into three 50-mL volumetric flasks, pipet 1.0, 2.0, and 3.0 mL of a nitrate ion (NO$_3$) standard solution. For the blank, add 1.0 mL of water to another 50-mL volumetric flask. To each flask, add approximately 5 mL of sulfuric acid, swirl, and cool in an ice bath. Add 2.0 mL of sample solution A to each flask, swirl again, and cool. Dilute to 50 mL with sulfuric acid, and shake well. Set a spectrophotometer at 410 nm, and using 5-cm cells, determine the absorbance of the standard solutions against the blank. A plot of absorbance versus concentration of nitrate should be linear, with a minimum absorbance of 0.20 for the solution containing 0.03 mg of nitrate.

Sensitivity to Formaldehyde. Into three 50-mL volumetric flasks, pipet 1.0, 2.0, and 3.0 mL of formaldehyde II standard solution. For the blank, add 1.0 mL of water to another 50-mL volumetric flask. To each flask, add 5.0 mL of sample solution A, stopper, and shake. Dilute to the mark with sulfuric acid and let stand in a hot water bath (70–80 °C) for 15 min. Cool in an ice bath to room temperature, fill to volume with sulfuric acid, stopper, and shake. Set a spectrophotometer at 580 nm, and using 1-cm cells, determine the absorbance of the standard solutions against the blank. A plot of absorbance versus concentration of formaldehyde should be linear, with a minimum absorbance of 0.30 for the solution containing 0.03 mg of formaldehyde.

Citric Acid, Anhydrous, and Citric Acid, Monohydrate
2-Hydroxy-1,2,3-propanetricarboxylic Acid

$$CH_2-\overset{\overset{O}{\|}}{C}-OH$$
$$HO-\overset{\overset{CH_2-\overset{\overset{O}{\|}}{C}-OH}{|}}{\underset{\underset{CH_2-\overset{\overset{O}{\|}}{C}-OH}{|}}{C}}-\overset{\overset{O}{\|}}{C}-OH \quad \cdot H_2O$$

$C_6H_8O_7$ (anhydrous)	Formula Wt 192.13	CAS No. 77-92-9
$C_6H_8O_7 \cdot H_2O$ (monohydrate)	Formula Wt 210.14	CAS No. 5949-29-1

Note: This reagent is available as either the anhydrous or monohydrate form. The label should identify the specific form.

GENERAL DESCRIPTION

Typical appearance: colorless or white solid

Analytical use: sequestering agent for trace metals; determination of citrate-soluble phosphorus pentoxide

Change in state (approximate): melting point, 100 °C (monohydrate), 153 °C (anhydrous)

Aqueous solubility: 165 g in 100 mL at 20 °C

SPECIFICATIONS

Assay (anhydrous)	$\geq$99.5% $C_6H_8O_7$
Assay (monohydrate)	99.0–102.0% $C_6H_8O_7 \cdot H_2O$

Maximum Allowable

Insoluble matter	0.005%
Residue after ignition	0.02%
Chloride (Cl)	0.001%

Oxalate (C_2O_4)...Passes test
Phosphate (PO_4)..0.001%
Sulfate (SO_4) 0.002%
Iron (Fe)...3 ppm
Lead (Pb)...2 ppm
Substances carbonizable by hot sulfuric acid ....Passes test

TESTS

Assay. (By acid–base titrimetry). Weigh accurately about 2.7 g, dissolve in about 40 mL of water, and add 0.15 mL of phenolphthalein indicator solution. Titrate with 1 N sodium hydroxide volumetric solution to a pink color that persists for 5 min. One milliliter of 1 N sodium hydroxide corresponds to 0.06404 g of $C_6H_8O_7$ or 0.07005 g of $C_6H_8O_7 \cdot H_2O$.

$$\% \ C_6H_8O_7 = \frac{(mL \times N \ NaOH) \times 6.404}{Sample \ wt \ (g)}$$

$$\% \ C_6H_8O_7 \cdot H_2O = \frac{(mL \times N \ NaOH) \times 7.005}{Sample \ wt \ (g)}$$

Insoluble Matter. (Page 25). Use 20 g dissolved in 150 mL of water.

Residue after Ignition. (Page 26). Ignite 5.0 g. Use 2 mL of sulfuric acid to moisten the residue.

Chloride. (Page 35). Use 1.0 g.

Oxalate. Dissolve 5 g in 25 mL of water and add 2 mL of 10% aqueous calcium acetate solution. No turbidity or precipitate should appear after standing 4 h. (Limit about 0.05%)

Phosphate. (Page 40, Method 1). Mix 2.0 g with 0.5 g of magnesium nitrate in a platinum dish, and ignite. Dissolve the residue in 5 mL of water, add 5 mL of nitric acid, and evaporate to dryness. Dissolve the residue in 25 mL of approximately 0.5 N sulfuric acid, and continue as described.

Sulfate. To 3.0 g, add 0.2–0.5 mg of ammonium metavanadate and 10 mL of nitric acid. Digest in a covered beaker on a hot plate ($\approx$100 °C) until the reaction ceases, remove the cover, and evaporate to dryness. Add 10 mL of nitric acid, and repeat the digestion and evaporation. Add 5 mL of dilute hydrochloric acid (1 + 1), and evaporate to dryness. Dissolve the residue in 4 mL of water plus 1 mL of dilute hydrochloric acid (1 + 19), filter if necessary through a small filter, wash with two 2-mL portions of water, and dilute with water to 10 mL. Add 1 mL of 12% barium chloride reagent solution. Any turbidity should not exceed that produced by 0.06 mg of sulfate ion (SO_4) treated in the same manner as the sample. Compare 10 min after adding the barium chloride to the sample and standard solutions.

Iron. (Page 38, Method 1). Use 3.3 g.

Lead. Dissolve 1.5 g in 10 mL of water, adjust the pH to between 9 and 10 (using a pH meter) with ammonium hydroxide, and dilute with water to 15 mL. Prepare a control containing 0.003 mg of lead and the amount of ammonium hydroxide required to neutralize the sample, and evaporate it to dryness on a hot plate ($\approx$100 °C). Moisten the residue from the control with 0.1 mL of nitric acid, dissolve in 5 mL of water, and add 5 mL of ammonium citrate (lead-free) reagent solution. Adjust the pH to between 9 and 10 (using a pH meter) with ammonium hydroxide, and dilute with water to 15 mL. Transfer the sample solution and the control solution to separatory funnels, add to each 5 mL of dithizone extraction reagent solution (in chloroform), and shake vigorously for 2 min. Let the layers separate, draw off the dithizone–chloroform layers into clean, dry comparison tubes, and dilute to 10 mL with chloroform. Any red color producing a purplish hue in the chloroform from the sample solution should not exceed that in the chloroform from the control.

Substances Carbonizable by Hot Sulfuric Acid (Tartrates, etc.). Carefully powder about 1.0 g of the sample, and take precautions to prevent contamination during the powdering. Mix 0.30 g with 10 mL of sulfuric acid in a test tube previously rinsed with the acid. Heat the mixture at 110 °C (in a brine bath) for 30 min, and keep the test tube covered during the heating. Any color should not exceed that produced in a color standard of the following composition: 1 part of cobalt chloride reagent solution, 2.4 parts of ferric chloride reagent solution, 0.4 parts of cupric sulfate reagent solution, and 6.2 parts of water.

Cobalt Acetate Tetrahydrate
Cobalt(II) Acetate Tetrahydrate; Cobaltous Acetate Tetrahydrate

$Co(CH_3COO)_2 \cdot 4H_2O$ Formula Wt 249.08 CAS No. 6147-53-1

GENERAL DESCRIPTION
Typical appearance: intense red solid
Analytical use: catalyst for oxidation and esterification
Change in state (approximate): on heating, becomes anhydrous by 140 °C
Aqueous solubility: soluble

SPECIFICATIONS

Assay .... 98.0–102.0% $Co(CH_3COO)_2 \cdot 4H_2O$

	Maximum Allowable
Insoluble matter	0.01%
Chloride (Cl)	0.002%
Nitrate (NO_3)	0.01%
Sulfate (SO_4)	0.005%
Calcium (Ca)	0.005%
Copper (Cu)	0.002%
Iron (Fe)	0.001%
Lead (Pb)	0.001%
Magnesium (Mg)	0.005%
Nickel (Ni)	0.1%
Potassium (K)	0.01%
Sodium (Na)	0.05%
Zinc (Zn)	0.01%

TESTS

Assay. (By complexometric titration of cobalt). Weigh accurately about 1.0 g, and transfer to a 400-mL beaker with the aid of 200 mL of water. Add 0.2 g of ascorbic acid and about 100 mg of murexide indicator mixture. Titrate with 0.1 M EDTA volumetric solution to a color change from yellow to violet that persists for at least 30 s. One milliliter of 0.1 M EDTA corresponds to 0.02491 g of $Co(CH_3COO)_2 \cdot 4H_2O$.

$$\% \ Co(CH_3COO)_2 \cdot 4H_2O = \frac{(mL \times M \ EDTA) \times 24.91}{Sample \ wt \ (g)}$$

Insoluble Matter. (Page 25). Use 10.0 g dissolved in 100 mL of water containing 1 mL of glacial acetic acid. Reserve the filtrate without washings for the test for sulfate.

Chloride. (Page 35). Use 0.5 g. The comparison is best made by the general method for chloride in colored solutions on page 35.

Nitrate. (Page 38, Method 1). Dissolve 1.0 g in 10 mL of water. Add this solution in small portions and with constant stirring to 10 mL of 10% sodium hydroxide reagent solution. Digest in a covered beaker on a hot plate ($\approx$100 °C) for 15 min. Cool, dilute with water to 20 mL, and filter.

> *Sample Solution A.* To 4 mL of the preceding filtrate, add 6 mL of water. Dilute to 50 mL with brucine sulfate reagent solution.

> *Control Solution B.* To 4 mL of the preceding filtrate, add 4 mL of water and 2 mL of nitrate ion (NO_3) standard solution. Dilute to 50 mL with brucine sulfate reagent solution.

Continue with the procedure, starting with the preparation of blank solution C.

Sulfate. (Page 40, Method 1). Use 1/10 of the filtrate from the insoluble matter test. Allow 30 min for turbidity to develop.

Calcium, Copper, Iron, Lead, Magnesium, Nickel, Potassium, Sodium, and Zinc. (By flame AAS, page 63).

> *Sample Stock Solution.* Dissolve 25.0 g in water in a 250-mL volumetric flask, add 2.5 mL of nitric acid, and dilute with water to volume (1 mL = 0.10 g).

Element	Wavelength (nm)	Sample Wt (g)	Standard Added (mg)	Flame Type*	Background Correction
Ca	422.7	0.80	0.02; 0.04	N/A	No
Cu	324.8	2.0	0.02; 0.04	A/A	Yes
Fe	248.3	2.50	0.025; 0.05	A/A	Yes
Pb	217.0	2.0	0.02; 0.04	A/A	Yes
Mg	285.2	0.20	0.005; 0.01	A/A	Yes
Ni	232.0	0.10	0.05; 0.10	A/A	Yes
K	766.5	0.20	0.01; 0.02	A/A	No
Na	589.0	0.04	0.01; 0.02	A/A	No
Zn	213.9	0.20	0.01; 0.02	A/A	Yes

*A/A is air/acetylene; N/A is nitrous oxide/acetylene.

Cobalt Chloride Hexahydrate

Cobalt(II) Chloride Hexahydrate; Cobaltous Chloride Hexahydrate

$CoCl_2 \cdot 6H_2O$ Formula Wt 237.93   CAS No. 7791-13-1

GENERAL DESCRIPTION

Typical appearance: pink to red solid

Analytical use: APHA color standard

Change in state (approximate): melting point, 86 °C

Aqueous solubility: 93 g in 100 mL at 20 °C

SPECIFICATIONS

Assay . . . . .98.0–102.0% $CoCl_2 \cdot 6H_2O$

	Maximum Allowable
Insoluble matter	0.01%
Nitrate (NO_3)	0.01%
Sulfate (SO_4)	0.01%
Calcium (Ca)	0.005%
Copper (Cu)	0.002%
Iron (Fe)	0.005%
Magnesium (Mg)	0.005%
Nickel (Ni)	0.1%
Potassium (K)	0.01%
Sodium (Na)	0.05%
Zinc (Zn)	0.03%

TESTS

Assay. (By complexometric titration of cobalt). Weigh, to the nearest 0.1 mg, 1.0 g of sample, and dissolve in 200 mL of water. Add about 0.2 g of ascorbic acid and 5 g of ammonium acetate, followed by 10 mL of ammonium hydroxide. Add about 0.1 g of murexide indicator mixture, and titrate with 0.1 M EDTA volumetric solution to the violet end point. The color change must persist for 30 s. One milliliter of 0.1 M EDTA corresponds to 0.02379 g of $CoCl_2 \cdot 6H_2O$.

$$\% \ CoCl_2 \cdot 6H_2O = \frac{(mL \times M \ EDTA) \times 23.79}{Sample \ wt \ (g)}$$

Insoluble Matter. (Page 25). Use 10.0 g dissolved in 100 mL of water.

Nitrate. (Page 38, Method 1). Dissolve 1.0 g in 10 mL of water. Add this solution in small portions and with constant stirring to 10 mL of 10% sodium hydroxide reagent solution. Digest in a covered beaker on a hot plate ($\approx$100 °C) for 15 min. Cool, dilute with water to 20 mL, and filter.

> *Sample Solution A.* To 4 mL of the preceding filtrate, add 6 mL of water. Dilute to 50 mL with brucine sulfate reagent solution.

Control Solution B. To 4 mL of the preceding filtrate, add 4 mL of water and 2 mL of nitrate ion (NO_3) standard solution. Dilute to 50 mL with brucine sulfate reagent solution.

Continue with the procedure, starting with preparation of blank solution C.

Sulfate. (Page 40, Method 1). Use 1/20 of the filtrate from the test for insoluble matter. Allow 30 min for turbidity to develop.

Calcium, Copper, Iron, Magnesium, Nickel, Potassium, Sodium, and Zinc.
(By flame AAS, page 63).

Sample Stock Solution. Dissolve 20.0 g of sample in water, add 2 mL of nitric acid, and dilute to the mark with water in a 200-mL volumetric flask (1 mL = 0.10 g).

Element	Wavelength (nm)	Sample Wt (g)	Standard Added (mg)	Flame Type*	Background Correction
Ca	422.7	0.80	0.02; 0.04	N/A	No
Cu	324.8	2.0	0.04; 0.08	A/A	Yes
Fe	248.3	2.0	0.05; 0.10	A/A	Yes
Mg	285.2	0.20	0.005; 0.01	A/A	Yes
Ni	232.0	0.10	0.05; 0.10	A/A	Yes
K	766.5	0.20	0.01; 0.02	A/A	No
Na	589.0	0.04	0.01; 0.02	A/A	No
Zn	213.9	0.10	0.01; 0.02	A/A	Yes

*A/A is air/acetylene; N/A is nitrous oxide/acetylene.

Cobalt Nitrate Hexahydrate
Cobalt(II) Nitrate Hexahydrate; Cobaltous Nitrate Hexahydrate

$Co(NO_3)_2 \cdot 6H_2O$ **Formula Wt 291.03**   **CAS No. 10026-22-9**

GENERAL DESCRIPTION
Typical appearance: red to plum-colored solid
Analytical use: preparation of cobalt standard solutions
Change in state (approximate): melting point, 55 °C
Aqueous solubility: 159 g in 100 mL at 20 °C

SPECIFICATIONS
Assay . . . . 98.0–102.0% $Co(NO_3)_2 \cdot 6H_2O$

Maximum Allowable

Insoluble matter 0.01%
Chloride (Cl). 0.002%
Sulfate (SO_4) 0.005%
Calcium (Ca) 0.005%

Copper (Cu)..0.002%
Iron (Fe)..0.001%
Lead (Pb) 0.002%
Magnesium (Mg)..0.005%
Nickel (Ni)...0.15%
Potassium (K) 0.01%
Sodium (Na) 0.05%
Zinc (Zn)..0.01%

TESTS

Assay. (By complexometric titration of cobalt). Weigh, to the nearest 0.1 mg, 1.0 g of sample, and dissolve in 200 mL of water. Add about 0.2 g of ascorbic acid and 5 g of ammonium acetate, followed by 10 mL of ammonium hydroxide. Add about 0.1 g of murexide indicator mixture, and titrate with 0.1 M EDTA volumetric solution to the violet end point. The color change must persist for 30 s. One milliliter of 0.1 M EDTA corresponds to 0.02910 g of $Co(NO_3)_2 \cdot 6H_2O$.

$$\% \ Co(NO_3)_2 \cdot 6H_2O = \frac{(mL \times M \ EDTA) \times 29.10}{Sample \ wt \ (g)}$$

Insoluble Matter. (Page 25). Use 10 g dissolved in 100 mL of water.

Chloride. Dissolve 0.50 g in 15 mL of water plus 1 mL of nitric acid, filter if necessary through a small chloride-free filter, and add 1 mL of silver nitrate reagent solution. Any turbidity should not exceed that produced by 0.01 mg of chloride in an equal volume of solution containing the quantities of reagents used in the test. The comparison is best made by the general method for chloride in colored solutions, page 35.

Sulfate. (Page 40, Method 2). Use 4 mL of dilute hydrochloric acid (1 + 1). Evaporate twice.

Calcium, Copper, Iron, Lead, Magnesium, Nickel, Potassium, Sodium, and Zinc. (By flame AAS, page 63).

Sample Stock Solution. Dissolve 20.0 g of sample in water, add 2 mL of nitric acid, and dilute to the mark with water in a 200-mL volumetric flask (1 mL = 0.1 g).

Element	Wavelength (nm)	Sample Wt (g)	Standard Added (mg)	Flame Type*	Background Correction
Ca	422.7	0.80	0.02; 0.04	N/A	No
Cu	324.8	2.0	0.04; 0.08	A/A	Yes
Fe	248.3	2.0	0.02; 0.04	A/A	Yes
Pb	217.0	2.0	0.04; 0.08	A/A	Yes
Mg	285.2	0.20	0.005; 0.01	A/A	Yes
Ni	232.0	0.10	0.075; 0.15	A/A	Yes
K	766.5	0.20	0.01; 0.02	A/A	No
Na	589.0	0.04	0.01; 0.02	A/A	No
Zn	213.9	0.10	0.01; 0.02	A/A	Yes

*A/A is air/acetylene; N/A is nitrous oxide/acetylene.

Congo Red

Direct Red 28, C.I. 22120

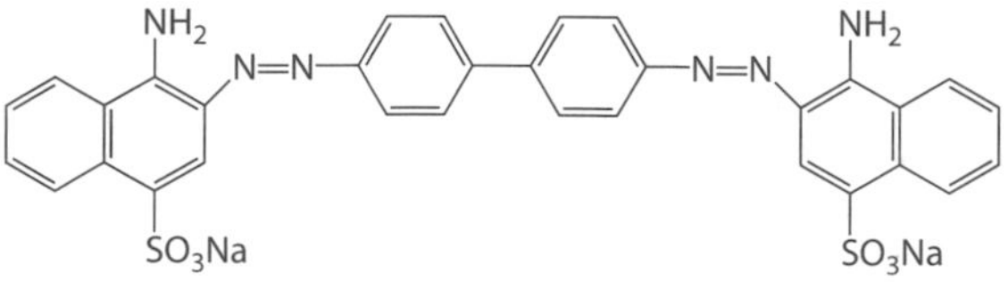

$C_{32}H_{22}N_6O_6S_2Na_2$ **Formula Wt 696.68**   **CAS No. 573-58-0**

GENERAL DESCRIPTION

Typical appearance: brownish-red solid
Analytical use: pH indicator; nitrogen analysis
Aqueous solubility: soluble

SPECIFICATIONS

Clarity of solution . . . . Passes test
Visual transition interval . . . . pH 3.0 (blue) to pH 5.2 (red)

Maximum Allowable

Loss on drying 3.0%

TESTS

Clarity of Solution. Dissolve 0.10 g in 100 mL of water. Not more than a trace of turbidity or insoluble matter should remain. Reserve the solution for the test for visual transition interval.

Visual Transition Interval. Dissolve 1.0 g of potassium chloride in 100 mL of water. Adjust the pH of the solution to 3.0 (using a pH meter) with 0.01 N hydrochloric acid. Add 0.15 mL of the 0.1% solution reserved from the clarity of solution test. The solution should be blue. Titrate with 0.01 N sodium hydroxide to pH 4.0. The solution should be violet. Continue the titration to pH 5.2. The solution should be red.

Loss on Drying. (Page 26). Use 1 g. Dry at 105 °C.

Copper

Cu   **Atomic Wt 63.55**   **CAS No. 7440-50-8**

GENERAL DESCRIPTION

Typical appearance: bright, reddish metal
Analytical use: catalyst; absorption of oxygen

SPECIFICATIONS

Assay . . . . ≥99.90% Cu

Maximum Allowable

Insoluble in dilute nitric acid 0.02%

Antimony and tin (as Sn) 0.01%

Arsenic (As) . . . . 5 ppm

Iron (Fe). 0.005%

Lead (Pb) 0.005%

Manganese (Mn) 0.001%

Silver (Ag). 0.002%

Phosphorus (P). 0.001%

TESTS

Assay. (By electrolytic determination of copper content). Accurately weigh 5.0050–5.0070 g of the sample, and transfer this weighed sample to a tall-form lipless beaker provided with a close-fitting cover glass. Add 42 mL of a solution that contains 10 mL of sulfuric acid plus 7 mL of nitric acid and 25 mL of water, cover, and allow to stand until the active reactions have subsided. Heat at 80–90 °C until the copper is completely dissolved and brown nitrous fumes are completely expelled. Wash down the cover glass and the sides of the beaker, and dilute the solution sufficiently with water so that it will cover a cathode cylinder. Insert a tared cathode and an anode, and electrolyze at a current density of about 0.6 A per dm^2 for about 16 h. If electrolysis of the copper is not completed, as indicated by copper plating on a new surface of the cathode stem when the level of the solution is raised, continue the electrolysis until all of the copper is deposited.

When all of the copper is deposited, slowly lower the beaker, without interrupting the current, and wash the electrodes continuously with a stream of water from a wash bottle while removing the beaker. Reserve this solution for preparing sample solution A. Rinse the electrodes by raising and lowering a beaker of water around them. Remove the beaker of rinse water, discontinue the current, and remove the electrodes. Wash the cathode in acetone, and dry it in an oven at 110 °C for 3–5 min. Cool, weigh, and calculate the percentage of copper from the weight of the deposit. This value may have to be corrected for a small amount of undeposited copper as determined in preparing sample solution A.

Sample Solution A. Place the anode in a 150-mL beaker, and add 10 mL of dilute nitric acid (1 + 1) containing 0.20 mL of alcohol. Heat for 5 min on a hot plate (≈100 °C), rotating the anode so that any deposit is dissolved in the dilute nitric acid. Remove the anode, and wash it with a stream of water. Add this solution and washings to the electrolyte from which the copper was deposited. Evaporate the electrolyte to fumes of sulfur trioxide, and continue until the volume is reduced to approximately 5 mL. Cool, add 50 mL of water, cool, and transfer the solution to a 100-mL volumetric flask. Dilute with water to 100 mL, and mix (1 mL = 0.05 g). To 20 mL of sample solution A (1-g sample), add an excess of ammonium hydroxide. If the presence of copper is indicated by a blue color, determine the amount colorimetrically, calculate the amount in sample solution A, and add to the amount determined electrolytically. Retain the remainder of sample solution A for succeeding tests.

Insoluble in Dilute Nitric Acid. In a well-ventilated fume hood, cautiously dissolve 30.0 g in 200 mL of dilute nitric acid (1 + 1), and dilute with water to 225 mL. Filter through a tared, preconditioned filtering crucible, and reserve the filtrate separate from the washings. Wash thoroughly with hot water, and dry at 105 °C.

Antimony and Tin. Dissolve 4.0 g in 35 mL of dilute nitric acid (1 + 1) in a 150-mL beaker, and evaporate to between 6 and 10 mL. Dilute to 30 mL with water, and digest on a hot plate ($\approx$100 °C) for 15 min. Any insoluble residue should not be greater than that obtained when a control solution containing 0.4 mg of tin dissolved in 35 mL of dilute nitric acid (1 + 1) is treated exactly like the 4.0 g of the sample.

Arsenic. (Page 34). Dissolve 0.60 g in 20 mL of dilute nitric acid (1 + 1) in a 250-mL beaker. Add 6 mL of sulfuric acid, evaporate in a hood to dense fumes of sulfur trioxide, and continue the fuming for 15 min. Cool, cautiously add 15 mL of water to dissolve the salts, and fume again in the same way. Repeat the addition of water and the fuming. Cautiously transfer the solution to a generator flask, using a total of 30 mL of water. Add 20 mL of hydrochloric acid containing 3 g of stannous chloride dihydrate. Cool in an ice bath, add 15 g of zinc shot, connect the scrubber–absorber assembly, and allow the hydrogen evolution and the color development to proceed for 45 min, keeping the flask in the ice bath. Any red color in the silver diethyldithiocarbamate solution of the sample should not exceed that in a standard containing 0.003 mg of arsenic ion (As) similarly treated.

Iron, Lead, Manganese, and Silver. (By flame AAS, page 63).

Sample Stock Solution. Dissolve 10.0 g of sample in 80 mL of (1 + 1) nitric acid. Boil a few minutes to remove the oxides of nitrogen, cool, and dilute with water to the mark in a 100-mL volumetric flask (1 mL = 0.1 g).

Element	Wavelength (nm)	Sample Wt (g)	Standard Added (mg)	Flame Type*	Background Correction
Fe	248.3	1.0	0.05; 0.10	A/A	Yes
Pb	217.0	1.0	0.05; 0.10	A/A	Yes
Mn	279.5	1.0	0.01; 0.02	A/A	Yes
Ag	328.1	1.0	0.02; 0.04	A/A	Yes

*A/A is air/acetylene.

Phosphorus. Dilute 10 mL of sample solution A (0.5-g sample) to 25 mL, add 1 mL of ammonium molybdate reagent solution and 1 mL of 4-(methylamino)phenol sulfate reagent solution, and allow to stand at room temperature for 2 h. Any blue color should not exceed that produced by 1.5 mL of phosphate standard solution (0.005 mg of P) in 25 mL of approximately 0.5 N sulfuric acid when treated with the reagents used with the 25-mL dilution of the 10 mL of sample solution A.

Crystal Violet
Hexamethylpararosaniline Chloride

$C_{25}H_{30}ClN_3$ **Formula Wt 407.98**   **CAS No. 548-62-9**

GENERAL DESCRIPTION

Typical appearance: dark green powder or greenish glistening pieces with metallic luster
Analytical use: indicator for copper salts
Aqueous solubility: soluble

SPECIFICATIONS

Assay . . . . $\geq$90.0% $C_{25}H_{30}ClN_3$
Sensitivity as indicator . . . . Passes test
Loss on drying . . . . $\leq$7.5%
Absorbance characteristics . . . . Passes test

TESTS

Assay. (By spectrophotometry). Weigh accurately 50.0 mg of sample previously dried for 4 h at 110 °C, dissolve in about 100 mL of water in a 250-mL volumetric flask, and dilute with water to volume. Dilute 10 mL of this solution with water to 1 L, and determine the absorbance in a 1-cm cell at the wavelength of maximum absorption at about 590 nm against water as the blank. Reserve the remainder of the solution for the determination of absorbance characteristics.

$$\% \text{ Crystal violet} = \text{Absorbance at 590 nm} \times 209$$

Sensitivity as Indicator.

Solution A. Dissolve 100 mg of sample in 10 mL of glacial acetic acid.

Solution B. Dilute 0.30 mL of 70% perchloric acid with dioxane to 25 mL.

To 50 mL of glacial acetic acid, add 0.10 mL of solution A. The purplish blue color of the solution is changed to bluish-green by the addition of not more than 0.05 mL of solution B.

Loss on Drying. Weigh accurately 1.0 g, and dry for 4 h at 110 °C.

Absorbance Characteristics. Using the solution retained from the assay, determine the absorbance in a 1-cm cell at $A - 15$ nm and at $A + 15$ nm, against water as the blank. The ratio of the absorbance at $A - 15$ nm to the absorbance at $A + 15$ nm should be between 0.98 and 1.9 (A is the wavelength of maximum absorption obtained in the assay).

Cupric Acetate Monohydrate

Copper(II) Acetate Monohydrate

$(CH_3COO)_2Cu \cdot H_2O$ Formula Wt 199.65   CAS No. 6046-93-1

GENERAL DESCRIPTION

Typical appearance: blue-green or dark green solid

Analytical use: preparation of Fehling's reagent for sugars

Change in state (approximate): melting point, 115 °C

Aqueous solubility: 7.2 g in 100 mL at 20 °C

SPECIFICATIONS

Assay . . . . 98.0–102.0% $(CH_3COO)_2Cu \cdot H_2O$

Maximum Allowable

Insoluble matter 0.01%

Chloride (Cl). 0.003%

Sulfate (SO_4) 0.01%

Calcium (Ca) 0.005%

Iron (Fe). 0.002%

Nickel (Ni). 0.01%

Potassium (K) 0.01%

Sodium (Na). 0.05%

TESTS

Assay. (By iodometric titration of oxidative capacity of Cu^{II}). Weigh accurately about 0.8 g, and dissolve in 50 mL of water. Add 4 mL of glacial acetic acid, 1 mL of 10% sulfuric acid, and 3 g of potassium iodide. Swirl. Titrate the liberated iodine with 0.1 N sodium thiosulfate, adding about 2 g of potassium thiocyanate and 3 mL of starch indicator solution near the end of the titration. One milliliter of 0.1 N sodium thiosulfate corresponds to 0.01996 g of $(CH_3COO)_2Cu \cdot H_2O$.

$$\% \ (CH_3COO)_2Cu \cdot H_2O = \frac{(mL \times N \ Na_2S_2O_3) \times 19.96}{Sample \ wt \ (g)}$$

Insoluble Matter. (Page 25). Use 10 g dissolved in 150 mL of water containing 5 mL of glacial acetic acid. Omit heating. Reserve the filtrate without washings for the sulfate test.

Chloride. Dissolve 0.50 g in 15 mL of water, filter if necessary through a small chloride-free filter, and add 1 mL of nitric acid and 1 mL of silver nitrate reagent solution. Any turbidity should not exceed that produced by 0.015 mg of chloride ion (Cl) in an equal volume of solution containing the quantities of reagents used in the test. The comparison is best made by the general method for chloride in colored solutions, page 35.

Sulfate. (Page 40, Method 1). Use 1/20 of the filtrate from the test for insoluble matter. Allow 30 min for turbidity to form.

Calcium, Iron, Nickel, Potassium, and Sodium. (By flame AAS, page 63).

Sample Stock Solution A. Dissolve by warming 25.0 g of sample in a mixture of 60 mL of water and 15 mL of nitric acid. Cool to room temperature, transfer to a 100-mL volumetric flask, and dilute to the mark with water (1 mL = 0.25 g).

Sample Stock Solution B. Dilute 4.0 mL of sample stock solution A with water to 100 mL in a volumetric flask (1 mL = 0.01 g).

Element	Wavelength (nm)	Sample Wt (g)	Standard Added (mg)	Flame Type*	Background Correction
Ca	422.7	2.0	0.05; 0.10	N/A	No
Fe	248.3	5.0	0.05; 0.10	A/A	Yes
Ni	232.0	2.0	0.10; 0.20	A/A	Yes
K	766.5	0.20	0.02; 0.04	A/A	No
Na	589.0	0.04	0.01; 0.02	A/A	No

*A/A is air/acetylene; N/A is nitrous oxide/acetylene.

Cupric Chloride Dihydrate

Copper(II) Chloride Dihydrate

$CuCl_2 \cdot 2H_2O$ **Formula Wt 170.48**   **CAS No. 10125-13-0**

GENERAL DESCRIPTION

Typical appearance: blue or blue-green solid
Analytical use: preparation of copper standard solutions; test for molybdenum
Change in state (approximate): melting point, 100 °C
Aqueous solubility: 121 g in 100 mL at 16 °C

SPECIFICATIONS

Assay . . . .$\geq$99.0% $CuCl_2 \cdot 2H_2O$

Maximum Allowable

Insoluble matter 0.01%
Nitrate (NO_3) 0.015%
Sulfate (SO_4) 0.005%
Calcium (Ca) 0.005%
Iron (Fe) 0.005%
Nickel (Ni) 0.01%
Potassium (K) 0.01%
Sodium (Na) 0.02%

TESTS

Assay. (By iodometric titration of oxidative capacity of Cu^{II}). Weigh accurately about 0.7 g, transfer to a glass-stoppered conical flask, and dissolve in 100 mL of water. Add 2 mL of glacial acetic acid and 3 g of potassium iodide. Stopper and swirl. Titrate the liberated iodine with 0.1 N sodium thiosulfate volumetric solution, adding about 2.0 g of potassium thiocyanate and 3 mL of starch indicator solution near the end of the titration. One milliliter of 0.1 N sodium thiosulfate corresponds to 0.01705 g of $CuCl_2 \cdot 2H_2O$.

$$\% \ CuCl_2 \cdot 2H_2O = \frac{(mL \times N \ Na_2S_2O_3) \times 17.05}{Sample \ wt \ (g)}$$

Insoluble Matter. (Page 25). Use 10 g in 100 mL of dilute hydrochloric acid (1 + 50). Omit heating. Retain the filtrate for the test for sulfate.

Nitrate. (Page 38, Method 1).

Sample Stock Solution. In a 150-mL beaker, dissolve 1.0 g in 10 mL of water, add the solution with stirring to 50 mL of 10% sodium hydroxide reagent solution, and digest on a hot plate ($\approx$100 °C) for 15 min. Cool, and filter through a retentive filter. Neutralize the filtrate with sulfuric acid, and dilute with water to 100 mL.

For sample solution A, dilute 30.0 mL of sample stock solution to 50 mL with brucine sulfate reagent solution. For control solution B, add 10.0 mL of sample stock solution and 0.03 mg of nitrate ion (NO_3), and dilute to 50 mL with brucine sulfate reagent solution. Continue with the procedure, starting with the preparation of blank solution C.

Sulfate. Heat the filtrate from the test for insoluble matter to boiling, add 10 mL of 12% barium chloride reagent solution, digest in a covered beaker on a hot plate ($\approx$100 °C) for 2 h, and allow to stand for at least 8 h. If a precipitate is formed, filter, wash thoroughly, and ignite it. The weight of the precipitate should not be 0.0012 g more than the weight obtained in a complete blank test.

Calcium, Iron, Nickel, Potassium, and Sodium. (By flame AAS, page 63).

Sample Stock Solution. Dissolve 10.0 g of sample in water, and add 5 mL of nitric acid. Transfer to a 100-mL volumetric flask, and dilute to the mark with water (1 mL = 0.10 g).

Element	Wavelength (nm)	Sample Wt (g)	Standard Added (mg)	Flame Type*	Background Correction
Ca	422.7	2.0	0.05; 0.10	N/A	No
Fe	248.3	2.0	0.05; 0.10	A/A	Yes
Ni	232.0	2.0	0.10; 0.20	A/A	No
K	766.5	0.10	0.01; 0.02	A/A	No
Na	589.0	0.10	0.01; 0.02	A/A	No

*A/A is air/acetylene; N/A is nitrous oxide/acetylene.

Cupric Nitrate Hydrate
Copper(II) Nitrate Hydrate

$Cu(NO_3)_2 \cdot 2.5H_2O$	**Formula Wt 232.59**	**CAS No. 19004-19-4**
$Cu(NO_3)_2 \cdot 3H_2O$	**Formula Wt 241.60**	**CAS No. 10031-43-3**

Note: This reagent is available containing either 2.5 or 3 molecules of water.

GENERAL DESCRIPTION

Typical appearance: hygroscopic blue solid
Analytical use: preparation of copper standard solutions
Change in state (approximate): melting point, 112–115 °C
Aqueous solubility: 140 g in 100 mL at 0 °C

SPECIFICATIONS

Assay . . . . 98.0–102.0% $Cu(NO_3)_2 \cdot 3H_2O$

Maximum Allowable

Insoluble matter	0.01%
Chloride (Cl)	0.002%
Sulfate (SO_4)	0.01%
Calcium (Ca)	0.005%
Iron (Fe).	0.005%
Lead (Pb)	0.001%
Nickel (Ni).	0.01%
Potassium (K)	0.005%
Sodium (Na)	0.01%

TESTS

Assay. (By iodometric titration of oxidative capacity of Cu^{II}). Weigh accurately about 1.0 g, dissolve in 5 mL of hydrochloric acid in a 100-mL beaker, and evaporate to dryness on a steam bath. Add 2 mL of sulfuric acid, and heat on a hot plate ($\approx$100 °C) to dense fumes of sulfur trioxide. Cool, add 5 mL of water and 5 mL of hydrochloric acid, and again evaporate to dryness on a hot plate ($\approx$100 °C). Dissolve the residue in water, transfer to a conical flask, and dilute with water to about 75 mL. Add 2 mL of glacial acetic acid and 3 g of potassium iodide. Titrate the liberated iodine with 0.1 N sodium thiosulfate volumetric solution, adding about 2 g of potassium thiocyanate and 3 mL of starch indicator solution near the end of the titration. Correct for a complete blank. One milliliter of 0.1 N sodium thiosulfate corresponds to 0.02326 g of $Cu(NO_3)_2 \cdot 2.5H_2O$ or 0.02416 g of $Cu(NO_3)_2 \cdot 3H_2O$.

$$\% \; Cu(NO_3)_2 \cdot 2.5H_2O = \frac{(mL \times N \; Na_2S_2O_3) \times 23.26}{Sample \; wt \; (g)}$$

$$\% \; Cu(NO_3)_2 \cdot 3H_2O = \frac{(mL \times N \; Na_2S_2O_3) \times 24.16}{Sample \; wt \; (g)}$$

Insoluble Matter. (Page 25). Use 10.0 g dissolved in 100 mL of dilute nitric acid (1 + 200). Omit heating.

Chloride. Dissolve 0.50 g in 15 mL of dilute nitric acid (1 + 15), filter if necessary through a small chloride-free filter, and add 1 mL of silver nitrate reagent solution. Any turbidity should not exceed that produced by 0.01 mg of chloride ion (Cl) in an equal volume of solution containing the quantities of reagents used in the test. The comparison is best made by the general method for chloride in colored solution, page 35.

Sulfate. (Page 41, Method 2). Use 1.5 mL of hot dilute hydrochloric acid (1 + 1). Do two evaporations. Allow 30 min for turbidity to form.

Calcium, Iron, Lead, Nickel, Potassium, and Sodium. (By flame AAS, page 63).

Sample Stock Solution. Dissolve 25.0 g of sample in 50 mL of water and 10 mL of nitric acid. Transfer to a 100-mL volumetric flask, and dilute to the mark with water (1 mL = 0.25 g).

Element	Wavelength (nm)	Sample Wt (g)	Standard Added (mg)	Flame Type*	Background Correction
Ca	422.7	2.0	0.05; 0.10	N/A	No
Fe	248.3	2.0	0.10; 0.20	A/A	Yes
Pb	217.0	5.0	0.05; 0.10	A/A	Yes
Ni	232.0	2.0	0.10; 0.20	A/A	Yes
K	766.5	0.20	0.01; 0.02	A/A	No
Na	589.0	0.20	0.01; 0.02	A/A	No

*A/A is air/acetylene; N/A is nitrous oxide/acetylene.

Cupric Oxide, Powdered
Copper(II) Oxide
CuO   **Formula Wt 79.55**   **CAS No. 1317-38-0**

GENERAL DESCRIPTION
Typical appearance: black or brownish-black powder
Analytical use: source for oxygen
Change in state (approximate): melting point, 1026 °C
Aqueous solubility: insoluble

SPECIFICATIONS
Assay . . . . ≥99.0% CuO

Maximum Allowable

Insoluble in dilute hydrochloric acid. 0.02%
Carbon compounds (as C) 0.01%
Chloride (Cl). 0.005%
Nitrogen compounds (as N). 0.002%
Sulfate (SO$_4$) 0.02%
Calcium (Ca) 0.01%
Iron (Fe). 0.05%
Potassium (K) 0.02%
Sodium (Na). 0.05%

TESTS
Assay. (By complexometric titration of copper). Weigh, to the nearest 0.1 mg, 0.2 g of sample. Dissolve in a mixture of 20 mL of water and 5 mL of nitric acid, and boil for a few minutes to dissolve the sample completely. Cool, dilute with 50 mL of water in a 400-mL beaker, and add 50 mL of alcohol and 50.0 mL of 0.1 M EDTA volumetric solution. Then

adjust the pH to about 5 (using a pH meter) with ammonium acetate–acetic acid buffer solution, add 0.2 mL of PAN indicator solution, and titrate with 0.1 M cupric sulfate volumetric solution while maintaining the pH at 5 by the addition of more buffer during the titration. The color change is from green to deep blue. Run a complete blank. One milliliter of 0.1 M EDTA corresponds to 0.007954 of CuO.

$$\% \text{ CuO} = \frac{\{[\text{mL (blank)} - \text{mL (sample)}] \times \text{M EDTA}\} \times 7.954}{\text{Sample wt (g)}}$$

Insoluble in Dilute Hydrochloric Acid. Dissolve 5.0 g by warming on a hot plate ($\approx$100 °C) with 30 mL of dilute hydrochloric acid (2 + 1). Dilute with about 100 mL of water, heat to boiling, and digest in a covered beaker on a hot plate ($\approx$100 °C) for 1 h. Filter through a tared filtering crucible, wash thoroughly, and dry at 105 °C.

Carbon Compounds. Ignite 1.2 g in a stream of carbon dioxide-free air or oxygen, and pass the effluent gases into 20 mL of 2.5% ammonium hydroxide reagent solution plus 2 mL of water. Prepare a standard of 20 mL of the 2.5% ammonium hydroxide and 1.06 mg of sodium carbonate (0.12 mg of C). Add 2 mL of 12% barium chloride reagent solution to each, and compare promptly. Any turbidity in the solution of the sample should not exceed that in the standard.

Chloride. For the sample solution, shake 1.0 g with 25 mL of dilute nitric acid (1 + 3) for 10 min, filter through a chloride-free filter, and dilute with water to 50 mL. To 10 mL of the sample solution, add 5 mL of water and 1 mL of silver nitrate reagent solution. Any turbidity should not exceed that produced by 0.01 mg of chloride ion (Cl) in an equal volume of solution containing 1 mL of nitric acid and 1 mL of silver nitrate reagent solution. The comparison is best made by the general method for chloride in colored solutions, page 35.

Nitrogen Compounds. (Page 39). Use 3.0 g dissolved in 30 mL of freshly boiled 10% sodium hydroxide. For the standard, use 0.06 mg of nitrogen.

Sulfate. (Page 41, Method 2). Use 1.5 mL of dilute aqua regia (1 volume nitric acid + 4 volumes hydrochloric acid + 6 volumes water). Allow 30 min for turbidity to form.

Calcium, Iron, Potassium, and Sodium. (By flame AAS, page 63).

    Sample Stock Solution A. Dissolve 4.0 g of sample by heating with 30 mL of dilute hydrochloric acid (2 + 1). Cool to room temperature, transfer to a 100-mL volumetric flask, and dilute to the mark with water (1 mL = 0.04 g).

    Sample Stock Solution B. Dilute 10.0 mL of sample stock solution A with water to 50 mL in a volumetric flask (1 mL = 0.008 g).

Element	Wavelength (nm)	Sample Wt (g)	Standard Added (mg)	Flame Type*	Background Correction
Ca	422.7	0.40	0.02; 0.04	N/A	No
Fe	248.3	0.40	0.10; 0.20	A/A	Yes
K	766.5	0.20	0.02; 0.04	A/A	No
Na	589.0	0.04	0.01; 0.02	A/A	No

*A/A is air/acetylene; N/A is nitrous oxide/acetylene.

Cupric Oxide, Wire
Copper(II) Oxide, Wire
CuO   Formula Wt 79.55   CAS No. 1317-38-0

GENERAL DESCRIPTION
Typical appearance: black, wirelike solid
Analytical use: determination of nitrogen
Change in state (approximate): melting point, 1026 °C
Aqueous solubility: insoluble

SPECIFICATIONS
                                                         Maximum Allowable
Carbon compounds (as C) 0.002%
Nitrogen compounds (as N). 0.002%
Sulfate (SO_4) 0.012%

TESTS

Carbon Compounds. Ignite 6.0 g in a stream of carbon dioxide-free air or oxygen, and pass the effluent gases into 20 mL of 2.5% ammonium hydroxide reagent solution plus 2 mL of water. Prepare a standard of 20 mL of the 2.5% ammonium hydroxide and 1.06 mg of sodium carbonate (0.12 mg of C). Add 2 mL of 12% barium chloride reagent solution to each, and compare promptly. Any turbidity in the solution representing the sample should not exceed that in the standard.

Nitrogen Compounds. (Page 39). Use 3.0 g dissolved in 30 mL of freshly boiled 10% sodium hydroxide. For the standard, use 0.06 mg of nitrogen ion (N).

Sulfate. (Page 41, Method 2). Use 2 mL of dilute aqua regia (1 volume nitric acid + 4 volumes hydrochloric acid + 6 volumes water). Allow 30 min for turbidity to form.

Cupric Sulfate Pentahydrate
Copper(II) Sulfate Pentahydrate
$CuSO_4 \cdot 5H_2O$ Formula Wt 249.68   CAS No. 7758-99-8

GENERAL DESCRIPTION
Typical appearance: blue solid
Analytical use: source of Cu^{II} ion
Change in state (approximate): melting point, 110 °C with loss of 4 water molecules
Aqueous solubility: 33 g in 100 mL at 20 °C

SPECIFICATIONS

Assay . . . . 98.0–102.0% $CuSO_4 \cdot 5H_2O$

Maximum Allowable

Insoluble matter 0.005%
Chloride (Cl) 0.001%
Nitrogen compounds (as N) 0.002%
Calcium (Ca) 0.005%
Iron (Fe). 0.003%
Nickel (Ni). 0.005%
Potassium (K) 0.01%
Sodium (Na) 0.02%

TESTS

Assay. (By iodometric titration of oxidative capacity of Cu^{II}). Weigh accurately about 1.0 g, dissolve in water, transfer to a conical flask, and dilute with water to about 50 mL. Add 4 mL of glacial acetic acid, 1 mL of 10% sulfuric acid, and 3 g of potassium iodide. Titrate the liberated iodine with 0.1 N sodium thiosulfate volumetric solution, adding about 2 g of potassium thiocyanate and 3 mL of starch indicator solution near the end of the titration. Correct for a blank. One milliliter of 0.1 N sodium thiosulfate corresponds to 0.02497 g of $CuSO_4 \cdot 5H_2O$.

$$\% \ CuSO_4 \cdot 5H_2O = \frac{(mL \times N \ Na_2S_2O_3) \times 24.97}{Sample \ wt \ (g)}$$

Insoluble Matter. (Page 25). Use 20 g dissolved in 200 mL of dilute sulfuric acid (1 + 40). Omit heating.

Chloride. Dissolve 1.0 g in 15 mL of water, filter if necessary through a small chloride-free filter, and add 1 mL of nitric acid and 1 mL of silver nitrate reagent solution. Any turbidity should not exceed that produced by 0.01 mg of chloride ion (Cl) in an equal volume of solution containing the quantities of reagents used in the test. The comparison is best made by the general method for chloride in colored solutions, page 35.

Nitrogen Compounds. (Page 39). Use 1.0 g. For the standard, use 0.02 mg of nitrogen.

Calcium, Iron, Nickel, Potassium, and Sodium. (By flame AAS, page 63).

Sample Stock Solution. Dissolve 20.0 g of sample in water, and add 5 mL of nitric acid. Transfer to a 100-mL volumetric flask, and dilute to the mark with water (1 mL = 0.20 g).

Element	Wavelength (nm)	Sample Wt (g)	Standard Added (mg)	Flame Type*	Background Correction
Ca	422.7	0.80	0.02; 0.04	N/A	No
Fe	248.3	4.0	0.12; 0.24	A/A	Yes
Ni	232.0	4.0	0.10; 0.20	A/A	Yes
K	766.5	0.40	0.02; 0.04	A/A	No
Na	589.0	0.05	0.005; 0.01	A/A	No

*A/A is air/acetylene; N/A is nitrous oxide/acetylene.

Cuprous Chloride
Copper(I) Chloride
CuCl   Formula Wt 99.00   CAS No. 7758-89-6

GENERAL DESCRIPTION

Typical appearance: white crystalline solid; may oxidize to green in air
Analytical use: absorption of carbon monoxide in gas analysis
Change in state (approximate): melting point, 430 °C
Aqueous solubility: 1.52 g in 100 mL at 25 °C

SPECIFICATIONS

Assay . . . . ≥90.0% CuCl

Maximum Allowable

Insoluble in acid. 0.02%
Sulfate (SO₄) 0.1%
Calcium (Ca) 0.01%
Iron (Fe). 0.005%
Potassium (K) 0.02%
Sodium (Na). 0.05%

TESTS

Assay. (By iodometric titration of reductive capacity of Cu^I). Weigh accurately 0.4 g and dissolve in the cold in 30 mL of ferric ammonium sulfate solution (10.0 g of ferric ammonium sulfate dissolved in 100 mL of 20% hydrochloric acid reagent solution). Add 5 mL of phosphoric acid, dilute with 200 mL of water, and titrate with 0.1 N potassium permanganate. The solution should be prepared and titrated in a flask in which the air is replaced with carbon dioxide. A stream of carbon dioxide should be passed through the flask during the titration. One milliliter of 0.1 N permanganate corresponds to 0.009900 g of CuCl.

$$\% \, CuCl = \frac{(mL \times N \, KMnO_4) \times 9.900}{Sample \, wt \, (g)}$$

Insoluble in Acid. Heat 5.0 g with 50 mL of dilute hydrochloric acid (1 + 1), and add nitric acid in small portions until the sample is dissolved. Dilute with 50 mL of water. If insoluble matter is present, filter through a tared filtering crucible, wash thoroughly, and dry at 105 °C. Retain the filtrate and washings for the test for sulfate.

Sulfate. (Page 40, Method 1). Dilute the combined filtrate and washings from the test for insoluble in acid with water to 200 mL. Use 2 mL of this solution. Allow 30 min for turbidity to form.

Calcium, Iron, Potassium, and Sodium. (By flame AAS, page 63).

Sample Stock Solution. To 10.0 g of sample, add 45 mL of dilute hydrochloric acid (1 + 1), and heat on a hot plate (≈100 °C). Cautiously add 5 mL of nitric acid, and keep warming and stirring until dissolved. Cool to room temperature, transfer to a 100-mL volumetric flask, and dilute to the mark with water (1 mL = 0.10 g).

Element	Wavelength (nm)	Sample Wt (g)	Standard Added (mg)	Flame Type*	Background Correction
Ca	422.7	0.20	0.02; 0.04	N/A	No
Fe	248.3	1.0	0.05; 0.10	A/A	Yes
K	766.5	0.20	0.02; 0.04	A/A	No
Na	589.0	0.02	0.005; 0.01	A/A	No

*A/A is air/acetylene; N/A is nitrous oxide/acetylene.

Cyclohexane

C_6H_{12} **Formula Wt 84.16**   **CAS No. 110-82-7**

GENERAL DESCRIPTION

Typical appearance: clear, colorless liquid
Analytical use: organic solvent
Change in state (approximate): boiling point, 80–81 °C
Aqueous solubility: virtually insoluble
Density: 0.78

SPECIFICATIONS

General Use

Assay . . . . ≥99.0% C_6H_{12}

Maximum Allowable

Color (APHA) 10
Residue after evaporation 0.002%
Substances darkened by sulfuric acid . . . . Passes test
Water (H$_2$O) 0.02%

Specific Use

Ultraviolet Spectrophotometry

Wavelength (nm)	Absorbance (AU)
300–400	0.01
260	0.02
250	0.03
240	0.08
230	0.20
220	0.50
210	1.00

TESTS

Assay. Analyze the sample by gas chromatography using the general parameters cited on page 80. The following specific conditions are also required.

Column: Type I, methyl silicone

Measure the area under all peaks, and calculate the cyclohexane content in area percent. Correct for water content.

Color (APHA). (Page 43).

Residue after Evaporation. (Page 25). Evaporate 100.0 g (129 mL) to dryness in a tared, preconditioned dish on a hot plate ($\approx$100 °C), and dry the residue at 105 °C for 30 min.

Substances Darkened by Sulfuric Acid. Shake 25 mL with 15 mL of sulfuric acid for 15–20 s, and allow to separate. Neither the cyclohexane nor the acid should be darkened.

Water. (Page 31, Method 2). Use 200 μL (160 mg) of the sample.

Ultraviolet Spectrophotometry. Use the procedure on page 86 to determine the absorbance.

Cyclohexanone

$C_6H_{10}O$ **Formula Wt 98.14**   **CAS No. 108-94-1**

GENERAL DESCRIPTION

Typical appearance: clear, oily liquid
Analytical use: organic solvent
Change in state (approximate): boiling point, 154–156 °C
Aqueous solubility: 5 g in 100 mL at 30 °C
Density: 0.95
pK_a: −6.8

SPECIFICATIONS

Assay . . . . $\geq$99.0% $C_6H_{10}O$

Maximum Allowable

Color (APHA) 10
Residue after evaporation 0.05%
Water (H$_2$O) 0.05%

TESTS

Assay. Analyze the sample by gas chromatography using the parameters cited on page 80. The following specific conditions are also required.

Column: Type I, methyl silicone

Measure the area under all peaks and calculate the cyclohexanone content in area percent. Correct for water content.

Color (APHA). (Page 43).

Residue after Evaporation. (Page 25). Evaporate 100 g (105 mL) to dryness in a tared, preconditioned dish on a hot plate (≈100 °C), and dry the residue at 105 °C for 30 min.

Water. (Page 31, Method 2). Use 30 mL (24 g) of the sample and a methanol-free system to prevent ketal formation with liberation of water.

(1,2-Cyclohexylenedinitrilo)tetraacetic Acid
CDTA

$C_{14}H_{22}N_2O_8 \cdot H_2O$ (free acid)	**Formula Wt 364.35**	**CAS No. 125572-95-4**
$C_{14}H_{20}N_2O_8Na_2 \cdot 3H_2O$ (disodium salt)	**Formula Wt 444.35**	**CAS No. 57137-35-6**

Note: This specification applies to both the free acid and disodium salt forms of this reagent. The residue after ignition test does not apply to the disodium salt form.

GENERAL DESCRIPTION

Typical appearance: white solid
Analytical use: chelating agent
Aqueous solubility: slightly soluble

SPECIFICATIONS

Assay (free acid)..97.5–100.5% $C_{14}H_{22}N_2O_8 \cdot H_2O$
Assay (disodium salt).................................≥94.0% $C_{14}H_{20}N_2O_8Na_2 \cdot 3H_2O$

Maximum Allowable

Residue after ignition 0.2%
Heavy metals (as Pb) 0.001%
Iron (Fe).. 0.005%

TESTS

Assay. (By complexometry). For the free acid, weigh accurately about 0.75 g, and dissolve it in 50 mL of 0.1 N sodium hydroxide in a 250-mL beaker. For the disodium salt, weigh accurately about 0.95 g, and dissolve in 50 mL of water in a 250-mL beaker. Add 25 mL of water, 10 mL of ammonium acetate buffer solution, and 0.15 mL of phenolphthalein indicator solution. If necessary, neutralize the solution to the pink color of phenolphthalein by adding dropwise 10% ammonium hydroxide reagent solution. Add glacial acetic acid dropwise to a pH of 5.8 (using a pH meter). Add 0.10 mL of xylenol orange indicator solution. Titrate with 0.1 M zinc chloride volumetric solution to the change from yellow to magenta purple that remains for at least 1 min. One milliliter of 0.1 M zinc chloride corresponds to 0.03644 g of $C_{14}H_{22}N_2O_8 \cdot H_2O$ and 0.04444 g of $C_{14}H_{20}N_2O_8Na_2 \cdot 3H_2O$.

$$\% \ C_{14}H_{22}N_2O_8 \cdot H_2O = \frac{(mL \times M \ ZnCl_2) \times 36.44}{\text{Sample wt (g)}}$$

$$\% \ C_{14}H_{20}N_2O_8Na_2 \cdot 3H_2O = \frac{(mL \times M \ ZnCl_2) \times 44.44}{\text{Sample wt (g)}}$$

Residue after Ignition. (Page 26). Use 3.0 g.

For the Determination of Heavy Metals and Iron:

Sample Solution A. Ignite 3.0 g thoroughly, and heat in an oven at 600 °C until most of the carbon is volatilized. Cool, add 0.15 mL of nitric acid, and heat at 600 °C until all the carbon is volatilized. Dissolve the residue in 2 mL of dilute hydrochloric acid (1 + 1), digest in a covered dish on a hot plate (≈100 °C) for 10 min, remove the cover, and evaporate to dryness. Dissolve in 1 mL of 1 N acetic acid and 20 mL of hot water, digest for 5 min, cool, and dilute with water to 30 mL (1 mL = 0.1 g).

Heavy Metals. (Page 36, Method 1). Dilute 20 mL of sample solution A (2-g sample) with water to 25 mL.

Iron. (Page 38, Method 1). Use 2.0 mL of sample solution A (0.2-g sample).

Devarda's Alloy
Decarda Metal

CAS No. 8049-11-4

GENERAL DESCRIPTION
Typical appearance: gray solid
Analytical use: reducing agent for determination of nitrogen by Kjeldahl analysis

SPECIFICATIONS
Aluminum (Al) (nominal) 44–46%
Copper (Cu) (nominal) 49–51%
Zinc (Zn) (nominal) 4–6%

Maximum Allowable

Nitrogen (N) 0.005%

TESTS

Aluminum, Copper, and Zinc. (By flame AAS, page 63).

Sample Stock Solution A. Weigh exactly 5.000 g, and in a fume hood, suspend in 25 mL of hydrochloric acid in a 1-L volumetric flask. After 1 h, add cautiously 25 mL of nitric acid in 5-mL increments. If a vigorous reaction with red-brown fumes takes place, immediately add 5–10 mL portions of water to quench the reaction. Once the alloy is in solution, cautiously add 500 mL of water, and cool. Dilute to the mark with water, and mix well (1 mL = 0.0050 g).

Sample Stock Solution B. Pipet 50.0 mL of sample stock solution A into a 1-L volumetric flask, and dilute to volume with 1% hydrochloric acid solution (1 mL = 0.00025 g).

Sample Stock Solution C. Pipet 50.0 mL of sample stock solution B to a 250-mL volumetric flask. Dilute to the mark with water, and mix well (1 mL = 0.00005 g).

Note: For aluminum determination, add 1.0 mL of 5% potassium chloride to the sample and standards. A reagent blank of equal amount of reagents is recommended.

Element	Wavelength (nm)	Sample Wt (g)	Standard Added (mg)	Flame Type*	Background Correction
Al	396.3	0.0005	0.15; 0.30	N/A	Yes
Cu	324.8	0.00005	0.015; 0.03	A/A	Yes
Zn	213.9	0.0005	0.015; 0.03	A/A	Yes

*A/A is air/acetylene; N/A is nitrous oxide/acetylene.

Note: Compensate for any variation on the summation of the components that deviate from 100%.

Nitrogen. Weigh accurately about 7.0 g of sample, and place in a Kjeldahl flask. Add slowly and cautiously 40 mL of hydrochloric acid with agitation. Cool, add 100 mL of 30% sodium hydroxide solution, and immediately connect to a Kjeldahl distillation apparatus, the tip of which is immersed in 10 mL of 0.02 N hydrochloric acid. Distill 75 mL into the 0.02 N hydrochloric acid solution, and titrate with 0.02 N sodium hydroxide solution to a methyl red end point. Not more than 0.9 mL of 0.02 N hydrochloric acid is consumed in excess of that used when a 2.0-g sample is treated in the same manner.

1,2-Dichloroethane

Ethylene Dichloride

CH_2ClCH_2Cl **Formula Wt 98.96** **CAS No. 107-06-2**

Suitable for general use or in ultraviolet spectrophotometry. Product labeling shall designate the uses for which suitability is represented on the basis of meeting the relevant speci-

fications and tests. The ultraviolet spectrophotometry specifications include all of the specifications for general use.

GENERAL DESCRIPTION

Typical appearance: clear, colorless liquid
Analytical use: organic solvent
Change in state (approximate): boiling point, 83–85 °C
Aqueous solubility: 0.9 g in 100 mL at 30 °C
Density: 1.26

SPECIFICATIONS

General Use

Assay . . . . $\geq$99.0% CH_2ClCH_2Cl

	Maximum Allowable
Color (APHA)	10
Residue after evaporation	0.002%
Titrable acid	0.0003 meq/g
Water (H_2O)	0.03%

Specific Use

Ultraviolet Spectrophotometry

Wavelength (nm)	Absorbance (AU)
255–400	0.01
250	0.02
245	0.05
240	0.10
235	0.20
230	0.50
228	1.00

TESTS

Assay. Analyze the sample by gas chromatography using the general parameters cited on page 80. The following specific conditions are also required.

 Column: Type I, methyl silicone

Measure the area under all peaks, and calculate the 1,2-dichloroethane content in area percent. Correct for water content.

Color (APHA). (Page 43).

Residue after Evaporation. (Page 25). Evaporate 100 g (82 mL) to dryness in a tared, preconditioned dish on a hot plate ($\approx$100 °C), and dry the residue at 105 °C for 30 min.

Titrable Acid.

 Note: Take special care during the addition of the sample and titration to avoid contamination from carbon dioxide.

To 25 mL of alcohol in a 100-mL glass-stoppered flask, add 0.10 mL of phenolphthalein indicator solution and 0.01 N sodium hydroxide until a faint pink color persists after shaking for 30 s. Add 31 g (25 mL) of sample, mix well, and titrate with 0.01 N sodium hydroxide until the pink color is restored. Not more than 0.93 mL of 0.01 N sodium hydroxide should be required.

Water. (Page 31, Method 2). Use 100 μL (125 mg) of the sample.

Ultraviolet Spectrophotometry. Use the procedure on page 86 to determine the absorbance.

2′,7′-Dichlorofluorescein

2′,7′-Dichloro-3′,6′-dihydroxyspiro[isobenzofuran-1(3*H*),9′-[9*H*]xanthen]-3-one;
2′,7′-Dichloro-3,6-fluorandiol

$C_{20}H_{10}Cl_2O_5$ **Formula Wt 401.20**   **CAS No. 76-54-0**

GENERAL DESCRIPTION

Typical appearance: orange to red-brown solid
Analytical use: argentimetric titration indicator

SPECIFICATIONS

Clarity of alcohol solution . . . . Passes test
Suitability as absorption indicator . . . . Passes test

TESTS

Clarity of Alcohol Solution. Dissolve 100 mg in 100 mL of 70% alcohol. The solution should be clear and greenish yellow. Reserve the solution for the test for suitability as absorption indicator.

Suitability as Absorption Indicator. Prepare a 0.10 M solution of sodium chloride by dissolving 0.58 g of sodium chloride in water, and diluting with water to 100.0 mL. Dilute 10.0 mL of the 0.10 M sodium chloride with 30 mL of water, add 0.5 mL of the solution reserved from the preceding test, and titrate with 0.10 N silver nitrate, protecting the titration vessel from direct sunlight. The color change from greenish yellow to rose should require between 9.9 and 10.1 mL of 0.10 N silver nitrate.

2,6-Dichloroindophenol Sodium Salt

2,6-Dichloro-*N*-[(4-hydroxyphenyl)imino]-2,5-cyclohexadien-1-one, Sodium Salt

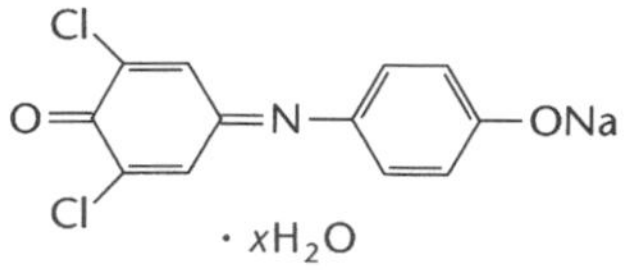

· xH$_2$O

O:C$_6$H$_2$Cl$_2$:NC$_6$H$_4$ONa   **Formula Wt 290.08**   CAS No. 620-45-1

GENERAL DESCRIPTION

Typical appearance: dark green solid
Analytical use: indicator
Aqueous solubility: slightly soluble

SPECIFICATIONS

	Maximum Allowable
Loss on drying	12.0%
Interfering dyes	Passes test

TESTS

Loss on Drying. Weigh accurately about 1.0 g, and dry at 120 °C to constant weight.

Interfering Dyes.

> **Solution A.** Dissolve 50 mg of the sample and 42 mg of sodium bicarbonate in water, and dilute with water to 200 mL. Filter through a dry filter, rejecting the first 20 mL of the filtrate.

> **Solution B.** Dissolve 50 mg of ascorbic acid in 50 mL of a solution composed of 1.5 g of metaphosphoric acid plus 4 mL of glacial acetic acid, and dilute to 50 mL with water.

To 15 mL of solution A, add 2.5 mL of solution B. The mixture should become nearly colorless, with no trace of red or blue.

Dichloromethane

Methylene Chloride

CH$_2$Cl$_2$ **Formula Wt 84.93**   CAS No. 75-09-2

Suitable for general use or in high-performance liquid chromatography, extraction–concentration analysis, or ultraviolet spectrophotometry. Product labeling shall designate the uses for which suitability is represented on the basis of meeting the relevant specifications and tests. The ultraviolet spectrophotometry and liquid chromatography suitability specifications include all of the specifications for general use. The extraction–concentration suitability specifications include only the general use specification for color.

GENERAL DESCRIPTION

Typical appearance: clear, colorless liquid

Analytical use: extraction solvent

Change in state (approximate): boiling point, 40 °C

Aqueous solubility: 1.3 g in 100 mL at 25 °C

Density: 1.33

SPECIFICATIONS

General Use

Assay . . . . $\geq$99.5% CH_2Cl_2

	Maximum Allowable
Color (APHA)	10
Residue after evaporation	0.002%
Titrable acid	0.0003 meq/g
Free halogens	Passes test
Water (H_2O)	0.02%

Specific Use

Ultraviolet Spectrophotometry

Wavelength (nm)	Absorbance (AU)
340–400	0.01
260	0.04
250	0.10
240	0.35
235	1.0

Liquid Chromatography Suitability

Absorbance . . . . Passes test

Extraction–Concentration Suitability

Absorbance . . . . Passes test

GC–FID . . . . Passes test

GC–ECD . . . . Passes test

TESTS

Assay. Analyze the sample by gas chromatography using the parameters cited on page 80. The following specific conditions are also required.

> *Column:* Type I, methyl silicone

Measure the area under all peaks, and calculate the dichloromethane content in area percent. Correct for water content.

Color (APHA). (Page 43).

Residue after Evaporation. (Page 25). Evaporate 100 g (76 mL) in a tared, preconditioned dish on a hot plate ($\approx$100 °C), and dry the residue at 105 °C for 30 min.

Titrable Acid.

> *Note:* Take special care during the addition of the sample and the titration to avoid contamination from carbon dioxide.

To 25 mL of alcohol in a 100-mL glass-stoppered flask, add 0.10 mL of phenolphthalein indicator solution and 0.01 N sodium hydroxide solution until a pink color persists for at least 30 s after vigorous shaking. Add 25 mL (33 g) of sample from a pipet, and mix thoroughly with the neutralized alcohol. If no pink color remains, titrate with 0.01 N sodium hydroxide to the end point where the pink color persists for at least 30 s. Not more than 0.90 mL of 0.01 N sodium hydroxide should be required.

Free Halogens. Shake 10 mL for 2 min with 10 mL of water to which 0.10 mL of 10% potassium iodide reagent solution has been added and allow to separate. The lower layer should not show a violet tint.

Water. (Page 31, Method 2). Use 250 μL (330 mg) of the sample.

Ultraviolet Spectrophotometry. Use the procedure on page 86 to determine the absorbance.

Liquid Chromatography Suitability. Use the procedure on page 86 to determine the absorbance.

Extraction–Concentration Suitability. Analyze the sample, using the general procedure cited on page 85. In the sample preparation step, exchange the sample with ACS-grade hexanes suitable for extraction–concentration. Use the procedure on page 86 to determine the absorbance.

Diethanolamine
2,2′-Iminodiethanol

(HOCH$_2$CH$_2$)$_2$NH	**Formula Wt 105.14**	**CAS No. 111-42-2**

GENERAL DESCRIPTION

Typical appearance: clear, viscous liquid
Analytical use: buffer formulation
Change in state (approximate): boiling point, 268–269 °C
Aqueous solubility: miscible
Density: 1.09

SPECIFICATIONS

Assay . . . . ≥98.5% (HOCH$_2$CH$_2$)$_2$NH
Apparent equivalent weight 104.0–106.0

Maximum Allowable
Color (APHA) 15

Residue after ignition 0.005%
2-Aminoethanol 1.0%
Triethanolamine 1.0%
Water (H_2O) 0.15%

TESTS

Assay, 2-Aminoethanol, and Triethanolamine. Analyze the sample by gas chromatography using the general parameters cited on page 80. The following specific conditions are also required.

> *Column:* Type I, methyl silicone

Make a 10% (w/w) mixture of the sample with ethyl alcohol.

Measure the area under all peaks, and calculate the diethanolamine, 2-aminoethanol, and triethanolamine content in area percent. Correct for water content and ethyl alcohol.

Apparent Equivalent Weight. To 50 mL of water in a 250-mL conical flask, add 0.3 mL of mixed indicator solution (described below), and neutralize by adding 0.1 N hydrochloric acid just to disappearance of the green color. From a suitable weighing pipet, introduce 1.8–2.0 g of sample, weighing to within 0.5 mg. Mix and titrate with standard 0.5 N hydrochloric acid to disappearance of the green color.

$$\text{Apparent equivalent weight} = \frac{\text{Sample wt (g)} \times 1000}{\text{mL} \times \text{N HCl}}$$

> *Mixed Indicator Solution.* Prepare separate solutions of bromocresol green and methyl red in methanol (0.1 g/100 mL). When needed, mix freshly each day 5 parts of the bromocresol green solution with 1 part of the methyl red solution.

Color (APHA). (Page 43).

Residue after Ignition. (Page 26). Evaporate 50.0 g (45 mL) to dryness in a tared, preconditioned dish. Add 0.10 mL of sulfuric acid, again evaporate to dryness, and finally ignite at 600 ± 25 °C.

Water. (Page 31, Method 2). Use 0.1 g of the sample.

Diethylamine
N-Ethylethanamine

$C_4H_{11}N$ **Formula Wt 73.14**   **CAS No. 109-89-7**

GENERAL DESCRIPTION
Typical appearance: clear liquid
Analytical use: buffer formulation

Change in state (approximate): boiling point, 56 °C
Aqueous solubility: very soluble
Density: 0.71
pK_a: 11.0

SPECIFICATIONS

Assay . . . . ≥99.0% $C_4H_{11}N$

Maximum Allowable

Color (APHA) 20
Monoethylamine 0.5%
Triethylamine 0.5%
Water (H_2O) 0.1%

TESTS

Assay, Monoethylamine, and Triethylamine. Analyze the sample by gas chromatography using the general parameters cited on page 80. The following specific conditions
are also required.

   Column: Type I, methyl silicone

Measure the area under each peak, and calculate the diethylamine, monoethylamine, and
triethylamine content in area percent. Correct for water content.

Color (APHA). (Page 43).

Water. (Page 31, Method 2). Use 50 µL (35 mg) of the sample.

4-(Dimethylamino)benzaldehyde

(CH₃)₂NC₆H₄CHO   **Formula Wt 149.19**   **CAS No. 100-10-7**

GENERAL DESCRIPTION

Typical appearance: white to pale yellow solid
Analytical use: derivatizing agent
Aqueous solubility: very slightly soluble

SPECIFICATIONS

Melting point . . . . 73–75 °C

Maximum Allowable

Solubility in alcohol . . . . Passes test
Color (APHA) of alcohol solution 60
Solubility in hydrochloric acid . . . . Passes test
Color of hydrochloric acid solution. . . . . Passes test
Residue after ignition 0.1%

TESTS

Melting Point. (Page 45).

Solubility in Alcohol. Dissolve 1.0 g in 25 mL of alcohol. The solution should be clear.

Color (APHA) of Alcohol Solution. (Page 43). Dissolve 4.0 g in alcohol, and dilute with alcohol to 100 mL.

Solubility in Hydrochloric Acid. Dissolve 1.0 g in 20 mL of dilute hydrochloric acid (1 + 10). Dissolution should be complete; the solution, clear. Reserve the solution for the test for color of hydrochloric acid solution.

Color of Hydrochloric Acid Solution. Transfer a portion of the solution prepared in the test for solubility in hydrochloric acid into a 1-cm curette. Measure the absorbance at 450 nm vs. water. The absorbance should not exceed 0.1 AU.

Residue after Ignition. (Page 26). Ignite 1.0 g.

N,N-Dimethylformamide

HCON(CH$_3$)$_2$ **Formula Wt 73.09**   **CAS No. 68-12-2**

Suitable for general use or in ultraviolet spectrophotometry. Product labeling shall designate the uses for which suitability is represented on the basis of meeting the relevant specifications and tests. The ultraviolet spectrophotometry specifications include all of the specifications for general use.

GENERAL DESCRIPTION

Typical appearance: clear, colorless to slightly yellow liquid
Analytical use: organic solvent
Change in state (approximate): boiling point, 152–154 °C
Aqueous solubility: soluble
Density: 0.94
pK_a: 0.0

SPECIFICATIONS
General Use
Assay . . . . $\geq$99.8% HCON(CH$_3$)$_2$

Maximum Allowable

Color (APHA) 15
Residue after evaporation 0.005%
Titrable base . . . . 0.003 meq/g
Titrable acid. . . . . 0.0005 meq/g
Water (H$_2$O) 0.15%

Specific Use
Ultraviolet Spectrophotometry

Wavelength (nm)	Absorbance (AU)
340–400	0.01
310	0.05
295	0.10
275	0.30
270	1.00

TESTS

Assay. Analyze the sample by gas chromatography using the general parameters cited on page 80. The following specific conditions are also required.

   Column: Type I, methyl silicone

Measure the area under all peaks, and calculate the *N,N*-dimethylformamide content in area percent. Correct for water content.

Color (APHA). (Page 43).

Residue after Evaporation. (Page 25). Evaporate 20.0 g (21 mL) to dryness in a tared, preconditioned dish on a hot plate ($\approx$100 °C), and dry the residue at 105 °C for 30 min.

> *Note:* In the tests for titrable base and titrable acid, take care to minimize exposure of the sample to the atmosphere because of rapid absorption of carbon dioxide. Flush the flask with nitrogen during the titrations.

Titrable Base. Mix 10.0 g (10.6 mL) with 25 mL of carbon dioxide-free water in a glass-stoppered flask, and add 0.05 mL of methyl red indicator solution. If the solution becomes yellow, titrate with 0.01 N hydrochloric acid until a red color appears. Not more than 3.0 mL of the hydrochloric acid solution should be required.

Titrable Acid. Quickly add 19.0 g (20 mL) of the sample from a graduated cylinder to a 125-mL conical flask. Add 0.10 mL of thymol blue indicator solution (described below). The color of the solution should be yellow. Titrate with 0.01 N sodium methoxide in methanol solution (described below) to a blue end point. Not more than 1.0 mL should be consumed in the titration.

> ***Sodium Methoxide in Methanol Solution, 0.01 N.*** Dissolve 0.540 g of sodium methylate in dry methanol, and dilute with the methanol to 1 L in a volumetric flask. Standardize by titrating 40 mL of 0.01 N sulfuric acid, using thymol blue indicator.

Thymol Blue Indicator Solution. Dissolve 0.30 g of thymol blue in 100 mL of dry methanol.

Water. (Page 31, Method 2). Use 25.0 μL (24.0 mg) of the sample.

Ultraviolet Spectrophotometry. Use the procedure on page 86 to determine the absorbance.

Dimethylglyoxime
2,3-Butanedione Dioxime

$CH_3C:NOHC:NOHCH_3$	**Formula Wt 116.12**	**CAS No. 95-45-4**

GENERAL DESCRIPTION
Typical appearance: white, crystalline solid
Analytical use: nickel-specific complexing reagent; determination of palladium
Change in state (approximate): melting point, 238–241 °C, with decomposition
Aqueous solubility: 0.06 g in 100 mL at 20 °C

SPECIFICATIONS
Melting point . . . . .Approximately 240 °C
Suitability for nickel determination . . . . Passes test

Maximum Allowable
Insoluble in alcohol 0.05%
Residue after ignition 0.05%

TESTS

Melting Point. (Page 45).

Suitability for Nickel Determination. Dissolve 0.665 g of nickel sulfate hexahydrate in water, and dilute with water to 50 mL. Dilute 20 mL of this solution with water to 100 mL, heat to boiling, and add 0.25 g of the dimethylglyoxime in 25 mL of alcohol. Add dilute ammonium hydroxide (1 + 4) dropwise until the solution is alkaline to litmus, cool, and filter. To the filtrate, add 1 mL of the nickel sulfate solution and heat to boiling. A substantial precipitate of red nickel dimethylglyoxime should appear.

Insoluble in Alcohol. Gently boil 2.0 g with 100 mL of alcohol under a reflux condenser until no more dissolves. Filter through a tared, preconditioned filtering crucible, wash with 50 mL of alcohol in small portions, and dry at 105 °C.

Residue after Ignition. (Page 26). Ignite 2.0 g.

2,9-Dimethyl-1,10-phenanthroline
Neocuproine

$C_{14}H_{12}N_2$ Formula Wt 208.27   CAS No. 484-11-7

Note: This reagent is available with water of hydration between anhydrous and hemihydrate. It loses water easily, and either form is acceptable.

GENERAL DESCRIPTION

Typical appearance: white solid
Analytical use: spectrophotometric determination of copper
Change in state (approximate): hemihydrate melting point, 159–160 °C

SPECIFICATIONS

Clarity of solution . . . . Passes test
Sensitivity to copper . . . . Passes test

Maximum Allowable

Loss on drying 6.0%

TESTS

Clarity of Solution. Dissolve 0.30 g in 10 mL of water by adding hydrochloric acid dropwise until solution is complete. The solution should be clear with not more than a faint yellow color. Save the solution for the sensitivity to copper test.

Sensitivity to Copper. Dilute the sample reserved from the clarity of solution test to 100 mL with water. In a 25-mL flask, add a standard containing 0.005 mg of copper, 1 mL of hydroxylamine hydrochloride reagent solution, and 2 mL of 10% sodium acetate reagent solution. Add 0.1 mL of the prepared sample solution. A yellow-orange color should develop with an absorption maximum at 454 nm.

Loss on Drying. Accurately weigh 1.0 g in a preconditioned weighing bottle, and dry at 100 °C for 1 h. Cool, and weigh.

Dimethyl Sulfoxide
Methyl Sulfoxide; Sulfinylbismethane
$(CH_3)_2SO$ Formula Wt 78.13   CAS No. 67-68-5

Suitable for general use or in ultraviolet spectrophotometry. Product labeling shall designate the uses for which suitability is represented on the basis of meeting the relevant speci-

fications and tests. The ultraviolet spectrophotometry specifications include all of the specifications for general use.

GENERAL DESCRIPTION

Typical appearance: clear, colorless liquid
Analytical use: determination of organic volatile impurities; solvent
Change in state (approximate): boiling point, 188–190 °C
Aqueous solubility: very soluble
Density: 1.10
pK_a: 31.3

SPECIFICATIONS

General Use

Assay . . . . $\geq$99.9% $(CH_3)_2SO$

Maximum Allowable

Residue after evaporation 0.01%
Titrable acid . . . . 0.001 meq/g
Water (H_2O) 0.1%

Specific Use

Ultraviolet Spectrophotometry

Wavelength (nm)	Absorbance (AU)
350–400	0.01
330	0.02
310	0.06
290	0.18
270	0.40

TESTS

Assay. Analyze the sample by gas chromatography using the general parameters cited on page 80. The following specific conditions are also required.

Column: Type III, polyethylene glycol

Measure the area under all peaks, and calculate the dimethylsulfoxide content in area percent. Correct for water content.

Residue after Evaporation. (Page 25). Evaporate 100.0 g (91 mL) to dryness in a tared, preconditioned dish on a hot plate.

Titrable Acid. Dissolve 50.0 g in 100 mL of water, and add 0.1 mL of phenolphthalein indicator solution. If the solution remains colorless, titrate with 0.01 N sodium hydroxide until a pink color appears. Not more than 5.0 mL of 0.01 N sodium hydroxide should be consumed.

Water. (Page 31, Method 2). Use 50 µL (55 mg) of the sample.

Ultraviolet Spectrophotometry. Use the procedure on page 86 to determine the absorbance.

Dioxane

1,4-Dioxane

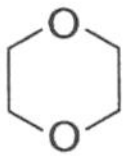

C$_4$H$_8$O$_2$ **Formula Wt 88.11**   **CAS No. 123-91-1**

Caution: Dioxane tends to form explosive peroxides, especially when anhydrous. It should not be allowed to evaporate to dryness unless the absence of peroxides has been shown.

Note: Dioxane usually contains a stabilizer. If a stabilizer is present, its identity and quantity must be stated on the label.

GENERAL DESCRIPTION

Typical appearance: clear liquid
Analytical use: organic solvent
Change in state (approximate): boiling point, 100–102 °C
Aqueous solubility: miscible
Density: 1.03
pK_a: −2.9

SPECIFICATIONS

Assay . . . . ≥99.0% C$_4$H$_8$O$_2$
Freezing point . . . . Not below 11.0 °C

Maximum Allowable

Color (APHA) 20
Peroxide (as H$_2$O$_2$) 0.005%
Residue after evaporation 0.005%
Titrable acid. . . . . 0.0016 meq/g
Carbonyl (as HCHO) 0.01%
Water (H$_2$O) 0.05%

TESTS

Assay. Analyze the sample by gas chromatography using the general parameters cited on page 80. The following specific conditions are also required.

Column: Type I, methyl silicone

Measure the area under all peaks, and calculate the dioxane content in area percent. Correct for water content.

Freezing Point.

Apparatus. For the sample container, use a test tube (25 × 100 mm) supported by a cork in a water-tight glass cylinder (50 × 110 mm). Mount the cylinder in a water bath that provides at least a 37-mm layer of water surrounding the sides and bottom of the cylinder. Center an accurate thermometer with 0.1 °C subdivisions in the test tube, and mount a thermometer with 1 °C subdivisions in the water bath. Use a wire about 30 cm long with a loop at the bottom as the stirrer.

Pour sufficient sample into the test tube to make a 50-mm column. Assemble the apparatus with the bulb of the 0.1 °C thermometer immersed halfway between the top and bottom of the sample in the test tube. Fill the water bath with a mixture of ice and water to within 12 mm of the top, and adjust the temperature to 0 °C by the addition of more ice, if necessary. Stir the sample continuously during the test by moving the wire loop up and down throughout the entire depth of the sample at a regular rate of 20 cycles per min. Record the reading of the 0.1 °C thermometer every 30 s. Anticipate that the temperature will fall gradually at first, then become constant for 1–2 min at the freezing point, and finally fall gradually again. The freezing point is the average of 4 consecutive readings that lie within a range of 0.2 °C.

Color (APHA). (Page 43).

Peroxide. Dilute 10 mL of the dioxane with water to 50 mL, and use 5.0 mL of the solution (1.0-g sample). For the standard, dilute a solution containing 0.05 mg of hydrogen peroxide standard with water to 5.0 mL. Add 5.0 mL of titanium tetrachloride reagent solution to each, and mix. After standing for 5 ± 1 min, any yellow color in the solution of the sample should not exceed that in the standard. The color intensities may be determined with a spectrophotometer in 1-cm cells at a wavelength of 410 nm.

Caution: If peroxide is present, do not perform the test for residue after evaporation.

Residue after Evaporation. (Page 25). Evaporate 20.0 g (19.0 mL) to dryness in a tared, preconditioned platinum dish on a hot plate ($\approx$100 °C), and dry the residue at 105 °C for 30 min.

Titrable Acid. Mix 31.0 g (30 mL) with 30 mL of carbon dioxide-free water, and titrate with 0.01 N sodium hydroxide, using phenolphthalein as the indicator. Not more than 5.0 mL of the titrant is required.

Carbonyl. To 1.0 mL of sample in a polarograph cell, add 2.0 mL of water, 5.0 mL of pH 6.5 buffer solution, 1.0 mL of 0.20% Triton X-100 reagent solution, and 1.0 mL of 2.0% hydrazine sulfate reagent solution. Prepare a similar mixture in which 1.0 mL of the water is replaced by 1.0 mL of a water solution containing 0.10 mg of formaldehyde. Deaerate each solution with nitrogen for 10 min, and record the polarograms from –0.6 to –1.6 V versus SCE at a sensitivity of 20 μA full scale. The total diffusion current for the sample should not be greater than one-half the diffusion current for sample plus formaldehyde.

The approximate half-wave potential for the formaldehyde hydrazone is –1.06 V.

Water. (Page 31, Method 2). Use 100 μL (103 mg) of the sample.

Diphenylamine
N-Phenylbenzeneamine

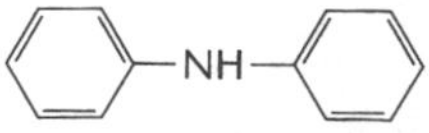

$(C_6H_5)_2NH$ Formula Wt 169.22   CAS No. 122-39-4

GENERAL DESCRIPTION

Typical appearance: white or almost white solid, darkening on exposure to light

Analytical use: detection of nitrate, chlorate, and other oxidizers; indicator for redox titrations

Aqueous solubility: 0.03 g in 100 mL at 25 °C

pK_a: 0.9

SPECIFICATIONS

Melting point . . . . 52.5–54.0 °C
Sensitivity to nitrate . . . . Passes test

Maximum Allowable

Solubility in alcohol . . . . Passes test
Residue after ignition. 0.03%
Nitrate (NO_3) . . . . Passes test

TESTS

Melting Point. (Page 45).

Sensitivity to Nitrate. To 8 mL of the solution retained from the test for nitrate, add 0.01 mg of nitrate ion (NO_3) and allow to stand at 60 °C for 5 min. A blue color should be produced.

Solubility in Alcohol. Dissolve 1.0 g in 50 mL of alcohol. The solution should be clear and colorless.

Residue after Ignition. (Page 26). Ignite 4.0 g. Moisten the char with 2 mL of sulfuric acid.

Nitrate. To 20 mL of water, add 60 mL of nitrate-free sulfuric acid, adjust the temperature to about 60 °C, and add 0.5 mL of hydrochloric acid and 0.0100 g of the sample. No blue color should be produced in 5 min. Retain the solution for the test for sensitivity to nitrate.

Diphenylaminesulfonic Acid Sodium Salt

Sodium Diphenylaminesulfonate

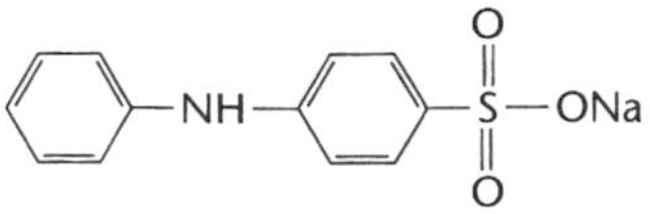

$C_6H_5NHC_6H_4SO_3Na$ Formula Wt 271.27   CAS No. 6152-67-6

GENERAL DESCRIPTION

Typical appearance: off-white solid

Analytical use: indicator for redox titrations

Aqueous solubility: very soluble

SPECIFICATION

Sensitivity as indicator . . . . Passes test

TEST

Sensitivity as Indicator. Dissolve 0.15 g in 100 mL of water. To 25 mL of water in a 50-mL test tube, add 10 mL of 4 N sulfuric acid reagent solution, 5 mL of phosphoric acid, 0.05 mL of 0.01 N ferrous ammonium sulfate, and 0.05 mL of the sample solution. The addition of 0.10 mL of 0.01 N potassium dichromate should produce a violet color. The color should be completely discharged by the addition of 0.10 mL of 0.01 N ferrous ammonium sulfate.

Diphenylcarbazone Compound with s-Diphenylcarbazide

(Phenylazo)formic Acid 2-Phenylhydrazide Compound with
1,5-Diphenylcarbohydrazide; "s-Diphenylcarbazone"

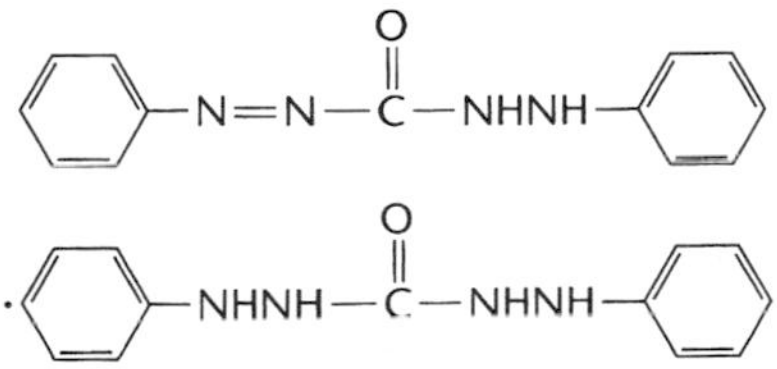

CAS No. 538-62-5 with 140-22-7

GENERAL DESCRIPTION

Typical appearance: bulky, orange crystalline solid

Analytical use: indicator

Change in state (approximate): melting point, 154 –161 °C
Aqueous solubility: insoluble

SPECIFICATIONS

Sensitivity. . . . . Passes test

Maximum Allowable

Residue after ignition. 0.1%

Solubility in acetone. . . . . Passes test

TESTS

Sensitivity. Dissolve 0.135 g of mercuric chloride in 500 mL of water. To 20 mL of this solution, add 0.25 mL of the solution reserved from the test for solubility in acetone. A violet color should develop.

Residue after Ignition. (Page 26). Ignite 2.0 g. Moisten the char with 2 mL of sulfuric acid.

Solubility in Acetone. Dissolve 0.2 g in 25 mL of acetone. The sample should dissolve completely, and the solution should be clear and red. Reserve this solution for the sensitivity test.

1,5-Diphenylcarbohydrazide

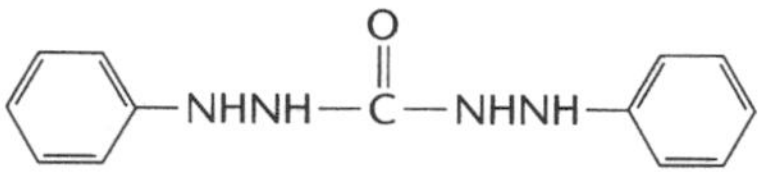

C₆H₅NHNHCONHNHC₆H₅   **Formula Wt 242.28**   **CAS No. 140-22-7**

GENERAL DESCRIPTION

Typical appearance: white or almost white crystalline powder; slowly becomes pink on exposure to air

Analytical use: indicator for chromate

Change in state (approximate): melting point, 175 °C

Aqueous solubility: very slightly soluble

SPECIFICATIONS

Melting point . . . . 173–176 °C

Sensitivity to chromate . . . . Passes test

Maximum Allowable

Solubility in aqueous acetone. . . . . Passes test

Residue after ignition. 0.05%

TESTS

Melting Point. (Page 45).

Sensitivity to Chromate. Dissolve 0.25 g in a mixture of 50 mL of acetone and 50 mL of water. To each of two 50-mL volumetric flasks, transfer 20 mL of water and 10 mL of 1 N sulfuric acid. To one flask, add 5.8 mL of chromate ion (CrO_4) standard solution, and to both, add 2.0 mL of the sample solution, mix, and allow to stand for 1 min. To both flasks, add 5 mL of 4 M monobasic sodium phosphate, dilute with water to the mark, and allow to stand for 5 min. Determine the absorbance of the solution containing chromate in a 1-cm cell at 540 nm against the other solution in a similar matched cell set at zero absorbance as the blank. The absorbance should be not less than 0.42.

Solubility in Aqueous Acetone. Mix 250 mL of acetone and 50 mL of water, and adjust the pH to 6.0 ± 0.2 (using a pH meter) with 0.02 M acetic acid or 0.02 M sodium hydroxide. Dissolve 10.0 g in the mixture without heating. Dissolution should be complete, and the solution clear, free of any red color, and not more than faintly yellow.

Residue after Ignition. (Page 26). Ignite 2.0 g. Moisten the char with 1 mL of sulfuric acid.

Dithizone
(Phenylazo)thioformic Acid 2-Phenylhydrazide; Diphenylthiocarbazone

$C_6H_5NHNHCSN{:}NC_6H_5$ Formula Wt 256.33   CAS No. 60-10-6

GENERAL DESCRIPTION

Typical appearance: black or purple-black crystalline solid
Analytical use: indicator for determination of cadmium, lead, and mercury
Change in state (approximate): melting point, 168 °C; decomposes
Aqueous solubility: insoluble

SPECIFICATIONS

Assay . . . . ≥85.0% $C_6H_5NHNHCSN{:}NC_6H_5$
Ratio of absorbances . . . . ≥1.55

Maximum Allowable

Residue after ignition 0.3%
Heavy metals (as Pb) 0.002%

TESTS

Assay. (By spectrophotometry). Weigh accurately 10.0 mg, transfer to a 100-mL volumetric flask, dissolve in about 75 mL of chloroform, and dilute to volume with chloroform. Dilute 5.0 mL of this solution to 100.0 mL with chloroform, and determine the absorbance of this solution in 1.00-cm cells at 620 nm, using chloroform as the blank. Reserve the remainder of the solution for determining the ratio of absorbances. Determine the molar absorptivity by dividing the absorbance by 1.95×10^{-5} (the molar concentration of the dithizone), and calculate the assay value as follows:

$$\% \text{ Dithizone} = \frac{\text{Molar absorptivity at 620 nm}}{34,600} \times 100$$

Ratio of Absorbances. Determine the molar absorptivity at 450 nm of the solution prepared in the test for assay, using the method described in the test. The ratio of the value at 620 nm to that at 450 nm should not be less than 1.55.

Residue after Ignition. (Page 26). Ignite 1.0 g. Moisten the char with 1 mL of nitric acid and 1 mL of sulfuric acid.

Heavy Metals. (Page 36, Method 2). Use 1.0 g.

Dodecyl Alcohol
Lauryl Alcohol
$CH_3(CH_2)_{11}OH$ **Formula Wt 186.34**   **CAS No. 112-53-8**

GENERAL DESCRIPTION
Typical appearance: off-white solid
Analytical use: esterification agent
Change in state (approximate): melting point, 24 °C; boiling point, 259 °C
Aqueous solubility: insoluble

SPECIFICATIONS
Assay . . . . ≥98.0% $CH_3(CH_2)_{11}OH$
Maximum Allowable
Color (APHA) 10
Water (H₂O) 0.05%

TESTS

Assay. Analyze the sample by gas chromatography by using the general parameters cited on page 80. The following specific conditions are also required.

> **Sample Solution.** Accurately weigh and transfer 0.10 g of the dodecyl alcohol into a 50-mL volumetric flask. Dissolve, and dilute to volume with hexanes. Mix well, and stopper.

Column: Type 1, methyl silicone

Column Temperature: 80–230 °C programmed at 10 °C/min, 5 min hold at 230 °C

Injector Temperature: 220 °C

Detector Temperature: 250 °C

Sample Size: 0.2 μL

Carrier Gas: Helium at 6 mL/min

Detector: Flame ionization

Measure the area under all the peaks after the solvent peaks, and calculate the dodecyl alcohol content in area percent. Relative retention for dodecyl alcohol to hexanes is approximately 8.8. Correct for water content.

Color (APHA). (Page 43).

Water. (Page 31, Method 2). Use 1.0 mL (0.832 g) of the sample.

Eosin Y
2′,4′,5′,7′-Tetrabromofluorescein, Disodium Salt

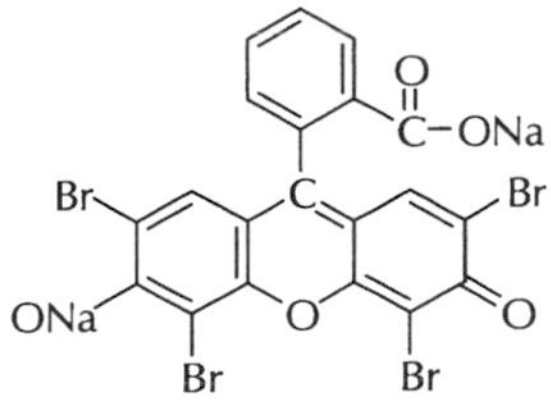

$C_{20}H_6Br_4Na_2O_5$ **Formula Wt 691.85**   **CAS No. 17372-87-1**

GENERAL DESCRIPTION

Typical appearance: red solid with a bluish tinge or brownish-red solid
Analytical use: indicator for argentimetric titrations
Aqueous solubility: freely soluble

SPECIFICATIONS

Clarity of aqueous solution . . . . Passes test
Suitability as adsorption indicator . . . . Passes test

TESTS

Clarity of Aqueous Solution. Dissolve 0.05 g in 100 mL of water. The solution should be clear and complete. Reserve the solution for the suitability test.

Suitability as Adsorption Indicator. Prepare a 0.10 N solution of sodium chloride by dissolving 0.58 g of sodium chloride in water, and dilute to 100.0 mL. Dilute 10.0 mL of the 0.10 N sodium chloride with 30.0 mL of methanol and 10.0 mL of acetic acid. Add 0.15 mL of the solution reserved from the preceding test, and with vigorous stirring, titrate with 0.1 N silver nitrate volumetric solution, protecting the titration vessel from direct sunlight. The color change from orange to bright pink should require between 9.9 and 10.1 mL of the silver nitrate solution.

Eriochrome Black T

3-Hydroxy-4-[(1-hydroxy-2-naphthalenyl)azo]-7-nitro-1-naphthalenesulfonic Acid, Monosodium Salt; C.I. Mordant Black 11

| $C_{20}H_{12}N_3NaO_7S$ | **Formula Wt 461.38** | **CAS No. 1787-61-7** |

GENERAL DESCRIPTION

Typical appearance: brownish-black powder
Analytical use: indicator for complexometric titrations

SPECIFICATIONS

Clarity of solution . . . . Passes test
Suitability as complexometric indicator . . . . Passes test

TESTS

Clarity of Solution. Dissolve 0.1 g in 100 mL of water, warming if necessary. Not more than a faint trace of turbidity or insoluble matter should remain. Reserve the solution for the test for suitability.

Suitability as Complexometric Indicator. Add 10 mL of pH 10 ammoniacal buffer solution to 100 mL of water. Add 0.5 mL of the solution reserved from the test for clarity of solution and 10.0 mL of magnesium ion I (Mg) standard solution. The color of the solution should be red-violet. Add 0.10 mL of 0.1 M EDTA volumetric solution. The color of the solution should change to blue.

Ethyl Acetate

$$CH_3-\overset{\overset{\displaystyle O}{\|}}{C}-OCH_2CH_3$$

$CH_3COOCH_2CH_3$	Formula Wt 88.11	CAS No. 141-78-6

Suitable for general use or in ultraviolet spectrophotometry. Product labeling shall designate the uses for which suitability is represented on the basis of meeting the relevant specifications and tests. The ultraviolet spectrophotometry specifications include all of the specifications for general use.

GENERAL DESCRIPTION

Typical appearance: colorless, clear liquid

Analytical use: chromatography solvent; extraction medium

Change in state (approximate): boiling point, 77 °C

Aqueous solubility: 7.9 g in 100 mL at 20 °C

Density: 0.90

SPECIFICATIONS

General Use

Assay . . . . ≥99.5% $CH_3COOCH_2CH_3$

Maximum Allowable

Color (APHA). 10

Residue after evaporation 0.003%

Water (H_2O) 0.2%

Titrable acid . . . . 0.0009 meq/g

Substances darkened by sulfuric acid . . . . Passes test

Specific Use

Ultraviolet Spectrophotometry

Wavelength (nm)	Absorbance (AU)
330–400	0.01
275	0.05
263	0.10
257	0.50
255	1.00

TESTS

Assay. Analyze the sample by gas chromatography using the general parameters cited on page 80. The following specific conditions are also required.

Column: Type I, methyl silicone

Measure the area under all peaks, and calculate the ethyl acetate content in area percent. Correct for water content.

Color (APHA). (Page 43).

Residue after Evaporation. (Page 25). Evaporate 40.0 g (45.0 mL) to dryness in a tared, preconditioned dish on a hot plate ($\approx$100 °C), and dry the residue at 105 °C for 30 min.

Water. (Page 31, Method 1). Use 20 mL (18 g) of the sample.

$$\% \text{ Water in sample} = \frac{100 \times \text{Karl Fischer titrant (mL)} \times F \text{(mg/mL)}}{\text{Sample wt (mg)}}$$

where F = water (mg) / Karl Fischer reagent (mL).

Titrable Acid. To 10 mL of ethyl alcohol, add 0.10 mL of phenolphthalein indicator solution, and neutralize with 0.01 N sodium hydroxide. Add 9.0 g (10.0 mL) of the sample, mix gently, and titrate with 0.01 N sodium hydroxide until the pink color persists for 15 s. Not more than 0.80 mL should be required.

Substances Darkened by Sulfuric Acid. Superimpose 5 mL of ethyl acetate upon 5 mL of sulfuric acid. No dark coloration should be produced at the zone of impact.

Ultraviolet Spectrophotometry. Use the procedure on page 86 to determine the absorbance.

Ethyl Alcohol
Ethanol

CH_3CH_2OH	**Formula Wt 46.07**	**CAS No. 64-17-5**

Suitable for general use or in ultraviolet spectrophotometry. Product labeling shall designate the uses for which suitability is represented on the basis of meeting the relevant specifications and tests. The ultraviolet spectrophotometry specifications include all of the specifications for general use.

GENERAL DESCRIPTION

Typical appearance: clear, colorless liquid

Analytical use: extraction medium; chromatography solvent; preparation of indicator solutions

Change in state (approximate): boiling point, 78 °C

Aqueous solubility: miscible

Density: 0.81

SPECIFICATIONS

General Use

Assay . . . . $\geq$95.0% CH_3CH_2OH

	Maximum Allowable
Color (APHA)..	10
Solubility in water...	Passes test
Residue after evaporation	0.001%
Acetone, isopropyl alcohol	Passes test
Titrable acid	0.0005 meq/g
Titrable base...	0.0002 meq/g
Methanol (CH_3OH)	0.1%
Substances darkened by sulfuric acid	Passes test
Substances reducing permanganate..................................	Passes test

Specific Use

Ultraviolet Spectrophotometry

Wavelength (nm)	Absorbance (AU)
270–400	0.01
240...	0.05
230...	0.15
220...	0.25
210...	0.40

TESTS

Assay. Analyze the sample by gas chromatography using the general parameters cited on page 80. The following specific conditions are also required.

    Column: Type I, methyl silicone

    Detector: TCD

Measure the area under all peaks, and calculate the ethyl alcohol content in area percent.

Color (APHA). (Page 43).

Solubility in Water. Mix 15 mL with 45 mL of water, and allow to stand for 1 h. The mixture should be as clear as an equal volume of water.

Residue after Evaporation. (Page 25). Evaporate 100.0 g (124 mL) to dryness in a tared, preconditioned dish on a hot plate (≈100 °C), and dry the residue at 105 °C for 30 min.

Acetone, Isopropyl Alcohol. Dilute 1.0 mL with 1 mL of water. Add 1 mL of a saturated solution of disodium hydrogen phosphate and 3 mL of a saturated solution of potassium permanganate. Warm the mixture to between 45 and 50 °C, and allow to stand until the permanganate color is discharged. Add 3 mL of 10% sodium hydroxide reagent solution, and filter, without washing, through glass. Prepare a control containing 1.2 mL of water, 1 mL of the saturated solution of disodium hydrogen phosphate, 3 mL of the saturated solution of potassium permanganate, 3 mL of 10% sodium hydroxide reagent solution, and 0.8 mL of acetone ion [$(CH_3)_2CO$] standard solution. To each solution, add 1 mL of 1% furfural solution (described below), and allow to stand for 10 min. To 1.0 mL of each solution, add 3 mL of hydrochloric acid. Any pink color produced in the solution of the sample should not exceed that in the control. (Limit about 0.001% acetone, 0.003% isopropyl alcohol)

Furfural Solution, 1%. Dissolve 1.0 g of furfural in 100 mL of reagent-grade ethyl alcohol.

Titrable Acid. Add 10 mL to 25 mL of water in a glass-stoppered flask, and add 0.15 mL of phenolphthalein indicator solution. Add 0.01 N sodium hydroxide until a slight pink color persists after shaking for 30 s. Add 25 mL of the sample, mix well, and titrate with 0.01 N sodium hydroxide until the pink color is restored. Not more than 1.0 mL of the sodium hydroxide solution should be required.

Titrable Base. Dilute 25 mL with 25 mL of water, and add 0.15 mL of methyl red indicator solution. Not more than 0.40 mL of 0.01 N sulfuric acid should be required to produce a pink color.

Methanol. Dilute 5 mL with water to 100 mL. To 1 mL of the solution add 0.2 mL of dilute phosphoric acid (1 + 9) and 0.25 mL of a 5% potassium permanganate reagent solution. Allow to stand for 15 min, add 0.3 mL of 10% sodium bisulfite solution, and shake until colorless. Add slowly 5 mL of ice-cold 80% sulfuric acid (3 volumes of acid plus 1 volume of water), keeping the mixture cool during the addition. Add 0.1 mL of 1% chromotropic acid reagent solution, mix, and digest on a hot plate ($\approx$100 °C) for 20 min. Any violet color should not exceed that produced by 0.04 mg of methanol in 1 mL of water treated in the same way as the 1 mL of the diluted sample.

Substances Darkened by Sulfuric Acid. Cool 10 mL of sulfuric acid, contained in a small conical flask, to 10 °C. Add dropwise, with constant agitation, 10 mL of the sample, while keeping the temperature of the mixture below 20 °C. The resulting solution should have no more color than either of the two liquids before mixing.

Substances Reducing Permanganate. Cool 20 mL to 15 °C, add 0.10 mL of 0.1 N potassium permanganate, and allow to stand at 15 °C for 5 min. The pink color should not be entirely discharged.

Ultraviolet Spectrophotometry. Use the procedure on page 86 to determine the absorbance.

Ethyl Alcohol, Absolute
Ethanol, Absolute
CH$_3$CH$_2$OH   Formula Wt 46.07   CAS No. 64-17-5

GENERAL DESCRIPTION
Typical appearance: colorless, clear liquid
Analytical use: extraction medium; chromatographic reagent
Change in state (approximate): boiling point, 78 °C
Aqueous solubility: miscible
Density: 0.79

SPECIFICATIONS

Assay . . . . ≥99.5% CH_3CH_2OH by volume
(about 99.2% by weight)

Maximum Allowable

Water (H_2O)	0.2%
Color (APHA)	10
Solubility in water	Passes test
Residue after evaporation	0.001%
Acetone, isopropyl alcohol	Passes test
Titrable acid	0.0005 meq/g
Titrable base	0.0002 meq/g
Methanol (CH_3OH)	0.1%
Substances darkened by sulfuric acid	Passes test
Substances reducing permanganate	Passes test

TESTS

Water. (Page 31, Method 2). Use 2 mL (1.58 g).

Assay and Other Tests. These are the same as for ethyl alcohol, page 308.

Ethyl Ether
Diethyl Ether
$(CH_3CH_2)_2O$ **Formula Wt 74.12**   **CAS No. 60-29-7**

Caution: Ethyl ether tends to form explosive peroxides, especially when anhydrous. It should not be allowed to evaporate to dryness unless the absence of peroxides has been shown. The presence of water or appropriate reducing agents lessens peroxide formation.

Note: Ethyl ether normally contains about 2% of alcohol and 0.5% of water as stabilizers.

GENERAL DESCRIPTION

Typical appearance: clear, colorless liquid
Analytical use: solvent; reagent in synthesis; extractant (hormones) from plant and animal tissues
Change in state (approximate): boiling point, 34 °C
Aqueous solubility: 7 g in 100 mL at 20 °C
Density: 0.71

SPECIFICATIONS

Assay . . . . ≥98.0% $(CH_3CH_2)_2O$

Maximum Allowable

Color (APHA) 10

Peroxide (as H_2O_2)...1 ppm
Residue after evaporation 0.001%
Titrable acid..0.0002 meq/g
Carbonyl (as HCHO) 0.001%

TESTS

Assay. Assay the sample by gas chromatography using the general parameters cited on page 80. The following specific conditions are also required.

Column: Type I, methyl silicone

Measure the area under all peaks, and calculate the ethyl ether content in area percent. The assay value is the combination of ethyl ether and added stabilizers.

Color (APHA). (Page 43).

Peroxide. To 35.0 g (50 mL) of the ethyl ether in a separatory funnel, add 5.0 mL of titanium tetrachloride reagent solution. Shake vigorously, allow the layers to separate, and drain the lower layer into a 25-mL glass-stoppered graduated cylinder. For the standard, transfer 5.0 mL of titanium tetrachloride reagent solution to a similar graduated cylinder, and add a volume of solution containing 0.035 mg of hydrogen peroxide ion (H_2O_2) standard solution. Dilute both solutions with water to 10.0 mL, and mix. Allow the mixture to stand at least 30 min. Any yellow color in the solution of the sample should not exceed that in the standard. The color intensities may be determined with a spectrophotometer in 1-cm cells at a wavelength of 410 nm.

> *Note:* Fresh ethyl ether should meet this test, but after storage for several months, peroxide may be formed.

> *Caution:* **If peroxide is present, do not perform the test for residue after evaporation.**

Residue after Evaporation. (Page 25). Evaporate 100.0 g (141 mL) to dryness in a tared, preconditioned dish on a water bath, and dry the residue at 105 °C for 30 min.

Titrable Acid.

> *Note:* Great care should be taken in the test during the addition of the sample and the titration to avoid contamination from carbon dioxide.

To 10 mL of water in a glass-stoppered flask, add 0.10 mL of bromthymol blue indicator solution and 0.01 N sodium hydroxide until a blue color persists after vigorous shaking. Add 25 mL of sample from a pipet, and shake briskly to mix the two layers. If no blue color remains, titrate with 0.01 N sodium hydroxide until the blue color is restored and persists for several minutes. Not more than 0.30 mL of 0.01 N sodium hydroxide should be required.

Carbonyl. To 5.0 g (7.0 mL) of sample in a 30-mL separatory funnel, add 5.0 mL of pH 6.5 buffer solution, 1.0 mL of 2.0% hydrazine sulfate reagent solution, 1.0 mL of 0.20% Triton X-100 reagent solution, and 3.0 mL of water. Prepare a similar mixture in which part of the water is replaced by an equal volume of aqueous solution containing 0.05 mg of formaldehyde. Shake each funnel well to mix the contents, allow the phases to separate, and draw off each lower layer into a polarographic cell. Deaerate with nitrogen for 10 min. Record the polarogram from -0.6 to -1.6 V versus SCE at a sensitivity of 5 μA full scale. The total diffusion current for the sample should not be greater than one-half the diffusion current for sample plus aldehyde.

Approximate half-wave potentials for the hydrazones of known carbonyl compounds are: formaldehyde, -1.06 V; acetaldehyde, -1.21 V; and acetone, -1.4 V.

Ethyl Ether, Anhydrous
Diethyl Ether, Anhydrous
$(CH_3CH_2)_2O$ **Formula Wt 74.12** **CAS No. 60-29-7**

> *Caution:* Ethyl ether tends to form explosive peroxides, especially when anhydrous. It should not be allowed to evaporate to dryness or near dryness unless the absence of peroxides has been shown. The formation of peroxides is more rapid in ethyl ether kept in containers that have been opened and partly emptied. Some ethyl ether may contain a stabilizer. If it does, the amount and type should be marked on the label.

GENERAL DESCRIPTION

Typical appearance: clear, colorless liquid
Analytical use: solvent; reagent in synthesis; extractant (hormones) from plant and animal tissues
Change in state (approximate): boiling point, 34 °C
Aqueous solubility: 7 g in 100 mL at 20 °C
Density: 0.71

SPECIFICATIONS

Assay . . . . ≥99.0% $(CH_3CH_2)_2O$

Maximum Allowable

Color (APHA) 10
Peroxide (as H_2O_2) . . . . 1 ppm
Residue after evaporation 0.001%
Titrable acid . . . . 0.0002 meq/g
Carbonyl (as HCHO) 0.001%
Alcohol (CH_3CH_2OH) . . . . Passes test
Water (H_2O) 0.03%

TESTS

Assay. Analyze the sample by gas chromatography using the general parameters cited on page 80. The following specific conditions are also required.

Column: Type I, methyl silicone

Measure the area under all peaks and calculate the ethyl ether content in area percent. Correct for water content.

Color (APHA). (Page 43).

Peroxide.

To 35.0 g (50 mL) of the ethyl ether in a separatory funnel, add 5.0 mL of titanium tetrachloride reagent solution. Shake vigorously, allow the layers to separate, and drain the lower layer into a 25-mL glass-stoppered graduated cylinder. For the standard, transfer 5.0 mL of titanium tetrachloride reagent solution to a similar graduated cylinder, and add a volume of solution containing 0.035 mg of hydrogen peroxide ion (H_2O_2) standard solution. Dilute both solutions with water to 10.0 mL, and mix. Allow the mixture to stand at least 30 min. Any yellow color in the solution of the sample should not exceed that in the standard. The color intensities may be determined with a spectrophotometer in 1-cm cells at a wavelength of 410 nm.

> *Note:* Fresh ethyl ether should meet this test, but after storage for several months, peroxide may be formed.

> *Caution:* **If peroxide is present, do not perform the test for residue after evaporation.**

Residue after Evaporation.

(Page 25). Evaporate 100.0 g (141 mL) to dryness in a tared, preconditioned dish on a water bath, and dry the residue at 105 °C for 30 min.

Titrable Acid.

> *Note:* Great care should be taken in the test during the addition of the sample and the titration to avoid contamination from carbon dioxide.

To 10 mL of water in a glass-stoppered flask, add 0.10 mL of bromthymol blue indicator solution and 0.01 N sodium hydroxide until a blue color persists after vigorous shaking. Add 25 mL of sample from a pipet, and shake briskly to mix the two layers. If no blue color remains, titrate with 0.01 N sodium hydroxide until the blue color is restored and persists for several minutes. Not more than 0.30 mL of 0.01 N sodium hydroxide should be required.

Carbonyl.

To 5.0 g (7.0 mL) of sample in a 30-mL separatory funnel add 5.0 mL of pH 6.5 buffer solution, 1.0 mL of 2.0% hydrazine sulfate reagent solution, 1.0 mL of 0.20% Triton X-100 reagent solution, and 3.0 mL of water. Prepare a similar mixture in which part of the water is replaced by an equal volume of aqueous solution containing 0.05 mg of formaldehyde. Shake each funnel well to mix the contents, allow the phases to separate, and draw off each lower layer into a polarographic cell. Deaerate with nitrogen for 10 min. Record the polarogram from –0.6 to –1.6 V versus SCE at a sensitivity of 5 µA full scale. The total diffusion current for the sample should not be greater than one-half the diffusion current for sample plus aldehyde.

Approximate half-wave potentials for the hydrazones of known carbonyl compounds are: formaldehyde, –1.06 V; acetaldehyde, –1.21 V; and acetone, –1.4 V.

Alcohol.

Transfer 100 mL to a separatory funnel, and shake with five successive portions—20 mL, 10 mL, 10 mL, 5 mL, and 5 mL, respectively—of distilled water at about 25

°C. Shake each portion for 2 min, and separate the water layer carefully. Finally, pour the combined water extract from one flask to another six times to ensure minimum contamination with ether. Transfer 1 mL of the water extract with a pipet to a comparison tube, and add 4 mL of water. For a standard take 5 mL of a solution of 0.2 mL of absolute alcohol in 1 L of water. Add 10 mL of nitrochromic acid solution (described below) to each solution, mix, and allow to stand for 1 h. At the end of this time, the color of the sample solution should show no more change from yellow to green or blue than is shown by the standard. The test and the standard must be kept at the same temperature. (Limit about 0.05%)

> **Nitrochromic Acid Solution.** Mix 1 volume of 5% potassium chromate solution with 133 volumes of water and 66 volumes of colorless nitric acid. This reagent should not be used if more than 1 month old.

Water. (Page 31, Method 2). Use 0.2 mL (0.14 g) of the sample.

(Ethylenedinitrilo)tetraacetic Acid

N,N,N′,N′-Ethylenediaminetetraacetic Acid; *N,N′*-1,2-Ethanediylbis[*N*-carboxymethyl]glycine; EDTA, Free Acid

$C_{10}H_{16}N_2O_8$ **Formula Wt 292.25**   **CAS No. 60-00-4**

GENERAL DESCRIPTION

Typical appearance: white solid
Analytical use: chelating agent
Change in state (approximate): melting point, 220 °C, with decomposition
Aqueous solubility: 0.05 g in 100 mL at 25 °C

SPECIFICATIONS

Assay . . . . .99.4–100.6% $C_{10}H_{16}N_2O_8$

	Maximum Allowable
Insoluble in dilute ammonium hydroxide	0.005%
Residue after ignition	0.2%
Nitrilotriacetic acid [(HOCOCH$_4$)$_3$N]	0.1%
Calcium (Ca)	0.001%
Magnesium (Mg)	5 ppm
Heavy metals (as Pb)	0.001%
Iron (Fe)	0.005%

TESTS

Assay. (By complexometric titration). Weigh accurately about 4.0 g, and transfer to a 250-mL volumetric flask. Dissolve in 25 mL of 1 N sodium hydroxide volumetric solution, dilute to volume with water, and mix thoroughly. Fill a 50-mL buret with this sample solution. Weigh accurately about 0.2 g of calcium carbonate, chelometric standard, that has been dried in an oven at 300 °C for 3 h and cooled in a charged desiccator for 2 h, and transfer to a 400-mL beaker. Cover the beaker with a watch glass. Add 2 mL of 10% hydrochloric acid reagent solution from a pipet placed between the lip of the beaker and the watch glass. Swirl the beaker to aid dissolution. With water, wash down the inner wall of the beaker, the outer surface of the pipet, and the watch glass. Dilute to about 100 mL with water. While stirring, add from the buret about 30 mL of the sample solution. Add 15 mL of 1 N sodium hydroxide volumetric solution and 300 mg of hydroxy naphthol blue indicator mixture. Continue the titration immediately with the sample solution to a blue color.

$$\% \ C_{10}H_{16}N_2O_8 = \frac{W_c \times 73,000}{V \times W_s}$$

where W_c = the weight, in g, of calcium carbonate; V = the volume, in mL, of the sample solution required in the titration; and W_s = the weight, in g, of the sample in aliquot.

Insoluble in Dilute Ammonium Hydroxide. To a 20.0-g sample, add 190 mL of water and 0.10 mL of methyl red indicator solution. Slowly add ammonium hydroxide, keeping the solution acidic, until all the sample is dissolved. Warm if necessary. Heat to boiling, and digest in a covered beaker on a hot plate ($\approx$100 °C) for 1 h. Filter through a tared filtering crucible, wash thoroughly, and dry at 105 °C.

Residue after Ignition. (Page 26). Ignite 3.0 g. Moisten the char with 1 mL of sulfuric acid.

Nitrilotriacetic Acid. For a stock solution, transfer 10.0 g of the sample to a 100-mL volumetric flask. Dissolve in 87 mL of potassium hydroxide solution (described below), dilute with water to volume, and mix.

> **Potassium Hydroxide Solution.** Dissolve 20.0 g of potassium hydroxide in 200 mL of water.

> **Sample Solution A.** Dilute 10 mL of the stock solution with water to 100 mL in a volumetric flask.

> **Control Solution B.** Add 1.0 mL (10 mg) of nitrilotriacetic acid ion [(HOCOCH$_2$)$_3$N] standard solution to 10 mL of the stock solution, and dilute with water to 100 mL in a volumetric flask.

Transfer 20 mL of sample solution A to a 150-mL beaker and 20 mL of control solution B to a second 150-mL beaker. Add to each beaker 1 mL of potassium hydroxide solution (described above) and 2 mL of ammonium nitrate reagent solution. Add approximately 0.05 g of Eriochrome Black T indicator mixture, and titrate with a cadmium nitrate reagent solution to a red end point.

Sample Solution C. Transfer 20 mL of sample solution A to a 100-mL volumetric flask.

Control Solution D. Transfer 20 mL of control solution B to a 100-mL volumetric flask.

Add to each flask a volume of the cadmium nitrate reagent solution equal to the volume determined by the titration plus 0.05 mL in excess. Add 1.5 mL of potassium hydroxide solution (described above), 10 mL of ammonium nitrate reagent solution, and 0.5 mL of methyl red indicator solution. Dilute each solution with water to 100 mL, and mix.

Transfer a portion of control solution D to an H-type polarographic cell equipped with a saturated calomel electrode, and deaerate with nitrogen for 10 min. Record the polarogram from −0.6 to −1.2 V versus SCE at a sensitivity of 0.006, μA/mm. Repeat with sample solution C. The diffusion current in sample solution C should not exceed 0.1 times the difference in diffusion current between control solution D and sample solution C.

Calcium and Magnesium.   (By flame AAS, page 63).

Sample Stock Solution. Dissolve 5.0 g of sample in 50 mL of water and 5 mL of ammonium hydroxide in a 150-mL beaker, and heat if necessary. Transfer to a 100-mL volumetric flask, and dilute to the mark with water (1 mL = 0.05 g).

Element	Wavelength (nm)	Sample Wt (g)	Standard Added (mg)	Flame Type*	Background Correction
Ca	422.7	1.0	0.01; 0.02	N/A	No
Mg	285.2	1.0	0.005; 0.01	A/A	Yes

*A/A is air/acetylene; N/A is nitrous oxide/acetylene.

For the Determination of Heavy Metals and Iron

Sample Solution E. Char 3.0 g thoroughly, and heat in an oven at 500 °C until most of the carbon is volatilized. Cool, add 0.15 mL of nitric acid, and heat at 500 °C until all of the carbon is volatilized. Dissolve the residue in 2 mL of dilute hydrochloric acid (1 + 1), digest in a covered crucible on a hot plate (≈100 °C) for 10 min, remove the cover, and evaporate to dryness. Dissolve in 1 mL of 1 N acetic acid and 20 mL of hot water, digest for 5 min, cool, and dilute with water to 30 mL (1 mL = 0.1 g).

Heavy Metals.   (Page 36, Method 1). Dilute 20 mL of sample solution E (2.0-g sample) with water to 25 mL.

Iron.   (Page 38, Method 1). Use 2.0 mL of sample solution E (0.2-g sample).

(Ethylenedinitrilo)tetraacetic Acid, Disodium Salt Dihydrate

N,N,N',N'-Ethylenediaminetetraacetic Acid, Disodium Salt, Dihydrate;
N,N'-1,2-Ethanediylbis[*N*-carboxymethyl]glycine, Disodium Salt, Dihydrate;
EDTA (Dihydrate)

$C_{10}H_{14}N_2O_8Na_2 \cdot 2H_2O$ **Formula Wt 372.24**   **CAS No. 6381-92-6**

GENERAL DESCRIPTION

Typical appearance: white solid
Analytical use: chelating agent; sequestering agent
Aqueous solubility: 11 g in 100 mL at 20 °C

SPECIFICATIONS

Assay . . . . 99.0–101.0% $C_{10}H_{14}N_2O_8Na_2 \cdot 2H_2O$
pH of a 5% solution at 25.0 °C 4.0–6.0

	Maximum Allowable
Insoluble matter	0.005%
Nitrilotriacetic acid [(HOCOCH₂)₂N]	0.1%
Heavy metals (as Pb)	0.005%
Iron (Fe)	0.01%

TESTS

Assay. (By complexometric titration). Weigh accurately about 5.0 g of sample, and transfer to a 250-mL volumetric flask. Dissolve in water, dilute to volume with water, and mix thoroughly. Fill a 50-mL buret with this sample solution. Weigh accurately about 0.2 g of calcium carbonate, chelometric standard, that has been dried in an oven at 300 °C for 3 h and cooled in a charged desiccator for 2 h, and transfer to a 400-mL beaker. Cover the beaker with a watch glass. Add 2 mL of 10% hydrochloric acid reagent solution from a pipet placed between the lip of the beaker and the watch glass. Swirl the beaker to aid dissolution. With water, wash down the inner wall of the beaker, the outer surface of the pipet, and the watch glass. Dilute to about 100 mL with water. While stirring, add from the buret about 30 mL of the sample solution. Add 15 mL of 1 M sodium hydroxide and 300 mg of hydroxy naphthol blue indicator mixture. Continue the titration immediately with the sample solution to a blue color.

$$\% \; C_{10}H_{14}N_2O_8Na_2 \cdot 2H_2O = \frac{W_c \times 92{,}980}{V \times W_s}$$

where W_c = the weight, in g, of calcium carbonate; V = the volume, in mL, of the sample solution required for titration; and W_s = the weight, in g, of the sample in aliquot.

pH of a 5% Solution at 25.0 °C. (Page 49).

Insoluble Matter. (Page 25). Use 20.0 g dissolved in 200 mL of hot water.

Nitrilotriacetic Acid. For a stock solution, transfer 10.0 g of the sample to a 100-mL volumetric flask. Dissolve in 40 mL of potassium hydroxide solution (described below), dilute with water to volume, and mix.

> **Potassium Hydroxide Solution.** Dissolve 20.0 g of potassium hydroxide in 200 mL of water.

> **Sample Solution A.** Dilute 10 mL of the stock solution with water to 100 mL in a volumetric flask.

> **Control Solution B.** Add 1.0 mL (10 mg) of nitrilotriacetic acid ion $[(HOCOCH_2)_3N]$ standard solution to 10 mL of the stock solution, and dilute with water to 100 mL in a volumetric flask.

Transfer 20 mL of sample solution A to a 150-mL beaker and 20 mL of control solution B to a second 150-mL beaker. Add to each beaker 1 mL of potassium hydroxide solution (described above) and 2 mL of ammonium nitrate reagent solution. Add approximately 0.05 g of Eriochrome Black T indicator mixture, and titrate with cadmium nitrate reagent solution to a red end point.

> **Sample Solution C.** Transfer 20 mL of sample solution A to a 100-mL volumetric flask.

> **Control Solution D.** Transfer 20 mL of control solution B to a 100-mL volumetric flask.

Add to each flask a volume of the cadmium nitrate reagent solution equal to the volume determined by the titration plus 0.05 mL in excess. Add 1.5 mL of potassium hydroxide solution (described above), 10 mL of ammonium nitrate reagent solution, and 0.5 mL of methyl red indicator solution. Dilute each solution with water to 100 mL, and mix.

Transfer a portion of control solution D to an H-type polarographic cell equipped with a saturated calomel electrode, and deaerate with nitrogen for 10 min. Record the polarogram from –0.6 to –1.2 V versus SCE at a sensitivity of 0.006 μA/mm. Repeat with sample solution C. The diffusion current in sample solution C should not exceed 0.1 times the difference in diffusion current between control solution D and sample solution C.

For the Determination of Heavy Metals and Iron

> **Sample Solution E.** To 1.0 g, add 1 mL of sulfuric acid, heat cautiously until the sample is charred, and ignite in an oven at 500 °C until most of the carbon is volatilized. Cool, add 1 mL of nitric acid, heat until the acid is evaporated, and ignite again at 500 °C until all of the carbon is volatilized. Cool, add 4 mL of dilute hydrochloric acid (1 + 1), digest on a hot plate (≈100 °C) for 10 min, and evaporate to dryness. Add 10 mL of aqueous 10% ammonium acetate solution, and digest on a hot plate (≈100 °C) for 30 min. Filter, wash thoroughly, and dilute with water to 50 mL (1 mL = 0.02 g).

Heavy Metals. (Page 36, Method 1). Dilute 20 mL of sample solution E (0.4-g sample) with water to 25 mL.

Iron. (Page 38, Method 1). Use 5.0 mL of sample solution E (0.1-g sample).

(Ethylenedinitrilo)tetraacetic Acid, Magnesium Disodium Salt

Disodium Magnesium EDTA; EDTA (Magnesium Disodium Salt)

$(NaO_2CCH_2)_2NC_2H_4N(CH_2CO_2)_2Mg$ **Formula Wt 358.52**   **CAS No. 14402-88-1**

GENERAL DESCRIPTION

Typical appearance: white solid

Analytical use: chelating agent

SPECIFICATIONS

Clarity of solution . . . . Passes test
Stoichiometry. . . . . Passes test

TESTS

Clarity of Solution. Dissolve 5 g in 100 mL of water. The solution should be clear and colorless with no more than a faint trace of turbidity.

Stoichiometry. To the solution from the clarity of solution test, add 0.1 mL of calmagite indicator solution and 5 mL of pH 10 ammoniacal buffer solution. The solution should be a light blue color with a red tint. Add 0.10 mL of 0.01 M EDTA. The color should be blue. Add 0.20 mL of 0.01 M calcium standard solution (described below), and the solution should change to red.

> **Calcium Standard Solution, 0.01 M.** Dissolve 0.100 g of calcium carbonate in 2 mL of 10% hydrochloric acid in a 100-mL volumetric flask. Dilute to the mark with water.

Ferric Ammonium Sulfate Dodecahydrate

Ammonium Iron(III) Sulfate Dodecahydrate; Ammonium Ferric Sulfate Dodecahydrate

$Fe(SO_4)_2 NH_4 \cdot 12H_2O$ **Formula Wt 482.20**   **CAS No. 7783-83-7**

GENERAL DESCRIPTION

Typical appearance: violet solid

Analytical use: indicator for argentimetric titrations

Change in state (approximate): melting point, 37 °C
Aqueous solubility: soluble

SPECIFICATIONS

Assay . . . . 98.5–102.0% $Fe(SO_4)_2NH_4 \cdot 12(H_2O)$

	Maximum Allowable
Insoluble matter	0.01%
Chloride (Cl)	0.001%
Nitrate (NO_3)	0.01%
Calcium (Ca)	0.01%
Copper (Cu)	0.003%
Magnesium (Mg)	0.005%
Potassium (K)	0.005%
Sodium (Na)	0.02%
Zinc (Zn)	0.003%
Ferrous iron (Fe^{++})	Passes test

TESTS

Assay. (By iodometric titration of oxidative capacity of Fe^{III}). Weigh accurately about 1.8 g and transfer to an iodine flask with 50 mL of water. Add 3 mL of hydrochloric acid and 3.0 g of potassium iodide. Stopper, swirl, and allow to stand in the dark for 30 min. Dilute with 100 mL of water and titrate the liberated iodine with 0.1 N sodium thiosulfate volumetric solution, adding 3 mL of starch indicator solution near the end of the titration. Correct for a blank. One milliliter of 0.1 N sodium thiosulfate corresponds to 0.04822 g of $Fe(SO_4)_2NH_4 \cdot 12H_2O$.

$$\% \ Fe(SO_4)_2NH_4 \cdot 12H_2O = \frac{(mL \times N \ Na_2S_2O_3) \times 48.22}{\text{Sample wt (g)}}$$

Insoluble Matter. (Page 25). Use 10.0 g of sample dissolved in 100 mL of dilute hydrochloric acid (1 + 99).

Chloride. Dissolve 4.0 g in 40 mL of dilute nitric acid (1 + 9), filter if necessary through a chloride-free filter, and dilute with water to 60 mL. For the sample, use 30 mL of the solution. For the control, add 0.01 mg of chloride ion (Cl) and 1 mL of nitric acid to 15 mL of the solution, and dilute with water to 30 mL. To each solution, add 1 mL of silver nitrate reagent solution. Any turbidity in the solution of the sample should not exceed that in the control.

Nitrate. (Page 38, Method 1). Dissolve 2.0 g in 1 mL of water by swirling in a test tube. Add 10 mL of dilute ammonium hydroxide (1 + 1), mix well, and filter through a dry filter paper. Do not wash the precipitate.

> **Sample Solution A.** Add 1.0 mL of the filtrate to 2 mL of water, dilute to 50 mL with brucine sulfate reagent solution, and mix.

> **Control Solution B.** Add 1.0 mL of the filtrate to 2 mL of the nitrate ion (NO_3) standard solution, dilute to 50 mL with brucine sulfate reagent solution, and mix.

Continue with the procedure, starting with the preparation of blank solution C.

Calcium, Copper, Magnesium, Potassium, Sodium, and Zinc. (By flame AAS, page 63).

> ***Sample Stock Solution.*** Dissolve 10.0 g of sample in 90 mL of nitric acid (1 + 19) in a 100-mL volumetric flask, and dilute to the mark with water (1 mL = 0.10 g).

Element	Wavelength (nm)	Sample Wt (g)	Standard Added (mg)	Flame Type*	Background Correction
Ca	422.7	1.00	0.05; 0.10	N/A	No
Cu	324.7	1.0	0.03; 0.06	A/A	Yes
Mg	285.2	0.10	0.005; 0.01	A/A	Yes
K	766.5	1.00	0.025; 0.05	A/A	No
Na	589.0	0.10	0.01; 0.02	A/A	No
Zn	213.9	0.20	0.005; 0.01	A/A	Yes

*A/A is air/acetylene; N/A is nitrous oxide/acetylene.

Ferrous Iron. Dissolve 1.0 g in a mixture of 20 mL of water and 1 mL of hydrochloric acid, and add 0.05 mL of a freshly prepared 5% potassium ferricyanide reagent solution in a 25-mL glass-stoppered flask. Stopper immediately. No blue or green color should be produced in 1 min. (Limit about 0.001%)

Ferric Chloride Hexahydrate

Iron(III) Chloride Hexahydrate

$FeCl_3 \cdot 6H_2O$ **Formula Wt 270.30**   **CAS No. 10025-77-1**

GENERAL DESCRIPTION

Typical appearance: brownish-yellow deliquescent solid
Analytical use: clinical reagent (amino acids in urine)
Change in state (approximate): melting point, 37 °C
Aqueous solubility: 92 g in 100 mL

SPECIFICATIONS

Assay . . . . 97.0–102.0% $FeCl_3 \cdot 6H_2O$

Maximum Allowable

Insoluble matter 0.01%
Nitrate (NO_3) 0.01%
Sulfate (SO_4) 0.01%
Phosphorus compounds (as PO_4) 0.01%
Calcium (Ca) 0.01%
Copper (Cu) 0.003%
Magnesium (Mg) 0.005%
Potassium (K) 0.005%

Sodium (Na) 0.05%

Zinc (Zn). 0.003%

Ferrous iron (Fe^{++}). 0.002%

TESTS

Assay. (By iodometric titration of oxidative capacity of Fe^{III}). Weigh accurately about 1.0 g, and transfer to an iodine flask with 50 mL of water. Add 3 mL of hydrochloric acid and 3.0 g of potassium iodide. Stopper, swirl, and allow to stand in the dark for 30 min. Dilute with 100 mL of water, and titrate the liberated iodine with 0.1 N sodium thiosulfate, adding 3 mL of starch indicator solution near the end of the titration. Correct for a blank. One milliliter of 0.1 N sodium thiosulfate corresponds to 0.02703 g of FeCl$_3 \cdot$ 6H$_2$O.

$$\% \text{ FeCl}_3 \cdot 6\text{H}_2\text{O} = \frac{(\text{mL} \times \text{N Na}_2\text{S}_2\text{O}_3) \times 27.03}{\text{Sample wt (g)}}$$

Insoluble Matter. (Page 25). Use 10.0 g dissolved in 50 mL of dilute hydrochloric acid (1 + 49).

For the Determination of Nitrate and Sulfate

Sample Solution S. In a 150-mL beaker, dissolve 10.0 g in 75 mL of water, heat to boiling, and pour slowly with constant stirring into 200 mL of dilute ammonium hydroxide (1 + 3). Filter through a folded filter while still hot, wash with hot water until the volume of the filtrate plus washings measures 300 mL, and thoroughly mix the solution (1 mL = 0.033 g).

Nitrate. (Page 38, Method 1).

Sample Solution A. Add 42 mL of brucine sulfate reagent solution and 2 mL of water to 6 mL of sample solution S (0.2-g sample).

Control Solution B. Add 42 mL of brucine sulfate reagent solution and 2 mL of nitrate ion (NO$_3$) standard solution to 6 mL of sample solution S.

Blank Solution C. Add 42 mL of brucine sulfate reagent solution to 8 mL of water.

Continue with the procedure as described.

Sulfate. (Page 40, Method 1). Evaporate 15 mL of sample solution S to 10 mL. Allow 30 min for turbidity to form.

Phosphorus Compounds. In a 250-mL beaker, dissolve 5.0 g in 40 mL of dilute nitric acid (1 + 1), and evaporate on a hot plate ($\approx$100 °C) to a syrupy residue. Dissolve the residue in about 80 mL of water, dilute with water to 100 mL, and dilute 20 mL of this solution with water to 100 mL. For the test solution, dilute 10 mL (0.1-g sample) with water to 70 mL. Prepare a standard containing 0.01 mg of phosphate ion (PO$_4$) in 70 mL of water. To each solution, add 5 mL of 10% ammonium molybdate reagent solution, and adjust the pH to 1.8 (using a pH meter) by adding dilute hydrochloric acid (1 + 1) or dilute ammonium hydroxide (1 + 1). Cautiously heat the solutions to boiling but do not boil, and cool to room temperature. If a precipitate forms, it will dissolve when the solution is acidified in the next step. To each solution, add 10 mL of hydrochloric acid, and dilute each with water to 100 mL. Transfer the solutions to separatory funnels, add 35 mL of diethyl ether to each, shake vigorously, and allow to separate. Draw off and discard the aqueous phases. Wash the ether phases twice with 10-mL portions of dilute hydrochloric acid (1 + 9), discarding the washings each time. To the washed ether phases, add 10 mL of dilute hydrochloric acid (1 + 9) to which has just been added 0.2 mL of a freshly prepared 2% stannous chloride reagent solution (in hydrochloric acid). Shake the solutions, and allow the phases to separate. Any blue color in the ether phase from the solution of the sample should not exceed that in the ether phase from the standard.

Calcium, Copper, Magnesium, Potassium, Sodium, and Zinc. (By flame AAS, page 63).

> **Sample Stock Solution A.** Dissolve 10.0 g of sample with water in a 100-mL volumetric flask, and dilute to the mark with water (1 mL = 0.10 g).

> **Sample Stock Solution B.** Transfer 5.0 mL (0.50 g) of sample stock solution A to a 100-mL volumetric flask, dilute to the mark with water, and mix (1 mL = 0.005 g).

Element	Wavelength (nm)	Sample Wt (g)	Standard Added (mg)	Flame Type*	Background Correction
Ca	422.7	1.0	0.05; 0.10	N/A	No
Cu	324.7	1.0	0.05; 0.10	A/A	Yes
Mg	285.2	0.20	0.005; 0.01	A/A	Yes
K	766.5	1.0	0.025; 0.05	A/A	No
Na	589.0	0.04	0.01; 0.02	A/A	No
Zn	213.9	0.20	0.005; 0.01	A/A	Yes

*A/A is air/acetylene; N/A is nitrous oxide/acetylene.

Ferrous Iron. Dissolve 0.5 g in 20 mL of dilute hydrochloric acid (1 + 19) in a 25-mL glass-stoppered flask, and add 0.05 mL of freshly prepared 5% potassium ferricyanide reagent solution. Stopper immediately. No blue or green color should be produced in 1 min. (Limit about 0.002%)

Ferric Nitrate Nonahydrate
Iron(III) Nitrate Nonahydrate
Fe(NO$_3$)$_3$ · 9H$_2$O   **Formula Wt 404.00**   **CAS No. 7782-61-8**

GENERAL DESCRIPTION

Typical appearance: pale mauve solid
Analytical use: determination of phosphate
Change in state (approximate): melting point, 47 °C
Aqueous solubility: very soluble

SPECIFICATIONS

Assay . . . . .98.0–101.0% $Fe(NO_3)_3 \cdot 9H_2O$

Maximum Allowable

Insoluble matter 0.005%
Chloride (Cl) . . . . ?5 ppm
Sulfate (SO_4) 0.01%
Calcium (Ca) 0.01%
Magnesium (Mg). 0.005%
Potassium (K) 0.005%
Sodium (Na) 0.05%

TESTS

Assay. (By complexometric titration of iron). Weigh accurately about 1.6 g, and transfer to a 400-mL beaker. Dissolve with 200 mL of water, and add 2 mL of 20% hydrochloric acid. From a 50-mL buret, add 30.0 mL of 0.1 M EDTA volumetric solution. Heat the solution to 40–50 °C, and adjust the pH to 2.5 ± 0.2 with 10% sodium acetate reagent solution. Add 0.25 mL of Variamine Blue B indicator solution, and continue the addition of 0.1 M EDTA until burgundy changes to canary yellow. One milliliter of 0.1 M EDTA corresponds to 0.04040 g of $Fe(NO_3)_3 \cdot 9H_2O$.

$$\% \ Fe(NO_3)_3 \cdot 9H_2O = \frac{(mL \times M \ EDTA) \times 40.40}{Sample \ wt \ (g)}$$

Insoluble Matter. (Page 25). Use 20.0 g dissolved in 200 mL of dilute nitric acid (1 + 99).

Chloride. Dissolve 2.0 g in 20 mL of water, filter if necessary through a chloride-free filter, and add 1 mL of nitric acid plus 1 mL of phosphoric acid and 1 mL of silver nitrate reagent solution. Any turbidity should not exceed that produced by 0.01 mg of chloride ion (Cl) in an equal volume of solution containing the quantities of reagents used in the test.

Sulfate.

    Sample Solution. Dissolve 5.0 g in 70 mL of water, heat to boiling, and pour slowly with constant stirring into 100 mL of dilute ammonium hydroxide (1 + 6). Filter through a folded filter while still hot, and wash with hot water until the filtrate (sample solution) measures 250 mL. Mix thoroughly (1 mL = 0.02 g).

To 25 mL (0.5-g sample) of sample solution, add 5 mL of hydrochloric acid and 10 mL of nitric acid. For the standard, add 0.05 mg of sulfate ion (SO_4) and 1.5 mL of ammonium hydroxide to 25 mL of water, and add 5 mL of hydrochloric acid and 10 mL of nitric acid. Digest each solution in a covered beaker on a hot plate ($\approx$100 °C) until reaction ceases, uncover, and evaporate to dryness. Dissolve the residues in 4 mL of water plus 1 mL of dilute hydrochloric acid (1 + 19), filter if necessary through a small filter, wash with two 2-mL portions of water, and dilute with water to 10 mL. To each solution, add 1 mL of 12% barium chloride reagent solution. Any turbidity in the solution of the sample should not exceed that in the standard. Compare 10 min after adding the barium chloride to the sample and standard solutions.

Calcium, Magnesium, Potassium, and Sodium. (By flame AAS, page 63).

> *Sample Stock Solution A.* Dissolve 10.0 g of sample in 75 mL of water, transfer to a 100-mL volumetric flask, and dilute to the mark with water (1 mL = 0.10 g).

> *Sample Stock Solution B.* Transfer 5.0 mL (0.50 g) of sample stock solution A to a 100-mL volumetric flask, dilute to the mark with water, and mix (1 mL = 0.005 g).

Element	Wavelength (nm)	Sample Wt (g)	Standard Added (mg)	Flame Type*	Background Correction
Ca	422.7	1.00	0.05; 0.10	N/A	No
Mg	285.2	0.20	0.005; 0.01	A/A	Yes
K	766.5	1.00	0.025; 0.05	A/A	No
Na	589.0	0.04	0.01; 0.02	A/A	No

*A/A is air/acetylene; N/A is nitrous oxide/acetylene.

Ferrous Ammonium Sulfate Hexahydrate

Ammonium Ferrous Sulfate Hexahydrate; Ammonium Iron(II) Sulfate Hexahydrate
$Fe(NH_4)_2(SO_4)_2 \cdot 6H_2O$ **Formula Wt 392.14**   **CAS No. 7783-85-9**

Note: Due to inherent oxidation, ferric ion (Fe^{3+}) content may increase on storage.

GENERAL DESCRIPTION

Typical appearance: pale blue-green solid
Analytical use: analytical standard
Aqueous solubility: soluble

SPECIFICATIONS

Assay . . . . .98.5–101.5% $Fe(NH_4)_2(SO_4)_2 \cdot 6H_2O$

Maximum Allowable

Insoluble matter 0.01%
Phosphate (PO_4). 0.003%
Calcium (Ca) 0.005%
Copper (Cu). 0.003%
Magnesium (Mg) 0.002%
Manganese (Mn) 0.01%
Potassium (K) 0.002%
Sodium (Na). 0.02%
Zinc (Zn) 0.003%
Ferric ion (Fe^{3+}). 0.01%

TESTS

Assay. (By titration of reductive capacity of Fe^{II}). Weigh accurately 1.6 g, and dissolve in a mixture of 100 mL of water and 3 mL of sulfuric acid that has previously been freed from oxygen by bubbling with an inert gas. Titrate with 0.1 N potassium permanganate to a permanent faint pink end point. One milliliter of 0.1 N permanganate corresponds to 0.03921 g of $Fe(NH_4)_2(SO_4)_2 \cdot 6H_2O$.

$$\% \ Fe(NH_4)_2(SO_4)_2 \cdot 6H_2O = \frac{(mL \times N \ KMnO_4) \times 39.21}{Sample \ wt \ (g)}$$

Insoluble Matter. (Page 25). Use 10.0 g dissolved in 100 mL of freshly boiled dilute sulfuric acid (1 + 99).

Phosphate. Dissolve 1.0 g in 10 mL of water, and add 2 mL of nitric acid. Boil to oxidize the iron and expel the excess gases. Dilute with water to 90 mL. Dilute 30 mL with water to 80 mL, add 5 mL of 10% ammonium molybdate reagent solution, and adjust the pH to 1.8 (using a pH meter) with dilute ammonium hydroxide (1 + 1). Prepare a standard containing 0.01 mg of phosphate ion (PO_4), the residue from evaporation of the amount of ammonium hydroxide used in adjusting the pH of the sample, and 5 mL of the ammonium molybdate solution in 85 mL of water. Adjust the pH to 1.8 (using a pH meter) with dilute hydrochloric acid (1 + 9). Cautiously heat the solutions to boiling, but do not boil, and cool to room temperature. If a precipitate forms, it will dissolve when acidified in the next operation. Add 10 mL of hydrochloric acid, and dilute with water to 100 mL. Transfer the solutions to separatory funnels, add 35 mL of diethyl ether to each, shake vigorously, and allow to separate. Draw off and discard the aqueous phase. Wash the diethyl ether phase twice with 10-mL portions of dilute hydrochloric acid (1 + 9), discarding the washings each time. Add 10 mL of dilute hydrochloric acid (1 + 9) to which has just been added 0.2 mL of a freshly prepared 2% stannous chloride reagent solution (in hydrochloric acid), shake, and allow to separate. Any blue color in the ether extract from the sample should not exceed that in the ether extract from the standard.

Calcium, Copper, Magnesium, Manganese, Potassium, Sodium, and Zinc.
(By flame AAS, page 63).

Sample Stock Solution. Dissolve 20.0 g of sample in 80 mL of water, transfer to a 100-mL volumetric flask, dilute to the mark with water, and mix (1 mL = 0.20 g).

Element	Wavelength (nm)	Sample Wt (g)	Standard Added (mg)	Flame Type*	Background Correction
Ca	422.7	2.0	0.05; 0.10	N/A	No
Cu	324.7	1.0	0.03; 0.06	A/A	Yes
Mg	285.2	1.0	0.01; 0.02	A/A	Yes
Mn	279.5	0.20	0.02; 0.04	A/A	Yes
K	766.5	2.0	0.02; 0.04	A/A	No
Na	589.0	0.10	0.01; 0.02	A/A	No
Zn	213.9	0.20	0.005; 0.01	A/A	Yes

*A/A is air/acetylene; N/A is nitrous oxide/acetylene.

Ferric Ion. Place 0.20 g of the sample and 0.5 g of sodium bicarbonate in a dry 125-mL glass-stoppered conical flask. For the control, place 0.10 g of the sample and 0.5 g of sodium bicarbonate in a similar flask. To each flask, add 94 mL of a freshly boiled and cooled solution of dilute sulfuric acid (1 + 25), and loosely stopper the flasks until effervescence stops. To the control, add 0.01 mg of ferric ion (Fe^{3+}). Add 6 mL of ammonium thiocyanate reagent solution to each flask. Any red color in the solution of the sample should not exceed that in the control.

Ferrous Sulfate Heptahydrate
Iron(II) Sulfate Heptahydrate
$FeSO_4 \cdot 7H_2O$ Formula Wt 278.01   CAS No. 7782-63-0

Note: Due to inherent oxidation, ferric ion (Fe^{3+}) content may increase on storage.

GENERAL DESCRIPTION

Typical appearance: green or bluish-green solid
Analytical use: quantitative analysis of nitrates
Change in state (approximate): melting point, forms monohydrate at 65 °C
Aqueous solubility: 48.6 g in 100 mL at 50 °C

SPECIFICATIONS

Assay . . . . ≥99.0% $FeSO_4 \cdot 7H_2O$

	Maximum Allowable
Insoluble matter	0.01%
Chloride (Cl)	0.001%
Phosphate (PO_4)	0.001%
Calcium (Ca)	0.005%
Copper (Cu)	0.005%
Magnesium (Mg)	0.002%
Manganese (Mn)	0.05%
Potassium (K)	0.002%
Sodium (Na)	0.02%
Zinc (Zn)	0.005%
Ferric ion (Fe^{3+})	0.1%

TESTS

Assay. (By titration of reductive capacity of Fe^{II}). Weigh accurately about 1.0 g, and dissolve in a mixture of 100 mL of water and 3 mL of sulfuric acid that has previously been freed from oxygen by bubbling with an inert gas. Titrate with 0.1 N potassium permanganate volumetric solution to a permanent faint pink end point. One milliliter of 0.1 N potassium permanganate corresponds to 0.02780 g of $FeSO_4 \cdot 7H_2O$.

$$\% \ FeSO_4 \cdot 7H_2O = \frac{(mL \times N \ KMnO_4) \times 27.80}{Sample \ wt \ (g)}$$

Insoluble Matter. (Page 25). Use 10.0 g dissolved in 100 mL of freshly boiled dilute sulfuric acid (1 + 99).

Chloride. Dissolve 1.0 g in 10 mL of water, and add slowly 2 mL of nitric acid. After the evolution of nitrogen oxides has ceased, dilute with water to 15 mL, filter if necessary through a small chloride-free filter, and add 1 mL of silver nitrate reagent solution. Any turbidity should not exceed that produced by 0.01 mg of chloride ion (Cl) in an equal volume of solution containing the quantities of reagents used in the test. The comparison is best made by the general method for chloride in colored solutions, page 35.

Phosphate. (Page 40, Method 2). Dissolve 1.0 g in 10 mL of dilute nitric acid (1 + 5), and boil to expel the excess gases. Dilute the solution with water to 80 mL, add 5 mL of 10% ammonium molybdate reagent solution, and adjust the pH to 1.8 (using a pH meter) with dilute ammonium hydroxide (1 + 1). Cautiously heat the solution to boiling but do not boil, and cool to room temperature. If a precipitate forms, it will dissolve in the next operation. Add 10 mL of hydrochloric acid, dilute with water to 100 mL, and continue as described.

Prepare a standard containing 0.01 mg of phosphate ion (PO_4), the residue from evaporation of the amount of ammonium hydroxide used in adjusting the pH of the sample, and 5 mL of the ammonium molybdate reagent solution in 85 mL of water. Adjust the pH to 1.8 (using a pH meter) with dilute hydrochloric acid (1 + 9). Cautiously heat the solutions to boiling but do not boil; cool to room temperature. Add 10 mL of hydrochloric acid, and dilute with water to 100 mL.

Transfer the solutions to separatory funnels, add 35 mL of diethyl ether to each, shake vigorously, and allow to separate. Draw off and discard the aqueous phase. Wash the diethyl ether phase twice with 100-mL portions of dilute hydrochloric acid (1 + 9), discarding the washings each time. Add 10 mL of dilute hydrochloric acid (1 + 9) to which has just been added 0.2 mL of a freshly prepared 2% stannous chloride reagent solution (in hydrochloric acid), shake, and allow to separate. Any blue color in the ether extract from the sample should not exceed that in the ether extract from the standard.

Calcium, Copper, Magnesium, Manganese, Potassium, Sodium, and Zinc. (By flame AAS, page 63).

Sample Stock Solution. Dissolve 20.0 g of sample in 80 mL of water, transfer to a 100-mL volumetric flask, dilute to the mark with water, and mix (1 mL = 0.20 g).

Element	Wavelength (nm)	Sample Wt (g)	Standard Added (mg)	Flame Type*	Background Correction
Ca	422.7	2.0	0.05; 0.10	N/A	No
Cu	324.7	1.0	0.05; 0.10	A/A	Yes
Mg	285.2	0.50	0.005; 0.01	A/A	Yes
Mn	279.5	0.10	0.025; 0.05	A/A	Yes
K	766.5	2.0	0.02; 0.04	A/A	No
Na	589.0	0.10	0.01; 0.02	A/A	No
Zn	213.9	0.10	0.005; 0.01	A/A	Yes

*A/A is air/acetylene; N/A is nitrous oxide/acetylene.

Ferric Ion. Place 0.20 g of the sample and 0.5 g of sodium bicarbonate in a dry 125-mL glass-stoppered conical flask. For the control, place 0.10 g of the sample and 0.5 g of sodium bicarbonate in a similar flask. To each flask, add 94 mL of a freshly boiled and cooled solution of dilute sulfuric acid (1 + 25), and loosely stopper the flasks until effervescence stops. To the control, add 0.1 mg of ferric ion (Fe^{3+}). Add 6 mL of ammonium thiocyanate reagent solution to each flask. Any red color in the solution of the sample should not exceed that in the control.

Formaldehyde Solution

HCHO	Formula Wt 30.03	CAS No. 50-00-0

Note: This reagent contains 10–15% of methanol as a stabilizer.

GENERAL DESCRIPTION

Typical appearance: colorless solution
Analytical use: reducing agent; to prevent polymerization
Change in state (approximate): boiling point, 96 °C
Aqueous solubility: miscible
Density: 1.08

SPECIFICATIONS

Assay . . . . .36.5–38.0% HCHO

Maximum Allowable

Color (APHA) 10
Residue after ignition. 0.005%
Titrable acid. . . . . .0.006 meq/g
Chloride (Cl). . . . . .5 ppm
Sulfate (SO_4) 0.002%
Heavy metals (as Pb). . . . . .5 ppm
Iron (Fe). . . . . .5 ppm

TESTS

Note: The assay procedure specified herein is intended to determine both formaldehyde and its paraformaldehyde polymerization product, which may form upon standing. A different procedure (see, for example, *Reagent Chemicals*, 5th ed.) should be used if only the formaldehyde content is desired.

Assay. (By bisulfite addition and acid–base titrimetry). Transfer a glass-stoppered weighing vial containing 2.0 g of sample, accurately weighed, to 100 mL of sodium sulfite solution in a 500-mL conical flask. Avoid getting any of the sample on the wall of the flask. Add 50 mL of 1 N sodium hydroxide and 0.15 mL of thymolphthalein indicator solution, and then titrate to a colorless end point with 1 N sulfuric acid. Correct for a reagent blank and any significant quantity of acid (as HCOOH) in the sample. One milliliter of 1 N sodium hydroxide corresponds to 0.03003 g of HCHO.

$$\% \text{ HCHO} = \frac{(\text{mL} \times \text{N NaOH}) \times 3.003}{\text{Sample wt (g)}}$$

Color (APHA). (Page 43).

Residue after Ignition. (Page 26). Evaporate 20.0 g (20 mL) to dryness in a tared, preconditioned dish on a hot plate ($\approx$100 °C), add 0.05 mL of sulfuric acid to the residue, and ignite gently to volatilize the excess sulfuric acid. Finally, ignite at 600 ± 25 °C for 15 min.

Titrable Acid. To 10 mL in a flask containing 20 mL of water, add 0.15 mL of bromthymol blue indicator solution, and titrate with 0.1 N sodium hydroxide. Not more than 0.60 mL of 0.1 N sodium hydroxide should be required.

Chloride. (Page 35). Dilute 2.0 mL with 20 mL of water.

Sulfate. (Page 40, Method 1). Use 2.5 g (2.5 mL).

Heavy Metals. (Page 36, Method 1). To 4.0 mL, add about 10 mg of sodium carbonate, evaporate to dryness, and heat gently to volatilize any organic matter. Dissolve the residue in about 20 mL of water, and dilute with water to 25 mL.

Iron. (Page 38, Method 2). Dilute 2.0 mL with water to 15 mL. Adjust the pH to 5 with 2.5% ammonium hydroxide reagent solution after addition of the 1,10-phenanthroline reagent solution.

Formamide

HCONH$_2$ Formula Wt 45.04   CAS No. 75-12-7

GENERAL DESCRIPTION

Typical appearance: slightly viscous, clear, colorless liquid
Analytical use: ionizing agent
Change in state (approximate): boiling point, 210 °C
Aqueous solubility: miscible
Density: 1.13

SPECIFICATIONS

Assay . . . . $\geq$99.5% HCONH$_2$
Freezing point . . . . 2.0–3.0 °C

Maximum Allowable

Color (APHA). 10

TESTS

Assay. Analyze the sample by gas chromatography using the general parameters cited on page 80. The following specific conditions are also required.

   Column: Type I, methyl silicone

Measure the area under all peaks, and calculate the formamide content in area percent.

Freezing Point. Place 15 mL in a test tube (20 × 150 mm) in which is centered an accurate thermometer. The sample tube is centered by corks in an outer tube about 38 × 200 mm. Cool the whole apparatus, without stirring, in a bath of shaved or crushed ice with enough water to wet the outer tube. When the temperature is about 0 °C, stir to start the freezing, and read the thermometer every 30 s. The temperature that remains constant for 1–2 min is the freezing point.

Color (APHA). (Page 43).

Formic Acid, 96%

HCOOH   Formula Wt 46.03   CAS No. 64-18-6

Caution: Slow decomposition of this reagent may produce pressure in the bottle. Loosen cap occasionally to vent the gas.

GENERAL DESCRIPTION

Typical appearance: clear, colorless liquid
Analytical use: chemical synthesis
Change in state (approximate): boiling point, 100 °C
Aqueous solubility: miscible
Density: 1.22
pK_a: 3.7

SPECIFICATIONS

Assay . . . . ≥96.0% HCOOH

Maximum Allowable

Color (APHA)	15
Dilution test.	Passes test
Residue after evaporation	0.003%
Acetic acid (CH_3COOH)	0.4%
Ammonium (NH_4)	0.005%
Chloride (Cl).	0.001%
Sulfate (SO_4)	0.003%
Sulfite (SO_3)	Passes test
Heavy metals (as Pb).	0.001%
Iron (Fe).	0.001%

TESTS

Assay. (By acid–base titrimetry). Tare a small glass-stoppered conical flask containing 15 mL of water. Quickly introduce 1.0–1.5 mL of the sample, and weigh. Dilute with water to about 50 mL and titrate with 1 N sodium hydroxide volumetric solution, using 0.15 mL of phenolphthalein indicator. One milliliter of 1 N sodium hydroxide corresponds to 0.04603 g of HCOOH.

$$\% \text{ HCOOH} = \frac{(\text{mL} \times \text{N NaOH}) \times 4.603}{\text{Sample wt (g)}}$$

Color (APHA). (Page 43).

Dilution Test. Dilute 1 volume of the acid with 3 volumes of water. No turbidity should be observed within 1 h.

Residue after Evaporation. (Page 25). Evaporate 50.0 g (42 mL) to dryness in a tared, preconditioned dish on a hot plate ($\approx$100 °C), and dry the residue at 105 °C for 30 min.

Acetic Acid. Analyze the sample by gas chromatography as described on page 80. The following specific conditions are satisfactory.

> *Column:* 1.8 m × 64 mm, stainless steel, packed with 80/100-mesh Porapak Q
>
> *Column Temperature:* 150 °C
>
> *Injection Port Temperature:* 150 °C
>
> *Detector Temperature:* 175 °C
>
> *Detector Current:* 200 mA
>
> *Carrier Gas:* Helium at 40 mL/min
>
> *Sample Size:* 5 μL
>
> *Detector:* Thermal conductivity
>
> *Approximate Retention Times (min):* Water, 1.2; formic acid, 3.5; acetic acid, 8.1

Measure the area under all peaks, and calculate the acetic acid content in area percent.

Ammonium. (By differential pulse polarography, page 53). Dilute 10.0 g (8.2 mL) of sample to 50 mL with water; use 5.0 mL in each test, neutralizing with 3.2 mL of ammonia-free 6 N sodium hydroxide reagent solution. For the standard, use 0.05 mg of ammonium ion (NH_4).

Chloride. (Page 35). Use 1.0 g (0.83 mL) sample.

Sulfate. (Page 40, Method 3). Use 1.7 g (1.4 mL) sample.

Sulfite. Dilute 25 mL with 25 mL of water; add 0.10 mL of 0.1 N iodine solution. The mixture should retain a distinct yellow color.

Heavy Metals. (Page 36, Method 1). To 2 g (1.6 mL) in a beaker, add 1.0 mL of 1% sodium carbonate reagent solution, and evaporate to dryness. Dissolve the residue in about 20 mL of water, and dilute with water to 25 mL.

Iron. (Page 38, Method 1). To 12.0 g (5.0 mL) in a beaker, add 1.0 mL of 1% sodium carbonate reagent solution, and evaporate to dryness. Dissolve the residue in 6 mL of hydrochloric acid, dilute with water to 60 mL, and use 10 mL of this solution without further acidification.

Formic Acid, 88%

HCOOH   **Formula Wt 46.03**   **CAS No. 64-18-6**

GENERAL DESCRIPTION

Typical appearance: clear, colorless liquid
Analytical use: chemical synthesis
Change in state (approximate): boiling point, 100 °C
Aqueous solubility: miscible
Density: 1.22
pK_a: 3.7

SPECIFICATIONS

Assay . . . . ≥88.0% HCOOH

Maximum Allowable

Color (APHA) 15
Dilution test. . . . . .Passes test
Residue after evaporation 0.002%
Acetic acid (CH_3COOH) 0.4%
Ammonium (NH_4) 0.005%
Chloride (Cl). 0.001%
Sulfate (SO_4) 0.002%
Sulfite (SO_3) . . . . .Passes test
Heavy metals (as Pb). . . . . .5 ppm
Iron (Fe). . . . . .5 ppm

TESTS

Assay and other tests except sulfate, heavy metals, and iron are the same as for Formic Acid, 96%, page 332.

Sulfate. (Page 41, Method 3). Use 2.5-g sample.

Heavy Metals. (Page 36, Method 1). Use 4.0-g sample.

Iron. (Page 38, Method 1). Use 12.0-g sample.

2-Furancarboxyaldehyde

2-Furaldehyde; Furfural

$C_5H_4O_2$ **Formula Wt 96.08**   **CAS No. 98-01-1**

GENERAL DESCRIPTION

Typical appearance: oily, colorless liquid; may become yellow, then brown, on exposure to air
Analytical use: catalyst; general reagent

Change in state (approximate): boiling point, 161 °C
Aqueous solubility: 9 g in 100 mL at 20 °C
Density: 1.16

SPECIFICATIONS

Assay . . . . $\geq$98.0% $C_5H_4O_2$

Maximum Allowable

Titrable acid . . . . 0.02 meq/g
Residue after evaporation 0.5%

TESTS

Assay. Analyze the sample by gas chromatography using the general parameters cited on page 80. The following specific conditions are also required.

   Column: Type I, methyl silicone

Measure the area under all peaks, and calculate the 2-furancarboxyaldehyde content in area percent.

Titrable Acid. To 300 mL of water in an Erlenmeyer flask, add 0.15 mL of phenolphthalein indicator solution, and titrate with 0.1 N sodium hydroxide to a faint pink color that is stable for about 1 min. Add 10.0 g (9.0 mL) of sample to the flask, mix well, and titrate with 0.1 N sodium hydroxide to the original faint pink color. Not more than 2.0 mL of 0.1 N sodium hydroxide should be consumed.

Residue after Evaporation. (Page 25). Evaporate 25.0 g (22 mL) to dryness in a tared, preconditioned dish on a hot plate ($\approx$100 °C), and dry the residue at 105 °C for 30 min.

Gallic Acid

$C_6H_2(OH)_3COOH$ (anhydrous)   **Formula Wt 170.12   CAS No. 149-91-7**
$C_6H_2(OH)_3COOH \cdot H_2O$ (monohydrate)   **Formula Wt 188.14   CAS No. 5995-86-8**

Note: This reagent is available in both monohydrate and anhydrous forms. The label should identify the degree of hydration.

GENERAL DESCRIPTION

Typical appearance: colorless solid
Analytical use: testing for free mineral acids, dihydroxyacetone, and alkaloids

Change in state (approximate): sublimes at 210 °C, giving a stable form with melting point 258–265 °C and an unstable form with a melting point 225–230 °C

Aqueous solubility: 1 g in 87 mL

SPECIFICATIONS

Assay . . . . $\geq$98.0% $C_6H_2(OH)_3COOH$
or $C_6H_2(OH)_3COOH \cdot H_2O$

Maximum Allowable

Insoluble matter 0.01%
Residue after ignition. 0.05%
Sulfate (SO_4) 0.02%

TESTS

Assay. (By acid–base titrimetry). Weigh accurately about 0.7 g, transfer to a beaker, add 100 mL of water, and stir to dissolve. When dissolution is complete, titrate potentiometrically with 0.1 N sodium hydroxide. Correct for a blank determination. One milliliter of 0.1 N sodium hydroxide corresponds to 0.017012 g of $C_6H_2(OH)_3COOH$ or 0.018814 g of $C_6H_2(OH)_3COOH \cdot H_2O$.

$$\% \ C_6H_2(OH)_3COOH = \frac{(mL \times N \ NaOH) \times 17.012}{Sample \ wt \ (g)}$$

$$\% \ C_6H_2(OH)_3COOH \cdot H_2O = \frac{(mL \times N \ NaOH) \times 18.814}{Sample \ wt \ (g)}$$

Insoluble Matter. (Page 25). Use 10.0 g dissolved in 300 mL of water.

Residue after Ignition. (Page 26). Ignite 4.0 g.

Sulfate. (Page 40, Method 1). Dissolve 2.5 g in 50 mL of hot water, cool in ice water while stirring, and filter. Use 5 mL.

D-Glucose, Anhydrous

Dextrose, Anhydrous

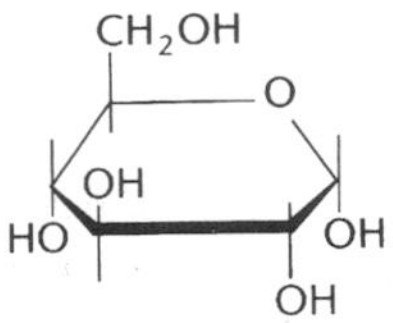

CH₂OH(CHOH)₄CHO   **Formula Wt 180.16**   **CAS No. 50-99-7**

GENERAL DESCRIPTION

Typical appearance: white solid
Analytical use: analytical standard

Change in state (approximate): melting point, 146 °C
Aqueous solubility: 91 g in 100 mL at 25 °C

SPECIFICATIONS

Specific rotation [α]$_D^{25°}$ . . . . +52.5° to +53.0°

Maximum Allowable

Insoluble matter 0.005%
Loss on drying 0.2%
Residue after ignition 0.02%
Titrable acid . . . . 0.002 meq/g
Chloride (Cl) 0.01%
Sulfate and sulfite (as SO$_4$) 0.005%
Starch . . . . Passes test
Heavy metals (as Pb) . . . . 5 ppm
Iron (Fe). . . . . 5 ppm

TESTS

Specific Rotation. (Page 46). Weigh accurately about 10 g, and dissolve in 90 mL of water in a 100-mL volumetric flask. Add 0.2 mL of ammonium hydroxide, and dilute with water to volume at 25 °C. Observe the optical rotation in a polarimeter at 25 °C using sodium light, and calculate the specific rotation. It should not be less than +52.5° nor more than +53.0°.

Insoluble Matter. (Page 25). Use 20 g dissolved in 150 mL of water.

Loss on Drying. Weigh accurately about 1 g, and dry in a preconditioned weighing bottle at 105 °C for 6 h.

Residue after Ignition. (Page 26). In a preconditioned dish, ignite 5.0 g until charred. Moisten the char with 2 mL of sulfuric acid.

Titrable Acid. To 100 mL of carbon dioxide-free water, add 0.15 mL of phenolphthalein indicator solution and 0.02 N sodium hydroxide until a pink color is produced. Dissolve 10.0 g of the sample in this solution, and titrate with 0.01 N sodium hydroxide to the same end point. Not more than 2.5 mL should be required.

Chloride. (Page 35). Dissolve 0.50 g in 50 mL of water, and use 10 mL of this solution.

Sulfate and Sulfite. (Page 40, Method 1). After sample dissolution, add 1 mL of bromine water, and boil.

Starch. Dissolve 1.0 g in 10 mL of water and add 0.05 mL of 0.1 N iodine solution. No blue color should appear.

Heavy Metals. (Page 36, Method 1). Dissolve 6.0 g in about 20 mL of water, and dilute with water to 30 mL. Use 25 mL to prepare the sample solution, and use the remaining 5.0 mL to prepare the control solution.

Iron. (Page 38, Method 1). Use 2.0 g.

Glycerol

1,2,3-Propanetriol

$$HOCH_2\underset{\underset{\displaystyle OH}{|}}{CH}CH_2OH$$

CH₂OHCHOHCH₂OH   **Formula Wt 92.09**   **CAS No. 56-81-5**

GENERAL DESCRIPTION

Typical appearance: clear, viscous liquid
Analytical use: titration reagent; tests for heavy metals
Change in state (approximate): boiling point, 290 °C
Aqueous solubility: miscible
Density: 1.26

SPECIFICATIONS

Assay (by volume) . . . . ≥99.5% $C_3H_5(OH)_3$

Maximum Allowable

Color (APHA) 10
Residue after ignition. 0.005%
Neutrality . . . . .Passes test
Chlorinated compounds (as Cl) 0.003%
Sulfate (SO_4) 0.001%
Acrolein and glucose . . . . .Passes test
Fatty acid esters (as butyric acid) 0.05%
Substances darkened by sulfuric acid . . . . .Passes test
Heavy metals (as Pb). . . . . .2 ppm
Water (H_2O) 0.5%

TESTS

Assay. Analyze the sample by gas chromatography using the general parameters cited on page 80. The following specific conditions are also required.

Column: Type II, mixed cyano, phenyl methyl silicone

Measure the area under all peaks, and calculate the glycerol content in area percent. Correct for water content.

Color (APHA). (Page 43).

Residue after Ignition. (Page 26). Heat 20 g in a tared, preconditioned open dish, ignite the vapors, and when the glycerol has been entirely consumed, ignite at 600 ± 25 °C for 15 min.

Neutrality. A 10% aqueous solution of glycerol should not affect the color of either red or blue litmus paper in 1 min.

Chlorinated Compounds. Transfer 5.0 g of the glycerol to a dry 100-mL round-bottom flask fitted with a ground joint. Add 15 mL of morpholine, connect a matching standard taper reflux condenser, and reflux the mixture gently for 3 h. Rinse the condenser with 10 mL of water, collecting the washing in the flask. Cool. Cautiously acidify the solution with nitric acid, add 1 mL of silver nitrate reagent solution, and dilute with water to 50 mL. Any turbidity should not exceed that produced by 0.15 mg of chloride ion (Cl) in an equal volume of solution containing the quantities of reagents used in the test, omitting the refluxing.

Sulfate. (Page 40, Method 1). Use 5.0 g.

Acrolein and Glucose. Heat a mixture of 5 mL of the glycerol and 5 mL of 10% potassium hydroxide solution at 60 °C for 5 min. No yellow color should develop.

Fatty Acid Esters. To 40 g in a 250-mL conical flask, add 50 mL of hot, freshly boiled water. Add 10.0 mL of 0.1 N sodium hydroxide, cover with a loosely fitting pear-shaped bulb, and digest on a hot plate ($\approx$100 °C) for 45 min. Cool, and titrate the excess alkali with 0.1 N hydrochloric acid, using 0.15 mL of bromthymol blue indicator solution and titrating to a bluish green end point. Run a blank with 50 mL of the same water and 10.0 mL of 0.1 N sodium hydroxide solution, heating for the same length of time and titrating to the same end point. The difference between the volumes of acid used in the titration of the blank and of the sample should be less than 2.3 mL.

Substances Darkened by Sulfuric Acid. Vigorously shake 5 mL of glycerol with 5 mL of 94.5–95.5% sulfuric acid in a glass-stoppered 25-mL cylinder for 1 min and allow the liquid to stand for 1 h. The liquid should not be darker than a standard made up of 0.4 mL of cobaltous chloride color solution (described below), 3.0 mL of ferric chloride color solution (described below), and 6.6 mL of water.

> **Ferric Chloride Color Solution.** Dissolve 1.80 g of ferric chloride hexahydrate in 40 mL of dilute hydrochloric acid (1 + 39) (1 mL = 45.0 mg $FeCl_3 \cdot 6H_2O$).

> **Cobaltous Chloride Color Solution.** Dissolve 2.38 g of cobaltous chloride hexahydrate in 40 mL of dilute hydrochloric acid (1 + 39) (1 mL = 59.5 mg $CoCl_2 \cdot 6H_2O$).

Heavy Metals. (Page 36, Method 1). Dilute 13 g (10 mL) with water to 36 mL. Use 32 mL to prepare the sample solution, and use the remaining 4.0 mL to prepare the control solution.

Water. (Page 31, Method 1). Use 4.0 mL (5.0 g) of sample.

Glycine
Aminoacetic Acid
H_2NCH_2COOH Formula Wt 75.07   CAS No. 56-40-6

GENERAL DESCRIPTION
Typical appearance: white solid
Analytical use: buffer
Change in state (approximate): decomposes at 233 °C
Aqueous solubility: 25.0 g in 100 mL at 25 °C

SPECIFICATIONS
Assay . . . . ≥98.5% H_2NCH_2COOH
Maximum Allowable
Residue after ignition. 0.1%
Heavy metals (as Pb). 0.002%
Chloride (Cl). 0.005%
Sulfate (SO_4) 0.005%
Ammonium (NH_4) 0.005%
Substances darkened by sulfuric acid . . . . .Passes test
Hydrolyzable substances . . . . .Passes test

TESTS

Assay. (Total alkalinity by nonaqueous titration). Weigh accurately 0.25 g, and dissolve in 100 mL of glacial acetic acid. Add 0.5 mL of crystal violet indicator solution, and titrate with 0.1 N perchloric acid in glacial acetic acid volumetric solution to a green end point. Correct for a blank. One milliliter of 0.1 N perchloric acid corresponds to 0.007507 g of H_2NCH_2COOH.

$$\% \ H_2NCH_2COOH = \frac{\{[mL \ (sample) - mL \ (blank)] \times N \ HClO_4\} \times 7.507}{Sample \ wt \ (g)}$$

Residue after Ignition. (Page 26). Use 2.0 g.

Heavy Metals. (Page 36, Method 1). Use 1.0 g.

Chloride. (Page 35). Use 0.2 g.

Sulfate. (Page 40, Method 1). Use 1.0 g.

Ammonium. (By colorimetry, page 32). Use 0.2 g. For the standard, use 0.01 mg of ammonium ion (NH_4) treated exactly as the sample.

Substances Darkened by Sulfuric Acid. Dissolve 0.5 g in 5 mL of sulfuric acid. Allow the solution to stand for 15 min. The color of the sample solution should not exceed that of an equal volume of the acid.

Hydrolyzable Substances. Boil 10 mL of a 1 in 10 solution of glycine for 1 min, and set aside for 2 h. The solution, when compared to an equal volume of an untreated sample solution, should appear as clear and as mobile.

Gold Chloride
Tetrachloroauric(III) Acid Trihydrate
$HAuCl_4 \cdot 3H_2O$ **Formula Wt 393.83**   **CAS No. 16961-25-4**

GENERAL DESCRIPTION
Typical appearance: yellow to reddish-yellow solid
Analytical use: preparation of gold standard solutions; graphite-furnace AAS
Change in state (approximate): melting point, 254 °C
Aqueous solubility: appreciable (>10%)

SPECIFICATIONS
Assay . . . . 49.0% Au

Maximum Allowable

Insoluble in ether 0.1%
Alkalis and other metals (as sulfates) 0.2%

TESTS

Assay. (Gold content by gravimetry). Evaporate the solution from the test for insoluble in ether to dryness or a syrupy residue. Dissolve the residue in 100 mL of dilute hydrochloric acid (1 + 19), disregarding any undissolved material. Add 10 mL of sulfurous acid, and digest on a hot plate (≈100 °C) until the precipitate is well coagulated. Filter, wash with dilute hydrochloric acid (1 + 99), and ignite in a tared, preconditioned crucible. Retain the filtrate and washings. The weight of the residue should be not less than 49.0% of the weight of the sample.

$$\% \text{ Gold} = \frac{\text{Residue wt (g)} \times 100}{\text{Sample wt (g)}}$$

Note: Sulfurous acid assay must be >6% as SO_2. If not, adjust the volume accordingly.

Insoluble in Ether. Weigh accurately about 1 g. Add 10 mL of diethyl ether, allow to stand for 10 min with occasional stirring, filter through a tared filtering crucible of fine porosity, wash with small portions of ether, and dry at 105 °C. Retain the filtrate and washings.

Alkalis and Other Metals (as Sulfates). Evaporate the filtrate and washings reserved from the assay in a tared, preconditioned dish to dryness. Ignite the residue cautiously, and finally ignite at 600 ± 25 °C for 15 min.

Hexamethylenetetramine

C₆H₁₂N₄   **Formula Wt 140.19**   **CAS No. 100-97-0**

GENERAL DESCRIPTION

Typical appearance: white solid
Analytical use: detection of metals
Change in state (approximate): sublimes at about 263 °C with slight decomposition
Aqueous solubility: 67 g in 100 mL at 20 °C
pK_a: 6.4

SPECIFICATIONS

Assay (dried basis)..≥99.0% C₆H₁₂N₄

Maximum Allowable

Loss on drying 2.0%
Residue after ignition..0.1%
Heavy metals (as Pb)..0.001%

TESTS

Assay. Weigh accurately about 1 g of previously dried sample, and transfer to a 100-mL beaker. Add 40.0 mL of 1 N sulfuric acid volumetric solution, and boil gently for at least 1 h to remove formaldehyde. Cool, add 20 mL of water, and titrate the excess acid with 1 N sodium hydroxide volumetric solution, using 0.15 mL of methyl red indicator solution. One milliliter of 1 N sulfuric acid corresponds to 0.035057 g of $C_6H_{12}N_4$.

$$\% \ C_6H_{12}N_4 = \frac{[(40.0 \times N \ H_2SO_4) - (mL \times N \ NaOH)] \times 3.5057}{\text{Sample wt (g)}}$$

Loss on Drying. Accurately weigh about 1 g of sample, and dry at 105 °C over phosphorus pentoxide for 4 h. Save the dried sample for the assay determination.

Residue after Ignition. (Page 26). Ignite 1–2 g, accurately weighed.

Heavy Metals. (Page 36, Method 1). Dissolve 2 g in 10 mL of water, add 2 mL of 3 N hydrochloric acid, and dilute with water to 25 mL. Proceed as directed, except use glacial acetic acid to adjust the pH.

Hexanes

Suitable for general use or in high-performance liquid chromatography, extraction–concentration analysis, or ultraviolet spectrophotometry. Product labeling shall designate the uses for which suitability is represented on the basis of meeting the relevant specifications and tests. The ultraviolet spectrophotometry and liquid chromatography suitability specifications include all of the specifications for general use. The extraction–concentration suitability specifications include only the general use specification for color.

Note: This reagent is generally a mixture of several isomers of hexane (C_6H_{14}), predominantly *n*-hexane, 2-methylpentane, and 3-methylpentane, plus methylcyclopentane (C_6H_{12}).

GENERAL DESCRIPTION

Typical appearance: colorless liquid

Analytical use: determining refractive index of minerals; extraction solvent

Density: 0.66

SPECIFICATIONS

General Use

Assay . . . . ≥98.5% hexanes (sum of 5 isomers, total hexanes, plus methylcyclopentane)

Maximum Allowable

Color (APHA). 10
Residue after evaporation 0.001%
Water-soluble titrable acid. . . . . 0.0003 meq/g
Sulfur compounds (as S) 0.005%
Thiophene . . . . Passes test

Specific Use

Ultraviolet Spectrophotometry

Wavelength (nm)	Absorbance (AU)
280–400	0.01
250	0.02
240	0.04
230	0.10
220	0.20
210	0.30
195	1.0

Liquid Chromatography Suitability

Absorbance. . . . . Passes test

Extraction–Concentration Suitability

Absorbance. . . . . Passes test
GC–FID . . . . Passes test
GC–ECD . . . . Passes test

TESTS

Assay. Analyze the sample by gas chromatography using the general parameters cited on page 80. The following specific conditions are also required.

Column: Type I, methyl silicone

Measure the area under all peaks, and calculate the hexanes and methylcyclopentane in area percent. The retention times are (min): 2-methylpentane, 7.5; 3-methylpentane, 8.5; *n*-hexane, 9.2; methylcyclopentane, 10.2.

Color (APHA). (Page 43).

Residue after Evaporation. (Page 25). Evaporate 100 g (145 mL) to dryness in a tared, preconditioned dish on a hot plate ($\approx$100 °C), and dry the residue at 105 °C for 30 min.

Water-Soluble Titrable Acid. To 30 g (44 mL) in a separatory funnel, add 50 mL of water, and shake vigorously for 2 min. Allow the layers to separate, draw off the aqueous layer, and add 0.15 mL of phenolphthalein indicator solution to the aqueous layer. Not more than 1.0 mL of 0.01 N sodium hydroxide should be required to produce a pink color.

Sulfur Compounds. To 30 mL of 0.5 N potassium hydroxide in methanol in a conical flask, add 5.5 g (8 mL) of the sample, and boil the mixture gently for 30 min under a reflux condenser, avoiding the use of a rubber stopper or connection. Detach the condenser, dilute with 50 mL of water, and heat on a hot plate ($\approx$100 °C) until the hexanes and methanol are evaporated. Add 50 mL of bromine water, heat for an additional 15 min, and transfer to a beaker. Neutralize with dilute hydrochloric acid (1 + 3), add an excess of 1 mL of the acid, evaporate to about 50 mL, and filter if necessary. Heat the filtrate to boiling, add 5 mL of 12% barium chloride reagent solution, digest in a covered beaker on the hot plate ($\approx$100 °C) for 2 h, and allow to stand for at least 8 h. If a precipitate is formed, filter, wash thoroughly, and ignite. Correct for the weight obtained in a complete blank test.

Thiophene. To 25 mL in a glass-stoppered flask, add 15 mL of sulfuric acid to which has been added a few milligrams of isatin, shake the mixture for 15–20 s, and allow to stand for 1 h. The acid layer should not be blue or green.

Ultraviolet Spectrophotometry. Use the procedure on page 86 to determine the absorbance.

Liquid Chromatography Suitability. Use the procedure on page 86 to determine the absorbance.

Extraction–Concentration Suitability. Analyze the sample by using the general procedure cited on page 85. Use the procedure on page 86 to determine the absorbance.

Hydrazine Sulfate

$(NH_2)_2 \cdot H_2SO_4$ Formula Wt 130.12   CAS No. 10034-93-2

GENERAL DESCRIPTION

Typical appearance: colorless solid

Analytical use: gravimetric determination of nickel, cobalt, and cadmium

Change in state (approximate): melting point, 254 °C

Aqueous solubility: 3.4 g in 100 mL at 25 °C

SPECIFICATIONS

Assay . . . . $\geq$99.0% $(NH_2)_2 \cdot H_2SO_4$

Maximum Allowable

Insoluble matter 0.005%

Residue after ignition 0.05%

Chloride (Cl) 0.005%

Heavy metals (as Pb) 0.002%

Iron (Fe). 0.001%

TESTS

Assay. (By titration of reducing power). Weigh accurately about 1 g, dissolve in water in a 500-mL volumetric flask, and dilute to the mark with water. Take 50.0 mL, dissolve 1 g of sodium bicarbonate, and add 50.0 mL of 0.1 N iodine volumetric solution. Titrate the excess iodine with 0.1 N sodium thiosulfate volumetric solution, adding 3 mL of starch indicator solution near the end point. One milliliter of 0.1 N iodine corresponds to 0.003253 g of $(NH_2)_2 \cdot H_2SO_4$.

$$\% \ (NH_2)_2 \cdot H_2SO_4 = \frac{[(50.0 \times N\,I_2) - (mL \times N\,Na_2S_2O_3)] \times 3.253}{\text{Sample wt (g)} / 10}$$

Insoluble Matter. (Page 25). Use 20 g dissolved in 300 mL of water.

Residue after Ignition. (Page 26). Ignite 2.0 g. Reserve the residue for the test for iron.

Chloride. (Page 35). Dissolve 1.0 g in water, dilute with water to 100 mL, and use 20 mL of this solution.

Heavy Metals. (Page 36, Method 1). Dissolve 2.0 g in 40 mL of warm water. Use 30 mL to prepare the sample solution, and use the remaining 10 mL to prepare the control solution.

Iron. (Page 38, Method 1). To the residue after ignition, add 3 mL of dilute hydrochloric acid (1 + 1) and 0.10 mL of nitric acid, cover with a watch glass, and digest on a hot plate ($\approx$100 °C) for 15–20 min. Remove the watch glass, and evaporate to dryness. Dissolve the residue in a mixture of 4 mL of hydrochloric acid and 10 mL of water, and dilute with water to 100 mL. Use 50 mL of this solution without further acidification.

Hydriodic Acid, 47%

(with stabilizer)

HI   Formula Wt 127.91   CAS No. 10034-85-2

Note: To avoid danger of explosions, this acid should be distilled only in an inert atmosphere. The reagent may have a slight yellow color. It contains hypophosphorus acid as a stabilizer.

GENERAL DESCRIPTION

Typical appearance: colorless liquid when freshly distilled; rapidly becomes yellow to reddish-brown

Analytical use: determination of methoxyl

Aqueous solubility: miscible

Density: 1.5

SPECIFICATIONS

Assay . . . . $\geq$47.0% HI

Maximum Allowable

Chloride and bromide (as Cl) 0.05%

Sulfate (SO_4) 0.005%

Stabilizer (H_3PO_2) 1.5%

Heavy metals (as Pb) 0.001%

Iron (Fe) 0.001%

TESTS

Assay. (By acid–base titrimetry) Tare a glass-stoppered flask containing about 25 mL of water. Under the water surface, quickly add about 3 mL of the sample, then stopper, and weigh accurately to the nearest 0.1 mg. Dilute to about 75 mL, add 0.15 mL of phenolphthalein indicator solution, and titrate with 1 N sodium hydroxide volumetric solution. One milliliter of 1 N sodium hydroxide corresponds to 0.12791 g of HI.

$$\% \text{ HI} = \frac{(\text{mL} \times \text{NaOH}) \times 12.79}{\text{Sample wt (g)}}$$

Chloride and Bromide. Dilute 1 g (0.67 mL) with water to 100 mL in a 250-mL conical flask. Add 1 mL of hydrogen peroxide and 1 mL of phosphoric acid. Heat to boiling, and boil gently until all the iodine is expelled and the solution is colorless. Cool, wash down the sides of the flask, and add 0.5 mL of hydrogen peroxide. If an iodine color develops, boil until the solution is colorless and for 10 min longer. If no color develops, boil for 10 min. Dilute with water to 100 mL, take a 2-mL portion of the solution, and dilute with water to 23 mL. Prepare a standard containing 0.01 mg of chloride ion (Cl) in 23 mL of water. Add 1 mL of nitric acid and 1 mL of silver nitrate reagent solution to each. Any turbidity produced in the solution of the sample should not exceed that in the standard.

Sulfate. For sample, dilute 2.0 mL (3.0 g) to 25 mL with water. For control, take 0.7 mL (1.0 g) of sample, and dilute to 25 mL with water. Neutralize both with (1 + 1) ammonium hydroxide, and add 1 mL of (1 + 19) hydrochloric acid to each. Heat to boil, and cool both. To the control solution, add 0.10 mg of sulfate ion (SO_4) and dilute both solutions to 25 mL, add 1 mL of 12% barium chloride reagent solution. Any turbidity in sample after 1 h should not exceed that of the control solution.

For the Determination of Stabilizer, Heavy Metals, and Iron

Sample Solution A. To 5 g (3.4 mL), add 10 mL of water, 7 mL of nitric acid, and 9 mL of hydrochloric acid. Evaporate to dryness on a hot plate ($\approx$100 °C), dissolve in water, and dilute with water to 50 mL in a volumetric flask (1 mL = 0.1 g).

Stabilizer. Dilute 1.0 mL of sample solution A (0.1-g sample) with water to 1 L. Dilute 4.8 mL of the solution with water to 85 mL. Add 5 mL of 10% ammonium molybdate reagent solution, and adjust the pH to 1.8 (using a pH meter) with dilute hydrochloric acid (1 + 9). Heat to boiling, and cool to room temperature. Add 10 mL of hydrochloric acid, and transfer to a separatory funnel. Add 35 mL of diethyl ether, and shake vigorously. Allow the layers to separate, and draw off and discard the aqueous layer. Wash the diethyl ether layer twice with 10-mL portions of dilute hydrochloric acid (1 + 9), drawing off and discarding the aqueous portion each time. Add 10 mL of dilute hydrochloric acid (1 + 9) to which has just been added 0.2 mL of a 2% stannous chloride reagent solution, and shake. Any blue color in the ether should not exceed that produced by 0.01 mg of phosphate ion (PO_4) treated exactly like the 4.8-mL portion of the sample.

Heavy Metals. (Page 36, Method 1). Dilute 20 mL of sample solution A (2-g sample) with water to 25 mL.

Iron. (Page 38, Method 1). Use 10 mL of sample solution A (1-g sample).

Hydriodic Acid, 55%

HI	Formula Wt 127.91	CAS No. 10034-85-2

GENERAL DESCRIPTION

Typical appearance: clear liquid; may darken with time
Analytical use: determination of methoxyl
Aqueous solubility: miscible
Density: 1.7

SPECIFICATIONS

Assay . . . . 55.0–58.0% HI

	Maximum Allowable
Free iodine (I_2)	0.75%
Residue after ignition	0.01%
Chloride and bromide (as Cl).	0.05%
Sulfate (SO_4)	0.005%
Phosphate (PO_4)	0.001%

Heavy metals (as Pb)... 0.001%
Iron (Fe).. 0.001%

TESTS

Assay. (By acid–base titrimetry). Tare a glass-stoppered flask containing about 25 mL of water. Under the water surface, quickly add about 3 mL of the sample, then stopper, and weigh accurately to the nearest 0.1 mg. Dilute to about 75 mL, add 0.15 mL of phenolphthalein indicator solution, and titrate with 1 N sodium hydroxide volumetric solution. One milliliter of 1 N sodium hydroxide corresponds to 0.12791 g of HI.

$$\% \text{ HI} = \frac{(\text{mL} \times \text{N NaOH}) \times 12.79}{\text{Sample wt (g)}}$$

Free Iodine. Weigh a glass-stoppered flask containing 15 mL of water. Add about 3 mL of sample, stopper, and reweigh. Dilute with 40 mL of water, and titrate with 0.1 N sodium thiosulfate volumetric solution to the disappearance of the yellow color. Not more than 0.60 mL of the sodium thiosulfate should be required per gram of the acid.

Residue after Ignition. (Page 26). Evaporate 10 g (6.0 mL) to dryness in a tared, preconditioned platinum crucible and ignite.

Chloride and Bromide. Dilute 1 g (0.6 mL) with water to 100 mL in a 250-mL conical flask. Add 1 mL of hydrogen peroxide and 1 mL of phosphoric acid. Heat to boiling, and boil gently until all the iodine is expelled and the solution is colorless. Cool, wash down the sides of the flask, and add 0.5 mL of hydrogen peroxide. If an iodine color develops, boil until the solution is colorless and for 10 min longer. If no color develops, boil for 10 min. Dilute with water to 100 mL, take a 2-mL portion, and dilute with water to 23 mL. Prepare a standard containing 0.01 mg of chloride ion (Cl) in 23 mL of water. Add 1 mL of nitric acid and 1 mL of silver nitrate reagent solution to each. Any turbidity produced in the solution of the sample should not exceed that in the standard.

Sulfate. For sample, dilute 2.0 mL (3.0 g) to 25 mL with water. For control, take 0.7 mL (1.0 g) of sample, and dilute to 25 mL with water. Neutralize both with (1 + 1) ammonium hydroxide, and add 1 mL of (1 + 19) hydrochloric acid to each. Heat to boiling, and cool both. To the control solution, add 0.10 mg of sulfate ion (SO_4), dilute both solutions to 35 mL, and add 1 mL of 12% barium chloride reagent solution. Any turbidity in sample after 1 h should not exceed that of the control solution.

For the Determination of Phosphate, Heavy Metals, and Iron

Sample Solution A. Evaporate 5 g (3.0 mL) of sample with 2 mL of nitric acid to dryness on a hot plate (≈ 100 °C). Re-evaporate with several portions of water until all the iodine has been volatilized. Warm the residue with 2 mL of 1 N hydrochloric acid, and take up with 40 mL of hot water. Cool, dilute with water to 50 mL, and mix well (1 mL = 0.1 g).

Phosphate. To 20 mL of sample solution A (2-g sample), add 2 mL of 25% sulfuric acid reagent solution, 10 mL of water, 1 mL of ammonium molybdate–sulfuric acid reagent solution, and 1 mL of 4-(triethylamino)phenol sulfate solution. Heat to 60 °C for 10 min. Any blue color should not exceed that produced by 0.02 mg of phosphate ion (PO$_4$) in an equal volume of solution containing the quantities of reagents used in the test.

Heavy Metals. (Page 36, Method 1). Dilute 20 mL of sample solution A (2-g sample) with water to 25 mL.

Iron. (Page 38, Method 1). Use 10 mL of sample solution A (1-g sample).

Hydrobromic Acid, 48%

HBr   **Formula Wt 80.91**   **CAS No. 10035-10-6**

GENERAL DESCRIPTION

Typical appearance: clear liquid, colorless to pale yellow when fresh; may become yellow to brown on storage
Analytical use: catalyst
Aqueous solubility: miscible
Density: 1.5
pK_a: −9

SPECIFICATIONS

Assay . . . . 47.0–49.0% HBr

Maximum Allowable

Residue after ignition 0.002%
Chloride (Cl) 0.05%
Iodide (I) 0.003%
Phosphate (PO$_4$) 0.001%
Sulfate and sulfite (as SO$_4$) 0.003%
Heavy metals (as Pb) . . . . 5 ppm
Iron (Fe) . . . . 1 ppm
Selenium (Se) . . . . 0.01 ppm

TESTS

Assay. (By acid–base titrimetry). Tare a glass-stoppered conical flask containing about 15 mL of water. Quickly add about 4 mL of the acid, and weigh. Dilute with water to 50 mL, and titrate with 1 N sodium hydroxide volumetric solution, using 0.15 mL of phenolphthalein indicator solution. One milliliter of 1 N sodium hydroxide corresponds to 0.08091 g of HBr.

$$\% \text{ HBr} = \frac{(\text{mL} \times \text{N NaOH}) \times 8.091}{\text{Sample wt (g)}}$$

Residue after Ignition. (Page 26). To 50 g (34 mL) in a tared, preconditioned dish (not platinum), add 0.05 mL of sulfuric acid, evaporate as far as possible on a hot plate ($\approx$100 °C), then heat gently to volatilize the solution. Finally, ignite at 600 ± 25 °C for 15 min.

Chloride. To 1 g (0.67 mL) in a 150-mL conical flask, add 50 mL of dilute nitric acid (1 + 3), and digest on a hot plate ($\approx$100 °C) until the solution is colorless. Wash down the sides of the flask with a little water, and digest for an additional 15 min. Cool, filter if necessary through a chloride-free filter, and dilute with water to 100 mL. Dilute 2.0 mL of the solution with water to 23 mL. Prepare a standard containing 0.01 mg of chloride ion (Cl) in 23 mL of water. Add 1 mL of nitric acid and 1 mL of silver nitrate reagent solution to each. Any turbidity produced in the solution of the sample should not exceed that in the standard.

Iodide. Dilute 6 g (4.0 mL) with 20 mL of water. Add 5 mL of chloroform, 0.2 mL of 10% ferric chloride solution, and 0.1 mL of dilute sulfuric acid (1 + 15), and mix gently in a separatory funnel. Draw off the chloroform layer into a comparison tube of about 1.5-cm diameter. No violet color should be observed on looking down through the chloroform.

Phosphate. (Page 40, Method 1). Add 5 mL of nitric acid to 6 g (4.0 mL) of the sample, and evaporate to dryness on a hot plate ($\approx$100 °C). Dissolve the residue in 75 mL of approximately 0.5 N sulfuric acid, and use 25 mL of the solution.

Sulfate and Sulfite. Dilute 2.7 mL (4.0 g) of the sample with water to 20 mL. Add 0.5 mL of bromine water, and gently heat to boil for a few minutes to expel excess bromine. Add 1.0 mL of 1% sodium carbonate reagent solution, and evaporate to dryness on a hot plate ($\approx$100 °C). Dissolve the residue in 1 mL of (1 + 19) hydrochloric acid and 20 mL of water. Filter through small, washed filter paper, add two 3-mL portions of water through, and dilute to 35 mL with water. For the control, take 0.12 mg of sulfate ion (SO_4) in 30 mL of water, add 1 mL of (1 + 19) hydrochloric acid, dilute to 35 mL with water, add 1 mL of 12% barium chloride reagent solution, and compare turbidity after 10 min. Sample turbidity should not exceed that of control solution.

Heavy Metals. (Page 36, Method 1). To 4 g (2.7 mL) in a beaker, add 1.0 mL of 1% sodium carbonate reagent solution, and evaporate to dryness on a hot plate ($\approx$100 °C). Dissolve the residue in about 20 mL of water, and dilute with water to 25 mL.

Iron. (Page 38, Method 1). To 10 g (6.6 mL), add about 10 mg of sodium carbonate, and evaporate to dryness on a hot plate ($\approx$100 °C). Dissolve the residue in 2 mL of hydrochloric acid, dilute with water to 50 mL, and use the solution without further acidification.

Selenium. Place 150 g (100 mL) in an all-glass distilling apparatus, add 2 mL of bromine, and distill off approximately 25 mL. Dilute the distillate with an equal volume of water, and decolorize with sulfurous acid. Add 0.1 g of hydroxylamine hydrochloride, and warm gently on a hot plate ($\approx$100 °C) for 30 min. Cool to room temperature, filter through a white glass-wool mat in a small Gooch crucible, without washing, and examine immediately. No pink or red color should be observed on the mat.

Hydrochloric Acid
HCl   **Formula Wt 36.46**   **CAS No. 7647-01-0**

GENERAL DESCRIPTION

Typical appearance: clear, fuming liquid
Analytical use: titrant; acidification; digestion
Aqueous solubility: miscible
Density: 1.18
pK_a: −8

SPECIFICATIONS

Assay . . . . 36.5–38.0% HCl

Maximum Allowable

Color (APHA) 10
Residue after ignition . . . . 5 ppm
Bromide (Br) 0.005%
Sulfate (SO_4) . . . . 1 ppm
Sulfite (SO_3) . . . . 1 ppm
Extractable organic substances . . . . Passes test (about 5 ppm)
Free chlorine (Cl) . . . . 1 ppm
Ammonium (NH_4) . . . . 3 ppm
Arsenic (As) . . . . 0.01 ppm
Heavy metals (as Pb) . . . . 1 ppm
Iron (Fe) . . . . 0.2 ppm

TESTS

Assay. (By acid–base titrimetry). Tare a glass-stoppered flask containing about 30 mL of water. Quickly add about 3 mL of the sample, stopper, and weigh accurately. Dilute to about 50 mL, add 0.15 mL of methyl orange indicator solution, and titrate with 1 N sodium hydroxide volumetric solution. One milliliter of 1 N sodium hydroxide corresponds to 0.03646 g of HCl.

$$\% \text{ HCl} = \frac{(\text{mL} \times \text{N NaOH}) \times 3.646}{\text{Sample wt (g)}}$$

Color (APHA). (Page 43).

Residue after Ignition. (Page 26). To 200 g (170 mL) in a tared, preconditioned platinum dish, add 0.05 mL of sulfuric acid, and evaporate as far as possible on a hot plate ($\approx$100 °C). Heat gently to volatilize the excess sulfuric acid, and ignite.

Bromide. Pipet 1 mL into 25 mL of water in a 100-mL beaker, add 2 mL of phenol red indicator solution, and titrate with 1 N sodium hydroxide volumetric solution. Add just sufficient titrant to change the yellow color to red. Transfer the solution to a 100-mL volumetric flask with about 25 mL of water. For the standard, add identical amounts of indicator and water to 0.06 mg of bromide ion (Br) in a 100-mL volumetric flask. To each flask, add 5 mL of acetate buffer solution (described below) and 2.0 mL of freshly prepared chloramine-T solution (described below). Mix, allow to stand for 20.0 min, and add 20 mL of 0.1 N sodium thiosulfate volumetric solution. Swirl, dilute with water to volume, and mix. Compare visually or spectrophotometrically at 590 nm in 1.0-cm cells. Any blue-violet color in the sample should not exceed that in the standard.

> *Acetate Buffer Solution.* Dissolve 68 g of sodium acetate and 30 mL of glacial acetic acid in a 1-L flask, and dilute to the mark with water.

> *Chloramine-T Solution.* Dissolve 0.125 g of chloramine-T trihydrate in 100 mL of water.

Sulfate. To 42 mL (50 g) of the sample, add 1 mL of 1% sodium carbonate reagent solution, and evaporate to dryness. Take up the residue with 10 mL of water, filter through washed filter paper, and wash with 5 mL of water. Add 1 mL of (1 + 9) hydrochloric acid, and dilute to 20 mL with water. For the control, use 0.05 mg of sulfate ion (SO_4) in 15 mL of water, add 1 mL (1 + 9) hydrochloric acid, and dilute to 20 mL with water. To both, add 1 mL of 12% barium chloride reagent solution. Compare after 10 min. Sample turbidity should not exceed that of the control solution.

Sulfite. Transfer 400 mL of freshly boiled and cooled water to each of two 500-mL conical flasks. Adjust the temperature of one flask (for the reagent blank) to 25 °C and the other flask (for the sample solution) to 15 °C. To each flask, add 3 mL of 10% potassium iodide reagent solution, 5 mL of hydrochloric acid, and 2 mL of starch indicator solution. To the sample flask, add 85 mL of hydrochloric acid sample, and adjust the temperature of the contents of the two flasks to within 1 °C of each other. Using a magnetic stirrer and a Teflon-coated stirring bar, titrate each with 0.01 N iodine until a faint permanent blue color is produced. Subtract the reagent blank titrant volume from the sample titrant volume to obtain the net sample titration. The net titration should not exceed 0.25 mL.

Extractable Organic Substances. Place 120 g (100 mL) of sample in each of two 125-mL separatory funnels. To one funnel, add 7.0 mL of 2,2,4-trimethylpentane as a sample blank. To the second funnel, add 5.0 mL of 2,2,4-trimethylpentane and 2.0 mL of internal standard solution (described below). Stopper the funnels, shake for 2 min, allow the layers to separate, and drain the hydrochloric acid layer. Prepare a standard solution by adding 2.0 mL of internal standard solution to 5.0 mL of chloroform standard solution (described below). (*Note:* 2,2,4-trimethylpentane suitable for extraction should be used; see page 708.)

> *Internal Standard Solution.* Add by syringe 30 mg (22.7 μL) of 1,1,1-trichloroethane to 100 mL of 2,2,4-trimethylpentane.

> *Chloroform Standard Solution.* Add by syringe 12 mg (8.1 μL) of chloroform to 100 mL of 2,2,4-trimethylpentane.

Analyze the sample blank and the sample 2,2,4-trimethylpentane layers and the standard solution by gas chromatography as described on page 80. With a flame ionization detector, most organic compounds can be detected at levels of 1 ppm or less. The use of an electron-capture or other selective detector would make it possible to detect chlorinated aliphatics or other selected compounds at much lower levels. The following parameters have given satisfactory results.

Column: Type I, methyl silicone, 30 m × 0.53 mm i.d., 5.0-μm film thickness

Column Temperature: 35 °C initial hold, 5 min, then programmed at 10 °C per min to 230 °C, final hold 21 min. Total run time, 45.5 min •

Injector Temperature: 230 °C

Detector Temperature: 250 °C

Carrier Gas: Helium, with flow rate 3–4 mL per min

Detector: Flame ionization, range 10^{-11} A

Sample Size: 0.5 μL. Adjust the attenuation to produce a peak height of at least 10% of full scale for the internal standard peak

Approximate Retention Times (min): Dichloromethane, 3.8; chloroform, 6.8; 1,2-dichloroethane, 7.7; 1,1,1-trichloroethane, 8.0; benzene, 8.6; carbon tetrachloride, 8.8; 2,2,4-trimethylpentane, 10–13; chlorobenzene, 13.8; 1,2,4-trichlorobenzene, 20.5; lindane, 37

Use the sample blank extraction to correct for impurities in the solvent or sample that interfere with the internal standard peak for 1,1,1-trichloroethane. The ratio of the total area of any peaks in the extraction (other than the internal standard and solvent) to the area of the internal standard peak in the sample should not exceed the ratio of the area of the chloroform peak to the area of the internal standard peak in the standard solution.

Free Chlorine. To 50 mL of freshly boiled and chilled water in a separatory funnel, cautiously add 50 mL of sample, and cool. For the standard, dilute 0.06 mg of freshly prepared chlorine ion (Cl_2) standard solution to 50 mL of freshly boiled and cooled water in a similar separatory funnel. Add 0.1 mL of 2% potassium iodide solution (prepared at time of testing) and 1 mL of carbon disulfide to each, mix for 5 s, and allow the phases to separate. The pink color of the carbon disulfide layer in the sample should not exceed that in the standard within 30 s.

Note: The free chlorine in this material will increase with time and storage conditions. The free chlorine test is applicable to freshly manufactured material.

Ammonium. (By differential pulse polarography, page 53). Use 3.3 g (2.8 mL) of sample and 5.5 mL of ammonia-free 6 N sodium hydroxide. For the standard, use 0.010 mg of ammonium ion (NH_4).

Arsenic. (Page 34). Mix 300 g (255 mL) with 10 mL of dilute sulfuric acid (1 + 1) in a generator flask. Add potassium chlorate crystals in small increments until a yellow color is produced, and evaporate on a hot plate just to fumes of sulfur trioxide, adding a few more crystals of potassium chlorate whenever the yellow color starts to fade. Cool, cautiously wash down the flask with 5–10 mL of water, and again heat just to fumes of sulfur trioxide. Cool, cautiously wash down the flask with 5–10 mL of water, and again heat just to fumes of sulfur trioxide. Cool, cautiously wash down the flask with 5–10 mL of water, and repeat the fuming. (The volume of sulfuric acid at this point should be about 4 mL.) Cool, cautiously dilute with water to 55 mL, and use this solution without further acidification. For the standard, use 0.003 mg of arsenic ion (As).

Heavy Metals. (Page 36, Method 1). To 20 g (17 mL) in a 150-mL beaker, add about 10 mg of sodium carbonate, and evaporate to dryness on a hot plate ($\approx$100 °C). Dissolve the residue in about 20 mL of water, and dilute with water to 25 mL.

Iron. (Page 38, Method 1). To 50 g (42 mL) in a porcelain or glass vessel, add about 10 mg of sodium carbonate and evaporate to dryness on a hot plate ($\approx$100 °C). Dissolve in 2 mL of hydrochloric acid, dilute with water to 50 mL, and use the solution without further acidification.

Hydrochloric Acid, Ultratrace
HCl   **Formula Wt 36.46**   **CAS No. 7647-01-0**

Suitable for use in ultratrace elemental analysis.

Note: The reagent must be packaged in a preleached Teflon bottle and used in a clean laboratory environment to maintain purity.

GENERAL DESCRIPTION
Typical appearance: clear, fuming liquid
Analytical use: trace metal analysis

SPECIFICATIONS
Assay 20–38%

	Maximum Allowable
Sulfate	1 ppm
Mercury (Hg)	1 ppb
Selenium (Se)	10 ppb
Aluminum (Al)	1 ppb
Barium (Ba)	1 ppb
Boron (B)	5 ppb
Cadmium (Cd)	1 ppb
Calcium (Ca)	1 ppb
Chromium (Cr)	1 ppb
Cobalt (Co)	1 ppb
Copper (Cu)	1 ppb
Iron (Fe)	5 ppb
Lead (Pb)	1 ppb
Lithium (Li)	1 ppb
Magnesium (Mg)	1 ppb

Manganese (Mn) . . . . 1 ppb
Molybdenum (Mo) . . . . 1 ppb
Potassium (K) . . . . 1 ppb
Silicon (Si) . . . . 5 ppb
Sodium (Na) . . . . 5 ppb
Strontium (Sr) . . . . 1 ppb
Tin (Sn) . . . . 1 ppb
Titanium (Ti) . . . . 1 ppb
Vanadium (V) . . . . 1 ppb
Zinc (Zn) . . . . 1 ppb
Zirconium (Zr) . . . . 1 ppb

TESTS

Assay. (By acid–base titrimetry). The assay of hydrochloric acid used for ultratrace metal analysis may vary from 20% to 38%, depending on the method of purification. Tare a glass-stoppered flask containing about 30 mL of water. Quickly add about 3 mL of the sample, stopper, and weigh accurately. Dilute to about 50 mL, add 0.15 mL of methyl orange indicator solution, and titrate with 1 N sodium hydroxide volumetric solution. One milliliter of 1 N sodium hydroxide corresponds to 0.03646 g of HCl.

$$\% \text{ HCl} = \frac{(\text{mL} \times \text{N NaOH}) \times 3.646}{\text{Sample wt (g)}}$$

Sulfate. See test for hydrochloric acid, page 352.

Mercury. (By CVAAS, page 65). To each of three 100-mL volumetric flasks containing about 35 mL of water, add 20.0 g of sample. To the second and third flasks, add mercury ion (Hg) standards of 10 ng (0.5 ppb) and 20 ng (1.0 ppb), respectively. Add 5 mL of nitric acid to all three flasks, and dilute to the mark with water. Mix. Zero the instrument with the blank, and determine the mercury content using a suitable mercury analyzer system (1.0 mL of 0.01 μg/mL Hg = 10 ng).

Selenium. (By HGAAS, page 66). To a set of three 100-mL volumetric flasks, transfer 50.0 g (42 mL) of sample. To two flasks, add selenium ion (Se) standards of 0.50 μg (10 ppb) and 0.10 μg (20 ppb), respectively. Dilute each to 100 mL with (1 + 1) hydrochloric acid. Mix. Prepare fresh working standard before use.

Trace Metals. Determine the aluminum, barium, boron, cadmium, calcium, chromium, cobalt, copper, iron, lead, lithium, magnesium, manganese, molybdenum, potassium, silicon, sodium, strontium, tin, titanium, vanadium, zinc, and zirconium by the ICP–OES method described on page 69.

Hydrofluoric Acid

HF   Formula Wt 20.01   CAS No. 7664-39-3

GENERAL DESCRIPTION

Typical appearance: clear liquid
Analytical use: determination of silicon dioxide
Aqueous solubility: miscible
Density: 1.15
pK_a: 3.1

SPECIFICATIONS

Assay . . . . 48.0–51.0% HF

Maximum Allowable

Fluosilicic acid (H_2SiF_6). 0.01%
Residue after ignition. . . . . 5 ppm
Chloride (Cl). . . . . 5 ppm
Phosphate (PO_4). . . . . 1 ppm
Sulfate and sulfite (as SO_4) . . . . 5 ppm
Arsenic (As) . . . . 0.05 ppm
Copper (Cu) . . . . 0.1 ppm
Iron (Fe). . . . . 1 ppm
Heavy metals (as Pb). . . . . 0.5 ppm

TESTS

Assay. (By acid–base titrimetry). Tare a stoppered plastic flask containing about 25 mL of water, and deliver about 1.5 mL of sample under the water surface, using a plastic pipet. Weigh accurately, wash down the sides, and dilute the solution to 50–60 mL with water. Add 0.15 mL of phenolphthalein indicator solution, and titrate with 1 N sodium hydroxide volumetric solution. One milliliter of 1 N sodium hydroxide corresponds to 0.02001 g of HF.

$$\% \text{ HF} = \frac{(\text{mL} \times \text{N NaOH}) \times 2.001}{\text{Sample wt (g)}}$$

Fluosilicic Acid. Weigh about 37.5 g (33 mL) into a large platinum dish. Add 2 g of potassium chloride and 3 mL of hydrochloric acid, and evaporate to dryness on a hot plate ($\approx 100\ ^\circ\text{C}$) in a fume hood. Wash down the sides of the dish with a small amount of water, add 3 mL of hydrochloric acid, and repeat the evaporation. Dissolve the residue in about 100 mL of water, cool to 0 °C, and add 0.15 mL of phenolphthalein indicator solution. Neutralize any free acid with 0.1 N sodium hydroxide solution to a colorless end point, keeping the temperature of the solution near 0 °C. Heat the solution to boiling, and titrate with 0.1 N sodium hydroxide solution. One milliliter of 0.1 N sodium hydroxide corresponds to 0.0036 g of fluosilicic acid.

$$\% \text{ H}_2\text{SiF}_6 = \frac{(\text{mL} \times \text{N NaOH}) \times 0.36}{\text{Sample wt (g)}}$$

Residue after Ignition. (Page 26). To 200 g (180 mL) in a tared, preconditioned platinum dish, add 0.05 mL of sulfuric acid, evaporate as far as possible on a hot plate (≈ 100 °C) in the hood, heat gently to volatilize the excess sulfuric acid, and ignite.

Chloride. Add 1.8 mL to 45 mL of water, filter if necessary through a chloride-free filter, and add 1 mL of nitric acid and 1 mL of silver nitrate reagent solution. Any turbidity should not exceed that produced by 0.01 mg of chloride ion (Cl) in an equal volume of solution containing the quantities of reagents used in the test.

For the Determination of Phosphate, Sulfate and Sulfite, Arsenic, Copper and Iron, and Heavy Metals

Sample Solution A. In a fume hood, to 500.0 g (425 mL) sample in a platinum or Teflon dish, add 10 mg of sodium carbonate, and evaporate to dryness on a hot plate (≈ 125 °C). Cool, add 5 mL of hydrochloric acid, evaporate to dryness. Cool, add 5 mL of water, evaporate to dryness. To the residue, add three 10-mL portions of (1 + 9) hydrochloric acid, transfer to a 50-mL volumetric flask, and dilute to the mark with water (1 mL = 10.0 g).

Phosphate. To 2.0 mL (20.0 g) of sample solution A, add 25 mL of approximately 0.5 N sulfuric acid, 1 mL of ammonium molybdate–sulfuric acid reagent solution, and 1 mL of 4-(methylamino)phenol sulfate reagent solution. Allow to stand for 2 h at room temperature. Any blue color should not exceed that produced by 0.02 mg of phosphate ion (PO_4) in an equal volume of solution containing the quantities of reagents used in the test.

Sulfate and Sulfite. To 1.0 mL (10.0 g) of sample solution A in a Teflon beaker, add 1 mL of 30% hydrogen peroxide. Evaporate to dryness on a hot plate (≈ 100 °C) in a fume hood, wash down the sides of the dish with a small volume of water, and add 3 mL of perchloric acid. Evaporate to about 1 mL, dilute with about 15 mL of water, and add 0.15 mL of phenolphthalein indicator solution. Neutralize with ammonium hydroxide, dilute with water to 20 mL, and add 2 mL of dilute hydrochloric acid (1 + 19) and 2 mL of 12% barium chloride reagent solution. Any turbidity should not exceed that produced by 0.1 mg of sulfate ion (SO_4) in an equal volume of solution containing the quantities of reagents used in the test. Compare 10 min after adding the barium chloride to the sample and standard solutions.

Arsenic. To 4.0 mL (40.0-g sample) of sample solution A in a platinum dish, add 5 mL of sulfuric acid and 5 mL of hydrochloric acid, swirling the beaker gently after each addition. Allow to stand at room temperature for 5 min. Evaporate in a hood in a sand bath to dense fumes of sulfur trioxide. Cool, wash down the sides of the beaker with 15–20 mL of water, and evaporate again in a sand bath to dense fumes of sulfur trioxide. Cool, and transfer to a 125-mL generator flask, rinsing the beaker thoroughly with water. Place the flask on a hot plate, evaporate to dense fumes of sulfur trioxide, add sulfuric acid if necessary to make a volume of about 4 mL, and cool. Add 50 mL of water, 2 mL of 16.5% potassium iodide solution, and 0.5 mL of 40% stannous chloride reagent solution, and mix.

Proceed as described in the general method for arsenic on page 34 under Procedure, starting with the second sentence, which begins "Allow the mixture to stand…" Any red color in the silver diethyldithiocarbamate solution of the sample should not exceed that in a standard containing 0.002 mg of arsenic ion (As).

Copper and Iron. (By flame AAS, page 63). To three 10.0-mL portions of sample solution A (100.0-g sample) in a 25-mL volumetric flask, add 5 mL of (1 +1) hydrochloric acid to each, and add standards as per the following table to second and third flasks, respectively.

Element	Wavelength (nm)	Sample Wt (g)	Standard Added (mg)	Flame Type*	Background Correction
Cu	324.8	100.0	0.01; 0.02	A/A	Yes
Fe	248.3	100.0	0.10; 0.20	A/A	Yes

*A/A is air/acetylene.

Heavy Metals. (Page 36, Method 1). Take 4.0 mL (40.0 g) of sample solution A, and dilute to 25 mL with water. Solution color should not exceed the control solution of 0.02 mg of lead.

Hydrofluoric Acid, Ultratrace
HF   **Formula Wt 20.01**   **CAS No. 7664-39-3**

Suitable for use in ultratrace elemental analysis.

Note: Reagent must be packaged in a preleached Teflon bottle and used in a clean laboratory environment to maintain purity.

GENERAL DESCRIPTION
Typical appearance: clear liquid
Analytical use: trace metal analysis
Aqueous solubility: miscible
pK_a: 3.1

SPECIFICATIONS
Assay 46–51.0%

Maximum Allowable
Chloride (Cl). . . . . .0.5 ppm
Phosphate (PO_4). . . . . .1 ppm
Sulfate and sulfite (SO_4). . . . . .1 ppm
Mercury (Hg) . . . . .1 ppb
Aluminum (Al) . . . . .1 ppb
Barium (Ba) . . . . .1 ppb
Boron (B). . . . . .5 ppb
Cadmium (Cd) . . . . .1 ppb
Calcium (Ca) . . . . .1 ppb
Chromium (Cr) . . . . .1 ppb
Cobalt (Co). . . . . .1 ppb

Copper (Cu).. 1 ppb
Iron (Fe)... 5 ppb
Lead (Pb) .... 1 ppb
Lithium (Li) .... 1 ppb
Magnesium (Mg)... 1 ppb
Manganese (Mn) .... 1 ppb
Molybdenum (Mo) .... 1 ppb
Potassium (K) .... 1 ppb
Silicon (Si) .... 5 ppb
Sodium (Na) .... 5 ppb
Strontium (Sr)... 1 ppb
Tin (Sn) .... 1 ppb
Titanium (Ti) .... 1 ppb
Vanadium (V) .... 1 ppb
Zinc (Zn).. 1 ppb
Zirconium (Zr) .... 1 ppb

TESTS

Assay. (By acid–base titrimetry). Tare a stoppered plastic flask containing about 25 mL of water, and deliver, using a plastic pipet, about 1.5 mL of sample under the water surface. Weigh accurately, wash down the sides, and dilute the solution to 50–60 mL with water. Add 0.15 mL of phenolphthalein indicator solution, and titrate with 1 N sodium hydroxide volumetric solution. One milliliter of 1 N sodium hydroxide corresponds to 0.02001 g of HF.

$$\% \text{ HF} = \frac{(\text{mL} \times \text{N NaOH}) \times 2.001}{\text{Sample wt (g)}}$$

Chloride. See test for hydrofluoric acid, page 357. Use 18 mL of sample, and add it to 45 mL of water.

Phosphate. See test for hydrofluoric acid, page 357.

Sulfate and Sulfite. See test for hydrofluoric acid, page 357. Use 5.0 mL (50.0 g) of sample solution A.

Mercury. (By CVAAS, page 65). To each of three polyethylene 100-mL volumetric containers, add about 30 mL of water and 10.0 g of sample. Cool the samples in an ice-water bath for about 15 min. Add slowly, with caution, 10 mL of ammonium hydroxide. Keeping the flasks in the ice-water bath for another 10 min, add 5 mL of nitric acid. Cool to room temperature. To the second and third flasks, add mercury ion (Hg) standards of 10 ng (1.0 ppb) and 20 ng (2.0 ppb), respectively. Dilute to the mark with water. Mix. Zero the instrument with the blank, and determine the mercury content using a suitable mercury analyzer system (1.0 mL of 0.01 μg/mL Hg = 10 ng).

Trace Metals. Determine the aluminum, barium, boron, cadmium, calcium, chromium, cobalt, copper, iron, lead, lithium, magnesium, manganese, molybdenum, potassium, silicon, sodium, strontium, tin, titanium, vanadium, zinc, and zirconium by the ICP–OES method described on page 69.

Hydrogen Peroxide

H_2O_2 **Formula Wt 34.01**   **CAS No. 7722-84-1**

Note: This reagent should be stored in a cool place in containers with a vent in the stopper. The assay requirement applies to material stored properly for a reasonable time.

GENERAL DESCRIPTION

Typical appearance: clear liquid
Analytical use: catalyst; oxidizing agent
Change in state (approximate): boiling point, 150 °C
Aqueous solubility: miscible
Density: 1.11

SPECIFICATIONS

Assay . . . . 29.0–32.0% H_2O_2

Maximum Allowable

Color (APHA) 10
Residue after evaporation 0.002%
Titrable acid. . . . . .0.0006 meq/g
Chloride (Cl). . . . . .3 ppm
Nitrate (NO_3) . . . . .2 ppm
Phosphate (PO_4). . . . . .2 ppm
Sulfate (SO_4) . . . . .5 ppm
Ammonium (NH_4) . . . . .5 ppm
Heavy metals (as Pb). . . . . .1 ppm
Iron (Fe). . . . . .0.5 ppm

TESTS

Assay. (By titration of the reductive capacity). Weigh accurately about 1 mL in a tared 100-mL volumetric flask, dilute to volume with water, and mix thoroughly. To 20.0 mL of this solution, add 20 mL of dilute sulfuric acid (1 + 15), and titrate with 0.1 N potassium permanganate volumetric solution. One milliliter of 0.1 N potassium permanganate corresponds to 0.0017 g of H_2O_2.

$$\% \ H_2O_2 = \frac{(mL \times N \ KMnO_4) \times 1.700}{Sample \ wt \ (g) \ / \ 5}$$

Color (APHA). (Page 43).

Residue after Evaporation. (Page 25). Evaporate 50 g (45 mL) to dryness in a tared, preconditioned porcelain or silica dish on a hot plate (≈100 °C), and dry the residue at 105 °C for 30 min.

Titrable Acid. Dilute 10 g (9 mL) with 90 mL of carbon dioxide-free water. Add 0.15 mL of methyl red indicator solution, and titrate with 0.01 N sodium hydroxide. The volume of sodium hydroxide solution consumed should not be more than 0.6 mL greater than the volume required for a blank test on 90 mL of the water used for dilution.

Chloride, Nitrate, Phosphate, and Sulfate. (By ion chromatography, page 87). Obtain three 100-mL beakers (blank, test, standard), and add 10 mL of water and 2.0 mg of sodium carbonate to each. Add 10.5 g (9.45 mL) of the sample to the standard and test. To the standard, add 0.02 mg each of chloride, nitrate, and phosphate ions and 0.05 mg of sulfate ion (SO_4). To the blank, add 0.5 g (0.45 mL) of the sample. Evaporate each to dryness on a low-temperature hot plate (≈ 100 °C). Dissolve the residues in a little water, and dilute each to 25 mL. Analyze 50-mL or 100-mL aliquots by ion chromatography, and measure the various peaks. The differences between the sample and the blank should not be greater than the corresponding differences between the standard and the sample.

Ammonium. Add 0.1 mL of sulfuric acid to 2 g (1.8 mL) of the sample, and evaporate on a hot plate (≈ 100 °C). Dissolve the residue in 45 mL of water and add 3 mL of 10% sodium hydroxide reagent solution and 2 mL of Nessler reagent. Any color should not exceed that produced by 0.01 mg of ammonium ion (NH_4) in an equal volume of solution containing the quantities of reagents used in the test.

Heavy Metals. (Page 36, Method 1). To 20 g (18 mL) in a 100-mL beaker, add about 10 mg of sodium chloride, and evaporate to dryness on a hot plate (≈ 100 °C). Dissolve the residue in about 20 mL of water, and dilute with water to 25 mL.

Iron. (Page 38, Method 1). To 20 g (18 mL) in a 100-mL beaker, add about 10 mg of sodium chloride, and evaporate to dryness on a hot plate (≈ 100 °C). Dissolve in 2 mL of hydrochloric acid, dilute with water to 50 mL, and use the solution without further acidification.

Hydrogen Peroxide, Ultratrace

H_2O_2	Formula Wt 34.01	CAS No. 7722-84-1

Suitable for use in ultratrace elemental analysis.

Note: Package in a precleaned polyethylene bottle and use in a clean laboratory environment to maintain purity.

GENERAL DESCRIPTION

Typical appearance: clear liquid
Analytical use: trace metal analysis
Change in state (approximate): boiling point, 150 °C
Density: 1.11

SPECIFICATIONS

Assay . . . . .25–35% H_2O_2

Maximum Allowable

Chloride (Cl). . . . . .3 ppm
Nitrate (NO_3) . . . . .2 ppm
Phosphate (PO_4). . . . . .2 ppm
Sulfate (SO_4) . . . . .5 ppm
Mercury (Hg). . . . . .1 ppb
Aluminum (Al) . . . . .1 ppb
Barium (Ba) . . . . .1 ppb
Boron (B). . . . . .5 ppb
Cadmium (Cd) . . . . .1 ppb
Calcium (Ca) . . . . .1 ppb
Chromium (Cr) . . . . .1 ppb
Cobalt (Co). . . . . .1 ppb
Copper (Cu) . . . . .1 ppb
Iron (Fe). . . . . .5 ppb
Lead (Pb). . . . . .1 ppb
Lithium (Li). . . . . .1 ppb
Magnesium (Mg) . . . . .1 ppb
Manganese (Mn) . . . . .1 ppb
Molybdenum (Mo) . . . . .1 ppb
Potassium (K) . . . . .1 ppb
Silicon (Si). . . . . .5 ppb
Sodium (Na). . . . . .5 ppb
Strontium (Sr) . . . . .1 ppb
Tin (Sn) . . . . .1 ppb
Titanium (Ti) . . . . .1 ppb
Vanadium (V) . . . . .1 ppb
Zinc (Zn) . . . . .1 ppb
Zirconium (Zr). . . . . .1 ppb

TESTS

Assay. (By titration of the reductive capacity). Weigh accurately about 1 mL of sample in a tared 100-mL volumetric flask containing about 20 mL of water. Dilute with water to the mark, and mix thoroughly. To 20.0 mL of this solution, add 20 mL of dilute sulfuric acid (1 + 15), and titrate with 0.1 N potassium permanganate volumetric solution. One milliliter of 0.1 N potassium permanganate corresponds to 0.0017 g of H_2O_2.

$$\% \ H_2O_2 = \frac{(mL \times N \ KMnO_4) \times 1.700}{Sample \ wt \ (g) \ / \ 5}$$

Chloride, Nitrate, Phosphate, and Sulfate. See test for hydrogen peroxide, page 361.

Mercury. (By CVAAS, page 65). To each of three 100-mL volumetric flasks containing about 35 mL of water, add 10.0 g of sample. To the second and third flasks, add mercury ion (Hg) standards of 10 ng (1.0 ppb) and 20 ng (2.0 ppb), respectively. Dilute to the mark with water. Mix. Zero the instrument with the blank, and determine the mercury content using a suitable mercury analyzer system (10 mL of 0.1 μg/mL Hg = 10 ng).

Trace Metals. Determine the aluminum, barium, boron, cadmium, calcium, chromium, cobalt, copper, iron, lead, lithium, magnesium, manganese, molybdenum, potassium, silicon, sodium, strontium, tin, titanium, vanadium, zinc, and zirconium by the ICP–OES method described on page 69.

Hydroxylamine Hydrochloride

$NH_2OH \cdot HCl$ **Formula Wt 69.49**   **CAS No. 5470-11-1**

GENERAL DESCRIPTION

Typical appearance: colorless, hygroscopic solid
Analytical use: determination of mercury
Change in state (approximate): melting point, 151 °C; decomposes above this temperature
Aqueous solubility: 85 g in 100 mL at 20 °C

SPECIFICATIONS

Assay . . . . $\geq$96.0% $NH_2OH \cdot HCl$

Maximum Allowable

Clarity of alcohol solution . . . . Passes test
Residue after ignition 0.05%
Titrable free acid . . . . 0.25 meq/g
Ammonium (NH_4) . . . . Passes test
Sulfur compounds (as SO_4) 0.005%
Heavy metals (as Pb) . . . . 5 ppm
Iron (Fe). . . . . 5 ppm

TESTS

Assay. (By titration of the reductive capacity). Weigh accurately about 1.5 g, previously dried for 24 h over magnesium perchlorate or phosphorus pentoxide, and transfer to a 250-mL volumetric flask. Dissolve in oxygen-free water, dilute to volume with oxygen-free water, and take 25.0 mL of the sample solution. Dissolve 5 g of ferric ammonium sulfate in 30 mL of oxygen-free dilute sulfuric acid (1 + 50), and add this solution to the sample solution. Protect the solution from oxygen in the air, boil gently for 5 min, and cool. Dilute with 150 mL of oxygen-free water, and titrate with 0.1 N potassium permanganate volumetric solution. One milliliter of 0.1 N potassium permanganate corresponds to 0.003474 g of $NH_2OH \cdot HCl$.

$$\% \ NH_2OH \cdot HCl = \frac{(mL \times N \ KMnO_4) \times 3.474}{\text{Sample wt (g) / 10}}$$

Clarity of Alcohol Solution. Dissolve 1.0 g in 25 mL of alcohol, and compare with an equal volume of alcohol. The sample solution should have no more color or turbidity than the alcohol. Retain the solution for the test for ammonium.

Residue after Ignition. (Page 26). Ignite 2.0 g. Retain the residue for the test for iron.

Titrable Free Acid. Dissolve 10.0 g in 50 mL of water, add 0.15 mL of bromphenol blue indicator solution, and titrate with 1 N sodium hydroxide volumetric solution to the blue end point. Not more than 2.5 mL of 1 N sodium hydroxide should be required.

Ammonium. To the solution obtained in the test for clarity of alcohol solution, add 1 mL of a solution of chloroplatinic acid (2.6 g in 20 mL). The solution should remain clear for 10 min.

Sulfur Compounds. Dissolve 2.0 g in 10 mL of water containing 10 mg of sodium carbonate and add 4 mL of nitric acid. Slowly add 12 mL of 10% hydrogen peroxide solution. Digest in a covered beaker until the reaction ceases, uncover, and evaporate to dryness on a hot plate ($\approx$100 °C). Dissolve the residue in 10 mL of water, add 2 mL of dilute hydrochloric acid (1 + 19), filter if necessary, and dilute with water to 20 mL. To 10 mL, add 1 mL of 12% barium chloride reagent solution. Any turbidity should not exceed that produced by 0.05 mg of sulfate ion (SO_4) in an equal volume of solution containing the quantities of reagents used in the test. Compare sample 10 min after adding the barium chloride to the standard.

Heavy Metals. (Page 36, Method 1). Dissolve 6.0 g in about 20 mL of water, and dilute with water to 30 mL. Use 25 mL to prepare the sample solution, and use the remaining 5.0 mL to prepare the control solution.

Iron. (Page 38, Method 1). To the residue from the test for residue after ignition, add 3 mL of dilute hydrochloric acid (1 + 1), cover the dish, and digest on a hot plate ($\approx$100 °C) for 15–20 min. Remove the cover, evaporate to dryness, dissolve the residue in 2 mL of hydrochloric acid, and dilute with water to 50 mL. Use the solution without further acidification.

Hydroxylamine Sulfate

Hydroxylammonium Sulfate

$(HONH_3)_2SO_4$ **Formula Wt 164.14**   **CAS No. 10039-54-0**

GENERAL DESCRIPTION

Typical appearance: white solid

Analytical use: to purify aldehydes and ketones; reagent for mercury and silver detection in water; reducing agent

Change in state (approximate): melting point, 170 °C
Aqueous solubility: freely soluble

SPECIFICATIONS

Assay . . . . ≥99.0% $(HONH_3)_2SO_4$

Maximum Allowable

Residue after ignition 0.05%
Chloride (Cl) 0.001%
Heavy metals (as Pb) . . . . 5 ppm
Copper (Cu). . . . . 5 ppm
Iron (Fe). . . . . 5 ppm

TESTS

Assay. (By titration of the reductive capacity). Weigh accurately about 0.16 g, and dissolve in 25 mL of water. Add a solution of 5 g of ferric ammonium sulfate in a mixture of 15 mL of 10% sulfuric acid and 20 mL of water. Boil gently for 5 min. Cool, dilute with 100 mL of water, and add 1 mL of phosphoric acid. Titrate with 0.1 N potassium permanganate volumetric solution to a permanent pink color. One milliliter of 0.1 N potassium permanganate corresponds to 0.00410 g of $(HONH_3)_2SO_4$.

$$\% \ (HONH_3)_2SO_4 = \frac{(mL \times N \ KMnO_4) \times 4.104}{Sample \ wt \ (g)}$$

Residue after Ignition. (Page 26). Ignite 2.0 g. Retain the residue for the test for iron.

Chloride. (Page 35). Use 1.0 g.

Heavy Metals. (Page 36, Method 1). Dissolve 5.0 g in about 20 mL of water, and dilute with water to 25.0 mL. Use 20.0 mL to prepare the sample solution, and use the remaining 5.0 mL to prepare the control solution.

Copper. (By flame AAS, page 63).

> **Sample Stock Solution.** Dissolve 20.0 g of sample in water, and transfer to a 100-mL volumetric flask. Add 5 mL of nitric acid, and dilute to the mark with water (1 mL = 0.2 g).

Element	Wavelength (nm)	Sample Wt (g)	Standard Added (mg)	Flame Type*	Background Correction
Cu	324.8	4.0	0.02; 0.04	A/A	Yes

*A/A is air/acetylene.

Iron. (Page 38, Method 1). To the residue after ignition, add 3 mL of dilute hydrochloric acid (1 + 1), cover the dish, and digest on a hot plate (≈100 °C) for 20 min. Remove the cover, evaporate to dryness, dissolve the residue in 2 mL of hydrochloric acid, and dilute with water to 50 mL. Use the solution without further acidification.

Hydroxy Naphthol Blue

1-(2-Naphtholazo-3,6-disulfonic acid)-2-naphthol-4-sulfonic Acid, Disodium Salt

CAS No. 165660-27-5

Note: This reagent is deposited on crystals of sodium chloride. The small blue crystals are freely water-soluble. In the pH range between 12 and 13, the solution of the indicator is reddish pink in the presence of calcium ion and deep blue in the presence of excess (ethylenedinitrilo)tetraacetate.

GENERAL DESCRIPTION

Typical appearance: dark blue solid
Analytical use: determination of calcium

SPECIFICATIONS

Suitability for calcium determination. . . . . Passes test

TEST

Suitability for Calcium Determination. Dissolve 0.3 g in 100 mL of water, add 10 mL of 1 N sodium hydroxide volumetric solution and 1.0 mL of calcium chloride solution (1 g in 200 mL), and dilute with water to 165 mL. The solution should be reddish pink. Add 1.0 mL of 0.05 M (ethylenedinitrilo)tetraacetic acid, disodium salt. The solution should be deep blue.

Imidazole

1,3-Diazole

NHCH:NCH:CH   **Formula Wt 68.08**   CAS No. 288-32-4

GENERAL DESCRIPTION

Typical appearance: solid
Analytical use: Karl Fischer reagent
Change in state (approximate): melting point, 90 °C
Aqueous solubility: freely soluble

SPECIFICATIONS

Assay . . . . ≥99% $C_3H_4N_2$
pH of a 5% solution at 25.0 °C 9.5–11.0

Maximum Allowable

Residue after ignition 0.1%
Iron (Fe) 0.001%
Water (H$_2$O) 0.2%

TESTS

Assay. (By acid–base titrimetry). Weigh accurately about 0.15 g, and dissolve in 25 mL of glacial acetic acid. Titrate with 0.1 N perchloric acid in acetic acid volumetric solution, using 0.1 mL of crystal violet indicator solution. One milliliter of 0.1 N perchloric acid corresponds to 0.006808 g of $C_3H_4N_2$. Correct for a blank.

$$\% \; C_3H_4N_2 = \frac{\{[mL \; (sample) - mL \; (blank)] \times N \; HClO_4\} \times 6.808}{Sample \; wt \; (g)}$$

pH of a 5% Solution at 25.0 °C. (Page 49).

Residue after Ignition. (Page 26). Use 1.0 g.

Iron. (By flame AAS, page 63).

> *Sample Stock Solution.* Dissolve 25.0 g of sample with water in a 100-mL volumetric flask, and dilute to the mark with water (1 mL = 0.25 g).

Element	Wavelength (nm)	Sample Wt (g)	Standard Added (mg)	Flame Type*	Background Correction
Fe	248.3	5.0	0.05; 0.10	A/A	No

*A/A is air/acetylene.

Water. (Page 31, Method 1). Use 2.0 g of sample.

Indigo Carmine
Acid Blue 74, C.I. 73015

$C_{16}H_8N_2Na_2O_8S_2$ **Formula Wt 466.37**   **CAS No. 860-22-0**

GENERAL DESCRIPTION

Typical appearance: dark blue solid
Analytical use: reagent for detection of nitrate
Aqueous solubility: 1 g in 100 mL at 25 °C

SPECIFICATIONS

Clarity of solution . . . . Passes test
Sensitivity to nitrate . . . . Passes test

TESTS

Clarity of Solution. Accurately weigh 1.0 g, and dissolve in 100 mL of water. The resulting solution should be clear, and no appreciable amount of insoluble residue should remain.

Sensitivity to Nitrate. To 7 mL of water, add 5 mg of sodium chloride. Add 3.0 mL of nitrate ion (NO_3) standard solution, and mix. Add 0.10 mL of 0.10% indigo carmine reagent solution (prepared using the sample) and 10 mL of sulfuric acid. The blue color should be completely discharged in 5 min.

Iodic Acid
Iodic(V) Acid
HIO_3   **Formula Wt 175.91**   **CAS No. 7782-68-5**

GENERAL DESCRIPTION

Typical appearance: white solid; darkens on exposure to light
Analytical use: titration standard
Change in state (approximate): melting point, 110 °C; with decomposition
Aqueous solubility: soluble

SPECIFICATIONS

Assay . . . . ≥99.5% HIO_3

Maximum Allowable

Insoluble matter 0.01%
Residue after ignition. 0.02%
Chloride and bromide (as Cl). 0.02%
Iodide (I) . . . . Passes test
Nitrogen compounds (as N). 0.1%
Sulfate (SO_4) 0.015%
Heavy metals (as Pb). 0.001%
Iron (Fe). 0.002%

TESTS

Assay. (By indirect iodometric titration of iodate). Weigh accurately about 0.5 g, transfer to a 250-mL volumetric flask, dissolve in water, dilute with water to volume, and mix thoroughly. Place a 50.0-mL aliquot of this solution in a glass-stoppered conical flask, and add 2 g of potassium iodide and 5 mL of 15% hydrochloric acid. Stopper, swirl, allow to stand in the dark for 10 min, and add 100 mL of cold water. Titrate the liberated iodine with 0.1 N sodium thiosulfate volumetric solution, adding 3 mL of starch indicator solution near the end of the titration. Correct for a blank. One milliliter of 0.1 N sodium thiosulfate corresponds to 0.002932 g of HIO_3.

$$\% \ HIO_3 = \frac{\{[mL \ (sample) - mL \ (blank)] \times N \ Na_2S_2O_3\} \times 2.932}{Sample \ wt \ (g) \ / \ 5}$$

Insoluble Matter. (Page 25). Use 10 g dissolved in 100 mL of water.

Residue after Ignition. (Page 26). Ignite 5.0 g.

Chloride and Bromide. Dissolve 1.0 g in 100 mL of water in a distilling flask. Add 1 mL of hydrogen peroxide and 1 mL of phosphoric acid, heat to boiling, and boil gently until all the iodine is expelled and the solution is colorless. Cool, wash down the sides of the flask, and add 0.5 mL of hydrogen peroxide. If an iodine color develops, boil until the solution is colorless and for 10 min longer. If no color develops, boil for 10 min, filter if necessary through a chloride-free filter, and dilute with water to 100 mL. Dilute 5.0 mL of this solution with 18 mL of water, and add 1 mL of nitric acid and 1 mL of silver nitrate reagent solution. Any turbidity should not exceed that produced by 0.01 mg of chloride ion (Cl) in an equal volume of solution containing the quantities of nitric acid and silver nitrate used in the test.

Iodide. Dissolve 1.0 g in 20 mL of water. Add 1 mL of chloroform and 0.5 mL of 1 N sulfuric acid volumetric solution, and mix. No violet color should appear in the chloroform within 1 min.

Nitrogen Compounds. Dissolve 0.50 g in water, and dilute with water to 200 mL. Transfer 40 mL to a flask connected through a spray trap to a condenser, the end of which dips beneath the surface of 10 mL of 0.1 N hydrochloric acid. Add to the flask 10 mL of 10% sodium hydroxide reagent solution and 0.5 g of aluminum wire in small pieces, allow to stand for 1 h, and slowly distill about 35 mL. To the distillate, add 1 mL of 10% sodium hydroxide reagent solution, dilute with water to 50 mL, and add 2 mL of Nessler reagent. Any color should not exceed that produced when a quantity of an ammonium salt containing 0.1 mg of nitrogen ion (N) is treated exactly like the sample.

For the Determination of Sulfate, Heavy Metals, and Iron

Sample Solution A. Dissolve 10 g in 20 mL of water, and add about 10 mg of sodium carbonate. Add 20 mL of hydrochloric acid, and evaporate to dryness on a hot plate (≈ 100 °C). Repeat the evaporation twice using 10 mL of hydrochloric acid.

Blank Solution B. Evaporate to dryness the quantities of hydrochloric acid and sodium carbonate used to prepare sample solution A.

Dissolve the residues in separate 20-mL portions of water, filter if necessary, and dilute each with water to 100 mL (1 mL of sample solution A = 0.1 g).

Sulfate. (Page 40, Method 1). Use 3.3 mL of sample solution A (0.33-g sample).

Heavy Metals. (Page 36, Method 1). Dilute 20 mL of sample solution A (2-g sample) to 25 mL. Use 20 mL of blank solution B to prepare the control solution.

Iron. (Page 38, Method 1). Use 5.0 mL of sample solution A (0.5-g sample).

Iodine

I_2 Formula Wt 253.81   CAS No. 7553-56-2

GENERAL DESCRIPTION

Typical appearance: slate gray to black-violet solid
Analytical use: iodometric standard
Change in state (approximate): melting point, 114 °C
Aqueous solubility: 0.03 g in 100 mL at 25 °C

SPECIFICATIONS

Assay . . . . ≥99.8% I_2

                                                                                          Maximum Allowable
Nonvolatile matter 0.01%
Chlorine and bromine (as Cl) 0.005%

TESTS

Assay. (By titration of oxidizing power). Weigh accurately about 0.5 g, and transfer to a glass-stoppered conical flask. Add 50 mL of water containing 3 g of potassium iodide and 1 mL of sulfuric acid. Stopper, swirl, and allow to stand for 2–3 min. Titrate the iodine with 0.1 N sodium thiosulfate volumetric solution, adding 3 mL of starch indicator solution near the end of the titration. One milliliter of 0.1 N sodium thiosulfate corresponds to 0.01269 g of I_2.

$$\% \, I_2 = \frac{(mL \times N \, Na_2S_2O_3) \times 12.69}{\text{Sample wt (g)}}$$

Nonvolatile Matter. Weigh accurately about 10 g, and transfer to a tared, preconditioned porcelain crucible or dish. Heat on a hot plate ($\approx$100 °C) until the sample is volatilized. Heat to 105 °C for 1 h, cool in a desiccator, and weigh.

Chlorine and Bromine. Add 1.0 g to 100 mL of hot water containing 0.6 g of hydrazine sulfate in a 250-mL Erlenmeyer flask. Heat on a steam bath until dissolution is effected. Cool, and neutralize with 1 N sodium hydroxide volumetric solution, using an external indicator. Add 2 mL of hydrogen peroxide and 1 mL of phosphoric acid. Heat to boiling, and boil gently until the solution is colorless. Cool, wash down the sides of the flask, and add 0.5 mL of hydrogen peroxide. If an iodine color develops, boil until the solution is colorless and for 10 min longer. If no color develops, boil for 10 min. Filter if necessary through a chloride-free filter, dilute with water to 100 mL, and take a 20-mL portion. For the standard, dilute a solution containing 0.01 mg of chloride ion (Cl) with water to 20 mL. Add 1 mL of nitric acid and 1 mL of silver nitrate reagent solution to each. Any turbidity in the solution of the sample should not exceed that in the standard.

Iodine Monochloride

ICl   Formula Wt 162.36   CAS No. 7790-99-0

Suitable for use in Wijs solution.

GENERAL DESCRIPTION

Typical appearance: black solid or reddish-brown liquid
Analytical use: determining iodine values of fats and oils
Change in state (approximate): melting point, α-form (stable), 27 °C; β-form, 14 °C

SPECIFICATIONS

Assay . . . . .The I/Cl ratio of the Wijs solution prepared
                                              from the reagent must be 1.10 ± 0.1.
Insoluble matter . . . . ≤0.005%

TESTS

Insoluble Matter.

Stock Solution. Dissolve 79.3 ± 0.1 g in glacial acetic acid, and dilute to 250 mL
with glacial acetic acid. Filter through a tared filtering crucible. Store the solution in a
dry, low-actinic glass bottle.

Wash the stock solution filter thoroughly with acetic acid, and dry at 105 °C.

Assay. (By selective titration of reducing and oxidizing power). Prepare Wijs solution
(1.4–1.5% ICl in acetic acid) by diluting 11.7 ± 0.1 mL of the stock solution with 217 mL of
glacial acetic acid. Determine the I/Cl ratio of this solution as follows:

Iodine. To 150 mL of saturated chlorine water in a 500-mL conical flask, add some
glass beads, and pipet 5 mL of the Wijs solution into the flask. Shake, heat to boiling,
and boil briskly for 10 min. Cool, add 30 mL of 2% sulfuric acid and 15 mL of 15%
potassium iodide solution, and mix well. Titrate immediately with 0.1 N sodium thiosulfate volumetric solution, adding 3 mL of starch indicator solution near the end
point.

Total Halogen. Transfer 150 mL of freshly boiled water to a dry 500-mL conical
flask. Add 15 mL of 15% potassium iodide solution, pipet 20 mL of Wijs solution into
the flask, and mix well. Titrate immediately with 0.1 N sodium thiosulfate volumetric
solution, adding 3 mL of starch indicator solution near the end point. Calculate the
halogen ratio as follows:

$$\frac{I}{Cl} = \frac{2A}{3B - 2A}$$

where A = volume, in mL, of 0.1 N sodium thiosulfate used to titrate iodine; and B =
volume, in mL, of 0.1 N sodium thiosulfate used to titrate total halogen (as iodine).

Iron, Low in Magnesium and Manganese

Fe   Atomic Wt 55.845   CAS No. 7439-89-6

GENERAL DESCRIPTION

Typical appearance: silver-white metal; gray or black solid
Analytical use: standard for elemental analysis
Change in state (approximate): melting point, 1535 °C

SPECIFICATIONS

	Maximum Allowable
Magnesium (Mg)	5 ppm
Manganese (Mn)	0.002%

TESTS

Magnesium and Manganese. (By flame AAS, page 63).

> **Sample Stock Solution.** In a fume hood, cautiously dissolve 10.0 g in 60 mL of dilute hydrochloric acid (1 + 1) and 20 mL of nitric acid in a covered 400-mL beaker. Substitute a ribbed cover glass, evaporate to dryness, and bake at moderate heat for 5 min. Add 30 mL of hydrochloric acid, heat gently until the salts are dissolved, and dilute with water to 100 mL in a volumetric flask (1 mL = 0.10 g).

Element	Wavelength (nm)	Sample Wt (g)	Standard Added (mg)	Flame Type*	Background Correction
Mg	285.2	1.0	0.005; 0.01	A/A	Yes
Mn	279.5	1.0	0.02; 0.04	A/A	Yes

*A/A is air/acetylene.

Isatin

2,3-Indolinedione

C$_8$H$_5$NO$_2$ Formula Wt 147.13   CAS No. 91-56-5

GENERAL DESCRIPTION

Typical appearance: orange solid
Analytical use: reagent for cuprous ions, mercaptans, and thiopene

Change in state (approximate): melting point, 203 °C (partial sublimation)
Aqueous solubility: soluble in boiling water

SPECIFICATIONS

Sensitivity to thiophene . . . . Passes test
Melting point . . . . 199–204 °C

Maximum Allowable

Alcohol-insoluble matter 0.01%
Residue after ignition 0.05%

TESTS

Sensitivity to Thiophene. Add 5.0 mL of freshly prepared (sample) isatin solution
(described below) to a dry, clean 50-mL porcelain crucible. Carefully overlay with 1.0 mL
of thiophene standard solution (described below). Allow to stand undisturbed for 1 h, and
compare with a blank containing 5.0 mL of isatin reagent solution and 1.0 mL of
thiophene-free hexanes (described below) in a 50-mL porcelain crucible. A bluish-green
color should develop as a halo or ring at the phase separation with the sample solution
when compared to the blank.

> **Solution A.** Dissolve 0.25 g of sample in 25 mL of sulfuric acid.

> **Solution B.** Dissolve 0.25 g of ferric chloride in 1 mL of water. Dilute to 50 mL with
> sulfuric acid, and allow the evolution of gas to cease.

> **Isatin Solution.** To 2.5 mL of solution A, add 5.0 mL of solution B, and dilute to 100
> mL with sulfuric acid.

> **Thiophene Standard Solution (0.001 mg of C_4H_4S in 1 mL).** Transfer 1.0 g (0.95
> mL) of thiophene to a 100-mL volumetric flask. Dilute to the mark with thiophene-
> free hexanes (described below), and mix. Transfer 1.0 mL of this solution to another
> 100-mL volumetric flask. Dilute to the mark with thiophene-free hexanes, and mix.
> Transfer 1.0 mL of this solution to a third 100-mL volumetric flask, dilute to the mark
> with thiophene-free hexanes, and mix.

> **Thiophene-Free Hexanes.** Add 350 mL of hexanes to a 500-mL separatory funnel,
> and add 50 mL of sulfuric acid. Shake, and allow the phases to separate. Extract the
> acid layer from the funnel. Add another 50 mL of sulfuric acid, and extract again. Dis-
> card the acid washings.

Melting Point. (Page 45).

Alcohol-Insoluble Matter. (Page 25). Dissolve 10.0 g in 250 mL of warm reagent alcohol. The solution should be orange-red. Do not digest the solution. Filter through a tared
filtering crucible, wash well with warm reagent alcohol, and dry at 105 °C.

Residue after Ignition. (Page 26). Gently ignite 2.0 g. Moisten the char with 1 mL of
sulfuric acid.

Isobutyl Alcohol
2-Methyl-1-propanol

$$CH_3$$
$$|$$
$$CH_3CHCH_2OH$$

(CH₃)₂CHCH₂OH   **Formula Wt 74.12**   **CAS No. 78-83-1**

Suitable for general use or in ultraviolet spectrophotometry. Product labeling shall designate the uses for which suitability is represented on the basis of meeting the relevant specifications and tests. The ultraviolet spectrophotometry specifications include all of the specifications for general use.

GENERAL DESCRIPTION

Typical appearance: colorless liquid
Analytical use: organic solvent
Change in state (approximate): boiling point, 108 °C
Aqueous solubility: soluble in 20 parts water
Density: 0.80

SPECIFICATIONS

General Use

Assay . . . . ≥99.0% (CH₃)₂CHCH₂OH
Solubility in water . . . . Passes test

Maximum Allowable

Color (APHA) 10
Residue after evaporation 0.001%
Titrable acid. . . . . .0.0005 meq/g
Water (H₂O) 0.1%
Carbonyl compounds . . . . .0.01% butyraldehyde
and 0.02% 2-butanone

Specific Use

Ultraviolet Spectrophotometry

Wavelength (nm)	Absorbance (AU)
400–330	0.01
260	0.05
250	0.07
240	0.25
230	1.00

Assay. Analyze the sample by gas chromatography using the parameters cited on page 80. The following specific conditions are also required.

   Column: Type I, methyl silicone

Measure the area under all peaks, and calculate the isobutyl alcohol content in area percent. Correct for water content.

Solubility in Water. Dilute 1 mL with 16 mL of water. The sample should dissolve completely, and the solution should be clear.

Color (APHA). (Page 43).

Residue after Evaporation. (Page 25). Evaporate 100 g (125 mL) to dryness in a tared, preconditioned dish on a hot plate ($\approx$100 °C), and dry the residue at 105 °C for 30 min.

Titrable Acid. To 60 g (75 mL) in a glass-stoppered flask, add 0.15 mL of phenolphthalein indicator solution, and shake for 1 min. No pink color should form. Then titrate the solution with 0.01 N methanolic potassium hydroxide volumetric solution to a faint pink color that persists for at least 15 s. Not more than 3.0 mL of 0.01 N potassium hydroxide should be consumed.

Water. (Page 31, Method 1). Use 50 mL (40 g) of the sample.

Carbonyl Compounds. (Page 54). Use 1.25 mL (1.0 g) of sample. For the standards, use 0.10 mg of butyraldehyde and 0.20 mg of 2-butanone.

Ultraviolet Spectrophotometry. Use the procedure on page 86 to determine the absorbance.

Isopentyl Alcohol
Isoamyl Alcohol; 3-Methyl-1-butanol

$$CH_3CHCH_2CH_2OH$$
with CH_3 branch

$(CH_3)_2CHCH_2CH_2OH$	Formula Wt 88.15	CAS No. 123-51-3

GENERAL DESCRIPTION

Typical appearance: clear liquid
Analytical use: organic solvent
Change in state (approximate): boiling point, 131 °C
Aqueous solubility: 2 g in 100 mL at 14 °C
Density: 0.81

SPECIFICATIONS

Assay . . . . $\geq$98.5% $C_5H_{11}OH$

Maximum Allowable

Water (H_2O) 0.5%
Titrable acid . . . . 0.002 meq/g
Residue after evaporation 0.003%
Acids and esters (as amyl acetate) 0.2%
Carbonyl compounds (as HCHO) 0.1%

TESTS

Assay. Analyze the sample by gas chromatography using the general parameters cited on page 80. The following specific conditions are also required.

Column: Type I, methyl silicone

Measure the area under all peaks, and calculate the isopentyl alcohol content in area percent. Correct for water content.

Water. (Page 31, Method 1). Use 25 mL (20 g) of the sample.

Titrable Acid. To 20 g (25 mL) of sample in a glass-stoppered flask, add 0.15 mL of phenolphthalein indicator solution and shake for 1 min. No pink color should form. Titrate the solution with 0.01 N methanolic potassium hydroxide volumetric solution to a faint pink color that persists for at least 15 s. Not more than 4.0 mL of the methanolic potassium hydroxide should be consumed.

Residue after Evaporation. (Page 25). Evaporate 40 g (50 mL) to dryness in a tared, preconditioned dish on a steam bath, and dry the residue at 105 °C for 30 min.

Acids and Esters. Dilute 100 mL of sample with 100 mL of anhydrous isopropyl alcohol, add 25.0 mL of 0.1 N methanolic potassium hydroxide volumetric solution, and heat gently under a reflux condenser for 10–15 min. Cool, and add 20 mL of carbon dioxide-free water through the reflux condenser. Detach the flask, add 0.15 mL of phenolphthalein indicator solution, and titrate the excess potassium hydroxide with 0.1 N hydrochloric acid to a colorless end point. The volume of 0.1 N potassium hydroxide consumed in the test should not be more than 12.3 mL greater than the volume consumed in a complete blank test.

Carbonyl Compounds. (Page 54). Use 0.25 g (0.30 mL) of sample. For the standard, use 0.25 mg of formaldehyde.

Isopropyl Alcohol
2-Propanol

$$CH_3\overset{\displaystyle OH}{\overset{|}{C}}HCH_3$$

$CH_3CHOHCH_3$ **Formula Wt 60.10**   **CAS No. 67-63-0**

Suitable for general use or in ultraviolet spectrophotometry. Product labeling shall designate the uses for which suitability is represented on the basis of meeting the relevant specifications and tests. The ultraviolet spectrophotometry specifications include all of the specifications for general use.

GENERAL DESCRIPTION

Typical appearance: clear liquid
Analytical use: organic solvent

Change in state (approximate): boiling point, 83 °C
Aqueous solubility: miscible
Density: 0.79

SPECIFICATIONS

General Use

Assay . . . . $\geq$99.5% $CH_3CHOHCH_3$
Solubility in water . . . . Passes test

	Maximum Allowable

Carbonyl compounds . . . . 0.002% propionaldehyde
                                                                  0.002% acetone
Color (APHA) 10
Residue after evaporation 0.001%
Water (H_2O) 0.2%
Titrable acid or base . . . . 0.0001 meq/g

Specific Use

Ultraviolet Spectrophotometry

Wavelength (nm)	Absorbance (AU)
400–330	0.01
300	0.02
275	0.03
260	0.04
245	0.08
230	0.20
220	0.40
210	1.00

TESTS

Assay. Analyze the sample by gas chromatography using the general parameters cited on page 80. The following specific conditions are also required.

Column: Type I, methyl silicone

Measure the area under all peaks, and calculate the isopropyl alcohol content in area percent. Correct for water content.

Solubility in Water. Mix 10 mL with 40 mL of water, and allow to stand 1 h. The solution should be as clear as an equal volume of water.

Carbonyl Compounds. (Page 54). Use 2.6 mL (2.0 g) of sample. Prepare separate standards with 0.04 mg of propionaldehyde and 0.04 mg of acetone.

Color (APHA). (Page 43).

Residue after Evaporation. (Page 25). Evaporate 100 g (128 mL) to dryness in a tared, preconditioned dish on a hot plate ($\approx$100 °C), and dry the residue at 105 °C for 30 min.

Water. (Page 31, Method 1). Use 10 mL (7.8 g) of the sample.

Titrable Acid or Base. Mix 10 mL with 10 mL of carbon dioxide-free water, and add 0.15 mL of bromthymol blue indicator solution. The solution should be neutral or require not more than 0.10 mL of 0.01 N hydrochloric acid or 0.10 mL of 0.01 N sodium hydroxide to render it so.

Ultraviolet Spectrophotometry. Use the procedure on page 86 to determine the absorbance.

Isopropyl Ether
Diisopropyl Ether

$$CH_3CH-O-CHCH_3$$

with CH_3 groups on each CH

$(C_3H_7)_2O$	**Formula Wt 102.18**	**CAS No. 108-20-3**

Caution: Isopropyl ether is a volatile and flammable liquid with a tendency to form an explosive peroxide. No evaporation or distillation should be performed unless peroxide is shown to be absent. Generally, a stabilizer is present to retard peroxide formation.

GENERAL DESCRIPTION

Typical appearance: clear liquid
Analytical use: organic solvent; extracting solvent
Change in state (approximate): boiling point, 68 °C
Density: 0.73

SPECIFICATIONS

Assay . . . . $\geq$99.0% $(C_3H_7)_2O$

Maximum Allowable

Color (APHA) 25
Peroxide (as $C_6H_{14}O_2$) 0.05%
Residue after evaporation 0.01%
Titrable acid . . . . .0.0007 meq/g

TESTS

Assay. Analyze the sample by gas chromatography using the general parameters cited on page 80. The following specific conditions are also required.

   Column: Type I, methyl silicone

Measure the area under all peaks, and calculate the isopropyl ether content in area percent.

Color (APHA). (Page 43).

Peroxide. Place 150 mL of 10% sulfuric acid in each of two 500-mL glass-stoppered Erlenmeyer flasks. Reserve one for a blank determination. Into the other, pipet 7.2 g (10 mL) of sample, and mix well. To each flask, add 0.15–0.25 mL of 1% ammonium molybdate solution and 15 mL of 10% potassium iodide reagent solution. Mix, and allow to stand in the dark for 15 min at room temperature. Titrate each with 0.05 N sodium thiosulfate until the yellow color begins to fade. Add 3 mL of starch indicator solution, and continue the titration just to disappearance of the blue color. Correct for a blank determination. One milliliter of 0.05 N sodium thiosulfate corresponds to 0.00296 g of $(C_3H_7)_2O_2$.

$$\% \ (C_3H_7)_2O_2 = \frac{(mL \times N \ NaS_2O_3) \times 5.91}{Sample \ wt \ (g)}$$

Residue after Evaporation. (Page 25). See the warning above. Evaporate 70 mL (50 g) of sample.

Titrable Acid. Titrate 36 g (50 mL) with 0.02 N methanolic potassium hydroxide solution, with 0.15 mL of phenolphthalein indicator solution, to a pink end point that remains for at least 15 s. Not more than 1.25 mL of the methanolic potassium hydroxide should be required.

Lactic Acid, 85%
2-Hydroxypropanoic Acid

$$\begin{array}{cc} OH & O \\ | & \| \\ CH_3CH\!-\!C\!-\!OH \end{array}$$

CH$_3$CHOHCOOH   **Formula Wt 90.08**   **CAS No. 50-21-5**

Note: The reagent generally available is a mixture of lactic acid, $CH_3CHOHCOOH$, and lactic acid lactate, $C_6H_{10}O_5$.

GENERAL DESCRIPTION

Typical appearance: clear, odorless, syrupy liquid
Analytical use: clinical standard
Change in state (approximate): boiling point, 122 °C
Aqueous solubility: miscible

SPECIFICATIONS

Assay . . . . 85.0–90.0% C$_3$H$_6$O$_3$
Substances darkened by sulfuric acid . . . . Passes test

	Maximum Allowable
Residue after ignition	0.02%
Chloride (Cl)	0.001%
Sulfate (SO$_4$)	0.002%
Heavy metals (as Pb)	5 ppm
Iron (Fe).	5 ppm

TESTS

Assay. (By indirect acid–base titrimetry). Weigh accurately 3 mL in a tared, glass-stoppered flask. Add 50.0 mL of 1 N sodium hydroxide volumetric solution, boil gently for 20 min, and titrate the excess alkali with 1 N sulfuric acid volumetric solution, using 0.15 mL of phenolphthalein indicator solution. Correct for a complete blank test. One milliliter of 1 N sodium hydroxide corresponds to 0.09008 g of $C_3H_6O_3$.

$$\% \ C_3H_6O_3 = \frac{[(mL \times N \ NaOH) - (mL \times N \ H_2SO_4)] \times 9.008}{Sample \ wt \ (g)}$$

Substances Darkened by Sulfuric Acid. Cool 5 mL of sulfuric acid to 15 °C, and transfer to a clean, dry test tube. Cool 5 mL of the sample to 15 °C, and add to the test tube in such a way as to form a distinct layer over the sulfuric acid. A dark color should not develop at the interface of the two layers within 10 min.

Residue after Ignition. (Page 26). In a tared, preconditioned crucible, ignite 10 g (8 mL), and moisten the char with 1 mL of sulfuric acid. Retain the final residue for the test for iron.

Chloride. (Page 35). Dilute 12 g (10 mL) with water to 60 mL, mix, and use 5 mL of this solution. Retain the remainder of the solution for the test for sulfate.

Sulfate. (Page 40, Method 1). Use 12.5 mL of the solution prepared in the test for chloride.

Heavy Metals. (Page 36, Method 2). Test a 4.0-g sample.

Iron. (Page 38, Method 1). To the residue from the test for residue after ignition, add 2 mL of water and 2 mL of hydrochloric acid, and evaporate to dryness on a hot plate (≈100 °C). Take up the residue in 1 mL of hydrochloric acid and 20 mL of water, filter if necessary, and dilute with water to 50 mL in a volumetric flask. Use 10 mL.

Lactose Monohydrate

$C_{12}H_{22}O_{11} \cdot H_2O$ **Formula Wt 360.32**   **CAS No. 64044-51-5**

GENERAL DESCRIPTION

Typical appearance: white solid

Analytical use: chromatographic adsorbent

Change in state (approximate): melting point, becomes anhydrous at 120 °C; at 201–202 °C when heated rapidly

Aqueous solubility: 40 g in 100 mL at 100 °C

SPECIFICATIONS

Water (H_2O) 4.0–6.0%

Maximum Allowable

Insoluble matter 0.005%
Residue after ignition 0.03%
Dextrose. . . . . Passes test
Sucrose. . . . . Passes test
Heavy metals (as Pb) . . . . 5 ppm
Iron (Fe). . . . . 5 ppm

TESTS

Water. (Page 31, Method 1). Use 2.0 g of the sample.

Insoluble Matter. (Page 25). Use 20 g dissolved in 150 mL of water.

Residue after Ignition. (Page 26). In a tared, preconditioned crucible, ignite 5.0 g. Moisten the char with 2 mL of sulfuric acid.

Dextrose. To 10 g, finely powdered, add 50 mL of 70% ethyl alcohol. Shake frequently for 30 min, and filter through a small, dry filter paper. Evaporate 10 mL of the filtrate to dryness on a hot plate ($\approx$100 °C), and reserve the remainder of the filtrate for the test for sucrose. Dissolve the residue in 5 mL of water, filter if necessary, and transfer the filtrate to a test tube. Add 5 mL of cupric acetate solution (described below), mix, immerse the test tube in a boiling-water bath for 3 min, and allow the solution to stand at room temperature for 25 min. No red precipitate should appear in the test tube.

> **Cupric Acetate Solution.** Dissolve 13.3 g of cupric acetate monohydrate in a mixture of 195 mL of water and 5 mL of glacial acetic acid.

Sucrose. Evaporate 25 mL of the filtrate reserved from the test for dextrose to dryness on a hot plate ($\approx$100 °C), and dissolve the residue in 9 mL of water. Add 1 mL of dilute hydrochloric acid (20 + 8) and 100 mg of resorcinol, transfer the mixture to a small test tube, and immerse the tube in a boiling-water bath for 8 min. The solution should remain colorless or acquire not more than a slight yellow coloration.

Heavy Metals. (Page 36, Method 1). Dissolve 6.0 g in about 20 mL of water, and dilute with water to 30 mL. Use 25 mL to prepare the sample solution, and use the remaining 5.0 mL to prepare the control solution.

Iron. (Page 38, Method 1). Use 2.0 g.

Lanthanum Chloride, Hydrated

LaCl$_3$ · (6–7)H$_2$O   **CAS No. 10099-58-8**

Note: This reagent is available in degrees of hydration ranging from 6 to 7 molecules of water.

GENERAL DESCRIPTION

Typical appearance: colorless or white solid
Analytical use: matrix modification
Change in state (approximate): melting point, 91 °C, with decomposition
Aqueous solubility: very soluble

SPECIFICATIONS

Assay . . . . .64.5–70.0% $LaCl_3$

Maximum Allowable

Insoluble matter 0.01%
Calcium (Ca) . . . . .10 ppm
Magnesium (Mg) . . . . .10 ppm

TESTS

Assay. (By gravimetric determination of La). Weigh accurately about 0.8 g, dissolve in 25 mL of water, filter any undissolved material, and wash with water. Add 100 mL of 5% oxalic acid solution, digest on a hot plate ($\approx$100 °C) until the supernatant solution is clear. Filter through a fine ashless filter paper, wash the precipitate with 10-mL portions of 1% oxalic acid solution, and dry. Ignite in a muffle furnace at 900 °C for 30 min. Cool and weigh as La_2O_3. One gram of lanthanum oxide corresponds to 1.506 g of $LaCl_3$.

$$\% \ LaCl_3 = \frac{\text{Residue wt (g) of } La_2O_3 \times 150.6}{\text{Sample wt (g)}}$$

Insoluble Matter. (Page 25). Use 20.0 g dissolved in 100 mL of water.

Calcium and Magnesium. (By flame AAS, page 63).

> **Sample Stock Solution.** Dissolve 5.0 g of sample in 40 mL of water, transfer to a 50-mL volumetric flask, and dilute to the mark with water (1 mL = 0.10 g).

Element	Wavelength (nm)	Sample Wt (g)	Standard Added (mg)	Flame Type*	Background Correction
Ca	422.7	1.0	0.01; 0.02	N/A	No
Mg	285.2	1.0	0.005; 0.01	A/A	Yes

*A/A is air/acetylene; N/A is nitrous oxide/acetylene.

Lead Acetate Trihydrate

Lead(II) Acetate Trihydrate

$(CH_3COO)_2Pb \cdot 3H_2O$ **Formula Wt 379.3**   **CAS No. 6080-56-4**

GENERAL DESCRIPTION

Typical appearance: white or colorless solid
Analytical use: detection of sulfide, chromate, and molybdenum trioxide

Change in state (approximate): melting point, 75 °C; decomposes at 200 °C
Aqueous solubility: 84 g in 100 mL at 40 °C

SPECIFICATIONS

Assay . . . . 99.0–103.0% $(CH_3COO)_2Pb \cdot 3H_2O$

Maximum Allowable

Insoluble matter 0.01%
Chloride (Cl) . . . . 5 ppm
Nitrate and nitrite (as NO_3). 0.005%
Calcium (Ca) 0.005%
Copper (Cu). 0.002%
Iron (Fe). 0.001%
Potassium (K) 0.005%
Sodium (Na) 0.01%

TESTS

Assay. (By complexometric titration of Pb). Weigh accurately about 1.5 g, transfer to a 400-mL beaker, and dissolve in 250 mL of water containing 0.15 mL of acetic acid. Add 15 mL of saturated hexamethylenetetramine reagent solution. Add a few milligrams of xylenol orange indicator mixture and titrate with 0.1 M EDTA volumetric solution to a yellow color. One milliliter of 0.1 M EDTA corresponds to 0.03793 g of $(CH_3COO)_2Pb \cdot 3H_2O$.

$$\% \ (CH_3COO)_2Pb \cdot 3H_2O = \frac{(mL \times M \ EDTA) \times 37.93}{Sample \ wt \ (g)}$$

Insoluble Matter. (Page 25). Use 10 g dissolved in 100 mL of dilute glacial acetic acid (1 + 99) free from carbon dioxide.

Chloride. (Page 35). Use 2.0 g.

Nitrate and Nitrite. (Page 38, Method 1).

> *Sample Solution A.* Dissolve 0.20 g in 1 mL of water. Add 20 mL of nitrate-free sulfuric acid (2 + 1), centrifuge, and decant the liquid portion into a 200-mm test tube. Dilute to 50 mL with brucine sulfate reagent solution.

> *Control Solution B.* Dissolve 0.20 g in 1 mL of nitrate ion (NO_3) standard solution. Add 20 mL of nitrate-free sulfuric acid (2 + 1), centrifuge, and decant the liquid portion into a 200-mm test tube. Dilute to 50 mL with brucine sulfate reagent solution.

Continue with the procedure, starting with the preparation of blank solution C.

Calcium, Copper, Iron, Potassium, and Sodium. (By flame AAS, page 63).

> *Sample Stock Solution A.* Dissolve 25.0 g in 75 mL of water and 5 mL of nitric acid. Transfer to a 100-mL volumetric flask, and dilute to the mark with water (1 mL = 0.25 g).

> *Sample Stock Solution B.* Transfer 4.0 mL of sample stock solution A to a 100-mL volumetric flask, and dilute to the mark with water (1 mL = 0.01 g).

Element	Wavelength (nm)	Sample Wt (g)	Standard Added (mg)	Flame Type*	Background Correction
Ca	422.7	1.0	0.025; 0.05	N/A	No
Cu	324.7	1.0	0.02; 0.04	A/A	Yes
Fe	248.3	5.0	0.05; 0.10	A/A	Yes
K	766.5	1.0	0.025; 0.05	A/A	No
Na	589.0	0.10	0.005; 0.01	A/A	No

*A/A is air/acetylene; N/A is nitrous oxide/acetylene.

Lead Carbonate

$PbCO_3$ (normal)	Formula Wt 267.20	CAS No. 598-63-0
$Pb(CO_3)_2Pb(OH)_2$ (basic)	Formula Wt 775.60	CAS No. 1344-36-1

Note: This specification applies to both the normal lead carbonate and the basic lead carbonate.

GENERAL DESCRIPTION

Typical appearance: heavy white solid
Analytical use: preparation of lead standard solutions
Change in state (approximate): decomposes at 400 °C
Aqueous solubility: insoluble

SPECIFICATIONS

	Maximum Allowable
Insoluble in dilute acetic acid	0.02%
Chloride (Cl)	0.002%
Nitrate and nitrite (as NO_3)	Passes test
Cadmium (Cd)	0.002%
Calcium (Ca)	0.01%
Iron (Fe)	0.005%
Potassium (K)	0.02%
Sodium (Na)	0.05%
Zinc (Zn)	0.003%

TESTS

Insoluble in Dilute Acetic Acid. Digest 5.0 g in 50 mL of water plus 7 mL of glacial acetic acid in a covered beaker on a steam bath for 1 h. Filter through a tared, preconditioned filtering crucible, wash thoroughly with dilute acetic acid (2 + 98), and dry at 105 °C.

Chloride. (Page 35). Dissolve 1.0 g in 20 mL of 10% nitric acid reagent solution, filter if necessary through a chloride-free filter, and dilute with water to 50 mL. Use 25 mL of this solution.

Nitrate and Nitrite. (Page 38, Method 1).

Sample Solution A. Dissolve 0.20 g in 1 mL of water and 8 mL of dilute acetic acid (1 + 7) by heating in a boiling-water bath. Add 20 mL of nitrate-free sulfuric acid (2 + 1), cool, centrifuge, and decant the liquid portion into a 200-mm test tube. Dilute to 50 mL with brucine sulfate reagent solution.

Control Solution B. Dissolve 0.20 g in 8 mL of dilute acetic acid (1 + 7) and 1 mL of nitrate ion (NO_3) standard solution by heating in a boiling-water bath. Add 20 mL of nitrate-free sulfuric acid (2 + 1), cool, centrifuge, and decant the liquid portion into a 200-mm test tube. Dilute to 50 mL with brucine sulfate reagent solution.

Continue with the procedure, starting with the preparation of blank solution C. (Limit about 0.005%)

Cadmium, Calcium, Iron, Potassium, Sodium, and Zinc. (By flame AAS, page 63).

Sample Stock Solution. Dissolve 10.0 g in 50 mL of water and sufficient nitric acid to achieve complete dissolution. Transfer the solution to a 100-mL volumetric flask, and dilute to the mark with water (1 mL = 0.1 g).

Element	Wavelength (nm)	Sample Wt (g)	Standard Added (mg)	Flame Type*	Background Correction
Cd	228.8	1.0	0.02; 0.04	A/A	Yes
Ca	422.7	0.40	0.02; 0.04	N/A	No
Fe	248.3	1.0	0.05; 0.10	A/A	Yes
K	766.5	0.20	0.02; 0.04	A/A	No
Na	589.0	0.04	0.01; 0.02	A/A	No
Zn	213.9	0.40	0.01; 0.02	A/A	Yes

*A/A is air/acetylene; N/A is nitrous oxide/acetylene.

Lead Chromate
$PbCrO_4$ **Formula Wt 323.19**   **CAS No. 7758-97-6**

GENERAL DESCRIPTION
Typical appearance: yellow or orange-yellow solid
Analytical use: chemical analysis of organic substances
Change in state (approximate): melting point, 844 °C
Aqueous solubility: 0.2 mg per L of water

SPECIFICATIONS
Assay . . . . ≥98.0% $PbCrO_4$
Maximum Allowable
Soluble matter 0.15%
Carbon compounds (as C) 0.01%

TESTS

Assay. (By titration of oxidizing power of chromate). Weigh accurately about 0.5 g of powdered sample, and dissolve by warming with 40 mL of 10% sodium hydroxide reagent solution in a glass-stoppered conical flask. Add 2 g of potassium iodide; when it is dissolved, dilute with 80 mL of water and 15 mL of hydrochloric acid. Stopper, swirl, and allow to stand in the dark for 5 min. Titrate the liberated iodine with 0.1 N sodium thiosulfate volumetric solution, adding 3 mL of starch indicator solution near the end of the titration. One milliliter of 0.1 N sodium thiosulfate corresponds to 0.01077 g of $PbCrO_4$.

$$\% \ PbCrO_4 = \frac{(mL \times N \ Na_2S_2O_3) \times 10.77}{Sample \ wt \ (g)}$$

Soluble Matter. Boil 0.50 g of the powdered sample for 5 min with 100 mL of dilute acetic acid $(1 + 20)$, stirring well during the heating. Cool, and filter. Evaporate 25 mL of the filtrate to dryness in a tared, preconditioned dish on a hot plate ($\approx$100 °C), and dry at 105 °C. Heat 50 mL of the filtrate with 2 g of the powdered sample for 5 min at 80–90 °C, cool, dilute with water to 50 mL, and filter. Evaporate 25 mL of this filtrate in a tared, preconditioned dish on a hot plate ($\approx$100 °C), and dry at 105 °C. The difference between the weights of the two residues should not be more than 0.0015 g.

Carbon Compounds. This procedure uses induction heating of the sample and a conductometric measurement of evolved carbon dioxide.

Transfer 1.0 g of sample and one tin capsule to a combustion crucible. Add 1.5 g each of iron and copper accelerator. Ignite in an induction furnace in a stream of carbon dioxide-free oxygen until combustion is complete. Pass the effluent gas through an approximately 0.1% solution of reagent-grade barium hydroxide. Measure the change in conductance, in mhos, using a suitable cell and detection system. Subtract the reading from a blank, and calculate the carbon content of the sample from a calibration curve.

Prepare the calibration curve as follows: transfer 0, 10, 25, 75, and 100 μL of carbon standard solution (1 μL = 4 μg of carbon) to tin capsules. Evaporate the solutions to dryness in the capsules at 80 °C. Transfer the capsules to combustion crucibles, add the accelerators, ignite, and measure the conductance under the same conditions as described for running the sample. Subtract the reading of the blank from that of each of the standards, and plot the calibration curve as mhos versus micrograms of carbon.

Lead Dioxide

Lead(IV) Oxide

PbO$_2$ **Formula Wt 239.20**   **CAS No. 1309-60-0**

GENERAL DESCRIPTION

Typical appearance: dark brown solid
Analytical use: oxidizing agent

Change in state (approximate): melting point, 290 °C, with decomposition

Aqueous solubility: insoluble

SPECIFICATIONS

Assay . . . . ≥97.0% PbO_2

Maximum Allowable

Acid-insoluble matter 0.2%

Carbon compounds (as C) 0.04%

Chloride (Cl) 0.002%

Nitrate (NO_3) 0.02%

Sulfate (SO_4) 0.05%

Manganese (Mn) . . . . 5 ppm

Calcium (Ca) 0.02%

Copper (Cu) 0.05%

Potassium (K) 0.05%

Sodium (Na) 0.1%

TESTS

Assay. (By titration of oxidizing power). Dissolve 25 g of sodium chloride in 100 mL of water in a stoppered flask. Add 2 g of potassium iodide and 20 mL of hydrochloric acid. Add, accurately weighed, about 0.25 g of finely powdered sample, and stir until the sample is dissolved. Titrate the liberated iodine with 0.1 N sodium thiosulfate volumetric solution, adding 3 mL of starch indicator solution near the end of the titration. One milliliter of 0.1 N sodium thiosulfate corresponds to 0.01196 g of PbO_2.

$$\% \ PbO_2 = \frac{(mL \times N \ Na_2S_2O_3) \times 11.96}{Sample \ wt \ (g)}$$

Acid-Insoluble Matter. To 1.0 g, add 25 mL of water and 3 mL of nitric acid. Add carefully with stirring 5 mL of 30% hydrogen peroxide or more, if needed, to dissolve all the lead dioxide. Filter through a tared, preconditioned filtering crucible, wash thoroughly, and dry at 105 °C.

Carbon Compounds. This procedure uses induction heating of the sample and a conductometric measurement of the evolved carbon dioxide.

Transfer 1.0 g and one tin capsule to a combustion crucible. Add 1.5 g each of iron and copper accelerator. Ignite in an induction furnace in a stream of carbon dioxide-free oxygen until combustion is complete. Pass the effluent gas through an approximately 0.1% solution of barium hydroxide. Measure the change in conductance, in mhos, by using a suitable cell and detection system. Subtract the reading from a blank, and calculate the carbon content of the sample from a calibration curve.

Prepare the calibration curve as follows: transfer 0, 10, 25, 75, and 100 µL of carbon standard solution (1 µL = 4 µg of carbon) to tin capsules. Evaporate the solutions to dryness in the capsules at 80 °C. Transfer the capsules to combustion crucibles, add the accelerators, ignite, and measure the conductance under the same conditions as described for running the sample. Subtract the reading of the blank from that of each of the standards, and plot the calibration curve as mhos versus micrograms of carbon.

Chloride. (Page 35). Dissolve 2.0 g in 20 mL of water plus 2 mL of glacial acetic acid and 2 mL of 30% hydrogen peroxide. Filter through a chloride-free filter, dilute with water to 100 mL, and use 25 mL of this solution.

Nitrate. (Page 38, Method 1). Dissolve 0.25 g in a test tube containing 10 mL of water, 1 mL of glacial acetic acid, and 0.5 mL of 30% hydrogen peroxide. Pass sulfur dioxide through the solution to precipitate the lead, filter, wash with two 2-mL portions of water, and dilute the filtrate with water to 20 mL.

> **Sample Solution A.** To 4.0 mL of the filtrate in a test tube, add 1 mL of water, dilute to 50 mL with brucine sulfate reagent solution, and mix.

> **Control Solution B.** To 4.0 mL of the filtrate in a test tube, add 1 mL of nitrate ion (NO_3) standard solution, dilute to 50 mL with brucine sulfate reagent solution, and mix.

Continue with the procedure, starting with the preparation of blank solution C.

Sulfate. Decompose 2.0 g with 10 mL of hydrochloric acid and 3 mL of nitric acid, and evaporate to dryness on a hot plate (≈ 100 °C). Add 25 mL of water and 2 g of sodium carbonate, and digest on a hot plate (≈ 100 °C) for several hours, with occasional stirring. Dilute with water to 100 mL, and filter. Neutralize the filtrate with hydrochloric acid and add an excess of 1 mL of the acid. Heat to boiling, add 5 mL of 12% barium chloride reagent solution, digest in a covered beaker on the hot plate for 2 h, and allow to stand for at least 8 h. Filter, using a fine ashless paper, wash thoroughly, and ignite. Correct for the weight obtained in a complete blank test.

$$\% \ SO_4 = \frac{[\text{Residue wt (sample) (g)} - \text{Residue wt (blank) (g)}] \times 100}{\text{Sample wt (g)}}$$

Manganese. Add 10 g to 15 mL of dilute nitric acid (1 + 1). Add carefully, with stirring, about 10 mL of 30% hydrogen peroxide—or more, if necessary—to dissolve all of the lead dioxide. Evaporate nearly to dryness, cool, add 20 mL of dilute sulfuric acid (1 + 1), and evaporate to dense fumes of sulfur trioxide. Cool, cautiously dilute to 100 mL with water, allow to settle, and filter. Dilute 20 mL of the filtrate with water to 50 mL, add 10 mL of nitric acid and 5 mL of phosphoric acid, and boil gently for 5 min. Cool slightly, add 0.25 g of potassium periodate, and again boil gently for 5 min. Any pink color should not exceed that produced by 0.01 mg of manganese in 50 mL of dilute sulfuric acid (1 + 9) treated exactly like the 50 mL of filtrate.

Calcium, Copper, Potassium, and Sodium. (By flame AAS, page 63).

> **Sample Stock Solution.** Dissolve by warming 10.0 g of sample with a mixture of 20 mL of water and 15 mL of nitric acid. Cool to room temperature, transfer to a 100-mL volumetric flask, and dilute to the mark with water (1 mL = 0.1 g).

Element	Wavelength (nm)	Sample Wt (g)	Standard Added (mg)	Flame Type*	Background Correction
Ca	422.7	0.40	0.04; 0.08	N/A	No
Cu	324.8	0.10	0.025; 0.05	A/A	Yes
K	766.5	0.10	0.025; 0.05	A/A	No
Na	589.0	0.02	0.01; 0.02	A/A	No

*A/A is air/acetylene; N/A is nitrous oxide/acetylene.

Lead Monoxide
Lead(II) Oxide; Litharge
PbO   **Formula Wt 223.19**   **CAS No. 1317-36-8**

GENERAL DESCRIPTION
Typical appearance: yellow or orange solid
Analytical use: assays of gold and silver ores
Change in state (approximate): melting point, 888 °C
Aqueous solubility: insoluble

SPECIFICATIONS
Assay . . . . ≥99.0% PbO

Maximum Allowable

Insoluble in dilute acetic acid 0.02%
Chloride (Cl) 0.002%
Nitrate (NO_3) 0.01%
Calcium (Ca) 0.005%
Copper (Cu) 0.005%
Iron (Fe) 0.002%
Potassium (K) 0.005%
Silver (Ag) . . . . 5 ppm
Sodium (Na) 0.02%

TESTS

Assay. (By complexometric titration of Pb). Weigh accurately about 0.8 g, transfer to a 400-mL beaker, and dissolve by warming with a mixture of 1 mL of acetic acid and 10 mL of water. After solution is complete, dilute to 250 mL with water and add 15 mL of saturated hexamethylenetetramine reagent solution and 50 mg of xylenol orange indicator mixture. Titrate with 0.1 M EDTA volumetric solution to a color change of purple-red to lemon yellow. One milliliter of 0.1 M EDTA corresponds to 0.02232 g of PbO.

$$\% \, PbO = \frac{(mL \times M \, EDTA) \times 22.32}{Sample \, wt \, (g)}$$

Insoluble in Dilute Acetic Acid. Heat 5.0 g with a mixture of 50 mL of water and 15 mL of acetic acid until solution is complete. Filter through a tared, preconditioned sintered-glass crucible, wash thoroughly with hot water, dry at 105 °C for 1 h, cool, and weigh.

Chloride. (Page 35). Use 0.5 g.

Nitrate. (Page 38, Method 1). Dissolve 0.5 g in a mixture of 10 mL of water and 1 mL of acetic acid. Add 5 mL of 25% sulfuric acid reagent solution, mix well, and allow to stand for 15 min, mixing occasionally. Filter, wash with 5 mL of water, and dilute the filtrate to 25 mL.

Sample Solution A. Dilute 5.0 mL of the nitrate-test filtrate to 50 mL with brucine sulfate reagent solution.

Control Solution B. To 5.0 mL of the nitrate-test filtrate, add 1.0 mL of nitrate ion (NO_3) standard solution. Dilute to 50 mL with brucine sulfate reagent solution.

Continue with the procedure, starting with the preparation of blank solution C.

Calcium, Copper, Iron, Potassium, Silver, and Sodium. (By flame AAS, page 63).

Sample Stock Solution. Dissolve by warming 20.0 g of sample with a mixture of 50 mL of water and 30 mL of nitric acid. Cool to room temperature, transfer to a 100-mL volumetric flask, and dilute to the mark with water (1 mL = 0.2 g).

Element	Wavelength (nm)	Sample Wt (g)	Standard Added (mg)	Flame Type*	Background Correction
Ca	422.7	0.80	0.02; 0.04	N/A	No
Cu	324.8	0.80	0.02; 0.04	A/A	Yes
Fe	248.3	4.0	0.08; 0.16	A/A	Yes
K	766.5	0.80	0.02; 0.04	A/A	No
Ag	328.1	4.0	0.02; 0.04	A/A	Yes
Na	589.0	0.10	0.01; 0.02	A/A	No

*A/A is air/acetylene; N/A is nitrous oxide/acetylene.

Lead Nitrate

Lead(II) Nitrate

$Pb(NO_3)_2$ **Formula Wt 331.21**   **CAS No. 10099-74-8**

GENERAL DESCRIPTION

Typical appearance: white solid
Analytical use: titrant for complexometric titration
Change in state (approximate): melting point, 470 °C, with decomposition
Aqueous solubility: 56.5 g in 100 mL at 20 °C

SPECIFICATIONS

Assay . . . . ≥99.0% $Pb(NO_3)_2$

Maximum Allowable

Insoluble matter 0.005%
Chloride (Cl). 0.001%
Calcium (Ca) 0.005%
Copper (Cu) 0.002%
Iron (Fe). 0.001%
Potassium (K) 0.005%
Sodium (Na). 0.02%

TESTS

Assay. (By complexometric titration of Pb). Weigh accurately about 1.3 g, transfer to a 400-mL beaker, and dissolve in 250 mL of water containing 0.15 mL of acetic acid. Add 15 mL of saturated hexamethylenetetramine reagent solution. Add a few milligrams of xylenol orange indicator mixture, and titrate with 0.1 M EDTA volumetric solution to a yellow color. One milliliter of 0.1 M EDTA corresponds to 0.03312 g of $Pb(NO_3)_2$.

$$\% \, Pb(NO_3)_2 = \frac{(mL \times M \, EDTA) \times 33.12}{Sample \, wt \, (g)}$$

Insoluble Matter. (Page 25). Use 20 g dissolved in 200 mL of dilute nitric acid (1 + 99).

Chloride. (Page 35). Use 1.0 g.

Calcium, Copper, Iron, Potassium, and Sodium. (By flame AAS, page 63).

    Sample Stock Solution. Dissolve 20.0 g of sample with 60 mL of water, and add 10 mL of nitric acid. Transfer to a 100-mL volumetric flask, and dilute to the mark with water (1 mL = 0.2 g).

Element	Wavelength (nm)	Sample Wt (g)	Standard Added (mg)	Flame Type*	Background Correction
Ca	422.7	1.0	0.05; 0.10	N/A	No
Cu	324.7	1.0	0.02; 0.04	A/A	Yes
Fe	248.3	4.0	0.04; 0.08	A/A	Yes
K	766.5	1.0	0.025; 0.05	A/A	No
Na	589.0	0.10	0.01; 0.02	A/A	No

*A/A is air/acetylene; N/A is nitrous oxide/acetylene.

Lead Perchlorate Trihydrate
$Pb(ClO_4)_2 \cdot 3H_2O$ **Formula Wt 460.15**   **CAS No. 13637-76-8**

GENERAL DESCRIPTION

Typical appearance: white solid
Analytical use: titrant for complexometric titrations
Change in state (approximate): melting point, 88 °C
Aqueous solubility: very soluble

SPECIFICATIONS

Assay . . . . 97.0%–102.0% $Pb(ClO_4)_2 \cdot 3H_2O$
pH of a 5% solution at 25.0 °C 3.0–5.0

                                                         Maximum Allowable
Insoluble matter 0.005%
Chloride (Cl) 0.01%
Iron (Fe). 0.001%

TESTS

Assay. (By complexometric titration of lead). While protecting from moisture, weigh accurately about 1.8 g, transfer to a 400-mL beaker, and dissolve in 250 mL of water. Add 15 mL of a saturated hexamethylenetetramine reagent solution and a few milligrams of xylenol orange indicator mixture, and titrate with 0.1 M EDTA volumetric solution to a yellow color. A precipitate that dissolves readily during the titration will form. One milliliter of 0.1 M EDTA corresponds to 0.04602 g of $Pb(ClO_4)_2 \cdot 3H_2O$.

$$\% \ Pb(ClO_4)_2 \cdot 3H_2O = \frac{(mL \times M \ EDTA) \times 46.02}{Sample \ wt \ (g)}$$

pH of a 5% Solution at 25.0 °C. (Page 49).

Insoluble Matter. (Page 25). Use 20 g dissolved in water.

Chloride. (Page 35). Use 1 g in 100 mL, and take a 10-mL aliquot.

Iron. (Page 38, Method 2). Use 1 g, and filter after addition of hydroxylamine hydrochloride reagent solution. Wash in the funnel and use the filtrate.

Lead Subacetate
$Pb(C_2H_3O_2)_2 \cdot 2Pb(OH)_2$ **Formula Wt 807.72**   **CAS No. 1335-32-6**

GENERAL DESCRIPTION
Typical appearance: white solid
Analytical use: analysis of sugars
Aqueous solubility: soluble in 16 parts of cold water with alkaline reaction

SPECIFICATIONS
Basic lead (PbO) . . . . ≥33.0%

Maximum Allowable

Loss on drying 1.5%
Insoluble in dilute acetic acid 0.02%
Insoluble in water 1.0%
Chloride (Cl). 0.003%
Nitrate (NO₃) 0.003%
Calcium (Ca) 0.01%
Copper (Cu) 0.002%
Iron (Fe). 0.002%
Potassium (K) 0.02%
Sodium (Na). 0.05%

TESTS

Basic Lead. (By indirect acidimetry). Weigh accurately about 5 g of sample, and dissolve in 100 mL of carbon dioxide-free water in a 500-mL volumetric flask. Add 50.0 mL of 1 N acetic acid volumetric solution and 100 mL of carbon dioxide-free 3% solution of sodium oxalate. Mix thoroughly, dilute to volume with carbon dioxide-free water, and allow the precipitate to settle. Decant and titrate 100.0 mL of the clear supernatant liquid with 1 N sodium hydroxide volumetric solution, using 0.15 mL of phenolphthalein indicator solution. Each milliliter of 1 N acetic acid consumed is equivalent to 0.1116 g of PbO.

$$\% \text{ PbO} = \frac{[(\text{mL} \times \text{N CH}_3\text{COOH}) - (\text{mL} \times 5 \times \text{N NaOH})] \times 11.16}{\text{Sample wt (g)}}$$

Loss on Drying. Weigh accurately about 0.5 g in a tared, preconditioned dish or crucible, and dry at 105 °C for 2 h. Cool, and reweigh.

Insoluble in Dilute Acetic Acid. Dissolve 5.0 g in 100 mL of dilute acetic acid (1 + 19), warming if necessary to effect complete dissolution. Filter through a tared, preconditioned filtering crucible, wash with dilute acetic acid (1 + 19) until the washings are no longer darkened by hydrogen sulfide, and dry at 105 °C.

Insoluble in Water. Agitate 1.0 g, just until dissolved, in a small stoppered flask with 50 mL of carbon dioxide-free water, and filter at once through a tared, preconditioned filtering crucible. Wash with carbon dioxide-free water, and dry at 105 °C.

Chloride. (Page 35). Dissolve 1.0 g in 30 mL of dilute nitric acid (1 + 9). Use 10 mL of this solution.

Nitrate. (Page 38, Method 1).

Sample Solution A. Suspend 0.50 g in 3 mL of water. Dilute to 50 mL with brucine sulfate reagent solution.

Control Solution B. Suspend 0.50 g in 1.5 mL of water and 1.5 mL of nitrate ion (NO_3) standard solution. Dilute to 50 mL with brucine sulfate reagent solution.

Blank Solution C. Use 50 mL of brucine sulfate reagent solution.

Heat sample solution A, control solution B, and blank solution C in a preheated (boiling) water bath until the lead subacetate in A and B is dissolved, then heat for 10 min more. Cool rapidly in an ice bath to room temperature, centrifuge sample solution A and control solution B for 10 min, and decant the liquids into 200-mm test tubes. Set a spectrophotometer at 410 nm, and using 1-cm cells, adjust the instrument to read zero absorbance with blank solution C in the light path, then determine the absorbance of sample solution A. Adjust the instrument to read 0 absorbance with sample solution A in the light path and determine the absorbance of control solution B. The absorbance of sample solution A should not exceed that of control solution B.

Calcium, Copper, Iron, Potassium, and Sodium. (By flame AAS, page 63).

Sample Stock Solution. Dissolve 20.0 g of sample in a mixture of 50 mL of water and 10 mL of nitric acid, warming to achieve complete dissolution. Cool to room temperature, transfer to a 100-mL volumetric flask, and dilute to the mark with water (1 mL = 0.2 g).

Element	Wavelength (nm)	Sample Wt (g)	Standard Added (mg)	Flame Type*	Background Correction
Ca	422.7	1.0	0.05; 0.10	N/A	No
Cu	324.8	1.0	0.02; 0.04	A/A	Yes
Fe	248.3	4.0	0.08; 0.16	A/A	Yes
K	766.5	0.20	0.02; 0.04	A/A	No
Na	589.0	0.04	0.01; 0.02	A/A	No

*A/A is air/acetylene; N/A is nitrous oxide/acetylene.

Lithium Carbonate

Li_2CO_3 **Formula Wt 73.89**   **CAS No. 554-13-2**

Note: The formula weight of this reagent is likely to deviate from the value cited, since the natural distribution of ^6Li and ^7Li isotopes is often altered in current sources of lithium compounds.

GENERAL DESCRIPTION

Typical appearance: white solid
Analytical use: matrix modifier
Change in state (approximate): melting point, 720 °C
Aqueous solubility: 0.7 g in 100 mL at 100 °C

SPECIFICATIONS

Assay . . . . ≥99.0% Li_2CO_3

Maximum Allowable

Insoluble in dilute hydrochloric acid. 0.01%
Chloride (Cl). 0.005%
Nitrate (NO_3) . . . . 5 ppm
Sulfur compounds (as SO_4) 0.2%
Heavy metals (as Pb). 0.002%
Iron (Fe). 0.002%
Calcium (Ca) 0.01%
Potassium (K) 0.01%
Sodium (Na). 0.1%

TESTS

Assay. (By indirect acid–base titrimetry for carbonate). Weigh accurately about 1.3 g of sample, add 50 mL of water, and then add cautiously 50.0 mL of 1 N hydrochloric acid volumetric solution. Boil gently to expel carbon dioxide, cool, add 0.15 mL of methyl red indicator solution, and titrate the excess hydrochloric acid with 1 N sodium hydroxide volumetric solution. One milliliter of 1 N hydrochloric acid corresponds to 0.03694 g of Li_2CO_3.

$$\% \, Li_2CO_3 = \frac{[(50.0 \times N \, HCl) - (mL \times N \, NaOH)] \times 3.694}{Sample \, wt \, (g)}$$

Insoluble in Dilute Hydrochloric Acid. Suspend 10 g in 75 mL of water, then add cautiously and slowly 25 mL of hydrochloric acid. Heat to boiling, and digest in a covered beaker on a hot plate (≈100 °C) for 1 h. Filter through a tared, preconditioned filtering crucible, wash thoroughly with hot water, and dry at 105 °C.

Chloride. (Page 35). Dissolve 2.0 g in 25 mL of dilute nitric acid (1 + 5), dilute with water to 50 mL, and use 5.0 mL of this solution.

Nitrate. (Page 38, Method 1).

Sample Solution A. Add 2.0 g to 4 mL of water in a test tube. Add brucine sulfate reagent solution, dropwise, until the sample is completely dissolved. Then dilute to 50 mL with the same reagent solution, and mix.

Control Solution B. Add 2.0 g to 3 mL of water and 1 mL of nitrate ion (NO_3) standard solution. Add brucine sulfate reagent solution, dropwise, until the sample is completely dissolved. Then dilute to 50 mL with the same reagent solution, and mix.

Continue with the procedure, starting with the preparation of blank solution C.

Sulfur Compounds. Suspend 0.10 g in 10 mL of water, and cautiously neutralize with dilute hydrochloric acid (1 + 9). Add 0.10 mL of bromine water, and boil gently until the bromine is completely expelled. Add 2 mL of dilute hydrochloric acid (1 + 19), filter, and dilute with water to 40 mL. To 10 mL, add 1 mL of 12% barium chloride reagent solution. Any turbidity should not exceed that produced by 0.05 mg of sulfate ion (SO_4) in an equal volume of solution containing 1 mL of dilute hydrochloric acid (1 + 19) and 1 mL of 12% barium chloride reagent solution. Compare 10 min after adding the barium chloride to the sample and standard solutions.

Heavy Metals. (Page 36, Method 1). Suspend 1.5 g in 10 mL of water, and cautiously add 10 mL of hydrochloric acid. Evaporate the solution to a moist residue on a hot plate (≈100 °C), dissolve the residue in 20 mL of water, and dilute with water to 25 mL. For the control, suspend 0.5 g in 10 mL of water, add 0.02 mg of lead, and treat exactly as the 1.5 g of sample.

Iron. (Page 38, Method 1). Suspend 0.5 g in 30 mL of water, cautiously add 3 mL of hydrochloric acid, and dilute with water to 50 mL. Use this solution without further acidification. Use only 2 mL of hydrochloric acid for the control solution.

Calcium, Potassium, and Sodium. (By flame AAS, page 63).

Sample Stock Solution. Suspend 5 g of sample in 50 mL of water, and cautiously add 25 mL of hydrochloric acid. Transfer the solution to a 100-mL volumetric flask, and dilute to the mark with water (1 mL = 0.05 g).

Element	Wavelength (nm)	Sample Wt (g)	Standard Added (mg)	Flame Type*	Background Correction
Ca	422.7	0.10	0.02; 0.04	N/A	No
K	766.5	1.0	0.05; 0.10	A/A	No
Na	589.0	0.01	0.01; 0.02	A/A	No

*A/A is air/acetylene; N/A is nitrous oxide/acetylene.

Lithium Chloride

LiCl	Formula Wt 42.39	CAS No. 7447-41-8

Note: The formula weight of this reagent is likely to deviate from the value cited, since the natural distribution of ^6Li and ^7Li isotopes is often altered in current sources of lithium compounds.

GENERAL DESCRIPTION

Typical appearance: white solid
Analytical use: matrix modifier
Change in state (approximate): melting point, 613 °C
Aqueous solubility: 81 g in 100 mL at 20 °C

SPECIFICATIONS

Assay . . . . ≥99% LiCl

Maximum Allowable

Insoluble matter 0.01%
Titrable base . . . . 0.008 meq/g
Loss on drying 1.0%
Nitrate (NO_3) 0.001%
Sulfate (SO_4) 0.01%
Barium (Ba) 0.003%
Heavy metals (as Pb). 0.002%
Iron (Fe). 0.001%
Calcium (Ca) 0.01%
Potassium (K) 0.01%
Sodium (Na). 0.20%

TESTS

Assay. (By indirect argentimetric titration of chloride). Weigh accurately about 0.15 g, dissolve in about 100 mL of water, and add 1 mL of dichlorofluorescein indicator solution. Titrate with 0.1 N silver nitrate volumetric solution to a pink end point. One milliliter of 0.1 N silver nitrate corresponds to 0.004239 g of LiCl.

$$\% \text{ LiCl} = \frac{(\text{mL} \times \text{N AgNO}_3) \times 4.239}{\text{Sample wt (g)}}$$

Insoluble Matter. (Page 25). Use 10.0 g dissolved in 150 mL of water.

Titrable Base. Transfer 10.0 g to a 150-mL beaker. Dissolve in 100 mL of carbon dioxide-free water, add 0.15 mL of methyl red indicator solution, and titrate with 0.01 N hydrochloric acid. Correct for a blank determination. Not more than 8.0 mL of 0.01 N hydrochloric acid should be required to produce a pink color.

Loss on Drying. Weigh accurately 1.0 g in a tared, preconditioned dish or crucible, and dry for 2 h at 105 °C.

Nitrate. Mix 1.0 g with 5 mL of phenoldisulfonic acid reagent solution, and cautiously dilute with water to 25 mL. Cautiously add ammonium hydroxide until the maximum yellow color develops. Any yellow color should not exceed that produced in a standard, treated similarly, containing 0.01 mg of added nitrate ion (NO_3).

Sulfate. (Page 40, Method 1). Use 0.5 g.

Barium. Dissolve 3.3 g in 25 mL of water, filter if necessary, and add 2 mL of potassium dichromate reagent solution. Add ammonium hydroxide until the orange color dissipates and the yellow color persists. Any turbidity produced within 15 min should not exceed that in a standard containing 0.10 mg of added barium.

Heavy Metals. (Page 36, Method 1). Dissolve 2.0 g in 40 mL of water. Use 30 mL to prepare the sample solution, and use the remaining 10 mL to prepare the control solution.

Iron. (Page 38, Method 1). Use 1.0 g.

Calcium, Potassium, and Sodium. (By flame AAS, page 63).

> **Sample Stock Solution A.** Dissolve 10.0 g of sample with water in a 100-mL volumetric flask, and dilute to the mark with water (1 mL = 0.1 g).

> **Sample Stock Solution B.** Pipet 10.0 mL of sample stock solution A into a 100-mL volumetric flask, and dilute to the mark with water (1 mL = 0.01 g).

Element	Wavelength (nm)	Sample Wt (g)	Standard Added (mg)	Flame Type*	Background Correction
Ca	422.7	0.50	0.05; 0.10	N/A	No
K	766.5	0.50	0.05; 0.10	A/A	No
Na	589.0	0.005	0.01; 0.02	A/A	No

*A/A is air/acetylene; N/A is nitrous oxide/acetylene.

Lithium Hydroxide Monohydrate
LiOH · H₂O   **Formula Wt 41.96**   **CAS No. 1310-66-3**

Note: The formula weight of this reagent is likely to deviate from the value cited, since the natural distribution of ^6Li and ^7Li is often altered in current sources of lithium compounds.

GENERAL DESCRIPTION
Typical appearance: white solid
Analytical use: catalyst for alkyd resins
Aqueous solubility: 26.8 g in 100 mL at 80 °C

SPECIFICATIONS
Assay . . . . ≥98.0% LiOH · H₂O

	Maximum Allowable
Lithium carbonate.	2.0%
Insoluble matter	0.01%
Chloride (Cl).	0.01%
Sulfate (SO_4)	0.05%
Heavy metals (as Pb).	0.002%
Iron (Fe).	0.002%

TESTS

Assay and Lithium Carbonate. (By acidimetry). Weigh accurately about 1.6 g, and dissolve in 50 mL of carbon dioxide-free water. Add 5 mL of 12% barium chloride reagent solution, stopper, and allow to stand for five minutes. Add 0.15 mL of phenolphthalein indicator solution, and titrate with 1 N hydrochloric acid volumetric solution just to a colorless end point (A mL). Add 0.15 mL of bromphenol blue indicator solution, and titrate with 0.1 N hydrochloric acid to a yellow end point (B mL). Calculate the assay using (A − 0.1 B) mL. One milliliter of 1 N hydrochloric acid corresponds to 0.04196 g of LiOH · H_2O.

$$\% \text{ LiOH} \cdot \text{H}_2\text{O} = \frac{[(A - 0.1\ B)\ \text{mL} \times \text{N HCl})] \times 4.196}{\text{Sample wt (g)}}$$

Calculate the carbonate content using B mL. One milliliter of 0.1 N hydrochloric acid corresponds to 0.007389 g of Li_2CO_3.

$$\% \text{ Li}_2\text{CO}_3 = \frac{(B\ \text{mL} \times \text{N HCl}) \times 0.7389}{\text{Sample wt (g)}}$$

Insoluble Matter. (Page 25). Use 10.0 g in 100 mL of water. Do not boil the sample.

Chloride. (Page 35). Use 0.1 g.

Sulfate. (Page 40, Method 1). Use 0.10 g, and add hydrochloric acid (1 + 1) until just acid. Then proceed as directed in the general procedure. For the standard, add the same amount of hydrochloric acid (1 + 1) as used for the sample.

Heavy Metals. (Page 36, Method 1). Use 1.0 g dissolved in 10 mL of water. Neutralize with hydrochloric acid (1 + 1), and dilute to 25 mL.

Iron. (Page 38, Method 1). Use 0.5 g in 10 mL of water, add 3 mL of hydrochloric acid, and dilute to 40 mL.

Lithium Metaborate

$LiBO_2$ **Formula Wt 49.75**   **CAS No. 13453-69-5**

Note: The formula weight of this reagent is likely to deviate from the value cited, since the natural distribution of ^{6}Li and ^{7}Li isotopes is often altered in current sources of lithium compounds.

GENERAL DESCRIPTION

Typical appearance: white solid

Analytical use: flux

Change in state (approximate): melting point, 845 °C

Aqueous solubility: moderate (1–10%)

SPECIFICATIONS

Assay . . . . 98.0–102.0% $LiBO_2$

Bulk density. . . . . ≥0.25 g/mL

	Maximum Allowable
Insoluble matter	0.01%
Loss on fusion at 950 °C	2.0%
Phosphorus compounds (as PO_4)	0.004%
Silicon (Si)	0.01%
Aluminum (Al)	0.001%
Calcium (Ca)	0.01%
Iron (Fe).	0.002%
Magnesium (Mg).	5 ppm
Potassium (K)	0.005%
Sodium (Na)	0.005%
Heavy metals (as Pb)	0.001%

TESTS

Assay. (By acidimetry). Weigh accurately about 1 g, dissolve in 80 mL of water in a 125-mL conical flask, and heat to boiling. Cool to room temperature, add 0.15 mL of 0.1% bromphenol blue indicator solution, and titrate with 1 N hydrochloric acid volumetric solution until the first appearance of yellow.

$$\% \ LiBO_2 = \frac{(mL \times N \ HCl) \times 497.5}{Sample \ wt \ (g) \times (100 - L)}$$

where L = % loss on fusion at 950 °C.

Bulk Density. Weigh a dry 50-mL glass-stoppered cylinder, calibrated to contain, then add approximately 50 mL of sample. Stopper, and reweigh to the nearest 0.1 g. Grasp the cylinder above its base, and from a height of 1 inch, lower the base sharply against a No. 12 rubber stopper. Level the surface of the powder with a gentle back and forth motion of the cylinder in mid-air. Record the volume. Repeat two more times, starting with the rotation of the cylinder. Use the mean of the three volume readings to calculate the apparent density.

Insoluble Matter. (Page 25). Use 10 g dissolved in 400 mL of water.

Loss on Fusion at 950 °C. Heat 1 g, accurately weighed, in a preconditioned platinum crucible or dish at 950 °C for 15 min. Cool in a desiccator.

Phosphorus Compounds. Dissolve 5.0 g in 40 mL of dilute nitric acid (1 + 1), and evaporate on a steam bath to a syrupy residue. Dissolve the residue in about 80 mL of water, dilute with water to 100 mL, and dilute 20 mL of this solution with water to 100 mL. For the test, dilute 10 mL (0.1-g sample) with water to 70 mL. Prepare a standard containing 0.01 mg of phosphate ion (PO_4) in 70 mL of water. To each solution add 5 mL of ammonium molybdate reagent solution, and adjust the pH to 1.8 (using a pH meter) by adding dilute hydrochloric acid (1 + 1) or dilute ammonium hydroxide (1 + 1). Cautiously heat the solutions to near boiling but do not boil, and cool to room temperature. If a precipitate forms, it will dissolve when the solution is acidified in the next steps. To each solution, add 10 mL of hydrochloric acid, and dilute each with water to 100 mL. Transfer the solutions to separatory funnels, add 35 mL of ethyl ether to each, shake vigorously, and allow to separate. Draw off and discard the aqueous phases. Wash the ether phases twice with 10-mL portions of dilute hydrochloric acid (1 + 9), and discard the washings each time. To the washed ether phases, add 10 mL of dilute hydrochloric acid (1 + 9), to which has been added 0.2 mL of a freshly prepared 2% stannous chloride reagent solution. Shake the solutions, and allow the phases to separate. Any blue color in the ether phase from the solution of the sample should not exceed that in the ether phase from the standard.

Silicon. Fuse 1.00 g at 950 °C for 15 min in a platinum crucible. Remove the crucible from the furnace, swirl to spread the melt around the sides of the crucible, and cool. Add a small magnetic stirring bar to the crucible, stir while adding 6.0 mL of 3 N hydrochloric acid, and stir until the melt disintegrates. Transfer the slurry with 50 mL of water to a 150-mL beaker. Adjust the pH to 2.0 ± 0.2 (using a pH meter) with 0.06 N hydrochloric acid (or dilute ammonium hydroxide). The solution should be clear at the final pH adjustment. Add 2.0 mL of ammonium molybdate reagent solution, mix, and let stand for 10 min. Add 4.0 mL of tartaric acid reagent solution, then 1.0 mL of reducing solution (described below). Mix the solution after each addition. Finally, dilute the solution with water to 100 mL. After 30 min, the blue color should not exceed that produced by 0.21 mg of silica in an equal volume of solution containing 10.0 mL of 0.06 N hydrochloric acid, 50 mL of water, and the quantities of reagents used in the test. If desired, the absorbances can be measured with a spectrophotometer at 650 nm, using water as a reference.

> **Reducing Solution.** Dissolve 0.7 g of sodium sulfite in 20 mL of water, then dissolve in this solution 0.15 g of 4-amino-3-hydroxy-1-naphthalene sulfonic acid. Add with constant stirring a solution of 9 g of sodium bisulfite in 80 mL of water. Filter the resulting solution into a plastic bottle. Discard after 3 days.

Aluminum. Add 25 mL of methanol and 1.8 mL of hydrochloric acid to 1.0 g in a 100-mL plastic beaker. Evaporate to dryness on a hot plate ($\approx$100 °C), add 15 mL of methanol and 0.3 mL of hydrochloric acid, and evaporate again to dryness. Add another 15 mL of methanol and 0.3 mL of hydrochloric acid, repeat the evaporation, and dissolve the residue in 15.0 mL of 0.06 N hydrochloric acid. Transfer the solution of the sample to a 100-mL volumetric flask with 45 mL of water. For the blank and the standard, respectively, add 15.0 mL of 0.06 N hydrochloric acid to two 100-mL volumetric flasks containing 0 and 10 μg of aluminum, then add 45 mL of water. To each of the three flasks, add 5.0 mL of lithium chloride reagent solution, 1.0 mL of hydroxylamine hydrochloride reagent solution, and 4.0 mL of ascorbic acid reagent solution. Mix, and let stand for 5 min. Add 10.0 mL of sodium acetate–acetic acid buffer solution (described below), mix, and let stand for 10 min. Add 5.0 mL of alizarin red S reagent solution, and dilute with water to 100 mL. The final pII should be 4.6 ± 0.1 (using a pH meter). After 1 h, measure the absorbance of the sample and the standard at 500 nm in 5-cm cells with the blank as the reference. The absorbance of the sample should not exceed that of the standard.

Sodium Acetate–Acetic Acid Buffer. Dissolve 100 g of sodium acetate trihydrate in water. Add 30 mL of glacial acetic acid, and dilute with water to 500 mL. Filter if necessary.

Calcium, Iron, Magnesium, Potassium, and Sodium. (By flame AAS, page 63).

Sample Stock Solution and Blank Solution. Weigh 10.0 g of sample into a 2-L beaker, add 400 mL of methanol and 5 mL of hydrochloric acid, and evaporate to dryness. Repeat the evaporation, using 200 mL of methanol and 2 mL of hydrochloric acid, then add 100 mL of methanol and 1 mL of hydrochloric acid, and perform a third evaporation. Prepare a blank in the same manner, using the quantities of reagents used for preparation of the sample. Dissolve each residue in 50 mL of water and 1 mL of hydrochloric acid, transfer to a 100-mL volumetric flask, and dilute to the mark with water (1 mL = 0.1 g).

Element	Wavelength (nm)	Sample Wt (g)	Standard Added (mg)	Flame Type*	Background Correction
Ca	422.7	0.20	0.02; 0.04	N/A	No
Fe	248.3	1.0	0.02; 0.04	A/A	Yes
Mg	285.2	1.0	0.005; 0.01	A/A	Yes
K	766.5	1.0	0.05; 0.10	A/A	No
Na	589.0	0.10	0.005; 0.01	A/A	No

*A/A is air/acetylene; N/A is nitrous oxide/acetylene.

Heavy Metals. Dilute 20 mL of glacial acetic acid with water to 80 mL, add 5.0 g of sample while stirring, and heat to about 80 °C to dissolve. Prepare a control with 1.0 g of sample, 5.0 mL of glacial acetic acid, and 0.04 mg of lead in a volume of 80 mL. The pH of the solutions should be between 3 and 4. Keep the solutions at 80 °C, add 10 mL of freshly prepared hydrogen sulfide water to each, and mix. Transfer to 100-mL Nessler tubes. Any color in the solution of the sample should not exceed that in the control. (If the solutions are permitted to cool, a white crystalline precipitate forms.)

Lithium Perchlorate

LiClO$_4$ (anhydrous)	**Formula Wt 106.39**	**CAS No. 7791-03-9**
LiClO$_4 \cdot$ 3H$_2$O (trihydrate)	**Formula Wt 160.45**	**CAS No. 13453-78-6**

Note: This specification includes the anhydrous form and the trihydrate form. The formula weight of this reagent is likely to deviate from the value cited, since the natural distribution of ^{6}Li and ^{7}Li isotopes is often altered in current sources of lithium compounds.

GENERAL DESCRIPTION

Typical appearance: solid
Analytical use: oxidizing agent
Change in state (approximate): melting point, 236 °C

SPECIFICATIONS

Assay . . . . ≥95.0% $LiClO_4$ (anhydrous)
63.0–68.0% $LiClO_4 \cdot 3H_2O$ (trihydrate)
pH of a 5% solution at 25.0 °C 6.0–7.5

Maximum Allowable

Insoluble matter 0.005%
Chloride (Cl). 0.003%
Sulfate (SO_4) 0.001%
Heavy metals (as Pb). . . . . .5 ppm
Iron (Fe). . . . . .5 ppm

TESTS

Assay. (By argentimetric titration of chloride after reduction). Weigh accurately about 0.4 g (anhydrous) or 0.6 g (trihydrate) into a platinum dish or crucible. Add 3 g of sodium carbonate. Heat gently at first, then to fusion, until a clear melt is obtained. Leach with minimal 10% nitric acid, and boil gently to expel carbon dioxide. Cool to room temperature, add 10 mL of 30% ammonium acetate solution, and titrate with 0.1 N silver nitrate volumetric solution, using dichlorofluorescein indicator solution to a pink end point. One milliliter of 0.1 N silver nitrate corresponds to 0.01064 g of $LiClO_4$, and 0.01604 g of $LiClO_4 \cdot 3H_2O$.

$$\% \ LiClO_4 = \frac{(mL \times N \ AgNO_3) \times 10.64}{Sample \ wt \ (g)}$$

$$\% \ LiClO_4 \cdot 3H_2O = \frac{(mL \times N \ AgNO_3) \times 16.04}{Sample \ wt \ (g)}$$

pH of a 5% Solution at 25.0 °C. (Page 49).

Insoluble Matter. (Page 25). Use 20 g dissolved in 200 mL of water.

Chloride. (Page 35). Use 0.33 g.

Sulfate. Dissolve 40 g in 300 mL of water, add 2 mL of hydrochloric acid, filter, and heat to boiling. Add 5 mL of 12% barium chloride reagent solution, digest in a covered beaker on a hot plate (≈100 °C) for 2 h, and allow to stand for at least 8 h. If a precipitate is formed, filter through a fine ashless paper, wash thoroughly, and ignite. Correct for the weight obtained in a complete blank test.

$$\% \ SO_4 = \frac{[Residue \ wt \ (sample) \ (g) - Residue \ wt \ (blank) \ (g)] \times 100}{Sample \ wt \ (g)}$$

Heavy Metals. (Page 36, Method 1). Dissolve 6.0 g in about 20 mL of water and dilute with water to 30 mL. Use 25 mL to prepare the sample solution, and use the remaining 5.0 mL to prepare the control solution.

Iron. Dissolve 1.0 g in 20 mL of water. Add 1 mL of hydroxylamine hydrochloride reagent solution, 4 mL of 1,10-phenanthroline reagent solution, and 1 mL of 10% sodium acetate reagent solution, and mix. Any red color should not exceed that produced by 0.005 mg of iron ion (Fe) in an equal volume of solution containing the quantities of reagents used in the test. Compare 1 h after adding the reagents to the sample and standard solutions.

Lithium Sulfate Monohydrate

$Li_2SO_4 \cdot H_2O$ **Formula Wt 127.96**   **CAS No. 10102-25-7**

Note: The formula weight of this reagent is likely to deviate from the value cited, since the natural distribution of ^6Li and ^7Li isotopes is often altered in current sources of lithium compounds.

GENERAL DESCRIPTION

Typical appearance: white solid
Analytical use: preparation of lithium standard solutions
Change in state (approximate): dehydrates at 130 °C; anhydrous melts at 860 °C
Aqueous solubility: 35 g in 100 mL at 25 °C

SPECIFICATIONS

Assay (dried basis). . . . . ≥99.0% Li_2SO_4
Loss on drying 13.0–15.0%

	Maximum Allowable
Insoluble matter	0.01%
Chloride (Cl)	0.002%
Nitrate (NO_3)	0.001%
Heavy metals (as Pb)	0.001%
Iron (Fe).	0.001%
Potassium (K)	0.05%
Sodium (Na)	0.05%

TESTS

Assay. (By indirect acid–base titrimetry). Weigh, to the nearest 0.1 mg, 0.3 g of previously dried sample (*see* loss on drying test), and dissolve in 50 mL of water. Pass the solution through a cation-exchange column (see page 29) at the rate of about 5 mL/min, and collect the eluate in a 500-mL titration flask. Then wash the resin in the column with water at a rate of about 10 mL/min, and collect in the same titration flask. Add 0.15 mL of phenolphthalein indicator solution to the flask, and titrate with 0.1 N sodium hydroxide. Continue the titration, as the washing proceeds, until 50 mL of eluate requires no further titration. One milliliter of 0.1 N sodium hydroxide corresponds to 0.06398 g of $Li_2SO_4 \cdot H_2O$.

$$\% \, Li_2SO_4 \cdot H_2O = \frac{(mL \times N \, NaOH) \times 63.98}{Sample \, wt \, (g)}$$

Loss on Drying. Weigh accurately about 1 g, and dry at 150 °C to constant weight. (The theoretical loss for $Li_2SO_4 \cdot H_2O$ is 14.08%.) Save the dried sample for the assay.

Insoluble Matter. (Page 25). Use 10 g dissolved in 100 mL of water.

Chloride. (Page 35). Use 0.50 g.

Nitrate. (Page 38, Method 1).

> *Sample Solution A.* Dissolve 1.0 g in 5 mL of water by heating in a boiling-water bath. Dilute to 50 mL with brucine sulfate reagent solution.

> *Control Solution B.* Dissolve 1.0 g in 4 mL of water and 1 mL of nitrate ion (NO_3) standard solution by heating in a boiling-water bath. Dilute to 50 mL with brucine sulfate reagent solution.

Continue with the procedure, starting with the preparation of blank solution C.

Heavy Metals. (Page 36, Method 1). Dissolve 4.0 g in 40 mL of water. Use 30 mL to prepare the sample solution, and use the remaining 10 mL to prepare the control solution.

Iron. (Page 38, Method 1). Use 1.0 g.

Potassium and Sodium. (By flame AAS, page 63).

> *Sample Stock Solution.* Dissolve 1.0 g of sample with water in a 100-mL volumetric flask, and dilute to the mark with water (1 mL = 0.01 g).

Element	Wavelength (nm)	Sample Wt (g)	Standard Added (mg)	Flame Type*	Background Correction
K	766.5	0.10	0.05; 0.10	A/A	No
Na	589.0	0.02	0.01; 0.02	A/A	No

*A/A is air/acetylene.

Lithium Tetraborate

$Li_2B_4O_7$ **Formula Wt 169.12**   **CAS No. 12007-60-2**

Suitable for use in X-ray spectroscopy.

Note: The formula weight of this reagent is likely to deviate from the value cited, since the natural distribution of ^6Li and ^7Li isotopes is often altered in current sources of lithium compounds.

GENERAL DESCRIPTION

Typical appearance: white solid
Analytical use: flux
Aqueous solubility: sparingly soluble

SPECIFICATIONS

Assay . . . . .98.0–102.0% $Li_2B_4O_7$
Bulk density. . . . . ≥0.25 g/mL

Maximum Allowable

Insoluble matter 0.01%
Loss on fusion at 950 °C 2.0%
Phosphorus compounds (as PO_4) 0.004%
Silicon (Si) 0.01%
Aluminum (Al) 0.001%
Calcium (Ca) 0.01%
Iron (Fe). 0.001%
Magnesium (Mg). . . . . 5 ppm
Potassium (K) 0.005%
Sodium (Na) 0.005%
Heavy metals (as Pb) 0.001%

TESTS

Assay. (By acidimetry). Weigh accurately about 1 g, dissolve in 80 mL of water in a 125-mL conical flask, and heat to boiling. Cool to room temperature, add 0.15 mL of 0.1% bromphenol blue indicator solution, and titrate with 1 N hydrochloric acid volumetric solution until the first appearance of yellow. One milliliter of 1 N hydrochloric acid corresponds to 0.08456 g of $Li_2B_4O_7$.

$$\% \, Li_2B_4O_7 = \frac{(mL \times N \, HCl) \times 845.6}{Sample \, wt \, (g) \times (100 - L)}$$

where L = % loss on fusion at 950 °C.

Bulk Density. Weigh a dry 50-mL glass-stoppered cylinder, calibrated to contain, then add approximately 50 mL of sample. Stopper, and reweigh to the nearest 0.1 g. Grasp the cylinder above its base, and from a height of 1 inch, lower the base sharply against a No. 12 rubber stopper. Level the surface of the powder with a gentle back and forth motion of the cylinder in mid-air. Record the volume. Repeat two more times, starting with the rotation of the cylinder. Use the mean of the three volume readings to calculate the apparent density.

Insoluble Matter. (Page 25). Use 10 g dissolved in 400 mL of water.

Loss on Fusion at 950 °C. Heat 1 g, accurately weighed, in a preconditioned platinum crucible or dish at 950 °C for 15 min.

Phosphorus Compounds. Dissolve 5.0 g in 40 mL of dilute nitric acid (1 + 1), and evaporate on a steam bath to a syrupy residue. Dissolve the residue in about 80 mL of water, dilute with water to 100 mL, and dilute 20 mL of this solution with water to 100 mL. For the test, dilute 10 mL (0.1-g sample) with water to 70 mL. Prepare a standard containing 0.01 mg of phosphate ion (PO_4) in 70 mL of water. To each solution, add 5 mL of ammonium molybdate reagent solution, and adjust the pH to 1.8 (using a pH meter) by adding dilute hydrochloric acid (1 + 1) or dilute ammonium hydroxide (1 + 1). Cautiously heat the solutions to near boiling but do not boil, then cool to room temperature. If a precipitate forms, it will dissolve when the solution is acidified in the next steps. To each solution, add 10 mL of hydrochloric acid, and dilute each with water to 100 mL. Transfer the solutions to separatory funnels, add 35 mL of ethyl ether to each, shake vigorously, and allow to separate. Draw off and discard the aqueous phases. Wash the ether phases twice with 10-mL portions of dilute hydrochloric acid (1 + 9), and discard the washings each time. To the washed ether phases, add 10 mL of dilute hydrochloric acid (1 + 9), to which has been added 0.2 mL of a freshly prepared 2% stannous chloride reagent solution. Shake the solutions, and allow the phases to separate. Any blue color in the ether phase from the solution of the sample should not exceed that in the ether phase from the standard.

Silicon. Fuse 1.00 g at 950 °C for 15 min in a 15-mL platinum crucible. Remove the crucible from the furnace, swirl to spread the melt around the sides of the crucible, and cool. Add a small magnetic stirring bar to the crucible, stir while adding 6.0 mL of 3 N hydrochloric acid, and stir until the melt disintegrates. Transfer the slurry with 50 mL of water to a 150-mL beaker. Adjust the pH to 2.0 ± 0.2 (using a pH meter) with 0.06 N hydrochloric acid (or dilute ammonium hydroxide). The solution should be clear at the final pH adjustment. Add 2.0 mL of ammonium molybdate solution (described below), mix, and let stand for 10 min. Add 4.0 mL of tartaric acid reagent solution, then 1.0 mL of reducing solution (described below). Mix the solution after each addition. Finally, dilute the solution with water to 100 mL. After 30 min, the blue color should not exceed that produced by 0.21 mg of silica in an equal volume of solution containing 10.0 mL of 0.06 N hydrochloric acid, 50 mL of water, and the quantities of reagents used in the test. If desired, the absorbances can be measured with a spectrophotometer at 650 nm, using water as a reference.

> ***Ammonium Molybdate Solution.*** Dissolve 7.5 g of ammonium molybdate tetrahydrate in 75 mL of water. Add 10.0 mL of 18 N sulfuric acid, dilute with water to 100 mL, filter, and store in a plastic bottle.

> ***Reducing Solution.*** Dissolve 0.7 g of sodium sulfite in 20 mL of water, then dissolve into this solution 0.15 g of 4-amino-3-hydroxy-1-naphthalene sulfonic acid. Add with constant stirring a solution of 9 g of sodium bisulfite in 80 mL of water. Filter the resulting solution into a plastic bottle. Discard after 3 days.

Aluminum. Add 25 mL of methanol and 1.8 mL of hydrochloric acid to 1.0 g in a 100-mL plastic beaker. Evaporate to dryness on a hot plate (≈100 °C), add 15 mL of methanol and 0.3 mL of hydrochloric acid, and evaporate again to dryness. Add another 15 mL of methanol and 0.3 mL of hydrochloric acid, repeat the evaporation, and dissolve the residue in 15.0 mL of 0.06 N hydrochloric acid. Transfer the solution of the sample to a 100-mL volumetric flask with 45 mL of water. For the blank and the standard, respectively, add 15.0 mL of 0.06 N hydrochloric acid to two 100-mL volumetric flasks containing 0 and 10 μg of aluminum, then add 45 mL of water. To each of the three flasks add 5.0 g of lithium chloride reagent solution, 1.0 mL of hydroxylamine hydrochloride reagent solution, and 4.0 mL of ascorbic acid reagent solution. Mix and let stand for 5 min. Add 10.0 mL of sodium acetate–acetic acid buffer solution (described below), mix, and let stand for 10 min. Add 5.0 mL of alizarin red S reagent solution, and dilute with water to 100 mL. The final pH should be 4.6 ± 0.1 (using a pH meter). After 1 h, measure the absorbance of the sample and the standard at 500 nm in 5-cm cells with the blank as the reference. The absorbance of the sample should not exceed that of the standard.

> **Sodium Acetate–Acetic Acid Buffer Solution.** Dissolve 100 g of sodium acetate trihydrate in water. Add 30 mL of glacial acetic acid, and dilute with water to 500 mL. Filter if necessary.

Calcium, Iron, Magnesium, Potassium, and Sodium. (By flame AAS, page 63).

> **Sample Stock Solution and Blank Solution.** Weigh 10 g of sample into a 2-L beaker, add 400 mL of methanol and 5 mL of hydrochloric acid, and evaporate to dryness. Repeat the evaporation using 200 mL of methanol and 2 mL of hydrochloric acid, then add 100 mL of methanol and 1 mL of hydrochloric acid, and perform a third evaporation. Prepare a blank in the same manner, using the quantities of reagents used for preparation of the sample. Dissolve each residue in 50 mL of water and 1 mL of hydrochloric acid, transfer to a 100-mL volumetric flask, and dilute to the mark with water (1 mL = 0.1 g).

Element	Wavelength (nm)	Sample Wt (g)	Standard Added (mg)	Flame Type*	Background Correction
Ca	422.7	0.20	0.02; 0.04	N/A	No
Fe	248.3	1.0	0.01; 0.02	A/A	Yes
Mg	285.2	1.0	0.005; 0.01	A/A	Yes
K	766.5	1.0	0.05; 0.10	A/A	No
Na	589.0	0.10	0.005; 0.01	A/A	No

*A/A is air/acetylene; N/A is nitrous oxide/acetylene.

Heavy Metals. Dilute 20 mL of glacial acetic acid with water to 80 mL, add 5.0 g of sample while stirring, and heat to about 80 °C to dissolve. Prepare a control with 1.0 g of sample, 5.0 mL of glacial acetic acid, and 0.04 mg of lead in a volume of 80 mL. The pH of the solutions should be between 3 and 4. Keep the solutions at 80 °C, add 10 mL of freshly prepared hydrogen sulfide water to each, and mix. Transfer to 100-mL Nessler tubes. Any color in the solution of the sample should not exceed that in the control. (If the solutions are permitted to cool, a white crystalline precipitate forms.)

Litmus Paper

Note: When litmus paper is used in a solution, the length taken of a 0.6-cm-wide strip should not exceed 0.5 cm.

GENERAL DESCRIPTION

Analytical use: pH indicator

SPECIFICATIONS

Maximum Allowable

Ash . . . . .0.4 mg per strip (about 3 cm²)
Phosphate (PO₄). . . . . .Passes test
Rosin acids (for blue paper only) . . . . .Passes test
Sensitivity. . . . . .Passes test

TESTS

Ash. Carefully ignite 10 strips in a tared, preconditioned crucible, and finally ignite at 600 ± 25 °C for 15 min.

Phosphate. (Page 40, Method 1). Cut five strips into small pieces, mix with 0.5 g of magnesium nitrate in a porcelain crucible, and ignite. To the ignited residue, add 5 mL of nitric acid, and evaporate to dryness. Dissolve in 25 mL of approximately 0.5 N sulfuric acid, and continue as described.

Rosin Acids. Immerse a strip of blue paper in a solution of 0.10 g of silver nitrate in 50 mL of water. The color of the paper should not change in 30 s.

Sensitivity. Drop the pieces from six strips of paper into 100 mL of test solution in a beaker, and stir continuously. Blue paper should change color in 45 s in 0.0005 N acid. Red paper should change color in 30 s in 0.0005 N alkali.

Magnesium Acetate Tetrahydrate

(CH₃COO)₂Mg · 4H₂O   **Formula Wt 214.45**   **CAS No. 16674-78-5**

GENERAL DESCRIPTION

Typical appearance: colorless, deliquescent solid
Analytical use: buffer; detection of sodium
Change in state (approximate): melting point, 80 °C
Aqueous solubility: very soluble

SPECIFICATIONS

Assay . . . . 98.0–102.0% (CH₃COO)₂Mg · 4H₂O

Maximum Allowable

Insoluble matter 0.005%
Sulfate (SO₄) 0.005%
Barium (Ba) 0.001%
Chloride (Cl). 0.001%
Iron (Fe) . . . . .5 ppm
Calcium (Ca) 0.01%
Manganese (Mn) 0.001%
Potassium (K) 0.005%

Sodium (Na) 0.005%
Strontium (Sr).. 0.005%
Heavy metals (as Pb) .... 5 ppm

TESTS

Assay. (By complexometric titration of Mg). Weigh accurately about 0.85 g, and transfer to a 250-mL beaker with 50 mL of water. Add 5 mL of pH 10 ammoniacal buffer solution and about 50 mg of Eriochrome Black T indicator mixture. Titrate with 0.1 M EDTA volumetric solution to a clear blue color (and disappearance of the last trace of red). One milliliter of 0.1 M EDTA corresponds to 0.02145 g of $(CH_3COO)_2Mg \cdot 4H_2O$.

$$\% \ (CH_3COO)_2Mg \cdot 4H_2O = \frac{(mL \times M \ EDTA) \times 21.45}{Sample \ wt \ (g)}$$

Insoluble Matter. (Page 25). Use 20 g dissolved in 200 mL of dilute acetic acid (1 + 99). Retain the filtrate, without washings, for the test for sulfate.

Sulfate. (Page 40, Method 1). Use 1/20 of the filtrate from the test for insoluble matter. Allow 30 min for turbidity to form.

Barium. Dissolve 6.0 g in 25 mL of dilute nitric acid (1 + 1). For the control, dissolve 1.0 g in 25 mL of dilute nitric acid (1 + 1), and add 0.05 mg of barium. Evaporate each solution on a hot plate ($\approx$100 °C) to a syrupy consistency, and add 20 mL of water. If the solutions are not clear, add 10% nitric acid until clear. Add 1 mL of 1 N acetic acid and 2 mL of potassium dichromate reagent solution to each, and adjust the pH to 7 (using a pH meter) with 10% ammonium hydroxide reagent solution. Add 25 mL of methanol to each, and stir vigorously. Any turbidity in the solution of the sample should not exceed that in the control.

For the Determination of Chloride and Iron

Sample Solution A. Dissolve 10 g in water, filter if necessary through a chloride-free filter, and dilute with water to 100 mL (1 mL = 0.10 g).

Chloride. (Page 35). Use 10 mL of sample solution A (1-g sample).

Iron. (Page 38, Method 1). Use 20 mL of sample solution A (2.0-g sample).

Calcium, Manganese, Potassium, Sodium, and Strontium. (By flame AAS, page 63).

Sample Stock Solution. Dissolve 10.0 g of sample in 75 mL of water, transfer to a 100-mL volumetric flask, and dilute to the mark with water (1 mL = 0.10 g).

Element	Wavelength (nm)	Sample Wt (g)	Standard Added (mg)	Flame Type*	Background Correction
Ca	422.7	0.20	0.02; 0.04	N/A	No
Mn	279.5	2.0	0.02; 0.04	A/A	Yes
K	766.5	1.0	0.05; 0.10	A/A	No
Na	589.0	0.20	0.01; 0.02	A/A	No
Sr	460.7	1.0	0.05; 0.10	N/A	No

*A/A is air/acetylene; N/A is nitrous oxide/acetylene.

Heavy Metals. (Page 36, Method 1). Dissolve 6.0 g in 20 mL of water, and dilute with water to 30 mL. Use 25 mL to prepare the sample solution, and use the remaining 5.0 mL to prepare the control solution.

Magnesium Chloride Hexahydrate

$MgCl_2 \cdot 6H_2O$ **Formula Wt 203.30**   **CAS No. 7791-18-6**

GENERAL DESCRIPTION

Typical appearance: colorless, deliquescent solid

Analytical use: adjustment of ion strength

Change in state (approximate): melting point, 115 °C, with decomposition

Aqueous solubility: 167 g in 100 mL

SPECIFICATIONS

Assay . . . . 99.0–102.0% $MgCl_2 \cdot 6H_2O$

Maximum Allowable

Insoluble matter 0.005%

Nitrate (NO_3) 0.001%

Phosphate (PO_4). . . . . 5 ppm

Sulfate (SO_4) 0.002%

Ammonium (NH_4) 0.002%

Barium (Ba) 0.005%

Calcium (Ca) 0.01%

Manganese (Mn) . . . . 5 ppm

Potassium (K) 0.005%

Sodium (Na). 0.005%

Strontium (Sr) 0.005%

Heavy metals (as Pb). . . . . 5 ppm

Iron (Fe). . . . . 5 ppm

TESTS

Assay. (By complexometric titration of Mg). Weigh accurately about 0.8 g, and transfer to a 250-mL beaker with 50 mL of water. Add 5 mL of pH 10 ammoniacal buffer solution and about 50 mg of Eriochrome Black T indicator mixture. Titrate with 0.1 M EDTA volumetric solution to a clear blue color (and disappearance of the last trace of red). One milliliter of 0.1 M EDTA corresponds to 0.02033 g of $MgCl_2 \cdot 6H_2O$.

$$\% \, MgCl_2 \cdot 6H_2O = \frac{(mL \times M \, EDTA) \times 20.33}{Sample \, wt \, (g)}$$

Insoluble Matter. (Page 25). Use 20 g dissolved in 200 mL of water.

Nitrate. (Page 38, Method 1).

> **Sample Solution A.** Dissolve 1.0 g in 3 mL of water by heating in a boiling-water bath. Dilute to 50 mL with brucine sulfate reagent solution.

> **Control Solution B.** Dissolve 1.0 g in 2 mL of water and 1 mL of nitrate ion (NO_3) standard solution by heating in a boiling-water bath. Dilute to 50 mL with brucine sulfate reagent solution.

Continue with the procedure, starting with the preparation of blank solution C.

Phosphate. (Page 40, Method 1). Dissolve 4.0 g in 25 mL of approximately 0.5 N sulfuric acid, and continue as described.

Sulfate. (Page 40, Method 1). Use 2.5 g of sample.

Ammonium. Dissolve 1.0 g in 90 mL of water and add 10 mL of freshly boiled 10% sodium hydroxide reagent solution. Allow to settle, decant 50 mL, and add 2 mL of Nessler reagent. Any color should not exceed that produced by 0.01 mg of ammonium ion (NH_4) in an equal volume of solution containing 5 mL of 10% sodium hydroxide reagent solution and 2 mL of Nessler reagent.

Barium. Dissolve 1.5 g in 20 mL of water. For the control, dissolve 0.5 g in 15 mL of water, add 0.05 mg of barium ion (Ba), and dilute with water to 20 mL. To each solution, add 1 mL of 1 N acetic acid and 2 mL of potassium dichromate reagent solution. Adjust the pH of the control and sample solutions to 7 (using a pH meter) with 10% ammonium hydroxide reagent solution, add 25 mL of methanol, and stir vigorously. Any turbidity in the solution of the sample should not exceed that in the control.

Calcium, Manganese, Potassium, Sodium, and Strontium. (By flame AAS, page 63).

> **Sample Stock Solution.** Dissolve 10.0 g of sample with water in a 100-mL volumetric flask, and dilute to the mark with water (1 mL = 0.10 g).

Element	Wavelength (nm)	Sample Wt (g)	Standard Added (mg)	Flame Type*	Background Correction
Ca	422.7	0.20	0.02; 0.04	N/A	No
Mn	279.5	2.0	0.01; 0.02	A/A	Yes
K	766.5	1.0	0.05; 0.10	A/A	No
Na	589.0	0.20	0.01; 0.02	A/A	No
Sr	460.7	1.0	0.05; 0.10	N/A	No

*A/A is air/acetylene; N/A is nitrous oxide/acetylene.

Heavy Metals. (Page 36, Method 1). Dissolve 6.0 g in 20 mL of water, and dilute with water to 30 mL. Use 25 mL to prepare the sample solution, and use the remaining 5.0 mL to prepare the control solution.

Iron. (Page 38, Method 1). Use 2.0 g.

Magnesium Nitrate Hexahydrate

$Mg(NO_3)_2 \cdot 6H_2O$ Formula Wt 256.41   CAS No. 13446-18-9

GENERAL DESCRIPTION

Typical appearance: colorless or white solid
Analytical use: preparation of standards
Change in state (approximate): melting point, 89 °C
Aqueous solubility: 125 g in 100 mL at 25 °C

SPECIFICATIONS

Assay . . . . 98.0–102.0% $Mg(NO_3)_2 \cdot 6H_2O$
pH of a 5% solution at 25.0 °C 5.0–8.2

Maximum Allowable

Insoluble matter 0.005%
Chloride (Cl). 0.001%
Phosphate (PO_4). . . . . 5 ppm
Sulfate (SO_4) 0.005%
Ammonium (NH_4) 0.003%
Barium (Ba) 0.005%
Calcium (Ca) 0.01%
Manganese (Mn) . . . . 5 ppm
Potassium (K) 0.005%
Sodium (Na). 0.005%
Strontium (Sr) 0.005%
Heavy metals (as Pb). . . . . 5 ppm
Iron (Fe). . . . . 5 ppm

TESTS

Assay. (By complexometric titration of Mg). Weigh accurately about 1 g, and transfer to a 250-mL beaker with 50 mL of water. Add 5 mL of pH 10 ammoniacal buffer solution and about 50 mg of Eriochrome Black T indicator mixture. Titrate with 0.1 M EDTA volumetric solution to a clear blue color (and disappearance of the last trace of red). One milliliter of 0.1 M EDTA corresponds to 0.02564 g of $Mg(NO_3)_2 \cdot 6H_2O$.

$$\% \ Mg(NO_3)_2 \cdot 6H_2O = \frac{(mL \times M \ EDTA) \times 25.64}{Sample \ wt \ (g)}$$

pH of a 5% Solution at 25.0 °C. (Page 49).

Insoluble Matter. (Page 25). Use 20 g dissolved in 200 mL of water.

Chloride. (Page 35). Use 1.0 g.

Phosphate. (Page 40, Method 1). Dissolve 4.0 g in 25 mL of approximately 0.5 N sulfuric acid, and continue as described.

Sulfate. (Page 41, Method 2). Use 10 mL of dilute hydrochloric acid (1 + 1), do two evaporations, and allow 30 min for turbidity to form.

Ammonium. Dissolve 1.0 g in 100 mL of water, add 10 mL of freshly boiled 10% sodium hydroxide reagent solution, and dilute with water to 150 mL. Allow the precipitate to settle. Filter through sintered-glass or porous porcelain filter. To 50 mL of the filtrate, add 2 mL of Nessler reagent. Any color should not exceed that produced by 0.01 mg of ammonium ion (NH_4) in an equal volume of solution containing the quantities of reagents used in the test.

Barium. Dissolve 1.5 g in 20 mL of water. For the control, dissolve 0.5 g in 15 mL of water, add 0.05 mg of barium ion (Ba), and dilute with water to 20 mL. To each solution, add 1 mL of 1 N acetic acid and 2 mL of potassium dichromate reagent solution. Adjust the pH of the control and sample solutions to 7 (using a pH meter) with 10% ammonium hydroxide reagent solution, add 25 mL of methanol, and stir vigorously. Any turbidity in the solution of the sample should not exceed that in the control.

Calcium, Manganese, Potassium, Sodium, and Strontium. (By flame AAS, page 63).

Sample Stock Solution. Dissolve 10.0 g of sample with water in a 100-mL volumetric flask, and dilute to the mark with water (1 mL = 0.10 g).

Element	Wavelength (nm)	Sample Wt (g)	Standard Added (mg)	Flame Type*	Background Correction
Ca	422.7	0.20	0.02; 0.04	N/A	No
Mn	279.5	2.0	0.01; 0.02	A/A	Yes
K	766.5	1.0	0.05; 0.10	A/A	No
Na	589.0	0.20	0.01; 0.02	A/A	No
Sr	460.7	1.0	0.05; 0.10	A/A	No

*A/A is air/acetylene; N/A is nitrous oxide/acetylene.

Heavy Metals. (Page 36, Method 1). Dissolve 6.0 g in 20 mL of water, and dilute with water to 30 mL. Use 25 mL to prepare the sample solution, and use the remaining 5.0 mL to prepare the control solution.

Iron. (Page 38, Method 1). Use 2.0 g.

Magnesium Oxide

MgO   **Formula Wt 40.30**   **CAS No. 1309-48-4**

GENERAL DESCRIPTION

Typical appearance: white solid

Analytical use: absorbant for colorants prior to determination; preparation of Eschka's reagent

Change in state (approximate): melting point, 2800 °C

Aqueous solubility: practically insoluble

SPECIFICATIONS

Assay (dried basis). . . . . ≥95.0% MgO

Maximum Allowable

Insoluble in dilute hydrochloric acid. 0.02%
Water-soluble substances 0.4%
Loss on ignition 2.0%
Chloride (Cl). 0.01%
Nitrate (NO_3) 0.005%
Sulfate and sulfite (as SO_4) 0.02%
Barium (Ba) 0.005%
Calcium (Ca) 0.05%
Manganese (Mn) . . . . .5 ppm
Potassium (K) 0.005%
Sodium (Na). 0.5%
Strontium (Sr) 0.005%
Heavy metals (as Pb). 0.003%
Iron (Fe). 0.01%

TESTS

Assay. (By complexometric titration of Mg). Weigh accurately about 0.5 g of the sample, previously ignited at 600 ± 25 °C. Dissolve cautiously in water and a slight excess of hydrochloric acid, transfer to a 250-mL volumetric flask, dilute to the mark with water, and mix thoroughly. Pipet 50.0 mL of this solution into a 250-mL beaker, add 50 mL of water and 0.10 mL of methyl red indicator solution, and neutralize with 10% sodium hydroxide reagent solution. Add 15 mL of pH 10 ammoniacal buffer solution, about 0.1 g of ascorbic acid, and a few milligrams of Eriochrome Black T indicator mixture. Titrate with 0.1 M EDTA volumetric solution to a blue end point. One milliliter of 0.1 M EDTA corresponds to 0.004030 g of MgO.

$$\% \text{ MgO} = \frac{(\text{mL} \times \text{M EDTA}) \times 4.03}{\text{Sample wt (g)} / 5}$$

Insoluble in Dilute Hydrochloric Acid. Dissolve 5.0 g in 125 mL of dilute hydrochloric acid (1 + 4), heat to boiling, and boil gently for 5 min. Digest the solution in a covered beaker on a hot plate (≈100 °C) for 1 h. Filter through a tared, preconditioned filtering crucible, wash thoroughly, and dry at 105 °C.

Water-Soluble Substances. Suspend 2.0 g in 50 mL of water, heat to boiling, and filter while hot. Evaporate 25 mL of the clear filtrate to dryness in a tared, preconditioned dish. Moisten the residue with 0.10 mL of sulfuric acid, ignite gently to remove the excess acid, and finally ignite at 600 ± 25 °C for 15 min.

Loss on Ignition. Weigh accurately about 0.25 g in a tared, preconditioned covered platinum crucible. Ignite at 600 ± 25 °C for 15 min. Cool, and reweigh.

Chloride. (Page 35). Use 0.10 g dissolved in 10 mL of dilute nitric acid (1 + 9).

Nitrate. (Page 38, Method 1).

Sample Solution A. Suspend 0.20 g in 3 mL of water. Dilute to 50 mL with brucine sulfate reagent solution.

Control Solution B. Suspend 0.20 g in 2 mL of water and 1 mL of nitrate ion (NO_3) standard solution. Dilute to 50 mL with brucine sulfate reagent solution.

Blank Solution C. Use 50 mL of brucine sulfate reagent solution.

Heat sample solution A, control solution B, and blank solution C in a preheated (boiling) water bath until the magnesium oxide is dissolved in A and B, then heat for 10 min more. Continue with the procedure as described.

Sulfate and Sulfite. Dissolve 0.50 g in (1 + 1) hydrochloric acid, and evaporate to dryness on a hot plate ($\approx$100 °C). Cool, dissolve the residue in 5 mL of water, and evaporate to dryness. Cool, and dissolve the residue in 20 mL of water. Filter through a small, washed filter paper. Add two 3-mL portions of water through the filter paper, and dilute with water to 30 mL. For control, take 0.10 mg of sulfate ion (SO_4) in 30 mL of water. To both, add 1 mL of (1 + 19) hydrochloric acid and 1 mL of 12% barium chloride reagent solution. After 30 min, sample turbidity should not exceed that of the control solution.

Barium, Calcium, Manganese, Potassium, Sodium, and Strontium. (By flame AAS, page 63).

Sample Stock Solution. Cautiously dissolve 20.0 g of sample with dilute nitric acid (1 + 3), and digest in a covered beaker on a hot plate ($\approx$100 °C) for 15 min. Cool, transfer to a 100-mL volumetric flask, and dilute to the mark with water (1 mL = 0.20 g).

Element	Wavelength (nm)	Sample Wt (g)	Standard Added (mg)	Flame Type*	Background Correction
Ba	553.6	4.0	0.10; 0.20	N/A	No
Ca	422.7	0.20	0.05; 0.10	N/A	No
Mn	279.5	4.0	0.02; 0.04	A/A	Yes
K	766.5	1.0	0.05; 0.10	A/A	No
Na	589.0	0.002	0.01; 0.02	A/A	No
Sr	460.7	1.0	0.05; 0.10	N/A	No

*A/A is air/acetylene; N/A is nitrous oxide/acetylene.

Heavy Metals. (Page 36, Method 1). Dissolve 5.0 g in 50 mL of dilute hydrochloric acid (1 + 1), heat if necessary to obtain complete dissolution, and dilute with water to 150 mL. Evaporate 30 mL to dryness on a hot plate ($\approx$100 °C), dissolve the residue in 20 mL of water, and dilute with water to 25 mL. For the control, add 0.02 mg of lead to 10 mL of the sample solution, evaporate to dryness on a hot plate ($\approx$100 °C), dissolve the residue in 20 mL of water, and dilute with water to 25 mL.

Iron. (Page 38, Method 1). Add 1.0 g to 50 mL of dilute hydrochloric acid (1 + 1), and boil gently for 5 min. Cool, dilute with water to 50 mL, and use 5 mL of this solution without further acidification.

Magnesium Perchlorate, Desiccant

$Mg(ClO_4)_2$ Formula Wt 223.21   CAS No. 10034-81-8

GENERAL DESCRIPTION

Typical appearance: very hygroscopic white solid
Analytical use: drying agent
Change in state (approximate): decomposes above 250 °C
Aqueous solubility: 99 g in 100 mL at 25 °C

SPECIFICATIONS

Suitability for moisture absorption . . . . Passes test

Maximum Allowable

Titrable free acid . . . . 0.005 meq/g
Titrable base . . . . 0.025 meq/g
Loss on drying 8%

TESTS

Suitability for Moisture Absorption. Weigh about 2 g in a tared, 50-mL weighing bottle having a diameter of 50 mm and a height of 30 mm. Place the bottle, with cover removed, for at least 16 h in a closed container in which the atmosphere possesses a relative humidity of 85%, maintained by equilibrium with sulfuric acid having a specific gravity of 1.16. The increase in weight should not be less than 25%.

Titrable Free Acid and Titrable Base. Dissolve 2.0 g in 25 mL of water, and add 0.15 mL of methyl orange indicator solution. Any red color produced should be changed to yellow by the addition of not more than 1.0 mL of 0.01 N sodium hydroxide (titrable free acid). If a yellow color is produced, not more than 5.0 mL of 0.01 N hydrochloric acid should be required to produce a red color (titrable base).

Loss on Drying.

    Caution: Do not dry the sample in the presence of any organic material.

Weigh accurately about 1 g, and dry at 190 ± 10 °C to constant weight. Calculate the loss in weight as percent water.

Magnesium Sulfate Heptahydrate

$MgSO_4 \cdot 7H_2O$ Formula Wt 246.47   CAS No. 10034-99-8

GENERAL DESCRIPTION

Typical appearance: colorless efflorescent solid
Analytical use: preparation of magnesium standard solutions

Change in state (approximate): loses 4 H_2O at 70–80 °C; loses 5 H_2O at 100 °C; loses 6 H_2O at 120 °C; loses last H_2O at 250 °C

Aqueous solubility: 72 g in 100 mL at 20 °C

SPECIFICATIONS

Assay . . . . 98.0–102.0% $MgSO_4 \cdot 7H_2O$
pH of a 5% solution at 25.0 °C 5.0–8.2

	Maximum Allowable
Insoluble matter	0.005%
Chloride (Cl)	5 ppm
Nitrate (NO_3)	0.002%
Ammonium (NH_4)	0.002%
Calcium (Ca)	0.02%
Manganese (Mn)	5 ppm
Potassium (K)	0.005%
Sodium (Na)	0.005%
Strontium (Sr)	0.005%
Heavy metals (as Pb)	5 ppm
Iron (Fe)	5 ppm

TESTS

Assay. (By complexometric titration of Mg). Weigh accurately about 1 g, and transfer to a 250-mL beaker with 50 mL of water. Add 5 mL of pH 10 ammoniacal buffer solution and about 50 mg of Eriochrome Black T indicator mixture. Titrate with 0.1 M EDTA volumetric solution to a clear blue color (and disappearance of the last trace of red). One milliliter of 0.1 M EDTA corresponds to 0.02465 g of $MgSO_4 \cdot 7H_2O$.

$$\% \; MgSO_4 \cdot 7H_2O = \frac{(mL \times M \; EDTA) \times 24.65}{Sample \; wt \; (g)}$$

pH of a 5% Solution at 25.0 °C. (Page 49).

Insoluble Matter. (Page 25). Use 20 g dissolved in 200 mL of water.

Chloride. (Page 35). Use 2.0 g.

Nitrate. (Page 38, Method 1).

> **Sample Solution A.** Dissolve 1.0 g in 5 mL of water by heating in a water bath. Dilute to 50 mL with brucine sulfate reagent solution.

> **Control Solution B.** Dissolve 1.0 g in 3 mL of water and 2 mL of nitrate ion (NO_3) standard solution by heating in a boiling water bath. Dilute to 50 mL with brucine sulfate reagent solution.

Continue with the procedure, starting with the preparation of blank solution C.

Ammonium. Dissolve 2.0 g in 90 mL of water, add 10 mL of 10% sodium hydroxide reagent solution and allow to settle. Draw off 25 mL of the clear solution, dilute with water to 50 mL, and add 2 mL of Nessler reagent. Any color should not exceed that produced by

0.01 mg of ammonium ion (NH_4) in an equal volume of solution containing 2.5 mL of 10% sodium hydroxide reagent solution and 2 mL of Nessler reagent.

Calcium, Manganese, Potassium, Sodium, and Strontium. (By flame AAS, page 63).

> **Sample Stock Solution.** Dissolve 10.0 g of sample with water in a 100-mL volumetric flask, and dilute to the mark with water (1 mL = 0.10 g).

Element	Wavelength (nm)	Sample Wt (g)	Standard Added (mg)	Flame Type*	Background Correction
Ca	422.7	0.20	0.04; 0.08	N/A	No
Mn	279.5	2.0	0.01; 0.02	A/A	Yes
K	766.5	0.50	0.025; 0.05	A/A	No
Na	589.0	0.20	0.01; 0.02	A/A	No
Sr	460.7	2.0	0.10; 0.20	N/A	No

*A/A is air/acetylene; N/A is nitrous oxide/acetylene.

Heavy Metals. (Page 36, Method 1). Dissolve 6.0 g in about 20 mL of water, and dilute with water to 30 mL. Use 25 mL to prepare the sample solution, and use the remaining 5.0 mL to prepare the control solution.

Iron. (Page 38, Method 1). Use 2.0 g.

Manganese Chloride Tetrahydrate

Manganese(II) Chloride Tetrahydrate; Manganous Chloride Tetrahydrate
$MnCl_2 \cdot 4H_2O$ Formula Wt 197.91   CAS No. 13446-34-9

GENERAL DESCRIPTION

Typical appearance: pink, slightly deliquescent solid
Analytical use: preparation of standard solutions
Change in state (approximate): melting point, 58 °C; dehydrates completely, 198 °C
Aqueous solubility: 150 g in 100 mL at 20 °C

SPECIFICATIONS

Assay . . . . 98.0–101.0% $MnCl_2 \cdot 4H_2O$
pH of a 5% solution at 25.0 °C 3.5–6.0

Maximum Allowable

Insoluble matter 0.005%
Sulfate (SO_4) 0.005%
Calcium (Ca) 0.005%
Magnesium (Mg) 0.005%
Potassium (K) 0.01%

Sodium (Na) 0.05%
Zinc (Zn).. 0.005%
Heavy metals (as Pb) .... 5 ppm
Iron (Fe).. 5 ppm

TESTS

Assay. (By complexometric titration of Mn). Weigh accurately about 0.8 g, transfer to a 500-mL beaker, and dissolve in 200 mL of water. Add a few milligrams of ascorbic acid to prevent oxidation. From a buret, add about 30 mL of 0.1 M EDTA volumetric solution. Add 10 mL of pH 10 ammoniacal buffer reagent solution and about 50 mg of Eriochrome Black T indicator mixture. Continue the titration with 0.1 M EDTA volumetric solution to a clear blue color (and disappearance of the last trace of red). One milliliter of 0.1 M EDTA corresponds to 0.01979 g of $MnCl_2 \cdot 4H_2O$.

$$\% \ MnCl_2 \cdot 4H_2O = \frac{(mL \times M \ EDTA) \times 19.79}{\text{Sample wt (g)}}$$

pH of a 5% Solution at 25.0 °C. (Page 49).

Insoluble Matter. (Page 25). Use 20 g dissolved in 150 mL of water.

Sulfate. Dissolve 10 g in 100 mL of water, add 1 mL of hydrochloric acid, filter, and heat the filtrate to boiling. Add 10 mL of 12% barium chloride reagent solution, digest in a covered beaker on a hot plate ($\approx$100 °C) for 2 h, and allow to stand for at least 8 h. If a precipitate is formed, filter through a fine ashless paper, wash thoroughly, and ignite. Correct for the weight obtained in a complete blank test.

$$\% \ SO_4 = \frac{[\text{Residue wt (sample) (g)} - \text{Residue wt (blank) (g)}] \times 100}{\text{Sample wt (g)}}$$

Calcium, Magnesium, Potassium, Sodium, and Zinc. (By flame AAS, page 63).

Sample Stock Solution. Dissolve 5.0 g in water in a 100-mL volumetric flask, add 0.5 mL of hydrochloric acid, and dilute with water to the mark (1 mL = 0.05 g).

Element	Wavelength (nm)	Sample Wt (g)	Standard Added (mg)	Flame Type*	Background Correction
Ca	422.7	0.80	0.02; 0.04	N/A	No
Mg	285.2	0.20	0.005; 0.01	A/A	Yes
K	766.5	0.20	0.01; 0.02	A/A	No
Na	589.0	0.04	0.01; 0.02	A/A	No
Zn	213.9	0.20	0.01; 0.02	A/A	Yes

*A/A is air acetylene; N/A is nitrous oxide/acetylene.

Heavy Metals. (Page 36, Method 1). Dissolve 6.0 g in about 20 mL of water, and dilute with water to 30 mL. Use 25 mL to prepare the sample solution, and use the remaining 5.0 mL to prepare the control solution. Adjust the pH with ammonium acetate solution.

Iron. (Page 38, Method 1). Use 2.0 g.

Manganese Sulfate Monohydrate

Manganese(II) Sulfate Monohydrate; Manganous Sulfate Monohydrate

$MnSO_4 \cdot H_2O$ Formula Wt 169.02   CAS No. 10034-96-5

GENERAL DESCRIPTION

Typical appearance: pink or almost white solid

Analytical use: preparation of manganese standard solutions

Change in state (approximate): dehydrates completely at 400 °C

Aqueous solubility: 100 g in 100 mL at 20 °C

SPECIFICATIONS

Assay . . . . 98.0–101.0% $MnSO_4 \cdot H_2O$
Loss on ignition 10.0–12.0%
Substances reducing permanganate . . . . Passes test

	Maximum Allowable
Insoluble matter	0.01%
Chloride (Cl)	0.005%
Calcium (Ca)	0.005%
Magnesium (Mg)	0.005%
Nickel (Ni)	0.02%
Potassium (K)	0.01%
Sodium (Na)	0.05%
Zinc (Zn)	0.005%
Heavy metals (as Pb)	0.002%
Iron (Fe)	0.002%

TESTS

Assay. (By complexometric titration of Mn). Weigh accurately about 0.7 g, transfer to a 500-mL beaker, and dissolve in 200 mL of water. Add a few milligrams of ascorbic acid to prevent oxidation. From a buret, add about 30 mL of 0.1 M EDTA volumetric solution, followed by 10 mL of pH 10 ammoniacal buffer reagent solution and about 50 mg of Eriochrome Black T indicator mixture. Continue the titration with 0.1 M EDTA to a clear blue color. One milliliter of 0.1 M EDTA corresponds to 0.01690 g of $MnSO_4 \cdot H_2O$.

$$\% \, MnSO_4 \cdot H_2O = \frac{(mL \times M \, EDTA) \times 16.90}{Sample \, wt \, (g)}$$

Loss on Ignition. Weigh accurately about 2 g. Ignite to constant weight at 400–500 °C.

Substances Reducing Permanganate. Dissolve 7.5 g in 200 mL of water containing 3 mL of sulfuric acid and 3 mL of phosphoric acid. To this solution, add 0.10 mL of 0.1 N potassium permanganate volumetric solution in excess of the amount required to produce a pink color in 200 mL of water containing 3 mL of sulfuric acid and 3 mL of phosphoric acid, and allow to stand for 1 min. The pink color should not be entirely discharged.

Insoluble Matter. (Page 25). Use 10 g dissolved in 130 mL of water.

Chloride. (Page 35). Dissolve 1.0 g in 100 mL of water, and use 20 mL of this solution.

Calcium, Magnesium, Nickel, Potassium, Sodium, and Zinc. (By flame AAS, page 63).

>   **Sample Stock Solution.** Dissolve 10.0 g in water in a 100-mL volumetric flask, add 0.5 mL of hydrochloric acid, and dilute with water to the mark (1 mL = 0.10 g).

Element	Wavelength (nm)	Sample Wt (g)	Standard Added (mg)	Flame Type*	Background Correction
Ca	422.7	0.80	0.02; 0.04	N/A	No
Mg	285.2	0.20	0.005; 0.01	A/A	Yes
Ni	232.0	0.50	0.05; 0.10	A/A	Yes
K	766.5	0.20	0.01; 0.02	A/A	No
Na	589.0	0.02	0.005; 0.01	A/A	No
Zn	213.9	0.20	0.005; 0.01	A/A	Yes

*A/A is air/acetylene; N/A is nitrous oxide/acetylene.

Heavy Metals. (Page 36, Method 1). Dissolve 1.0 g in about 20 mL of water, and dilute with water to 25 mL.

Iron. (Page 38, Method 1). Dissolve 1.0 g in water, dilute with water to 50 mL, and use 25 mL of this solution without further acidification.

Mannitol

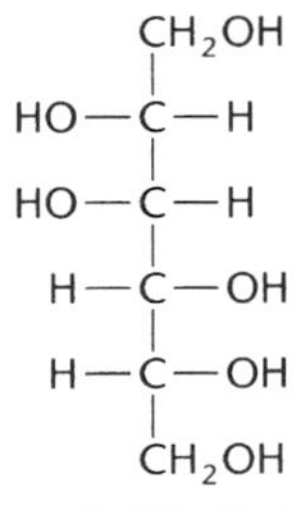

$$
\begin{array}{c}
CH_2OH \\
HO-C-H \\
HO-C-H \\
H-C-OH \\
H-C-OH \\
CH_2OH
\end{array}
$$

HOCH$_2$(CHOH)$_4$CH$_2$OH   **Formula Wt 182.17**   **CAS No. 69-65-8**

GENERAL DESCRIPTION

Typical appearance: white solid
Analytical use: titrimetric determination of boric acid
Change in state (approximate): melting point, 167 °C
Aqueous solubility: 18 g in 100 mL at 20 °C

SPECIFICATIONS

Specific rotation [α]$_D^{25°}$ .... +23.3° to +24.3°
Reducing sugars .... Passes test

	Maximum Allowable
Insoluble matter	0.01%
Loss on drying	0.05%
Residue after ignition	0.01%
Titrable acid	0.0008 meq/g
Heavy metals (as Pb)	5 ppm

TESTS

Specific Rotation. (Page 46). Weigh accurately about 10 g, transfer to a 100-mL volumetric flask, and add 12.8 g of sodium borate. Dissolve the mixture in sufficient water to make about 90 mL of solution, allow to stand with occasional shaking for 1 h, dilute with water to volume at 25 °C, and mix. Observe the optical rotation in a polarimeter at 25 °C using sodium D-line, and calculate the specific rotation.

Reducing Sugars. Add 1 mL of a saturated solution of the mannitol (about 0.2 g per mL) to 5 mL of Benedict's solution on a hot plate ($\approx$100 °C), and heat for 5 min. No more than a slight precipitate should form.

Insoluble Matter. (Page 25). Use 20 g dissolved in 100 mL of water.

Loss on Drying. Weigh accurately 5.0 g in a tared, preconditioned dish or crucible, and dry at 105 °C for 2 h.

Residue after Ignition. (Page 26). Ignite 10 g. Moisten the char with 1 mL of sulfuric acid.

Titrable Acid. To 100 mL of carbon dioxide-free water, add 0.15 mL of phenolphthalein indicator solution and 0.01 N sodium hydroxide until a pink color is produced. Dissolve 10 g of the sample in this solution, and titrate with 0.01 N sodium hydroxide to the same end point. Not more than 0.8 mL should be required.

Heavy Metals. (Page 36, Method 1). Dissolve 4.0 g in about 25 mL of water, and dilute with water to 30 mL.

Mercuric Acetate

Mercury(II) Acetate

(CH₃COO)₂Hg   **Formula Wt 318.68**   **CAS No. 1600-27-7**

GENERAL DESCRIPTION

Typical appearance: white solid with a faint odor of acetic acid; turns slightly yellow on exposure to light

Analytical use: determination of nitrate in chromium compounds

Change in state (approximate): melting point, 179 °C, with decomposition

Aqueous solubility: 40 g in 100 mL at 25 °C

SPECIFICATIONS

Assay . . . . ≥98.0% (CH$_3$COO)$_2$Hg

Maximum Allowable

Insoluble matter 0.01%
Nitrate (NO$_3$) 0.005%
Residue after reduction 0.02%
Chloride (Cl) 0.005%
Sulfate (SO$_4$) 0.005%
Other heavy metals (as Pb) 0.002%
Iron (Fe). 0.001%
Mercurous mercury (as Hg) 0.4%

TESTS

Assay. (By thiocyanate precipitation titration of mercuric ion). Weigh accurately about 0.7 g of sample, and dissolve in 50 mL of dilute nitric acid (1 + 19). Add 2 mL of ferric ammonium sulfate indicator solution, and titrate with 0.1 N ammonium thiocyanate volumetric solution. One milliliter of 0.1 N ammonium thiocyanate corresponds to 0.01593 g of (CH$_3$COO)$_2$Hg.

$$\% \text{ (CH}_3\text{COO)}_2\text{Hg} = \frac{(\text{mL} \times \text{N NH}_4\text{SCN}) \times 15.93}{\text{Sample wt (g)}}$$

Insoluble Matter. (Page 25). Use 10.0 g dissolved in 100 mL of dilute acetic acid (5 + 95). Also use dilute acetic acid (1 + 99) as the wash.

Nitrate.

> **Sample Solution A.** Dissolve 0.50 g completely in 3.0 mL of water in a dry test tube. Put the tube in an ice bath, and immediately, but slowly, add 7 mL of cold chromotropic acid reagent solution while swirling. Keep the tube in the ice bath an additional 2–3 min, remove, and let stand for 30 min, swirling occasionally.

> **Control Solution B.** Dissolve 0.50 g in 0.5 mL of water plus 2.5 mL of nitrate ion (NO$_3$) standard solution in a dry test tube. Put the tube in an ice bath, and immediately, but slowly, add 7 mL of cold chromotropic acid reagent solution while stirring. Keep the tube in the ice bath an additional 2–3 min, remove, and let stand for 30 min, stirring occasionally.

> **Blank Solution C.** Add 7 mL of cold chromotropic acid reagent solution to 3.0 mL of water in a dry test tube immersed in an ice bath. Remove and let stand for 30 min.

Transfer sample solution A and control solution B to dry 15-mL centrifuge tubes, and centrifuge until the supernatant liquid is clear. Set a spectrophotometer at 405 nm, and using 1-cm cells, adjust the instrument to read zero absorbance with blank solution C in the light path, then determine the absorbance of sample solution A. Adjust the instrument to read zero absorbance with sample solution A in the light path, and determine the absorbance of control solution B. The absorbance of sample solution A versus blank solution C should not exceed that of control solution B versus sample solution A.

For the Determination of Residue after Reduction, Chloride, Sulfate, Other Heavy Metals, and Iron

Sample Solution D. Dissolve 10.0 g in 15 mL of water, 2 mL of nitric acid, and 10 mL of 96% formic acid. Digest under total reflux until all the mercury is reduced to metal and the solution is clear. Cool, filter through a thoroughly washed filter paper, and wash with a small quantity of water. Dilute the filtrate and washings with water to 100 mL in a volumetric flask (1 mL = 0.1 g).

Residue after Reduction. Evaporate 50 mL (5-g sample) of sample solution D plus 0.10 mL of sulfuric acid to dryness in a tared, preconditioned dish in a well-ventilated hood. Continue heating until the excess sulfuric acid has been volatilized. Finally, ignite at 600 ± 25 °C for 15 min. Correct for the weight obtained in a complete blank test.

Chloride. (Page 35). Use 2.0 mL of sample solution D (0.2-g sample).

Sulfate. (Page 41, Method 3). Use 10 mL of sample solution D (1-g sample).

Other Heavy Metals. (Page 36, Method 1). Add 1.0 mL of 1% sodium carbonate reagent solution to 10 mL of sample solution D (1-g sample), and evaporate to dryness. Dissolve the residue in about 20 mL of water, filter, and dilute with water to 25 mL.

Iron. (Page 38, Method 1). To 10 mL of sample solution D (1-g sample), add about 10 mg of sodium carbonate, and evaporate to dryness. Dissolve the residue in 2 mL of hydrochloric acid, dilute with water to 50 mL, and use the solution without further acidification.

Mercurous Mercury. Dissolve 5.0 g in 100 mL of a 12.5% solution of potassium iodide. Add 5.0 mL of 0.1 N iodine and 2 mL of 1 N hydrochloric acid. Allow to stand, protected from the light, for 1 h with frequent agitation. Titrate the iodine with 0.1 N sodium thiosulfate volumetric solution, adding 3 mL of starch indicator solution near the end point, and correct for a complete blank. Not more than 1.0 mL of 0.1 N iodine should be consumed.

Mercuric Bromide
Mercury(II) Bromide

$HgBr_2$	Formula Wt 360.40	CAS No. 7789-47-1

GENERAL DESCRIPTION
Typical appearance: white or very faint yellow solid
Analytical use: colorimetric detection of arsenic
Change in state (approximate): melting point, 237 °C
Aqueous solubility: 0.5 g in 100 mL at 20 °C

SPECIFICATIONS
Maximum Allowable

Residue after reduction	0.02%
Insoluble in methanol	0.05%
Chloride (Cl)	0.25%

TESTS

Residue after Reduction. Dissolve 5.0 g in 10 mL of water plus 10 mL of ammonium hydroxide. Add 40 mL of formic acid (96%), and reflux until all the mercury is reduced to metal. Cool, filter through a well-washed filter paper, and wash with a small quantity of water. Add 0.10 mL of sulfuric acid to the combined filtrate and washings. Evaporate in a tared, preconditioned dish in a well-ventilated hood. Continue heating until the excess sulfuric acid has been volatilized. Finally, ignite at 600 ± 25 °C for 15 min. Correct for the weight obtained in a complete blank test.

Insoluble in Methanol. Dissolve 2.0 g in 30 mL of methanol. Filter through a tared, preconditioned filtering crucible, wash with 100 mL of methanol, and dry at 105 °C.

Chloride. To 0.70 g, add 20 mL of water, 5 mL of 10% sodium hydroxide reagent solution, and 4 mL of 30% hydrogen peroxide. Digest on a hot plate ($\approx$100 °C) until all reaction ceases. Cool, filter through a chloride-free filter, wash with water, and dilute the filtrate and washings with water to 100 mL. To 5.0 mL in a small conical flask, add 20 mL of water and 1.5 mL of ammonium carbonate solution (described below). Add, with agitation, 5 mL of silver nitrate reagent solution, and allow to stand, with frequent agitation, for 10 min. Filter through a chloride-free filter, wash with water, and dilute the filtrate and washings with water to 100 mL. To 25 mL, add 0.5 mL of nitric acid. Any turbidity should not exceed that in a control obtained by treating 0.20 g of sample and 1.25 mg of added chloride ion (Cl) in 20 mL of water exactly like the solution of the sample.

> *Ammonium Carbonate Solution.* In a 100-mL volumetric flask, add 20 g of ammonium carbonate and 20 mL of 10% ammonium hydroxide reagent solution. Fill to the mark with water.

Mercuric Chloride
Mercury(II) Chloride

$HgCl_2$ Formula Wt 271.50   CAS No. 7487-94-7

GENERAL DESCRIPTION

Typical appearance: white or colorless solid
Analytical use: preparation of standard solutions; amalgamation of zinc for Jones reductor
Change in state (approximate): melting point, 277 °C; sublimes substantially at 100 °C
Aqueous solubility: 6.5 g in 100 mL at 20 °C

SPECIFICATIONS

Assay . . . . $\geq$99.5% $HgCl_2$
Solution in ethyl ether . . . . Passes test

Maximum Allowable

Residue after reduction 0.02%
Iron (Fe). 0.002%

TESTS

Assay. (By complexometric titration of Hg^{II}). Weigh accurately about 0.7 g, and dissolve completely in 250 mL of water. From a buret, add 20 mL of 0.1 M EDTA volumetric solution, then add 10 mL of a solution containing 2 g of silver nitrate. Add 2 g of ammonium nitrate, 20 mL of a saturated hexamethylenetetramine reagent solution, and a few milligrams of xylenol orange indicator mixture. Continue the titration with 0.1 M EDTA to a yellow end point. One milliliter of 0.1 M EDTA corresponds to 0.02715 g of $HgCl_2$.

$$\% \ HgCl_2 = \frac{(mL \times M \ EDTA) \times 27.15}{Sample \ wt \ (g)}$$

Solution in Ethyl Ether. Dissolve 2.0 g in 60 mL of ethyl ether in a stoppered flask. Shake the flask to agitate the sample and ether, but do not expose the contents to atmospheric moisture. Not more than a faint trace of insoluble residue should remain.

Residue after Reduction. Suspend 5.0 g in 10 mL of water plus 10 mL of ammonium hydroxide. Add 40 mL of formic acid (96%), and reflux until all the mercury is reduced to metal. Cool, filter through a well-washed filter paper, and wash with a small quantity of water. Add 0.10 mL of sulfuric acid to the combined filtrate and washings. Evaporate in a tared, preconditioned dish in a well-ventilated hood. Continue heating until the excess sulfuric acid has been volatilized. Finally, ignite at 600 ± 25 °C for 15 min. Correct for the weight obtained in a complete blank test. Retain residue for the determination of iron.

Iron. (Page 38, Method 1). To the residue obtained in the test for residue after reduction, add 3 mL of dilute hydrochloric acid (1 + 1), cover the dish, and digest on a hot plate (≈ 100 °C) for 15–20 min. Remove the cover, and evaporate to dryness. Dissolve the residue in 10 mL of dilute hydrochloric acid (1 + 9), dilute with water to 50 mL, and use 5.0 mL of this solution.

Mercuric Iodide, Red

Mercury(II) Iodide, Red

HgI_2   Formula Wt 454.40   CAS No. 7774-29-0

GENERAL DESCRIPTION

Typical appearance: red solid
Analytical use: preparation of Nessler's reagent
Change in state (approximate): melting point, 259 °C
Aqueous solubility: insoluble

SPECIFICATIONS

Assay (dried basis). . . . . ≥99.0% HgI_2
Solubility in potassium iodide solution . . . . Passes test

Maximum Allowable

Mercurous mercury (as Hg) 0.1%

Soluble mercury salts (as Hg) 0.05%

TESTS

Assay. (By iodometry). Dry about 1 g of sample over magnesium perchlorate for at least 8 h. Weigh accurately about 0.5 g of the dried sample, place in a glass-stoppered conical flask, and add 30 mL of hydrochloric acid and 20 mL of water. Rotate the flask until the mercuric iodide is dissolved, add 5 mL of chloroform, and titrate the solution with 0.05 M potassium iodate volumetric solution until the iodine color is discharged from the aqueous layer. Stopper the flask, stir well for 30 s, then continue the titration, stirring vigorously after each addition of the potassium iodate until the chloroform is free of iodine color. One milliliter of 0.05 M potassium iodate corresponds to 0.02272 g of HgI_2.

$$\% \, HgI_2 = \frac{(mL \times M \, KIO_3) \times 45.44}{Sample \ wt \ (g)}$$

Solubility in Potassium Iodide Solution. Dissolve 10.0 g of the sample in 100 mL of 10% potassium iodide reagent solution in a glass-stoppered flask. A complete, or practically complete, solution results. Retain the solution for the test for mercurous mercury.

Mercurous Mercury. To the solution reserved from the test for solubility in potassium iodide solution, add 5.0 mL of 0.1 N iodine and 3 mL of 1 N hydrochloric acid. Allow to stand in a dark place for 1 h with frequent agitation. Titrate the excess iodine with 0.1 N sodium thiosulfate volumetric solution, adding 3 mL of starch indicator solution near the end point, and correct for a complete blank. Not more than 0.50 mL of the 0.1 N iodine should be consumed.

Soluble Mercury Salts. Shake 1.0 g of the sample with 20 mL of water for 2 min, and filter. Dilute 10 mL of the filtrate with water to 40 mL, and add 10 mL of freshly prepared hydrogen sulfide water. Any color should not exceed that produced by 0.25 mg of mercury ion (Hg) in an equal volume of solution containing the quantities of reagents used in the test.

Mercuric Nitrate Monohydrate or Dihydrate
Mercury(II) Nitrate Monohydrate or Dihydrate

$Hg(NO_3)_2 \cdot H_2O$ (monohydrate) Formula Wt 342.62   CAS No. 7783-34-8

$Hg(NO_3)_2 \cdot 2H_2O$ (dihydrate)   Formula Wt 360.63   CAS No. 22852-67-1

Note: This reagent is available as either the mono- or dihydrate. The label should identify the degree of hydration.

GENERAL DESCRIPTION

Typical appearance: white solid; may become colored on exposure to light

Analytical use: preparation of standards

Change in state (approximate): melting point, 70 °C
Aqueous solubility: decomposes

SPECIFICATIONS

Assay . . . . ≥98.0% $Hg(NO_3)_2 \cdot H_2O$
or $Hg(NO_3)_2 \cdot 2H_2O$

Maximum Allowable

Residue after reduction. 0.01%
Chloride (Cl). 0.002%
Sulfate (SO_4) 0.002%
Iron (Fe). 0.001%

TESTS

Assay. (By thiocyanate precipitation titration of Hg^{II}). Weigh accurately about 0.6 g of sample, and transfer with 100 mL of water to a 250-mL flask. Add 5 mL of nitric acid and 2 mL of ferric ammonium sulfate indicator solution. Cool to 15 °C, and titrate with 0.1 N ammonium thiocyanate volumetric solution to a permanent reddish brown color. One milliliter of 0.1 N ammonium thiocyanate corresponds to 0.01713 g of $Hg(NO_3)_2 \cdot H_2O$ or 0.01803 g of $Hg(NO_3)_2 \cdot 2H_2O$.

$$\% \ Hg(NO_3)_2 \cdot H_2O = \frac{(mL \times N \ NH_4SCN) \times 17.13}{Sample \ wt \ (g)}$$

$$\% \ Hg(NO_3)_2 \cdot 2H_2O = \frac{(mL \times N \ NH_4SCN) \times 18.03}{Sample \ wt \ (g)}$$

Residue after Reduction. Dissolve 10 g in 15 mL of water, 2 mL of nitric acid, and 10 mL of formic acid (96%). Digest under total reflux until all the mercury is reduced to metal and the solution is clear. Cool, filter through a well-washed filter paper, and wash with a small quantity of water. Add 0.10 mL of sulfuric acid to the combined filtrate and washings. Evaporate in a tared, preconditioned dish in a well-ventilated hood. Continue heating until the excess sulfuric acid has been volatilized. Finally, ignite at 600 ± 25 °C for 15 min. Correct for the weight obtained in a complete blank test. Retain the residue for the test for iron.

Chloride. Dissolve 2.0 g in 50 mL of water and 2 mL of formic acid (96%). Add dropwise 10% sodium hydroxide reagent solution until a small amount of permanent precipitate is formed. Digest under a reflux condenser until the mercury is all reduced to the metal and the supernatant liquid is clear. Cool, filter through a chloride-free filter, wash thoroughly, and dilute the filtrate with water to 100 mL. To 50 mL of the dilution, add 1 mL of nitric acid and 1 mL of silver nitrate reagent solution. Any turbidity should not exceed that produced by 0.02 mg of chloride ion (Cl) in an equal volume of solution containing the quantities of reagents used in the test.

Sulfate. Dissolve 5.0 g in 50 mL of water and 5 mL of formic acid. Digest under a reflux condenser until the mercury is all reduced to the metal and the supernatant liquid is clear. Cool, filter, and wash thoroughly. To the combined filtrate and washings, add 20 mg of sodium carbonate, and evaporate to dryness on a hot plate (≈100 °C). Dissolve the residue in 10 mL of water and 1 mL of dilute hydrochloric acid (1 + 19), and filter if necessary. To the filtrate, add 1 mL of 12% barium chloride reagent solution. Any turbidity should not exceed that produced by 0.1 mg of sulfate ion (SO_4) in an equal volume of solution containing the quantities of reagents used in the test. Compare 10 min after adding the barium chloride to the sample and standard solutions.

Iron. (Page 38, Method 1). To the residue obtained in the test for residue after reduction, add 3 mL of dilute hydrochloric acid (1 + 1), cover with a watch glass, and digest on a hot plate ($\approx$100 °C) for 20 min. Remove the cover, and evaporate to dryness. Dissolve the residue in 2 mL of dilute hydrochloric acid (1 + 1), add about 40 mL of water, filter if necessary, and dilute with water to 100 mL. Use 10 mL of this solution.

Mercuric Oxide, Red
Mercury(II) Oxide, Red

HgO	Formula Wt 216.59	CAS No. 21908-53-2

GENERAL DESCRIPTION

Typical appearance: red to orange-red solid

Analytical use: catalyst

Change in state (approximate): decomposes into mercury and oxygen at 500 °C

Aqueous solubility: practically insoluble

SPECIFICATIONS

Assay . . . . $\geq$99.0% HgO

Maximum Allowable

Insoluble in dilute hydrochloric acid 0.03%

Residue after reduction 0.025%

Sulfate (SO_4) 0.015%

Chloride (Cl) 0.025%

Nitrogen compounds (as N) 0.005%

Iron (Fe). 0.005%

TESTS

Assay. (By complexometric titration of Hg^{II}). Weigh accurately about 0.8 g of sample. Dissolve in 3.5 mL of nitric acid, dilute to about 100 mL with water, and add 50 mg of xylenol orange indicator mixture. Add saturated hexamethylenetetramine reagent solution until the color changes to purple, followed by 3–4 mL in excess, then titrate with 0.1 M EDTA volumetric solution until the color changes to yellow. One milliliter of 0.1 M EDTA corresponds to 0.02166 g of HgO.

$$\% \, HgO = \frac{(mL \times M \, EDTA) \times 21.66}{Sample \, wt \, (g)}$$

Insoluble in Dilute Hydrochloric Acid. Dissolve 4.0 g in 40 mL of dilute hydrochloric acid (1 + 3), heat to boiling, and digest in a covered beaker on a hot plate ($\approx$100 °C) for 1 h. Filter through a tared, preconditioned filtering crucible, wash well with water, and dry at 105 °C.

For the Determination of Residue after Reduction and Sulfate

Sample Solution A. Dissolve 10.0 g in 15 mL of water, 2 mL of nitric acid, and 10 mL of 96% formic acid. Digest under total reflux until all the mercury is reduced to metal and the solution is clear. Cool, filter through a well-washed filter paper, and wash with a small quantity of water. Dilute the filtrate and washings with water to 200 mL in a volumetric flask (1 mL = 0.05 g).

Residue after Reduction. In a well-ventilated fume hood, evaporate 80 mL (4-g sample) of sample solution A plus 0.10 mL of sulfuric acid to dryness in a tared, preconditioned dish. Continue heating until the excess sulfuric acid has been volatilized. Finally, ignite at 600 ± 25 °C for 15 min. Correct for the weight obtained in a complete blank test. Retain the residue for the test for iron.

Sulfate. (Page 41, Method 3). Use 6.7 mL of sample solution A (0.33-g sample).

Chloride. Dissolve 1.0 g in 50 mL of water plus 1 mL of formic acid (96%). Add dropwise 10% sodium hyroxide reagent solution until a small amount of permanent precipitate is formed. Digest under a reflux condenser until all the mercury is reduced to metal and the solution is clear. Cool, filter through a chloride-free filter, and dilute with water to 100 mL. Dilute 4.0 mL of this solution with water to 20 mL, and add 1 mL of nitric acid and 1 mL of silver nitrate reagent solution. Any turbidity should not exceed that produced by 0.01 mg of chloride ion (Cl) in an equal volume of solution containing the quantities of reagents used in the test.

Nitrogen Compounds. (Page 39). Dissolve 0.2 g in 5 mL of dilute hydrochloric acid (1 + 1). For the standard, use 0.01 mg of nitrogen ion (N) and 5 mL of the dilute acid.

Iron. (Page 38, Method 1). To the residue obtained in the test for residue after reduction, add 3 mL of dilute hydrochloric acid (1 + 1), cover with a watch glass, and digest on a hot plate ($\approx$100 °C) for 20 min. Remove the cover, and evaporate to dryness. Dissolve in 2 mL of dilute hydrochloric acid (1 + 1), add about 40 mL of water, filter if necessary, and dilute with water to 100 mL. Use 5 mL of this solution.

Mercuric Oxide, Yellow
Mercury(II) Oxide, Yellow

HgO	**Formula Wt 216.59**	**CAS No. 21908-53-2**

GENERAL DESCRIPTION

Typical appearance: yellow or yellow-orange solid
Analytical use: catalyst

Change in state (approximate): decomposes into mercury and oxygen at 500 °C
Aqueous solubility: practically insoluble

SPECIFICATIONS

Assay . . . . ≥99.0% HgO

Maximum Allowable

Insoluble in dilute hydrochloric acid 0.03%
Residue after reduction 0.05%
Sulfate (SO$_4$) 0.01%
Chloride (Cl) 0.025%
Nitrogen compounds (as N) 0.005%
Iron (Fe). 0.003%

TESTS

Assay. (By complexometric titration of Hg^{II}). Weigh accurately about 0.8 g of sample. Dissolve in 3.5 mL of nitric acid, dilute to about 100 mL with water, and add 50 mg of xylenol orange indicator mixture. Add saturated hexamethylenetetramine reagent solution until the color changes to purple, followed by 3–4 mL in excess, then titrate with 0.1 M EDTA volumetric solution until the color changes to yellow. One milliliter of 0.1 M EDTA corresponds to 0.02166 g of HgO.

$$\% \text{ HgO} = \frac{(\text{mL} \times \text{M EDTA}) \times 21.66}{\text{Sample wt (g)}}$$

Insoluble in Dilute Hydrochloric Acid. Dissolve 3.0 g in 30 mL of dilute hydrochloric acid (1 + 3), heat to boiling, and digest in a covered beaker on a hot plate (≈100 °C) for 1 h. Filter through a tared, preconditioned filtering crucible, wash thoroughly, and dry at 105 °C.

For the Determination of Residue after Reduction and Sulfate

Sample Solution A. Dissolve 10.0 g in 15 mL of water, 2 mL of nitric acid, and 10 mL of 96% formic acid. Digest under total reflux until all the mercury is reduced to metal and the solution is clear. Cool, filter through a well-washed filter paper, and wash with a small quantity of water. Dilute the filtrate and washings with water to 200 mL in a volumetric flask (1 mL = 0.05 g).

Residue after Reduction. Evaporate 60 mL (3 g) of sample solution A plus 0.10 mL of sulfuric acid to dryness in a tared, preconditioned dish in a well-ventilated hood. Continue heating until the excess sulfuric acid has been volatilized. Finally, ignite at 600 ± 25 °C for 15 min. Correct for the weight obtained in a complete blank test. Retain the residue for the test for iron.

Sulfate. (Page 41, Method 3). Use 10 mL of sample solution A (0.5-g sample).

Chloride. Dissolve 1.0 g in 50 mL of water and 1 mL of 96% formic acid. Add dropwise 10% sodium hydroxide reagent solution until a small amount of permanent precipitate is formed. Digest under a reflux condenser until all the mercury is reduced to metal and the solution is clear. Cool, filter through a chloride-free filter, and dilute with water to 100 mL.

Dilute 4.0 mL of this solution with water to 20 mL, and add 1 mL of nitric acid and 1 mL of silver nitrate reagent solution. Any turbidity should not exceed that produced by 0.01 mg of chloride ion (Cl) in an equal volume of solution containing the quantities of reagents used in the test.

Nitrogen Compounds. (Page 39). Dissolve 0.2 g in 5 mL of dilute hydrochloric acid (1 + 1). For the standard, use 0.01 mg of nitrogen ion (N) and 5 mL of the dilute acid.

Iron. (Page 38, Method 1). To the residue obtained in the test for residue after reduction, add 1 mL of hydrochloric acid, 0.15 mL of nitric acid, and about 10 mg of sodium carbonate, cover, and digest on a hot plate ($\approx$100 °C) for 15–20 min. Uncover, and evaporate to dryness. Dissolve the residue in 9 mL of hydrochloric acid, and dilute with water to 90 mL. Use 10 mL of this solution and 1 mL of additional hydrochloric acid.

Mercuric Sulfate
Mercury(II) Sulfate

HgSO₄	**Formula Wt 296.65**	**CAS No. 7783-35-9**

$HgSO_4$ with Formula Wt 296.65 and CAS No. 7783-35-9.

GENERAL DESCRIPTION
Typical appearance: white solid
Analytical use: determination of chemical oxygen demand; catalyst for Kjeldahl method
Aqueous solubility: decomposes into basic salt and sulfuric acid

SPECIFICATIONS
Assay . . . . $\geq$98.0% $HgSO_4$

Maximum Allowable

Residue after reduction.	0.02%
Chloride (Cl).	0.003%
Nitrate (NO₃)	Passes test
Iron (Fe).	0.005%
Mercurous mercury (as Hg)	0.15%

TESTS

Assay. (By thiocyanate precipitation titration of Hg^{II}). Weigh accurately about 0.5 g, and dissolve in 50 mL of dilute nitric acid (1 + 1). Add 1 mL of 10% ferric nitrate solution, and titrate with 0.1 N ammonium thiocyanate volumetric solution to a permanent reddish-brown color. One milliliter of 0.1 N ammonium thiocyanate corresponds to 0.01483 g of $HgSO_4$.

$$\% \; HgSO_4 = \frac{(mL \times N \; NH_4SCN) \times 14.83}{Sample \; wt \; (g)}$$

Residue after Reduction. Dissolve 5.0 g in 50 mL of water plus 10 mL of formic acid (96%), and add 25 mL of 30% ammonium hydroxide. Digest under total reflux until all the mercury is reduced to metal. Cool, filter through a thoroughly washed filter paper, and wash with a small quantity of water. Add 0.1 mL of sulfuric acid to the combined filtrate and washings, and evaporate in a tared, preconditioned dish in a well-ventilated hood. Continue heating until the excess sulfuric acid has been volatilized. Finally, ignite at 600 ± 25 °C for 15 min. Correct for the weight obtained in a complete blank test. Retain the residue for the test for iron.

Chloride. Dissolve 1.0 g in 50 mL of water plus 1 mL of formic acid (96%). Add dropwise 10% sodium hydroxide reagent solution until a small amount of permanent precipitate is formed. Digest under total reflux until all the mercury is reduced to metal and the solution is clear. Cool, filter through a chloride-free filter, and dilute to 90 mL. To 30 mL of this solution, add 1 mL of nitric acid and 1 mL of silver nitrate reagent solution. Any turbidity should not exceed that produced by 0.01 mg of chloride ion (Cl) in an equal volume of solution containing the quantities of reagents used in the test.

Nitrate. Disperse 1.0 g in 9 mL of water, add 1 mL of sodium chloride solution (1 in 200), mix, and add 0.1 mL of indigo carmine reagent solution, followed by 10 mL of sulfuric acid. The blue color of the clear solution should not be discharged entirely within 5 min. (Limit about 0.003%)

Iron. (Page 38, Method 1). To the residue obtained in the test for residue after reduction, add 3 mL of dilute hydrochloric acid (1 + 1), cover with a watch glass, and digest on a hot plate ($\approx$100 °C) for 20 min. Remove the watch glass, and evaporate to dryness. Take up the residue in a mixture of 1 mL of dilute hydrochloric acid (1 + 1) and 30 mL of water, filter if necessary, and dilute with water to 100 mL. Use 4.0 mL of the solution.

Mercurous Mercury. Transfer 5.0 g to a glass-stoppered flask. Add 100 mL of 15% potassium iodide solution, 5.00 mL of 0.1 N iodine, and 3 mL of 1 N hydrochloric acid. Allow to stand in the dark, with frequent agitation, for 1 h. Titrate the excess iodine with 0.1 N sodium thiosulfate volumetric solution, adding 3 mL of starch indicator solution near the end point. Correct for a blank determination. Not more than 0.38 mL of the 0.1 N iodine should be consumed.

Mercurous Chloride
Mercury(I) Chloride

Hg_2Cl_2 **Formula Wt 472.09**   **CAS No. 10112-91-1**

GENERAL DESCRIPTION

Typical appearance: white solid

Analytical use: preparation of standard electrodes

Change in state (approximate): slowly decomposes by sunlight into mercuric chloride and metallic mercury; sublimes at 400–500 °C without melting

Aqueous solubility: practically insoluble

SPECIFICATIONS

Assay . . . . ≥99.5% Hg_2Cl_2

Maximum Allowable

Residue after reduction. 0.02%
Mercuric chloride ($HgCl_2$) 0.01%
Sulfate (SO_4) 0.01%

TESTS

Assay. (By iodometric titration of mercury). Weigh accurately about 0.9 g, and transfer to a 250-mL glass-stoppered conical flask. Add 50.0 mL of 0.1 N iodine and 2 g of potassium iodide. Stopper, and swirl until the precipitate redissolves. Titrate the excess of iodine with 0.1 N sodium thiosulfate volumetric solution, adding 3 mL of starch indicator solution near the end point. One milliliter of 0.1 N iodine corresponds to 0.02360 g of Hg_2Cl_2.

$$\% \ Hg_2Cl_2 = \frac{[(50.0 \times N \ I_2) - (mL \times N \ Na_2S_2O_3)] \times 23.60}{Sample \ wt \ (g)}$$

Residue after Reduction. Dissolve 5.0 g in 10 mL of water plus 10 mL of ammonium hydroxide. Add 40 mL of formic acid (96%), and reflux until all the mercury is reduced to metal. Cool, filter through a thoroughly washed filter paper, and wash with a small quantity of water. Add 0.10 mL of sulfuric acid to the combined filtrate and washings. Evaporate in a tared, preconditioned dish in a well-ventilated hood. Continue heating until the excess sulfuric acid has been volatilized. Ignite at 600 ± 25 °C for 15 min. Correct for the weight obtained in a complete blank test.

Mercuric Chloride. Shake 1.0 g with 10 mL of ethyl alcohol for 5 min, and filter. To 5 mL of the filtrate, add 0.10 mL of hydrochloric acid and 5 mL of freshly prepared hydrogen sulfide water. Any darkening should not be more than that produced by 0.05 mg of mercuric chloride ($HgCl_2$) in a mixture of 5 mL of alcohol, 0.10 mL of hydrochloric acid, and 5 mL of hydrogen sulfide water.

Sulfate. (Page 41, Method 3). Digest for 10 min with 10 mL of dilute hydrochloric acid (1 + 1), and filter.

Mercury

Hg   **Atomic Wt 200.59**   **CAS No. 7439-97-6**

GENERAL DESCRIPTION

Typical appearance: silver-white liquid metal
Analytical use: electrodes in polarography
Change in state (approximate): boiling point, 357 °C
Aqueous solubility: insoluble

SPECIFICATIONS

Appearance.. Passes test

Maximum Allowable

Nonvolatile matter .... 5 ppm

TESTS

Appearance. The mercury should have a bright mirrorlike surface, free from film or scum. It should pour freely from a thoroughly clean, dry glass container without leaving any mercury adhering to the glass.

Nonvolatile Matter.

> *Note:* In this case, because of environmental considerations, the manufacturer- or supplier-certified value may be accepted, provided company or organization policy allows this.

Perform all tests in a fume hood. Transfer 200 g to a tared, preconditioned boat of 25–30-mL capacity. Place the boat in the approximate center of a glass combustion tube, about 350 mm long and 35 mm in diameter, and mount in a horizontal position. Fit the tube at one end with a removable closure. Bend the exit end downward at 90° to the horizontal, and draw it out so as to fit it into a one-hole rubber stopper held in the mouth of a suction flask. Cool the suction flask in an ice bath. Place the boat and sample in the tube, close the entrance end, and reduce the pressure to less than 15 mm of mercury. Heat gently until all the mercury has distilled from a quiet surface (without ebullition). Remove the boat, heat it in a muffle furnace at 600 °C for 15 min, and cool.

Metalphthalein

Phthalein Purple; *o*-Cresolphthalein complexone

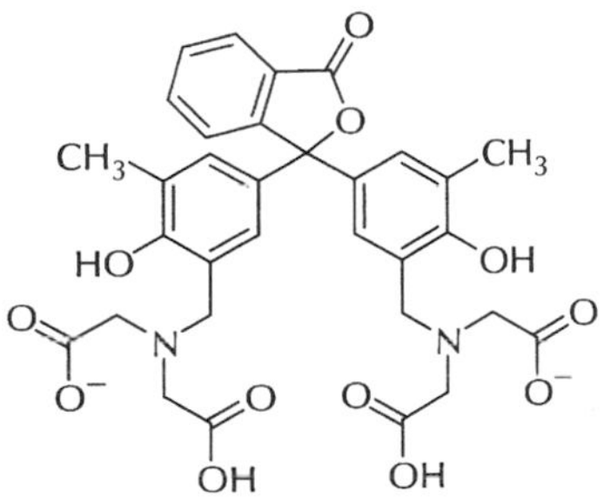

$C_{32}H_{32}N_2O_{12}$ **Formula Wt 636.60**   **CAS No. 2411-89-4**

GENERAL DESCRIPTION

Typical appearance: light pink to tan crystalline solid

Analytical use: indicator

SPECIFICATIONS

Clarity of solution . . . . Passes test
Suitability as a mixed indicator for complexometry . . . . Passes test

TESTS

Clarity of Solution. Dissolve 0.18 g in 100 mL of water containing 0.5 mL of ammonium hydroxide. Not more than a trace of turbidity or insoluble matter should remain.

Suitability as a Mixed Indicator for Complexometry. Using the test sample, prepare metalphthalein-screened indicator solution. Mix 50 mL of water and 50 mL of alcohol. Add 10.0 mL of 0.1 M EDTA volumetric solution, 10 mL of ammonium hydroxide, and 0.3 mL of the metalphthalein-screened indicator solution. Titrate with 0.1 M barium nitrate reagent solution. The color change is from gray-green to magenta. Add 0.05 mL of 0.1 M EDTA; the color should return to gray-green.

Methanol
Methyl Alcohol

CH_3OH Formula Wt 32.04   CAS No. 67-56-1

Suitable for general use or in high-performance liquid chromatography, extraction–concentration analysis, or ultraviolet spectrophotometry. Product labeling shall designate the uses for which suitability is represented based on meeting the relevant specifications and tests. The ultraviolet spectrophotometry and liquid chromatography suitability specifications include all of the specifications for general use. The extraction–concentration suitability specifications include only the general use specification for color.

GENERAL DESCRIPTION

Typical appearance: clear, colorless liquid
Analytical use: organic solvent
Change in state (approximate): boiling point, 65 °C
Aqueous solubility: miscible
Density: 0.79

SPECIFICATIONS

General Use

Assay . . . . $\geq$99.8% CH_3OH
Substances darkened by sulfuric acid . . . . Passes test
Substances reducing permanganate . . . . Passes test
Solubility in water . . . . Passes test

                                                            Maximum Allowable
Color (APHA) 10

Water (H_2O) 0.1%
Residue after evaporation 0.001%
Carbonyl compounds . . . . 0.001% each of acetone,
formaldehyde, and acetaldehyde
Titrable acid . . . . 0.0003 meq/g
Titrable base . . . . 0.0002 meq/g

Specific Use

Ultraviolet Spectrophotometry

Wavelength (nm)	Absorbance (AU)
280–400	0.01
260	0.04
240	0.10
230	0.20
220	0.40
210	0.80
205	1.00

Liquid Chromatography Suitability

Absorbance . . . . Passes test
Gradient elution . . . . Passes test

Extraction–Concentration Suitability

Absorbance . . . . Passes test
GC–FID . . . . Passes test
GC–ECD . . . . Passes test

TESTS

Assay. Analyze the sample by gas chromatography using the general parameters cited on page 80. The following specific conditions are also required.

Column: Type I, methyl silicone

Measure the area under all peaks, and calculate the methanol content in area percent. Correct for water content.

Substances Darkened by Sulfuric Acid. Cool 10 mL of sulfuric acid contained in a small conical flask to 10 °C, and add dropwise with constant agitation 10 mL of the sample, keeping the temperature of the mixture below 20 °C. The mixture should have no more color than the acid or methanol before mixing.

Substances Reducing Permanganate. Cool 20 mL to 15 °C, add 0.10 mL of 0.1 N potassium permanganate volumetric solution, and allow to stand at 15 °C for 15 min. The pink color should not be entirely discharged.

Solubility in Water. Dilute 15 mL with 45 mL of water, mix, and allow to stand for 1 h. The solution should be as clear as an equal volume of water.

Color (APHA). (Page 43).

Water. (Page 31, Method 1). Use 25 mL (19.8 g) of the sample.

Residue after Evaporation. (Page 25). Evaporate 100 g (125 mL) in a tared, preconditioned dish on a hot plate ($\approx$100 °C), and dry the residue at 105 °C for 30 min.

Carbonyl Compounds. (Page 54). Use 5.0 g (6.4 mL) of sample. For the standard, use 0.05 mg each of acetaldehyde, acetone, and formaldehyde.

Titrable Acid. To 25 mL of water and 10 mL of ethyl alcohol in a glass-stoppered flask, add 0.50 mL of phenolphthalein indicator solution and 0.01 N sodium hydroxide until a slight pink color persists after shaking for 30 s. Add 15 g (19 mL) of sample, mix well, and titrate with 0.01 N sodium hydroxide until the pink color is restored. Not more than 0.50 mL should be required.

> *Note:* Special care should be taken during the addition of the sample and titration to avoid contamination from carbon dioxide.

Titrable Base. Dilute 22.6 g (28.6 mL) with 25 mL of water and add 0.15 mL of methyl red indicator solution. Not more than 0.40 mL of 0.01 N hydrochloric acid should be required to produce a pink color.

Ultraviolet Spectrophotometry. Use the procedure on page 86 to determine the absorbance.

Liquid Chromatography Suitability. Analyze the sample, using the gradient elution procedure cited on page 84. Use the procedure on page 86 to determine the absorbance.

Extraction–Concentration Suitability. Analyze the sample using the general procedure cited on page 85. Use the procedure on page 86 to determine the absorbance.

2-Methoxyethanol
Ethylene Glycol Monomethyl Ether
$CH_3OCH_2CH_2OH$ Formula Wt 76.10   CAS No. 109-86-4

GENERAL DESCRIPTION
Typical appearance: clear liquid
Analytical use: solvent
Change in state (approximate): boiling point, 124 °C
Aqueous solubility: miscible
Density: 0.97

SPECIFICATIONS
Assay . . . . $\geq$99.3% $CH_3OCH_2CH_2OH$

Maximum Allowable

Color (APHA) 10
Titrable acid. . . . . .0.002 meq/g
Water (H_2O) 0.1%

TESTS

Assay. Analyze the sample by gas chromatography using the general parameters cited on page 80. The following specific conditions are also required.

Column: Type III, polyethylene glycol

Measure the area under all peaks, and calculate the 2-methoxyethanol content in area percent. Correct for water content.

Color (APHA). (Page 43).

Titrable Acid. To 50 mL of water in a conical flask, add 0.15 mL of phenol red indicator solution, and adjust to a pink color with 0.01 N sodium hydroxide. Add 25 g (26 mL) of sample, and titrate with 0.01 N sodium hydroxide to the same color. Not more than 5.0 mL should be required.

Water. (Page 31, Method 1). Use 25 mL (24 g) of sample.

4-(Methylamino)phenol Sulfate

$$\left[\begin{array}{c} \text{OH} \\ \text{NHCH}_3 \end{array}\right]_2 \cdot \text{H}_2\text{SO}_4$$

$(CH_3NHC_6H_4OH)_2 \cdot H_2SO_4$ **Formula Wt 344.38**   **CAS No. 55-55-0**

GENERAL DESCRIPTION

Typical appearance: solid; discolors in air

Analytical use: determination of phosphate

Change in state (approximate): melting point, 260 °C, with decomposition

Aqueous solubility: soluble in 20 parts cold, 6 parts boiling water

SPECIFICATIONS

Assay 99.0–101.5%
Residue after ignition . . . . ≤0.1%
Suitability for determination of phosphate . . . . Passes test

TESTS

Assay. (By titration of reductive capacity). Weigh about 250 mg to the nearest 0.1 mg. Transfer to a 500-mL conical flask containing 100 mL of water and 10 mL of 0.1 N sulfuric acid. Dissolve, add 0.15 mL of 0.025 M ferroin indicator solution, and titrate with 0.1 N ceric ammonium sulfate volumetric solution to a light green color that persists for 15 s. One milliliter of 0.1 N ceric ammonium sulfate corresponds to 0.00861 g of $(CH_3NHC_6H_4OH)_2 \cdot H_2SO_4$.

$$\% \ (CH_3NHC_6H_4OH)_2 \cdot H_2SO_4 = \frac{[mL \times N \ (NH_4)_4Ce(SO_4)_4] \times 8.61}{Sample \ wt \ (g)}$$

Residue after Ignition. (Page 26). Ignite 5.0 g.

Suitability for Determination of Phosphate. Dissolve 2.0 g in 100 mL of water. To 10 mL of this solution add 90 mL of water and 20 g of sodium bisulfite, dissolve, and mix. Transfer 1 mL of this solution to each of two solutions containing 25 mL of 0.5 N sulfuric acid and 1 mL of ammonium molybdate–sulfuric acid reagent solution. For the control, add 0.005 mg of phosphate ion (PO_4) to one of the solutions. For the sample, use the remaining solution. Allow the solutions to stand at room temperature for 2 h. The control should turn perceptibly darker blue than the sample.

Methyl Orange
4-(Dimethylamino)azobenzenesulfonic Acid, Sodium Salt; 4-[[4-(Dimethylamino)-phenyl]azo]benzenesulfonic Acid, Sodium Salt; C.I. Acid Orange 52

$$(CH_3)_2N\!-\!\!\bigcirc\!\!-\!N\!=\!N\!-\!\!\bigcirc\!\!-\!SO_3Na$$

C₁₄H₁₄N₃NaO₃S   **Formula Wt 327.33**   **CAS No. 547-58-0**

GENERAL DESCRIPTION

Typical appearance: orange-yellow solid
Analytical use: pH indicator
Aqueous solubility: soluble in 500 parts water; more soluble in hot water

SPECIFICATIONS

Clarity of solution . . . . Passes test
Visual transition interval . . . . From pH 3.2 (pink) to pH 4.4 (yellow)

TESTS

Clarity of Solution. Dissolve 0.1 g in 100 mL of water. Not more than a faint trace of turbidity or insoluble matter should remain. Reserve the solution for the test for visual transition interval.

Visual Transition Interval. Dissolve 1 g of potassium chloride in 100 mL of water. Adjust the pH of the solution to 3.0 (using a pH meter) with 0.01 N hydrochloric acid. Add 0.15 mL of the 0.1% solution reserved from the test for clarity of solution. The solution should have a definite pink color. Titrate the solution with 0.01 N sodium hydroxide to a pH of 3.2 (using a pH meter). A small amount of yellow color should appear, producing a pinkish orange color. Continue the titration to pH 4.4 (using a pH meter); the solution should have a definite yellow color.

4-Methyl-2-pentanone
Methyl Isobutyl Ketone

$$CH_3 - \overset{\overset{\displaystyle O}{\|}}{C}CH_2\overset{\overset{\displaystyle CH_3}{|}}{C}HCH_3$$

(CH₃)₂CHCH₂COCH₃   **Formula Wt 100.16**   **CAS No. 108-10-1**

GENERAL DESCRIPTION

Typical appearance: clear, colorless liquid
Analytical use: solvent; extraction solvent
Change in state (approximate): boiling point, 118 °C
Aqueous solubility: moderately soluble
Density: 0.8

SPECIFICATIONS

Assay . . . . ≥98.5% (CH₃)₂CHCH₂COCH₃

Maximum Allowable

Color (APHA) 15
Residue after evaporation 0.005%
Titrable acid . . . . 0.002 meq/g
Water (H₂O) 0.1%

TESTS

Assay. Analyze the sample by gas chromatography using the general parameters cited on page 80. The following specific conditions are also required.

   Column: Type I, methyl silicone

Measure the area under all peaks, and calculate the 4-methyl-2-pentanone content in area percent. Correct for water content.

Color (APHA). (Page 43).

Residue after Evaporation. (Page 25). Evaporate 100 g (125 mL) to dryness in a tared, preconditioned dish on a hot plate (≈100 °C), and dry the residue at 125 °C for 30 min.

Titrable Acid. To 25 mL of ethyl alcohol in a glass-stoppered flask, add 0.50 mL of phenolphthalein indicator solution. Add 0.01 N sodium hydroxide solution until a slight pink color persists after shaking for 30 s. Add 25 mL of the sample, mix, and titrate with 0.01 N sodium hydroxide until the pink color is restored. Not more than 4.0 mL of the sodium hydroxide solution should be required.

   Note: Great care should be taken in the test during the addition of the sample and the titration to avoid contamination from carbon dioxide.

Water. (Page 31, Method 1). Use 25 mL (20 g) of the sample and a methanol-free system to prevent ketal formation with liberation of water.

1-Methyl-2-pyrrolidone
N-Methyl Pyrrolidone

C_5H_9NO	Formula Wt 99.13	CAS No. 872-50-4

Suitable for general use or in ultraviolet spectrophotometry. Product labeling shall desig-nate the uses for which suitability is represented on the basis of meeting the relevant speci-fications and tests. The ultraviolet spectrophotometry specifications include all of the spec-ifications for general use.

GENERAL DESCRIPTION

Typical appearance: clear liquid

Analytical use: solvent

Change in state (approximate): boiling point, 202 °C

Aqueous solubility: miscible

Density: 1.03

SPECIFICATIONS

General Use

Assay . . . . ≥99.0% C_5H_9NO

Maximum Allowable

Color (APHA) 50

Water (H_2O) 0.05%

Free amines (as CH_3NH_2) 0.01%

Chloride (Cl) . . . . 1 ppm

Specific Use

Ultraviolet Spectrophotometry

Wavelength (nm)	Absorbance (AU)
400	0.01
350	0.01
300	0.05
285	0.15
275	1.00

TESTS

Assay. Analyze the sample by gas chromatography by using the general parameters cited on page 80. The following specific conditions are also required.

Column: Type I, methyl silicone

Measure the area under all the peaks, and calculate the 1-methyl-2-pyrrolidone content in area percent. Correct for water content.

Color (APHA). (Page 43).

Water. (Page 31, Method 1). Use 24.3 mL (25 g) of the sample and a methanol-free system to prevent ketal formation with liberation of water.

Free Amines. Accurately weigh a 10.0-g sample in a tared, glass-stoppered flask. Add 100 mL of water and 0.1 mL of methyl red indicator solution. Titrate with standardized 0.01 N hydrochloric acid to the first permanent pink color. No more than 3.2 mL should be required. One milliliter of 0.01 N hydrochloric acid corresponds to 0.0003106 g of CH_3NH_2.

$$\% \ CH_3NH_2 = \frac{(mL \times N \ HCl) \times 3.106}{Sample \ wt \ (g)}$$

Chloride. Dilute 24.3 mL (25 g) of sample to the mark in a 100-mL volumetric flask with deionized water. Transfer 10.0 mL of sample solution to each of two 50-mL Nessler tubes and 30.0 mL to each of two additional Nessler tubes. Add 0.5 mL of chloride ion (Cl) standard solution to one tube containing 10 mL of sample solution (making a 0.5-ppm standard), and add 1.0 mL of chloride standard solution to the other tube, making a 1.0-ppm standard. Also add 0.5 mL of chlorine standard to one tube containing 30 mL (making a spiked sample). Dilute all tubes to about 45 mL with deionized water, and mix. Add 1 mL of nitric acid and 1 mL of 0.1 N silver nitrate volumetric solution to each tube. Dilute to 50 mL, and mix thoroughly. After 10 min, observe the opalescence. Any opalescence in the sample must be no greater than that in the 0.5-ppm standard, and any opalescence in the spiked sample must be no greater than that in the 1-ppm standard.

Ultraviolet Spectrophotometry. Use the procedure on page 86 to determine the absorbance.

Methyl Red
2-[4-(Dimethylamino)phenylazo]benzoic Acid; C.I. Acid Red 2

$C_{15}H_{15}N_3O_2$ (free acid)	**Formula Wt 269.30**	**CAS No. 493-52-7**
$C_{15}H_{14}N_3O_2Na$ (sodium salt)	**Formula Wt 291.28**	**CAS No. 845-10-3**

Note: Three forms of this indicator are available: the compound named above (I), the sodium salt of the compound (II), and the hydrochloride of the compound (III). Compound I must meet the specifications for melting point (range), clarity of alcohol solution, and visual transition interval. This form is recommended for nonaqueous titrations, particularly when an aprotic solvent is used. Compound II must meet the specifications for clarity of alcohol solution, clarity of aqueous solution, and visual transition interval. This form is the choice for titrations in aqueous media and is also suitable for nonaqueous titrations where the medium is an amphiprotic solvent. Compound III must meet the specifications for clarity of alcohol solution and visual transition interval. This form can be used as an indicator for titrations in aqueous media and amphiprotic solvents.

GENERAL DESCRIPTION

Typical appearance: red to brown solid
Analytical use: pH indicator

SPECIFICATIONS

Melting point . . . . 179–182 °C (I)
Clarity of alcohol solution. . . . . Passes test (I, II, III)
Clarity of aqueous solution . . . . Passes test (II)
Visual transition interval . . . . From pH 4.2 (pink) to pH 6.2 (yellow) (I,II,III)

TESTS

Melting Point. (Page 45).

Clarity of Alcohol and Aqueous Solutions. Dissolve 0.1 g of II in 100 mL of water. For all three forms (I, II, and III), dissolve 0.1 g in 100 mL of alcohol. Not more than a faint trace of turbidity or insoluble matter should remain. Reserve the alcohol solutions for the test for visual transition interval.

Visual Transition Interval. Dissolve 1 g of potassium chloride in 100 mL of water. Adjust the pH of the solution to 4.2 (using a pH meter) with 0.01 N hydrochloric acid. Add 0.15 mL of the 0.1% solution reserved from the test for clarity of alcohol and aqueous solutions. The color of the solution should be pink. Titrate the solution with 0.01 N sodium hydroxide to a pH of 5.5 (using a pH meter). The solution should be orange. Continue the titration to a pH of 6.2. The solution should be yellow.

Methyl *tert*-Butyl Ether

$CH_3OC(CH_3)_3$ **Formula Wt 88.14**   **CAS No. 1634-04-4**

Suitable for general use or in ultraviolet spectrophotometry or extraction–concentration analysis. Product labeling shall designate the uses for which suitability is represented on the basis of meeting the relevant specifications and tests. The ultraviolet spectrophotometry specifications include all of the specifications for general use. The extraction–concentration suitability specifications include only the general use specification for color.

GENERAL DESCRIPTION

Typical appearance: clear liquid
Analytical use: organic solvent
Change in state (approximate): boiling point, 55 °C

Aqueous solubility: 4.8 g in 100 g

Density: 0.74

SPECIFICATIONS

General Use

Assay . . . . $\geq$99.0% $C_5H_{12}O$

Maximum Allowable

Color (APHA). 10

Peroxide (as H_2O_2). . . . . 1 ppm

Residue after evaporation 0.001%

Water (H_2O) 0.05%

Specific Use

Ultraviolet Spectrophotometry

Wavelength (nm)	Absorbance (AU)
350	0.01
300	0.01
250	0.10
225	0.50
210	1.0

Extraction–Concentration Suitability

Absorbance. . . . . Passes test

GC–FID . . . . Passes test

GC–ECD . . . . Passes test

TESTS

Assay. Analyze the sample by gas chromatography by using the general parameters cited on page 80. The following specific conditions are also required.

Column: Type I, methyl silicone

Measure the area under all the peaks, and calculate the methyl *tert*-butyl ether content in area percent. Correct for water content.

Color (APHA). (Page 43).

Peroxide. To 47 mL (35 g) of sample in a separatory funnel, add 5.0 mL of titanium tetrachloride reagent solution. Shake vigorously, let separate, and drain the lower layer into a 25-mL glass-stoppered, graduated cylinder. For the standard, transfer 5.0 mL of titanium tetrachloride reagent solution to a similar graduated cylinder, and add 0.035 mg of freshly prepared hydrogen peroxide ion (H_2O_2) standard solution. Dilute both solutions with water to 10.0 mL, and mix. Let mixtures stand for 30 min. Any yellow color in the sample solution should not exceed that in the standard. The color intensities may be determined with a spectrophotometer in 1-cm cells at a wavelength of 410 nm.

Caution: If the sample fails the peroxide test, do not perform the test for residue after evaporation.

Residue after Evaporation. (Page 25). Evaporate 136 mL (100 g) to dryness in a tared, preconditioned platinum dish on a hot plate ($\approx$100 °C), and dry the residue at 105 °C for 30 min. The weight of the residue should not exceed 0.001 g.

Water. (Page 31, Method 1). Use 25 mL (18.4 g) of the sample.

Ultraviolet Spectrophotometry. Use the procedure on page 86 to determine the absorbance.

Extraction–Concentration Suitability. Analyze the sample by using the general procedure cited on page 85. Use the procedure on page 86 to determine the absorbance.

Methylthymol Blue, Sodium Salt

3,3′-Bis[*N*,*N*-di(carboxymethyl)-aminomethyl]thymolsulfonphthalein, Sodium Salt

$C_{37}H_{40}N_2O_{13}Na_4S$	Formula Wt 844.74	CAS No. 1945-77-3

GENERAL DESCRIPTION

Typical appearance: green or brown solid
Analytical use: pH indicator

SPECIFICATIONS

Clarity of solution . . . . Passes test
Suitability as a metal indicator . . . . Passes test
Visual transition interval . . . . From pH 6.5 (yellow) to pH 8.5 (blue)

TESTS

Clarity of Solution. Dissolve 0.1 g in 100 mL of water. Not more than a trace of turbidity or insoluble matter should remain. Reserve this solution for the test for visual transition interval.

Suitability as a Metal Indicator. Mix 50 mL of water with 5 mL of diethylamine. Add 25–50 mg of methylthymol blue indicator mixture and 10.0 mL of 0.1 M EDTA volumetric solution. The solution is colorless or gray. Titrate with 0.1 M strontium chloride solution (described below). The color change is from colorless-gray to blue. Add 0.05 mL of 0.1 M EDTA. The color should return to the colorless-gray.

> **Strontium Chloride Solution, 0.1 M.** Dissolve 0.666 g of strontium chloride hexahydrate in water, and dilute to 25.0 mL in a volumetric flask.

Visual Transition Interval. Dissolve 1.0 g of potassium chloride in 100 mL of water. Adjust the pH of the solution to 6.5 (using a pH meter) with 0.01 N hydrochloric acid or 0.01 N sodium hydroxide. Add 0.15 mL of the 0.1% solution reserved from the test for clarity of solution. The color of solution should be yellow. Titrate the solution with 0.01 N sodium hydroxide to a pH of 8.5 (using a pH meter). The color of the solution should be blue.

Molybdenum Trioxide
Molybdenum(VI) Oxide; Molybdic Acid Anhydride

MoO$_3$	Formula Wt 143.94	CAS No. 1313-27-5

GENERAL DESCRIPTION

Typical appearance: white, slightly yellow, green, or gray solid
Analytical use: trace metal analysis
Change in state (approximate): melting point, 795 °C
Aqueous solubility: very slightly soluble

SPECIFICATIONS

Assay	≥99.5% MoO$_3$

Maximum Allowable

Insoluble in dilute ammonium hydroxide	0.01%
Chloride (Cl)	0.002%
Nitrate (NO$_3$)	Passes test
Arsenate, phosphate, and silicate (as SiO$_2$)	0.001%
Phosphate (PO$_4$)	5 ppm
Sulfate (SO$_4$)	0.02%
Ammonium (NH$_4$)	0.002%
Heavy metals (as Pb)	0.005%

TESTS

Assay. (By complexometric titration of molybdenum). Weigh accurately about 0.6 g, and dissolve in 100 mL of 2.5% ammonium hydroxide reagent solution. Adjust the pH of the solution to 4.0 (using a pH meter) with 10% nitric acid reagent solution. Add saturated hexamethylenetetramine reagent solution to a pH of 5–6. Heat the solution to 60 °C, and titrate with 0.1 M lead nitrate volumetric solution. Using 0.1% PAR indicator solution, titrate from yellow color to the first permanent pink end point. One milliliter of 0.1 N lead nitrate corresponds to 0.01439 g of MoO$_3$.

$$\% \, MoO_3 = \frac{[mL \times M \, Pb(NO_3)_2] \times 14.39}{Sample \, wt \, (g)}$$

Insoluble in Dilute Ammonium Hydroxide. Dissolve 10 g in 120 mL of dilute ammonium hydroxide (1 + 6), heat to boiling, and digest in a covered beaker on a hot plate ($\approx$100 °C) for 2 h. Filter through a tared, preconditioned filtering crucible, wash thoroughly, and dry at 105 °C.

Chloride. Dissolve 1.0 g in 4 mL of ammonium hydroxide, heat on a hot plate ($\approx 100\ °C$) until completely dissolved, and evaporate to dryness. Dissolve the residue in 20 mL of water, filter if necessary through a chloride-free filter, add 4 mL of nitric acid, and dilute with water to 30 mL. To 15 mL, add 1 mL of silver nitrate reagent solution. Any turbidity should not exceed that produced by 0.01 mg of chloride ion (Cl) in an equal volume of solution containing the quantities of reagents used in the test. The comparison is best made by the general method for chloride in colored solutions, page 35.

Nitrate. Grind 1.0 g with 9 mL of water and 1 mL of sodium chloride solution containing 5 mg of sodium chloride. Add 0.20 mL of indigo carmine reagent solution and 10 mL of sulfuric acid. The blue color should not be completely discharged in 5 min. (Limit about 0.003%)

Arsenate, Phosphate, and Silicate. Dissolve 2.5 g in 60 mL of water and enough silica-free ammonium hydroxide to effect dissolution (about 4 mL) in a platinum dish. For the control, dissolve 0.5 g in 60 mL of water and enough silica-free ammonium hydroxide to effect dissolution and add 0.02 mg of silica (SiO_2). Heat on a hot plate ($\approx 100\ °C$) until the solutions are neutral as determined with an external indicator (at least 30 min). Cool, and adjust the pH to between 3 and 4 with an external indicator. Transfer to beakers, and dilute with water to 80 mL. Add enough bromine water to impart a distinct yellow color to the solutions. Adjust the pH of each solution to 1.8 with dilute hydrochloric acid (1 + 9) (using a pH meter). Heat just to boiling, and allow to cool to room temperature. (If a precipitate forms, it will dissolve when the solution is acidified in the next operation.) Add 10 mL of hydrochloric acid, and dilute with water to 100 mL. Transfer the solutions to separatory funnels, and add 30 mL of 4-methyl-2-pentanone and 1 mL of butyl alcohol. Shake vigorously, and allow to separate. Draw off and discard the aqueous phase. Wash the ketone phase three times with 10-mL portions of dilute hydrochloric acid (1 + 99), discarding the aqueous phase each time. To the washed ketone phase, add 10 mL of dilute hydrochloric acid (1 + 99), to which has just been added 0.2 mL of a freshly prepared 2% stannous chloride reagent solution. Any blue color in the solution of the sample should not exceed that in the control.

Phosphate. (Page 40, Method 3).

Sulfate. Boil 1.0 g with 15 mL of dilute nitric acid (1 + 2) for 5 min. Cool thoroughly, dilute with water to 50 mL, mix well, and filter. Evaporate 10 mL of the filtrate to dryness on a hot plate ($\approx 100\ °C$). Dissolve the residue in 4 mL of water plus 1 mL of dilute hydrochloric acid (1 + 19), filter if necessary through a small filter, wash with two 2-mL portions of water, and dilute with water to 10 mL. Add 1 mL of 12% barium chloride reagent solution. Any turbidity should not exceed that produced by 0.04 mg of sulfate ion (SO_4) in an equal volume of solution containing the quantities of reagents used in the test. Compare 10 min after adding the barium chloride to the sample and standard solutions.

Ammonium. Dissolve 1.0 g in 15 mL of freshly boiled 10% sodium hydroxide reagent solution, and dilute with water to 100 mL. To 50 mL, add 2 mL of Nessler reagent. Any color should not exceed that produced by 0.01 mg of ammonium ion (NH_4) in an equal volume of solution containing 1.5 mL of 10% sodium hydroxide reagent solution and 2 mL of Nessler reagent.

Heavy Metals. Dissolve 1.0 g in 15 mL of 10% sodium hydroxide reagent solution, add 2 mL of ammonium hydroxide, and dilute with water to 40 mL. For the control, add 0.025 mg of lead ion (Pb) to 10 mL of the solution, and dilute with water to 40 mL. For the sample, dilute the remaining solution with water to 40 mL. Add 5 mL of freshly prepared hydrogen sulfide water to each, and mix. Any color in the solution of the sample should not exceed that in the control.

Molybdic Acid, 85%

CAS No. 7782-91-4

Note: This reagent consists largely of an ammonium molybdate.

GENERAL DESCRIPTION

Typical appearance: white or slightly yellowish solid
Analytical use: determination of phosphate
Aqueous solubility: slightly soluble

SPECIFICATIONS

Assay . . . . $\geq$85.0% MoO_3

Maximum Allowable

Insoluble in dilute ammonium hydroxide 0.01%
Chloride (Cl) 0.002%
Arsenate, phosphate, and silicate (as SiO_2). 0.001%
Phosphate (PO_4) . . . . 5 ppm
Sulfate (SO_4) 0.2%
Heavy metals (as Pb) 0.003%

TESTS

Assay. (By complexometric titration of molybdenum). Weigh accurately about 0.6 g, and dissolve in 100 mL of 2.5% ammonium hydroxide reagent solution. Adjust the pH of the solution to 4.0 (using a pH meter) with 10% nitric acid reagent solution. Add saturated hexamethylenetetramine reagent solution to a pH of 5–6. Heat the solution to 60 °C, and titrate with 0.1 M lead nitrate volumetric solution. Using 0.1% PAR indicator solution, titrate from yellow to the first permanent pink end point. One milliliter of 0.1 N lead nitrate corresponds to 0.014394 g of MoO_3.

$$\% \, MoO_3 = \frac{[mL \times M \, Pb(NO_3)_2] \times 14.394}{Sample \, wt \, (g)}$$

Insoluble in Dilute Ammonium Hydroxide. Dissolve 10 g in 100 mL of dilute ammonium hydroxide (1 + 9), heat to boiling, and digest in a covered beaker on a hot plate ($\approx$100 °C) for 2 h. Filter through a tared, preconditioned filtering crucible, wash thoroughly, and dry at 105 °C.

Chloride. Dissolve 1.0 g in 4 mL of ammonium hydroxide, heat on a hot plate ($\approx$100 °C) until completely dissolved, and evaporate to dryness. Dissolve the residue in 20 mL of water, add 4 mL of nitric acid, and dilute with water to 30 mL. To 15 mL, add 1 mL of silver nitrate reagent solution. Any turbidity should not exceed that produced by 0.01 mg of chloride ion (Cl) in an equal volume of solution containing the quantities of reagents used in the test. The comparison is best made by the general method for chloride in colored solutions, page 35.

Arsenate, Phosphate, and Silicate. Dissolve 2.5 g in 60 mL of water and enough silica-free ammonium hydroxide to effect dissolution (about 4 mL) in a platinum dish. For the control, dissolve 0.5 g in 60 mL of water and enough silica-free ammonium hydroxide to effect dissolution, and add 0.02 mg of silica (SiO_2). Heat on a hot plate ($\approx$100 °C) until the solutions are neutral as determined with an external indicator (at least 30 min). Cool, and adjust the pH to between 3 and 4 with an external indicator. Transfer to beakers, and dilute with water to 80 mL. Add enough bromine water to impart a distinct yellow color to the solutions. Adjust the pH of each solution to 1.8 with dilute hydrochloric acid (1 + 9) (using a pH meter). Heat just to boiling, and allow to cool to room temperature. (If a precipitate forms, it will dissolve when the solution is acidified in the next operation.) Add 10 mL of hydrochloric acid, and dilute with water to 100 mL. Transfer the solutions to separatory funnels, and add 30 mL of 4-methyl-2-pentanone and 1 mL of butyl alcohol. Shake vigorously, and allow to separate. Draw off and discard the aqueous phase. Wash the ketone phase three times with 10-mL portions of dilute hydrochloric acid (1 + 99), discarding the aqueous phase each time. To the washed ketone phase, add 10 mL of dilute hydrochloric acid (1 + 99), to which has just been added 0.2 mL of a freshly prepared 2% stannous chloride reagent solution. Any blue color in the solution of the sample should not exceed that in the control.

Phosphate. (Page 40, Method 3). Use a 20-g sample.

Sulfate. Boil 1.0 g with 15 mL of dilute nitric acid (1 + 2) for 5 min. Cool thoroughly, dilute with water to 100 mL, and filter. Evaporate 2 mL of the filtrate to dryness on a hot plate ($\approx$100 °C), dissolve the residue in 4 mL of water plus 1 mL of dilute hydrochloric acid (1 + 19), and filter if necessary through a small filter. Wash with two 2-mL portions of water, dilute with water to 10 mL, and add 1 mL of 12% barium chloride reagent solution. Any turbidity should not exceed that produced by 0.04 mg of sulfate ion (SO_4) in an equal volume of solution containing the quantities of reagents used in the test. Compare 10 min after adding the barium chloride to the sample and standard solutions.

Heavy Metals. Dissolve 1.3 g in 15 mL of 10% sodium hydroxide reagent solution, add 2 mL of ammonium hydroxide, and dilute with water to 40 mL. For the control, add 0.02 mg of lead ion (Pb) to 10 mL of the solution, and dilute with water to 40 mL. For the sample, dilute the remaining portion with water to 40 mL. Add 5 mL of freshly prepared hydrogen sulfide water to each, and mix. Any color in the solution of the sample should not exceed that in the control.

Morpholine

C₄H₉NO   **Formula Wt 87.12**   **CAS No. 110-91-8**

GENERAL DESCRIPTION

Typical appearance: clear liquid
Analytical use: determination of chlorinated compounds in glycerol
Change in state (approximate): boiling point, 130 °C
Aqueous solubility: miscible with evolution of some heat
Density: 1.0
pK_a: 8.4

SPECIFICATIONS

Assay . . . . ≥99.0% C₄H₉NO
Color (APHA) . . . . ≤15
Boiling range . . . . 126.0–130.0 °C

TESTS

Assay. (By acid–base titrimetry). Add 0.30–0.40 mL of mixed indicator solution (described below) to 50 mL of water in a 250-mL flask, and neutralize by the dropwise addition of 0.1 N hydrochloric acid just to the disappearance of the green color. Weigh accurately about 3.0 g of the sample into the flask, and swirl to effect complete dissolution. Titrate with 1 N hydrochloric acid volumetric solution to the disappearance of the green color. One milliliter of 1.0 N hydrochloric acid corresponds to 0.04356 g of C₄H₉NO.

$$\% \ C_4H_9NO= \frac{(mL \times N \ HCl) \times 8.71}{Sample \ wt \ (g)}$$

Mixed Indicator Solution. Prepare separate 0.1% solutions of bromocresol green in methanol and the sodium salt of methyl red in water. Mix 5 parts by volume of the bromocresol green solution with 1 part of the methyl red solution.

Color (APHA). (Page 43).

Boiling Range. Distill 100 mL by the method referenced on page 42. The corrected initial boiling point should not lie below 126 °C, and the corrected dry point should not be above 130 °C. The barometric pressure correction for morpholine is +0.045 °C per mm under 760 mm and –0.045 °C per mm above 760 mm. (The boiling point for pure morpholine at 760-mm mercury is 128.2 °C.)

Murexide
5,5′-Nitrilobarbituric Acid Monoammonium Salt

| C$_8$H$_8$N$_6$O$_6$ (anhydrous) | Formula Wt 284.20 | CAS No. 3051-09-0 |
| C$_8$H$_8$N$_6$O$_6$ · H$_2$O (monohydrate) | Formula Wt 302.22 | CAS No. 6032-80-0 |

Note: This material is available in anhydrous and monohydrate forms. The tests are identical for both forms.

GENERAL DESCRIPTION

Typical appearance: purple-red solid with green metallic luster
Analytical use: indicator
Aqueous solubility: sparingly soluble in cold water; moreso in hot

SPECIFICATIONS

Suitability as a complexometric indicator. . . . . .Passes test

TESTS

Suitability as a Complexometric Indicator. Dissolve 0.1 g in 100 mL of water. Add 1 mL to 100 mL of water, then add 10 mL of pH 10 ammoniacal buffer solution, and divide into two equal portions, A and B. Titrate portion A with nickel ion (Ni) standard solution until the blue component of the color is discharged. The color of the solution should change from purple to red. Use portion B as a control. The purple color is restored by adding 0.1 mL of 0.1 M EDTA volumetric solution. Portion A should match portion B in appearance.

Naphthol Green B
Acid Green 1

| C$_{30}$H$_{15}$FeN$_3$Na$_3$O$_{15}$S$_3$ | Formula Wt 878.46 | CAS No. 19381-50-1 |

GENERAL DESCRIPTION

Typical appearance: dark green to black solid
Analytical use: indicator

SPECIFICATIONS

Clarity of solution . . . . Passes test
Suitability as a mixed indicator for complexometry . . . . Passes test

TESTS

Clarity of Solution. Dissolve 0.02 g in 100 mL of water that contains 0.5 mL of ammonium hydroxide. Not more than a trace of turbidity or insoluble matter should remain.

Suitability as a Mixed Indicator for Complexometry. Using the test sample, prepare metalphthalein-screened indicator solution. Mix 50 mL of water and 50 mL of ethyl alcohol. Add 10.0 mL of 0.1 M EDTA volumetric solution, 10 mL of ammonium hydroxide, and 0.3 mL of the metalphthalein-screened indicator solution. Titrate with 0.1 M barium nitrate reagent solution. The color change is from gray-green to magenta. Add 0.05 mL of 0.1 M EDTA. The color should return to a gray-green.

N-(1-Naphthyl)ethylenediamine Dihydrochloride

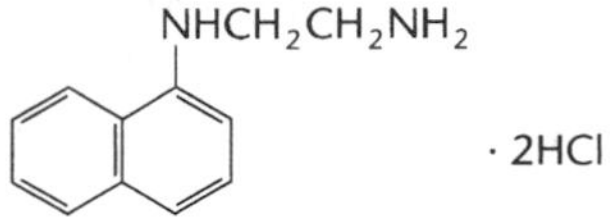

NHCH$_2$CH$_2$NH$_2$

· 2HCl

| C$_{12}$H$_{14}$N$_2$ · 2HCl | Formula Wt 259.18 | CAS No. 1465-25-4 |

GENERAL DESCRIPTION

Typical appearance: white solid; becoming pink on exposure to light
Analytical use: determination of sulfanilamide
Aqueous solubility: slightly soluble cold; readily soluble hot

SPECIFICATIONS

Sensitivity to sulfanilamide . . . . Passes test
Solubility . . . . Passes test

Maximum Allowable

Water (H$_2$O) 5%

TESTS

Sensitivity to Sulfanilamide. Dissolve 0.1 g of sample in 100 mL of water to make the sample solution. Then, in two different Nessler tubes, dilute 1.0 and 2.5 mL of sulfanilamide solution (described below) with water to 100 mL. For the blank, omit the sulfanilamide from a third tube, and fill with water to 100 mL. To each tube, add 8 mL of trichloroacetic acid solution and 1 mL of sodium nitrite solution (both described below). After 3 min, add 1 mL of ammonium sulfamate solution (described below), then after 2 additional min, add 1 mL of sample solution. The tubes containing the sulfanilamide should turn pink to red, the higher concentration producing the deeper color.

Sulfanilamide Solution. Dissolve 0.02 g of sulfanilamide in 100 mL of water.

Trichloroacetic Acid Solution. Dissolve 15 g of trichloroacetic acid in 100 mL of water.

Sodium Nitrite Solution. Dissolve 0.1 g of sodium nitrite in 100 mL of water.

Ammonium Sulfamate Solution. Dissolve 0.5 g of ammonium sulfamate in 100 mL of water.

Solubility. A solution of 1 g in 50 mL of water is clear, and dissolution is complete.

Water. (Page 31, Method 1). Use 3.0 g of the sample.

Neutral Red

Basic Red 5, C.I. 50040; 2-Amino-8-(dimethylamino)toluphenazine Hydrochloride

$C_{15}H_{17}ClN_4$ **Formula Wt 288.78**   **CAS No. 553-24-2**

GENERAL DESCRIPTION

Typical appearance: dark green solid
Analytical use: indicator for nitrogen determination; pH indicator
Aqueous solubility: soluble

SPECIFICATIONS

Clarity of solution . . . . Passes test
Visual transition interval . . . . pH 6.8 (red) to pH 8.0 (yellow)

TESTS

Clarity of Solution. Accurately weigh 0.1 g, and dissolve in 100 mL of water. The resulting solution should be clear, and no visible amount of insoluble residue remains. Reserve the solution for the visual transition interval test.

Visual Transition Interval. Dissolve 1 g of potassium chloride in 100 mL of water. Adjust the pH of the solution to 6.0, using a pH meter, with 0.01 N hydrochloric acid or 0.1 N sodium hydroxide. Add 0.15 mL of the 0.1% solution reserved from the test for clarity of solution. Titrate the solution with 0.01 N sodium hydroxide to pH 6.8 (using a pH meter).

The color of the solution should be red. Continue the titration to a pH of 8.0. The solution should be yellow.

Nickel Chloride Hexahydrate
Nickel(II) Chloride Hexahydrate; Nickelous Chloride Hexahydrate

$NiCl_2 \cdot 6H_2O$	Formula Wt 237.69	CAS No. 7791-20-0

GENERAL DESCRIPTION

Typical appearance: green solid
Analytical use: electroplating
Aqueous solubility: 254 g in 100 mL at 20 °C

SPECIFICATIONS

Assay . . . . ≥97% $NiCl_2 \cdot 6H_2O$

Maximum Allowable

Insoluble matter	0.005%
Nitrogen compounds (as N)	0.005%
Sulfate (SO₄)	0.005%
Calcium (Ca)	0.01%
Cobalt (Co)	0.1%
Copper (Cu)	0.002%
Iron (Fe)	0.002%
Lead (Pb)	0.005%
Magnesium (Mg)	0.01%
Manganese (Mn)	0.002%
Potassium (K)	0.01%
Sodium (Na)	0.05%
Zinc (Zn)	0.01%

TESTS

Assay. (By complexometric titration of nickel). Weigh accurately 1 g, transfer to a 400-mL beaker, and dissolve in 250 mL of water. Add 10 mL of pH 10 ammoniacal buffer solution and about 50 mg of murexide indicator mixture. Titrate immediately with 0.1 M EDTA volumetric solution to a purple color. One milliliter of 0.1 M EDTA corresponds to 0.02377 g of $NiCl_2 \cdot 6H_2O$.

$$\% \, NiCl_2 \cdot 6H_2O = \frac{(mL \times M \, EDTA) \times 23.77}{Sample \, wt \, (g)}$$

Insoluble Matter. (Page 25). Dissolve 20 g in 150 mL of water. Retain the filtrate for the test for sulfate.

Nitrogen Compounds. (Page 39). Use 0.4 g. For the standard, use 0.02 mg of nitrogen.

Sulfate. (Page 40, Method 1). Use 7.5 mL (1.0 g) of the filtrate retained from the test for insoluble matter. Allow 30 min for turbidity to develop.

Calcium, Cobalt, Copper, Iron, Lead, Magnesium, Manganese, Potassium, Sodium, and Zinc. (By flame AAS, page 63).

Sample Stock Solution. Dissolve 20.0 g of sample in 75 mL of water and 2 mL of nitric acid. Transfer to a 100-mL volumetric flask, and dilute to the mark with water (1 mL = 0.20 g).

Element	Wavelength (nm)	Sample Wt (g)	Standard Added (mg)	Flame Type*	Background Correction
Ca	422.7	2.0	0.05; 0.10	N/A	No
Co	240.7	2.0	0.02; 0.04	A/A	Yes
Cu	324.8	2.0	0.05; 0.10	A/A	Yes
Fe	248.3	4.0	0.04; 0.08	A/A	Yes
Pb	217.0	2.0	0.10; 0.20	A/A	Yes
Mg	285.2	0.20	0.005; 0.01	A/A	Yes
Mn	279.5	2.0	0.02; 0.04	A/A	Yes
K	766.5	0.20	0.01; 0.02	A/A	No
Na	589.0	0.10	0.025; 0.05	A/A	No
Zn	213.9	2.0	0.05; 0.10	A/A	Yes

*A/A is air/acetylene; N/A is nitrous oxide/acetylene.

Nickel Nitrate Hexahydrate
Nickel(II) Nitrate Hexahydrate; Nickelous Nitrate Hexahydrate

$Ni(NO_3)_2 \cdot 6H_2O$ **Formula Wt 290.81**   **CAS No. 13478-00-7**

GENERAL DESCRIPTION

Typical appearance: green solid
Analytical use: electroplating; standard preparation
Change in state (approximate): melting point, 140 °C (release of crystalline water)
Aqueous solubility: about 238 g in 100 mL at 0 °C

SPECIFICATIONS

Assay . . . . 98.0–102.0% $Ni(NO_3)_2 \cdot 6H_2O$

Maximum Allowable

Insoluble matter 0.005%
Chloride (Cl) 0.002%
Sulfate (SO₄) 0.002%
Calcium (Ca) 0.01%
Cobalt (Co) 0.1%
Copper (Cu) . . . . 5 ppm
Iron (Fe) 0.001%

Lead (Pb) 0.005%
Magnesium (Mg). 0.01%
Manganese (Mn) 0.002%
Potassium (K) 0.01%
Sodium (Na) 0.05%
Zinc (Zn). 0.01%

TESTS

Assay. (By complexometric titration of nickel). Weigh accurately 1 g, transfer to a 400-mL beaker, and dissolve in 250 mL of water. Add 10 mL of pH 10 ammoniacal buffer solution and about 50 mg of murexide indicator mixture. Titrate immediately with 0.1 M EDTA volumetric solution to a purple color. One milliliter of 0.1 M EDTA corresponds to 0.02908 g of $Ni(NO_3)_2 \cdot 6H_2O$.

$$\% \ Ni(NO_3)_2 \cdot 6H_2O = \frac{(mL \times M \ EDTA) \times 29.08}{Sample \ wt \ (g)}$$

Insoluble Matter. (Page 25). Dissolve 20 g in 200 mL of water. Retain the filtrate for the test for sulfate.

Chloride. Dissolve 0.5 g in 15 mL of water, filter if necessary through a small, chloride-free filter, and add 1 mL of nitric acid and 1 mL of silver nitrate reagent solution. Any turbidity should not exceed that produced by 0.01 mg of chloride ion (Cl) in an equal volume of solution containing the quantities of reagents used in the test. The comparison is best made by the general method for chloride in colored solutions, page 35.

Sulfate. (Page 40, Method 1). Use 25 mL (2.5 g) of the filtrate retained from the test for insoluble matter. Allow 30 min for turbidity to develop.

Calcium, Cobalt, Copper, Iron, Lead, Magnesium, Manganese, Potassium, Sodium, and Zinc. (By flame AAS, page 63).

Sample Stock Solution. Dissolve 20.0 g of sample in 75 mL of water and 2 mL of nitric acid. Transfer to a 100-mL volumetric flask, and dilute to the mark with water (1 mL = 0.20 g).

Element	Wavelength (nm)	Sample Wt (g)	Standard Added (mg)	Flame Type*	Background Correction
Ca	422.7	2.0	0.05; 0.10	N/A	No
Co	240.7	2.0	0.02; 0.04	A/A	Yes
Cu	324.8	2.0	0.05; 0.10	A/A	Yes
Fe	248.3	4.0	0.04; 0.08	A/A	Yes
Pb	217.0	2.0	0.10; 0.20	A/A	Yes
Mg	285.2	0.20	0.005; 0.01	A/A	Yes
Mn	279.5	2.0	0.02; 0.04	A/A	Yes
K	766.5	0.20	0.01; 0.02	A/A	No
Na	589.0	0.10	0.25; 0.50	A/A	No
Zn	213.9	2.0	0.05; 0.10	A/A	Yes

*A/A is air/acetylene; N/A is nitrous oxide/acetylene.

Nickel Sulfate Hexahydrate

Nickel(II) Sulfate Hexahydrate; Nickelous Sulfate Hexahydrate

$NiSO_4 \cdot 6H_2O$ Formula Wt 262.85   CAS No. 10101-97-0

GENERAL DESCRIPTION

Typical appearance: blue to green solid

Analytical use: source of nickel in standards

Change in state (approximate): melting point, loses 5 H_2O at 100 °C; greenish-yellow anhydrous salt forms at 280 °C

Aqueous solubility: soluble in 1.4 parts water

SPECIFICATIONS

Assay . . . . 98.0–102.0% $NiSO_4 \cdot 6H_2O$

Maximum Allowable

Insoluble matter 0.005%
Chloride (Cl). 0.001%
Nitrogen compounds (as N). 0.002%
Calcium (Ca) 0.005%
Cobalt (Co). 0.002%
Copper (Cu) 0.005%
Iron (Fe). 0.001%
Magnesium (Mg) 0.005%
Manganese (Mn) 0.002%
Potassium (K) 0.01%
Sodium (Na). 0.05%

TESTS

Assay. (By complexometric titration of nickel). Weigh accurately about 1 g, transfer to a 400-mL beaker, and dissolve in 250 mL of water. Add 10 mL of pH 10 ammoniacal buffer solution and about 50 mg of murexide indicator mixture. Titrate immediately with 0.1 M EDTA volumetric solution to a purple color. One milliliter of 0.1 M EDTA corresponds to 0.02628 g of $NiSO_4 \cdot 6H_2O$.

$$\% \, NiSO_4 \cdot 6H_2O = \frac{(mL \times M \, EDTA) \times 26.28}{Sample \, wt \, (g)}$$

Insoluble Matter. (Page 25). Use 20 g dissolved in 150 mL of water.

Chloride. Dissolve 1.0 g in 15 mL of water, filter if necessary through a small chloride-free filter, and add 1 mL of nitric acid and 1 mL of silver nitrate reagent solution. Any turbidity should not exceed that produced by 0.01 mg of chloride ion (Cl) in an equal volume of solution containing the quantities of reagents used in the test. The comparison is best made by the general method for chloride in colored solutions, page 35.

Nitrogen Compounds. (Page 39). Use 1.0 g. For the standard, use 0.02 mg of nitrogen ion (N).

Calcium, Cobalt, Copper, Iron, Magnesium, Manganese, Potassium, and Sodium. (By flame AAS, page 63).

> **Sample Stock Solution.** Dissolve 20.0 g of sample in 75 mL of water and 2 mL of nitric acid. Transfer to a 100-mL volumetric flask, and dilute to the mark with water (1 mL = 0.20 g).

Element	Wavelength (nm)	Sample Wt (g)	Standard Added (mg)	Flame Type*	Background Correction
Ca	422.7	2.0	0.05; 0.10	N/A	No
Co	240.7	2.0	0.02; 0.04	A/A	Yes
Cu	324.8	2.0	0.05; 0.10	A/A	Yes
Fe	248.3	4.0	0.04; 0.08	A/A	Yes
Mg	285.2	0.20	0.005; 0.01	A/A	Yes
Mn	279.5	2.0	0.02; 0.04	A/A	Yes
K	766.5	0.20	0.01; 0.02	A/A	No
Na	589.0	0.04	0.01; 0.02	A/A	No

*A/A is air/acetylene; N/A is nitrous oxide/acetylene.

Ninhydrin

1*H*-Indene-1,2,3-trione Monohydrate; 1,2,3-Triketohydrindene Monohydrate

$C_9H_6O_4$ **Formula Wt 178.14**   **CAS No. 485-47-2**

GENERAL DESCRIPTION

Typical appearance: white to brownish-white crystals
Analytical use: determination of amino acids
Change in state (approximate): melting point, 250 °C, with decomposition
Aqueous solubility: 2 g in 100 mL

SPECIFICATIONS

Identification and melting point . . . . Passes test
Solubility . . . . Passes test
Sensitivity to amino acids. . . . . Passes test

TESTS

Identification and Melting Point. A sample heated above 150 °C becomes purple and melts with decomposition at approximately 250 °C.

Solubility. A solution of 0.10 g of sample in 10 mL of water is clear and pale yellow in color. Dissolution is complete. Retain the solution for the test for sensitivity to amino acids.

Sensitivity to Amino Acids. Dissolve 0.10 g of β-alanine in 10 mL of warm water, and heat to boiling for 5 min. Add 0.50 mL of the ninhydrin solution reserved from the solubility test, and continue heating. A rose color starts to develop within a minute and intensifies to a deep violet color on continued heating.

Nitric Acid

HNO_3 **Formula Wt 63.01**   **CAS No. 7697-37-2**

Note: This material may darken during storage as a result of photochemical reactions.

GENERAL DESCRIPTION

Typical appearance: clear, colorless liquid, fuming in moist air; with a characteristic odor
Analytical use: solvent; acidifier; trace metal standard matrix
Change in state (approximate): boiling point, 83 °C
Aqueous solubility: very soluble
Density: 1.50
pK_a: −2.3

SPECIFICATIONS

Appearance . . . . Colorless and free from suspended matter or sediment
Assay . . . . 68.0–70.0% HNO_3

	Maximum Allowable
Color (APHA)	10
Residue after ignition	5 ppm
Chloride (Cl)	0.5 ppm
Sulfate (SO_4)	1 ppm
Arsenic (As)	0.01 ppm
Heavy metals (as Pb)	0.2 ppm
Iron (Fe)	0.2 ppm

TESTS

Appearance. Mix the acid in the original container, transfer 10 mL to a test tube (20 mm × 150 mm), and compare with distilled water in a similar tube. The liquids should be equally clear and free from suspended matter.

Assay. (By acid–base titrimetry). Tare a small glass-stoppered flask containing about 15 mL of water. Quickly pipet about 2 mL of the sample under the surface, stopper, cool, and weigh accurately. Dilute with about 40 mL of water, add 0.15 mL of methyl orange indicator solution, and titrate with 1 N sodium hydroxide volumetric solution. One milliliter of 1 N sodium hydroxide corresponds to 0.06301 g of HNO₃.

$$\% \ \mathrm{HNO_3} = \frac{(\mathrm{mL \times N \ NaOH}) \times 6.301}{\mathrm{Sample \ wt \ (g)}}$$

Color (APHA). (Page 43).

Residue after Ignition. To 200 g (140 mL) in a tared, preconditioned dish add 0.05 mL of sulfuric acid, evaporate as far as possible on low heat ($\approx$100 °C), and heat gently to volatilize the excess sulfuric acid. Finally, ignite at 600 ± 25 °C for 15 min.

Chloride. Dilute 20 g (14 mL) with 10 mL of water, and add 1 mL of silver nitrate reagent solution. Prepare a standard containing 0.01 mg of chloride ion (Cl) in 20 mL of water, and add 1 mL of silver nitrate reagent solution. Evaporate the solutions to dryness on a hot plate ($\approx$100 °C). Dissolve the residues in 0.5 mL of ammonium hydroxide, dilute with water to 20 mL, and add 1.5 mL of nitric acid. Any turbidity in the solution of the sample should not exceed that in the standard.

Sulfate. (Page 41, Method 3). Use 50 g (35 mL) of sample.

Arsenic. (Page 34). To 300 g (210 mL) in a 250-mL beaker, add 5 mL of sulfuric acid, and evaporate on a hot plate ($\approx$100 °C) to a volume of about 5 mL. Transfer the beaker to a hot plate in a fume hood, and heat to dense fumes of sulfur trioxide. Cool, cautiously wash down the beaker with about 10 mL of water, evaporate again to dense fumes of sulfur trioxide, and continue the fuming for 15 min. Cool, wash down the beaker with about 10 mL of water, and repeat the fuming. Cool, cautiously wash the solution into a generator flask with sufficient water to make 35 mL, add 0.5 g of hydrazine sulfate, and proceed as described. For the standard, use 0.003 mg of arsenic ion (As).

Heavy Metals. (Page 36, Method 1). To 100 g (70 mL) in a beaker, add 1 mL of 1% sodium carbonate reagent solution, evaporate to dryness on a hot plate ($\approx$100 °C), dissolve the residue in about 20 mL of water, and dilute with water to 25 mL.

Iron. (Page 38, Method 1). Evaporate 50 g (35 mL) to dryness, dissolve the residue in 2 mL of hydrochloric acid, dilute with water to 50 mL, and use the solution without further acidification.

Nitric Acid, 90%

HNO₃	Formula Wt 63.01	CAS No. 7697-37-2

GENERAL DESCRIPTION

Typical appearance: clear, yellow, fuming liquid
Analytical use: oxidizing solvent
Aqueous solubility: miscible

Density: 1.52

pK_a: −2.3

SPECIFICATIONS

Assay . . . . ≥90.0% HNO_3
Dilution test. . . . . Passes test

	Maximum Allowable

Residue after ignition. 0.002%
Dissolved oxides . . . . .0.1% as N_2O_3
Chloride (Cl). . . . . .0.7 ppm
Sulfate (SO_4) . . . . .5 ppm
Arsenic (As) . . . . .0.3 ppm
Heavy metals (as Pb). . . . . .5 ppm
Iron (Fe). . . . . .2 ppm

TESTS

Assay. (By acid–base titrimetry). Tare a small glass-stoppered flask containing about 15 mL of water. Quickly pipet about 2 mL of the sample under the surface, stopper, cool, and weigh accurately. Dilute with about 40 mL of water, add 0.15 mL of methyl orange indicator solution, and titrate with 1 N sodium hydroxide volumetric solution. One milliliter of 1 N sodium hydroxide corresponds to 0.06301 g of HNO_3.

$$\% \ HNO_3 = \frac{(mL \times N \ NaOH) \times 6.301}{Sample \ wt \ (g)}$$

Dilution Test. Dilute 1 volume of acid with 3 volumes of water. Mix, and allow to stand for 1 h. No turbidity or precipitate should be observed.

Residue after Ignition. To 50 g (33 mL) in a tared, preconditioned dish add 0.10 mL of sulfuric acid, evaporate as far as possible on low heat (≈100 °C), and heat gently to volatilize the excess sulfuric acid. Finally, ignite at 600 ± 25 °C for 15 min.

Dissolved Oxides. Dilute 10 g (6.6 mL) to 150 mL with cold water in a casserole. Add 20 mL of dilute sulfuric acid (1 + 4), and titrate with 0.1 N potassium permanganate volumetric solution, rapidly at first, then slowly until the pink color lasts 3 min. Not more than 5.0 mL of the potassium permanganate solution should be required. One milliliter of 0.1 N potassium permanganate solution corresponds to 0.0019 g of N_2O_3.

$$\% \ N_2O_3 = \frac{(mL \times N \ KMnO_4) \times 1.9}{Sample \ wt \ (g)}$$

Chloride. Dilute 15 g (10 mL) with 10 mL of water, and add 1 mL of silver nitrate reagent solution. Prepare a standard of 0.01 mg of chloride ion (Cl) in 20 mL of water and add 1 mL of silver nitrate reagent solution. Evaporate the solutions to dryness on a hot plate (≈100 °C). Dissolve the residues in 0.5 mL of ammonium hydroxide, dilute with water to 20 mL, and add 1.5 mL of nitric acid. Any turbidity in the solution of the sample should not exceed that in the standard.

Sulfate. (Page 41, Method 3). Use 10.0 g (6.6 mL) of sample.

Arsenic. (Page 34). To 10 g (6.6 mL) in a 150-mL beaker, add 5 mL of sulfuric acid, and evaporate on a hot plate ($\approx$100 °C) to a volume of about 5 mL. Transfer the beaker to a hot plate in a hood, and heat to dense fumes of sulfur trioxide. Cool, cautiously wash down the sides of the beaker with about 10 mL of water, evaporate again to dense fumes of sulfur trioxide, and continue the fuming for 15 min. Cool, cautiously wash down the sides of the beaker with about 10 mL of water, and repeat the fuming. Cool, cautiously wash the solution into an arsine generator flask with 35 mL of water, add 0.5 g of hydrazine sulfate, and proceed as described. For the standard, use 0.003 mg of arsenic ion (As).

Heavy Metals. (Page 36, Method 1). To 4 g (2.7 mL), add 1 mL of 1% sodium carbonate reagent solution, evaporate to dryness on a hot plate ($\approx$100 °C), dissolve the residue in about 20 mL of water, and dilute with water to 25 mL.

Iron. (Page 38, Method 1). Evaporate 5 g (3.3 mL) to dryness, dissolve the residue in 2 mL of hydrochloric acid, dilute with water to 50 mL, and use the solution without further acidification.

Nitric Acid, Ultratrace

HNO_3	Formula Wt 63.01	CAS No. 7697-37-2

Suitable for use in ultratrace elemental analysis.

Note: This material may darken during storage as a result of photochemical reactions. Reagent must be packaged in a preleached Teflon bottle and used in a clean laboratory environment to maintain purity.

GENERAL DESCRIPTION

Typical appearance: clear, colorless liquid; fuming in moist air with a characteristic odor
Analytical use: analysis for trace metal content
Change in state (approximate): boiling point, 83 °C
Aqueous solubility: miscible
Density: 1.50
pK_a: −2.3

SPECIFICATIONS

Assay . . . . 65–70% HNO_3

Maximum Allowable

Chloride (Cl)	0.5 ppm
Sulfate (SO$_4$)	1 ppm
Mercury (Hg)	1 ppb
Aluminum (Al)	1 ppb

Barium (Ba) ....1 ppb
Boron (B)...5 ppb
Cadmium (Cd) ....1 ppb
Calcium (Ca) ....1 ppb
Chromium (Cr) ....1 ppb
Cobalt (Co)..1 ppb
Copper (Cu)...1 ppb
Iron (Fe)..5 ppb
Lead (Pb)...1 ppb
Lithium (Li)..1 ppb
Magnesium (Mg) ....1 ppb
Manganese (Mn) ....1 ppb
Molybdenum (Mo) ....1 ppb
Potassium (K) ....1 ppb
Silicon (Si)...5 ppb
Sodium (Na)...5 ppb
Strontium (Sr) ....1 ppb
Tin (Sn) ....1 ppb
Titanium (Ti) ....1 ppb
Vanadium (V) ....1 ppb
Zinc (Zn) ....1 ppb
Zirconium (Zr)...1 ppb

TESTS

Assay. (By acid–base titrimetry). Tare a small, glass-stoppered flask containing about 15 mL of water. Quickly pipet about 2 mL of the sample under the surface, stopper, and weigh accurately. Dilute with about 40 mL of water, add 0.15 mL of methyl orange indicator solution, and titrate with 1 N sodium hydroxide volumetric solution. One milliliter of 1 N sodium hydroxide corresponds to 0.06301 g of HNO_3.

$$\% \, HNO_3 = \frac{(mL \times N \, NaOH) \times 6.301}{Sample \, wt \, (g)}$$

Chloride. See test for nitric acid, page 461.

Sulfate. (Page 41, Method 3). Use a 50-g sample.

Mercury. (By CVAAS, page 65). To each of three 100-mL volumetric flasks containing about 35 mL of water, add 20.0 g of sample. To the second and third flasks, add mercury ion (Hg) standards of 10 ng (0.5 ppb) and 20 ng (1.0 ppb), respectively. Dilute to the mark with water. Mix. Zero the instrument with the blank, and determine the mercury content using a suitable mercury analyzer system (1.0 mL of 0.01 μg/mL Hg = 10 ng).

Trace Metals. Determine the aluminum, barium, boron, cadmium, calcium, chromium, cobalt, copper, iron, lead, lithium, magnesium, manganese, molybdenum, potassium, silicon, sodium, strontium, tin, titanium, vanadium, zinc, and zirconium by the ICP–OES method described on page 69.

Nitrilotriacetic Acid

N,N-Bis(carboxymethyl)glycine

$$HO-\underset{\underset{O}{\|}}{C}-CH_2NCH_2-\underset{\underset{O}{\|}}{C}-OH$$
$$CH_2-\underset{\underset{O}{\|}}{C}-OH$$

$C_6H_9O_6N$	Formula Wt 191.14	CAS No. 139-13-9

GENERAL DESCRIPTION

Typical appearance: white to off-white solid
Analytical use: sequestration of metals; chelometric analysis
Aqueous solubility: 1.28 g in 1 L at 22.5 °C

SPECIFICATIONS

Assay . . . . $\geq$98.0% $C_6H_9O_6N$
Clarity of solution . . . . Passes test

TESTS

Assay (by Alkalimetry) and Clarity of Solution. To check for clarity of solution, weigh accurately about 0.35 g, transfer to a beaker, and dissolve in 100 mL of water. The sample should dissolve completely, with heating and stirring if necessary, to produce a clear solution.

For the assay, add 0.15 mL of phenolphthalein indicator solution, and titrate with 0.1 N sodium hydroxide. One milliliter of 0.1 N sodium hydroxide corresponds to 0.006372 g of $C_6H_9O_6N$.

$$\% \ C_6H_9O_6N = \frac{(mL \times N \ NaOH) \times 6.37}{Sample \ wt \ (g)}$$

Nitrobenzene

NO_2

$C_6H_5NO_2$	Formula Wt 123.11	CAS No. 98-95-3

GENERAL DESCRIPTION

Typical appearance: pale yellow liquid
Analytical use: solvent; anticoagulant for titrations

Change in state (approximate): boiling point, 211 °C
Aqueous solubility: 0.19 g in 100 mL at 20 °C
Density: 1.21

SPECIFICATIONS

Assay . . . . ≥99.0% $C_6H_5NO_2$

Maximum Allowable

Residue after evaporation 0.005%
Water-soluble titrable acid . . . . 0.0005 meq/g
Chloride (Cl) . . . . 5 ppm

TESTS

Assay. Analyze the sample by gas chromatography using the general parameters cited on page 80. The following specific conditions are also required.

  Column: Type I, methyl silicone

Measure the area under all peaks, and calculate the nitrobenzene content in area percent. Correct for water content.

Residue after Evaporation. (Page 25). Evaporate 20 g (16.7 mL) to dryness in a tared, preconditioned dish on a hot plate (≈100 °C), and dry the residue at 105 °C for 30 min.

Water-Soluble Titrable Acid. Shake 25 g (21 mL) with 60 mL of water for 1 min, and allow the layers to separate. Draw off and discard the nitrobenzene. Titrate 48 mL of the water extract with 0.01 N sodium hydroxide, using 0.15 mL of bromphenol blue indicator solution. Not more than 1.0 mL of 0.01 N sodium hydroxide should be required.

Chloride. Dilute the remaining 12 mL of the water extract from the test for acidity with water to 25 mL, filter through a chloride-free filter, and dilute 10 mL of the filtrate with water to 20 mL. Add 1 mL of nitric acid and 1 mL of silver nitrate reagent solution. Any turbidity should not exceed that produced by 0.01 mg of chloride ion (Cl) in an equal volume of solution containing the quantities of reagents used in the test.

Nitromethane

CH_3NO_2   **Formula Wt 61.04**   **CAS No. 75-52-5**

GENERAL DESCRIPTION

Typical appearance: clear, oily liquid
Analytical use: solvent
Change in state (approximate): boiling point, 101 °C
Aqueous solubility: slightly soluble
Density: 1.13

SPECIFICATIONS

Assay . . . . ≥95% CH_3NO_2

Maximum Allowable

Color (APHA) 10
Water (H_2O) 0.05%

TESTS

Assay. Analyze the sample by gas chromatography using the general parameters cited on page 80. The following specific conditions are also required.

Column: Type III, polyethylene glycol

Measure the area under all peaks, and calculate the nitromethane content in area percent. Correct for water content.

Color (APHA). (Page 43).

Water. (Page 31, Method 2). Use 100 μL (100 ng) of the sample.

1-Octanol

n-Octyl Alcohol

$C_8H_{18}O$ Formula Wt 130.23 CAS No. 111-87-5

GENERAL DESCRIPTION

Typical appearance: clear, colorless liquid; penetrating odor
Analytical use: solvent; determinatin of TSCA partition coefficients
Change in state (approximate): boiling point, 194–195 °C
Aqueous solubility: insoluble
Density: 0.83

SPECIFICATIONS

General Use

Assay . . . . ≥99% $C_8H_{18}O$

Maximum Allowable

Color (APHA) 10
Residue after evaporation 0.004%
Titrable acid . . . . 0.0002 meq/g

Specific Use

Ultraviolet Spectrophotometry

Wavelength (nm)	Absorbance (AU)
280–400	0.01
260	0.05
240	0.10
230	0.5
220	1.0

TESTS

Assay. Analyze the sample by gas chromatography using the general parameters cited on page 80. The following specific conditions are also required.

Column: Type I, methyl silicone

Measure the area under all peaks and calculate the 1-octanol content in area percent. Correct for water content.

Color (APHA). (Page 43).

Residue after Evaporation. (Page 25). Evaporate 100 g (120.5 mL) to dryness in a tared, preconditioned platinum dish on a hot plate (≈ 100 °C) in a well-ventilated hood, and dry the residue at 105 °C for 30 min.

Titrable Acid. To 25 mL of ethyl alcohol, add 0.15 mL of phenolphthalein indicator solution, and neutralize the solution with either 0.01 N sodium hydroxide or 0.01 N hydrochloric acid. Add 21.0 g (25 mL) of sample, and titrate with 0.01 N sodium hydroxide to a pink end point. Not more than 0.4 mL should be required.

Ultraviolet Spectrophotometry. Use the procedure on page 86 to determine the absorbance.

2-Octanol
Methyl Hexyl Carbinol; 2-Capryl Alcohol

$C_8H_{18}O$	**Formula Wt 130.23**	**CAS No. 123-96-6**

GENERAL DESCRIPTION

Typical appearance: clear, oily liquid
Analytical use: solvent
Change in state (approximate): boiling point, 179 °C
Aqueous solubility: 0.096 mL in 100 mL
Density: 0.82

SPECIFICATIONS

Assay . . . . $\geq$99% $C_8H_{18}O$

Maximum Allowable

Color (APHA) 10
Residue after evaporation 0.004%
Titrable acid . . . . 0.0002 meq/g

TESTS

Assay. Analyze the sample by gas chromatography using the general parameters cited on page 80. The following specific conditions are also required.

Column: Type I, methyl silicone

Measure the area under each peak, and calculate the 2-octanol content in area percent.

Color (APHA). (Page 43).

Residue after Evaporation. (Page 25). Evaporate 100 g (122 mL) to dryness in a tared, preconditioned platinum dish on a hot plate ($\approx$100 °C) in a fume hood, and dry the residue at 105 °C for 30 min.

Titrable Acid. To 25 mL of ethyl alcohol, add 0.15 mL of phenolphthalein indicator solution, and neutralize the solution with either 0.01 N sodium hydroxide or 0.01 N hydrochloric acid. Add 20.0 g (24.4 mL) of sample, and titrate with 0.01 N sodium hydroxide to a pink end point. Not more than 0.4 mL should be required.

Osmium Tetroxide
Osmic Acid
OsO_4 **Formula Wt 254.23**   CAS No. 20816-12-0

Caution: Osmium tetroxide vapors are poisonous and highly irritating to the eyes and respiratory membranes. Use only in a well-ventilated fume hood.

GENERAL DESCRIPTION

Typical appearance: pale yellow solid
Analytical use: catalyst in redox titrations
Change in state (approximate): melting point, 41 °C
Aqueous solubility: 7.24 g in 100 g at 25 °C

SPECIFICATIONS

Assay . . . . $\geq$98.0% OsO_4
Clarity of solution . . . . Passes test
Identification . . . . Passes test

TESTS

Clarity of Solution. Accurately weigh about 0.20 g, and add to 5 mL of reagent alcohol. The resulting solution is clear and not more than slightly yellow, and no appreciable amount of insoluble residue remains. Reserve the solution for the assay test.

Assay and Identification. (Assay by iodometry). To the solution from the clarity of solution test, add 5–7 mL of 1 N sodium hydroxide volumetric solution in a glass-stoppered flask. A dark red precipitate is produced (identification). Add 3 g of potassium iodide, mix to dissolve, slowly add 40 mL of 25% sulfuric acid, stopper, and let stand in a dark place for 20 min. Titrate the liberated iodine with 0.1 N sodium thiosulfate volumetric solution, using starch indicator solution as an external indicator (the reduced osmium solution is green, and the end point may be difficult to determine with an internal indicator). Correct for a blank. One milliliter of 0.1 N sodium thiosulfate corresponds to 0.006355 g of OsO_4.

$$\% \text{ OsO}_4 = \frac{\{[\text{mL (sample)} - \text{mL (blank)}] \times N \text{ Na}_2\text{S}_2\text{O}_3\} \times 6.355}{\text{Sample wt (g)}}$$

Oxalic Acid Dihydrate
Ethanedioic Acid Dihydrate

$$\underset{\displaystyle HO-C-C-OH \cdot 2H_2O}{\overset{\displaystyle O \quad O}{\overset{\|\quad\|}{}}}$$

$H_2C_2O_4 \cdot 2H_2O$	**Formula Wt 126.07**	**CAS No. 6153-56-6**

GENERAL DESCRIPTION

Typical appearance: white or colorless solid

Analytical use: reducing agent; precipitant for certain metals

Change in state (approximate): melting point, hydrated acid, 101 °C; anhydrous acid, 189 °C

Aqueous solubility: 14 g in 100 mL at 20 °C

Density: 1.65

pK_a: 1.3

SPECIFICATIONS

Assay . . . . 99.5–102.5% $H_2C_2O_4 \cdot 2H_2O$
Substances darkened by hot sulfuric acid . . . . Passes test

Maximum Allowable

Insoluble matter 0.005%
Residue after ignition. 0.01%
Chloride (Cl). 0.002%
Sulfate (SO_4) 0.005%
Calcium (Ca) 0.001%
Nitrogen compounds (as N). 0.001%
Heavy metals (as Pb). . . . . 5 ppm
Iron (Fe). . . . . 2 ppm

TESTS

Assay. (By titration of reductive capacity). Weigh accurately 0.25 g, and add 50 mL of water and 5 mL of sulfuric acid. Titrate slowly with 0.1 N potassium permanganate volumetric solution until about 25 mL has been added. Heat the solution to about 70 °C, and complete the titration to a permanent faint pink color. One milliliter of 0.1 N potassium permanganate corresponds to 0.006303 g of $H_2C_2O_4 \cdot 2H_2O$.

$$\% \ H_2C_2O_4 \cdot 2H_2O = \frac{(mL \times N \ KMnO_4) \times 6.303}{Sample \ wt \ (g)}$$

Substances Darkened by Hot Sulfuric Acid. Dissolve 2.0 g in 20 mL of sulfuric acid, and heat at 150 °C for 30 min. Any color should not exceed that produced in a color standard of the following composition: 0.2 mL of cobalt chloride reagent solution, 1.0 mL of ferric chloride reagent solution, and 0.2 mL of cupric sulfate reagent solution, in 20 mL of water.

Insoluble Matter. (Page 25). Use 20.0 g dissolved in 250 mL of water.

Residue after Ignition. (Page 26). Dry 10.0 g in a tared, preconditioned crucible or dish for 1 h at 105 °C, slowly ignite, and continue as described. Retain the residue for the test for iron.

Chloride. (Page 35). Dissolve 0.5 g in 20 mL of nitric acid.

Sulfate. Digest 1.0 g of sample with 1 mL of 1% sodium carbonate reagent solution, 3 mL of 30% hydrogen peroxide, and 0.5 mL of nitric acid in a covered beaker on a hot plate ($\approx$100 °C) until reaction ceases. Remove the cover, and evaporate to dryness. Dissolve the residue in 6 mL of dilute hydrochloric acid (1 + 1), and again evaporate to dryness. Dissolve the residue in 4 mL of water plus 1 mL of dilute hydrochloric acid (1 + 19), filter if necessary through a small filter, wash the precipitate with two 2-mL portions of water, and dilute with water to 10 mL. Add 1 mL of 12% barium chloride reagent solution. Any turbidity should not exceed that produced when a solution containing 0.05 mg of sulfate ion (SO_4) is treated exactly like the 1.0 g of sample. Compare 10 min after adding the barium chloride to the sample and standard solutions.

Calcium. (By flame AAS, page 63).

> **Sample Stock Solution.** Suspend 10.0 g of sample in 50 mL of nitric acid (1 + 24), and add 25 mL of 30% hydrogen peroxide. Cover, and digest on a hot plate ($\approx$100 °C) until the reaction ceases. Add 5 mL more of the hydrogen peroxide, and when the reaction ceases, uncover and evaporate to dryness. Dissolve the residue in 80 mL of nitric acid (1 + 99), transfer to a 100-mL volumetric flask, and dilute to the mark with nitric acid (1 + 99) (1 mL = 0.10 g).

Element	Wavelength (nm)	Sample Wt (g)	Standard Added (mg)	Flame Type*	Background Correction
Ca	422.7	2.0	0.02; 0.04	N/A	No

*N/A is nitrous oxide/acetylene.

Nitrogen Compounds. (Page 39). Use 1.0 g. For the standard, use 0.01 mg of nitrogen ion (N).

Heavy Metals. (Page 36, Method 1). Place 4.0 g in a covered beaker, and add 1 mL of 1% sodium carbonate reagent solution, 5 mL of 30% hydrogen peroxide, and 0.5 mL of nitric acid. Digest on a hot plate ($\approx$100 °C) until the reaction ceases, remove the cover, and evaporate to dryness. Digest the residue with 0.5 mL of hydrochloric acid and 0.1 mL of nitric acid in a covered beaker for 15 min, remove the cover, and again evaporate to dryness. Dissolve the residue in about 20 mL of water, and dilute with water to 25 mL. For the standard solution, treat 0.02 mg of lead exactly as the 4.0 g of sample.

Iron. (Page 38, Method 1). To the residue after ignition, add 3 mL of dilute hydrochloric acid (1 + 1), cover with a watch glass, and digest on a hot plate ($\approx$100 °C) for 15 min. Remove the cover, and evaporate to dryness. Dissolve the residue in 2 mL of dilute hydrochloric acid (1 + 1), add about 40 mL of water, filter if necessary, and dilute with water to 50 mL. Use 25.0 mL (5.0 g) of the solution.

Palladium Chloride
Palladium(II) Chloride

$PdCl_2$	**Formula Wt 177.33**	**CAS No. 7647-10-1**

GENERAL DESCRIPTION
Typical appearance: red solid
Analytical use: catalyst; carbon monoxide detection
Change in state (approximate): melting point, 678–680 °C

SPECIFICATIONS
Assay (dried basis)..58.5–60.5% Pd

TEST

Assay. (By gravimetry, as palladium). Dry 0.4 g of sample at 110 °C for 2 h. Accurately weigh 0.3 g of the cooled and dried sample in a 250-mL beaker. Add 10 mL of 10% hydrochloric acid, and gently warm on a hot plate ($\approx$100 °C) to dissolve the sample. Dilute to 75 mL with water. Add 100 mL of 1% dimethylglyoxime reagent solution, heat on a hot plate for 1 h. Filter through fine-porosity ashless filter paper, and wash the precipitate with 50 mL of hot water. Carefully place the filter paper containing the precipitate in a preconditioned and tared porcelain crucible with lid. Heat in an oven for 1 h at 150 °C. Finally, ignite in a muffle furnace at 600 ± 25 °C for 15 min. Cool, and weigh. The weight of the precipitate should be within limits.

$$\% \text{ Pd} = \frac{\text{Wt of precipitate (g)} \times 100}{\text{Sample wt (g)}}$$

PAN
1-(2-Pyridylazo)-2-naphthol

C₁₅H₁₁N₃O $\text{C}_{15}\text{H}_{11}\text{N}_3\text{O}$ **Formula Wt 249.27**   **CAS No. 85-85-8**

GENERAL DESCRIPTION
Typical appearance: orange-red to brown crystals
Analytical use: complexometric indicator
Change in state (approximate): melting point, 128–141°C

SPECIFICATIONS
Suitability as a complexometric indicator. . . . . Passes test

TESTS

Suitability as a Complexometric Indicator. Weigh 0.01 g, and dilute to 10 mL with reagent alcohol. Mix 50 mL of water, 50 mL of alcohol, 10.0 mL of 0.1 M EDTA volumetric solution, and 5 mL of ammonium acetate–acetic acid buffer solution. Add 0.2 mL of the PAN indicator solution, and titrate with 0.1 M cupric sulfate volumetric solution. The color change is from blue-green to deep blue. Add 0.05 mL of 0.1 M EDTA. The color should return to blue-green.

PAR
4-(2′-Pyridylazo)resorcinol

C₁₁H₉N₃O₂ (free acid)   **Formula Wt 215.21**   **CAS No. 1141-59-9**
C₁₁H₈N₃NaO₂ (monosodium salt)   **Formula Wt 237.21**   **CAS No. 16593-81-0**

Note: This specification includes both the free acid and the salt forms of the indicator.

GENERAL DESCRIPTION

Typical appearance: orange solid (free acid form)

Analytical use: complexometric indicator

SPECIFICATIONS

Clarity of solution . . . . .Passes test

Suitability as a complexometric indicator (Pb) . . . . .Passes test

TESTS

Clarity of Solution. If the indicator is the salt form, weigh 0.10 g of sample, and dissolve in 100 mL of water. If the indicator is the acid form, dissolve 0.1 g in 1 N sodium hydroxide volumetric solution, and dilute to 100 mL, maintaining the pH at 8.9. Not more than a trace of turbidity or insoluble matter should remain. Reserve the solution for the test for suitability as a complexometric indicator.

Suitability as a Complexometric Indicator. Add 10.0 mL of 0.1 M EDTA volumetric solution to 100 mL of water. Adjust the pH of the solution to 4.0 with 1% nitric acid. Add saturated hexamethylenetetramine reagent solution to a pH of 5–6. Add 0.2 mL of the 0.1% solution reserved from the clarity of solution test. Heat the solution to 60 °C, and titrate with 0.1 M lead nitrate volumetric solution. The color of the solution should change from yellow to pink. Add back 0.05 mL of 0.1 M EDTA. The solution should return to the original yellow color.

Pararosaniline Hydrochloride

4-[(4-Aminophenyl)(-4-imino-2,5-cyclohexadien-1-ylidene)methyl]benzenamine Monohydrochloride

$C_{19}H_{17}N_3 \cdot HCl$ **Formula Wt 323.82**   CAS No. 569-61-9

GENERAL DESCRIPTION

Typical appearance: green crystalline solid

Analytical use: determination of sulfur dioxide

Change in state (approximate): melting point, 268–270 °C

SPECIFICATIONS

Assay. . . . . ≥99.0% $C_{19}H_{18}N_3Cl$
Suitability for SO_2 determination:
    Absorbance of reagent blank. . . . . ≤0.170
    Calibration (μg/mL). 0.706–0.786

TESTS

Assay. (By spectrophotometry at 540 nm).

> ***Sample Stock Solution.*** Weigh accurately 0.200 g of sample. Dissolve completely by shaking with 1 N hydrochloric acid in a 100-mL volumetric flask, and dilute to volume (1 mL = 0.002 g).

Dilute 1.0 mL of sample stock solution with water to the mark in a 100-mL volumetric flask. Pipet 5.0 mL into a 50-mL volumetric flask, add 5 mL of 1 M sodium acetate–acetic acid buffer solution (described below), and dilute with water to volume. Allow to stand for 1 h at room temperature, and determine the absorbance at 540 nm with a spectrophotometer, using 1-cm cells at a slit width of 0.04 mm.

$$\% \ C_{19}H_{18}N_3Cl = \frac{A \times 21.30}{W}$$

where A = absorbance and W = weight, in grams, of sample in 1.0 mL of sample stock solution.

> ***Sodium Acetate–Acetic Acid Buffer Solution, 1 M.*** Dissolve 13.61 g of sodium acetate trihydrate in water in a 100-mL volumetric flask. Add 5.7 mL of glacial acetic acid, and dilute to volume with water.

Suitability for SO_2 Determination.

> *Note:* All absorbance measurements using this procedure are temperature-dependent (0.015 AU/°C) and should be performed at 22 °C.

Absorbance of Reagent Blank. Pipet 10 mL of 0.04 M potassium tetrachloromercurate and 1 mL of 0.6% sulfamic acid solution (both described below) into a 25-mL volumetric flask. Allow to stand for 10 min, then add from a pipet 2.0 mL of 0.2% formaldehyde solution (described below) and 5.0 mL of pararosaniline reagent (described below). Dilute to volume with water, allow to stand for 30 min, and determine the absorbance at 548 nm in 1-cm cells, using water as the reference.

Calibration Curve Slope. Pipet 1.0 mL of sulfite standard solution (described below) into a 25-mL volumetric flask. Add 9 mL of 0.04 M potassium tetrachloromercurate and 1 mL of 0.6% sulfamic acid, allow to stand for 10 min, then add from a pipet 2.0 mL of 0.2% formaldehyde solution and 5.0 mL of pararosaniline reagent. Dilute to volume with water, allow to stand for 30 min, and determine the absorbance at 548 nm in 1-cm cells, using water as the reference.

$$\text{Absorbance (g/mL)} = (A_s - A_b) / W$$

where A_s = absorbance of sample; A_b = absorbance of reagent blank; and W = weight, in micrograms, of SO_2 in 1 mL.

Reagents for Suitability for SO_2 Determination

All six reagents cited should be freshly prepared.

Formaldehyde Solution, 0.2%. Dilute 0.5 mL of 36–38% formaldehyde with water to 100 mL. Prepare this solution daily.

Pararosaniline Reagent. Pipet 20.0 mL of sample stock solution of pararosaniline (described above in the assay) into a 250-mL volumetric flask. Add an additional 0.2 mL of stock for each percent the pararosaniline assay is below 100%. Add 25 mL of 3 M phosphoric acid, and dilute to volume with water.

Potassium Tetrachloromercurate, 0.04 M. Dissolve 10.86 g of mercuric chloride in water in a 1-L volumetric flask. Add 5.96 g of potassium chloride and 0.066 g of EDTA, dissolve, and dilute with water to volume. The pH of this reagent should not be less than 5.2.

Sulfamic Acid Solution, 0.6%. Dissolve 0.6 g of sulfamic acid in 100 mL of water. This solution can be kept for a few days if protected from air.

Sulfite Standard Solution. Immediately after its standardization, pipet a 2-mL aliquot of sulfite stock solution (described below) into a 100-mL volumetric flask, and dilute to volume with 0.04 M potassium tetrachloromercurate. This solution is stable for 30 days if stored at 5 °C.

Sulfite Stock Solution. Dissolve 0.40 g of sodium sulfite or 0.30 g of sodium metabisulfite in 500 mL of deaerated water (purged with high-purity nitrogen gas or by boiling). This solution contains 320–400 μg/mL as sulfur dioxide. The actual concentration is determined by adding excess iodine solution and back-titrating with sodium thiosulfate solution that has been standardized against potassium iodate or potassium dichromate. Sulfite solutions are unstable.

Standardize as follows: For blank solution A, in the back-titration, transfer 25 mL of water to a 500-mL iodine flask, and add from a pipet 50.0 mL of 0.01 N iodine. For sample solution B, in a second 500-mL iodine flask, add successively from a pipet 25.0 mL of sulfite stock solution and 50.0 mL of 0.01 N iodine. Stopper the flasks, allow to react for 5 min, and titrate each with 0.01 N sodium thiosulfate to a pale yellow color. Add 5 mL of starch indicator solution, and continue the titration to the disappearance of the blue color. Calculate the concentration of sulfur dioxide as follows:

$$\% \ SO_2 \ (\mu g/mL) = \frac{\{[mL \ (blank) - mL \ (sample)] \times N \ Na_2S_2O_3\} \times 32.030}{Sample \ volume \ (mL)}$$

1-Pentanol
n-Amyl Alcohol

CH₃(CH₂)₃CH₂OH   **Formula Wt 88.15**   **CAS No. 71-41-0**

GENERAL DESCRIPTION

Typical appearance: liquid, with mild characteristic odor

Analytical use: solvent

Change in state (approximate): boiling point, 138 °C

Aqueous solubility: 2.7 g in 100 mL at 22 °C

Density: 0.81

SPECIFICATIONS

Assay . . . . ≥98.0% CH₃(CH₂)₃CH₂OH

Maximum Allowable

Water (H₂O) 0.5%

Color (APHA) 30

Residue after evaporation 0.003%

Acids and esters . . . . 0.075 meq/g

Carbonyl compounds (as HCHO) 0.1%

TESTS

Assay. Analyze the sample by gas chromatography using the parameters cited on page 80. The following specific conditions are also required.

   Column: Type I, methyl silicone

Measure the area under all peaks, and calculate the 1-pentanol content in area percent. Correct for water content.

Water. (Page 31, Method 1). Use 25 mL (20 g) of the sample.

Color (APHA). (Page 43).

Residue after Evaporation. (Page 25). Use 40.0 g (50 mL) of sample, evaporate to dryness on a hot plate (≈100 °C), and complete drying at 105 °C.

Acids and Esters. Dilute 8.1 g (10 mL) with 20 mL of ethyl alcohol, add 10 mL of 0.1 N sodium hydroxide, and reflux gently for 10 min. Cool, add 0.15 mL of phenolphthalein indicator solution, and titrate the excess sodium hydroxide with 0.1 N hydrochloric acid. Run a reagent blank. The net volume of 0.1 N sodium hydroxide consumed by the sample should not exceed 6.1 mL.

Carbonyl Compounds. (Page 54). Use 0.25 g (0.30 mL) of sample. For the standard, use 0.25 mg of formaldehyde.

Perchloric Acid

HClO₄   **Formula Wt 100.46**   **CAS No. 7601-90-3**

Note: This specification applies to perchloric acid with nominal concentrations of 50%, 60%, or 70%.

GENERAL DESCRIPTION

Typical appearance: clear, colorless liquid

Analytical use: solvent for metals and ores; standard acid; oxidizer; nonaqueous titrimetry; wet ashing

pK_a: < -10

SPECIFICATIONS

Assay . . . . 48.0–50.0% $HClO_4$
60.0–62.0% $HClO_4$
69.0–72.0% $HClO_4$

	Maximum Allowable
Color (APHA)	10
Residue after ignition.	0.003%
Silicate and phosphate (as SiO_2).	5 ppm
Chloride (Cl).	0.001%
Nitrogen compounds (as N).	0.001%
Sulfate (SO_4)	0.001%
Heavy metals (as Pb).	1 ppm
Iron (Fe).	1 ppm

TESTS

Note: All tests are written for 70% concentration. The densities of the 60% acid (1.54 g/mL) and the 50% acid (1.40 g/mL) are less than the density of the 70% reagent (1.67 g/mL), and proper allowance should be made for this when measured volumes are taken for relevant tests.

Assay. (By alkalimetric titration). Weigh accurately about 3 mL in a suitable flask, and dilute with water to 60 mL. Add 0.15 mL of phenolphthalein indicator solution, and titrate with 1 N sodium hydroxide volumetric solution. One milliliter of 1 N sodium hydroxide corresponds to 0.1005 g of $HClO_4$.

$$\% \ HClO_4 = \frac{(mL \times N\ NaOH) \times 10.05}{Sample\ wt\ (g)}$$

Color (APHA). (Page 43).

Residue after Ignition. (Page 26). To 67 g (42 mL) in a tared, preconditioned dish, add 1 mL of nitric acid, evaporate to dryness, and ignite at 600 ± 25 °C for 15 min.

Silicate and Phosphate. Dilute 2.0 g (1.2 mL) with 50 mL of water in a platinum dish, add 1.0 g of sodium carbonate, and digest on a hot plate (≈100 °C) for 15–30 min. Cool, add

5 mL of a 10% ammonium molybdate reagent solution, acidify with hydrochloric acid to about pH 4 (using indicator paper), and transfer to a beaker. Adjust the pH to 1.8 (using a pH meter) with dilute hydrochloric acid (1 + 9). Dilute with water to 90 mL, heat just to boiling, and cool to room temperature. Add 10 mL of hydrochloric acid, and dilute with water to 100 mL. Transfer to a separatory funnel, add 50 mL of 4-methyl-2-pentanone, shake vigorously, and allow to separate. Draw off and discard the aqueous phase. Wash the ketone layer three times with 20-mL portions of dilute hydrochloric acid (1 + 99), drawing off and discarding the aqueous layer each time. Add 10 mL of dilute hydrochloric acid (1 + 99), to which has just been added 0.2 mL of a freshly prepared 2% stannous chloride reagent solution. Shake vigorously, and allow to separate. Any blue color should not exceed that produced by 0.01 mg of silica and 1 g of sodium carbonate, treated exactly like the sample.

Chloride. (Page 35). Dilute 1 g (0.6 mL) with 25 mL of water.

Nitrogen Compounds. (Page 39). Use 5 g (3.0 mL). For the standard, use 0.05 mg of nitrogen ion (N).

Sulfate. Dilute 32.0 g (20 mL) with water to 160 mL, and neutralize with ammonium hydroxide, using litmus paper as an indicator. Add 1 mL of hydrochloric acid, heat to boiling, add 5 mL of 12% barium chloride reagent solution, digest in a covered beaker on a hot plate ($\approx$100 °C) for 2 h, and allow to stand for at least 8 h. If a precipitate is formed, filter, wash thoroughly, and ignite. The weight of the precipitate should not be more than 0.0008 g greater than the weight obtained from a complete blank test, including the evaporated residue from a volume of ammonium hydroxide equal to that used to neutralize the sample.

For the Determination of Heavy Metals and Iron

Sample Solution A. To 40.0 g (25 mL), add 1 mL of nitric acid and 1 mL of 1% sodium carbonate reagent solution. Evaporate to dryness on a hot plate in a well-ventilated hood, dissolve the residue in 2 mL of 1 N acetic acid, and dilute with water to 40 mL (1 mL = 1.0 g).

Heavy Metals. (Page 36, Method 1). Dilute 20 mL of sample solution A (20-g sample) with water to 25 mL.

Iron. (Page 38, Method 1). Use 10 mL of sample solution A (10-g sample).

Perchloric Acid, Ultratrace
$HClO_4$ Formula Wt 100.46   CAS No. 7601-90-3

Suitable for use in ultratrace elemental analysis.

Note: This reagent must be packaged in a preleached Teflon bottle and used in a clean laboratory environment to maintain purity.

GENERAL DESCRIPTION

Typical appearance: clear, colorless liquid
Analytical use: solvent in trace metal analysis; oxidizer
pK_a: < -10

SPECIFICATIONS

Assay . . . . .65–72% $HClO_4$

	Maximum Allowable
Chloride (Cl)	3 ppm
Silicate and phosphate (as SiO_2)	5 ppm
Sulfate (SO_4)	5 ppm
Mercury (Hg)	1 ppb
Aluminum (Al)	1 ppb
Barium (Ba)	1 ppb
Boron (B)	5 ppb
Cadmium (Cd)	1 ppb
Calcium (Ca)	1 ppb
Chromium (Cr)	1 ppb
Cobalt (Co)	1 ppb
Copper (Cu)	1 ppb
Iron (Fe)	5 ppb
Lead (Pb)	1 ppb
Lithium (Li)	1 ppb
Magnesium (Mg)	1 ppb
Manganese (Mn)	1 ppb
Molybdenum (Mo)	1 ppb
Potassium (K)	1 ppb
Silicon (Si)	5 ppb
Sodium (Na)	5 ppb
Strontium (Sr)	1 ppb
Tin (Sn)	1 ppb
Titanium (Ti)	1 ppb
Vanadium (V)	1 ppb
Zinc (Zn)	1 ppb
Zirconium (Zr)	1 ppb

TESTS

Assay. (By alkalimetric titration). Weigh accurately about 3 mL, dilute with water to 60 mL, add 0.15 mL of phenolphthalein indicator solution, and titrate with 1 N sodium hydroxide volumetric solution. One milliliter of 1 N sodium hydroxide corresponds to 0.1005 g of $HClO_4$.

$$\% \ HClO_4 = \frac{(mL \times N \ NaOH) \times 10.05}{Sample \ wt \ (g)}$$

Chloride. (Page 35). Dilute 3.3 g with 25 mL of water.

Silicate and Phosphate. Use test for perchloric acid, page 478.

Sulfate. Use test for perchloric acid, page 479. Precipitate weight <0.0004 g.

Mercury. (By CVAAS, page 65). To each of three 100-mL volumetric flasks containing about 35 mL of water, add 10.0 g of sample. To the second and third flasks, add mercury ion (Hg) standards of 10 ng (1.0 ppb) and 20 ng (2.0 ppb), respectively. Dilute to the mark with water. Mix. Zero the instrument with the blank, and determine the mercury content using a suitable mercury analyzer system (1.0 mL of 0.01 μg/mL of Hg = 10 ng).

Trace Metals. Determine the aluminum, barium, boron, cadmium, calcium, chromium, cobalt, copper, iron, lead, lithium, magnesium, manganese, molybdenum, potassium, silicon, sodium, strontium, tin, titanium, vanadium, zinc, and zirconium by the ICP–OES method described on page 69.

Perchloroethylene
Tetrachloroethylene

$Cl_2C:CCl_2$	**Formula Wt 165.85**	**CAS No. 127-18-4**

Note: Perchloroethylene usually contains a stabilizer. If a stabilizer is present, its identity and quantity must be stated on the label.

GENERAL DESCRIPTION

Typical appearance: clear liquid with a characteristic odor
Analytical use: solvent
Change in state (approximate): boiling point, 121 °C
Aqueous solubility: insoluble
Density: 1.62

SPECIFICATIONS

General Use

Assay . . . . ≥99.0% C_2Cl_4

Maximum Allowable

Color (APHA) 10
Residue after evaporation . . . . 5 ppm
Water (H_2O) 0.05%

Specific Use

Ultraviolet Spectrophotometry

Wavelength (nm)	Absorbance (AU)
400	0.03
350	0.05
305	0.2
300	1.0

TESTS

Assay. Analyze the sample by gas chromatography using the general parameters cited on page 80. The following specific conditions are also required.

Column: Type I, methyl silicone

Measure the area under all the peaks, and calculate the perchloroethylene content in area percent. Correct for water content. The assay value is the combination of perchloroethylene and added stabilizers.

Color (APHA). (Page 43).

Residue after Evaporation. (Page 25). Evaporate 124 mL (200 g) to dryness in a tared, preconditioned dish on a hot plate ($\approx$100 °C), and dry the residue at 105 °C for 30 min. The weight of the residue should not exceed 0.001 g.

Water. (Page 31, Method 2). Use 1.0 mL (1.6 g) of the sample.

Ultraviolet Spectrophotometry. Use the procedure on page 86 to determine the absorbance.

Periodic Acid
para-Periodic Acid

H_5IO_6 Formula Wt 227.96   CAS No. 10450-60-9

GENERAL DESCRIPTION

Typical appearance: hygroscopic, colorless solid; may become pale yellow on exposure to air
Analytical use: oxidizer; determination of manganese
Change in state (approximate): melting point, 122 °C; decomposes at 130 –140 °C
Aqueous solubility: very soluble

SPECIFICATIONS

Assay . . . . 99.0–101.0% H_5IO_6

	Maximum Allowable
Insoluble matter	0.01%
Residue after ignition	0.01%
Other halogens (as Cl)	0.01%
Sulfate (SO_4)	0.01%
Heavy metals (as Pb)	0.005%
Iron (Fe)	0.003%

TESTS

Assay. (By titration of oxidizing power). Weigh accurately about 0.12 g, transfer to an iodine flask, and dissolve in about 50 mL of water. Add 1 mL of sulfuric acid and 3 g of potassium iodide, swirl, and let stand in the dark for 10 min. Titrate the liberated iodine with 0.1 N sodium thiosulfate volumetric solution, adding starch indicator solution near the end of the titration. One milliliter of 0.1 N thiosulfate corresponds to 0.002849 g of H_5IO_6.

$$\% \ H_5IO_6 = \frac{(mL \times N \ Na_2S_2O_3) \times 2.849}{Sample \ wt \ (g)}$$

Insoluble Matter. (Page 25). Use 10.0 g dissolved in 100 mL of water.

Residue after Ignition. (Page 26). Use a 10.0-g sample.

Other Halogens. Dissolve 1.0 g in 20 mL of water with 1 mL of phosphoric acid. Add 1 mL of hydrogen peroxide, and boil until the iodine is expelled. Wash down the sides with water, add 1 mL of hydrogen peroxide, and boil. If iodine color forms, repeat the treatment until no more iodine is evolved. Boil an additional 10 min and dilute to 100 mL. Dilute 10 mL to 23 mL, and add 1 mL of nitric acid and 1 mL of silver nitrate reagent solution. The turbidity should not exceed that produced by 0.01 mg of chloride ion in an equal volume containing the quantities of all reagents used in the test.

For the Determination of Sulfate, Heavy Metals, and Iron

Sample Solution A. Dissolve 10.0 g in 20 mL of water. Add 1 mL of 1% sodium carbonate reagent solution and 20 mL of hydrochloric acid, and evaporate to dryness on a hot plate ($\approx$100 °C). Repeat the evaporation twice, using 10 mL of hydrochloric acid. Dissolve the residue in 20 mL of water and dilute to 100 mL (1 mL = 0.10 g).

Blank Solution B. Use the quantity of hydrochloric acid added to sample solution A and 10 mg of sodium carbonate. Evaporate to dryness, dissolve in 20 mL of water, and dilute to 100 mL.

Sulfate. (Page 40, Method 1). Use 5 mL of sample solution A (0.5-g sample).

Heavy Metals. (Page 36, Method 1). Dilute 4 mL of sample solution A (0.4-g sample) to 25 mL. Use 4 mL of blank solution B to prepare the standard.

Iron. (Page 38, Method 2). Use 3.3 mL of sample solution A (0.33-g sample). For the standard, use 3.3 mL of blank solution B.

Petroleum Ether
Ligroin

CAS No. 8032-32-4

Suitable for general use or in extraction–concentration analysis. Product labeling shall designate the uses for which suitability is represented on the basis of meeting the relevant specifications and tests. The extraction–concentration suitability specifications include only the general use specification for color.

GENERAL DESCRIPTION

Typical appearance: clear liquid with a characteristic odor
Analytical use: solvent

SPECIFICATIONS

General Use

Color (APHA) . . . . ≤10
Boiling range. . . . . .35–60 °C
Residue after evaporation . . . . ≤0.001%
Acidity. . . . . Passes test

Specific Use

Extraction–Concentration Suitability

GC–FID . . . . Passes test
GC–ECD . . . . Passes test

TESTS

Color (APHA). (Page 43).

Boiling Range. Distill 100 mL by the method referenced on page 42. The corrected initial boiling point should not be below 35 °C, and the corrected dry point should not be above 60 °C. The barometric pressure correction for petroleum ether is +0.039 °C per mm under 760 mm and –0.039 °C per mm above 760 mm.

Residue after Evaporation. (Page 25). Evaporate 100 g (150 mL) to dryness in a tared, preconditioned dish at 70–80 °C, and dry the residue at 105 °C for 30 min.

Acidity. Thoroughly shake 10 mL with 5 mL of water for 2 min and allow to separate. The aqueous layer should not turn blue litmus paper red in 15 s.

Extraction–Concentration Suitability. Analyze the sample by using the general procedure cited on page 85.

1,10-Phenanthroline Monohydrate

$C_{12}H_8N_2 \cdot H_2O$ **Formula Wt 198.22**   **CAS No. 5144-89-8**

GENERAL DESCRIPTION

Typical appearance: white solid; may become cream-colored with storage
Analytical use: colorimetric indicator; determination of iron; base for redox indicator

Change in state (approximate): melting point, 90–93 °C
Aqueous solubility: 0.3 g in 100 mL

SPECIFICATIONS

Suitability as redox indicator. . . . . Passes test
Suitability for determining iron . . . . Passes test

TESTS

Sample Solution A. Dissolve 0.10 g in water, and dilute with water to 100 mL.

Suitability as Redox Indicator. To 2.5 mL of sample solution A, add 0.1 mL of 0.1 N ferrous ammonium sulfate solution and dilute to 50 mL with 12 N sulfuric acid. A distinct red color should be produced. Add 0.1 mL of 0.1 N potassium dichromate solution. The red color should be discharged.

Suitability for Determining Iron. To 20 mL of water, add 0.001 mg of iron and 1 mL of hydroxylamine hydrochloride reagent solution. For the control, add 1 mL of hydroxyl-amine hydrochloride reagent solution to 20 mL of water. To each solution, add 2.5 mL of sample solution A, and allow to stand for 1 h. The color in the solution containing the added iron should be definitely pink compared with the control. The control containing no added iron should not show any turbidity or color.

Phenol

C_6H_5OH	Formula Wt 94.11	CAS No. 108-95-2

Note: Phenol that conforms to this specification may contain a stabilizer. If a stabilizer is present, its presence should be stated on the label.

GENERAL DESCRIPTION

Typical appearance: colorless hygroscopic solid; may become pink on exposure to air
Analytical use: pH indicator
Change in state (approximate): when free from water and cresol, it congeals at 41 °C and melts at 43 °C; boiling point, 182 °C
Aqueous solubility: 8.2 g in 100 mL at 15 °C; very soluble in hot water
Density: 1.07
pK_a: 9.9

SPECIFICATIONS

Assay . . . . $\geq$99.0% C_6H_5OH
Freezing point . . . . Not below 40.5 °C (dry basis)
Clarity of solution . . . . Passes test

Maximum Allowable

Residue after evaporation 0.05%
Water (H_2O) 0.5%

TESTS

Assay. (By iodometry after bromination of hydroxyl groups). Weigh accurately about 1.5 g, and dissolve in sufficient water to make exactly 1 L. Transfer 25 mL into a 500-mL glass-stoppered iodine flask, add 30 mL of 0.1 N bromine solution and 5 mL of hydrochloric acid, and immediately stopper the flask. Shake frequently during one-half hour, and allow to stand for 15 min. Then quickly add 10 mL of 10% potassium iodide reagent solution, being careful that no bromine escapes, and stopper immediately. Shake, add 1 mL of chloroform, and titrate the liberated iodine (which represents the excess of bromine) with 0.1 N sodium thiosulfate volumetric solution, using 3 mL of starch indicator solution. One milliliter of 0.1 N bromine corresponds to 0.001569 g of C_6H_5OH.

$$\% \ C_6H_5OH = \frac{[(mL \times N \ Br_2) - (mL \times N \ Na_2S_2O_3)] \times 1.569}{Sample \ wt \ (g) \ / \ 40}$$

Freezing Point. Dry the sample by adding 15 g of molecular sieve zeolite having 4A-size pores to 100 g of molten sample in a 400-mL conical flask. (The zeolite may be in the form of a powder or of cylindrical granules about 3 mm in diameter.) Stopper the flask loosely, and maintain the mixture at 55–60 °C for 20 min with frequent stirring. Quickly transfer a portion of the dried sample to a jacketed test tube, filling the tube to a depth of 8 cm. Insert a suitable thermometer, graduated in 0.1 °C units, and stir until the temperature falls to 10 °C above the expected freezing point. Immerse the tube in a water bath that is adjusted to about 4 °C below the expected freezing point and clamp in a vertical position. Insert a ring-type stirrer around the thermometer that is clamped to place the bottom of the bulb about 0.5 cm above the bottom of the test tube. Stir the sample continuously until crystallization occurs and the temperature does not change more than 0.05 °C in 2 min or, if supercooling occurs, the temperature rises rapidly to maximum accompanied by rapid crystallization. Record the constant temperature or the maximum temperature as the freezing point.

Clarity of Solution. Dissolve 5.0 g in 100 mL of water. The solution should be free of turbidity and insoluble residue.

Residue after Evaporation. (Page 25). Evaporate 10.0 g to dryness in a tared, preconditioned dish on a hot plate ($\approx$100 °C), and dry the residue at 105 °C for 30 min.

Water. (Page 31, Method 1). Use 25.0 g of molten sample.

Phenol Red

Phenolsulfonphthalein; 4,4'-(3*H*-2,1-Benzoxathiol-3-ylidene)bisphenol *S,S*-Dioxide

$C_{19}H_{14}O_5S$ (sultone)	Formula Wt 354.38	CAS No. 143-74-8
$C_{19}H_{13}O_5SNa$ (sodium salt)	Formula Wt 376.36	CAS No. 34487-61-1

Note: This standard applies to both the sultone and the salt forms of this indicator.

GENERAL DESCRIPTION

Typical appearance: bright red to dark red solid
Analytical use: indicator for acidimetry
Aqueous solubility: 1 g in 1300 mL

SPECIFICATIONS

Clarity of solution . . . . Passes test
Visual transition interval. . . . . From pH 6.8 (yellow) to pH 8.2 (red)

TESTS

Clarity of Solution. If the indicator is the sultone form, dissolve 0.1 g in 100 mL of reagent alcohol. If the indicator is a salt form, dissolve 0.1 g in 100 mL of water. Not more than a faint trace of turbidity or insoluble matter should remain. Reserve the solution for the test for visual transition interval.

Visual Transition Interval. Dissolve 1.0 g of potassium chloride in 100 mL of water. Adjust the pH of the solution to 6.8 (using a pH meter) with 0.01 N acid or alkali. Add 0.15 mL of the 0.1% solution reserved from the test for clarity of solution. The color of the solution should be yellow, with not more than a faint trace of green color. Titrate the solution with 0.01 N sodium hydroxide to a pH of 7.0 (using a pH meter). The color of the solution should be orange. Continue the titration to pH 8.2. The color of the solution should be red.

Phenolphthalein

3,3-Bis(4-hydroxyphenyl)-1(3*H*)-isobenzofuranone;
3,3-Bis(*p*-hydroxyphenyl)phthalide

$C_{20}H_{14}O_4$	Formula Wt 318.32	CAS No. 77-09-8

GENERAL DESCRIPTION

Typical appearance: white or yellowish-white solid
Analytical use: indicator for acidimetric titrations
Change in state (approximate): melting point, 258–262 °C
Aqueous solubility: almost insoluble
Density: 1.3
pK_a: 9.7

SPECIFICATIONS

Clarity of alcohol solution. . . . . Passes test

Visual transition interval . . . . From pH 8.0 (colorless) to pH 10 (red)

TESTS

Clarity of Alcohol Solution. Dissolve 1.0 g in 100 mL of ethyl alcohol. Not more than a faint trace of turbidity or insoluble matter should remain. Reserve the solution for the test for visual transition interval.

Visual Transition Interval. Dissolve 1.0 g of potassium chloride in 100 mL of water. Adjust the pH of the solution to 8.0 (using a pH meter) with 0.01 N acid or base. Add 0.15 mL of the 1% solution reserved from the test for clarity of alcohol solution. The solution should be colorless. Titrate the solution with 0.01 N sodium hydroxide until the pH is 8.2 (using a pH meter). The solution should have a pale pink color. Continue the titration until the pH is 8.6 (using a pH meter). The solution should have a definite pink color.

Phosphomolybdic Acid
Molybdophosphoric Acid

CAS No. 51429-74-4

GENERAL DESCRIPTION

Typical appearance: yellow solid

Analytical use: test for alkaloids

Change in state (approximate): melting point, 79–90 °C

Aqueous solubility: very soluble

SPECIFICATIONS
Maximum Allowable

Insoluble matter 0.01%

Chloride (Cl). 0.02%

Sulfate (SO_4) 0.025%

Ammonium (NH_4) 0.01%

Calcium (Ca) 0.02%

Heavy metals (as Pb). 0.005%

Iron (Fe). 0.005%

TESTS

Insoluble Matter. (Page 25). Use 10 g dissolved in 250 mL of water.

Chloride. Dissolve 1.0 g in 50 mL of water, add 1 mL of nitric acid, filter through a chloride-free filter, and divide into two equal portions. For the control, add 0.5 mL of silver nitrate reagent solution to one-half of the solution, let stand for 10 min, filter, and add 0.1 mg of chloride ion (Cl). For the sample, add 0.5 mL of silver nitrate reagent solution to the remaining half. Any turbidity in the solution of the sample should not exceed that in the control.

Sulfate. (Page 40, Method 1).

Ammonium. (By colorimetry, page 32). Dilute the distillate from a 1.0-g sample with water to 100 mL; 20 mL of the dilution shows not more than 0.02 mg of ammonium ion (NH_4).

Calcium. (By flame AAS, page 63).

Sample Stock Solution. Dissolve 1.0 g of sample in 50 mL of water in a 100-mL volumetric flask, and dilute to the mark with water (1 mL = 0.01 g).

Element	Wavelength (nm)	Sample Wt (g)	Standard Added (mg)	Flame Type*	Background Correction
Ca	422.7	0.10	0.02; 0.04	N/A	No

*N/A is nitrous oxide/acetylene.

Heavy Metals. Dissolve 3.0 g in 30 mL of water, filter, neutralize the filtrate to litmus with 10% sodium hydroxide reagent solution, and add 5 mL of 1 N sodium hydroxide in excess. Heat the solution to boiling to destroy any blue color. For the standard, boil a solution containing 5 mL of 1 N sodium hydroxide and 0.15 mg of lead in 30 mL of water. Cool the solutions to about 10 °C, and to each add 4.5 mL of 1 N hydrochloric acid and 10 mL of 10% (*w/v*) sodium sulfide solution. Any color in the solution of the sample should not exceed that in the standard.

Iron. Dissolve 5.0 g in 30 mL of water, make alkaline with ammonium hydroxide, and add 0.25 to 0.30 mL of bromine water. Boil for a few minutes, filter, wash the precipitate several times, and discard the filtrate and washings. Pass 25 mL of dilute hydrochloric acid (1 + 1) through the filter to dissolve any precipitate, wash thoroughly, and dilute with water to 50 mL. Dilute 20 mL of this solution with water to 45 mL, and add 2 mL of ammonium thiocyanate reagent solution. Any red color should not exceed that produced by 0.1 mg of ferric ion (Fe^{3+}) in an equal volume of solution containing 10 mL of dilute hydrochloric acid (1 + 1) and 2 mL of ammonium thiocyanate reagent solution.

Phosphoric Acid
Orthophosphoric Acid; Phosphoric(V) Acid
H_3PO_4 Formula Wt 98.00   CAS No. 7664-38-2

GENERAL DESCRIPTION

Typical appearance: clear, viscous liquid
Analytical use: various buffers
Aqueous solubility: miscible
Density: 1.68
pK_a: 2.1

SPECIFICATIONS

Assay . . . . $\geq$85.0% H_3PO_4

Maximum Allowable

Color (APHA)	10
Insoluble matter	0.001%
Chloride (Cl)?	3 ppm
Nitrate (NO_3)	5 ppm
Sulfate (SO_4)	0.003%
Volatile acids (as CH_3COOH)	0.001%
Antimony (Sb)	0.002%
Calcium (Ca)	0.002%
Magnesium (Mg)	0.002%
Potassium (K)	0.005%
Sodium (Na)	0.025%
Arsenic (As)	1 ppm
Heavy metals (as Pb)	0.001%
Iron (Fe)	0.003%
Manganese (Mn)	0.5 ppm
Reducing substances	Passes test

TESTS

Assay. (By acid–base titrimetry). Weigh accurately about 2.0 g, and dilute with 120 mL of water. Titrate potentiometrically with 1 N sodium hydroxide volumetric solution to the inflection point at a pH near 9, using the first-derivative method (page 27). One milliliter of 1 N sodium hydroxide corresponds to 0.04900 g of H_3PO_4.

$$\% \, H_3PO_4 = \frac{(mL \times N \, NaOH) \times 4.900}{Sample \, wt \, (g)}$$

Color (APHA). (Page 43).

Insoluble Matter. (Page 25). Dilute 100 g (60 mL) to 500 mL with water.

Chloride. (Page 35). Dilute 3.4 g (2.0 mL) with water to 25 mL, and add 0.5 mL of nitric acid.

Nitrate. (Page 38, Method 1).

>   **Sample Solution A.** Add 4.0 g (2.4 mL) to 2 mL of water, dilute to 50 mL with brucine sulfate reagent solution, and mix.

>   **Control Solution B.** Add 4.0 g (2.4 mL) to 2 mL of nitrate ion (NO_3) standard solution, dilute to 50 mL with brucine sulfate reagent solution, and mix.

Continue with the procedure, starting with the preparation of blank solution C.

Sulfate. (Page 40, Method 1). Use 1.7 g (1.0 mL), and neutralize with ammonium hydroxide. Allow 30 min for turbidity to form.

Volatile Acids. Dilute 64.0 g (37.5 mL) with 75 mL of freshly boiled and cooled water in a distilling flask provided with a spray trap, and distill off 50 mL. To the distillate, add 0.15 mL of phenolphthalein indicator solution, and titrate with 0.01 N sodium hydroxide. Not more than 1.0 mL of the 0.01 N sodium hydroxide should be required to produce a pink color.

Antimony, Calcium, Magnesium, Potassium, and Sodium. (By flame AAS, page 63).

> ***Sample Stock Solution.*** Dilute 6.0 mL (10 g) of sample with 20 mL of water in a 100-mL volumetric flask, and dilute to the mark with water (1 mL = 0.1 g). For antimony, use a 10.0-g sample and standard additions.

Element	Wavelength (nm)	Sample Wt (g)	Standard Added (mg)	Flame Type*	Background Correction
Sb	217.6	10.0	0.20; 0.40	A/A	Yes
Ca	422.7	1.0	0.02; 0.04	N/A	No
Mg	285.2	1.0	0.01; 0.02	A/A	Yes
K	766.5	1.0	0.04; 0.08	A/A	No
Na	589.0	0.10	0.02; 0.04	A/A	No

*A/A is air/acetylene; N/A is nitrous oxide/acetylene.

Arsenic. (Page 34). To an arsine generator flask, add 10.0 g (6 mL) of sample, 30 mL of water, and 10 mL of ferric ammonium sulfate solution (described below). Then add 2% potassium permanganate solution dropwise to a persistent pink color. Add 1.5 g of sodium chloride, and mix well until dissolution is complete. Heat nearly to boiling, remove from the heat, and add 1 mL of 40% stannous chloride reagent solution. Dilute to 60 mL, and cool to about 25 °C. For the standard, to a second arsine generator flask, add 0.01 mg of arsenic and 20 mL of dilute sulfuric acid (1 + 4), and treat as described for the sample. Assemble the test apparatus (shown on page 34), and continue according to the procedure, using 10.0 g of No. 20-mesh granulated zinc. After 30 min, transfer the two silver diethyldithiocarbamate-absorbing solutions to 1-cm cells, and read the absorbance at 540 nm against the stock silver diethyldithiocarbamate solution. The absorbance for the sample should not exceed that for the standard.

> ***Ferric Ammonium Sulfate Solution.*** Mix 8.4 g of ferric ammonium sulfate dodecahydrate, 0.2 mL of sulfuric acid, and 0.10 g of sodium chloride. Dissolve, and dilute with water to 100 mL.

Heavy Metals. (Page 36, Method 1). Dilute 4.0 g (2.4 mL) with water to 32 mL. Use 24 mL to prepare the sample solution, and use the remaining 8.0 mL to prepare the control solution. Adjust the pH to 4.5.

Iron. (Page 38, Method 2). Dilute 6.7 g (4.0 mL) to 100 mL. To 5.0 mL of the solution, add 8 mL of 0.5 N sodium acetate.

Manganese. Add 25.0 g (15 mL) to 100 mL of dilute sulfuric acid (1 + 9). For the control, add 5 g (3 mL) of sample and 0.01 mg of manganese to 100 mL of dilute sulfuric acid (1 + 9). To each solution, add 20 mL of nitric acid, heat to boiling, and boil gently for 5 min. Cool slightly, add 0.25 g of potassium periodate, and again boil for 5 min. Any red color in the solution of the sample should not exceed that in the control.

Reducing Substances. Add 5.0 mL of 0.1 N bromine solution to each of two 500-mL iodine determination flasks. To each flask, add 120 mL of cold, dilute sulfuric acid (1 + 59),

and to one of the flasks, add rapidly 17 g (10 mL) of the sample. Stopper the flasks, shake gently, seal with water, and allow to stand at room temperature for 10 min. Cool in an ice bath, add 10 mL of 10% potassium iodide reagent solution, and reseal with water. Shake gently to dissolve all fumes, and place in the ice bath for about 5 min. Wash the stoppers and inside walls of the flasks with water, and titrate each with 0.01 N sodium thiosulfate, adding 3 mL of starch indicator solution near the end of the titration. The difference between the two titrations should not exceed 1.0 mL of 0.01 N sodium thiosulfate.

Phosphoric Acid, Meta-
Vitreous Sodium Acid Metaphosphate

CAS No. 37267-86-0

Note: This reagent contains as a stabilizer a somewhat greater proportion of sodium meta-phosphate than that corresponding to the empirical formula $NaH(PO_3)_2$.

GENERAL DESCRIPTION

Typical appearance: transparent, glass-like solid or very silky masses; hygroscopic
Analytical use: acid digestion
Aqueous solubility: very slowly soluble in cold water, slowly changing to H_3PO_4
pK_a: 2.1

SPECIFICATIONS

Assay (HPO_3) 33.5–36.5%
Stabilizer ($NaPO_3$) 57.0–63.0%

Maximum Allowable

Nitrate (NO_3) 0.001%
Sulfate (SO_4) 0.005%
Chloride (Cl) 0.001%
Arsenic (As) . . . . 1 ppm
Heavy metals (as Pb) 0.005%
Iron (Fe) 0.005%
Substances reducing permanganate (as H_3PO_3) 0.02%

TESTS

Assay. (By alkalimetry). Weigh accurately about 5.0 g. In a suitable flask, dissolve in 400 mL of water, and titrate with 1 N sodium hydroxide volumetric solution to a pH of 4.4 (using a pH meter). One milliliter of 1 N sodium hydroxide corresponds to 0.07998 g of metaphosphoric acid (HPO_3).

$$\% \ HPO_3 = \frac{(mL \times N \ NaOH) \times 7.998}{Sample \ wt \ (g)}$$

Stabilizer. Weigh accurately 20.0 g, dissolve in water, and dilute with water to 1 L. To 50.0 mL of this solution in a 200-mL volumetric flask, add 50 mL of water and 50 mL of a

10% lead acetate reagent solution, dilute with water to 200 mL, mix thoroughly, and decant through a filter. Treat 100 mL of the clear filtrate with freshly prepared hydrogen sulfide water sufficiently to precipitate the excess of lead, filter, and wash with about 20 mL of water. To the filtrate, add 2 mL of sulfuric acid, evaporate to dryness in a tared, preconditioned dish, and ignite to constant weight at 600 ± 25 °C. Calculate the weight of $NaPO_3$ by multiplying the weight of sodium sulfate by the factor 1.4356.

$$Wt\ NaPO_3 = Wt\ NaSO_4 \times 1.4356$$

Nitrate. (Page 38, Method 1). For sample solution A, use 1.0 g in 4 mL of water. For control solution B, use 1.0 g and 1 mL of nitrate ion (NO_3) standard solution.

Sulfate. Dissolve 8.0 g in 180 mL of water, heat the solution to boiling, cool, filter, and heat again to boiling. Add 10 mL of 40% barium chloride reagent solution, digest in a covered beaker on a hot plate ($\approx$100 °C) for 2 h, and allow to stand for at least 8 h. If a precipitate is formed, filter, wash thoroughly, and ignite. The weight of the precipitate should not be more than 0.001 g greater than the weight obtained in a complete blank test, including filtration of a solution of 2 mL of hydrochloric acid in 200 mL of water.

For the Determination of Chloride, Arsenic, Heavy Metals, Iron, and Substances Reducing Permanganate

Sample Solution A. Dissolve 20 g in 150 mL of water, cool, and dilute with water to 200 mL (1 mL = 0.1 g).

Chloride. (Page 35). Dilute 10 mL of sample solution A (1-g sample) with water to 25 mL, and add 0.5 mL of nitric acid.

Arsenic. (Page 34). Dilute 30 mL of sample solution A (3-g sample) with water to 35 mL. For the standard, use 0.003 mg of arsenic.

Heavy Metals. To 50 mL of sample solution A (5-g sample), add 0.15 mL of phenolphthalein indicator solution, and neutralize with ammonium hydroxide. Adjust the pH of this solution to 4.5 (using a pH meter) with 1 N sulfuric acid and dilute with water to 250 mL. For the sample, use 40 mL of the solution. For the control, add 0.02 mg of lead to 20 mL of the solution, and dilute with water to 40 mL. Add 10 mL of freshly prepared hydrogen sulfide water to each, and mix. Any color in the solution of the sample should not exceed that in the control.

Iron. To 2.0 mL of sample solution A (0.2-g sample), add 5 mL of water and 3 mL of hydrochloric acid, boil gently for 10 min, and cool. Add 3.5 mL of 0.5 N sodium acetate, 6 mL of hydroxylamine hydrochloride reagent solution, and 4 mL of 1,10-phenanthroline reagent solution. Adjust the pH to between 4 and 6 with ammonium hydroxide. Any red color produced within 1 h should not exceed that produced by 0.01 mg of iron in an equal volume of solution containing the quantities of reagents used in the test.

Substances Reducing Permanganate. To 50 mL of sample solution A (5-g sample), add 10 mL of 10% sulfuric acid reagent solution, heat the solution to boiling, and titrate with 0.1 N potassium permanganate volumetric solution to a pink color that remains for 5 min. Not more than 0.25 mL of permanganate should be required.

Phosphoric Acid, Ultratrace

H_3PO_4 Formula Wt 98.00   CAS No. 7664-38-2

Suitable for use in ultratrace elemental analysis.

Note: This reagent must be packaged in a preleached Teflon bottle and used in a clean laboratory environment to maintain purity.

GENERAL DESCRIPTION

Typical appearance: clear, viscous liquid
Analytical use: trace metal analysis
Aqueous solubility: miscible
Density: 1.68
pK_a: 2.1

SPECIFICATIONS

Assay . . . . $\geq$85.0% H_3PO_4

Maximum Allowable

Chloride (Cl)	3 ppm
Nitrate (NO₃)	5 ppm
Sulfate (SO₄)	0.003%
Mercury (Hg)	5 ppb
Aluminum (Al)	50 ppb
Barium (Ba)	5 ppb
Boron (B)	5 ppb
Cadmium (Cd)	5 ppb
Calcium (Ca)	50 ppb
Chromium (Cr)	5 ppb
Cobalt (Co)	10 ppb
Copper (Cu)	5 ppb
Iron (Fe)	20 ppb
Lead (Pb)	10 ppb
Lithium (Li)	5 ppb
Magnesium (Mg)	10 ppb
Manganese (Mn)	5 ppb
Molybdenum (Mo)	5 ppb
Potassium (K)	5 ppb
Sodium (Na)	50 ppb
Strontium (Sr)	5 ppb
Tin (Sn)	10 ppb
Titanium (Ti)	5 ppb
Vanadium (V)	5 ppb
Zinc (Zn)	50 ppb
Zirconium (Zr)	5 ppb

TESTS

Assay. (By acid–base titrimetry). Weigh accurately about 2.0 g, and dilute with 120 mL of water. Titrate potentiometrically with 1 N sodium hydroxide volumetric solution to the inflection point at a pH near 9, using the first derivative method (page 27). One milliliter of 1 N sodium hydroxide corresponds to 0.04900 g of H_3PO_4.

$$\% \ H_3PO_4 = \frac{(mL \times N \ NaOH) \times 4.90}{Sample \ wt \ (g)}$$

Chloride. (Page 35). Dilute 3.0 g of sample with water to 25 mL, and add 0.5 mL of nitric acid.

Nitrate. (Page 38, Method 1). For sample solution A, use 2.0 g in 4 mL of water. For control solution B, use 1.0 g and 1 mL of nitrate ion (NO_3) standard solution.

Sulfate. Use test for phosphoric acid, page 490.

Mercury. (By CVAAS, page 65). To each of three 100-mL volumetric flasks containing about 35 mL of water, add 4.0 g of sample. To the second and third flasks, add mercury ion (Hg) standards of 10 ng (2.5 ppb) and 20 ng (5.0 ppb), respectively. Add 5 mL of nitric acid to all three flasks, and dilute to the mark with water. Mix. Zero the instrument with the blank, and determine the mercury content using a suitable mercury analyzer system (1.0 mL of 0.01 μg/mL Hg = 10 ng).

Trace Metals. Determine the aluminum, barium, boron, cadmium, calcium, chromium, cobalt, copper, iron, lead, lithium, magnesium, manganese, molybdenum, potassium, sodium, strontium, tin, titanium, vanadium, zinc, and zirconium by the ICP-OES method described on page 69.

Phosphorus Pentoxide
Phosphorus(V) Oxide

P_2O_5 **Formula Wt 141.94**   **CAS No. 1314-56-3**

GENERAL DESCRIPTION

Typical appearance: white to off-white, very hygroscopic solid
Analytical use: desiccant
Aqueous solubility: exothermic hydrolysis by water to form phosphoric acid

SPECIFICATIONS

Assay . . . . $\geq$98.0% P_2O_5

Maximum Allowable

Insoluble matter 0.02%
Phosphorus trioxide (P_2O_3) . . . . Passes test
Ammonium (NH_4) 0.01%
Heavy metals (as Pb) 0.01%

TESTS

Note: When making a solution, the phosphorus pentoxide must be added carefully and in small portions to water that has been cooled. Quiet dissolution may be achieved by allowing the sample to remain for at least 8 h in the uncovered weighing bottle placed in a covered beaker containing the necessary water. The pentoxide will absorb enough water from the air to dissolve without sputtering.

Assay. (By acid–base titrimetry). Weigh accurately about 1.0 g, carefully dissolve in 100 mL of water, and evaporate to about 25 mL. Cool, dilute to 120 mL, add 0.5 mL of thymolphthalein indicator solution, and titrate with 1 N sodium hydroxide volumetric solution to the first appearance of a permanent blue color. One milliliter of 1 N sodium hydroxide corresponds to 0.03549 g of P_2O_5.

$$\% \, P_2O_5 = \frac{(mL \times N \, NaOH) \times 3.549}{Sample \, wt \, (g)}$$

Insoluble Matter. Carefully dissolve 5.0 g in 40 mL of water, warming if necessary. Filter through a tared, preconditioned filtering crucible, and set aside the filtrate for sample solution A. Wash the residue thoroughly, and dry at 105 °C.

For the Determination of Phosphorus Trioxide, Ammonium, and Heavy Metals

Sample Solution A. Dilute the filtrate, without washings, obtained in the test for insoluble matter with water to 100 mL (1 mL = 0.05 g).

Phosphorus Trioxide. To 60 mL of sample solution A (3-g sample), add 0.20 mL of 0.1 N potassium permanganate volumetric solution. Heat to boiling, and allow to digest on a hot plate ($\approx$100 °C) for 10 min. The pink color should not be entirely discharged. (Limit about 0.02%)

Ammonium. Dilute 2.0 mL of sample solution A (0.1-g sample) with water to 40 mL, and add 10 mL of freshly boiled 10% sodium hydroxide reagent solution and 2 mL of Nessler reagent. Any color should not exceed that produced by 0.01 mg of ammonium ion (NH_4) in an equal volume of solution containing the quantities of reagents used in the test.

Heavy Metals. (Page 36, Method 1). Dilute 8.0 mL of sample solution A (0.4-g sample) with water to about 30 mL, boil for 5 min, cool, and dilute with water to 32 mL. Use 24 mL to prepare the sample solution, and use the remaining 8.0 mL to prepare the control solution.

Phthalic Acid
1,2-Benzenedicarboxylic Acid
C₆H₄(COOH)₂   **Formula Wt 166.13**   **CAS No. 88-99-3**

GENERAL DESCRIPTION

Typical appearance: white solid
Analytical use: buffers

Aqueous solubility: 0.6 g in 100 mL at 20 °C

pK_a: 3.0

SPECIFICATIONS

Assay . . . . ≥99.5% $C_6H_4(COOH)_2$

Maximum Allowable

Insoluble matter 0.05%
Residue after ignition (as SO_4) 0.02%
Chloride (Cl) 0.001%
Nitrate (NO_3) 0.005%
Sulfate (SO_4) 0.005%
Heavy metals (as Pb) 0.001%
Iron (Fe). 0.001%
Water (H_2O). 0.5%

TESTS

Assay. (By acid–base titrimetry). Weigh accurately about 2.8 g, transfer to a conical flask, cool, and add 50.0 mL of 1 N sodium hydroxide volumetric solution. Add 25 mL of water, and boil on a hot plate until dissolved. Add 0.15 mL of phenolphthalein indicator solution, and titrate the excess sodium hydroxide with 1 N hydrochloric acid volumetric solution. Perform a blank determination, and make any necessary correction. One milliliter of 1 N sodium hydroxide consumed corresponds to 0.08306 g of $C_6H_4(COOH)_2$.

$$\% \ C_6H_4(COOH)_2 = \frac{[(50.0 \times N \ NaOH) - (mL \times N \ HCl)] \times 8.306}{\text{Sample wt (g)}}$$

Insoluble Matter. (Page 25). Use 10.0 g dissolved in 250 mL of 10% sodium carbonate solution.

Residue after Ignition. (Page 26). Use 4.0 g. Retain the residue to prepare sample solution B for the tests for heavy metals and iron.

For the Determination of Chloride, Nitrate, and Sulfate

Sample Solution A. Dissolve 10.0 g in 100 mL of boiling water, and allow to cool to room temperature. The supernatant liquid after decantation from the crystallized phthalic acid is sample solution A (1 mL = 0.10 g).

Chloride. (Page 35). Dilute 10 mL of sample solution A (1.0-g sample) to 20 mL with water. Filter if necessary through a chloride-free filter. For the standard, use 0.01 mg of chloride ion (Cl) and dilute to 20 mL with water. To each, add 1 mL of nitric acid and 1 mL of silver nitrate reagent solution. Mix, allow to stand for 5 min protected from sunlight, and compare. The solutions can best be viewed visually against a black background. Any turbidity in the solution of the sample should not exceed that in the standard.

Nitrate. (Page 38, Method 1). Use 10 mL of sample solution A (1-g sample) and 0.05 mg of nitrate ion (NO_3).

Sulfate. (Page 40, Method 3). Add 1 mL of 1% sodium carbonate reagent solution to 10 mL of sample solution A (1-g sample), and evaporate to dryness on a hot plate (≈100 °C).

Redissolve in a minimum volume of water, and add 1 mL of dilute hydrochloric acid (1 + 19). If necessary, filter through a small filter, and wash with two 2-mL portions of water. Dilute to 10 mL, and add 1 mL of 12% barium chloride reagent solution. Any turbidity should not exceed that produced by 0.05 mg of sulfate ion (SO_4) in an equal volume of solution containing the quantities of reagents used in the test. Compare 30 min after adding the barium chloride to the sample and standard solutions.

For the Determination of Heavy Metals and Iron

Sample Solution B. To the residue after ignition, add 10 mL of water, and dilute to 20 mL (1 mL = 0.2 g).

Heavy Metals. (Page 36, Method 1). Use 10 mL of sample solution B (2.0-g sample).

Iron. (Page 38, Method 2). Use 5 mL of sample solution B (1.0-g sample).

Water. (Page 31, Method 1). Use 5.0 g.

Phthalic Anhydride
1,3-Isobenzofurandione

$C_6H_4(CO)_2O$ **Formula Wt 148.12**   **CAS No. 85-44-9**

GENERAL DESCRIPTION
Typical appearance: white solid
Analytical use: buffers
Change in state (approximate): melting point, 130 °C

SPECIFICATIONS
Assay . . . . .99.0–100.2% $C_8H_4O_3$
Melting point . . . . .Not more than 3 °C range, including 131 °C

Maximum Allowable

Residue after ignition. 0.01%
Chloride (Cl). 0.002%
Sulfate (SO_4) 0.003%
Heavy metals (as Pb). . . . . .5 ppm
Iron (Fe). . . . . .5 ppm

TESTS

Assay. (By acid–base titrimetry). Accurately weigh about 0.3 g into a 150-mL beaker containing 15.0 mL of 0.5 N morpholine methanolic solution (described below) and approximately 80 mL of methanol. Stir at least 30 min, and titrate potentiometrically with 1 N hydrochloric acid volumetric solution. Similarly, titrate a blank containing 15.0 mL of 0.5 N morpholine methanolic solution and 80 mL of methanol. Correct for a blank. One milliliter of 1 N hydrochloric acid corresponds to 0.1481 g of $C_6H_4(CO)_2O$.

$$\% \ C_6H_4(CO)_2O = \frac{\{[mL \ (blank) - mL \ (sample)] \times N \ HCl\} \times 14.81}{Sample \ wt \ (g)}$$

> *Morpholine Methanolic Solution, 0.5 N.* Dilute 4.4 mL (4.4 g) of morpholine to 100 mL with methanol.

Melting Point. (Page 45).

Residue after Ignition. (Page 26). Gently ignite 10.0 g in a tared, preconditioned crucible or dish. Ignite at 600 ± 25 °C for 15 min. (Reserve this residue for the test for iron.)

Chloride. Mix 0.50 g of sample with 0.5 g of sodium carbonate, and add 10–15 mL of water. Evaporate on a hot plate ($\approx$100 °C), and ignite until the mass is thoroughly charred, avoiding an excessively high temperature. When the sample is completely ignited, cool, and extract the fusion with 10 mL of water and 2 mL of nitric acid. Filter through chloride-free filter paper, wash with water to a volume of 20 mL, and add 1 mL of silver nitrate reagent solution. Any turbidity should not exceed that produced by 0.01 mg of chloride ion in an equal volume of solution containing 0.5 g of sodium carbonate, 2 mL of nitric acid, and 1 mL of silver nitrate reagent solution.

Sulfate. Mix 3.0 g of sample with 1.0 g of sodium carbonate in a platinum crucible, add 20 mL of water, and evaporate to dryness on a hot plate ($\approx$100 °C). Ignite completely, taking care to protect the fusion from the flame because of the presence of sulfur compounds in the natural gas. Cool, dissolve the residue in 20 mL of water, add 2 mL of 30% hydrogen peroxide, and boil for 5 min. Add 3 mL of hydrochloric acid and evaporate to dryness on the hot plate. Dissolve the residue in 10 mL of water, filter, and wash with 10 mL of water. To the filtrate, add 1 mL of 1 N hydrochloric acid and 2 mL of 12% barium chloride reagent solution. Any turbidity should not be greater than that of a standard prepared as follows: Evaporate 1 g of sodium carbonate, 2 mL of 30% hydrogen peroxide, and 3 mL of hydrochloric acid to dryness on a hot plate ($\approx$100 °C). Dissolve the residue and 0.09 mg of sulfate ion (SO_4) in sufficient water to make 20 mL, and add 1 mL of 1 N hydrochloric acid and 2 mL of 12% barium chloride reagent solution. Compare 10 min after adding the barium chloride to the sample and standard solutions.

Heavy Metals. (Page 36, Method 2). Use 4.0 g of sample.

Iron. (Page 38, Method 1). To the residue from the residue after ignition test, add 3 mL of hydrochloric acid, cover with a watch glass, and digest on a hot plate ($\approx$100 °C) for 15–20 min. Remove the cover, and evaporate to dryness. Dissolve the residue in 5 mL of hydrochloric acid, filter if necessary, and dilute with water to 100 mL. Use 20 mL of this solution.

Picric Acid

2,4,6-Trinitrophenol

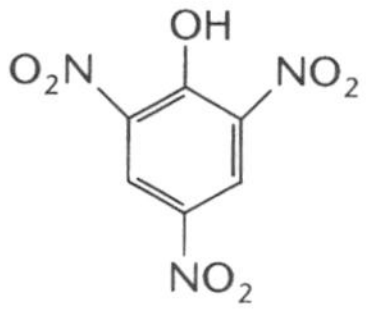

$(NO_2)_3C_6H_2OH$ **Formula Wt 229.11**   **CAS No. 88-89-1**

Caution: Trinitrophenol explodes when heated rapidly or when subjected to percussion. For safety in transportation, trinitrophenol is usually mixed with a minimum of 30% water.

GENERAL DESCRIPTION

Typical appearance: pale yellow solid
Analytical use: preparation of organic derivatives for identification
Aqueous solubility: 1 g in 100 mL
pK_a: 0.4

SPECIFICATIONS

Water (H_2O) . . . . ≥30%
Melting point (dried) . . . . 121–123 °C

Maximum Allowable

Insoluble and resinous matter 0.01%
Insoluble in toluene 0.1%
Sulfate (SO_4) 0.1%

TESTS

Water. Weigh accurately about 1.0 g, and dissolve in 200 mL of water. Titrate with 0.1 N sodium hydroxide, using 0.15 mL of phenolphthalein indicator solution. One milliliter of 0.1 N sodium hydroxide corresponds to 0.02291 g of $(NO_2)_3C_6H_2OH$.

$$\% \ (NO_2)_3C_6H_2OH = \frac{(mL \times N \ NaOH) \times 22.91}{Sample \ wt \ (g)}$$

Subtract the percent of picric acid from 100% to determine the water content.

Melting Point (Dried). Carefully transfer a small amount of dried sample (prepared in the test for insoluble in toluene) to a capillary tube, and run the melting point, using an oil immersion bath.

Insoluble and Resinous Matter. Dissolve a sample equivalent to 10 g on the dry basis in 500 mL of water, add 1 mL of sulfuric acid, and digest on a hot plate (≈100 °C) for 1 h. Filter through a tared, preconditioned filtering crucible, wash thoroughly, dry at 105 °C, and weigh. No resinous material should be apparent in the crucible.

Insoluble in Toluene. Dry 5–10 g at 70 °C and dissolve 4.0 g of the dried sample in 100 mL of warm toluene. (*Note:* Add water to the remaining dried sample before discarding it.) Filter off any insoluble material using a tared, preconditioned filtering crucible, wash thoroughly with warm toluene, and dry at 105 °C.

Sulfate. (Page 41, Method 2). Dissolve 0.5 g in 100 mL of water. Take 10 mL of this solution, and add 25 mL of nitric acid. Allow 30 min for turbidity to form.

Potassium Acetate

CH₃COOK   **Formula Wt 98.14**   **CAS No. 127-08-2**

GENERAL DESCRIPTION

Typical appearance: white or colorless solid
Analytical use: buffers
Change in state (approximate): melting point, 290 °C
Aqueous solubility: 253 g in 100 mL at 20 °C

SPECIFICATIONS

Assay . . . . ≥99.0% CH₃COOK
pH of a 5% solution at 25.0 °C 6.5–9.0

	Maximum Allowable
Insoluble matter	0.005%
Chloride (Cl)	0.003%
Phosphate (PO₄)	0.001%
Sulfate (SO₄)	0.002%
Heavy metals (as Pb)	5 ppm
Iron (Fe)	5 ppm
Calcium (Ca)	0.005%
Magnesium (Mg)	0.002%
Sodium (Na)	0.03%

TESTS

Assay. (By nonaqueous acid–base titrimetry). Weigh, to the nearest 0.1 mg, about 0.4 g, and dissolve in 5 mL of acetic anhydride and 25 mL of acetic acid. Add 0.15 mL of crystal violet indicator solution, and titrate in a closed system with 0.1 N perchloric acid in glacial acetic acid volumetric solution to a green end point. Perform a blank determination, and make any necessary correction. One milliliter of 0.1 N perchloric acid corresponds to 0.00981 g of CH₃COOK.

$$\% \ CH_3COOK = \frac{\{[mL \ (sample) - mL \ (blank)] \times N \ HClO_4\} \times 9.81}{Sample \ wt \ (g)}$$

pH of a 5% Solution at 25.0 °C. (Page 49).

Insoluble Matter. (Page 25). Use 20 g dissolved in 200 mL of water.

Chloride. (Page 35). Use 0.33 g.

Phosphate. (Page 40, Method 2). Dissolve 1.0 g in 10 mL of sulfuric acid, and evaporate to near dryness on a hot plate ($\approx$100 °C). Gently heat with a burner until the salt is dry. Add 10 mL of sulfuric acid, and repeat the evaporation and drying. Dissolve in 80 mL of water, add 0.5 g of ammonium molybdate, and adjust the pH to 1.8 (using a pH meter) with dilute hydrochloric acid (1 + 9). Heat to boiling, cool, add 10 mL of hydrochloric acid, and proceed as described. Carry along a standard containing 0.01 mg of phosphate ion (PO_4) treated in the same manner as the sample after the addition of 80 mL of water.

Sulfate. (Page 40, Method 1).

Heavy Metals. (Page 36, Method 1). Dissolve 6.0 g in about 10 mL of water, add 15 mL of 10% hydrochloric acid reagent solution, and dilute with water to 30 mL. Use 25 mL to prepare the sample solution, and use the remaining 5.0 mL to prepare the control solution.

Iron. (Page 38, Method 1). Use 2.0 g.

Calcium, Magnesium, and Sodium. (By flame AAS, page 63).

    Sample Stock Solution. Dissolve 5.0 g of sample in 80 mL of water. Transfer to a 100-mL volumetric flask, and dilute to the mark with water (1 mL = 0.05 g).

Element	Wavelength (nm)	Sample Wt (g)	Standard Added (mg)	Flame Type*	Background Correction
Ca	422.7	0.50	0.025; 0.05	N/A	No
Mg	285.2	0.50	0.01; 0.02	A/A	Yes
Na	589.0	0.03	0.01; 0.02	A/A	No

*A/A is air/acetylene; N/A is nitrous oxide/acetylene.

Potassium Antimony Tartrate Trihydrate
Antimony Potassium Tartrate Trihydrate
$K_2(C_4H_2O_6Sb)_2 \cdot 3H_2O$ **Formula Wt 667.87**   **CAS No. 11071-15-1**

GENERAL DESCRIPTION
Typical appearance: transparent, white solid
Analytical use: catalyst in phosphate analysis
Aqueous solubility: slightly soluble

SPECIFICATIONS

Assay . . . . 99.0–103.0% $K_2(C_4H_2O_6Sb)_2 \cdot 3H_2O$

Maximum Allowable

Titrable acid or base . . . . 0.020 meq/g
Loss on drying 2.7%
Arsenic (As) 0.015%

TESTS

Assay. (By iodometric titration of antimony). Weigh accurately about 0.5 g of sample, transfer to a 250-mL beaker, and dissolve in 50 mL of water. Add 5.0 g of potassium sodium tartrate tetrahydrate, 2 g of sodium borate decahydrate, and 3 mL of starch indicator solution. Titrate immediately with 0.1 N iodine volumetric solution to the production of a persistent blue color. One milliliter of 0.1 N iodine corresponds to 0.01670 g of $K_2(C_4H_2O_6Sb)_2 \cdot 3H_2O$.

$$\% \ K_2(C_4H_2O_6Sb)_2 \cdot 3H_2O = \frac{(mL \times N\ I_2) \times 16.70}{Sample\ wt\ (g)}$$

Titrable Acid or Base. Dissolve 1.0 g in 50 mL of carbon dioxide-free water, and titrate to a pH of 4.5 using 0.01 N hydrochloric acid or 0.01 N sodium hydroxide as required. Not more than 2.0 mL of either titrant should be required.

Loss on Drying. Weigh accurately 1.0 g, and dry in a tared, preconditioned dish at 105 °C to constant weight.

Arsenic. Dissolve 0.1 g in 5 mL of hydrochloric acid, and add 10 mL of a freshly prepared solution of 20 g of stannous chloride dihydrate in 30 mL of hydrochloric acid. Mix, transfer to a color-comparison tube, and allow to stand for 30 min. Viewed downward over a white surface, the color of the solution is not darker than that of a standard containing 0.015 g of arsenic in 100 mL of solution.

Potassium Bicarbonate
Potassium Hydrogen Carbonate
$KHCO_3$ **Formula Wt 100.12**   CAS No. 298-14-6

GENERAL DESCRIPTION
Typical appearance: colorless or white solid
Analytical use: buffer

SPECIFICATIONS
Assay (dried basis) . . . . 99.7–100.5% $KHCO_3$

	Maximum Allowable
Insoluble matter	0.01%
Chloride (Cl)	0.001%
Phosphate (PO_4)	5 ppm
Sulfur compounds (as SO_4)	0.003%
Ammonium (NH_4)	5 ppm
Heavy metals (as Pb)	5 ppm
Iron (Fe)	5 ppm
Calcium (Ca)	0.002%
Magnesium (Mg)	0.001%
Sodium (Na)	0.03%

TESTS

Assay. (By acid–base titrimetry). Weigh accurately 3.0 g, previously dried over sulfuric acid for 24 h, dissolve it in 50 mL of water, add 0.15 mL of methyl orange indicator solution, and titrate with 1 N hydrochloric acid volumetric solution. The potassium bicarbonate content calculated from the total alkalinity, as determined by the titration, should not be less than 99.7% nor more than 100.5% of the weight taken. One milliliter of 1 N hydrochloric acid corresponds to 0.1001 g of $KHCO_3$.

$$\% \ KHCO_3 = \frac{(mL \times N \ HCl) \times 10.01}{Sample \ wt \ (g)}$$

Insoluble Matter. (Page 25). Dissolve 10 g in 100 mL of hot water, heat to boiling, and boil gently for 2–3 min.

Chloride. (Page 35). Use 1.0 g.

Phosphate. (Page 40, Method 1). Dissolve 4.0 g in 20 mL of water, add 5 mL of hydrochloric acid, and evaporate to dryness on a hot plate ($\approx$100 °C). Dissolve the residue in 25 mL of approximately 0.5 N sulfuric acid, and proceed as described.

Sulfur Compounds. (Page 41, Method 4). Allow 30 min for turbidity to form.

Ammonium. Dissolve 2.0 g in 40 mL of water, and add 10 mL of 10% sodium hydroxide reagent solution and 2 mL of Nessler reagent solution. Any color should not exceed that produced by 0.01 mg of ammonium ion in an equal volume of solution containing the quantities of reagents used in the test.

Heavy Metals. (Page 36, Method 1). To 5.0 g in a 150-mL beaker, add 10 mL of water, mix, and add cautiously 10 mL of hydrochloric acid. Evaporate to dryness on a hot plate ($\approx$100 °C), dissolve the residue in about 20 mL of water, and dilute with water to 25 mL. For the control, add 0.02 mg of lead ion (Pb) to 1.0 g of sample, and treat exactly as the 5.0 g of sample.

Iron. (Page 38, Method 1). Dissolve 2.0 g in 20 mL of water, cautiously add 5 mL of hydrochloric acid, and dilute with water to 50 mL. Use the solution without further acidification. In the standard, use only 3 mL of hydrochloric acid.

Calcium, Magnesium, and Sodium. (By flame AAS, page 63).

Sample Stock Solution. Cautiously dissolve 5.0 g of sample in 30 mL of dilute hydrochloric acid (1 + 2). Cool to room temperature, transfer to a 100-mL volumetric flask, and dilute to the mark with water (1 mL = 0.05 g).

Element	Wavelength (nm)	Sample Wt (g)	Standard Added (mg)	Flame Type*	Background Correction
Ca	422.7	1.0	0.02; 0.04	N/A	No
Mg	285.2	1.0	0.01; 0.02	A/A	Yes
Na	589.0	0.05	0.015; 0.03	A/A	No

*A/A is air/acetylene; N/A is nitrous oxide/acetylene.

Potassium Bitartrate
Potassium Hydrogen Tartrate

KOCO(CHOH)₂COOH   **Formula Wt 188.18**   **CAS No. 868-14-4**

GENERAL DESCRIPTION

Typical appearance: white solid
Analytical use: buffering agent
Aqueous solubility: 1 g in 162 mL

SPECIFICATIONS

Assay . . . . 99.0–101.0% $C_4H_5O_6K$

Maximum Allowable

Insoluble matter in ammonium hydroxide 0.005%
Chloride (Cl) 0.005%
Ammonium (NH_4) 0.01%
Sulfate (SO_4) 0.01%
Calcium (Ca) 0.05%
Sodium (Na) 0.05%
Heavy metals (as Pb) 0.001%
Iron (Fe) 0.002%

TESTS

Assay. (By acid–base titrimetry). Weigh accurately about 0.7 g, and dissolve in 200 mL of water. Add 0.15 mL of phenolphthalein indicator solution, and titrate with 0.1 N sodium hydroxide to a pink end point. One milliliter of 0.1 N sodium hydroxide corresponds to 0.01882 g of $C_4H_5O_6K$.

$$\% \ C_4H_5O_6K = \frac{(mL \times N \ NaOH) \times 18.82}{Sample \ wt \ (g)}$$

Insoluble Matter in Ammonium Hydroxide. (Page 25). Use 20 g dissolved in a mixture of 100 mL of water and 30 mL of ammonium hydroxide.

Chloride. (Page 35). Use 0.20 g.

Ammonium. Dissolve 1.0 g in 200 mL of ammonia-free water in a 250-mL volumetric flask, and dilute to the mark with water. To 50.0 mL (0.2-g sample) of this solution, add 2 mL of freshly boiled 10% sodium hydroxide reagent solution, and mix. Add 2 mL of Nessler reagent, and mix again. Any color should not exceed that produced by 0.02 mg of ammonium ion (NH_4) in an equal volume of solution containing the quantities of reagents used in the test.

Sulfate. (Page 41, Method 3). Ignite 0.50 g, protected from sulfur and preferably in an electric muffle furnace, until nearly free of carbon. Boil the residue with 20 mL of water and 2 mL of hydrogen peroxide for 5 min. Add 10 mg of sodium carbonate and 5 mL of hydrochloric acid, and evaporate to dryness on a hot plate ($\approx$100 °C). Prepare a standard by evaporating 10 mg of sodium carbonate, 2 mL of hydrogen peroxide, 0.05 mg of sulfate ion (SO_4) standard solution, and 5 mL of hydrochloric acid to dryness on a hot plate ($\approx$100 °C). Dissolve the residue from each in a minimum amount of water and proceed as directed in the test beginning with "Add 1mL of dilute hydrochloric acid (1 + 19)…".

Calcium and Sodium. (By flame AAS, page 63).

> **Sample Stock Solution.** Dissolve 1.0 g of sample in 80 mL of (1 + 7) hydrochloric acid, and digest on a hot plate ($\approx$100 °C) for 20 min. Cool, transfer to a 100-mL volumetric flask, and dilute to the mark with water (1 mL = 0.01 g).

To each of four flasks, add 5 mL of 5% potassium chloride solution to suppress ionization interferences, dilute to the mark with water, mix well, and continue with the procedure.

Element	Wavelength (nm)	Sample Wt (g)	Standards Added (mg)	Flame Type*	Background Correction
Ca	422.7	0.10	0.025; 0.05	N/A	No
Na	589.0	0.10	0.025; 0.05	A/A	No

*A/A is air/acetylene, N/A is nitrous oxide/acetylene.

Heavy Metals. Ignite 5.0 g in a platinum dish at 500–600 °C. Dissolve the residue in 30 mL of water and 2 mL of hydrochloric acid, filter, dilute to 50 mL, and mix. For the sample, transfer 30 mL of the filtrate to a suitable Nessler tube. For the standard, add 0.02 mg of lead ion (Pb) to a 10-mL portion of the filtrate, also in a Nessler tube, dilute to 30 mL, and mix. Adjust the pH of the sample and standard solutions to between 3 and 4 (using a pH meter) with 1 N acetic acid or 10% ammonium hydroxide reagent solution. Dilute each solution to 40 mL, mix, add 10 mL of freshly prepared hydrogen sulfide water to each, and mix. Any brown color in the sample solution must not exceed that in the standard.

Iron. To 0.5 g, add 25 mL of water, 3 mL of hydrochloric acid, and 3 drops of nitric acid. Boil for 1 min. Cool, dilute to 50 mL, and add 3 mL of 30% ammonium thiocyanate reagent solution. Compare to an iron ion (Fe) standard solution treated exactly as the sample.

Potassium Bromate

| KBrO₃ | Formula Wt 167.00 | CAS No. 7758-01-2 |

$KBrO_3$ **Formula Wt 167.00**   **CAS No. 7758-01-2**

GENERAL DESCRIPTION

Typical appearance: colorless or white solid

Analytical use: oxidizing agent in acid solutions

Change in state (approximate): melting point, 434 °C; begins to decompose at 370 °C

Aqueous solubility: 6.9 g in 100 mL at 20 °C

SPECIFICATIONS

Assay (dried basis) . . . . ≥99.8% $KBrO_3$
pH of a 5% solution at 25.0 °C 5.0–9.0

Maximum Allowable

Insoluble matter 0.005%
Bromide (Br) . . . . Passes test
Sulfate (SO₄) 0.005%
Heavy metals (as Pb) . . . . 5 ppm
Iron (Fe) 0.002%
Sodium (Na) 0.01%

TESTS

Assay. (By oxidation–reduction titration of bromate). Dry a powdered sample to constant weight at 150 °C. Weigh accurately 1.0 g, dissolve in water, and transfer to a 250-mL volumetric flask. Dilute to volume with water, and mix thoroughly. Place a 25.0-mL aliquot of this solution in a glass-stoppered conical flask, and add 3 g of potassium iodide and 3 mL of hydrochloric acid. Stopper, swirl, and allow to stand for 5 min. Titrate the liberated iodine with 0.1 N sodium thiosulfate volumetric solution, adding 3 mL of starch indicator solution near the end of the titration. One milliliter of 0.1 N sodium thiosulfate corresponds to 0.002783 g of $KBrO_3$.

$$\% \ KBrO_3 = \frac{(mL \times N \ Na_2S_2O_3) \times 2.783}{\text{Sample wt (g) / 10}}$$

pH of a 5% Solution at 25.0 °C. (Page 49).

Insoluble Matter. (Page 25). Use 20 g dissolved in 150 mL of hot water.

Bromide. Dissolve 4.0 g in 80 mL of water, and divide the solution into two equal portions. Add 0.15 mL of 1 N sulfuric acid volumetric solution to one portion. At the end of 2 min, this portion should show no more yellow color than the portion to which no acid was added. (Limit about 0.05%)

Sulfate. (Page 41, Method 2). Use 6 mL of dilute hydrochloric acid (1 + 1). Do two evaporations.

Heavy Metals. (Page 36, Method 1). Dissolve 5.0 g in 20 mL of dilute hydrochloric acid (1 + 1). Evaporate the solution to dryness on a hot plate (≈100 °C). Add 10 mL more of dilute hydrochloric acid (1 + 1), and again evaporate to dryness. Dissolve in about 20 mL of water, and dilute with water to 25 mL. For the control, add 0.02 mg of lead ion (Pb) to 1.0 g of sample, and treat exactly as the 5.0 g of sample.

Iron. (Page 38, Method 1). Dissolve 1.0 g in 20 mL of dilute hydrochloric acid (1 + 1), and evaporate on a hot plate ($\approx$100 °C) to dryness. Add 10 mL of additional dilute hydrochloric acid (1 + 1), and again evaporate to dryness. Dissolve the residue in water, add 4 mL of hydrochloric acid, and dilute with water to 100 mL. Use 50 mL of the solution. For the control, use 0.02 mg of iron ion (Fe) carried through the same procedure as the 1.0 g of sample.

Sodium. (By flame AAS, page 63).

> **Sample Stock Solution.** Dissolve 1.0 g of sample in 25 mL of hydrochloric acid (1 + 3), and digest in a covered beaker on a hot plate ($\approx$100 °C) until the reaction ceases. Uncover, evaporate to dryness, add 10 mL of hydrochloric acid (1 + 3), and dilute with hydrochloric acid (1 + 99) to the mark in a 100-mL volumetric flask (1 mL = 0.01 g).

Element	Wavelength (nm)	Sample Wt (g)	Standard Added (mg)	Flame Type*	Background Correction
Na	589.0	0.10	0.01; 0.02	A/A	No

*A/A is air/acetylene.

Potassium Bromide

KBr   **Formula Wt 119.00**   **CAS No. 7758-02-3**

GENERAL DESCRIPTION

Typical appearance: colorless solid
Analytical use: analytical standard; redox reagent
Change in state (approximate): melting point, 730 °C
Aqueous solubility: 65 g in 100 mL at 20 °C
Density: 2.75

SPECIFICATIONS

Assay . . . . $\geq$99.0% KBr
pH of a 5% solution at 25.0 °C 5.0–8.8

Maximum Allowable

Insoluble matter 0.005%
Bromate (BrO_3) 0.001%
Iodate (IO_3). 0.001%
Chloride (Cl). 0.2%
Iodide (I) 0.001%
Sulfate (SO_4) 0.005%
Barium (Ba) 0.002%

Heavy metals (as Pb) . . . . 5 ppm
Iron (Fe). . . . . 5 ppm
Calcium (Ca) 0.002%
Magnesium (Mg). 0.001%
Sodium (Na) 0.02%

TESTS

Assay. (By argentimetric titration of bromide content). Weigh accurately about 0.5 g of sample, and dissolve in 50 mL of water. Add 50.0 mL of 0.1 N silver nitrate volumetric solution and 10 mL of dilute nitric acid (1 + 10), and titrate the excess silver nitrate with 0.1 N ammonium thiocyanate volumetric solution, using ferric ammonium sulfate indicator solution. One milliliter of 0.1 N silver nitrate corresponds to 0.0119 g of KBr.

$$\% \text{ KBr} = \frac{[(\text{mL} \times \text{N AgNO}_3) - (\text{mL} \times \text{N NH}_4\text{SCN})] \times 11.90}{\text{Sample wt (g)}}$$

pH of a 5% Solution at 25.0 °C. (Page 49).

Insoluble Matter. (Page 25). Use 20 g dissolved in 150 mL of water.

Bromate and Iodate. (Page 54). Use 10.0 g of sample in 25 mL of solution. For the standard, add 0.10 mg of bromate ion (BrO_3) and 0.10 mg of iodate ion (IO_3).

Chloride. Dissolve 0.50 g in 15 mL of dilute nitric acid (1 + 2) in a small flask. Add 3 mL of 30% hydrogen peroxide, and digest on a hot plate ($\approx$100 °C) until the solution is colorless. Wash down the sides of the flask with a little water, digest for an additional 15 min, cool, and dilute with water to 200 mL. Dilute 2.0 mL to 20 mL and add 1 mL of nitric acid and 1 mL of silver nitrate reagent solution. Any turbidity should not exceed that produced by 0.01 mg of chloride ion (Cl) in an equal volume of solution containing the quantities of reagents used in the test.

Iodide. Dissolve 2.5 g in 50 mL of water in a 100-mL volumetric flask. For the control, dissolve 0.5 g in a similar flask, and add 0.02 mg of iodide ion (I). Treat each solution as follows: Add 2 mL of bromine water, mix, and allow to stand for 5 min. Add 2 mL of sodium formate solution (25 g per 100 mL), and mix. Add 1 mL of 1 N sodium hydroxide volumetric solution or enough to bring the pH to 9 or greater, dilute to volume with water, and mix. Using differential pulse polarography, as described on page 54, starting with transfer of the solution to the polarographic cell, determine any iodate present. The wave for the sample, after correction for any iodate originally present, should not exceed that for the control (similarly corrected).

Sulfate. (Page 40, Method 1).

Barium. For the sample, dissolve 6.0 g in 15 mL of water. For the control, dissolve 1.0 g in 15 mL of water and add 0.1 mg of barium. To each solution, add 5 mL of acetic acid, 5 mL of 30% hydrogen peroxide, and 1 mL of hydrochloric acid. Digest in a covered beaker on a hot plate ($\approx$100 °C) until reaction ceases, uncover, and evaporate to dryness. Dissolve the residues in 15 mL of water, filter if necessary, and dilute with water to 23 mL. Add 2 mL of

10% potassium dichromate reagent solution, and add ammonium hydroxide until the orange color is just dissipated and the yellow color persists. Add 25 mL of methanol, stir vigorously, and allow to stand for 10 min. The turbidity in the solution of the sample should not exceed that in the control.

Heavy Metals. (Page 36, Method 1). Dissolve 6.0 g in about 20 mL of water, and dilute with water to 30 mL. Use 25 mL to prepare the sample solution, and use the remaining 5.0 mL to prepare the control solution.

Iron. (Page 38, Method 1). Use 2.0 g. In both sample and control solutions, use 4 mL of hydrochloric acid.

Calcium, Magnesium, and Sodium. (By flame AAS, page 63).

Sample Stock Solution. Dissolve 5.0 g of sample in 80 mL of water. Transfer to a 100-mL volumetric flask, and dilute to the mark with water (1 mL = 0.05 g).

Element	Wavelength (nm)	Sample Wt (g)	Standard Added (mg)	Flame Type*	Background Correction
Ca	422.7	1.0	0.02; 0.04	N/A	No
Mg	285.2	1.0	0.01; 0.02	A/A	Yes
Na	589.0	0.05	0.01; 0.02	A/A	No

*A/A is air/acetylene; N/A is nitrous oxide/acetylene.

Potassium Carbonate
Potassium Carbonate, Anhydrous

K_2CO_3 **Formula Wt 138.21**   **CAS No. 584-08-7**

GENERAL DESCRIPTION
Typical appearance: white, hygroscopic solid
Analytical use: drying agent for organic solvents; a base
Change in state (approximate): melting point, 891 °C
Aqueous solubility: 112 g in 100 mL at 20 °C

SPECIFICATIONS
Assay . . . . $\geq$99.0% K_2CO_3

Maximum Allowable

Insoluble matter 0.01%
Chloride (as Cl) 0.003%
Phosphate (PO_4). 0.001%
Silica (SiO_2). 0.005%
Sulfur compounds (as SO_4) 0.004%
Heavy metals (as Pb). . . . . 5 ppm

Iron (Fe). . . . . 5 ppm
Calcium (Ca) 0.005%
Magnesium (Mg). 0.002%
Sodium (Na) 0.02%

TESTS

Assay. (By acid–base titrimetry). Weigh, to the nearest 0.1 mg, about 3 g of sample, transfer to a 125-mL glass-stoppered flask, and dissolve with 50 mL of water. Add 0.15 mL of methyl orange indicator solution, and titrate with 1 N hydrochloric acid volumetric solution. One milliliter of 1 N hydrochloric acid corresponds to 0.0691 g of K_2CO_3.

$$\% \, K_2CO_3 = \frac{(mL \times N \, HCl) \times 6.91}{Sample \, wt \, (g)}$$

Insoluble Matter. (Page 25). Use 10 g dissolved in 100 mL of water.

Chloride. Dissolve 1.0 g in about 40 mL of water plus 6 mL of nitric acid. Filter through a chloride-free filter, wash, and dilute the filtrate with water to 60 mL. To 20 mL, add 1 mL of silver nitrate reagent solution. Any turbidity should not exceed that produced by 0.01 mg of chloride ion (Cl) in an equal volume of solution containing the quantities of reagents used in the test.

Phosphate. (Page 40, Method 2). Dissolve 1.0 g in 50 mL of water in a platinum dish, and digest on a hot plate ($\approx$100 °C) for 30 min. Cool, neutralize with dilute sulfuric acid (1 + 19) to a pH of about 4, and dilute with water to about 75 mL. Add 0.5 g of ammonium molybdate to the solution, and when it is dissolved, adjust the pH to 1.8 (using a pH meter) with dilute hydrochloric acid (1 + 9). Heat to boiling, cool, add 10 mL of hydrochloric acid, and dilute with water to 100 mL. Proceed as described. Carry along a standard containing 0.01 mg of phosphate ion (PO_4) and 0.05 mg of silica ion (SiO_2) in about 75 mL of water treated as the 75 mL of sample solution. Reserve the aqueous phase for the determination of silica.

Silica. Add 10 mL of hydrochloric acid to the solutions reserved from the determination of phosphate, and transfer to separatory funnels. Add 40 mL of butyl alcohol, shake vigorously, and allow to separate. Draw off and discard the aqueous phase. Wash the butyl alcohol three times with 20-mL portions of dilute hydrochloric acid (1 + 99), discarding the washings each time. Dilute each butyl alcohol solution to 50 mL with butyl alcohol, take 10 mL from each, and dilute each to 50 mL with butyl alcohol. Add 0.5 mL of a freshly prepared 2% stannous chloride reagent solution. The blue color in the extract from the sample should not exceed that in the control. If the butyl alcohol extracts are turbid, wash them with 10 mL of dilute hydrochloric acid (1 + 99).

Sulfur Compounds. (Page 41, Method 4). Cautiously acidify with hydrochloric acid. Allow 30 min for turbidity to form.

Heavy Metals. (Page 36, Method 1). To 5.0 g in a 150-mL beaker, add 10 mL of water, mix, and cautiously add 10 mL of hydrochloric acid. Evaporate the solution to dryness on a hot plate ($\approx$100 °C), dissolve the residue in about 20 mL of water, and dilute with water to 25 mL. For the control, add 0.02 mg of lead ion (Pb) to 1.0 g of sample, and treat exactly as the 5.0 g of sample.

Iron. (Page 38, Method 1). Dissolve 2.0 g in 20 mL of dilute hydrochloric acid (1 + 1), and evaporate to dryness on a hot plate ($\approx$100 °C). Dissolve the residue in 5 mL of dilute hydrochloric acid (1 + 1), and again evaporate to dryness. Dissolve the residue in 50 mL of dilute hydrochloric acid (1 + 24), and use the solution without further acidification. Use the residue from evaporation of 15 mL of hydrochloric acid to prepare the standard.

Calcium, Magnesium, and Sodium. (By flame AAS, page 63).

Sample Stock Solution. Dissolve 5.0 g of sample with water in a 100-mL volumetric flask, and dilute to the mark with water (1 mL = 0.05 g).

Element	Wavelength (nm)	Sample Wt (g)	Standard Added (mg)	Flame Type*	Background Correction
Ca	422.7	0.50	0.025; 0.05	N/A	No
Mg	285.2	0.50	0.01; 0.02	A/A	Yes
Na	589.0	0.05	0.01; 0.02	A/A	No

*A/A is air/acetylene; N/A is nitrous oxide/acetylene.

Potassium Carbonate Sesquihydrate

$K_2CO_3 \cdot 1.5H_2O$ **Formula Wt 165.23**   **CAS No. 6381-79-9**

GENERAL DESCRIPTION

Typical appearance: white solid
Analytical use: invert sugar test; buffer

SPECIFICATIONS

Assay . . . . 98.5–101.0% $K_2CO_3 \cdot 1.5H_2O$
Loss on heating at 285 °C 14.0–16.5%

	Maximum Allowable
Insoluble matter	0.01%
Chloride (as Cl)	0.003%
Phosphate (PO$_4$).	0.001%
Silica (SiO$_2$).	0.005%
Sulfur compounds (as SO$_4$)	0.004%
Heavy metals (as Pb).	5 ppm
Iron (Fe).	5 ppm
Calcium (Ca)	0.005%
Magnesium (Mg)	0.002%
Sodium (Na).	0.02%

TESTS

Assay. (By acidimetric titration of carbonate). Weigh accurately about 3.3 g, and dissolve in 50 mL of water. Add 0.15 mL of methyl orange indicator solution, and titrate with 1 N hydrochloric acid volumetric solution to the change from yellow to orange. One milliliter of 1 N hydrochloric acid corresponds to 0.08262 g of $K_2CO_3 \cdot 1.5H_2O$.

$$\% \ K_2CO_3 \cdot 1.5H_2O = \frac{(mL \times N \ HCl) \times 8.262}{Sample \ wt \ (g)}$$

Loss on Heating at 285 °C. Weigh accurately about 2.0 g, and heat in a tared, preconditioned container to constant weight at 270–300 °C.

Insoluble Matter. (Page 25). Use 10.0 g dissolved in 100 mL of water.

Chloride. Dissolve 1.0 g in about 40 mL of water plus 6 mL of nitric acid. Filter through a chloride-free filter, wash, and dilute the filtrate with water to 60 mL. To 20 mL, add 1 mL of silver nitrate reagent solution. Any turbidity should not exceed that produced by 0.01 mg of chloride ion (Cl) in an equal volume of solution containing the quantities of reagents used in the test.

Phosphate. (Page 40, Method 2). Dissolve 1.0 g in 50 mL of water in a platinum dish, and digest on a hot plate ($\approx$100 °C) for 30 min. Cool, neutralize with dilute sulfuric acid (1 + 19) to a pH of about 4, and dilute with water to about 75 mL. Add 0.5 g of ammonium molybdate to the solution, and when it is dissolved, adjust the pH to 1.8 (using a pH meter) with dilute hydrochloric acid (1 + 9). Heat the solution to boiling, cool to room temperature, add 10 mL of hydrochloric acid, and dilute with water to 100 mL. Proceed as described. Carry along a standard containing 0.01 mg of phosphate ion (PO_4) and 0.05 mg of silica ion (SiO_2) in about 75 mL of water treated as the 75 mL of sample solution. Reserve the aqueous phase for the determination of silica.

Silica. Add 10 mL of hydrochloric acid to the solutions reserved from the determination of phosphate, and transfer to separatory funnels. Add 40 mL of butyl alcohol, shake vigorously, and allow to separate. Draw off and discard the aqueous phase. Wash the butyl alcohol three times with 20-mL portions of dilute hydrochloric acid (1 + 99), discarding the washings each time. Dilute the butyl alcohol solutions to 50 mL with butyl alcohol, take 10 mL of each, and dilute each to 50 mL with butyl alcohol. Add 0.2 mL of a freshly prepared 2% stannous chloride reagent solution. The blue color in the extract from the sample should not exceed that in the control. If the butyl alcohol extracts are turbid, wash them with 10 mL of dilute hydrochloric acid (1 + 99).

Sulfur Compounds. (Page 41, Method 4). Cautiously acidify with hydrochloric acid. Allow 30 min for turbidity to form.

Heavy Metals. (Page 36, Method 1). To 5.0 g in a 150-mL beaker, add 10 mL of water, mix, and add cautiously 10 mL of hydrochloric acid. Evaporate the solution to dryness on a hot plate ($\approx$100 °C), dissolve the residue in about 20 mL of water, and dilute with water to 25 mL. For the control, add 0.02 mg of lead to 1.0 g of sample and treat exactly as the 5.0 g of sample.

Iron. (Page 38, Method 1). Dissolve 2.0 g in 20 mL of dilute hydrochloric acid $(1 + 1)$, and evaporate to dryness on a hot plate ($\approx$100 °C). Dissolve the residue in 5 mL of dilute hydrochloric acid $(1 + 1)$, and again evaporate to dryness. Dissolve the residue in 50 mL of dilute hydrochloric acid $(1 + 24)$, and use the solution without further acidification. Use the residue from evaporation of 15 mL of hydrochloric acid to prepare the standard.

Calcium, Magnesium, and Sodium. (By flame AAS, page 63).

Sample Stock Solution. Dissolve 5.0 g of sample with water in a 100-mL volumetric flask, and dilute to the mark with water (1 mL = 0.05 g).

Element	Wavelength (nm)	Sample Wt (g)	Standard Added (mg)	Flame Type*	Background Correction
Ca	422.7	0.50	0.025; 0.05	N/A	No
Mg	285.2	0.50	0.01; 0.02	A/A	Yes
Na	589.0	0.05	0.01; 0.02	A/A	No

*A/A is air/acetylene; N/A is nitrous oxide/acetylene.

Potassium Chlorate
KClO$_3$ **Formula Wt 122.55**   **CAS No. 3811-04-9**

GENERAL DESCRIPTION
Typical appearance: white or colorless solid
Analytical use: source of oxygen
Change in state (approximate): melting point, 368 °C
Aqueous solubility: 7 g in 100 mL at 20 °C
Density: 2.3

SPECIFICATIONS
Assay . . . . $\geq$99.0% KClO$_3$

	Maximum Allowable
Insoluble matter	0.005%
Bromate (BrO$_3$)	0.015%
Chloride (Cl)	0.001%
Sulfate (SO$_4$)	Passes test
Heavy metals (as Pb)	5 ppm
Iron (Fe)	3 ppm
Calcium (Ca)	0.002%
Magnesium (Mg)	0.002%
Sodium (Na)	0.01%

TESTS

Assay. (By oxidation–reduction titration of chlorate). Weigh, to the nearest 0.1 mg, about 0.05 g of sample, and dissolve it with 10 mL of water in a 250-mL glass-stoppered flask. Add 50.0 mL of 0.1 N ferrous ammonium sulfate volumetric solution, cover the flask to prevent exposure to the air, and boil for 10 min. Cool, add 10 mL of 10% manganese sulfate solution (described below) and 5 mL of phosphoric acid, and titrate the excess ferrous ammonium sulfate with 0.1 N potassium permanganate volumetric solution. Run a blank in the same manner. One milliliter of 0.1 N ferrous ammonium sulfate corresponds to 0.002043 g of $KClO_3$.

$$\% \ KClO_3 = \frac{\{[mL \ (blank) - mL \ (sample)] \times N \ KMnO_4\} \times 2.043}{Sample \ wt \ (g)}$$

Manganese Sulfate Solution, 10%. Dissolve 10.0 g of manganese sulfate monohydrate in 100 mL of water.

Insoluble Matter. (Page 25). Use 20 g dissolved in 250 mL of water.

Bromate. (Page 54). Use 1.0 g of sample and 0.5 g of calcium chloride dihydrate in 25 mL of solution. For the standard, add 0.15 mg of bromate ion (BrO_3). Also prepare a reagent blank.

Chloride. Dissolve 2.0 g in 40 mL of warm water, and filter if necessary through a chloride-free filter. Add 0.25 mL of nitric acid, free from lower oxides of nitrogen, and 1 mL of silver nitrate reagent solution. Any turbidity should not exceed that produced by 0.02 mg of chloride ion (Cl) in an equal volume of solution containing the quantities of reagents used in the test.

Sulfate. Dissolve 5.0 g in 150 mL of water, filter if necessary, and add 1 mL of 10% hydrochloric acid reagent solution and 5 mL of 12% barium chloride reagent solution. No precipitate should be produced on standing for at least 8 h. (Limit about 0.002%)

Heavy Metals. (Page 36, Method 1). Dissolve 5.0 g in 20 mL of dilute hydrochloric acid (1 + 1). Evaporate the solution to dryness on a hot plate ($\approx$100 °C), add 5 mL more of dilute hydrochloric acid (1 + 1), and again evaporate to dryness. Dissolve the residue in about 20 mL of water, and dilute with water to 25 mL. For the control, add 0.02 mg of lead ion (Pb) to 1.0 g of sample, and treat exactly as the 5.0 g of sample.

Iron. (Page 38, Method 1). Dissolve 3.3 g in 20 mL of dilute hydrochloric acid (1 + 1), and evaporate to dryness on a hot plate ($\approx$100 °C). Add 5 mL of dilute hydrochloric acid (1 + 1), and again evaporate to dryness. Dissolve in 50 mL of dilute hydrochloric acid (1 + 25), and use the solution without further acidification. In preparing the control, use the residue from evaporation of 15 mL of hydrochloric acid.

Calcium, Magnesium, and Sodium. (By flame AAS, page 63).

Sample Stock Solution. Dissolve 5.0 g in 25 mL of dilute hydrochloric acid (1 + 3), and digest in a covered beaker on a hot plate ($\approx$100 °C) until the reaction ceases. Uncover the beaker, and evaporate to dryness. Add 10 mL of dilute hydrochloric acid

(1 + 3), and again evaporate to dryness. Dissolve in dilute hydrochloric acid (1 + 99), and dilute to 100 mL with the hydrochloric acid (1 + 99) in a volumetric flask (1 mL = 0.05 g).

Element	Wavelength (nm)	Sample Wt (g)	Standard Added (mg)	Flame Type*	Background Correction
Ca	422.7	1.0	0.02; 0.04	N/A	No
Mg	285.2	0.20	0.004; 0.008	A/A	Yes
Na	589.0	0.20	0.01; 0.02	A/A	No

*A/A is air/acetylene; N/A is nitrous oxide/acetylene.

Potassium Chloride

KCl   **Formula Wt 74.55**   **CAS No. 7447-40-7**

GENERAL DESCRIPTION

Typical appearance: colorless or white solid
Analytical use: buffer solutions
Change in state (approximate): melting point, 773 °C
Aqueous solubility: 35 g in 100 mL at 20 °C

SPECIFICATIONS

Assay . . . . .99.0–100.5% KCl
pH of a 5% solution at 25.0 °C 5.4–8.6

	Maximum Allowable

Insoluble matter 0.005%
Iodide (I) 0.002%
Bromide (Br) 0.01%
Chlorate and nitrate (as NO_3). 0.003%
Phosphate (PO_4). . . . . .5 ppm
Sulfate (SO_4) 0.001%
Barium (Ba) . . . . .Passes test
Heavy metals (as Pb). . . . . .5 ppm
Iron (Fe). . . . . .3 ppm
Calcium (Ca) 0.002%
Magnesium (Mg) 0.001%
Sodium (Na). 0.005%

TESTS

Assay. (By argentimetric titration of chloride content). Weigh accurately about 0.3 g, and dissolve in 50 mL of water in a 250-mL glass-stoppered flask. Add 1 mL of dichlorofluorescein indicator solution. While stirring, mix and titrate with 0.1 N silver nitrate volumetric solution. The silver chloride should flocculate, and the mixture should change to pink. One milliliter of 0.1 N silver nitrate corresponds to 0.007455 g of KCl.

$$\% \text{ KCl} = \frac{(\text{mL} \times \text{N AgNO}_3) \times 7.455}{\text{Sample wt (g)}}$$

pH of a 5% Solution at 25.0 °C. (Page 49).

Insoluble Matter. (Page 25). Use 20 g dissolved in 150 mL of water.

Iodide. Dissolve 11 g in 50 mL of water. Prepare a control by dissolving 1 g of the sample, 0.2 mg of iodide ion (I), and 1 mg of bromide ion (Br) in 50 mL of water. To each solution, in a separatory funnel, add 2 mL of hydrochloric acid and 5 mL of ferric chloride reagent solution. Allow to stand for 5 min. Add 10 mL of chloroform, shake for 1 min, allow the phases to separate, and draw off the chloroform layer. Reserve the water solution for the test for bromide. Any violet color in the chloroform extract from the solution of the sample should not exceed that in the extract from the control.

Bromide. Dissolve 0.10 g of sample in 5.0 mL of water. Add 2.0 mL of pH 4.7 phenol red indicator solution, then add 1.0 mL of chloramine-T reagent solution, and stir immediately. After 2 min, add 0.15 mL of 0.1 N sodium thiosulfate volumetric solution, stir, and dilute to 10.0 mL with water. Simultaneously, set up a control solution by using 5.0 mL of potassium bromide (KBr) standard solution and adding the same quantity of reagents, carried out similarly in the same order as the sample solution, and dilute to 10.0 mL with water. The absorbance of the sample solution at 590 nm should not exceed that of the control solution, using water as the blank.

Chlorate and Nitrate.

Sample Solution A. Dissolve 0.50 g in 3 mL of water by heating in a boiling-water bath. Dilute to 50 mL with brucine sulfate reagent solution.

Control Solution B. Dissolve 0.50 g in 1.5 mL of water and 1.5 mL of nitrate ion (NO_3) standard solution by heating in a boiling-water bath. Dilute to 50 mL with brucine sulfate reagent solution.

Blank Solution C. Use 50 mL of brucine sulfate reagent solution.

Heat the three solutions in a preheated (boiling) water bath for 10 min. Cool rapidly in an ice bath to room temperature. Set a spectrophotometer at 410 nm, and using 1-cm cells, adjust the instrument to read 0 absorbance with blank solution C in the light path, then determine the absorbance of sample solution A. Adjust the instrument to read 0 absorbance with sample solution A in the light path and determine the absorbance of control solution B. The absorbance of sample solution A should not exceed that of control solution B.

Phosphate. (Page 40, Method 1). Dissolve 4.0 g in 25 mL of approximately 0.5 N sulfuric acid, and continue as described.

Sulfate. Dissolve 25.0 g in 200 mL of water, add 2 mL of hydrochloric acid, heat to boiling, add 10 mL of 12% barium chloride reagent solution, digest in a covered beaker on a hot plate ($\approx$100 °C) for 2 h, and allow to stand for at least 8 h. If a precipitate is formed, filter, wash thoroughly, and ignite. Correct for the weight of barium sulfate obtained in a complete blank test.

Barium. Dissolve 4.0 g in 20 mL of water, filter if necessary, and divide into two portions. To one portion, add 2 mL of 10% sulfuric acid reagent solution and to the other, 2 mL of water. The solutions should be equally clear at the end of 2 h. (Limit about 0.001%)

Heavy Metals. (Page 36, Method 1). Dissolve 6.0 g in about 20 mL of water, and dilute with water to 30 mL. Use 25 mL to prepare the sample solution, and use the remaining 5.0 mL to prepare the control solution.

Iron. (Page 38, Method 1). Use 3.3 g.

Calcium, Magnesium, and Sodium. (By flame AAS, page 63).

> *Sample Stock Solution.* Dissolve 5.0 g of sample in water, and dilute to the mark with water in a 100-mL volumetric flask (1 mL = 0.05 g).

Element	Wavelength (nm)	Sample Wt (g)	Standard Added (mg)	Flame Type*	Background Correction
Ca	422.7	1.0	0.02; 0.04	N/A	No
Mg	285.2	1.0	0.01; 0.02	A/A	Yes
Na	589.0	0.10	0.005; 0.01	A/A	No

*A/A is air/acetylene; N/A is nitrous oxide/acetylene.

Potassium Chromate

K_2CrO_4 **Formula Wt 194.19**   **CAS No. 7789-00-6**

GENERAL DESCRIPTION

Typical appearance: yellow solid
Analytical use: oxidizing agent
Change in state (approximate): melting point, 980 °C
Aqueous solubility: 62 g in 100 mL, cold

SPECIFICATIONS

Assay . . . . $\geq$99.0% K_2CrO_4
pH of a 5% solution at 25.0 °C 8.6–9.8

Maximum Allowable

Insoluble matter 0.005%
Chloride (Cl). 0.005%
Sulfate (SO_4) 0.03%
Calcium (Ca) 0.005%
Sodium (Na). 0.02%

TESTS

Assay. (By iodometric oxidation–reduction titration). Weigh, to the nearest 0.1 mg, about 0.25 g of sample, and dissolve in 200 mL of water in a 500-mL glass-stoppered iodine flask. Add 3 g of potassium iodide and 7 mL of hydrochloric acid, and allow to stand in the dark for 10 min. Titrate the liberated iodine with 0.1 N sodium thiosulfate volumetric solution, adding 5 mL of starch indicator solution near the end point. One milliliter of 0.1 N sodium thiosulfate corresponds to 0.006473 grams of K_2CrO_4.

$$\% \, K_2CrO_4 = \frac{(mL \times N \, Na_2S_2O_3) \times 6.473}{Sample \, wt \, (g)}$$

pH of a 5% Solution at 25.0 °C. (Page 49).

Insoluble Matter. (Page 25). Use 20 g dissolved in 150 mL of water.

Chloride. Dissolve 0.20 g in 10 mL of water, filter if necessary through a small chloride-free filter, and add 1 mL of ammonium hydroxide, 1 mL of silver nitrate reagent solution, and 2 mL of nitric acid. Any turbidity should not exceed that produced by 0.01 mg of chloride ion (Cl) in an equal volume of solution containing the quantities of reagents used in the test. The comparison is best made by the general method for chloride in colored solutions, page 35.

Sulfate. Dissolve 10.0 g of sample in 250 mL of water. Add 2.5 g of barium chloride in 15 mL of water and 5 mL of hydrochloric acid. Digest on low heat ($\approx$100 °C) for 2 h, and allow to stand at room temperature for 12 h. If any precipitate is formed, filter, wash thoroughly, and ignite in a platinum crucible. Fuse the ignited residue with 1 g of anhydrous sodium carbonate, extract the fused mass with water, and filter off the insoluble residue. Add 5 mL of hydrochloric acid to the filtrate, dilute to about 150 mL with water, heat the solution to boiling, and add 10 mL of 95% ethyl alcohol. Digest on low heat until reduction of chromate is complete as indicated by the change to a clear green or colorless solution. Neutralize the solution with ammonium hydroxide, filter off any chromic hydroxide precipitate, and dilute to 200 mL in a volumetric flask. Use 10.0 mL of this solution for the sample (0.5 g). Dilute to 30 mL with water. For the control, take 0.15 mg of sulfate ion (SO_4) in 30 mL of water. To each, add 1 mL of (1 + 19) hydrochloric acid and 1 mL of 12% barium chloride reagent solution. Compare after 10 min. Sample turbidity should not exceed that of the control solution.

Calcium and Sodium. (By flame AAS, page 63).

   Sample Stock Solution. Dissolve 5.0 g in water, and dilute to 100 mL in a volumetric flask with water (1 mL = 0.05 g).

Element	Wavelength (nm)	Sample Wt (g)	Standard Added (mg)	Flame Type*	Background Correction
Ca	422.7	0.50	0.025; 0.05	N/A	No
Na	589.0	0.05	0.01; 0.02	A/A	No

*A/A is air/acetylene; N/A is nitrous oxide/acetylene.

Potassium Cyanide

KCN   Formula Wt 65.12   CAS No. 151-50-8

GENERAL DESCRIPTION

Typical appearance: white solid
Analytical use: complexing agent in alkaline solutions
Change in state (approximate): melting point, 634 °C
Aqueous solubility: 72 g in 100 mL at 25 °C

SPECIFICATIONS

Assay . . . . ≥96.0% KCN

Maximum Allowable

Chloride (Cl) 0.5%
Phosphate (PO₄) 0.005%
Sulfate (SO₄) 0.04%
Sulfide (S) 0.003%
Thiocyanate (SCN) . . . . Passes test
Iron, total (as Fe) 0.03%
Lead (Pb) . . . . 2 ppm
Sodium (Na) 0.5%

TESTS

Assay. (By argentimetric titration of cyanide content). Weigh accurately about 0.5 g, and dissolve in 30 mL of water. Add 0.2 mL of 10% potassium iodide reagent solution and 1 mL of ammonium hydroxide, and titrate with 0.1 N silver nitrate volumetric solution to a slight yellowish permanent turbidity. One milliliter of 0.1 N silver nitrate corresponds to 0.01302 g of KCN.

$$\% \text{ KCN} = \frac{(\text{mL} \times \text{N AgNO}_3) \times 13.02}{\text{Sample wt (g)}}$$

For the Determination of Chloride, Phosphate, Sulfate, Sulfide, Thiocyanate, and Total Iron

Sample Solution A. Dissolve 10 g in water, and dilute with water to 200 mL. Filter, if necessary, under a well-ventilated fume hood into a dry flask (1 mL = 0.05 g).

Chloride. (Page 35). Dilute 1.0 mL of sample solution A with water to 50 mL. To 2.0 mL (0.002-g sample) of this solution, add 2 mL of 30% hydrogen peroxide, and allow to stand in a covered beaker until reaction ceases, then digest in the covered beaker on a hot plate (≈100 °C) for 20–30 min. Cool, and dilute with water to 25 mL.

Phosphate. (Page 40, Method 1). To 20 mL of sample solution A, add 2 mL of hydrochloric acid, and evaporate to dryness in a well-ventilated hood. Add 5 mL of dilute hydrochloric acid (1 + 1), and evaporate to dryness again. Dissolve in 50 mL of approximately 0.5 N sulfuric acid. To 20 mL of the solution, add 5 mL of approximately 0.5 N sulfuric acid, and continue as described.

Sulfate. (Page 41, Method 2). Use 2.5 mL of sample solution A plus 1 mL of dilute hydrochloric acid (1 + 1) in a hood.

Sulfide. To 20 mL of sample solution A (measured in a graduated cylinder) (1-g sample), add 0.15 mL of alkaline lead solution (made by adding 10% sodium hydroxide reagent solution to a 10% lead acetate reagent solution until the precipitate is redissolved). The color should not be darker than is produced by 0.03 mg of sulfide ion (S) in an equal volume of solution when treated with 0.15 mL of the alkaline lead solution.

Thiocyanate. To 20 mL of sample solution A (measured in a graduated cylinder) (1-g sample), add 4 mL of hydrochloric acid and 0.20 mL of 5% ferric chloride reagent solution. At the end of 5 min, the solution should show no reddish tint when compared with 20 mL of water to which have been added the quantities of hydrochloric acid and ferric chloride used in the test. (Limit about 0.02%)

Iron, Total. (Page 38, Method 1). Transfer 10 mL of sample solution A (0.5-g sample) to a platinum evaporating dish, add 3 mL of hydrochloric acid, and evaporate to dryness in a well-ventilated hood. Heat the residue at 650 °C for 30 min. Cool, add 2 mL of hydrochloric acid and 10 mL of water, and digest on a hot plate ($\approx$100 °C) until dissolution is complete. Dilute with water to 30 mL, and use 2.0 mL of this solution.

Lead. Dissolve 1.2 g in 10 mL of water in a separatory funnel. Add 5 mL of ammonium citrate reagent solution, 2 mL of hydroxylamine hydrochloride reagent solution for the dithizone test, and 0.10 mL of phenol red indicator solution, and make the solution alkaline if necessary by the addition of ammonium hydroxide. Add 5 mL of dithizone test solution, shake gently but well for 1 min, and allow the layers to separate. The intensity of the red color of the chloroform layer should be no greater than that of a control made with 0.002 mg of lead ion (Pb) and 0.2 g of the sample treated exactly like the solution of 1.2 g of sample in 10 mL of water.

Sodium. (By flame AAS, page 63).

    Sample Stock Solution. Dissolve 0.5 g of sample in a 100-mL volumetric flask, and dilute to the mark with water (1 mL = 0.005 g).

Element	Wavelength (nm)	Sample Wt (g)	Standard Added (mg)	Flame Type*	Background Correction
Na	589.0	0.01	0.05; 0.10	A/A	No

*A/A is air/acetylene.

Potassium Dichromate

$K_2Cr_2O_7$ Formula Wt 294.18   CAS No. 7778-50-9

GENERAL DESCRIPTION

Typical appearance: orange-red solid
Analytical use: oxidizing agent; redox standard
Change in state (approximate): melting point, 398 °C
Aqueous solubility: 12.3 g in 100 mL at 20 °C

SPECIFICATIONS

Assay . . . . $\geq$99.0% $K_2Cr_2O_7$

Maximum Allowable

Insoluble matter 0.005%
Loss on drying 0.05%
Chloride (Cl). 0.001%
Sulfate (SO_4) 0.005%
Calcium (Ca) 0.003%
Iron (Fe). 0.001%
Sodium (Na). 0.02%

TESTS

Assay. (By iodometric oxidation–reduction titration). Weigh, to the nearest 0.1 mg, about 0.2 g of sample, transfer to a 500-mL glass-stoppered iodine flask, and dissolve with 125 mL of water. Add 5 g of potassium iodide and 5 mL of 6 M hydrochloric acid with constant swirling. Carefully wash down the sides of the flask with water so that a layer of water is formed on top of the solution, stopper the flask, and allow to stand in the dark for 10 min. Add about 160 mL of water, and titrate with 0.1 N sodium thiosulfate volumetric solution, adding 5 mL of starch indicator solution near the end point. One milliliter of 0.1 N sodium thiosulfate corresponds to 0.004903 grams of $K_2Cr_2O_7$.

$$\% \ K_2Cr_2O_7 = \frac{(mL \times N \ Na_2S_2O_3) \times 4.903}{\text{Sample wt (g)}}$$

Insoluble Matter. (Page 25). Dissolve 20 g in 200 mL of water.

Loss on Drying. Weigh accurately about 2 g, and dry in a tared, preconditioned container to constant weight at 105 °C.

Chloride. Dissolve 1.0 g in 10 mL of water, filter if necessary through a small chloride-free filter, and add 1 mL of ammonium hydroxide and 1 mL of silver nitrate reagent solution. Prepare a standard containing 0.01 mg of chloride ion (Cl) in 10 mL of water, and add 1 mL of ammonium hydroxide and 1 mL of silver nitrate reagent solution. Add 2 mL of nitric acid to each. The comparison is best made by the general method for chloride in colored solutions, page 35.

Sulfate. Dissolve 10 g in 250 mL of water, filter if necessary, and heat to boiling. Add 25 mL of a solution containing 1 g of barium chloride and 2 mL of hydrochloric acid per 100 mL of solution. Digest in a covered beaker on a hot plate ($\approx$100 °C) for 2 h, and allow to stand for at least 8 h. If a precipitate is formed, filter, wash thoroughly, and ignite. Fuse the residue with 1 g of sodium carbonate. Extract the fused mass with water, and filter off the insoluble residue. Add 5 mL of hydrochloric acid to the filtrate, dilute with water to about 200 mL, heat to boiling, and add 10 mL of ethyl alcohol. Digest in a covered beaker on the hot plate ($\approx$100 °C) until reduction of chromate is complete, as indicated by the change to a clear green or colorless solution. Neutralize the solution with ammonium hydroxide, and add 2 mL of hydrochloric acid. Heat to boiling, and add 10 mL of 12% barium chloride reagent solution. Digest in a covered beaker on a hot plate ($\approx$100 °C) for 2 h, and allow to stand for at least 8 h. Filter, wash thoroughly, and ignite. Correct for the weight of barium sulfate obtained in a complete blank test. If the original precipitate of barium sulfate weighs less than the specification allows, the fusion with sodium carbonate is not necessary.

Calcium, Iron, and Sodium. (By flame AAS, page 63).

> **Sample Stock Solution for Calcium and Sodium.** Dissolve 5.0 g of sample in 20 mL of water, transfer the solution to a 100-mL volumetric flask, and dilute to the mark with water (1 mL = 0.05 g).

> **Sample Stock Solution for Iron.** Dissolve 10.0 g of sample in 20 mL of water, transfer the solution to a 100-mL volumetric flask, and dilute to the mark with water (1 mL = 0.10 g).

Element	Wavelength (nm)	Sample Wt (g)	Standard Added (mg)	Flame Type*	Background Correction
Ca	422.7	0.50	0.015; 0.03	N/A	No
Fe	248.3	2.0	0.04; 0.08	A/A	Yes
Na	589.0	0.05	0.01; 0.02	A/A	No

*A/A is air/acetylene; N/A is nitrous oxide/acetylene.

Potassium Ferricyanide

Potassium Hexacyanoferrate(III)

K$_3$Fe(CN)$_6$ **Formula Wt 329.25**   **CAS No. 13746-66-2**

GENERAL DESCRIPTION

Typical appearance: ruby-red solid
Analytical use: mild oxidizing agent
Aqueous solubility: 46 g in 100 mL at 20 °C

SPECIFICATIONS

Assay . . . . $\geq$99.0% K$_3$Fe(CN)$_6$

Maximum Allowable

Insoluble matter 0.005%
Chloride (Cl). 0.01%
Sulfate (SO$_4$) 0.01%
Ferro compounds . . . . 0.05% as ferrocyanide radical [Fe(CN)$_6$]$^{4-}$

TESTS

Assay. (By iodometric titration). Weigh accurately about 1.4 g, and dissolve in 50 mL of water in a glass-stoppered conical flask. Add 3 g of potassium iodide and 1 drop of glacial acetic acid. Then add 3.0 g of zinc sulfate heptahydrate in 20 mL of water. Stopper, swirl, and allow to stand in the dark for 30 min. Titrate the liberated iodine with 0.1 N sodium thiosulfate volumetric solution, adding 3 mL of starch indicator solution near the end of the titration. Correct for a blank. One milliliter of 0.1 N sodium thiosulfate corresponds to 0.03292 g of K$_3$Fe(CN)$_6$.

$$\% \; K_3Fe(CN)_6 = \frac{\{[mL \; (sample) - mL \; (blank)] \times N \; Na_2S_2O_3\} \times 32.92}{Sample \; wt \; (g)}$$

Insoluble Matter. (Page 25). Dissolve 20 g in 200 mL of water at room temperature, and filter promptly.

Chloride. Dissolve 2.0 g in 175 mL of water in a cylinder, add 2.5 g of cupric sulfate crystals dissolved in 25 mL of water, mix thoroughly, and allow the precipitate to settle. Filter the supernatant liquid, if necessary, through a chloride-free filter. To 10 mL of the clear solution, add 1 mL of nitric acid, 10 mL of water, and 1 mL of silver nitrate reagent solution. Any turbidity should not exceed that produced by 0.01 mg of chloride ion (Cl) in an equal volume of solution containing the quantities of nitric acid and silver nitrate used in the test and enough cupric sulfate to match the color in the test solution.

Sulfate. Dissolve 5.0 g in 100 mL of water without heating, filter promptly, and to 10 mL (0.5 g) of the filtrate, add 0.2 mL of glacial acetic acid and 5 mL of 12% barium chloride reagent solution. Any turbidity should not exceed that produced by a standard containing 0.05 mg of sulfate ion (SO$_4$) in an equal volume of solution containing 0.2 mL of glacial acetic acid and 5 mL of 12% barium chloride reagent solution. Compare 10 min after adding the barium chloride to the sample and standard solutions.

Ferro Compounds. Mix 400 mL of water with 10 mL of 25% sulfuric acid reagent solution, and add 0.1 N potassium permanganate volumetric solution until the pink color persists for 1 min. Dissolve 4.0 g of the sample in the solution, add 0.10 mL of the 0.1 N potassium permanganate, and stir. The solution should retain a pink tint in comparison with a blank prepared with the same quantities of potassium ferricyanide, water, and sulfuric acid.

Potassium Ferrocyanide Trihydrate

Potassium Hexacyanoferrate(II) Trihydrate

$K_4Fe(CN)_6 \cdot 3H_2O$ Formula Wt 422.39   CAS No. 14459-95-1

GENERAL DESCRIPTION

Typical appearance: pale yellow solid

Analytical use: moderately strong oxidizer when coupled with ferricyanide

Change in state (approximate): dehydrates partially from 60 °C and completely at 100 °C

Aqueous solubility: 28 g in 100 mL at 12 °C

SPECIFICATIONS

Assay . . . . .98.5–102.0% $K_4Fe(CN)_6 \cdot 3H_2O$

Maximum Allowable

Insoluble matter 0.005%

Chloride (Cl) 0.01%

Sulfate (SO_4) . . . . Passes test

TESTS

Assay. (By titration of reducing power). Weigh accurately about 1.6 g and dissolve in 250 mL of water. Add 30 mL of 25% sulfuric acid reagent solution, and titrate with 0.1 N potassium permanganate volumetric solution to a change from yellow to orange-yellow. One milliliter of 0.1 N potassium permanganate corresponds to 0.04224 g of $K_4Fe(CN)_6 \cdot 3H_2O$.

$$\% \ K_4Fe(CN)_6 \cdot 3H_2O = \frac{(mL \times N \ KMnO_4) \times 42.24}{Sample \ wt \ (g)}$$

Insoluble Matter. (Page 25). Use 20 g dissolved in 200 mL of water.

Chloride. Dissolve 2.0 g in 175 mL of water, filter if necessary through a chloride-free filter, add 2.5 g of cupric sulfate crystals dissolved in 25 mL of water, mix thoroughly, and allow the precipitate to settle in a cylinder. To 10 mL of the clear supernatant liquid, add 1 mL of nitric acid, 10 mL of water, and 1 mL of silver nitrate reagent solution. Any turbidity should not exceed that produced by 0.01 mg of chloride ion (Cl) in an equal volume of solution containing the quantities of nitric acid and silver nitrate used in the test and enough cupric sulfate to match the color in the test solution.

Sulfate. Dissolve 5.0 g in 100 mL of water without heating, filter promptly, and to 10 mL (0.5 g) of the filtrate, add 0.2 mL of glacial acetic acid and 5 mL of 12% barium chloride reagent solution. Any turbidity should not exceed that produced by a standard containing 0.05 mg of sulfate ion (SO_4) in an equal volume of solution containing 0.2 mL of glacial acetic acid and 5 mL of 12% barium chloride reagent solution. Compare 10 min after adding the barium chloride to the sample and standard solutions. (Limit about 0.01%)

Potassium Fluoride

KF   Formula Wt 58.10   CAS No. 7789-23-3

GENERAL DESCRIPTION

Typical appearance: white, hygroscopic solid
Analytical use: complexing agent; fluorination of organic compounds
Change in state (approximate): melting point, 855 °C
Aqueous solubility: 96 g in 100 mL at 21 °C

SPECIFICATIONS

Assay . . . . ≥99.0% KF

Maximum Allowable

Chloride (Cl). 0.005%
Titrable acid. . . . . .0.03 meq/g
Titrable base . . . . .0.01 meq/g
Potassium fluosilicate (K_2SiF_6). 0.1%
Sulfate (SO_4) 0.005%
Heavy metals (as Pb). 0.001%
Iron (Fe). 0.001%
Sodium (Na). 0.2%

TESTS

Assay. (By indirect acid–base titrimetry). Weigh accurately about 0.2 g of sample and dissolve in 50 mL of water in a polyethylene vessel. Pass the solution through the cation column at a rate of about 5 mL/min and collect the eluate in a 500-mL polyethylene titration flask. Then wash the resin in the column with water at a rate of about 10 mL/min, and collect in the same titration flask. Add 0.15 mL of phenolphthalein indicator solution to the flask and titrate with 0.1 N sodium hydroxide. Continue the titration as the washing proceeds until 50 mL of eluate requires no further titration. One milliliter of 0.1 N sodium hydroxide corresponds to 0.005810 g of KF.

$$\% \text{ KF} = \frac{(\text{mL} \times \text{N NaOH}) \times 5.810}{\text{Sample wt (g)}}$$

Chloride. (Page 35). Dissolve 1.0 g of sample and 0.2 g of boric acid in 80 mL of water, filter if necessary through a chloride-free filter, and dilute to 100 mL with water. Use 20 mL (0.2-g sample) of this solution.

Titrable Acid. Dissolve 5.0 g in 40 mL of carbon dioxide-free water in a platinum dish, add 10 mL of a freshly prepared saturated solution of potassium nitrate, cool the solution to 0 °C, and add 0.15 mL of phenolphthalein indicator solution. If a pink color is produced, omit the titration for titrable acid, and reserve the solution for titrable base. If no pink color is produced, titrate with 0.1 N sodium hydroxide until the pink color persists for 15 s while the temperature is near 0 °C. Not more than 1.50 mL of the 0.1 N sodium hydroxide should be required. Reserve the solution for the test for potassium fluosilicate.

Titrable Base. If a pink color was produced in the solution prepared for the test for titrable acid, add 0.1 N hydrochloric acid, stirring the liquid only gently, until the pink color is discharged. Not more than 0.50 mL of the 0.1 N hydrochloric acid should be required. Reserve the solution for the test for potassium fluosilicate.

Potassium Fluosilicate. Heat the solution reserved from the test for titrable acid or titrable base to boiling, and titrate while hot with 0.01 N sodium hydroxide to a permanent pink color. Not more than 9.0 mL of 0.01 N sodium hydroxide should be required.

Sulfate. (Page 41, Method 2). Evaporate 1.0 g to dryness with 2 mL of hydrochloric acid in a platinum dish. Repeat the evaporation four additional times.

Heavy Metals. (Page 36, Method 1). Use 2.0 g. Adjust the pH, using pH indicator paper.

Iron. Dissolve 0.50 g of sample in 20 mL of water in a 100-mL beaker. Prepare the standards containing 0.0025 and 0.005 mg of iron ion, respectively, and dilute each to 25 mL with water. To each solution, add in the order specified, 15 mL of aqueous 4% boric acid, 3 mL of hydrochloric acid, 30 to 50 mg of ammonium persulfate, and 3 mL of ammonium thiocyanate reagent solution, mixing after each addition. Any red color in the sample must not exceed that of the 0.005 mg standard solution (0.001%).

Sodium. (By flame AAS, page 63).

> **Sample Stock Solution.** Dissolve 0.10 g in 100 mL of water, using suitable plastic volumetric flasks (1 mL = 0.001 g).

Element	Wavelength (nm)	Sample Wt (g)	Standard Added (mg)	Flame Type*	Background Correction
Na	589.0	0.005	0.005; 0.01	A/A	No

*A/A is air/acetylene.

Potassium Hydrogen Diiodate
Potassium Bi-iodate

$KH(IO_3)_2$ **Formula Wt 389.91**   **CAS No. 13455-24-8**

GENERAL DESCRIPTION

Typical appearance: white solid
Analytical use: standardizing bases
Aqueous solubility: 1.3 g in 100 mL at 15 °C

SPECIFICATIONS

Assay (acid–base titrimetry) .... 99.95–100.05% $KH(IO_3)_2$
Assay (iodometric) ....99.9–100.1% $KH(IO_3)_2$

	Maximum Allowable
Insoluble matter	0.01%
Chloride (Cl)	0.02%
Sulfate (SO_4)	0.005%
Heavy metals (as Pb)	0.001%
Iron (Fe)	0.001%

TESTS

Note: Both assay procedures represent common uses of this reagent. The sample should be thoroughly crushed and dried for 1 h at 105 °C before use in either assay procedure.

Assay. (By acid–base titrimetery). Weigh in duplicate (to the nearest 0.1 mg) 1.980 ± 0.001 g of dried sample. Dissolve in 150 mL of carbon dioxide-free water. Add 0.610 g of NIST standard tris(hydroxymethyl)aminomethane (TRIS), stir, and titrate potentiometrically with a solution containing 1 mg of NIST standard TRIS per mL to a pH of 5.00. Calculate the assay value of the sample as follows:

$$\% \, KH(IO_3)_2 = \frac{[(V \times C) + A] \times F \times 38.991}{\text{Sample wt (g)} \times 121.14}$$

where V = volume, in mL, of TRIS titrant solution; C = concentration, in mg/mL, of TRIS titrant solution; A = weight, in mg, of solid TRIS added; and F = assay value of NIST standard TRIS ($\div 100$).

Assay. (By iodometric titration of oxidizing power). Weigh accurately about 0.13 g of dried sample, and dissolve 100 mL of water in a glass-stoppered conical flask. Add 3.0 g of potassium iodide and 1 mL of sulfuric acid. Stopper, swirl, and let stand for 30 min. Titrate the liberated iodine with 0.1 N sodium thiosulfate volumetric solution, and add 3 mL of starch indicator solution near the end of the titration. Correct for a blank. One milliliter of 0.1 N sodium thiosulfate corresponds to 0.003249 g of $KH(IO_3)_2$.

$$\% \, KH(IO_3)_2 = \frac{(\text{mL} \times N \, Na_2S_2O_3) \times 3.249}{\text{Sample wt (g)}}$$

Insoluble Matter. (Page 25). Use 10.0 g dissolved in 200 mL of water.

Chloride. Dissolve 1.0 g in 100 mL of water in a conical flask, filter if necessary through a chloride-free filter, and add 6 mL of hydrogen peroxide and 1 mL of phosphoric acid. Heat to boiling, and boil gently until all the iodine is expelled and the solution is colorless. Cool, wash down the sides of the flask, and add 0.5 mL of hydrogen peroxide. If an iodine color develops, boil until the solution is colorless and then for an additional 10 min. If no color develops, boil for 10 min. Dilute with water to 100 mL, and dilute a 10-mL portion to 25 mL. Prepare a standard containing 0.02 mg of chloride ion (Cl) and 0.6 mL of hydrogen peroxide in 23 mL of water. Add 1 mL of nitric acid and 1 mL of silver nitrate reagent solution to each. Any turbidity in the sample solution should not exceed that in the standard.

Sulfate. (Page 41, Method 2). Evaporate 1.0 g to dryness with 3 mL of hydrochloric acid, and repeat the evaporation three additional times. Allow 30 min for turbidity to form.

Heavy Metals. Treat 10.0 g with 10 mL of hydrochloric acid, and evaporate to dryness. Repeat the evaporation two additional times. Dissolve the residue in water, and add 1 mL of 1 N acetic acid and 5 mL of hydrogen sulfide water. For the standard, use 0.1 mg of lead ion (Pb) treated exactly as the sample. Any brown color produced in the sample solution within 5 min should not exceed that in the standard.

Iron. (Page 38, Method 2). Treat 1.0 g with 1 mL of 50% sulfuric acid and 5 mL of hydroxylamine hydrochloride reagent solution. Evaporate to expel the iodine. Dissolve the residue in 10 mL of water, and add 1 mL of hydroxylamine hydrochloride reagent solution.

Potassium Hydrogen Phthalate, Acidimetric Standard

Potassium Acid Phthalate; Potassium Biphthalate; Phthalic Acid, Monopotassium Salt

HOCOC$_6$H$_4$COOK | **Formula Wt 204.22** | **CAS No. 877-24-7**

Note: This reagent is satisfactory for use as a pH standard. For use as an acidimetric standard, this material should be lightly crushed and dried for 2 h at 120 °C to remove any absorbed moisture.

GENERAL DESCRIPTION

Typical appearance: white solid
Analytical use: acidimetric standard
Change in state (approximate): melting point, 295 °C, with decomposition
Aqueous solubility: 10 g in 100 mL at 25 °C
Density: 1.64

SPECIFICATIONS

Assay (dried basis)..99.95–100.05% C$_8$H$_5$O$_4$K
pH of a 0.05 M solution ....4.00–4.02 at 25.0 ± 0.2 °C

	Maximum Allowable
Insoluble matter	0.005%
Chlorine compounds (as Cl)	0.003%
Sulfur compounds (as S)	0.002%
Heavy metals (as Pb)	5 ppm
Iron (Fe)	5 ppm
Sodium (Na)	0.005%

TESTS

Assay. (By acid–base titrimetry). This is a comparative procedure in which the potassium hydrogen phthalate to be assayed is compared with NIST SRM Potassium Hydrogen Phthalate. Accurately weighed portions of anhydrous sodium carbonate, dissolved in water, are allowed to react with accurately weighed portions of the two potassium hydrogen phthalate samples of such size as to provide a small excess of the potassium hydrogen phthalate. The carbon dioxide is removed by boiling and the excess potassium hydrogen phthalate is determined by titration with carbonate-free sodium hydroxide using a potentiometric end point.

> *Caution:* Due to the small tolerance in assay limits for this acidimetric standard, extreme care must be observed in the weighing, transferring, and titrating operations in the following procedure. Strict adherence to the specified sample weights and final titration volumes is absolutely necessary. It is recommended that the titrations be run at least in duplicate on both the sample and the SRM. Duplicate values of M, the number of milliequivalents of total alkali required per gram of potassium hydrogen phthalate, obtained on the sample and on the SRM, should agree within 2 parts in 5000 to be acceptable for averaging.

Procedure. Place duplicate 1.300 ± 0.001-g portions of a uniform sample of anhydrous sodium carbonate in clean, dry weighing bottles. Dry at 285 °C for 3 h, stopper the bottles, and cool in a desiccator for at least 2 h. Open the weighing bottles momentarily to equalize the pressure, weigh each bottle and its contents to the nearest 0.1 mg, and transfer the entire contents of each bottle to a clean, dry 50-mL beaker. Store the beaker in a desiccator until ready for use, weigh each bottle again, and determine by difference, to the nearest 0.1 mg, the weight of each portion of sodium carbonate.

Transfer the portions of sodium carbonate to 500-mL titration flasks by pouring them through powder funnels. Rinse the beakers, funnels, and sides of the flasks thoroughly with several small portions of water. In the same manner, add to one of these flasks 5.100 ± 0.003 g of the potassium hydrogen phthalate to be assayed and to the other flask the same weight of the NIST SRM Potassium Hydrogen Phthalate—both of which have been previously crushed (not ground) in an agate or mullite mortar to approximately 100-mesh fineness, dried at 120 °C for 2 h, and cooled in a desiccator for at least 2 h. Dilute the contents of each titration flask with water to about 90 mL, swirl to dissolve the samples, stopper, and bubble carbon dioxide-free air through the solutions at a moderately fast rate during all subsequent operations. Boil the solutions gently for 20 min to complete the removal of carbon dioxide, taking care to prevent loss of solutions by spraying. Cool the solutions to room temperature in an ice bath, and dilute to 75–80 mL with carbon dioxide-free water. Titrate each solution with 0.02 N carbonate-free sodium hydroxide using a combination electrode (glass–calomel or glass–silver, silver chloride) for the measurement of E in millivolts or pH. The end point of the titration is determined by the second derivative method (page 28).

Calculations. Calculate the number of milliequivalents of total alkali (sodium carbonate plus sodium hydroxide) required to neutralize exactly 1 g of the sample of potassium hydrogen phthalate and of the NIST SRM by the following equations:

$$M_u = \frac{(B/0.052994) + (C \times D)}{W_u}$$

$$M_s = \frac{(B/0.052994) + (C \times D)}{W_s \times F}$$

where M_u = number of meq of total alkali required per g of potassium hydrogen phthalate being assayed; M_s = number of meq of total alkali required per g of NIST Standard Potassium Hydrogen Phthalate; B = weight, in g, of sodium carbonate; C = volume, in mL, of sodium hydroxide standard solution; D = normality of sodium hydroxide standard solution; W_u = weight, in g, of potassium hydrogen phthalate being assayed; W_s = weight, in g, of NIST SRM Potassium Hydrogen Phthalate; and F = assay value of NIST SRM in percent divided by 100.

Calculate the assay value of the sample of potassium hydrogen phthalate as follows:

$$A = (M_u \times 100) / M_s$$

where A is percent of potassium hydrogen phthalate in the sample assayed and M_u and M_s are defined as before.

pH of a 0.05 M Solution. Dissolve 1.021 g in 100 g of carbon dioxide- and ammonia-free water (or 1.012 g in a volume of 100 mL). Standardize the pH meter and electrodes at pH 4.00 at 25.0 ± 0.2 °C with 0.05 M NIST SRM Potassium Hydrogen Phthalate. Determine the pH of the solution by the method described on page 49 for pH of buffer standard solutions. The pH should be 4.00–4.02 at 25.0 ± 0.2 °C.

Insoluble Matter. (Page 25). Use 20 g dissolved in 200 mL of water.

Chlorine Compounds. Mix 1.0 g with 0.5 g of sodium carbonate in a platinum vessel, moisten the mixture, and ignite until thoroughly charred, avoiding an unduly high temperature. Treat the residue with 30 mL of water, cautiously add 3 mL of nitric acid, and filter through a chloride-free filter. Wash the residue with a few milliliters of hot water, dilute the filtrate with water to 50 mL, and to 25 mL of the filtrate add 1 mL of silver nitrate reagent solution. Any turbidity should not exceed that produced by 0.015 mg of chloride ion (Cl) in an equal volume of solution containing the quantities of reagents used in the test.

Sulfur Compounds. Mix 2.0 g with 1 g of sodium carbonate, add in small portions 15 mL of water, evaporate, and thoroughly ignite, taking care to protect the sample from the flame because of the presence of sulfur compounds in the natural gas. Treat the residue with 20 mL of water and 2 mL of 30% hydrogen peroxide, and heat on a hot plate ($\approx$100 °C) for 15 min. Add 5 mL of hydrochloric acid, and evaporate to dryness on the hot plate. Dissolve the residue in 10 mL of water, filter, wash with several portions of water, and dilute the filtrate with water to 25 mL. Add to this solution 0.5 mL of 1 N hydrochloric acid and 2 mL of 12% barium chloride reagent solution. Any turbidity should not exceed that in a standard prepared as follows: Treat 1 g of sodium carbonate with 2 mL of 30% hydrogen peroxide and 5 mL of hydrochloric acid, and evaporate to dryness on a hot plate ($\approx$100 °C). Dissolve the residue and 0.12 mg of sulfate ion (SO_4) in sufficient water to make 25 mL, and add 0.5 mL of 1 N hydrochloric acid and 2 mL of 12% barium chloride reagent solution.

Heavy Metals. (Page 36, Method 1). Dissolve 6.6 g in warm water, and dilute with water to 50 mL. Use 40 mL to prepare the sample solution, and use the remaining 10 mL to prepare the control solution.

Iron. (Page 38, Method 2). For the test, use 2.0 g, dissolve in 30 mL of water, and dilute to 50 mL.

Sodium. (By flame AAS, page 63).

> ***Sample Stock Solution.*** Dissolve 1.0 g of sample in water, and dilute to 100 mL in a volumetric flask with water (1 mL = 0.01 g).

Element	Wavelength (nm)	Sample Wt (g)	Standard Added (mg)	Flame Type*	Background Correction
Na	589.0	0.20	0.01; 0.02	A/A	No

*A/A is air/acetylene.

Potassium Hydrogen Sulfate, Fused
Potassium Pyrosulfate

CAS No. 7790-62-7

Note: This product is a mixture of potassium pyrosulfate, $K_2S_2O_7$, and potassium hydrogen sulfate, $KHSO_4$.

GENERAL DESCRIPTION

Typical appearance: colorless solid
Analytical use: fusion flux for ores and minerals
Change in state (approximate): melting point, 325 °C
Aqueous solubility: soluble

SPECIFICATIONS

Acidity (as H_2SO_4) 37.5–38.6%

Maximum Allowable

Water (H_2O) 2.5%
Insoluble matter 0.01%
Chloride (Cl). 0.002%
Phosphate (PO_4). 0.001%
Heavy metals (as Pb). 0.001%
Iron (Fe). 0.002%
Calcium (Ca) 0.002%
Magnesium (Mg) 0.001%
Sodium (Na). 0.01%

TESTS

Acidity. (By acid–base titrimetry). Weigh accurately about 4.0 g, dissolve in 50 mL of water, and titrate with 1 N sodium hydroxide volumetric solution, using methyl orange indicator solution. One milliliter of 1 N sodium hydroxide corresponds to 0.04904 g of H_2SO_4.

$$\% \ H_2SO_4 = \frac{(mL \times N \ NaOH) \times 4.904}{Sample \ wt \ (g)}$$

Water. Weigh accurately about 5 g, dissolve in water, and dilute with water to 250 mL. Evaporate 25.0 mL of the solution to dryness in a tared, preconditioned platinum vessel, and ignite to remove all sulfuric acid and leave potassium sulfate. Calculate the percentage of potassium sulfate (*B*). Calculate the percentage of sulfite (*A*) from the percentage of sulfuric acid obtained in the test for acidity. One percent of sulfuric acid is equivalent to 0.8163% of sulfite. Calculate the percentage of water (*C*) as the difference between 100% and the sum of the percentages of sulfite and potassium sulfate.

$$C = 100 - (A + B)$$

Insoluble Matter. (Page 25). Dissolve 20 g in 200 mL of water.

Chloride. (Page 35). Use 0.50 g.

Phosphate. Dissolve 2.0 g in 15 mL of dilute ammonium hydroxide (1 + 2), and evaporate to dryness. Dissolve the residue in 25 mL of approximately 0.5 N hydrochloric acid, add 1 mL of ammonium molybdate reagent solution and 1 mL of 4-(methylamino)phenol sulfate reagent solution, and allow to stand for 2 h at room temperature. Any blue color should not exceed that produced by 0.02 mg of phosphate ion (PO_4) in an equal volume of solution containing the quantities of reagents used in the test, including the residue from evaporation of 5 mL of ammonium hydroxide.

Heavy Metals. (Page 36, Method 1). Dissolve 4.0 g in about 20 mL of water, and dilute with water to 32 mL. Use 24 mL to prepare the sample solution, and use the remaining 8.0 mL to prepare the control solution.

Iron. (Page 38, Method 1). Dissolve 5.0 g in 80 mL of dilute hydrochloric acid (1 + 3), and boil gently for 10 min. Cool, dilute with water to 100 mL, and use 10 mL of the solution.

Calcium, Magnesium, and Sodium. (By flame AAS, page 63).

   Sample Stock Solution. Dissolve 5.0 g of sample with water in a 100-mL volumetric flask, and dilute to the mark with water (1 mL = 0.05 g).

Element	Wavelength (nm)	Sample Wt (g)	Standard Added (mg)	Flame Type*	Background Correction
Ca	422.7	1.0	0.02; 0.04	N/A	No
Mg	285.2	1.0	0.01; 0.02	A/A	Yes
Na	589.0	0.10	0.01; 0.02	A/A	No

*A/A is air/acetylene; N/A is nitrous oxide/acetylene.

Potassium Hydroxide

KOH   Formula Wt 56.11   CAS No. 1310-58-3

Note: Reagent potassium hydroxide usually contains 10–15% water.

GENERAL DESCRIPTION

Typical appearance: white, deliquescent solid
Analytical use: alkalimetric titrations
Change in state (approximate): melting point, 360 °C
Aqueous solubility: 11 g in 100 mL at 20 °C

SPECIFICATIONS

Assay and potassium carbonate. . . . . ≥85% KOH and ≤2.0% K_2CO_3

Maximum Allowable

Chloride (Cl). 0.01%
Nitrogen compounds (as N). 0.001%
Phosphate (PO_4). . . . . .5 ppm
Sulfate (SO_4) 0.003%
Heavy metals (as Ag) 0.001%
Iron (Fe). 0.001%
Nickel (Ni). 0.001%
Calcium (Ca) 0.005%
Magnesium (Mg) 0.002%
Sodium (Na). 0.05%

TESTS

Note: Special care must be taken in sampling to obtain a representative sample and to avoid absorption of water and carbon dioxide by the sample taken.

Assay and Potassium Carbonate. (By acid–base titrimetry). Weigh rapidly 35–40 g to within 0.1 g, dissolve, cool, and dilute to 1 L with water. Dilute 50.0 mL of the well-mixed solution to 200 mL with water, add 5 mL of 12% barium chloride reagent solution, shake, and allow to stand for a few minutes. Titrate with 1 N hydrochloric acid volumetric solution to the phenolphthalein end point. One milliliter of 1 N hydrochloric acid corresponds to 0.05611 g of KOH. Continue the titration with 1 N hydrochloric acid to the methyl orange end point to determine carbonate. One milliliter of 1 N hydrochloric acid corresponds to 0.06911 g of K_2CO_3.

$$\% \text{ KOH} = \frac{(\text{mL} \times \text{N HCl}) \times 5.611}{\text{Sample wt (g)} / 20}$$

$$\% \text{ } K_2CO_3 = \frac{(\text{mL} \times \text{N HCl}) \times 6.911}{\text{Sample wt (g)} / 20}$$

For the Determination of Chloride, Nitrogen Compounds, Phosphate, Sulfate, Heavy Metals, Iron, and Nickel

Sample Solution A. Dissolve 50.0 g in carbon dioxide- and ammonia-free water, cool, and dilute with water to 500 mL in a volumetric flask (1 mL = 0.10 g).

Chloride. (Page 35). Use 1.0 mL of sample solution A (0.1-g sample).

Nitrogen Compounds. (Page 39). Use 20 mL of sample solution A (2-g sample). For the control, use 10 mL of sample solution A and 0.01 mg of nitrogen ion (N).

Phosphate. (Page 40, Method 1). To 40 mL of sample solution A (4-g sample), add 10 mL of hydrochloric acid, and evaporate to dryness on a hot plate ($\approx$100 °C). Dissolve the residue in 25 mL of approximately 0.5 N sulfuric acid, and continue as described. For the standard, include the residue from evaporation of 5 mL of hydrochloric acid.

Sulfate. (Page 40, Method 1). Neutralize 16.7 mL of sample solution A, and evaporate to 10 mL. Allow 30 min for turbidity to form.

Heavy Metals. To 60 mL of sample solution A (6-g sample), cautiously add 15 mL of nitric acid. For the control, add 0.05 mg of silver to 10 mL of sample solution A, and cautiously add 15 mL of nitric acid. Evaporate both solutions to dryness over a low flame or on an electric hot plate. Dissolve each residue in about 20 mL of water, filter if necessary through a chloride-free filter, and dilute with water to 25 mL. Adjust the pH of the control and sample solutions to between 3 and 4 (using a pH meter) with 1 N acetic acid or 10% ammonium hydroxide reagent solution, dilute with water to 40 mL, and mix. Add 10 mL of freshly prepared hydrogen sulfide water to each, and mix. Any yellow-brown color in the solution of the sample should not exceed that in the control.

Iron. (Page 38, Method 1). Neutralize 10 mL of sample solution A (1-g sample) with hydrochloric acid, using phenolphthalein indicator. Add 2 mL in excess, and dilute with water to 50 mL. Use this solution. For the standard, use the residue obtained by evaporating the quantity of acid used to neutralize the sample solution.

Nickel. Dilute 20 mL of sample solution A (2-g sample) to 50 mL with water and neutralize with hydrochloric acid. Dilute with water to 85 mL and adjust the pH to 8 with ammonium hydroxide. Add 5 mL of bromine water, 5 mL of 1% dimethylglyoxime reagent solution, and 5 mL of 10% sodium hydroxide reagent solution. Any red color should not exceed that produced by 0.02 mg of nickel in an equal volume of solution containing the quantities of reagents used in the test.

Calcium, Magnesium, and Sodium. (By flame AAS, page 63).

> **Sample Stock Solution.** Using an ice-water bath, cautiously dissolve 10.0 g of sample in 50 mL of water, cool to about 15 °C, and slowly add 40 mL of nitric acid (1 + 1) with stirring. Cool to room temperature, transfer to a 100-mL volumetric flask, dilute to the mark with water, and mix (1 mL = 0.10 g).

Element	Wavelength (nm)	Sample Wt (g)	Standard Added (mg)	Flame Type*	Background Correction
Ca	422.7	2.0	0.05; 0.10	N/A	No
Mg	285.2	0.50	0.005; 0.01	A/A	Yes
Na	589.0	0.05	0.02; 0.04	A/A	No

*A/A is air/acetylene; N/A is nitrous oxide/acetylene.

Potassium Iodate

KIO_3 Formula Wt 214.00   CAS No. 7758-05-6

GENERAL DESCRIPTION

Typical appearance: colorless or white solid
Analytical use: oxidizing agent
Change in state (approximate): melting point, 560 °C, with partial decomposition
Aqueous solubility: 8.1 g in 100 mL at 20 °C

SPECIFICATIONS

Assay . . . . 99.4–100.4% KIO_3
pH of a 5% solution at 25.0 °C 5.0–8.0

	Maximum Allowable
Insoluble matter	0.005%
Chloride and bromide (as Cl).	0.01%
Iodide (I)	0.001%
Nitrogen compounds (as N).	0.005%
Sulfate (SO4)	0.005%
Heavy metals (as Pb).	5 ppm
Iron (Fe).	0.001%
Sodium (Na).	0.005%

TESTS

Assay. (By iodometric titration of oxidizing power). Weigh accurately about 0.14 g, and dissolve in 50 mL of water in a glass-stoppered conical flask. Add 3 g of potassium iodide and 1 mL of sulfuric acid. Stopper, swirl, and allow to stand for 3 min. Titrate the liberated iodine with 0.1 N sodium thiosulfate volumetric solution, adding 3 mL of starch indicator solution near the end of the titration. One milliliter of 0.1 N sodium thiosulfate corresponds to 0.003567 g of KIO_3.

$$\% \, KIO_3 = \frac{(mL \times N \, Na_2S_2O_3) \times 3.567}{Sample \, wt \, (g)}$$

pH of a 5% Solution at 25.0 °C. (Page 49).

Insoluble Matter. (Page 25). Use 20 g dissolved in 250 mL of water.

Chloride and Bromide. Dissolve 1.0 g in 100 mL of water in a flask, filter if necessary through a chloride-free filter, and add 6 mL of 30% hydrogen peroxide and 1 mL of phosphoric acid. Heat to boiling, and boil gently until all the iodine is expelled and the solution is colorless. Cool, wash down the sides of the flask, and add 0.5 mL of hydrogen peroxide. If an iodine color develops, boil until the solution is colorless and for 10 min longer. If no color develops, boil for 10 min. Dilute with water to 100 mL, take 10 mL, and dilute with water to 23 mL. Prepare a standard containing 0.01 mg of chloride ion (Cl) and 0.6 mL of hydrogen peroxide in 23 mL of water. Add 1 mL of nitric acid and 1 mL of silver nitrate reagent solution to each. Any turbidity in the solution of the sample should not exceed that in the standard.

Iodide. Dissolve 11 g in 160 mL of water, and add 1 g of citric acid and 5 mL of chloroform. Prepare a control solution containing 1 g of the sample, 1 g of citric acid, and 0.1 mg of iodide ion (I) in 160 mL of water; add 5 mL of chloroform. Shake vigorously and allow the chloroform to separate. Any pink color in the chloroform layer from the sample should not exceed that in the chloroform layer from the control.

Nitrogen Compounds. (Page 39). Dilute the distillate of a 1.0-g sample to 50 mL. Use 10 mL of the distillate. For the standard, use 0.01 mg of nitrogen ion (N).

Sulfate. (Page 41, Method 2). Evaporate 1.0 g to dryness with 3 mL of hydrochloric acid. Repeat the evaporation two more times. Allow 30 min for turbidity to form.

Heavy Metals. (Page 36, Method 1). Dissolve 7.0 g in 50 mL of water, heat to boiling, add 1.5 mL of dilute formic acid (1 + 9), and continue heating until the initial reaction ceases. Add dropwise 50 mL of dilute formic acid (1 + 9), and continue boiling until the solution appears to be free of iodine. Wash down the sides of the container, and boil again if the iodine color reappears. Repeat the washing down and boiling until the solution remains colorless. Evaporate to dryness on a hot plate ($\approx$100 °C). Dissolve the residue in about 20 mL of water, and dilute with water to 35 mL. Use 25 mL to prepare the sample solution, use 5.0 mL to prepare the control solution, and reserve the remaining 5.0 mL for the test for iron.

Iron. (Page 38, Method 2). Use 5.0 mL of the solution reserved from the test for heavy metals.

Sodium. (By flame AAS, page 63).

> *Sample Stock Solution.* Dissolve 1.0 g of sample in water, add 5 mL of nitric acid, transfer to a 100-mL volumetric flask, and dilute to the mark with water (1 mL = 0.01 g).

Element	Wavelength (nm)	Sample Wt (g)	Standard Added (mg)	Flame Type*	Background Correction
Na	589.0	0.20	0.005; 0.01	A/A	No

*A/A is air/acetylene.

Potassium Iodide

KI	**Formula Wt 166.00**	**CAS No. 7681-11-0**

GENERAL DESCRIPTION

Typical appearance: colorless or white solid; may become yellow when exposed to light and humidity

Analytical use: iodometric titrations; reducing agent
Change in state (approximate): melting point, 680 °C
Aqueous solubility: 145 g in 100 mL at 20 °C

SPECIFICATIONS

Assay . . . . ≥99.0% KI
pH of a 5% solution at 25.0 °C 6.0–9.2

	Maximum Allowable
Insoluble matter	0.005%
Loss on drying	0.2%
Chloride and bromide (as Cl)	0.01%
Iodate (IO_3)	3 ppm
Phosphate (PO_4)	0.001%
Sulfate (SO_4)	0.005%
Barium (Ba)	0.002%
Heavy metals (as Pb)	5 ppm
Iron (Fe)	3 ppm
Calcium (Ca)	0.002%
Magnesium (Mg)	0.001%
Sodium (Na)	0.005%

TESTS

Assay. (By titration of reductive power). Weigh, to the nearest 0.1 mg, about 0.5 g of sample, and dissolve with 20 mL of water in a 200-mL glass-stoppered titration flask. Add 30 mL of hydrochloric acid and 5 mL of chloroform, cool if necessary, and titrate with 0.05 M potassium iodate volumetric solution until the iodine color disappears from the aqueous layer. Stopper, shake vigorously for 30 s, and continue the titration, shaking vigorously after each addition of the iodate, until the iodine color in the chloroform is discharged. One milliliter of 0.05 M potassium iodate corresponds to 0.0166 g of KI.

$$\% \text{ KI} = \frac{(\text{mL} \times \text{M KIO}_3) \times 33.20}{\text{Sample wt (g)}}$$

pH of a 5% Solution at 25.0 °C. (Page 49).

Insoluble Matter. (Page 25). Use 20 g dissolved in 150 mL of water.

Loss on Drying. Weigh accurately about 2.0 g of sample in which the large crystals have been crushed, but not ground, with a mortar and pestle. Dry in a tared, preconditioned dish for 6 h at 150 °C.

Chloride and Bromide. Dissolve 1.0 g in 100 mL of water in a distilling flask. Add 1 mL of hydrogen peroxide and 1 mL of phosphoric acid, heat to boiling, and boil gently until all the iodine is expelled and the solution is colorless. Cool, wash down the sides of the flask, and add 0.5 mL of 30% hydrogen peroxide. If an iodine color develops, boil until the solution is colorless and then for an additional 10 min. If no color develops, boil for 10 min, filter if necessary through a chloride-free filter, and dilute with water to 100 mL. Dilute 10 mL with water to 23 mL, and add 1 mL of nitric acid and 1 mL of silver nitrate reagent solution. Any turbidity should not exceed that produced by 0.01 mg of chloride ion (Cl) in an equal volume of solution containing the quantities of nitric acid and silver nitrate used in the test.

Iodate. (Page 54). Use 10.0 g of sample in 25 mL of solution. For the standard, add 0.03 mg of iodate ion (IO_3).

Phosphate. (Page 40, Method 1). Dissolve 2.0 g in water, and add 10 mL of nitric acid and 5 mL of hydrochloric acid. Evaporate to dryness on a hot plate (≈ 100 °C). Dissolve the residue in 25 mL of approximately 0.5 N sulfuric acid and continue as described. For the standard, include the residue from evaporation of the quantities of acids used with the sample.

Sulfate. (Page 40, Method 1). Allow 30 min for turbidity to form.

Barium. Dissolve 3.0 g of the sample in 10 mL of dilute hydrochloric acid $(1 + 1)$, and add 5 mL of nitric acid. For the control, dissolve 0.5 g of sample and 0.05 mg of barium in 10 mL of dilute hydrochloric acid $(1 + 1)$, and add 5 mL of nitric acid. Evaporate both solutions to dryness. Dissolve the residues in 10 mL of dilute hydrochloric acid $(1 + 1)$, add 5 mL of nitric acid, and again evaporate to dryness. Dissolve each residue in water and dilute with water to 23 mL. To each solution, add 2 mL of potassium dichromate reagent solution and 10% ammonium hydroxide reagent solution until the orange color is just dissipated and the yellow color persists. To each solution, add with constant stirring 25 mL of methanol. Any turbidity in the solution of the sample should not exceed that in the control.

Heavy Metals. (Page 36, Method 1). Dissolve 6.0 g in water, and dilute with water to 30 mL. Use 25 mL to prepare the sample solution, and use the remaining 5.0 mL to prepare the control solution.

Iron. (Page 38, Method 2). Dissolve 3.3 g in 50 mL of water, and add 0.1 mL of ammonium hydroxide.

Calcium, Magnesium, and Sodium. (By flame AAS, page 63).

> **Sample Stock Solution.** Dissolve 5.0 g of sample in water in a 100-mL volumetric flask, and dilute to the mark with water (1 mL = 0.05 g).

Element	Wavelength (nm)	Sample Wt (g)	Standard Added (mg)	Flame Type*	Background Correction
Ca	422.7	1.0	0.02; 0.04	N/A	No
Mg	285.2	1.0	0.01; 0.02	A/A	Yes
Na	589.0	0.20	0.005; 0.01	A/A	No

*A/A is air/acetylene; N/A is nitrous oxide/acetylene.

Potassium Nitrate
KNO$_3$ Formula Wt 101.10   CAS No. 7757-79-1

GENERAL DESCRIPTION
Typical appearance: white solid
Analytical use: preparation of standard solutions
Change in state (approximate): melting point, 334 °C
Aqueous solubility: 32 g in 100 mL at 20 °C

SPECIFICATIONS
Assay . . . . ≥99.0% KNO$_3$
pH of a 5% solution at 25.0 °C 4.5–8.5

	Maximum Allowable
Insoluble matter	0.005%
Chloride (Cl)	0.002%
Iodate (IO$_3$)	5 ppm
Nitrite (NO$_2$)	0.001%
Phosphate (PO$_4$)	5 ppm
Sulfate (SO$_4$)	0.003%
Heavy metals (as Pb)	5 ppm
Iron (Fe)	3 ppm
Calcium (Ca)	0.005%
Magnesium (Mg)	0.002%
Sodium (Na)	0.005%

TESTS

Assay. (By acid–base titrimetry). Weigh, to the nearest 0.1 mg, about 0.4 g of sample, and dissolve in 100 mL of water. Pass the solution through a cation-exchange column (see page 29) with water at a rate of about 5 mL/min, and collect the eluate in a 500-mL titration flask. Then, wash the resin in a column at a rate of about 10 mL/min, and collect in the same titration flask. Add 0.15 mL of phenolphthalein to the flask, and titrate with 0.1 N sodium hydroxide. Continue the titration as the washing proceeds until 50 mL of the eluate requires no further titration. One milliliter of 0.1 N sodium hydroxide corresponds to 0.01011 g of KNO$_3$.

$$\% \ KNO_3 = \frac{(mL \times N \ NaOH) \times 10.11}{Sample \ wt \ (g)}$$

pH of a 5% Solution at 25.0 °C. (Page 49).

Insoluble Matter. (Page 25). Use 20 g dissolved in 150 mL of water.

Chloride. (Page 35). Use 0.50 g.

Iodate. (Page 54). Use 10.0 g of sample in 25 mL of solution. For the standard, add 0.05 mg of iodate ion (IO$_3$).

Nitrite. (Page 55). Use 1.0 g of sample and 0.01 mg of nitrite.

Phosphate. (Page 40, Method 1). Dissolve 4.0 g in 25 mL of approximately 0.5 N sulfuric acid, and continue as directed.

Sulfate. (Page 41, Method 2). Evaporate 1.7 g using 7 mL of dilute hydrochloric acid (1 + 1). Repeat the evaporation, and heat the last residue for 1 h at 120 °C.

Heavy Metals. (Page 36, Method 1). Dissolve 6.0 g in about 20 mL of water, and dilute with water to 30 mL. Use 25 mL to prepare the sample solution, and use the remaining 5.0 mL to prepare the control solution.

Iron. (Page 38, Method 1). Use 3.3 g.

Calcium, Magnesium, and Sodium. (By flame AAS, page 63).

  Sample Stock Solution. Dissolve 5.0 g of sample in water in a 100-mL volumetric flask, and dilute to the mark with water (1 mL = 0.05 g).

Element	Wavelength (nm)	Sample Wt (g)	Standard Added (mg)	Flame Type*	Background Correction
Ca	422.7	0.50	0.025; 0.05	N/A	No
Mg	285.2	0.50	0.01; 0.02	A/A	Yes
Na	589.0	0.20	0.005; 0.01	A/A	No

*A/A is air/acetylene; N/A is nitrous oxide/acetylene.

Potassium Nitrite

KNO_2 **Formula Wt 85.10**   **CAS No. 7758-09-0**

GENERAL DESCRIPTION

Typical appearance: white or slightly yellow solid

Analytical use: complexing agent

Change in state (approximate): melting point, 441 °C; decomposition starts at 350 °C

Aqueous solubility: 300 g in 100 mL at 20 °C

SPECIFICATIONS

Assay . . . . ≥96.0% KNO_2

Maximum Allowable

Insoluble matter 0.01%

Chloride (Cl) 0.03%

Sulfate (SO_4) 0.01%

Heavy metals (as Pb) 0.001%

Iron (Fe). 0.001%

Calcium (Ca) 0.005%

Magnesium (Mg). 0.002%

Sodium (Na) 0.5%

TESTS

Assay. (By oxidation–reduction titrimetry). Weigh accurately about 6 g, dissolve in water, and dilute with water to 1 L in a volumetric flask. Add 5 mL of sulfuric acid to 300 mL of water, and while the solution is still warm, add 0.1 N potassium permanganate volumetric solution until a faint pink color that persists for 2 min is produced. Then, add 40.0 mL of 0.1 N potassium permanganate, and mix gently. Add, slowly and with constant agitation, 25.0 mL of the sample solution, holding the tip of the pipet well below the surface of the liquid. Add 15.0 mL of 0.1 N ferrous ammonium sulfate volumetric solution, allow the solution to stand for 5 min, and titrate the excess of ferrous ammonium sulfate with 0.1 N potassium permanganate. Each milliliter of 0.1 N potassium permanganate consumed by the potassium nitrite corresponds to 0.004255 g of KNO_2.

$$\% \ KNO_2 = \frac{\{[(40.0 + V\,mL) \times N\ KMnO_4] - (15.0 \times N\ Fe(NH_4)_2(SO_4)_2\} \times 4.255}{Sample\ wt\ (g)\ /\ 40}$$

where V = the volume, in mL, of potassium permanganate used in the back-titration.

Insoluble Matter. (Page 25). Use 20 g dissolved in 200 mL of water.

Chloride. Dissolve 3.3 g in 20 mL of water, filter if necessary through a chloride-free filter, and dilute with water to 100 mL. Dilute 1.0 mL of this solution with 15 mL of water, and slowly add 1 mL of acetic acid. Heat to boiling, boil gently for 5 min, cool, and dilute with water to 20 mL. Add 1 mL of nitric acid and 1 mL of silver nitrate reagent solution. Any turbidity should not exceed that produced by 0.01 mg of chloride ion (Cl) in an equal volume of solution containing the quantities of reagents used in the test.

For the Determination of Sulfate, Heavy Metals, and Iron

Sample Solution A. Dissolve 10.0 g in water, and dilute with water to 100 mL (1 mL = 0.1 g).

Sulfate. (Page 41, Method 2). Use 5.0 mL of sample solution A (0.5-g sample) plus 1 mL of hydrochloric acid.

Heavy Metals. (Page 36, Method 1). To 30 mL of sample solution A (3-g sample), add 5 mL of hydrochloric acid, and evaporate to dryness on a hot plate ($\approx$100 °C). Add 5 mL of hydrochloric acid, and again evaporate to dryness. Dissolve in about 20 mL of water, and dilute with water to 25 mL. For the control, add 0.02 mg of lead to 10 mL of sample solution A (1-g sample), add 5 mL of hydrochloric acid, and treat exactly as the 30 mL of sample solution A.

Iron. (Page 38, Method 1). To 10 mL of sample solution A (1-g sample), add 5 mL of hydrochloric acid, and evaporate to dryness on a hot plate ($\approx$100 °C). Dissolve the residue in 15–20 mL of water, add 2 mL of hydrochloric acid, filter if necessary, and dilute with water to 50 mL. Use the solution without further acidification.

Calcium, Magnesium, and Sodium. (By flame AAS, page 63).

Sample Stock Solution A. Dissolve 0.40 g of sample in water in a 100-mL volumetric flask, and dilute to the mark with water (1 mL = 0.004 g).

Sample Stock Solution B. Dissolve 5.0 g of sample in 80 mL of water. Transfer to a 100-mL volumetric flask, and dilute to the mark with water (1 mL = 0.05 g).

Element	Wavelength (nm)	Sample Wt (g)	Standard Added (mg)	Flame Type*	Background Correction
Ca	422.7	0.50	0.025; 0.05	N/A	No
Mg	285.2	0.50	0.01; 0.02	A/A	Yes
Na	589.0	0.004	0.01; 0.02	A/A	No

*A/A is air/acetylene; N/A is nitrous oxide/acetylene.

Potassium Oxalate Monohydrate
Oxalic Acid, Potassium Salt, Monohydrate

$(COOK)_2 \cdot H_2O$ **Formula Wt 184.23** **CAS No. 6487-48-5**

GENERAL DESCRIPTION

Typical appearance: colorless solid
Analytical use: reducing agent
Change in state (approximate): dehydrates at 100 °C
Aqueous solubility: 36 in 100 mL at 20 °C

SPECIFICATIONS

Assay . . . . .98.5–101.0% $K_2C_2O_4 \cdot H_2O$
Substances darkened by hot sulfuric acid . . . . Passes test
Neutrality. . . . . Passes test

Maximum Allowable

Insoluble matter 0.01%
Chloride (Cl) 0.002%
Sulfate (SO_4) 0.01%
Ammonium (NH_4). 0.002%
Heavy metals (as Pb) 0.002%
Iron (Fe). 0.001%
Sodium (Na) 0.02%

TESTS

Assay. (By oxidation–reduction titrimetry). Weigh accurately about 1.8 g, and dissolve in water in a 250-mL volumetric flask. Dilute to volume, and mix. To a 50-mL aliquot, add 50 mL of water and 5 mL of sulfuric acid. Titrate slowly with 0.1 N potassium permanganate volumetric solution until about 25 mL has been added. Heat the solution to about 70 °C, and complete the titration to a permanent faint pink color. One milliliter of 0.1 N potassium permanganate consumed corresponds to 0.009212 g of $K_2C_2O_4 \cdot H_2O$.

$$\% \ K_2C_2O_4 \cdot H_2O = \frac{(mL \times N \ KMnO_4) \times 9.212}{\text{Sample wt (g) / 5}}$$

Substances Darkened by Hot Sulfuric Acid. Heat 1.0 g in a recently ignited test tube with 10 mL of sulfuric acid until the appearance of dense fumes. The acid when cooled should have no more color than a mixture of the following composition: 0.2 mL of cobalt chloride reagent solution, 0.3 mL of ferric chloride reagent solution, 0.3 mL of cupric sulfate reagent solution, and 9.2 mL of water.

Neutrality. Dissolve 2.0 g in 200 mL of water, and add 10.0 mL of 0.01 N oxalic acid and 0.20 mL of 1% phenolphthalein indicator solution. Boil the solution in a flask for 10 min, passing through it a stream of carbon dioxide-free air. Cool the solution rapidly to room temperature while keeping the flow of carbon dioxide-free air passing through it, and titrate with 0.01 N sodium hydroxide. Not less than 9.2 nor more than 10.5 mL of 0.01 N sodium hydroxide should be required to match the pink color produced in a buffer solution containing 0.20 mL of 1% phenolphthalein indicator solution. The buffer solution contains: 3.1 g of boric acid, 3.8 g of potassium chloride, and 5.90 mL of 1 N sodium hydroxide in 1 L of water.

Insoluble Matter. (Page 25). Use 10 g dissolved in 100 mL of water.

Chloride. (Page 35). Dissolve 2.0 g in water plus 10 mL of nitric acid, filter if necessary through a chloride-free filter, and dilute with water to 100 mL. To 25 mL of the solution, add 1 mL of silver nitrate reagent solution. Any turbidity should not exceed that produced by 0.01 mg of chloride ion (Cl) in an equal volume of solution containing the quantities of reagents used in the test.

Sulfate. (Page 41, Method 4).

Ammonium. Dissolve 1.0 g in 50 mL of ammonia-free water, and add 2 mL of Nessler reagent. Any color should not exceed that produced by 0.02 mg of ammonium ion (NH_4) in an equal volume of solution containing 2 mL of Nessler reagent.

Heavy Metals. (Page 36, Method 1). Mix 2.0 g with 2 mL of water, add 2 mL of nitric acid and 2 mL of 30% hydrogen peroxide, and digest in a covered beaker on a hot plate ($\approx$100 °C) until reaction ceases. Remove the cover, and evaporate to dryness. Dissolve the residue in 4 mL of dilute nitric acid (1 + 1), and again evaporate to dryness. Dissolve the residue in about 20 mL of water, dilute with water to 32 mL, and use 24 mL to prepare the sample solution. Use the remaining 8.0 mL, plus the residue from evaporation of 1 mL of 30% hydrogen peroxide and 3 mL of nitric acid to dryness, to prepare the control solution.

Iron. (Page 38, Method 1). Mix 2.0 g with 2 mL of water, add 2 mL of nitric acid and 2 mL of 30% hydrogen peroxide, and digest in a covered beaker on a hot plate ($\approx$100 °C) until reaction ceases. Remove the cover, and evaporate to dryness. Add 5 mL of hydrochloric acid, and again evaporate to dryness. Dissolve the residue in a few milliliters of water, add 4 mL of hydrochloric acid, and dilute with water to 100 mL. Use 50 mL without further acidification. For the control, use the residue from evaporating to dryness 1 mL of nitric acid, 1 mL of 30% hydrogen peroxide, and 2.5 mL of hydrochloric acid.

Sodium. (By flame AAS, page 63).

    Sample Stock Solution. Dissolve 1.0 g of sample in water, and dilute to 100 mL in a volumetric flask with water (1 mL = 0.01 g).

Element	Wavelength (nm)	Sample Wt (g)	Standard Added (mg)	Flame Type*	Background Correction
Na	589.0	0.05	0.01; 0.02	A/A	No

*A/A is air/acetylene.

Potassium Perchlorate
KClO$_4$ **Formula Wt 138.55** **CAS No. 7778-74-7**

GENERAL DESCRIPTION

Typical appearance: colorless or white solid
Analytical use: oxidizing agent
Aqueous solubility: 1.7 g in 100 mL at 20 °C

SPECIFICATIONS

Assay . . . . 99.0–100.5% KClO$_4$

Maximum Allowable

Insoluble matter 0.005%
Chloride (Cl) 0.003%
Sulfate (SO$_4$) 0.001%
Calcium (Ca) 0.005%
Sodium (Na) 0.02%
Heavy metals (as Pb) . . . . 5 ppm
Iron (Fe). . . . . 5 ppm

TESTS

Assay. (By argentimetric titration after decomposition to chloride). Weigh accurately about 0.28 g into a platinum crucible or dish. Add 3 g of sodium carbonate, and heat, gently at first, then to fusion until a clear melt is obtained. Leach with minimal 10% nitric acid, and boil gently to expel carbon dioxide. Cool to room temperature, add 10 mL of aqueous 30% ammonium acetate solution, and titrate with 0.1 N silver nitrate volumetric solution, using dichlorofluorescein indicator solution to a pink end point. One milliliter of 0.1 N silver nitrate corresponds to 0.01385 g of KClO$_4$.

$$\% \text{ KClO}_4 = \frac{(\text{mL} \times \text{N AgNO}_3) \times 13.85}{\text{Sample wt (g)}}$$

Insoluble Matter. (Page 25). Use 50 g dissolved in 600 mL of hot water.

Chloride. (Page 35). Dissolve 0.33 g in 20 mL of hot water, filter, and cool.

Sulfate. Dissolve 40 g in 400 mL of hot water, add 4 mL of hydrochloric acid, and heat the solution to boiling. Add 5 mL of 12% barium chloride reagent solution, digest in a covered beaker on a hot plate ($\approx$100 °C) for 2 h, and allow to stand for at least 8 h. Warm to dissolve any potassium perchlorate that may have crystallized. If a precipitate remains, filter, wash thoroughly, and ignite. The weight of the precipitate should not be more than 0.001 g greater than the weight obtained in a complete blank test.

Calcium and Sodium. (By flame AAS, page 63).

> **Sample Stock Solution.** Dissolve 5.0 g of sample in water, and dilute to 100 mL in a volumetric flask with water (1 mL = 0.05 g).

Element	Wavelength (nm)	Sample Wt (g)	Standard Added (mg)	Flame Type*	Background Correction
Ca	422.7	0.50	0.025; 0.05	N/A	No
Na	589.0	0.05	0.01; 0.02	A/A	No

*A/A is air/acetylene; N/A is nitrous oxide/acetylene.

Heavy Metals. (Page 36, Method 1). Dissolve 3.0 g in hot water, dilute with water to 60 mL, and use 50 mL to prepare the sample solution. For the control, add 0.01 mg of lead ion (Pb) to the remaining 10 mL, and dilute with water to 50 mL.

Iron. (Page 38, Method 1). Use 2.0 g.

Potassium Periodate
Potassium Metaperiodate

KIO_4 **Formula Wt 230.00**   **CAS No. 7790-21-8**

GENERAL DESCRIPTION

Typical appearance: white solid
Analytical use: oxidizing agent; determination of manganese
Change in state (approximate): melting point, 582 °C, with decomposition
Aqueous solubility: 0.42 g in 100 mL at 20 °C

SPECIFICATIONS

Assay (dried basis) . . . . 99.8–100.3% KIO_4

Maximum Allowable

Other halogens (as Cl) 0.01%
Manganese (Mn) . . . . 1 ppm

TESTS

Assay. (By titration of oxidative capacity). Weigh accurately about 1 g, previously dried over magnesium perchlorate or phosphorus pentoxide for 6 h. Dissolve in water, and dilute with water to 500 mL in a volumetric flask. To 50.0 mL in a glass-stoppered flask, add 10 g of potassium iodide and 10 mL of a cooled solution of 25% sulfuric acid reagent solution. Allow to stand for 5 min, add 100 mL of cold water, and titrate the liberated iodine with 0.1

N sodium thiosulfate volumetric solution, adding starch indicator solution near the end point. Carry out a complete blank test and make any necessary correction. One milliliter of 0.1 N sodium thiosulfate corresponds to 0.002875 g of KIO_4.

$$\% \ KIO_4 = \frac{(mL \times N \ Na_2S_2O_3) \times 2.875}{Sample \ wt \ (g) \ / \ 10}$$

Other Halogens. Dissolve 0.50 g in a mixture of 10 mL of water, and 16 mL of sulfurous acid and boil for 3 min. Cool, add 5 mL of ammonium hydroxide and 20 mL of 2.5% silver nitrate solution, and filter. Dilute the filtrate with water to 100 mL, and to 20 mL of this solution add 1.5 mL of nitric acid. Any turbidity should not exceed that produced by 0.01 mg of chloride ion (Cl) in an equal volume of solution containing 0.5 mL of ammonium hydroxide, 1.5 mL of nitric acid, and 1 mL of silver nitrate reagent solution.

Manganese. Add 5.5 g to 85 mL of 10% sulfuric acid reagent solution. For the control, add 0.5 g of the sample and 0.005 mg of manganese ion (Mn) to 85 mL of 10% sulfuric acid reagent solution. Add 10 mL of nitric acid and 5 mL of phosphoric acid to each, boil gently for 10 min, and cool. Any pink color in the solution of the sample should not exceed that in the control.

Potassium Permanganate
$KMnO_4$ Formula Wt 158.03   CAS No. 7722-64-7

GENERAL DESCRIPTION
Typical appearance: dark solid
Analytical use: oxidizing agent
Aqueous solubility: 6.4 g in 100 mL at 20 °C

SPECIFICATIONS
Assay . . . . $\geq$99.0% $KMnO_4$

	Maximum Allowable
Insoluble matter	0.2%
Chloride and chlorate (as Cl)	0.005%
Sulfate (SO_4)	0.02%

TESTS

Assay. (By titration of oxidizing power). Weigh, to the nearest 0.1 mg, about 1.5 g of sample. Transfer to a 500-mL volumetric flask, dissolve with water, then dilute to the mark with water, and mix thoroughly. Weigh accurately about 0.25 g of NIST SRM or primary standard sodium oxalate (dried at 105 °C), and dissolve in a 500-mL glass-stoppered flask. Add 15 mL of 25% sulfuric acid reagent solution, and then add from a buret 30 mL of 0.1 N potassium permanganate volumetric solution. Heat the solutions to about 70 °C, and complete the titration with 0.1 N potassium permanganate. One milliliter of 0.1 N potassium permanganate corresponds to 0.02358 g of $KMnO_4$.

$$\% \, KMnO_4 = \frac{(g \, Na_2C_2O_4) \times (Assay / 100) \times 23.58}{Sample \, wt \, (g) \times [mL \, (total) \, KMnO_4]}$$

Insoluble Matter. (Page 25). Dissolve 2.0 g in 150 mL of warm water on a hot plate (≈ 100 °C). Filter at once through a tared filtering crucible, wash thoroughly, and dry at 105 °C.

Chloride and Chlorate. Dissolve 1.0 g in 75 mL of water, and add 5 mL of nitric acid and 3 mL of 30% hydrogen peroxide. After reduction is complete, filter if necessary through a chloride-free filter, and dilute with water to 100 mL. To 20 mL of this solution, add 1 mL of silver nitrate reagent solution. Any turbidity should not exceed that produced by 0.01 mg of chloride ion (Cl) in an equal volume of solution containing the quantities of reagents used in the test.

Sulfate. Dissolve 5.0 g in 75 mL of water, and add 10 mL of reagent alcohol and 10 mL of hydrochloric acid. Heat until colorless, adding more alcohol and a little hydrochloric acid if necessary. Evaporate to dryness. Dissolve the residue in 100 mL of dilute hydrochloric acid (1 + 99), filter, and heat to boiling. Add 5 mL of 12% barium chloride reagent solution, digest in a covered beaker on a hot plate (≈ 100 °C) for 2 h, and allow to stand for at least 8 h. Filter, wash thoroughly, and ignite. The weight of the precipitate should not be more than 0.0024 g greater than the weight obtained in a complete blank test.

Potassium Permanganate, Low in Mercury

$KMnO_4$ Formula Wt 158.03   CAS No. 7722-64-7

GENERAL DESCRIPTION

Typical appearance: dark solid
Analytical use: oxidizing agent
Aqueous solubility: 6.4 g in 100 mL at 20 °C

SPECIFICATIONS

Assay . . . . $\geq$99.0% $KMnO_4$

	Maximum Allowable
Insoluble matter	0.2%
Chloride and chlorate (as Cl)	0.005%
Sulfate (SO_4)	0.02%
Mercury (Hg)	0.05 ppm

TESTS

Assay and other tests except mercury are the same as for potassium permanganate, page 547.

Mercury. For the sample, place 1.0 g in a 125-mL conical flask, and add 10 mL of water. For the standard, place 10 mL of water in a similar flask, and add 0.05 μg of mercury [0.5 mL of a solution prepared freshly by diluting 1.0 mL of mercury ion (Hg) standard solution with water to 500 mL]. Add 2 mL of 4% potassium permanganate solution and 2 mL of 25% sulfuric acid reagent solution to each flask. Heat on a hot plate ($\approx$100 °C) for 30 min, then cool to room temperature. Transfer the contents of the sample flask to the aeration vessel, making the total volume about 50 mL. Rinse the flask with 4 mL of 10% hydroxylamine hydrochloride reagent solution, and add this to the aeration vessel also. Complete the reduction of manganese dioxide and mercury with 2 mL of 10% stannous chloride solution. Treat the standard solution in the same way. Obtain the peak height for the sample and standard solutions by CVAAS (page 65). The peak height for the sample should not exceed that for the standard.

Potassium Peroxydisulfate
Potassium Persulfate
$K_2S_2O_8$ **Formula Wt 270.32**   **CAS No. 7727-21-1**

GENERAL DESCRIPTION
Typical appearance: colorless or white solid
Analytical use: oxidizing agent
Change in state (approximate): decomposed by heat or humidity
Aqueous solubility: 5 g in 100 mL at 20 °C

SPECIFICATIONS

Assay . . . . $\geq$99.0% $K_2S_2O_8$

Maximum Allowable

Insoluble matter	0.005%
Chlorine compounds (as Cl)	0.001%
Heavy metals (as Pb)	0.001%
Iron (Fe).	5 ppm
Manganese (Mn)	2 ppm

TESTS

Assay. (By titration of oxidizing power). Weigh accurately about 0.5 g, and add it to 50.0 mL of freshly prepared 0.1 N ferrous ammonium sulfate volumetric solution in a glass-stoppered flask. Stopper the flask, allow it to stand for 1 h with frequent shaking, add 5 mL of phosphoric acid, and titrate the excess ferrous ammonium sulfate with 0.1 N potassium permanganate volumetric solution. Add 5 mL of phosphoric acid to 50.0 mL of ferrous ammonium sulfate volumetric solution, and titrate with 0.1 N potassium permanganate volumetric solution. The difference in the volume of permanganate consumed in the two titrations is equivalent to the potassium peroxydisulfate. One milliliter of 0.1 N potassium permanganate corresponds to 0.01352 g of $K_2S_2O_8$.

$$\% \ K_2S_2O_8 = \frac{\{[mL \ (blank) - mL \ (sample)] \times N \ KMnO_4\} \times 13.52}{Sample \ wt \ (g)}$$

Insoluble Matter. (Page 25). Use 20 g dissolved in 200 mL of water.

Chlorine Compounds. Mix 1.0 g with 1 g of anhydrous sodium carbonate, and fuse in a porcelain dish in a muffle furnace. Cool, dissolve the residue in 20 mL of water, neutralize with nitric acid, and add an excess of 1 mL of the acid. Filter if necessary through a chloride-free filter, and add 1 mL of silver nitrate reagent solution. Any turbidity should not exceed that in a standard containing 0.01 mg of chloride ion (Cl) and 1 g of sodium carbonate fused and acidified as in the test.

Heavy Metals. (Page 36, Method 1). Dissolve 2.0 g in 20 mL of hot water, add 5 mL of hydrochloric acid, and evaporate to about 5 mL. Cool, and dilute with water to 25 mL. For the standard, use 0.02 mg of lead in 5 mL of hydrochloric acid, add 10 mL of water and 0.25 mL of sulfuric acid, evaporate to fumes of sulfur trioxide, cool, and dilute with water to 25 mL.

Iron. (Page 38, Method 1). To 2.0 g, add 20 mL of hot water and 5 mL of hydrochloric acid, and evaporate to dryness. Moisten the residue with 2 mL of hydrochloric acid, dilute with water to 50 mL, and use the solution without further acidification.

Manganese. To 5.0 g in a 150-mL beaker, add 5 mL of sulfuric acid and 5 mL of phosphoric acid. Heat to dissolve the sample, and continue heating to fumes of sulfur trioxide. Cool, dilute with water to 45 mL, add 10 mL of nitric acid, and boil for 5 min. Cool, add about 0.25 g of potassium periodate, and boil for 5 min. Cool, and compare with a standard prepared by treating 0.01 mg of manganese ion (Mn) in the same manner, using the same quantities of reagents (with no sample), except that the standard solution need not be fumed before adding the nitric acid. Any color in the solution of the sample should not exceed that in the standard.

Potassium Phosphate, Dibasic
Dipotassium Hydrogen Phosphate
K_2HPO_4 Formula Wt 174.18   CAS No. 7758-11-4

GENERAL DESCRIPTION

Typical appearance: white solid
Analytical use: buffer solutions
Aqueous solubility: very soluble

SPECIFICATIONS

Assay . . . . $\geq$98.0% K_2HPO_4
pH of a 5% solution at 25.0 °C 8.5–9.6

Maximum Allowable

Insoluble matter 0.01%
Loss on drying 1.0%
Chloride (Cl) 0.003%
Nitrogen compounds (as N) 0.001%
Sulfate (SO_4) 0.005%
Heavy metals (as Pb) . . . . 5 ppm
Iron (Fe). 0.001%
Sodium (Na) 0.05%

TESTS

Assay. (By indirect acid–base titrimetry). Weigh accurately about 7.0 g of sample, and dissolve in 50 mL of water and exactly 50.0 mL of 1 N hydrochloric acid volumetric solution. Titrate the excess acid with 1 N sodium hydroxide volumetric solution to the first inflection point at about pH 4, as measured with a pH meter. Calculate *A*, the equivalent of 1 N hydrochloric acid consumed by the sample. Continue the titration with 1 N sodium hydroxide to the second inflection point at about pH 8.8. Calculate *B*, the equivalent of 1 N sodium hydroxide required in the titration betwen the two inflection points.

$$A = [(50.0 \times N\ HCl) - (mL \times N\ NaOH)]$$

$$B = mL \times N\ NaOH$$

If *A* is equal to or less than *B*:

$$\% K_2HPO_4 = \frac{A \times 0.1742}{Sample\ wt\ (g)} \times 100$$

If *A* is greater than *B*:

$$\% K_2HPO_4 = \frac{(2B - A) \times 0.1742}{Sample\ wt\ (g)} \times 100$$

pH of a 5% Solution at 25.0 °C. (Page 49).

Insoluble Matter. (Page 25). Use 10 g dissolved in 100 mL of water.

Loss on Drying. Weigh accurately about 2 g. Dry in a tared, preconditioned vessel, at 105 °C to constant weight.

Chloride. (Page 35). Use 0.33 g.

Nitrogen Compounds. (Page 39). Use 1.0 g. For the standard, use 0.01 mg of nitrogen ion (N).

Sulfate. Dissolve 1.0 g in 15 mL of water, and add 4.0 mL of 10% hydrochloric acid (solution approximately pH 2). Filter through washed filter paper, and add 10 mL of water through the same filter. For the control, add 0.05 mg of sulfate ion (SO_4) in 20 mL of water, and add 1.0 mL of (1 + 19) hydrochloric acid. Dilute both sample and control solutions to

35 mL with water, add 5.0 mL of 40% barium chloride reagent solution to each, and mix. Observe turbidity after 10 min. Sample solution turbidity should not exceed that of the control solution.

Heavy Metals. (Page 36, Method 1). Dissolve 6.0 g in about 20 mL of water, add 8 mL of 4 N hydrochloric acid, and dilute with water to 30 mL. Use 25 mL to prepare the sample solution, and use the remaining 5.0 mL to prepare the control solution.

Iron. (Page 38, Method 2). Use 1.0 g.

Sodium. (By flame AAS, page 63).

>**Sample Stock Solution.** Dissolve 2.0 g of sample in a 100-mL volumetric flask and dilute to the mark with water (1 mL = 0.02 g).

Element	Wavelength (nm)	Sample Wt (g)	Standard Added (mg)	Flame Type*	Background Correction
Na	589.0	0.02	0.01; 0.02	A/A	No

*A/A is air/acetylene.

Potassium Phosphate, Monobasic
Potassium Dihydrogen Phosphate
KH_2PO_4 **Formula Wt 136.09**   **CAS No. 7778-77-0**

GENERAL DESCRIPTION
Typical appearance: white or colorless solid
Analytical use: buffer solutions
Change in state (approximate): melting point, 253 °C
Aqueous solubility: 33 g in 100 mL at 25 °C

SPECIFICATIONS
Assay . . . . ≥99.0% KH_2PO_4
pH of a 5% solution at 25.0 °C 4.1–4.5

	Maximum Allowable
Insoluble matter	0.01%
Loss on drying	0.2%
Chloride (Cl)	0.001%
Sulfate (SO_4)	0.003%
Heavy metals (as Pb)	0.001%
Iron (Fe)	0.002%
Sodium (Na)	0.005%

TESTS
Assay. (By acid–base titrimetry). Weigh accurately about 5.0 g of sample, and dissolve in 50 mL of water. Titrate with 1 N sodium hydroxide volumetric solution to the inflection

point at about pH 8.8, as measured with a pH meter. One milliliter of 1 N sodium hydroxide corresponds to 0.1361 g of KH_2PO_4.

$$\% \ KH_2PO_4 = \frac{(mL \times N \ NaOH) \times 13.61}{Sample \ wt \ (g)}$$

pH of a 5% Solution at 25.0 °C. (Page 49).

Insoluble Matter. (Page 25). Dissolve 10 g in 100 mL of water.

Loss on Drying. Weigh accurately about 2 g, and dry for 2 h at 105 °C.

Chloride. (Page 35). Use 1.0 g of sample and 2 mL of nitric acid.

Sulfate. Dissolve 2.0 g in 15 mL of water, and add 2.5 mL of 10% hydrochloric acid (solution approximately pH 2). Filter through washed filter paper, and add 10 mL of water through the same filter. For the control, add 0.06 mg of sulfate ion (SO_4) in 20 mL of water, and add 1.0 mL of (1 + 19) hydrochloric acid. Dilute both sample and control solutions to 35 mL with water, add 5.0 mL of 40% barium chloride solution to each, and mix. Observe turbidity after 10 min. Sample solution turbidity should not exceed that of the control solution.

Heavy Metals. (Page 36, Method 1). Dissolve 4.0 g in about 20 mL of water, and dilute with water to 32 mL. Use 24 mL to prepare the sample solution, and use the remaining 8.0 mL to prepare the control solution.

Iron. (Page 38, Method 2). Dissolve 1.0 g in water, dilute with water to 20 mL, and use 10 mL.

Sodium. (By flame AAS, page 63).

> **Sample Stock Solution.** Dissolve 1.0 g of sample in water, and dilute to 100 mL in a volumetric flask with water (1 mL = 0.01 g).

Element	Wavelength (nm)	Sample Wt (g)	Standard Added (mg)	Flame Type*	Background Correction
Na	589.0	0.20	0.01; 0.02	A/A	No

*A/A is air/acetylene.

Potassium Phosphate, Tribasic
Tripotassium Orthophosphate

K_3PO_4 (anhydrous)	**Formula Wt 212.27**	**CAS No. 7778-53-2**
$K_3PO_4 \cdot 7H_2O$ (heptahydrate)	**Formula Wt 338.38**	**CAS No. 22763-02-6**

Note: This reagent must be labeled to indicate whether it is the anhydrous or heptahydrate form.

GENERAL DESCRIPTION
Typical appearance: solid
Analytical use: buffering agent

SPECIFICATIONS

Assay (as-is basis). . . . . $\geq 98\%$ K_3PO_4 or $K_3PO_4 \cdot 7H_2O$

Maximum Allowable

Dibasic potassium phosphate (K_2HPO_4) 1%
Excess alkali (as KOH) 1%
Insoluble matter 0.01%
Chloride (Cl). 0.005%
Heavy metals (as Pb). 0.002%
Iron (Fe). 0.001%
Sulfate (SO_4) 0.005%
Sodium (Na). 0.1%

TESTS

Assay. (By indirect acid–base titrimetry). Weigh accurately about 4.0 g for the anhydrous salt and 6.5 g for the heptahydrate salt, and dissolve in 50 mL of water and exactly 50.0 mL of 1 N hydrochloric acid volumetric solution. Heat to boiling to expel carbon dioxide. While protecting the solution from absorbing carbon dioxide, cool, and titrate with 1 N sodium hydroxide volumetric solution to the inflection point at about pH 4, as measured with a pH meter. Calculate A, the equivalent of 1 N hydrochloric acid consumed by the sample. Continue the titration with 1 N sodium hydroxide to the inflection point at about pH 8.8. Calculate B, the equivalent of 1 N sodium hydroxide required in the titration between the two inflection points.

$$A = [(50.0 \times N\ HCl) - (mL \times N\ NaOH)]$$

$$B = mL \times N\ NaOH$$

If A is equal to or greater than $2B$, then

$$\% \ K_3PO_4 = \frac{B \times F}{\text{Sample wt (g)}} \times 100$$

If A is less than $2B$, then

$$\% \ K_3PO_4 = \frac{(A-B) \times F}{\text{Sample wt (g)}} \times 100$$

where F is 0.2123 for the anhydrous salt and 0.3384 for the heptahydrate salt.

Dibasic Potassium Phosphate and Excess Alkali. Calculate the amount of dibasic potassium phosphate as K_2HPO_4 or the amount of excess alkali as potassium hydroxide from the titration values obtained in the assay.

If A is less than $2B$, then

$$\% \ K_2HPO_4 = \frac{(2B-A) \times 0.1742}{\text{Sample wt (g)}} \times 100$$

If A is greater than $2B$, then

$$\% \ \text{Excess alkali (as KOH)} = \frac{(A-2B) \times 0.0561}{\text{Sample wt (g)}} \times 100$$

Insoluble Matter. (Page 25). Dissolve 10.0 g in 100 mL of water, add 0.10 mL of methyl red indicator solution, add hydrochloric acid until the solution is slightly acid to the indicator, and continue as described.

Chloride. (Page 35). Dissolve 0.20 g of sample in 10 mL of water, add 3 mL of nitric acid, and dilute to 20 mL.

Heavy Metals. (Page 36, Method 1). Dissolve 2.0 g in about 20 mL of water, and dilute with water to 32 mL. Use 24 mL to prepare the sample solution, and use the remaining 8.0 mL to prepare the control solution.

Iron. (Page 38, Method 2). Use 1.0 g.

Sulfate. Dissolve 1.0 g in 15 mL of water, and add 4.0 mL of 10% hydrochloric acid (solution pH 2). Filter through washed filter paper, and add 10 mL of water through the same filter. For the control, add 0.05 mg of sulfate ion (SO_4) in 20 mL of water, and then add 1.0 mL of dilute hydrochloric acid (1 + 19). Dilute both to 35 mL with water, add 5.0 mL of 40% barium chloride solution, and mix. Observe turbidity after 10 min. The turbidity of the sample solution should not exceed that of the control.

Sodium. (By flame AAS, page 63).

> **Sodium Stock Solution.** Dissolve 1.0 g of sample in water, and dilute to 100 mL in a volumetric flask with water (1 mL = 0.01 g).

Element	Wavelength (nm)	Sample Wt (g)	Standard Added (mg)	Flame Type*	Background Correction
Na	589.0	0.02	0.01;0.02	A/A	No

*A/A is air/acetylene.

Potassium Sodium Tartrate Tetrahydrate

2,3-Dihydroxybutanedioic Acid, Potassium Sodium Salt, Tetrahydrate

$$NaO - \overset{O}{\underset{\|}{C}} - \overset{OH}{\underset{|}{CH}} - \overset{OH}{\underset{|}{CH}} - \overset{O}{\underset{\|}{C}} - OK \cdot 4H_2O$$

$KNaC_4H_4O_6 \cdot 4H_2O$ **Formula Wt 282.22**   **CAS No. 6381-59-5**

GENERAL DESCRIPTION

Typical appearance: colorless or white solid
Analytical use: constituent of Fehling's solution
Change in state (approximate): melting point, 75 °C
Aqueous solubility: 111 g in 100 mL at 25 °C

SPECIFICATIONS

Assay . . . . 99.0–102.0% $KNaC_4H_4O_6 \cdot 4H_2O$
pH of a 5% solution at 25.0 °C 6.0–8.5

Maximum Allowable

Insoluble matter 0.005%
Chloride (Cl) 0.001%
Phosphate (PO_4) 0.002%
Sulfate (SO_4) 0.005%
Ammonium (NH_4) 0.002%
Calcium (Ca) 0.005%
Heavy metals (as Pb) . . . . 5 ppm
Iron (Fe) 0.001%

TESTS

Assay. (Total alkalinity by nonaqueous titration). Weigh accurately about 0.5 g, and mix with 3 mL of 96% formic acid to dissolve. Add 5 mL of acetic anhydride and 50 mL of glacial acetic acid. Titrate with 0.1 N perchloric acid in glacial acetic acid volumetric solution to a green end point with crystal violet indicator. One milliliter of 0.1 N perchloric acid corresponds to 0.01411 g of $KNaC_4H_4O_6 \cdot 4H_2O$.

$$\% \ KNaC_4H_4O_6 \cdot 4H_2O = \frac{(mL \times N \ HClO_4) \times 14.11}{Sample \ wt \ (g)}$$

pH of a 5% Solution at 25.0 °C. (Page 49).

Insoluble Matter. (Page 25). Use 20 g in 200 mL of water.

Chloride. (Page 35). Use 1.0 g.

Phosphate. (Page 40, Method 1). Ignite 1.0 g in a platinum dish. Dissolve the residue in 5 mL of water, add 5 mL of nitric acid, and evaporate to dryness. Dissolve the residue in 25 mL of approximately 0.5 N sulfuric acid, and continue as described.

Sulfate. To 5.0 g of sample, add 4 mL of hydrochloric acid and 8 mL of 30% hydrogen peroxide. Digest in a covered beaker on a hot plate (≈100 °C) until reaction ceases. Carefully rinse down the cover glass and sides of the beaker with about 5 mL of water. Add 4 mL of hydrochloric acid and 3 mL of 30% hydrogen peroxide, and evaporate to dryness. Add 10 mL of water, and evaporate to dryness. Dissolve the residue in 5 mL of 10% hydrochloric acid, and dilute to 25 mL with water (5 mL = 1 g). For the sample, take 15 mL, and for the control, take 5 mL of the sample solution. Add 0.10 mg of sulfate ion (SO_4) to each. Dilute both solutions to 25 mL, add 1 mL of 12% barium chloride reagent solution to each, and compare after 10 min. Sample solution turbidity should not exceed that of the control solution.

Ammonium. Dissolve 1.0 g in 50 mL of water. To 25 mL of the solution, add 2 mL of 10% sodium hydroxide reagent solution, dilute with water to 50 mL, and add 2 mL of Nessler reagent. Any color should not exceed that produced by 0.01 mg of ammonium ion (NH_4) in an equal volume of solution containing the quantities of reagents used in the test.

Calcium. (By flame AAS, page 63).

Sample Stock Solution. Dissolve 4.0 g, and dilute to 100 mL with water in a volumetric flask (1 mL − 0.04 g).

Element	Wavelength (nm)	Sample Wt (g)	Standard Added (mg)	Flame Type*	Background Correction
Ca	422.7	0.40	0.02; 0.04	N/A	No

*N/A is nitrous oxide/acetylene.

Heavy Metals. (Page 36, Method 1). Dissolve 8.0 g in about 20 mL of water, and dilute with water to 32 mL. Use 24 mL to prepare the sample solution, and use the remaining 8.0 mL to prepare the control solution.

Iron. (Page 38, Method 1). Use 1.0 g.

Potassium Sulfate

K_2SO_4 **Formula Wt 174.26**   **CAS No. 7778-80-5**

GENERAL DESCRIPTION

Typical appearance: white or colorless solid
Analytical use: Kjeldahl analysis
Change in state (approximate): melting point, 1069 °C
Aqueous solubility: 11 g in 100 mL at 20 °C

SPECIFICATIONS

Assay . . . . ≥99.0% K_2SO_4
pH of a 5% solution at 25.0 °C 5.5–8.5

Maximum Allowable

Insoluble matter 0.01%
Chloride (Cl) 0.001%
Nitrogen compounds (as N) . . . . 5 ppm
Heavy metals (as Pb) . . . . 5 ppm
Iron (Fe) . . . . 5 ppm
Calcium (Ca) 0.01%
Magnesium (Mg) 0.005%
Sodium (Na) 0.02%

TESTS

Assay. (By acid–base titrimetry). Weigh, to the nearest 0.1 mg, about 0.4 g of sample, and dissolve in 50 mL of water. Pass the solution through the cation-exchange column (see page 29) at a rate of about 5 mL/min, and collect the eluate in a 500-mL titration flask. Then, wash the resin in the column with water at a rate of about 10 mL/min, and collect in the same titration flask. Add 0.15 mL of phenolpthalein indicator solution to the flask, and titrate with 0.1 N sodium hydroxide. Continue the titration as the washing proceeds until

50 mL requires no further titration. One milliliter of 0.1 N sodium hydroxide corresponds to 0.008713 g of K_2SO_4.

$$\% \, K_2SO_4 = \frac{(mL \times N \, NaOH) \times 8.713}{Sample \, wt \, (g)}$$

pH of a 5% Solution at 25.0 °C. (Page 49).

Insoluble Matter. (Page 25). Use 10 g dissolved in 150 mL of water.

Chloride. (Page 35). Use 1.0 g.

Nitrogen Compounds. (Page 39). Use 2.0 g. For the standard, use 0.01 mg of nitrogen ion (N).

Heavy Metals. (Page 36, Method 1). Dissolve 6.0 g in water, and dilute with water to 54 mL. Use 45 mL to prepare the sample solution, and use the remaining 9.0 mL to prepare the control solution.

Iron. (Page 38, Method 1). Use 2.0 g.

Calcium, Magnesium, and Sodium. (By flame AAS, page 63).

> **Sample Stock Solution.** Dissolve 2.0 g of sample with water in a 100-mL volumetric flask, and dilute to the mark with water (1 mL = 0.02 g).

Element	Wavelength (nm)	Sample Wt (g)	Standard Added (mg)	Flame Type*	Background Correction
Ca	422.7	0.20	0.02; 0.04	N/A	No
Mg	285.2	0.20	0.01; 0.02	A/A	Yes
Na	589.0	0.10	0.01; 0.02	A/A	No

*A/A is air/acetylene; N/A is nitrous oxide/acetylene.

Potassium Thiocyanate
KSCN   **Formula Wt 97.18**   **CAS No. 333-20-0**

GENERAL DESCRIPTION
Typical appearance: colorless or white solid
Analytical use: analysis of silver ion; indirect determination of chloride, bromide, and iodide
Change in state (approximate): melting point, 173°C
Aqueous solubility: 217 g in 100 mL at 20 °C

SPECIFICATIONS
Assay . . . . ≥98.5% KSCN
pH of a 5% solution at 25.0 °C 5.3–8.7

	Maximum Allowable
Insoluble in water	0.005%
Chloride (Cl)	0.005%
Sulfate (SO$_4$)	0.005%
Ammonium (NH$_4$)	0.003%
Heavy metals (as Pb)	5 ppm
Iron (Fe)	2 ppm
Sodium (Na)	0.005%
Iodine-consuming substances	Passes test

TESTS

Caution: Cyanide gases are given off during ignition and with addition of acids. Use a well-ventilated hood.

Assay. (By argentimetric titration of thiocyanate content). Weigh, to the nearest 0.1 mg, about 7 g of sample, dissolve with 100 mL of water in a 1-L volumetric flask, dilute to the mark with water, and mix thoroughly. Transfer a 50.0-mL aliquot to a 250-mL glass-stoppered iodine-type flask, add 5 mL of 1 N nitric acid, then add with agitation exactly 50.0 mL of 0.1 N silver nitrate volumetric solution, and shake vigorously. Add 2 mL of ferric ammonium sulfate indicator solution, and chill in an ice bath to approximately 10 °C or lower. Titrate the excess silver nitrate with 0.1 N ammonium thiocyanate volumetric solution. Near the end point, shake after the addition of each drop. One milliliter of 0.1 N silver nitrate corresponds to 0.009718 g of KSCN.

$$\% \, KSCN = \frac{[(50.0 \times N \, AgNO_3) - (mL \times N \, NH_4SCN)] \times 9.718}{Sample \ wt \ (g) \ / \ 20}$$

pH of a 5% Solution at 25.0 °C. (Page 49).

Insoluble in Water. Dissolve 20 g in 200 mL of water, heat to boiling, and digest in a covered beaker on a hot plate ($\approx$100 °C) for 1 h. Filter through a tared, preconditioned filtering crucible, wash thoroughly, and dry at 105 °C.

Chloride. Dissolve 1.0 g in 20 mL of water, filter if necessary through a chloride-free filter, and add 10 mL of 25% sulfuric acid reagent solution and 7 mL of 30% hydrogen peroxide. Evaporate to 20 mL by boiling in a well-ventilated hood, add 15–20 mL of water, and evaporate again. Repeat until all the cyanide has been volatilized, cool, and dilute with water to 100 mL. To 20 mL of this solution, add 1 mL of nitric acid and 1 mL of silver nitrate reagent solution. Any turbidity should not exceed that produced by 0.01 mg of chloride ion (Cl) in an equal volume of solution containing the quantities of reagents used in the test.

Sulfate. (Page 40, Method 1). Allow 30 min for turbidity to form.

Ammonium. (By colorimetry, page 32). Dilute the distillate from a 1.0-g sample to 100 mL; 50 mL of the dilution should not show more than 0.015 mg of ammonium ion (NH$_4$).

Heavy Metals. (Page 36, Method 1). Dissolve 6.0 g in about 20 mL of water, and dilute with water to 30 mL. Use 25 mL to prepare the sample solution, and use the remaining 5.0 mL to prepare the control solution.

Iron. (Page 38, Method 2). Dissolve 10 g in 60 mL of water. To 30 mL of the solution, add 0.1 mL of hydrochloric acid, and continue as described.

Sodium. (By flame AAS, page 63).

> **Sample Stock Solution.** Dissolve 1.0 g of sample with water in a 100-mL volumetric flask, and dilute to the mark with water (1 mL = 0.01 g).

Element	Wavelength (nm)	Sample Wt (g)	Standard Added (mg)	Flame Type*	Background Correction
Na	589.0	0.10	0.005; 0.01	A/A	No

*A/A is air/acetylene.

Iodine-Consuming Substances. Dissolve 5.0 g in 50 mL of water and add 1.7 mL of 10% sulfuric acid reagent solution. Add 1 g of potassium iodide and 1 mL of starch indicator solution, and titrate with 0.1 N iodine volumetric solution. Not more than 1.0 mL of 0.1 N iodine solution should be required.

1,2-Propanediol
Propylene Glycol

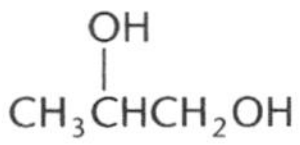

$$CH_3\overset{\displaystyle OH}{\overset{|}{C}}HCH_2OH$$

CH₃CHOHCH₂OH   **Formula Wt 76.09**   **CAS No. 57-55-6**

GENERAL DESCRIPTION

Typical appearance: viscous liquid
Analytical use: solvent
Change in state (approximate): boiling point, 188 °C
Aqueous solubility: miscible
Density: 1.04

SPECIFICATIONS

Assay . . . . ≥99.5% CH₃CHOHCH₂OH

Maximum Allowable

Color (APHA) 10
Residue after ignition 0.005%
Titrable acid . . . . .0.0005 meq/g
Chloride (Cl) . . . . .1 ppm
Water (H₂O) 0.2%

TESTS

Assay. Analyze the sample by gas chromatography using the general parameters cited on page 80. The following specific conditions are also required.

Column: Type III, polyethylene glycol

Measure the area under all peaks, and calculate the 1,2-propanediol content in area percent. Correct for water content.

Color (APHA). (Page 43).

Residue after Ignition. (Page 26). Use 50 g (48.5 mL).

Titrable Acid. Place 60 g (58 mL) in a 250 mL conical flask. Add 0.15 mL of phenol red indicator solution, and titrate with 0.01 N sodium hydroxide to a pink color. Not more than 3.0 mL should be required.

Chloride. (Page 35). Use 10 g (9.7 mL).

Water. (Page 31, Method 1). Use 20 mL (19.4 g).

Propionic Acid
Propanoic Acid
C₃H₆O₂ — $C_3H_6O_2$ **Formula Wt 74.08** **CAS No. 79-09-4**

GENERAL DESCRIPTION

Typical appearance: liquid
Analytical use: esterifying agent
Change in state (approximate): boiling point, 141 °C
Aqueous solubility: miscible
Density: 0.99

SPECIFICATIONS

Assay . . . . ≥99.5% CH_3CH_2COOH

Maximum Allowable
Color (APHA). 20
Residue after evaporation 0.01%
Readily oxidizable substances (as HCOOH) 0.10%
Heavy metals (as Pb) 0.001%
Carbonyl compounds [formaldehyde, acetone, or
    acetaldehyde plus propionaldehyde (as the latter)] 0.002%
Water (H₂O). 0.15%

TESTS

Assay. (By acid–base titrimetry). Weigh accurately about 3.0 g, and dissolve in 100 mL of water. Add 0.15 mL of phenolphthalein indicator solution, and titrate with 1 N sodium hydroxide volumetric solution to the first appearance of a faint pink end point that persists for at least 30 s. One milliliter of 1 N sodium hydroxide corresponds to 0.07408 g of CH_3CH_2COOH.

$$\% \ CH_3CH_2COOH = \frac{(mL \times N \ NaOH) \times 7.408}{Sample \ wt \ (g)}$$

Color (APHA). (Page 43).

Residue after Evaporation. (Page 25). Evaporate 20 g (20 mL) to dryness in a tared, preconditioned dish on a hot plate ($\approx$100 °C), and dry the residue at 105 °C for 30 min.

Readily Oxidizable Substances. Dissolve 7.5 g of sodium hydroxide in 50 mL of water, cool, add 3 mL of bromine water, stir until dissolved, and dilute to 1 L with water. Transfer 25.0 mL of this solution to a glass-stoppered conical flask containing 100 mL of water, and add 10 mL of a 1-in-5 solution of sodium acetate trihydrate and 10.0 mL of the sample. Allow to stand for 15 min, then add 1.0 g of potassium iodide dissolved in 5 mL of water, mix, and add 10 mL of hydrochloric acid. Titrate with 0.1 N sodium thiosulfate volumetric solution just to the disappearance of the brown color. Perform a blank determination. One milliliter of 0.1 N sodium thiosulfate corresponds to 0.004603 g of formic acid ($HCOOH$). The difference between the blank and sample titrations should not exceed 2.2 mL.

$$\% \ HCOOH = \frac{\{[mL \ (sample) - mL \ (blank)] \times N \ Na_2S_2O_3\} \times 4.603}{Sample \ wt \ (g)}$$

Heavy Metals. (Page 36, Method 1). Add about 25 mg of sodium carbonate to 3.0 g of the sample in a platinum dish, and evaporate to dryness on a hot plate ($\approx$100 °C). Dissolve the residue in 1 mL of dilute hydrochloric acid (1 + 19), and dilute with water to 30 mL. Use 25 mL to prepare the sample solution, and use the remaining 5 mL to prepare the control solution.

Carbonyl Compounds. (Page 54). Use 1.0-g (1.0-mL) samples diluted with 4 mL of water and neutralized to a pH between 6.5 and 6.8 with 10% sodium hydroxide reagent solution. For the standards, use 0.02 mg each of formaldehyde, propionaldehyde, and acetone. The peak potentials for acetaldehyde and propionaldehyde overlap at about −1.15 to −1.25 V.

Water. (Page 31, Method 1). Use 10.0 mL (9.9 g) of the sample.

n-Propyl Alcohol
1-Propanol
CH₃(CH₂)₂OH   **Formula Wt 60.10**   **CAS No. 71-23-8**

GENERAL DESCRIPTION
Typical appearance: clear liquid
Analytical use: solvent for resins and cellulose esters
Change in state (approximate): boiling point, 97 °C
Aqueous solubility: miscible
Density: 0.81

SPECIFICATIONS
Assay . . . . ≥99.5% CH₃(CH₂)₂OH
Solubility in water . . . . Passes test

Maximum Allowable

Color (APHA) 10
Residue after evaporation 0.001%
Titrable acid . . . . 0.0004 meq/g
Carbonyl compounds (as C₂H₅CHO) 0.03%
Ethyl alcohol (CH₃CH₂OH) 0.01%
Methanol (CH₃OH) 0.01%
Isopropyl alcohol (CH₃CHOHCH₃) 0.05%
Water (H₂O) 0.2%

TESTS

Assay, Ethyl Alcohol, Methanol, and Isopropyl Alcohol. Analyze the sample by gas chromatography using the general parameters cited on page 80. The following specific conditions are also required.

   Column: Type I, methyl silicone

   Detector: Flame ionization

Measure the area under all peaks, and calculate the area percent for ethyl alcohol, methanol, isopropyl alcohol, and *n*-propyl alcohol. Correct for water content.

Solubility in Water. Mix 10 mL with 40 mL of water, and allow to stand for 1 h. The solution should be as clear as an equal volume of water.

Color (APHA). (Page 43).

Residue after Evaporation. (Page 25). Evaporate 100 g (124 mL) to dryness in a tared, preconditioned dish on a hot plate (≈100 °C), and dry the residue at 105 °C for 30 min.

Titrable Acid. Add 10 mL of sample to 30 mL of carbon dioxide-free water in a glass-stoppered flask, and add 0.5 mL of phenolphthalein indicator solution. Add 0.01 N sodium hydroxide until a slight pink color persists after shaking for 30 s. Add 25 g (31 mL) of the sample, mix well, and titrate with 0.01 N sodium hydroxide until the pink color is restored. Not more than 1.0 mL of the sodium hydroxide solution should be required.

Carbonyl Compounds. (Page 54). Use 1.0 g (1.25 mL). For the standard, use 0.30 mg of propionaldehyde.

Water. (Page 31, Method 1). Use 10.0 mL (8.0 g) of the sample.

Pyridine

C₅H₅N C_5H_5N

Formula Wt 79.10

CAS No. 110-86-1

GENERAL DESCRIPTION
Typical appearance: clear liquid
Analytical use: solvent; complexing agent
Change in state (approximate): boiling point, 116 °C
Aqueous solubility: miscible
Density: 0.98

SPECIFICATIONS
Assay . . . . ≥99.0% C_5H_5N
Solubility in water . . . . Passes test

Maximum Allowable

Residue after evaporation 0.002%
Water (H₂O) 0.1%
Chloride (Cl). 0.001%
Sulfate (SO₄) 0.001%
Ammonia (NH₃) 0.002%
Copper (Cu) . . . . 5 ppm
Reducing substances . . . . Passes test

TESTS

Assay. Analyze the sample by gas chromatography using the general parameters cited on page 80. The following specific conditions are also required.

Column: Type I, methyl silicone

Measure the area under all peaks, and calculate the pyridine content in area percent. Correct for water content.

Solubility in Water. Dilute 10 mL with 90 mL of water. The solution should show no turbidity in 30 min.

Residue after Evaporation. (Page 25). Evaporate 50 g (51 mL) to dryness in a preconditioned dish on a hot plate ($\approx$100 °C), and dry the residue at 105 °C for 30 min. Save residue after weighing for the test for copper.

Water. (Page 31, Method 1). Use 20 mL (19.6 g) of the sample.

Chloride. (Page 35). Dilute 1.0 g (1.0 mL) with water to 25 mL.

Sulfate. (Page 41, Method 3).

Ammonia. Add 2 g (2.0 mL) to 10 mL of carbon dioxide-free water. Add 0.10 mL of phenolphthalein indicator solution. If a pink color develops, it should be discharged by not more than 0.24 mL of 0.01 N hydrochloric acid.

Copper. (By flame AAS, page 63). Add 5 mL of nitric acid to the residue from the test for residue after evaporation, and digest on a hot plate ($\approx$100 °C) for about 5 min. Cool, and add 25 mL of water. Transfer to a 50-mL volumetric flask, dilute to the mark with water, and mix (1 mL = 1.0 g).

Element	Wavelength (nm)	Sample Wt (g)	Standard Added (mg)	Flame Type*	Background Correction
Cu	324.8	5.0	0.025; 0.05	A/A	Yes

*A/A is air/acetylene.

Reducing Substances. To 5 g (5.0 mL), add 0.50 mL of 0.1 N potassium permanganate volumetric solution. The pink color should not be entirely discharged in 30 min. If a brown color interferes, the solution should be centrifuged or filtered through a sintered glass filter. If a pink color persists, the sample passes.

Pyrogallol
1,2,3-Trihydroxybenzene; 1,2,3-Benzenetriol

$C_6H_3(OH)_3$ Formula Wt 126.11   CAS No. 87-66-1

GENERAL DESCRIPTION

Typical appearance: white solid; may become grayish on exposure to light and air
Analytical use: complexing agent; reducing agent; alkaline solution indicator for gaseous oxygen
Change in state (approximate): melting point, 132 °C, sublimes when heated slowly
Aqueous solubility: 40 g in 100 mL

SPECIFICATIONS

Melting point . . . . 131.0–135.0 °C

Maximum Allowable

Residue after ignition. 0.005%
Chloride (Cl). 0.001%
Sulfate (SO$_4$) 0.005%
Heavy metals (as Pb). . . . . 5 ppm
Iron (Fe). 0.001%

TESTS

Melting Point. (Page 45).

Residue after Ignition. (Page 26). Ignite 20 g, and use 1 mL of sulfuric acid. Retain the residue for the test for iron.

Chloride. (Page 35). Dissolve 1.0 g in 20 mL of water, and add 3 mL of 10% nitric acid.

Sulfate. (Page 40, Method 1).

Heavy Metals. (Page 36, Method 2). Use 4.0 g.

Iron. (Page 38, Method 1). For the sample solution, take the residue retained from the test for residue after ignition, add 3 mL of hydrochloric acid and 1 mL of nitric acid, and evaporate to dryness in a dish on a hot plate ($\approx$100 °C). Take up the residue in 10 mL of water and 1 mL of dilute hydrochloric acid (1 + 19), filter if necessary, and wash to a volume of 40 mL (1 mL = 0.5 g). Dilute 20 mL of sample solution with water to 100 mL, and use 10 mL.

8-Quinolinol
8-Hydroxyquinoline; Oxine

C$_9$H$_7$NO   **Formula Wt 145.16**   **CAS No. 148-24-3**

GENERAL DESCRIPTION

Typical appearance: white solid
Analytical use: chelating agent for trace metal analysis
Change in state (approximate): melting point, 73–76 °C
Aqueous solubility: insoluble

SPECIFICATIONS

Melting point . . . . 72.5–74.0 °C
Suitability for magnesium determination. . . . . Passes test

	Maximum Allowable
Insoluble in alcohol	0.05%
Residue after ignition	0.05%
Sulfate (SO_4)	0.02%

TESTS

Melting Point. (Page 45).

Suitability for Magnesium Determination. Dissolve 2.5 g in 5 mL of glacial acetic acid, warming if necessary, and dilute with water to 100 mL. Add 3.5 mL of this solution to 50 mL of a solution containing 6 mg of magnesium ion (Mg). Heat to 80 °C, add with stirring 2 mL of ammonium hydroxide, allow to stand for 10 min, and filter. To the filtrate, which should be alkaline and yellow in color, add a solution containing 3 mg of magnesium ion (Mg), and heat to 80 °C. The characteristic yellow magnesium quinolate precipitate should form.

Insoluble in Alcohol. Dissolve 3.0 g in 40 mL of alcohol, filter through a tared, preconditioned filtering crucible, wash with 95% ethyl alcohol, and dry at 105 °C.

Residue after Ignition. (Page 26). Ignite 2.0 g. Use 1 mL of sulfuric acid.

Sulfate. Dissolve 2.0 g in 15 mL of methanol and 5 mL of hydrochloric acid (1 + 3). For control, dissolve 1.0 g in 15 mL of methanol and 5 mL of hydrochloric acid (1 + 3), and add 0.20 mg of sulfate ion (SO_4). Dilute sample and control to 40 mL with water, and add 1 mL of 12% barium chloride reagent solution to each. After 10 min, turbidity of the sample solution should not exceed that of the control solution.

Reagent Alcohol
Alcohol, Reagent

Note: This material is a denatured form of ethyl alcohol, approved for sale under U.S. regulations governing the stated description or equivalent, consisting of about 5 volumes of isopropyl alcohol and about 95 volumes of formula 3-A specially denatured alcohol (which consists of about 5 volumes of methanol and about 100 volumes of ethyl alcohol).

GENERAL DESCRIPTION

Typical appearance: colorless liquid
Analytical use: solvent
Density: 0.79

SPECIFICATIONS

Assay	Within 94.0–96.0% (v/v) methanol and ethyl alcohol and 4.0–6.0% (v/v) isopropyl alcohol

Maximum Allowable

Water (H$_2$O) 0.5%

Color (APHA) 10

Residue after evaporation 0.001%

TESTS

Assay. Analyze the sample by gas chromatography using the general parameters cited on page 80. The following specific conditions are also required.

Column: Type I, methyl silicone

Measure the area under all peaks, and calculate the ethyl alcohol plus methanol and isopropyl alcohol content, correcting for response factors determined by injections of suitably prepared standards. Correct for water content.

Water. (Page 31, Method 1). Use 10 mL (7.4 g) of the sample.

Color (APHA). (Page 43).

Residue after Evaporation. (Page 25). Evaporate 100 g (124 mL) to dryness in a tared, preconditioned dish on a hot plate ($\approx$100 °C), and dry the residue at 105 °C for 30 min.

Reinecke Salt

Ammonium Tetra(thiocyanato)diamminechromate(III) Monohydrate;
Ammonium Diamminetetrakis(thiocyanato-*N*)chromate(1–)

NH$_4$[Cr(SCN)$_4$(NH$_3$)$_2$] · H$_2$O   **Formula Wt 354.44**   **CAS No. 13573-16-5**

> *Caution:* Aqueous solutions of Reinecke salt decompose slowly with the evolution of hydrogen cyanide gas. Decomposition is rapid above 65 °C.

GENERAL DESCRIPTION

Typical appearance: dark red solid

Analytical use: precipitant for primary and secondary amines, proline, hydroxyproline, and certain amino acids

Aqueous solubility: soluble in hot water; sparingly in cold

SPECIFICATIONS

Assay . . . . $\geq$93.0% NH$_4$[Cr(SCN)$_4$(NH$_3$)$_2$] · H$_2$O

Insoluble in dilute hydrochloric acid . . . . $\leq$0.05%

Sensitivity . . . . Passes test

TESTS

Assay. (By gravimetric determination as mercury salt). Weigh accurately about 0.5 g, and dissolve in 100 mL of water containing 0.5 mL of hydrochloric acid. Add a hot solution containing 0.5 g of mercuric acetate dissolved in a mixture of 140 mL of water and 10 mL of hydrochloric acid. Digest on a hot plate ($\approx$100 °C) for 5 min, filter through a tared, pre-conditioned porcelain filtering crucible, and wash several times with hot water. The precipitate can be dried at 105 °C and weighed as $Hg[Cr(SCN)_4(NH_3)_2]_2$:

$$\% \text{ Reinecke salt} = \frac{\text{Precipitate wt (g)} \times 85.67}{\text{Sample wt (g)}}$$

Alternatively, the precipitate can be ignited in a hood and heated to constant weight at 600 °C as Cr_2O_3:

$$\% \text{ Reinecke salt} = \frac{\text{Wt of } Cr_2O_3 \text{ (g)} \times 466.3}{\text{Sample wt (g)}}$$

Insoluble in Dilute Hydrochloric Acid. Dissolve 2.0 g in 200 mL of dilute hydrochloric acid (1 + 99), heat to boiling, and digest in a covered beaker on a hot plate ($\approx$100 °C) for 1 h. Filter through a tared filtering crucible, wash thoroughly with water, and dry at 105 °C.

Sensitivity. Dissolve 50 mg in 10 mL of water. Add 0.2 mL of this solution to 1 mL of a solution of 10 mg of choline chloride in 20 mL of water, and shake gently. A distinct red precipitate should form within 10 s.

Salicylic Acid
2-Hydroxybenzoic Acid

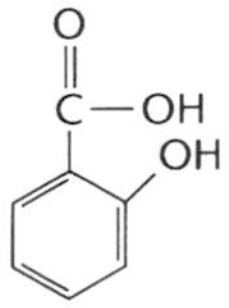

$C_7H_6O_3$ **Formula Wt 138.12**   **CAS No. 69-72-7**

GENERAL DESCRIPTION

Typical appearance: colorless or white solid

Analytical use: buffers

Change in state (approximate): sublimation temperature, 76 °C

Aqueous solubility: 0.2 g in 100 mL at 20 °C

pK_a: 3.0

SPECIFICATIONS

Assay . . . . $\geq$99.0% $C_7H_6O_3$

Melting point . . . . 158.0–161.0 °C

Maximum Allowable

Residue after ignition. 0.01%	
Chloride (Cl). 0.001%	
Sulfate (SO_4) 0.003%	
Heavy metals (as Pb). . . . . .5 ppm	
Iron (Fe). . . . . .2 ppm	
Substances darkened by sulfuric acid . . . . .Passes test	

TESTS

Assay. Analyze the sample by liquid chromatography using the general method described on page 84. The parameters cited have given satisfactory results.

Weigh 0.5 g of sample in a vial, and dissolve in 5 mL of acetonitrile plus 5 mL of mobile phase. Inject duplicate 10-μL aliquots into the chromatograph, and compare the peak areas with those obtained from standards under the same conditions.

Mobile Phase: Acetonitrile–0.01 M phosphoric acid (25/75) at 2 mL/min. Prepare 1 M phosphoric acid by diluting 7 mL of the concentrated (85%) acid to 100 mL with water. Mix 10 mL of this solution with 740 mL of water and 250 mL of acetonitrile. Deaerate under vacuum or with a sonic bath.

Column: Decyl (C-8 monomeric phase), 250 × 4.6 mm i.d., 5 μm, 9% C loading, endcapped.

Column Temperature: Ambient

Flow Rate: 2 mL/min

Sample Size: 10 μL

Detector: Ultraviolet at 280 nm

Approximate Retention Times (min): *p*-hydroxybenzoic   acid,   1.3;   *p*-hydroxy-isophthalic acid, 1.8; phenol, 3.0; salicylic acid, 4.1. Actual times may vary, depending upon the age or condition of the column.

Melting Point. (Page 45).

Residue after Ignition. (Page 26). Ignite 20 g in a preconditioned crucible. Use 1 mL of sulfuric acid. Retain the residue to prepare sample solution A for the determination of heavy metals and iron.

Chloride. (Page 35). Dissolve 1.0 g in 20 mL of 95% ethyl alcohol.

Sulfate. Mix 4.0 g with 1 g of sodium carbonate in a dish, add 30 mL of hot water in small portions, and evaporate to dryness. Ignite gently, taking care to protect the mixture from the flame because of the presence of sulfur compounds in the natural gas. To the residue, add 15 mL of water and 1 mL of 30% hydrogen peroxide, boil for 5 min, and add 2 mL of hydrochloric acid. Evaporate to dryness on a hot plate ($\approx$100 °C), cool, and add 10 mL of water. Filter, wash the filter with two 5-mL portions of water, and dilute the combined filtrate and washings with water to 25 mL. For the standard, evaporate 1 mL of 30% hydrogen peroxide, 2 mL of hydrochloric acid, and 1 g of sodium carbonate to dryness on the hot plate. Take up the residue in 10 mL of water, add 0.12 mg of sulfate ion (SO_4), and dilute with water to 25 mL. To both sample and standard solutions, add 0.5 mL of 1 N hydrochloric acid and 2 mL of 12% barium chloride reagent solution. Any turbidity in the solution of the sample should not exceed that in the standard. Compare 10 min after adding the barium chloride to the sample and standard solutions.

For the Determination of Heavy Metals and Iron

Sample Solution A. To the residue retained from the test for residue after ignition, add 2 mL of hydrochloric acid and 0.5 mL of nitric acid, and evaporate to dryness on a hot plate ($\approx$100 °C). Warm the residue with 1 mL of 1 N hydrochloric acid, add 30 mL of hot water, cool, and dilute with water to 40 mL (1 mL = 0.5 g).

Heavy Metals. (Page 36, Method 2). Use 8.0 mL of sample solution A (4-g sample).

Iron. (Page 38, Method 1). Use 10 mL of sample solution A (5-g sample).

Substances Darkened by Sulfuric Acid. Dissolve 0.5 g in 10 mL of sulfuric acid. The solution should have not more than a pale yellow color.

Silica Gel Desiccant

CAS No. 7631-86-9

GENERAL DESCRIPTION
Typical appearance: colorless or white solid
Analytical use: drying agent

SPECIFICATIONS
Suitability for moisture absorption. . . . . Passes test

TESTS

Suitability for Moisture Absorption. Weigh accurately about 5 g in a tared weighing dish. Place the dish for 24 h in a desiccator in which the atmosphere possesses a relative humidity of 80%, maintained by equilibrium with sulfuric acid having a specific gravity of 1.19 (27% H_2SO_4). The increase in weight should not be less than 27%.

Note: This test is adapted from Rosin, 1967.

Silicic Acid, Hydrated

$SiO_2 \cdot nH_2O$ CAS No. 1343-98-2

Note: This reagent is available in variable amounts of water up to 16%.

GENERAL DESCRIPTION

Typical appearance: white solid
Analytical use: sorbent for column chromatography
Aqueous solubility: insoluble

SPECIFICATIONS

Loss on ignition . . . . ≤16.0%

	Maximum Allowable
Chloride (Cl).	0.01%
Nonvolatile with hydrofluoric acid	0.40%
Sulfate (SO_4)	0.02%
Heavy metals (as Pb).	0.002%
Iron (Fe).	0.01%

TESTS

Loss on Ignition. (Page 25). Ignite 0.5 g, accurately weighed, gradually to a temperature of 600 ± 25°C, in a preconditioned platinum crucible. Cool in a desiccator, weigh, and calculate the percent loss on ignition. Retain the crucible and its contents for the nonvolatile with hydrofluoric acid test.

Chloride. (Page 35). Digest 1.0 g with 20 mL of water and 1 mL of nitric acid for 5 min. Filter through a chloride-free filter, wash the filter, and dilute with water to 100 mL. To 20 mL of this solution, add 1 mL of nitric acid and 1 mL of silver nitrate reagent solution. Any turbidity should not exceed that produced by 0.02 mg of chloride ion in an equal volume of solution containing the quantities of reagents used in the test.

Nonvolatile with Hydrofluoric Acid. To the crucible containing the residue from loss on ignition, add 0.5 mL of sulfuric acid and 10 g (8.8 mL) of hydrofluoric acid, and evaporate as far as possible on a hot plate (≈100 °C) in a fume hood. Heat gently to volatilize the excess sulfuric acid, and ignite at 600 °C for 1 h. Cool in a desiccator, weigh, and calculate the percent nonvolatile with hydrofluoric acid, based on the sample weight in the loss on ignition test.

$$\% \text{ Nonvolatile with HF} = \frac{\text{Residue wt (g)} \times 100}{\text{Sample wt (g)}}$$

For the Determination of Sulfate, Heavy Metals, and Iron

Sample Solution A. Weigh 5.0 g of the sample, and boil with a mixture of 8 mL of hydrochloric acid and 40 mL of water. Filter the residue, and wash with 10 mL of water. To the filtrate, add 20 mg of sodium carbonate, anhydrous, and evaporate to dryness on a hot plate ($\approx$100 °C). Dissolve the residue in 40 mL of water containing 0.05 mL of 10% hydrochloric acid reagent solution. Filter if necessary, and dilute with water to 50 mL (1 mL = 0.1 g).

Sulfate. (Page 40, Method 1). For the test, use 5.0 mL (0.50 g) of sample solution A, and dilute with water to 20 mL. For the standard, use 0.10 mg of sulfate ion (SO_4), and dilute with water to 20 mL.

Heavy Metals. (Page 36, Method 1). Use 10 mL (1.0 g) of sample solution A.

Iron. (Page 38, Method 1). Use 1.0 mL (0.10 g) of sample solution A and 3 mL of hydrochloric acid.

Silver Diethyldithiocarbamate

(C_2H_5)$_2$NCS$_2$Ag   Formula Wt 256.14   CAS No. 1470-61-7

Note: To enhance stability, storage below 8 °C is recommended.

GENERAL DESCRIPTION

Typical appearance: greenish-yellow solid
Analytical use: determination of arsenic
Change in state (approximate): melting point, 175 °C
Aqueous solubility: insoluble

SPECIFICATIONS

Solubility in pyridine . . . . Passes test
Suitability for determination of arsenic. . . . . Passes test

TESTS

Solubility in Pyridine. Transfer 1 g of sample to a 200-mL volumetric flask, and dilute to the mark with freshly distilled pyridine. The solution should be clear, bright yellow, and dissolution should be complete. (Save this solution for the suitability for determination of arsenic test.)

Suitability for Determination of Arsenic. Transfer 0.002 mg of arsenic ion (As) to an arsine generator flask and dilute with water to 35 mL. (For a description of the apparatus used in this test, see page 34.) Add 20 mL of dilute sulfuric acid (1 + 4), 2 mL of 16.5% potassium iodide reagent solution, and 0.5 mL of 40% stannous chloride dihydrate reagent solution in concentrated hydrochloric acid. Mix, and allow to stand for 30 min at room temperature. Pack the scrubber tube loosely with two pledgets of lead acetate cotton. Place 3.0 mL of sample solution, retained from the solubility in pyridine test, in the absorber tube. Quickly add 3.0 g of granulated (No. 20 mesh) zinc to the arsine generator flask, and immediately connect the flask to the scrubber–absorber assembly. Allow the evolution of hydrogen to proceed at room temperature for 45 min, swirling the solution every 10 min. Disconnect the absorber tube, and determine the absorbance of the solution in a 1.00-cm cell at 525 nm against a fresh portion of the sample solution in a similar matched cell set at zero absorbance as the reference liquid. The absorbance should not be less than 0.10.

Silver Nitrate
AgNO₃   **Formula Wt 169.87**   **CAS No. 7761-88-8**

GENERAL DESCRIPTION

Typical appearance: colorless or white solid; may become brown in the presence of impurities or light

Analytical use: titrimetric reagent

Change in state (approximate): melting point, 212 °C

Aqueous solubility: 216 g in 100 mL at 20 °C

SPECIFICATIONS

Assay . . . . ≥99.0% AgNO₃

Clarity of solution . . . . Passes test

	Maximum Allowable
Chloride (Cl).	5 ppm
Free acid	Passes test
Substances not precipitated by hydrochloric acid.	0.01%
Sulfate (SO₄)	0.002%
Copper (Cu)	2 ppm
Iron (Fe).	2 ppm
Lead (Pb).	0.001%

TESTS

Assay. (By argentimetric titrimetry). Weigh, to the nearest 0.1 mg, about 0.65 g of sample in a 250-mL Erlenmeyer flask. Dissolve in 100 mL of water, and add 5 mL of nitric acid. Add 2 mL of ferric ammonium sulfate indicator solution, and titrate with 0.1 N potassium thiocyanate volumetric solution to the first appearance of a reddish-brown end point. One milliliter of 0.1 N potassium thiocyanate corresponds to 0.01699 g of AgNO₃.

$$\% \text{ AgNO}_3 = \frac{(\text{mL} \times \text{N KSCN}) \times 16.99}{\text{Sample wt (g)}}$$

Clarity of Solution. Dissolve 20 g in 100 mL of water in a 250-mL conical flask. The solution should be clear, and no significant amount of insoluble matter, such as minute fibers and/or other particles of any shape or color, should be observed. Reserve the solution for the test for substances not precipitated by hydrochloric acid.

Chloride. Dissolve 2.0 g in 40 mL of water, add ammonium hydroxide dropwise until the precipitate first formed is redissolved, and dilute with water to 50 mL. Transfer the solution to a 100-mL platinum dish that becomes the cathode. Insert a rotating anode, and electro-lyze for 1 h, starting with a current of 1 ampere. Decant the solution, and evaporate to approximately 25 mL. Neutralize the solution to the phenolphthalein end point with nitric acid, and add 1 mL of nitric acid in excess and 1 mL of silver nitrate reagent solution. Any turbidity should not exceed that produced by 0.01 mg of chloride ion (Cl) in an equal vol-ume of solution containing 1 mL of nitric acid and 1 mL of silver nitrate reagent solution.

Free Acid. Dissolve 5.0 g in 50 mL of water, add 0.25 mL of aqueous 0.04% bromocresol green indicator solution, and mix well. The solution should be colored blue, not green or yellow.

Substances Not Precipitated by Hydrochloric Acid. Dilute the solution obtained in the test for clarity of solution to about 600 mL. Heat to boiling, and add hydrochloric acid to precipitate the silver completely (about 11 mL). Allow to stand for at least 8 h, and filter. Evaporate the filtrate to dryness, and add 0.15 mL of hydrochloric acid and 10 mL of water. Heat, filter, and wash with about 10 mL of water. Evaporate the filtrate to dryness in a tared, preconditioned dish or crucible, and dry at 105 °C. Correct for the weight obtained in a complete blank test. Reserve the residues for the preparation of sample solution A and blank solution B.

For the Determination of Sulfate, Copper, Iron, and Lead

Sample Solution A and Blank Solution B . To each of the residues remaining from the test for substances not precipitated by hydrochloric acid, add 3 mL of dilute hydrochloric acid (1 + 1), cover with a watch glass, and digest on a hot plate ($\approx$100 °C) for 15–20 min. Cool, and dilute with water to 100 mL. One is sample solution A (1 mL = 0.2 g); the other is blank solution B.

Sulfate. (Page 40, Method 1). Use 12.5 mL of sample solution A. Use 12.5 mL of blank solution B to prepare the standard solution.

Copper. Add a slight excess of ammonium hydroxide to 25 mL of sample solution A (5-g sample). For the standard, add 0.01 mg of copper ion (Cu) to 25 mL of blank solution B, and make slightly alkaline with ammonium hydroxide. Add 10 mL of 0.1% sodium diethyldithio-carbamate reagent solution to each. Any yellow color in the solution of the sample should not exceed that in the standard. Estimate the copper content to aid in the test for lead.

Iron. (Page 38, Method 1). Use 25 mL of sample solution A (5-g sample). Add the stand-ard to 25 mL of blank solution B.

Lead. Dilute 25 mL of sample solution A (5-g sample) to 35 mL. For the standard, add 0.05 mg of lead ion (Pb) and the amount of copper estimated to be present in the test for copper to 25 mL of blank solution B, and dilute with water to 35 mL. Adjust the pH of the standard and sample solutions to between 3 and 4 (using a pH meter) with 1 N acetic acid or 10% ammonium hydroxide reagent solution, dilute with water to 40 mL, and mix. Add 10 mL of freshly prepared hydrogen sulfide water to each, and mix. Any color in the solu-tion of the sample should not exceed that in the standard.

Silver Sulfate

Ag_2SO_4 Formula Wt 311.80   CAS No. 10294-26-5

GENERAL DESCRIPTION

Typical appearance: colorless or white solid; may darken on exposure to light

Analytical use: titrimetric reagent

Change in state (approximate): melting point, 657 °C

Aqueous solubility: 0.8 g in 100 mL at 20 °C

SPECIFICATIONS

Assay . . . . ≥98.0% Ag_2SO_4

Maximum Allowable

Insoluble matter and silver chloride 0.02%

Nitrate (NO_3) 0.001%

Substances not precipitated by hydrochloric acid. 0.03%

Iron (Fe). 0.001%

TESTS

Assay. (By argentimetric titrimetry). Weigh, to the nearest 0.1 mg, about 0.5 g of sample in a 250-mL Erlenmeyer flask. Dissolve in 50 mL of water and 5 mL of nitric acid. Add 2 mL of ferric ammonium sulfate indicator solution, and titrate with 0.1 N potassium thiocyanate to the first appearance of a reddish-brown end point. One milliliter of 0.1 N potassium thiocyanate corresponds to 0.01559 g of Ag_2SO_4.

$$\% \ Ag_2SO_4 = \frac{(mL \times N \ KSCN) \times 15.59}{Sample \ wt \ (g)}$$

Insoluble Matter and Silver Chloride. Add 5.0 g of the powdered salt to 500 mL of boiling water, and boil gently until the silver sulfate is dissolved. If any insoluble matter remains, filter while hot through a tared, preconditioned filtering crucible (retain the filtrate for the test for substances not precipitated by hydrochloric acid), wash thoroughly with hot water (discard the washings), and dry at 105 °C.

Nitrate. To 0.50 g of the powdered salt, add 2 mL of phenoldisulfonic acid reagent solution, and heat on a hot plate (≈100 °C) for 15 min. Cool, dilute with 20 mL of water, filter, and make alkaline with ammonium hydroxide. Any yellow color should not exceed that produced when a solution containing 0.005 mg of nitrate ion (NO_3) is evaporated to dryness and the residue is treated in the same manner as the sample.

Substances Not Precipitated by Hydrochloric Acid. Heat to boiling the filtrate obtained in the test for insoluble matter and silver chloride, add 5 mL of hydrochloric acid, and allow to stand for at least 8 h. Dilute with water to 500 mL, and filter. Evaporate 250 mL of the filtrate to dryness, and add 0.15 mL of hydrochloric acid and 10 mL of water. Heat and filter. Add 0.10 mL of sulfuric acid to the resulting filtrate, evaporate to dryness in a tared, preconditioned dish, and ignite at 600 ± 25 °C for 15 min. Correct for the weight obtained in a complete blank test. Retain the residue for the test for iron.

Iron. (Page 38, Method 1). To the residue obtained in the test for substances not precipitated by hydrochloric acid, add 3 mL of dilute hydrochloric acid (1 + 1), cover with a watch glass, and digest on a hot plate (≈100 °C) for 15–20 min. Remove the watch glass, and evaporate to dryness. Dissolve the residue in 35 mL of dilute hydrochloric acid (1 + 6), filter if necessary, and dilute with water to 50 mL. Use 20 mL of the solution without further acidification.

Soda Lime

CAS No. 8006-28-8

Note: Soda lime is a mixture of variable proportions of sodium hydroxide with calcium oxide or hydroxide. This reagent may or may not include an indicator.

GENERAL DESCRIPTION
Typical appearance: white or gray-white solid
Analytical use: to absorb carbon dioxide

SPECIFICATIONS
Carbon dioxide absorption capacity .... ≥19.0%

Maximum Allowable

Loss on drying 7%
Fines 1%

TESTS

Carbon Dioxide Absorption Capacity. Fill the lower transverse section of a U-shaped drying tube of about 15-mm internal diameter and 15-cm height with loosely packed glass wool. Place in one arm of the tube about 5 g of anhydrous calcium chloride, and accurately weigh the tube, and its contents. Into the other arm of the tube place 9.5–10.5 g of soda lime, and again weigh accurately. Insert stoppers in the open arms of the U-tube, and connect the side tube of the arm filled with soda lime to a calcium chloride drying tube, which in turn is connected to a suitable source of carbon dioxide. Pass the carbon dioxide though the U-tube at a rate of 75 mL/min for 30 min, accurately timed. Disconnect the U-tube, cool to room temperature, remove the stoppers, and weigh.

Loss on Drying. Weigh accurately about 10 g, and dry in a preconditioned crucible at 200 °C for 18 h.

Fines. Place 100 g on a clean U.S. No. 100 standard sieve nested in a receiving pan. Cover the sieve, and shake on a mechanical shaker for 5 min. The weight of the fine material in the receiving pan should not exceed 1.0 g.

Sodium

Na   Atomic Wt 22.99   CAS No. 7440-23-5

GENERAL DESCRIPTION

Typical appearance: light, silvery-white metal; may become gray upon oxidation
Analytical use: preparation of alkoxide titrants
Change in state (approximate): melting point, 98 °C

SPECIFICATIONS

Maximum Allowable

Chloride (Cl). 0.002%
Nitrate (NO_3) 0.003%
Phosphate (PO_4). . . . . 5 ppm
Sulfate (SO_4) 0.002%
Heavy metals (as Pb). . . . . 5 ppm
Iron (Fe). 0.001%

TESTS

Sample Solution A. If the metal contains any adhering oil or other foreign material, shave off a thin layer, and use only bright clean metal for the sample. Weigh 20 g, and cut it into small pieces. Cool about 100 mL of water in a beaker in an ice bath. Add the small pieces of sodium one at a time to the ice-cold water. Keep the solution cool, and do not add another piece until the preceding one has completely reacted and dissolved. If desired, a magnetic stirrer may be used to aid dissolution. After all the sample has been dissolved, cool, and dilute with water to 500 mL (1 mL = 0.04 g).

Chloride. (Page 35). Neutralize 12.5 mL of sample solution A (0.5-g sample) with nitric acid.

Nitrate. (Page 38, Method 1). Use 12.5 mL of sample solution A (0.5-g sample). For the standard, use 0.015 mg of nitrate ion (NO_3).

Phosphate. (Page 40, Method 1). To 100 mL of sample solution A (4-g sample), add 15 mL of hydrochloric acid, and evaporate to about 30 mL. Add 30 mL of hydrochloric acid, filter through a filtering crucible, and wash the precipitated sodium chloride twice with 5-mL portions of hydrochloric acid. Evaporate the filtrate and washings to dryness on a hot plate ($\approx$100 °C). Dissolve the residue in 25 mL of approximately 0.5 N sulfuric acid, and continue as described.

Sulfate. (Page 40, Method 1). Neutralize 62.5 mL of sample solution A (2.5-g sample), with hydrochloric acid, and evaporate to about 20 mL. Add 1 mL of dilute hydrochloric acid (1 + 19).

Heavy Metals. (Page 36, Method 1). To 125 mL of sample solution A, add 25 mL of hydrochloric acid, and evaporate to dryness on a hot plate ($\approx$100 °C). Dissolve the residue in about 20 mL of water, and dilute with water to 25 mL. For the control, add 0.02 mg of lead to 25 mL of sample solution A, and treat exactly as the 125 mL of sample solution A.

Iron. (Page 38, Method 1). Neutralize 25 mL of sample solution A (1-g sample) with hydrochloric acid, add 2 mL of hydrochloric acid in excess, and dilute with water to 50 mL. Use the solution without further acidification.

Sodium Acetate
Sodium Acetate, Anhydrous
CH₃COONa   **Formula Wt 82.03**   **CAS No. 127-09-3**

GENERAL DESCRIPTION

Typical appearance: white solid
Analytical use: buffers
Change in state (approximate): melting point, 324 °C
Aqueous solubility: 46.5 g in 100 mL at 20 °C

SPECIFICATIONS

Assay . . . . $\geq$99.0% $C_2H_3O_2Na$
pH of a 5% solution at 25.0 °C 7.0–9.2

	Maximum Allowable
Insoluble matter	0.01%
Loss on drying	1.0%
Chloride (Cl)	0.002%
Phosphate (PO₄)	0.001%
Sulfate (SO₄)	0.003%
Calcium (Ca)	0.005%
Magnesium (Mg)	0.002%
Heavy metals (as Pb)	0.001%
Iron (Fe)	0.001%

TESTS

Assay. (Total alkalinity by nonaqueous titration). Weigh accurately about 0.3 g of sample into a 125-mL flask, and dissolve in 50 mL of acetic acid and 5 mL of acetic anhydride. Prepare a blank using 50 mL of acetic acid and 5 mL of acetic anhydride. Allow to stand for 15 min. Add 0.10 mL of crystal violet indicator solution to both flasks, and titrate with 0.1 N perchloric acid in glacial acetic acid volumetric solution until the solution color changes from violet to emerald green. One milliliter of 0.1 N perchloric acid corresponds to 0.008203 g of CH_3COONa.

$$\% \ CH_3COONa = \frac{\{[mL \ (sample) - mL \ (blank)] \times N \ HClO_4\} \times 8.203}{Sample \ wt \ (g)}$$

pH of a 5% Solution at 25.0 °C. (Page 49).

Insoluble Matter. (Page 25). Use 20 g dissolved in 150 mL of water.

Loss on Drying. Weigh accurately about 2 g in a tared, preconditioned weighing bottle, and dry at 120 °C to constant weight.

Chloride. (Page 35). Use 0.50 g.

Phosphate. (Page 40, Method 2). Dissolve 2.0 g in 10 mL of nitric acid, and evaporate to dryness on a hot plate ($\approx$100 °C). Add 10 mL of nitric acid, and repeat the evaporation. Dissolve in 80 mL of water, add 0.5 g of ammonium molybdate, and adjust the pH to 1.8 (using a pH meter) with dilute hydrochloric acid (1 + 9). Heat to boiling, cool, add 10 mL of hydrochloric acid, and continue as described. Carry along a standard containing 0.02 mg of phosphate ion (PO_4) treated in the same manner as the sample after the addition of 80 mL of water.

Sulfate. (Page 40, Method 1). Allow 30 min for turbidity to form.

Calcium and Magnesium. (By flame AAS, page 63).

> **Sample Stock Solution.** Dissolve 5.0 g in 80 mL of water, and transfer to a 100-mL volumetric flask. Dilute to the mark with water (1 mL = 0.05 g).

Element	Wavelength (nm)	Sample Wt (g)	Standard Added (mg)	Flame Type*	Background Correction
Ca	422.7	0.50	0.025; 0.05	N/A	No
Mg	285.2	0.50	0.01; 0.02	A/A	Yes

*A/A is air/acetylene; N/A is nitrous oxide/acetylene.

Heavy Metals. (Page 36, Method 1). Dissolve 6.0 g in about 10 mL of water, add 15 mL of dilute hydrochloric acid (10%), and dilute with water to 60 mL. Use 25 mL to prepare the sample solution, and use 5.0 mL of the remaining solution to prepare the control solution.

Iron. (Page 38, Method 1). Dissolve 1.0 g in 50 mL of dilute hydrochloric acid (1 + 24), and use the solution without further acidification.

Sodium Acetate Trihydrate

$NaC_2H_3O_2 \cdot 3H_2O$ **Formula Wt 136.08**   **CAS No. 6131-90-4**

GENERAL DESCRIPTION

Typical appearance: colorless or white solid
Analytical use: buffers
Change in state (approximate): melting point, 58 °C; dehydrates at 120 °C
Aqueous solubility: 125 g in 100 mL at 20 °C

SPECIFICATIONS

Assay . . . . 99.0–101% $NaC_2H_3O_2 \cdot 3H_2O$
pH of a 5% solution at 25.0 °C 7.5–9.2
Substances reducing permanganate . . . . Passes test

Maximum Allowable

Insoluble matter 0.005%
Chloride (Cl) 0.001%
Phosphate (PO_4) . . . . 5 ppm
Sulfate (SO_4) 0.002%
Heavy metals (as Pb) . . . . 5 ppm
Iron (Fe) . . . . 5 ppm
Calcium (Ca) 0.005%
Magnesium (Mg) 0.002%
Potassium (K) 0.005%

TESTS

Assay. (Total alkalinity by nonaqueous titration). Weigh accurately about 0.5 g of sample into a 125-mL flask, and dissolve in 50 mL of acetic acid and 5 mL of acetic anhydride. Use a second flask containing 50 mL of acetic acid and 5 mL of acetic anhydride as a blank. Add 0.10 mL of crystal violet indicator solution to both flasks, and titrate with 0.1 N perchloric acid in glacial acetic acid volumetric solution until the solution color changes from violet to emerald green. One milliliter of 0.1 N perchloric acid corresponds to 0.01361 g of $CH_3COONa \cdot 3H_2O$.

$$\% \ CH_3COONa \cdot 3H_2O = \frac{\{[mL \ (sample) - mL \ (blank)] \times N \ HClO_4\} \times 13.61}{Sample \ wt \ (g)}$$

pH of a 5% Solution at 25.0 °C. (Page 49).

Substances Reducing Permanganate. Dissolve 5.0 g in 50 mL of water, and add 5 mL of 10% sulfuric acid reagent solution and 0.10 mL of 0.1 N potassium permanganate volumetric solution. The pink color should persist for at least 1 h.

Insoluble Matter. (Page 25). Use 20 g dissolved in 150 mL of water.

Chloride. (Page 35). Use 1.0 g.

Phosphate. Dissolve 2.0 g in 10 mL of nitric acid, and evaporate to dryness on a hot plate ($\approx$100 °C). Add 10 mL of nitric acid, and repeat the evaporation. Dissolve in 80 mL of water, add 0.5 g of ammonium molybdate tetrahydrate, and adjust the pH to 1.8 (using a pH meter) with dilute hydrochloric acid (1 + 9). Heat to boiling, cool, and add 10 mL of hydrochloric acid, and continue as described. Carry along a standard containing 0.01 mg of phosphate ion (PO_4) treated in the same manner as the sample after the addition of 80 mL of water.

Sulfate. (Page 40, Method 1). Allow 30 min for turbidity to form.

Heavy Metals. (Page 36, Method 1). Dissolve 6.0 g in about 10 mL of water, add 15 mL of dilute hydrochloric acid (10%), and dilute with water to 30 mL. Use 25 mL to prepare the sample solution, and use 5.0 mL of the remaining solution to prepare the control solution.

Iron. (Page 38, Method 1). Dissolve 2.0 g in 50 mL of dilute hydrochloric acid (1 + 24), and use the solution without further acidification.

Calcium, Magnesium, and Potassium. (By flame AAS, page 63).

    Sample Stock Solution. Dissolve 5.0 g of sample in water in a 100-mL volumetric flask, and dilute to the mark with water (1 mL = 0.05 g).

Element	Wavelength (nm)	Sample Wt (g)	Standard Added (mg)	Flame Type*	Background Correction
Ca	422.7	0.50	0.025; 0.05	N/A	No
Mg	285.2	0.50	0.01; 0.02	A/A	Yes
K	766.5	0.50	0.025; 0.05	A/A	No

*A/A is air/acetylene; N/A is nitrous oxide/acetylene.

Sodium Arsenate Heptahydrate
Disodium Hydrogen Arsenate Heptahydrate
$Na_2HAsO_4 \cdot 7H_2O$ **Formula Wt 312.01**   **CAS No. 10048-95-0**

GENERAL DESCRIPTION
Typical appearance: white solid
Analytical use: source of soluble arsenic
Change in state (approximate): melting point, 57 °C
Aqueous solubility: 67 g in 100 mL at 20 °C

SPECIFICATIONS
Assay . . . . 98.0–102.0% $Na_2HAsO_4 \cdot 7H_2O$

                                                               Maximum Allowable
Insoluble matter 0.005%
Arsenite (As_2O_3) 0.01%
Chloride (Cl). 0.001%
Nitrate (NO_3) 0.005%
Sulfate (SO_4) 0.01%
Heavy metals (as Pb). 0.002%
Iron (Fe). 0.001%

TESTS

Assay. (By titration of oxidizing power of arsenate). Weigh accurately about 0.55 g, and dissolve in 50 mL of water in a glass-stoppered conical flask. Heat to 80 °C, and add 10 mL of hydrochloric acid and 3 g of potassium iodide. Stopper the flask, swirl, and maintain at 80 °C for 15 min. Cool to room temperature, and titrate the liberated iodine with 0.1 N sodium thiosulfate volumetric solution, adding 3 mL of starch indicator solution near the end of the titration. One milliliter of 0.1 N sodium thiosulfate corresponds to 0.01560 g of $Na_2HAsO_4 \cdot 7H_2O$.

$$\% \text{ Na}_2\text{HAsO}_4 \cdot 7\text{H}_2\text{O} = \frac{(\text{mL} \times \text{N Na}_2\text{S}_2\text{O}_3) \times 15.60}{\text{Sample wt (g)}}$$

Insoluble Matter. (Page 25). Use 20 g dissolved in 200 mL of water.

Arsenite. Dissolve 10 g in 75 mL of water, make the solution just acidic to litmus paper with 10% sulfuric acid, and add 2 g of sodium bicarbonate. When dissolution is complete, add starch indicator solution, and titrate with 0.02 N iodine to a blue end point. Not more than 1.0 mL of 0.02 N iodine should be required.

Chloride. (Page 35). Use 1.0 g of sample and 5 mL of nitric acid.

Nitrate. (Page 38, Method 1).

> **Sample Solution A.** Dissolve 0.20 g in 3 mL of water by heating on a hot plate ($\approx$100 °C). Dilute to 50 mL with brucine sulfate reagent solution.

> **Control Solution B.** Dissolve 0.20 g in 2 mL of water and 1 mL of nitrate ion (NO_3) standard solution by heating on a hot plate ($\approx$100 °C). Dilute to 50 mL with brucine sulfate reagent solution.

Continue with the procedure, starting with the preparation of blank solution C.

Sulfate. Dissolve 10 g in 10 mL of water, add 5 mL of hydrochloric acid, and heat the solution to boiling. Add 5 mL of 12% barium chloride reagent solution, digest in a covered beaker on a hot plate ($\approx$100 °C) for 2 h, and allow to stand for at least 8 h. If any precipitate is formed, filter, wash thoroughly, and ignite. Correct for the weight obtained in a complete blank test.

Heavy Metals. (Page 36, Method 1). Dissolve 2.5 g in 20 mL of water in a small dish. Add 2 g of potassium iodide, 10 mL of hydrobromic acid, and 0.10 mL of sulfuric acid. Evaporate to dryness on a hot plate ($\approx$100 °C), wash down the sides of the dish with a few milliliters of water, add 5 mL of hydrochloric acid, and again evaporate to dryness on a hot plate ($\approx$100 °C). Dissolve the residue in a few milliliters of water, neutralize to litmus paper with 10% ammonium hydroxide reagent solution, and dilute with water to 50 mL. Use 30 mL to prepare the sample solution, and use 10 mL of the remaining solution to prepare the control solution.

Iron. (Page 38, Method 1). Use 1.0 g of sample and 5 mL of hydrochloric acid.

Sodium Azide

NaN₃   **Formula Wt 65.01**   **CAS No. 26628-22-8**

GENERAL DESCRIPTION

Typical appearance: colorless solid
Analytical use: reagent for chromium in water; reagent for metals
Aqueous solubility: highly soluble

SPECIFICATIONS

Assay . . . . ≥99.0% NaN$_3$

Maximum Allowable

Insoluble matter 0.05%

Loss on drying 0.1%

Titrable base . . . . .0.05 meq/g

TESTS

Assay. (By oxidation–reduction titrimetry). Weigh accurately 0.25 g of sample, and dissolve in 50 mL of water. In a fume hood, add 50.0 mL of 0.1 N ceric ammonium sulfate volumetric solution with stirring (nitrogen is evolved). Add 2 drops of nitric acid and 3 drops of 0.025 M ferroin indicator solution, and rinse the sides of the flask with water. Stir for 5 min. The sample can now be removed from the hood, and the excess ceric ammonium sulfate titrated with freshly standardized 0.1 N ferrous ammonium sulfate volumetric solution. Add 0.25 mL of sulfuric acid just before the end point. The color changes from yellow to light green to an orange-red end point. Perform a blank titration on 50.0 mL of ceric ammonium sulfate. One milliliter of 0.1 N ceric ammonium sulfate consumed corresponds to 0.006501 g of NaN$_3$.

$$\% \, NaN_3 = \frac{\{[mL \, (blank) - mL \, (sample)] \times N \, Fe(NH_4)_2(SO_4)_2 \cdot 6H_2O\} \times 6.501}{Sample \, wt \, (g)}$$

Insoluble Matter. (Page 25). Dissolve 2.0 g in 200 mL of water.

Loss on Drying. (Page 26). Weigh accurately 1.0 g, and dry at 105 °C to constant weight.

Titrable Base. To 50 mL of water, add 0.15 mL of phenolphthalein indicator solution, and neutralize the solution using 0.01 N sodium hydroxide. Add 0.2 g of sample, and dissolve. If a pink color persists, not more than 1.0 mL of 0.01 N hydrochloric acid should be required to discharge the color.

Sodium Bicarbonate

Sodium Hydrogen Carbonate

NaHCO$_3$ **Formula Wt 84.01**   **CAS No. 144-55-8**

GENERAL DESCRIPTION

Typical appearance: white solid

Analytical use: buffers; pH adjustment

Aqueous solubility: soluble in 10 parts water at 25 °C

SPECIFICATIONS

Assay (dried basis). . . . . .99.7–100.3% NaHCO$_3$

Maximum Allowable

Insoluble matter 0.015%
Chloride (Cl) 0.003%
Phosphate (PO$_4$) 0.001%
Sulfur compounds (as SO$_4$) 0.003%
Ammonium (NH$_4$). . . . . 5 ppm
Heavy metals (as Pb) . . . . 5 ppm
Iron (Fe). 0.001%
Calcium (Ca) 0.02%
Magnesium (Mg). 0.005%
Potassium (K) 0.005%

TESTS

Assay. (By acid–base titrimetry). Weigh accurately about 3 g, previously dried over indicating-type silica gel desiccant for 24 h, dissolve it in 50 mL of water, add 0.1 mL of methyl orange indicator solution, and titrate with 1 N hydrochloric acid volumetric solution. The sodium bicarbonate content calculated from the total alkalinity, as determined by the titration, should not be less than 99.7% nor more than 100.3% of the weight taken. One milliliter of 1 N hydrochloric acid corresponds to 0.08401 g of NaHCO$_3$.

$$\% \text{ NaHCO}_3 = \frac{(\text{mL} \times \text{N HCl}) \times 8.401}{\text{Sample wt (g)}}$$

Insoluble Matter. (Page 25). Use 10.0 g dissolved in 100 mL of hot water.

Chloride. (Page 35). Use 0.33 g, and neutralize with nitric acid.

Phosphate. (Page 40, Method 1). Dissolve 2.0 g in 15 mL of dilute hydrochloric acid (1 + 2), and evaporate to dryness on a hot plate ($\approx$100 °C). Dissolve the residue in 25 mL of approximately 0.5 N sulfuric acid, and continue as described.

Sulfur Compounds. Dissolve 2.0 g of sample in 20 mL of water, evaporate to 5 mL, add 1 mL of bromine water, and evaporate to dryness. Cool. Dissolve in 10 mL of 10% hydrochloric acid, and evaporate to dryness. Cool. Add 5 mL of 10% hydrochloric acid, and evaporate to dryness. Cool. Dissolve with 10 mL of water, and evaporate to dryness. Cool. Dissolve in 10 mL of water, adjust to approximately pH 2 with 10% hydrochloric acid or dilute ammonium hydroxide (1 + 3). Filter through a washed filter paper, wash with two 2-mL portions of water, and dilute with water to 30 mL. Add 1 mL of 12% barium chloride reagent solution.

For the control, take 0.06 mg of sulfate ion (SO$_4$) in 15 mL of water. Add 1 mL of dilute hydrochloric acid (1 + 19), dilute with water to 30 mL, and add 1 mL of 12% barium chloride reagent solution. Compare the sample and control solutions after 30 min. Sample solution turbidity should not exceed that of the control solution.

Ammonium. Dissolve 2.0 g in 40 mL of ammonia-free water, and add 10 mL of 10% sodium hydroxide reagent solution and 2 mL of Nessler reagent. Any color should not exceed that produced by 0.01 mg of ammonium ion (NH_4) in an equal volume of solution containing the quantities of reagents used in the test.

Heavy Metals. (Page 36, Method 1). To 5.0 g in a 150-mL beaker, add 10 mL of water, mix, and cautiously add 10 mL of hydrochloric acid. Evaporate to dryness on a hot plate ($\approx$100 °C), dissolve residue in about 20 mL of water, and dilute with water to 25 mL. For the control, add 0.02 mg of lead ion (Pb) to 1.0 g of sample, and treat exactly as the 5.0 g of sample.

Iron. (Page 38, Method 1). Dissolve 1.0 g in 30 mL of dilute hydrochloric acid (1 + 9), dilute with water to 50 mL, and use the solution without further acidification.

Calcium, Magnesium, and Potassium. (By flame AAS, page 63).

Sample Stock Solution. Dissolve 4.0 g of sample with water in a 100-mL volumetric flask, and dilute to the mark with water (1 mL = 0.04 g).

Element	Wavelength (nm)	Sample Wt (g)	Standard Added (mg)	Flame Type*	Background Correction
Ca	422.7	0.20	0.04; 0.08	N/A	No
Mg	285.2	0.20	0.01; 0.02	A/A	Yes
K	766.5	0.20	0.01; 0.02	A/A	No

*A/A is air/acetylene; N/A is nitrous oxide/acetylene.

Sodium Bismuthate

NaBiO₃   **Formula Wt 279.97**   **CAS No. 12232-99-4**

GENERAL DESCRIPTION

Typical appearance: yellow to brown solid
Analytical use: oxidizer; determination of manganese in iron and steel
Aqueous solubility: insoluble in cold water; decomposes in hot water

SPECIFICATIONS

Assay . . . . ≥80.0% NaBiO₃
Oxidizing efficiency . . . . ≥99.6%

Maximum Allowable
Chloride (Cl). 0.002%
Manganese (Mn) . . . . .5 ppm

TESTS

Assay. (By titration of oxidizing power). Weigh accurately about 0.7 g, place in a flask, add 50.0 mL of 0.1 N ferrous ammonium sulfate volumetric solution, and stopper the flask. Transfer 50.0 mL of the ferrous ammonium sulfate to another flask, and stopper the flask. Allow each flask to stand for 30 min, shaking frequently, and titrate the ferrous ammonium sulfate in each with 0.1 N potassium permanganate volumetric solution. The difference in the volume of permanganate consumed in the two titrations is equivalent to the sodium bismuthate. One milliliter of 0.1 N potassium permanganate corresponds to 0.01400 g of $NaBiO_3$.

$$\% \ NaBiO_3 = \frac{\{[mL \ (blank) - mL \ (sample)] \times N \ KMnO_4\} \times 14.00}{Sample \ wt \ (g)}$$

Chloride. Add 1.0 g to 25 mL of water, heat to boiling, and keep at the boiling temperature for 10 min. Dilute with water to 50 mL, and filter through a chloride-free filter. To 25 mL of the filtrate, add 0.15 mL of 30% hydrogen peroxide to clear the solution, and then add 1 mL of nitric acid and 1 mL of silver nitrate reagent solution. Any turbidity should not exceed that produced by 0.01 mg of chloride ion (Cl) in an equal volume of solution containing the quantities of reagents used in the test.

Manganese. Dissolve 2.0 g in 35 mL of dilute nitric acid (5 + 2), heat to boiling, and boil gently for 5 min. Prepare a standard containing 0.01 mg of manganese ion (Mn) in 35 mL of dilute nitric acid (5 + 2). To each, add 5 mL of sulfuric acid, 5 mL of phosphoric acid, and 0.5 mL of sulfurous acid. Boil gently to expel oxides of nitrogen, cool the solutions to 15 °C, and add 0.5 g of sodium bismuthate to each. Allow to stand for 5 min with occasional stirring, dilute each with 25 mL of water, and filter through a filter other than paper. Any pink color in the solution of the sample should not exceed that in the standard.

Oxidizing Efficiency.

Manganese Metal for Use as Oxidimetric Standard. Assay a selected lot of commercial high-purity electrolytic manganese, previously screened through a U.S. number 10 and retained on a U.S. number 20 screen, by determining the concentration of impurities. Evaluate metals at levels below 0.03% by the spectrographic semiquantitative method, which can be done with sufficient accuracy. Evaluate metals at higher concentrations by suitable quantitative methods; determine carbon and sulfur by classical combustion methods. Determine oxygen, hydrogen, and nitrogen by vacuum fusion analysis, employing a 25-g iron bath containing 2–3 g of tin. Place the sample in a tin capsule and drop it into the bath, which is held at 1500–1550 °C. Analyze up to three samples before discarding the bath. The assay of the manganese metal for use as an oxidimetric standard should not be less than 99.8% Mn. (Because of the tendency of the metal to react with oxygen, it must be stored in a tightly sealed container after the assay has been performed.)

To 0.20 g of oxidimetric standard manganese metal in a 1-L conical flask, add 15 mL of dilute nitric acid (1 + 3), and heat cautiously until the manganese is dissolved. Add 8 mL of 70% perchloric acid, and boil gently until the acid fumes strongly and manganese dioxide begins to separate. Cool, add 5 mL of water and 25 mL of dilute nitric acid (1 + 3), and boil for several minutes to expel free chlorine. Add sufficient sulfurous acid or sodium nitrite solution to just dissolve the manganese dioxide. Boil the solution to expel completely the oxides of nitrogen. Cool to room temperature, add 225 mL of colorless, dilute nitric acid (2 + 5) and sufficient water to bring the total volume to 250 mL, and cool to 10–15 °C.

Add 7 g of sodium bismuthate (weighed to the nearest 10 mg) to the flask, agitate briskly for 1 min, dilute with 250 mL of cold water (10–15 °C), and filter immediately through a fine-porosity fritted glass filter (pretreat the frit in hot nitric acid and then wash it free of acid with hot water). The filter can be washed free of manganese more readily if not allowed to run dry during the filtering and washing. Wash the filter with cold, freshly boiled dilute nitric acid (3 + 97) until the washings are entirely colorless, and immediately treat the filtrate and washings as directed in the next paragraph.

Add 8.5 g of ferrous ammonium sulfate heptahydrate (weighed to the nearest mg) to the filtered solution of permanganic acid. Stir briskly. As soon as reduction is complete and all the salt is dissolved, add 0.01 M 1,10-phenanthroline indicator solution, and titrate the excess of ferrous ion with 0.1 N potassium permanganate volumetric solution to a clear green color that persists for at least 30 s.

Determine the manganese equivalent of the ferrous ammonium sulfate heptahydrate by titrating 1.75 g of the salt with the 0.1 N potassium permanganate volumetric solution in 500 mL of cold, dilute nitric acid that has been pretreated with 2 g of sodium bismuthate under the conditions described.

Calculate the oxidizing efficiency of the sodium bismuthate as follows:

$$\text{Oxidizing efficiency} = \frac{(A - B) \times 0.00110 \times 100}{C}$$

where A = milliliters of exactly 0.1 N potassium permanganate volumetric solution equivalent to the ferrous sulfate added; B = milliliters of exactly 0.1 N potassium permanganate required to titrate the excess ferrous ions; and C = grams of manganese used, taking into account the assay of the metal.

Sodium Bisulfite

$NaHSO_3$ (sodium bisulfite)	Formula Wt 104.06	CAS No. 7631-90-5
$Na_2S_2O_5$ (sodium metabisulfite)	Formula Wt 190.11	CAS No. 7681-57-4

Note: This reagent is usually a mixture of sodium bisulfite and sodium metabisulfite.

GENERAL DESCRIPTION

Typical appearance: white solid
Analytical use: reducing agent; convenient source of sulfur dioxide
Aqueous solubility: soluble in 3.5 parts water

SPECIFICATIONS

Assay . . . . $\geq$58.5% SO_2

	Maximum Allowable
Insoluble matter	0.005%
Chloride (Cl)	0.02%
Heavy metals (as Pb)	0.001%
Iron (Fe)	0.002%

TESTS

Assay. (Titration of reducing power). Weigh accurately about 0.47 g, and add to a mixture of 100.0 mL of 0.1 N iodine volumetric solution and 5 mL of 10% hydrochloric acid solution. Swirl gently until the sample is dissolved completely. Titrate the excess of iodine with 0.1 N sodium thiosulfate volumetric solution, adding 3 mL of starch indicator solution near the end of the titration. One milliliter of 0.1 N iodine consumed corresponds to 0.003203 g of SO_2.

$$\% \ SO_2 = \frac{[(mL \times N \ I_2) - (mL \times N \ Na_2S_2O_3)] \times 3.203}{Sample \ wt \ (g)}$$

Insoluble Matter. (Page 25). Use 20 g dissolved in 200 mL of water.

Chloride. Dissolve 0.50 g in 100 mL of water, and transfer 10 mL of the well-mixed solution to a platinum dish. Add low-chloride 10% sodium hydroxide reagent solution until the solution is slightly alkaline to litmus paper, making note of the volume of sodium hydroxide added. Prepare a standard containing 0.01 mg of chloride ion (Cl) in 10 mL of water, and add the same volume of low-chloride 10% sodium hydroxide reagent solution as was added to the sample solution. To each solution, add dropwise 2 mL of 30% hydrogen peroxide, and allow to stand at room temperature for 10 min. Evaporate the solutions to dryness on a hot plate ($\approx$100 °C), dissolve the residues in 10 mL of water, and add 1 mL of nitric acid and 1 mL of silver nitrate reagent solution to each. Any turbidity in the solution of the sample should not exceed that of the standard.

Heavy Metals. (Page 36, Method 1). Dissolve 3.0 g in a solution of 15 mL of water and 8 mL of hydrochloric acid. Evaporate to dryness on a hot plate ($\approx$100 °C), dissolve the residue in about 20 mL of water, and dilute with water to 25 mL. For the control, add 0.02 mg of lead ion (Pb) to 1.0 g of sample, and treat exactly as the 3.0 g of sample.

Iron. (Page 38, Method 1). Dissolve 1.0 g in 10 mL of water, add 2 mL of hydrochloric acid, and evaporate to dryness on a hot plate ($\approx$100 °C). Dissolve the residue in a mixture of 5 mL of water and 2 mL of hydrochloric acid, and again evaporate to dryness. Dissolve the residue in 4 mL of hydrochloric acid, dilute with water to 100 mL, and use 50 mL of the solution without further acidification.

Sodium Borate Decahydrate
Borax, Sodium Tetraborate Decahydrate
$Na_2B_4O_7 \cdot 10H_2O$ **Formula Wt 381.37**   **CAS No. 1303-96-4**

GENERAL DESCRIPTION

Typical appearance: white solid
Analytical use: buffers; complexing or masking agent

Change in state (approximate): melting point, when rapidly heated, at 75 °C; loses 5 H_2O at 100 °C, 9 H_2O at 150 °C; becomes anhydrous at 320 °C
Aqueous solubility: 1 g dissolves in 16 mL

SPECIFICATIONS

Assay ....99.5–105.0% $Na_2B_4O_7 \cdot 10H_2O$
pH of a 0.01 M solution ....9.15–9.20 at 25.0 °C

	Maximum Allowable
Insoluble matter	0.005%
Chloride (Cl)	0.001%
Phosphate (PO_4)	0.001%
Sulfate (SO_4)	0.005%
Calcium (Ca)	0.005%
Heavy metals (as Pb)	0.001%
Iron (Fe)	5 ppm

TESTS

Assay. (By acid–base titrimetry). Weigh accurately about 1 g, and dissolve in 50 mL of water. Make slightly acidic to methyl red indicator with 1 N hydrochloric acid volumetric solution, cover with a watch glass, and boil gently for 2 min to expel any carbon dioxide. Cool, and adjust to the methyl red end point (pinkish yellow) with 1 N sodium hydroxide volumetric solution. Add phenolphthalein indicator solution and 8 g of mannitol, then titrate with 1 N sodium hydroxide through the development of a yellow color to a permanent pink end point. One milliliter of 1 N sodium hydroxide corresponds to 0.09536 g of $Na_2B_4O_7 \cdot 10H_2O$. The pH adjustment may be done potentiometrically. The end points are pH 5.4 and 8.5.

$$\% \ Na_2B_4O_7 \cdot 10H_2O = \frac{(mL \times N \ NaOH) \times 9.536}{Sample \ wt \ (g)}$$

pH of a 0.01 M Solution. Dissolve 0.381 g in 100 g of carbon dioxide- and ammonia-free water (or 0.380 g in a volume of 100 mL). Standardize the pH meter and electrode at pH 9.18 at 25 °C with 0.01 M NIST SRM Sodium Tetraborate Decahydrate prepared from NIST SRM. Determine the pH by the method described on page 49.

Insoluble Matter. (Page 25). Use 20 g dissolved in 300 mL of water.

Chloride. (Page 35). Use 1.0 g.

Phosphate. (Page 40, Method 1). Dissolve 2.0 g in 10 mL of warm water, add 2 mL of hydrochloric acid, and evaporate to dryness on a hot plate ($\approx$100 °C). Dissolve in 25 mL of approximately 0.5 N sulfuric acid, filter to remove any boric acid, and use the clear filtrate. Continue as described.

Sulfate. Dissolve 8.0 g in 120 mL of warm water plus 6 mL of hydrochloric acid. Filter, and wash with 30 mL of water. Heat to boiling, add 5 mL of 12% barium chloride reagent solution, digest in a covered beaker on a hot plate ($\approx$100 °C) for 2 h, and allow to stand for at least 8 h. Heat to dissolve any boric acid that may be crystallized. If a precipitate is formed, filter, wash thoroughly, and ignite. Correct for the weight obtained in a complete blank test.

Calcium. (By flame AAS, page 63).

> **Sample Stock Solution.** Dissolve 10.0 g of sample with water in a 200-mL volumetric flask, add 10 mL of hydrochloric acid, and dilute to the mark with water (1 mL = 0.05 g).

Element	Wavelength (nm)	Sample Wt (g)	Standard Added (mg)	Flame Type*	Background Correction
Ca	422.7	1.0	0.05; 0.10	N/A	No

*N/A is nitrous oxide/acetylene.

Heavy Metals. (Page 36, Method 1). Dissolve 4.0 g in 40 mL of hot water, add 5 mL of glacial acetic acid, and dilute with water to 48 mL. Use 36 mL to prepare the sample solution, and use the remaining 12 mL to prepare the control solution.

Iron. (Page 38, Method 1). Use 2.0 g of sample and 3 mL of hydrochloric acid.

Sodium Borohydride
Sodium Tetrahydroborate

NaBH₄   **Formula Wt 37.83**   **CAS No. 16940-66-2**

GENERAL DESCRIPTION

Typical appearance: hygroscopic solid
Analytical use: reducing agent
Change in state (approximate): decomposes at 400–500 °C
Aqueous solubility: 55% (w/w) at 25 °C

SPECIFICATIONS

Assay . . . . $\geq$ 98% NaBH₄

TESTS

Assay. (By titration of reductive capacity). Weigh accurately about 0.5 g of sample and dissolve in 125 mL of dilute sodium hydroxide solution (1 + 24) in a 250-mL volumetric flask. Dilute with the sodium hydroxide solution to volume, and mix. Pipet 10.0 mL of the sample solution into a 250-mL iodine flask, add 30.0 mL of 0.3 N potassium iodate volumetric solution, and mix. Add 2 g of potassium iodide, mix, and then add 10 mL of 10% sulfuric acid. Insert the stopper in the flask, and allow to stand in the dark for 3 min. Titrate with 0.1 N sodium thiosulfate volumetric solution to a colorless end point, adding 3 mL of starch indicator solution near the end point. One milliliter of 0.3 N potassium iodate consumed corresponds to 0.00473 g of NaBH₄.

$$\% \, NaBH_4 = \frac{[(30.0 \times N \, KIO_3) - (mL \times N \, Na_2S_2O_3)] \times 0.473}{Sample \, wt \, (g) \, / \, 25}$$

Sodium Bromide

NaBr   Formula Wt 102.89   CAS No. 7647-15-6

GENERAL DESCRIPTION

Typical appearance: white or colorless solid
Analytical use: standard in ion chromatography
Change in state (approximate): melting point, 755 °C
Aqueous solubility: 90 g in 100 mL at 20 °C

SPECIFICATIONS

Assay (corrected) . . . . ≥99.0% NaBr
pH of a 5% solution at 25.0 °C 5.0–8.8

Maximum Allowable

Insoluble matter 0.005%
Bromate (BrO$_3$) 0.001%
Chloride (Cl) 0.2%
Sulfate (SO$_4$) 0.002%
Barium (Ba) 0.002%
Heavy metals (as Pb) . . . . 5 ppm
Iron (Fe) . . . . 5 ppm
Calcium (Ca) 0.002%
Magnesium (Mg) 0.001%
Potassium (K) 0.1%

TESTS

Assay. (By argentimetric titration of bromide content). Weigh, to the nearest 0.1 mg, about 0.4 g of sample. Transfer to a 250-mL titration flask, and dissolve in 25 mL of water. Add slowly, while agitating, 50.0 mL of 0.1 N silver nitrate volumetric solution, then add 3 mL of nitric acid and 10 mL of benzyl alcohol, and shake vigorously. Add 2 mL of ferric ammonium sulfate indicator solution, and titrate the excess silver nitrate with 0.1 N ammonium thiocyanate volumetric solution. One milliliter of 0.1 N silver nitrate corresponds to 0.01029 g of NaBr (uncorrected).

$$\% \text{ NaBr (uncorrected)} = \frac{[(\text{mL} \times \text{N AgNO}_3) - (\text{mL} \times \text{N NH}_4\text{SCN})] \times 10.29}{\text{Sample wt (g)}}$$

$$\% \text{ NaBr (corrected)} = [\% \text{ NaBr (uncorrected)}] - (2.90 \times \% \text{ Cl})$$

pH of a 5% Solution at 25.0 °C. (Page 49).

Insoluble Matter. (Page 25). Use 20 g dissolved in 150 mL of water.

Bromate. (Page 54). Use 10.0 g of sample in 25 mL of solution. For the standard, add 0.10 mg of bromate ion (BrO$_3$).

Chloride. Dissolve 0.50 g in 15 mL of dilute nitric acid (1 + 2) in a small flask. Add 3 mL of 30% hydrogen peroxide, and digest on a hot plate ($\approx$100 °C) until the solution is colorless. Wash down the sides of the flask with a little water, digest for an additional 15 min, cool, and dilute with water to 200 mL. Dilute 2.0 mL with water to 20 mL, and add 1 mL of nitric acid and 1 mL of silver nitrate reagent solution. Any turbidity should not exceed that produced by 0.01 mg of chloride ion (Cl) in an equal volume of solution containing the quantities of reagents used in the test.

Sulfate. (Page 40, Method 1).

Barium. For the sample, dissolve 6.0 g in 15 mL of water. For the control, dissolve 1.0 g in 15 mL of water, and add 0.1 mg of barium ion (Ba). To each solution, add 5 mL of acetic acid, 5 mL of 30% hydrogen peroxide, and 1 mL of hydrochloric acid. Digest in a covered beaker on a hot plate ($\approx$100 °C) until reaction ceases, uncover, and evaporate to dryness. Dissolve the residues in 15 mL of water, filter if necessary, and dilute with water to 23 mL. Add 2 mL of 10% potassium dichromate reagent solution, and add ammonium hydroxide until the orange color is just dissipated and the yellow color persists. Add 25 mL of methanol, stir vigorously, and allow to stand for 10 min. Any turbidity in the solution of the sample should not exceed that in the control.

Heavy Metals. Dissolve 6.0 g in about 20 mL of water, and dilute with water to 30 mL. For the control, add 0.02 mg of lead ion (Pb) to 5.0 mL of the solution, and dilute with water to 25 mL. For the sample, use the remaining 25-mL portion. Adjust the pH of the control and sample solutions to between 3 and 4 (using a pH meter) with 1 N acetic acid or 10% ammonium hydroxide reagent solution, dilute with water to 40 mL, and mix. Add 10 mL of freshly prepared hydrogen sulfide water to each, and mix. Any color in the solution of the sample should not exceed that in the control.

Iron. Dissolve 2.0 g in 40 mL of water plus 2 mL of hydrochloric acid, and dilute with water to 50 mL. Add 30–50 mg of ammonium peroxydisulfate crystals and 3 mL of 30% ammonium thiocyanate reagent solution. Any red color should not exceed that produced by 0.01 mg of iron ion (Fe) in an equal volume of solution containing the quantities of reagents used in the test.

Calcium, Magnesium, and Potassium. (By flame AAS, page 63).

Sample Stock Solution. Dissolve 5.0 g of sample with water in a 100-mL volumetric flask, and dilute to the mark with water (1 mL = 0.05 g).

Element	Wavelength (nm)	Sample Wt (g)	Standard Added (mg)	Flame Type*	Background Correction
Ca	422.7	1.0	0.02, 0.04	N/A	No
Mg	285.2	1.0	0.01; 0.02	A/A	Yes
K	766.5	0.05	0.025; 0.05	A/A	No

*A/A is air/acetylene; N/A is nitrous oxide/acetylene.

Sodium Carbonate

Sodium Carbonate, Anhydrous

Na$_2$CO$_3$ Formula Wt 105.99   CAS No. 497-19-8

GENERAL DESCRIPTION

Typical appearance: white solid

Analytical use: buffers; pH adjustment

Change in state (approximate): melting point, 851 °C

Aqueous solubility: 21.5 g in 100 mL at 20 °C

SPECIFICATIONS

Assay (dried basis). . . . . ≥99.5% Na$_2$CO$_3$

Maximum Allowable

Insoluble matter 0.01%

Loss on heating at 285 °C 1.0%

Chloride (Cl). 0.001%

Phosphate (PO$_4$). 0.001%

Silica (SiO$_2$). 0.005%

Sulfur compounds (as SO$_4$) 0.003%

Heavy metals (as Pb). . . . . 5 ppm

Iron (Fe). . . . . 5 ppm

Calcium (Ca) 0.03%

Magnesium (Mg). 0.005%

Potassium (K) 0.005%

TESTS

Assay. (By acid–base titrimetry of carbonate). Weigh, to the nearest 0.1 mg, about 2 g of the dried sample from the test for loss on heating at 285 °C. Transfer to a 125-mL glass-stoppered flask, dissolve with 50 mL of water, add 0.10 mL of methyl orange indicator solution, and titrate with 1 N hydrochloric acid volumetric solution. One milliliter of 1 N hydrochloric acid corresponds to 0.053 g of Na$_2$CO$_3$.

$$\% \, Na_2CO_3 = \frac{(mL \times N \, HCl) \times 5.300}{Sample \, wt \, (g)}$$

Insoluble Matter. (Page 25). Use 10 g dissolved in 100 mL of water.

Loss on Heating at 285 °C. Crush sample, and accurately weigh 10 g in a preconditioned, low-form weighing bottle. Heat to constant weight at 270–300 °C. Retain this sample for assay determination.

Chloride. (Page 35). Use 1.0 g of sample and 2 mL of nitric acid.

Phosphate. (Page 40, Method 2). Dissolve 1.0 g in 50 mL of water in a platinum dish, and digest on a hot plate (≈100 °C) for 30 min. Cool, neutralize with dilute sulfuric acid (1 + 19) to a pH of about 4, and dilute with water to about 75 mL. Add 0.5 g of ammonium molybdate, and adjust the pH to 1.8 (using a pH meter) with dilute hydrochloric acid

(1 + 9). Heat to boiling, cool, add 10 mL of hydrochloric acid, and dilute with water to 100 mL. Continue as described, and concurrently prepare a standard containing 0.01 mg of phosphate ion (PO_4) and 0.05 mg of silica ion (SiO_2) in about 75 mL of water treated as the 75 mL of sample solution. Reserve the aqueous phase for the determination of silica.

Silica. Add 10 mL of hydrochloric acid to the solutions reserved from the determination of phosphate, and transfer to separatory funnels. Add 40 mL of butyl alcohol, shake vigorously, and allow to separate. Draw off and discard the aqueous phase. Wash the butyl alcohol three times with 20-mL portions of dilute hydrochloric acid (1 + 99), discarding the washings each time. Dilute each butyl alcohol solution with butyl alcohol to 50 mL, take 10 mL from each, and dilute each to 50 mL with butyl alcohol. Add 0.5 mL of a freshly prepared 2% stannous chloride reagent solution. The blue color in the extract from the sample should not exceed that in the standard. If the butyl alcohol extracts are turbid, wash with 10 mL of dilute hydrochloric acid (1 + 99).

Sulfur Compounds. To prepare the sample solution, dissolve 2.0 g of sample in 20 mL of water, evaporate to 5 mL, add 1 mL of bromine water, and evaporate to dryness. Cool. Dissolve in 10 mL of 10% hydrochloric acid, and evaporate to dryness. Cool. Add 5 mL of 10% hydrochloric acid, and evaporate to dryness. Cool. Dissolve with 10 mL of water and evaporate to dryness. Cool. Dissolve in 10 mL of water, adjust the pH to approximately pH 2 with 10% hydrochloric acid or dilute ammonium hydroxide (1 + 3). Filter through a washed, fine-porosity filter paper, wash with two 2-mL portions of water, and dilute with water to 20 mL.

To prepare the standard solution, take 6 mL of sulfate ion (SO_4) standard solution, and add 1 mL of dilute hydrochloric acid (1 + 19). Dilute with water to 20 mL.

To the sample and standard solutions, add 1 mL of 12% barium chloride reagent solution and allow to stand 30 min. Any turbidity in the sample solution must not exceed that of the standard.

Heavy Metals. (Page 36, Method 1). To 5.0 g in a 150-mL beaker, add 10 mL of water, mix, and cautiously add 10 mL of hydrochloric acid. Evaporate to dryness on a hot plate ($\approx$100 °C), dissolve the residue in about 20 mL of water, and dilute with water to 25 mL. For the control, add 0.02 mg of lead ion (Pb) to 1.0 g of sample, and treat exactly as the 5.0 g of sample.

Iron. (Page 38, Method 1). Dissolve 2.0 g in 50 mL of dilute hydrochloric acid (1 + 9). Use the solution without further acidification.

Calcium, Magnesium, and Potassium. (By flame AAS, page 63).

Sample Stock Solution. Dissolve 5.0 g of sample with water in a 100-mL volumetric flask, and dilute to the mark with water (1 mL = 0.05 g).

Element	Wavelength (nm)	Sample Wt (g)	Standard Added (mg)	Flame Type*	Background Correction
Ca	422.7	0.10	0.03; 0.06	N/A	No
Mg	285.2	0.10	0.005; 0.01	A/A	Yes
K	766.5	0.20	0.01; 0.02	A/A	No

*A/A is air/acetylene; N/A is nitrous oxide/acetylene.

Sodium Carbonate, Alkalimetric Standard

Na_2CO_3 **Formula Wt 105.99**   **CAS No. 497-19-8**

Note: For use as an alkalimetric standard, this reagent should be heated at 285 °C for 2 h.

GENERAL DESCRIPTION

Typical appearance: white solid
Analytical use: primary standard material for acid–base titrimetry
Change in state (approximate): melting point, 851 °C
Aqueous solubility: 21.5 g in 100 mL at 20 °C

SPECIFICATIONS

Assay (dried basis) . . . . 99.95–100.05% Na_2CO_3

Maximum Allowable

Insoluble matter 0.01%
Loss on heating at 285 °C 1.0%
Chloride (Cl) 0.001%
Phosphate (PO_4) 0.001%
Silica (SiO_2) 0.005%
Sulfur compounds (as SO_4) 0.003%
Heavy metals (as Pb) . . . . 5 ppm
Iron (Fe) . . . . 5 ppm
Calcium (Ca) 0.02%
Magnesium (Mg) 0.004%
Potassium (K) 0.005%

TESTS

Except for assay and calcium, magnesium, and potassium, other tests are the same as for sodium carbonate, page 594.

Assay. (By acid–base titrimetry of carbonate). Transfer a quantity of NIST SRM Potassium Hydrogen Phthalate to an agate or mullite mortar and grind to approximately 100-mesh fineness. Place 5.100 ± 0.003 g of the 100-mesh material in a weighing bottle, dry at 120 °C for 2 h, and cool in a desiccator for at least 2 h. Weigh the bottle and contents accurately, transfer the contents to the titration flask, and weigh the empty bottle to obtain the exact weight of the potassium hydrogen phthalate. Place 1.300 g of the alkalimetric standard sodium carbonate in a clean, dry weighing bottle. Heat at 285 °C for 2 h, cool in a desiccator for at least 2 h, and weigh the bottle and contents accurately. Transfer the contents to a clean, dry 50-mL beaker, and weigh the empty bottle to obtain the exact weight of the sodium carbonate.

Transfer the sodium carbonate to the titration flask by pouring it through a powder funnel. Rinse the beaker and funnel thoroughly with carbon dioxide-free water, using several small portions of the water to rinse the funnel and sides of the flask. Dilute the solution with carbon dioxide-free water to 90 mL, swirl to dissolve the sample, stopper the flask, and bubble carbon dioxide-free air through the solution during all subsequent operations. Boil the solution for 15 min, cool to room temperature in an ice bath, and dilute with carbon dioxide-free water to 75–80 mL. Titrate with 0.02 N sodium hydroxide, using a combination electrode (glass–calomel or glass–silver, silver chloride) for the measurement of E in millivolts or pH. The end point of the titration is determined by the second derivative method (page 28). Calculate the % Na_2CO_3 from the following formula:

$$\% \, Na_2CO_3 = \frac{25.949 \times F}{Sample \; wt \; (g)}$$

The calculation factor, F, is obtained as follows:

$$F = (A \times B) - (0.20422 \times C \times D)$$

where A = weight, in g, of the potassium hydrogen phthalate; B = assay value of the potassium hydrogen phthalate; C = normality of the sodium hydroxide solution, as determined against NIST Standard Potassium Hydrogen Phthalate, using the method in the NIST assay certificate; and D = volume, in mL, of sodium hydroxide consumed.

Calcium, Magnesium, and Potassium.   (By flame AAS, page 63).

Sample Stock Solution.   Dissolve 5.0 g of sample with water in a 100-mL volumetric flask, and dilute to the mark with water (1 mL = 0.05 g).

Element	Wavelength (nm)	Sample Wt (g)	Standard Added (mg)	Flame Type*	Background Correction
Ca	422.7	0.10	0.02; 0.04	N/A	No
Mg	285.2	0.10	0.004; 0.008	A/A	Yes
K	766.5	0.20	0.01; 0.02	A/A	No

*A/A is air/acetylene; N/A is nitrous oxide/acetylene.

Sodium Carbonate Monohydrate

$Na_2CO_3 \cdot H_2O$ **Formula Wt 124.00**   **CAS No. 5968-11-6**

GENERAL DESCRIPTION

Typical appearance: colorless or white solid
Analytical use: buffers; pH adjustment
Change in state (approximate): becomes anhydrous at 100 °C
Aqueous solubility: soluble in 3 parts of water

SPECIFICATIONS

Assay . . . . $\geq$99.5% $Na_2CO_3 \cdot H_2O$
Loss on drying 13.0–15.0%

	Maximum Allowable
Insoluble matter	0.01%
Chloride (Cl)	0.001%
Phosphate (PO_4)	5 ppm
Silica (SiO_2)	0.005%
Sulfur compounds (as SO_4)	0.004%
Heavy metals (as Pb)	5 ppm
Iron (Fe)	5 ppm
Calcium (Ca)	0.03%
Magnesium (Mg)	0.005%
Potassium (K)	0.005%

TESTS

Assay. (By acid–base titrimetry of carbonate). Weigh, to the nearest 0.1 mg, about 2.5 g of sample. Transfer to a 125-mL glass-stoppered flask, dissolve with 50 mL of water, add 0.10 mL of methyl orange indicator solution, and titrate with 1 N hydrochloric acid volumetric solution. One milliliter of 1 N hydrochloric acid corresponds to 0.0621 g of $Na_2CO_3 \cdot H_2O$.

$$\% \ Na_2CO_3 \cdot H_2O = \frac{(mL \times N \ HCl) \times 6.21}{Sample \ wt \ (g)}$$

Loss on Drying. Weigh accurately about 1 g, and dry in a preconditioned crucible at 150 °C to constant weight.

Insoluble Matter. (Page 25). Use 10 g dissolved in 100 mL of water.

Chloride. (Page 35). Use 1.0 g of sample and 2 mL of nitric acid.

Phosphate. (Page 40, Method 2). Dissolve 2.0 g in 50 mL of water in a platinum dish, and digest on a hot plate ($\approx$100 °C) for 30 min. Cool, neutralize with dilute sulfuric acid (1 + 19) to a pH of about 4, and dilute with water to about 75 mL. Add 0.5 g of ammonium molybdate, and adjust the pH to 1.8 (using a pH meter) with dilute hydrochloric acid (1 + 9). Heat to boiling, cool, add 10 mL of hydrochloric acid, and dilute with water to 100 mL. Continue as described, and concurrently prepare a standard containing 0.01 mg of phosphate ion (PO_4) and 0.10 mg of silica (SiO_2) in about 75 mL of water treated as the 75 mL of sample solution. Reserve the aqueous phase for the determination of silica.

Silica. Add 10 mL of hydrochloric acid to the solutions reserved from the determination of phosphate, and transfer to separatory funnels. Add 40 mL of butyl alcohol, shake vigorously, and allow to separate. Draw off and discard the aqueous phase. Wash the butyl alcohol three times with 20-mL portions of dilute hydrochloric acid (1 + 99), discarding the washings each time. Dilute each butyl alcohol solution with butyl alcohol to 50 mL, take 10 mL from each, and dilute each to 50 mL with butyl alcohol. Add 0.5 mL of a freshly prepared 2% stannous chloride reagent solution. The blue color in the extract from the sample should not exceed that in the standard. If the butyl alcohol extracts are turbid, wash with 10 mL of dilute hydrochloric acid (1 + 99).

Sulfur Compounds. To prepare the sample solution, dissolve 0.5 g of sample in 20 mL of water, evaporate to 5 mL, add 1 mL of bromine water, and evaporate to dryness. Cool.

Dissolve in 10 mL of 10% hydrochloric acid, and evaporate to dryness. Cool. Add 5 mL of 10% hydrochloric acid, and evaporate to dryness. Cool. Dissolve with 10 mL of water, and evaporate to dryness. Cool. Dissolve in 10 mL of water, and adjust the pH to approximately pH 2 with 10% hydrochloric acid or dilute ammonium hydroxide (1 + 3). Filter through a washed, fine-porosity filter paper, wash with two 2-mL portions of water, and dilute with water to 20 mL.

To prepare the standard solution, take 6 mL of sulfate ion (SO_4) standard solution, and add 1 mL of dilute hydrochloric acid (1 + 19). Dilute with water to 20 mL.

To sample and standard solutions, add 1 mL of 12% barium chloride reagent solution, and allow to stand 30 min. Any turbidity in the sample solution must not exceed that of the standard solution.

Heavy Metals. (Page 36, Method 1). To 5.0 g in a 150-mL beaker, add 10 mL of water, mix, and cautiously add 10 mL of hydrochloric acid. Evaporate to dryness on a hot plate (≈ 100 °C), dissolve the residue in about 20 mL of water, and dilute with water to 25 mL. For the control, add 0.02 mg of lead ion (Pb) to 1.0 g of sample, and treat exactly as the 5.0 g of sample.

Iron. (Page 38, Method 1). Dissolve 2.0 g in 50 mL of dilute hydrochloric acid (1 + 9). Use the solution without further acidification.

Calcium, Magnesium, and Potassium. (By flame AAS, page 63).

> **Sample Stock Solution.** Dissolve 5.0 g of sample with water in a 100-mL volumetric flask, and dilute to the mark with water (1 mL = 0.05 g).

Element	Wavelength (nm)	Sample Wt (g)	Standard Added (mg)	Flame Type*	Background Correction
Ca	422.7	0.10	0.03; 0.06	N/A	No
Mg	285.2	0.10	0.005; 0.01	A/A	Yes
K	766.5	0.20	0.01; 0.02	A/A	No

*A/A is air/acetylene; N/A is nitrous oxide/acetylene.

Sodium Chlorate

NaClO₃   **Formula Wt 106.44**   CAS No. 7775-09-9

GENERAL DESCRIPTION

Typical appearance: colorless or white solid
Analytical use: oxidimetric reagent
Change in state (approximate): melting point, 248 °C
Aqueous solubility: soluble in 1 mL cold, 0.5 mL hot

SPECIFICATIONS

Assay . . . . $\geq$99.0% $NaClO_3$

Maximum Allowable

Insoluble matter 0.005%
Bromate (BrO_3) 0.015%
Chloride (Cl). 0.005%
Sulfate (SO_4) 0.001%
Heavy metals (as Pb). 0.001%
Iron (Fe). . . . . .5 ppm
Calcium (Ca) 0.002%
Magnesium (Mg) 0.002%
Potassium (K) 0.01%

TESTS

Assay. (By argentimetric titration of chloride after reduction). Weigh accurately about 0.5 g of sample, transfer to a glass-stoppered conical flask, and dissolve in 100 mL of water. Add 25 mL of fresh 6% sulfurous acid and 5 mL of 25% nitric acid, and heat for 30 min to remove the excess sulfur dioxide. Cool, add 50.0 mL of 0.1 N silver nitrate volumetric solution and 10 mL of toluene, and shake vigorously. Titrate the excess silver nitrate with 0.1 N ammonium thiocyanate volumetric solution, using ferric ammonium sulfate indicator solution. One milliliter of 0.1 N silver nitrate corresponds to 0.01064 g of $NaClO_3$.

$$\% \ NaClO_3 = \frac{[(50.0 \times N \ AgNO_3) - (mL \times N \ NH_4SCN)] \times 10.64}{Sample \ wt \ (g)}$$

Insoluble Matter. (Page 25). Use 20 g dissolved in 250 mL of water.

Bromate. (Page 54). Use 1.0 g of sample and 0.5 g of calcium chloride dihydrate in 25 mL of solution. For the standard, add 0.15 mg of bromate ion (BrO_3). Also prepare a reagent blank.

Chloride. (Page 35). Use 0.2 g. Use only nitric acid that is free from lower oxides of nitrogen in the test.

Sulfate. (Page 40, Method 1). Use 5.0 g.

Heavy Metals. (Page 36, Method 1). Dissolve 2.5 g in 20 mL of dilute hydrochloric acid (1 + 1). Evaporate the solution to dryness on a hot plate ($\approx$100 °C), add 5 mL more of dilute hydrochloric acid (1 + 1), and again evaporate to dryness. Dissolve the residue in about 20 mL of water, and dilute with water to 25 mL. For the standard–control solution, add 0.02 mg of lead ion (Pb) to 0.5 g of sample, and treat exactly as the 2.5-g sample.

Iron. (Page 38, Method 1). Dissolve 2.0 g in 20 mL of dilute hydrochloric acid (1 + 1), and evaporate to dryness on a hot plate ($\approx$100 °C). Add 5 mL of dilute hydrochloric acid (1 + 1), and again evaporate to dryness. Dissolve the residue in 50 mL of dilute hydrochloric acid (1 + 25), and use the solution without further acidification. In preparing the standard, use the residue from evaporation of 15 mL of hydrochloric acid.

Calcium, Magnesium, and Potassium. (By flame AAS, page 63).

Sample Stock Solution. Dissolve 5.0 g in 25 mL of dilute hydrochloric acid (1 + 3), and digest in a covered beaker on a hot plate (≈100 °C) until the reaction ceases. Uncover the beaker, and evaporate to dryness. Add 10 mL of dilute hydrochloric acid (1 + 3), and again evaporate to dryness. Dissolve in dilute hydrochloric acid (1 + 99), and dilute to 100 mL with dilute hydrochloric acid (1 + 99) (1 mL = 0.05 g).

Element	Wavelength (nm)	Sample Wt (g)	Standard Added (mg)	Flame Type*	Background Correction
Ca	422.7	1.0	0.02; 0.04	N/A	No
Mg	285.2	0.20	0.004; 0.008	A/A	Yes
K	766.5	0.20	0.01; 0.02	A/A	No

*A/A is air/acetylene; N/A is nitrous oxide/acetylene.

Sodium Chloride

NaCl   **Formula Wt 58.44**   **CAS No. 7647-14-5**

GENERAL DESCRIPTION

Typical appearance: white solid
Analytical use: electrolyte; buffers; matrix modification
Change in state (approximate): melting point, 800 °C
Aqueous solubility: 36 g in 100 mL at 20 °C

SPECIFICATIONS

Assay . . . . ≥99.0% NaCl
pH of a 5% solution at 25.0 °C 5.0–9.0

Maximum Allowable

Insoluble matter 0.005%
Iodide (I) 0.002%
Bromide (Br) 0.01%
Chlorate and nitrate (as NO_3) 0.003%
Phosphate (PO_4) . . . . 5 ppm
Sulfate (SO_4) 0.004%
Barium (Ba) . . . . Passes test
Heavy metals (as Pb) . . . . 5 ppm
Iron (Fe). . . . . 2 ppm
Calcium (Ca) 0.002%
Magnesium (Mg) 0.001%
Potassium (K) 0.005%

TESTS

Assay. (By argentimetric titration of chloride content). Weigh accurately about 0.25 g, and dissolve with 50 mL of water in a 250-mL glass-stoppered flask. Add 1 mL of dichlorofluorescein indicator solution. While stirring, mix, and titrate with 0.1 N silver nitrate volumetric solution. The silver chloride flocculates, and the mixture changes to a pink color. One milliliter of 0.1 N silver nitrate corresponds to 0.005844 g of NaCl.

$$\% \, NaCl = \frac{(mL \times N \, AgNO_3) \times 5.844}{Sample \, wt \, (g)}$$

pH of a 5% Solution at 25.0 °C. (Page 49).

Insoluble Matter. (Page 25). Use 20 g dissolved in 200 mL of water.

Iodide. Dissolve 11 g in 50 mL of water. Prepare a control by dissolving 1 g of the sample, 0.2 mg of iodide ion (I), and 1.0 mg of bromide ion (Br) in 50 mL of water. To each solution, in a separatory funnel, add 2 mL of hydrochloric acid and 5 mL of ferric chloride reagent solution. Allow to stand for 5 min. Add 10 mL of chloroform, shake for 1 min, allow the chloroform to settle, and draw it off. Reserve the water solution for the test for bromide. Any violet color in the chloroform extract from the solution of the sample should not exceed that in the extract from the control.

Bromide. Dissolve 0.10 g of sample in 5.0 mL of water. Add 2.0 mL of pH 4.7 phenol red indicator solution, then add 1.0 mL of chloramine-T reagent solution, and stir immediately. After 2 min, add 0.15 mL of 0.1 N sodium thiosulfate volumetric solution, stir, and dilute to 10.0 mL with water. Simultaneously, set up a control solution by using 5.0 mL of potassium bromide (KBr) standard solution and adding the same quantity of reagents, carried out similarly in the same order as the sample solution, and dilute to 10.0 mL with water. The absorbance of the sample solution at 590 nm should not exceed that of the control solution, using water as the blank.

Chlorate and Nitrate. (Page 38, Method 1).

    Sample Solution A. Dissolve 0.50 g in 3 mL of water by heating in a boiling-water bath. Dilute to 50 mL with brucine sulfate reagent solution.

    Control Solution B. Dissolve 0.50 g in 1.5 mL of water and 1.5 mL of nitrate ion (NO_3) standard solution by heating in a boiling-water bath. Dilute to 50 mL with brucine sulfate reagent solution.

Continue with the procedure, starting with the preparation of blank solution C.

Phosphate. (Page 40, Method 1). Dissolve 4.0 g in 25 mL of approximately 0.5 N sulfuric acid, and continue as described.

Sulfate. Dissolve 3.0 g of the sample in 20 mL of water, and add 1.0 mL of 10% hydrochloric acid. For the control, take 1.0 g of the sample, dissolve in 20 mL of water, and add 0.08 mg of sulfate ion (SO_4) and 1.0 mL of 10% hydrochloric acid. Dilute both solutions to 35 mL, and add 1 mL of 12% barium chloride reagent solution to each. Turbidity of the sample solution after 10 min should not exceed that of the control solution.

Barium. Dissolve 4.0 g in 20 mL of water, filter if necessary, and divide into two equal portions. To one portion, add 2 mL of 10% sulfuric acid reagent solution, and to the other, 2 mL of water. The solutions should be equally clear at the end of 2 h.

Heavy Metals. (Page 36, Method 1). Dissolve 6.0 g in about 20 mL of water, and dilute with water to 30 mL. Use 25 mL to prepare the sample solution, and use the remaining 5.0 mL to prepare the control solution.

Iron. (Page 38, Method 1). Use 5.0 g.

Calcium, Magnesium, and Potassium. (By flame AAS, page 63).

> **Sample Stock Solution.** Dissolve 10.0 g of sample in 75 mL of water, transfer to a 100-mL volumetric flask, and dilute to the mark with water (1 mL = 0.10 g).

Element	Wavelength (nm)	Sample Wt (g)	Standard Added (mg)	Flame Type*	Background Correction
Ca	422.7	1.0	0.02; 0.04	N/A	No
Mg	285.2	1.0	0.01; 0.02	A/A	Yes
K	766.5	1.0	0.025; 0.05	A/A	No

*A/A is air/acetylene; N/A is nitrous oxide/acetylene.

Sodium Citrate Dihydrate
2-Hydroxy-1,2,3-propanetricarboxylic Acid, Trisodium Salt, Dihydrate

$Na_3C_6H_5O_7 \cdot 2H_2O$ **Formula Wt 294.10**   **CAS No. 6132-04-3**

GENERAL DESCRIPTION

Typical appearance: colorless or white solid
Analytical use: sequestering agent to remove trace metals; buffers
Change in state (approximate): becomes anhydrous at 150 °C
Aqueous solubility: 57 g in 100 mL at 25 °C

SPECIFICATIONS

Assay . . . . ≥99.0% $Na_3C_6H_5O_7 \cdot 2H_2O$
pH of a 5% solution at 25.0 °C 7.0–9.0

	Maximum Allowable

Insoluble matter 0.005%
Chloride (Cl) 0.003%
Sulfate (SO₄) 0.005%
Ammonia (NH₃) 0.003%
Calcium (Ca) 0.005%
Heavy metals (as Pb) . . . . 5 ppm
Iron (Fe). . . . . 5 ppm

TESTS

Assay. (Total alkalinity by nonaqueous titration). Weigh accurately about 0.35 g, and transfer to a 250-mL beaker. Dissolve in 100 mL of glacial acetic acid, add 5 mL of acetic anyhdride, stir until dissolution is complete, and titrate with 0.1 N perchloric acid in glacial acetic acid volumetric solution, determining the end point potentiometrically. Correct for a reagent blank. One milliliter of 0.1 N perchloric acid corresponds to 0.009803 g of $Na_3C_6H_5O_7 \cdot 2H_2O$.

$$\% \ Na_3C_6H_5O_7 \cdot 2H_2O = \frac{(mL \times N \ HClO_4) \times 9.803}{Sample \ wt \ (g)}$$

pH of a 5% Solution at 25.0 °C. (Page 49).

Insoluble Matter. (Page 25). Use 20.0 g dissolved in 200 mL of water.

Chloride. (Page 35). Use 0.33 g.

Sulfate. Drive off any moisture by heating a sample in a dish on a hot plate. Then, carefully ignite 1 g in an electric muffle furnace until nearly free of carbon. Boil the residue with 10 mL of water and 0.5 mL of 30% hydrogen peroxide for 5 min. Add 5 mg of sodium carbonate and 1 mL of hydrochloric acid, and evaporate on a hot plate ($\approx$100 °C) to dryness. Dissolve the residue in 4 mL of hot water to which has been added 1 mL of dilute hydrochloric acid (1 + 19), filter through a small filter, wash with two 2-mL portions of water, and dilute the filtrate to 10 mL. Add 1 mL of 12% barium chloride reagent solution, and mix well. Any turbidity produced should not be greater than that in a standard prepared as described below and to which the 12% barium chloride reagent solution is added at the same time as it is added to the sample solution. To 10 mL of water, add 5 mg of sodium carbonate, 1 mL of hydrochloric acid, 0.5 mL of 30% hydrogen peroxide, and 0.05 mg of sulfate ion (SO_4), and evaporate on a hot plate ($\approx$100 °C) to dryness. Dissolve the residue in 9 mL of water, and add 1 mL of dilute hydrochloric acid (1 + 19). If necessary, adjust with water to the same volume as the sample solution, add 1 mL of 12% barium chloride reagent solution, and mix well.

Ammonia. Dissolve 1 g in 50 mL of ammonia-free water, and add 2 mL of Nessler reagent. Any color should not exceed that produced by 0.03 mg of ammonia ion (NH_3) in an equal volume of solution containing 2 mL of Nessler reagent.

Calcium. (By flame AAS, page 63).

>   *Sample Stock Solution.* Dissolve 10.0 g of sample with water in a 100-mL volumetric flask, add 5 mL of hydrochloric acid, and dilute to the mark with water (1 mL = 0.10 g).

Element	Wavelength (nm)	Sample Wt (g)	Standard Added (mg)	Flame Type*	Background Correction
Ca	422.7	1.0	0.05; 0.10	N/A	No

*N/A is nitrous oxide/acetylene.

Heavy Metals. (Page 36, Method 1). Dissolve 6.0 g in water, add 7.5 mL of dilute hydrochloric acid (1 + 1), and dilute with water to 42 mL. Use 35 mL to prepare the sample solution, and use the remaining 7.0 mL to prepare the control solution.

Iron. (Page 38, Method 1). Use 2.0 g.

Sodium Cobaltinitrite

Trisodium Hexakis(nitrito-*N*)cobaltate(3–); Sodium Hexanitritocobaltate(III)

$Na_3Co(NO_2)_6$ **Formula Wt 403.94** **CAS No. 13600-98-1**

GENERAL DESCRIPTION

Typical appearance: orange-yellow solid
Analytical use: detection of potassium
Aqueous solubility: very soluble

SPECIFICATIONS

Insoluble matter . . . . ≤0.02%
Suitability for determination of potassium . . . . Passes test

TESTS

Insoluble Matter. (Page 25). Use 5.0 g dissolved in 25 mL of dilute acetic acid (1 + 25), and allow to stand in a covered beaker for at least 8 h.

Suitability for Determination of Potassium. Dissolve 1.583 g of potassium chloride in water, and dilute with water to 500 mL. To 10.0 mL of this solution (20 mg of K_2O) in a 50-mL beaker, add 2 mL of 1 N nitric acid and 8 mL of water. Dissolve 5 g of the sodium cobaltinitrite in water, dilute with water to 25 mL, and filter. Cool both solutions to approximately 20 °C, and add 10 mL of the sodium cobaltinitrite solution to the potassium chloride solution. Allow to stand for 2 h, and filter through a sintered-glass crucible that has been washed with ethyl alcohol, dried at 105 °C, and weighed. Wash the precipitate with cobaltinitrite wash solution (described below), finally wash with 5–10 mL of ethyl alcohol, and dry at 105 °C. The weight of the potassium sodium cobaltinitrite precipitate $[K_2NaCo(NO_2)_6 \cdot H_2O]$ should be between 0.0945 and 0.0985 g.

> **Cobaltinitrite Wash Solution.** (0.01 N nitric acid saturated with potassium sodium cobaltinitrite). Transfer 75–100 mg of potassium sodium cobaltinitrite to a 125-mL glass-stoppered flask. Add 50 mL of 0.01 N nitric acid, shake on a mechanical shaker for 1 h, and filter through a fine-porosity sintered-glass funnel. The filtrate is cobaltinitrite wash solution.

> *Note:* If no potassium sodium cobaltinitrite is available, a sufficient amount may be prepared by following the suitability for determination of potassium test, but substituting 0.01 N nitric acid in place of the cobaltinitrite wash solution. When it is possible to follow the suitability procedure as written, the analytical precipitate of potassium sodium cobaltinitrite so obtained may be used to prepare the wash solution.

Sodium Cyanide

NaCN   **Formula Wt 49.01**   **CAS No. 143-33-9**

GENERAL DESCRIPTION

Typical appearance: white solid
Analytical use: electroplating analysis; complexing agent
Change in state (approximate): melting point, 563 °C
Aqueous solubility: 58 g in 100 mL

SPECIFICATIONS

Assay . . . . ≥95.0% NaCN

Maximum Allowable

Phosphate (PO$_4$). 0.02%
Chloride (Cl). 0.15%
Sulfate (SO$_4$) 0.05%
Sulfide (S). 0.005%
Thiocyanate (SCN). 0.02%
Iron, total (as Fe). 0.005%
Lead (Pb). . . . . .5 ppm

TESTS

Assay. (By argentimetric titration of cyanide content). Weigh accurately about 0.4 g, and dissolve in 30 mL of water. Add 0.20 mL of 10% potassium iodide reagent solution and 1 mL of ammonium hydroxide, and titrate with 0.1 N silver nitrate volumetric solution until a slight permanent yellowish turbidity forms. One milliliter of 0.1 N silver nitrate corresponds to 0.009802 g of NaCN.

$$\% \text{ NaCN} = \frac{(\text{mL} \times \text{N AgNO}_3) \times 9.802}{\text{Sample wt (g)}}$$

Phosphate. (Page 40, Method 1). Dissolve 1.0 g in 5 mL of water in a dish, add 2 mL of hydrochloric acid, and evaporate to dryness in a well-ventilated hood. Add 5 mL of dilute hydrochloric acid (1 + 1), and evaporate to dryness again. Dissolve the residue in 50 mL of approximately 0.5 N sulfuric acid. To 5.0 mL of the solution, add 20 mL of approximately 0.5 N sulfuric acid, and continue as described.

For the Determination of Chloride, Sulfate, Sulfide, Thiocyanate, and Total Iron

Sample Solution A. Dissolve 10.0 g in water, filter if necessary through a chloride-free filter, and dilute with water to 200 mL (1 mL = 0.05 g).

Chloride. Dilute 1.0 mL of sample solution A (0.05-g sample) with water to 50 mL. To 10 mL of this solution in a beaker, add 5 mL of 30% hydrogen peroxide, and cover the beaker until the reaction ceases. Digest in the covered beaker on a hot plate (≈100 °C) for about 30 min, cool, and neutralize with nitric acid. Dilute with water to 25 mL, and add 1 mL of nitric acid and 1 mL of silver nitrate reagent solution. The turbidity should not exceed that produced by 0.015 mg of chloride ion (Cl) in an equal volume of solution containing the quantities of reagents used in the test.

Sulfate. (Page 41, Method 2). Use 2 mL of sample solution A (0.1-g sample) plus 1 mL of hydrochloric acid in a well-ventilated hood.

Sulfide. To 10 mL of sample solution A (measured in a graduated cylinder) (0.5-g sample), add 10 mL of water and 0.15 mL of alkaline lead solution (made by adding 10% sodium hydroxide reagent solution to a 10% lead acetate reagent solution until the precipitate first formed is redissolved). The color should not be darker than that produced by 0.025 mg of sulfide ion (S) in an equal volume of solution when treated with 0.15 mL of the alkaline lead solution.

Thiocyanate. To 20 mL (measured in a graduated cylinder) of sample solution A (1-g sample), add 4 mL of hydrochloric acid and 0.20 mL of ferric chloride reagent solution. At the end of 5 min, the solution should show no reddish tint when compared with 20 mL of water to which have been added the quantities of hydrochloric acid and ferric chloride used in the test.

Iron, Total. (Page 38, Method 1). Transfer 4.0 mL of sample solution A (0.2-g sample) to a platinum dish, add 3 mL of hydrochloric acid, and evaporate to dryness in a well-ventilated hood. Heat the residue at 650 °C for 30 min, cool, and add 2 mL of hydrochloric acid and 10 mL of water. Cover with a watch glass, digest on a hot plate ($\approx$100 °C) until dissolution is complete, and use the solution without further acidification.

Lead. Dissolve 0.50 g in 10 mL of water in a separatory funnel. Add 5 mL of ammonium citrate reagent solution (lead-free), 2 mL of hydroxylamine hydrochloride reagent solution for dithizone test, and 0.10 mL of phenol red indicator solution, and make the solution alkaline if necessary by adding ammonium hydroxide. Add 5 mL of dithizone extraction solution, shake gently but well for 1 min, and allow the layers to separate. The intensity of the red color of the chloroform layer should be no greater than that of a control made with 0.002 mg of lead ion (Pb) and 0.1 g of the sample dissolved in 10 mL of water and treated exactly like the solution of 0.50 g of sample in 10 mL of water.

Sodium Dichromate Dihydrate

$Na_2Cr_2O_7 \cdot 2H_2O$ **Formula Wt 298.00**   **CAS No. 7789-12-0**

GENERAL DESCRIPTION

Typical appearance: orange solid

Analytical use: preparation of titrant in redox titrations

Change in state (approximate): becomes anhydrous on prolonged heating at 100 °C; melting point, anhydrous salt, 357 °C

Aqueous solubility: 180 g in 100 mL

SPECIFICATIONS

Assay . . . . 99.5–100.5% $Na_2Cr_2O_7 \cdot 2H_2O$

Maximum Allowable

Insoluble matter 0.005%
Chloride (Cl) 0.005%
Sulfate (SO_4) 0.01%
Calcium (Ca) 0.003%
Magnesium (Mg) 0.005%
Potassium (K) 0.01%
Aluminum (Al) 0.002%

TESTS

Assay. (By titration of oxidizing power). Weigh accurately about 0.2 g, and dissolve in 200 mL of water. Add 7 mL of hydrochloric acid and 3 g of potassium iodide, mix, and allow to stand in the dark for 10 min. Titrate the liberated iodine with 0.1 N sodium thiosulfate volumetric solution, using starch indicator solution near the end point, to a greenish blue color. One milliliter of 0.1 N sodium thiosulfate corresponds to 0.004967 g of $Na_2Cr_2O_7 \cdot 2H_2O$.

$$\% \ Na_2Cr_2O_7 \cdot 2H_2O = \frac{(mL \times N \ Na_2S_2O_3) \times 4.967}{\text{Sample wt (g)}}$$

Insoluble Matter. (Page 25). Dissolve 20.0 g in 200 mL of water.

Chloride. Dissolve 0.2 g in 10 mL of water, filter if necessary through a chloride-free filter, and add 1 mL of ammonium hydroxide and 1 mL of silver nitrate reagent solution. Prepare a standard containing 0.01 mg of chloride ion (Cl) in 10 mL of water, and add 1 mL of ammonium hydroxide and 1 mL of silver nitrate reagent solution. Add 2 mL of nitric acid to each. The comparison is best made by the general method for chloride in colored solutions, page 35.

Sulfate. Dissolve 10 g in 250 mL of water, filter if necessary, and heat to boiling in a dish. Add 25 mL of a solution containing 1 g of barium chloride and 2 mL of hydrochloric acid in 100 mL of solution. Digest in a covered beaker on a hot plate ($\approx 100 \ °C$) for 2 h, and allow to stand for at least 8 h. If a precipitate forms, filter, wash thoroughly, and ignite. Fuse the residue with 1 g of sodium carbonate. Extract the fused mass with water, and filter off the insoluble residue. Add 5 mL of hydrochloric acid to the filtrate, dilute with water to about 200 mL, heat to boiling, and add 10 mL of ethyl alcohol. Digest in a covered beaker on the hot plate until the reduction of chromate is complete, as indicated by the change to a clear green or colorless solution. Neutralize the solution with ammonium hydroxide, and add 2 mL of hydrochloric acid. Heat to boiling, add 10 mL of 12% barium chloride reagent solution, digest in a covered beaker on a hot plate ($\approx 100 \ °C$) for 2 h, and allow to stand for at least 8 h. Filter, wash thoroughly, and ignite. Correct for the weight obtained in a complete blank test. If the original precipitate of barium sulfate weighs less than the requirement permits, the fusion with sodium carbonate is not necessary.

Calcium, Magnesium, and Potassium. (By flame AAS, page 63).

> *Sample Stock Solution.* Dissolve 5.0 g in 50 mL of water, transfer to a 100-mL volumetric flask, and dilute to the mark with water (1 mL = 0.05 g).

Element	Wavelength (nm)	Sample Wt (g)	Standard Added (mg)	Flame Type*	Background Correction
Ca	422.7	1.0	0.03; 0.06	N/A	No
Mg	285.2	0.20	0.005; 0.01	A/A	Yes
K	766.5	0.20	0.01; 0.02	N/A	No

*A/A is air/acetylene; N/A is nitrous oxide/acetylene.

Aluminum. Dissolve 20.0 g in 140 mL of water, filter, and add 5 mL of glacial acetic acid to the filtrate. Make alkaline with ammonium hydroxide. Digest for 2 h on a hot plate ($\approx$100 °C), filter, wash, ignite, and weigh.

Sodium Diethyldithiocarbamate Trihydrate

$(CH_3CH_2)_2NCS_2Na \cdot 3H_2O$ **Formula Wt 225.31**   **CAS No. 20624-25-3**

GENERAL DESCRIPTION

Typical appearance: white or colorless solid
Analytical use: reagent for copper
Change in state (approximate): melting point, 94–96 °C after drying
Aqueous solubility: freely soluble with gradual decomposition

SPECIFICATIONS

Solubility in water . . . . Passes test
Sodium (as Na_2SO_4) 30.5%–32.5%
Sensitivity to copper . . . . Passes test

TESTS

Solubility in Water. Dissolve 1 g in 50 mL of water. There should be no undissolved residue, and the solution should be substantially clear. Retain the solution for the test for sensitivity.

> *Note:* Because of inherent instability, this reagent may be expected to react with oxygen in the atmosphere to form insoluble oxidation products. After storage for some time, the reagent may fail to meet the requirement for solubility in water.

Sodium. Weigh accurately 2 g in a tared, preconditioned dish or crucible. Moisten the sample with concentrated sulfuric acid, and ignite cautiously to char the material. Finally, ignite to constant weight at 600 ± 25 °C.

Sensitivity to Copper.

> ***Solution A.*** Dilute 5 mL of the solution retained from the test for solubility in water to 100 mL with water.

For the sample, prepare a solution containing 0.002 mg of copper ion (Cu) in 100 mL of dilute ammonium hydroxide (1 + 99). For the control, use 100 mL of dilute ammonium hydroxide (1 + 99). Transfer the solutions to separatory funnels and add to each 10 mL of solution A and 10 mL of isopentyl alcohol. Shake for about 1 min and allow to separate (about 30 min). The isopentyl alcohol separated from the solution containing the 0.002 mg of copper should show a distinct yellow color when compared with the isopentyl alcohol from the control.

Sodium Dodecyl Sulfate
SDS; Sodium Lauryl Sulfate

$CH_3(CH_2)_{11}SO_4Na$ **Formula Wt 288.38**   **CAS No. 151-21-3**

GENERAL DESCRIPTION
Typical appearance: white solid
Analytical use: wetting agent; electrophoretic separation of proteins and lipids
Aqueous solubility: soluble

SPECIFICATIONS
Assay (acidimetric, on dried basis) . . . . $\geq$99.0% $C_{12}H_{25}NaO_4S$
Fatty alcohols (as $C_{12}H_{25}OH$) . . . . $\geq$96.0%

	Maximum Allowable
Absorbance (3% w/v)	0.1 AU
Loss on drying	1.0%
Titrable base	0.06 meq/g
Heavy metals (as Pb)	0.002%
Sodium chloride and sodium sulfate	8.0% total
Unsulfated alcohols	4.0%

TESTS

Assay. Weigh accurately about 5.0 g, transfer to a 500-mL round-bottom flask equipped with a ground-glass joint. Add 25.0 mL of 1 N sulfuric acid, and attach a condenser. Using a heating mantle and a magnetic stirrer, heat cautiously and gently at about 100 °C until foaming ceases (about 4 h). Boil under reflux for 3–4 h or more after foaming ceases. Cool, and wash the condenser with 30 mL of ethanol and then 30 mL of water. Titrate with 1 N sodium hydroxide volumetric solution, using 0.15 mL of phenolphthalein indicator solution. Perform a complete blank. One milliliter of 1 N sodium hydroxide corresponds to 0.2884 g $C_{12}H_{25}NaO_4S$. Correct the assay value for loss on drying.

$$\% \ C_{12}H_{25}NaO_4S = \frac{\{[mL \ (blank) - mL \ (sample)] \times N \ NaOH\} \times 28.84}{Sample \ wt \ (g) \times [(100 \quad \% \ LOD)/100]}$$

Fatty Alcohols.

Sample Preparation. Using a 150-mL glass-stoppered conical flask, warm 25 mL of distilled water on a hot plate ($\approx$100 °C). Add approximately 2 g of sample and dissolve. Add 5 mL of hydrochloric acid, connect a cold finger condenser, and let reflux for 2 h. When hydrolysis is complete and the molten fatty alcohol has separated as a completely clear liquid, cool the flask to 0 °C by using an ice bath. (*Note:* This should be done with minimum disturbance to the separated fatty alcohol. It is desirable to solidify the fatty alcohol in one piece without increasing the surface area or allowing occluded water to be introduced.) Prepare a funnel with a suitable filter paper, and transfer the fatty alcohol and solution to the funnel. Wash the solid fatty alcohol with cold water. The fatty alcohol may be broken up and transferred to the filter. However, this should only be done after the first washing. Using a dry filter paper, blot the fatty alcohol dry.

Chromatographic Conditions.

Column: Type I, methyl silicone

Column Temperature: 80 to 230 °C, programmed at 10 °C/min, hold 5 min at 230 °C

Injector Temperature: 220 °C

Detector Temperature: 250 °C

Sample Size: 0.2 µL

Carrier Gas: Helium at 6 mL/min

Detector: Flame ionization

Standard Solution. Accurately weigh 0.10 g of pure dodecyl alcohol, and transfer to a 50-mL volumetric flask. Dissolve and dilute to volume with hexanes. Mix well, and stopper.

Sample Solution. Accurately weigh 0.10 g of the dried fatty alcohol from the sample preparation, and transfer to a 50-mL volumetric flask. Dissolve, and dilute to volume with hexanes. Mix well, and stopper.

Consecutively inject the standard and sample solutions into a prepared gas chromatograph. Record the chromatogram for the duration of the temperature-programmed run. Electronically integrate the area under the peaks of interest. Determine the amount of dodecyl alcohol (C_{12}) in the sample by a conventional area normalization calculation. Correct for assay content of dodecyl alcohol.

Absorbance. Determine the absorbance of 3.0% w/v aqueous solution in 10-mm silica or quartz cuvettes over the range 220 to 350 nm against water as the reference. The absorbance does not exceed 0.1.

Loss on Drying. Weigh accurately about 2 g, and in a preconditioned crucible dry at 105 °C to constant weight.

Titrable Base. Dissolve 1.0 g in 100 mL of water, add 0.15 mL phenol red indicator solution, and titrate with 0.1 N hydrochloric acid. Not more than 0.60 mL is required for neutralization.

Heavy Metals. (Page 36, Method 2). Use 1.0 g.

Sodium Chloride. Accurately weigh about 5 g, and dissolve in about 50 mL of water. Neutralize the solution with 10% nitric acid using litmus paper as the indicator, add 2 mL of potassium chromate reagent solution, and titrate with 0.1 N silver nitrate volumetric solution. One milliliter of 0.1 N silver nitrate is equivalent to 5.844 mg of NaCl.

Sodium Sulfate. Transfer about 1 g of sodium dodecyl sulfate, accurately weighed, to a 250-mL beaker, add 35 mL of water, and warm to dissolve. To the warm solution, add 2.0 mL of 1 N nitric acid, mix, and add 50 mL of ethyl alcohol. Heat the solution to boiling, and slowly add 20 mL of 0.1 N lead nitrate volumetric solution while stirring. Cover the beaker, simmer for 5 min, and let settle. If the supernatant liquid is hazy, let stand for 10 min, heat to boiling, and decant as much liquid as possible through a 9-cm filter paper (Whatman No. 41 or equivalent). Wash four times by decantation, each time using 50 mL of 50% ethyl alcohol, and bring the mixture to a boil. Finally, transfer the filter paper to the original beaker, and immediately add 30 mL of water, 20.0 mL of 0.05 M EDTA, and 2 mL of pH 10 ammoniacal buffer reagent solution. Warm to dissolve the precipitate, add 0.3 g of Eriochrome Black T indicator mixture, and titrate with 0.05 M zinc sulfate. One milliliter of 0.05 M EDTA is equivalent to 7.102 mg of Na_2SO_4.

Unsulfated Alcohols. Dissolve about 10 g, accurately weighed, in 100 mL of water, and add 100 mL of ethyl alcohol. Transfer the solution to a separator, and extract with three 50-mL portions of solvent hexane. If an emulsion forms, sodium chloride may be added to promote separation of the two layers. Wash the combined solvent hexane extracts with three 50-mL portions of water, and dry with anhydrous sodium sulfate. Filter the solvent hexane extract into a tared beaker, evaporate on a steam bath until the hexane is completely volatilized, dry the residue at 105 °C for 30 min, cool, and weigh. The weight of the residue is not more than 4.0% of the weight of the sodium dodecyl sulfate taken.

Sodium Fluoride

NaF   Formula Wt 41.99   CAS No. 7681-49-4

GENERAL DESCRIPTION

Typical appearance: white solid
Analytical use: sequestering agent
Change in state (approximate): melting point, 990 °C
Aqueous solubility: 4.3 g in 100 mL at 25 °C

SPECIFICATIONS

Assay . . . . ≥99% NaF

Maximum Allowable

Insoluble matter 0.02%
Loss on drying 0.3%
Chloride (Cl) 0.005%
Titrable acid . . . . 0.03 meq/g
Titrable base . . . . 0.01 meq/g
Sodium fluosilicate (Na_2SiF_6) 0.1%
Sulfate (SO_4) 0.03%
Sulfite (SO_2) 0.005%
Heavy metals (as Pb) 0.003%
Iron (Fe) 0.003%
Potassium (K) 0.02%

TESTS

Assay. (By indirect acid–base titrimetry). Weigh, to the nearest 0.1 mg, about 1.6 g of sample into a 100-mL volumetric flask. Dissolve, and dilute to the mark with water. Dilute 10.0 mL of the diluted sample with water to 50 mL. Pass the solution through a cation-exchange column (see page 29) at a rate of about 5 mL/min, and collect the eluate in a 500-mL titration flask. Then wash the resin in the column with water at a rate of about 10 mL/min, and collect in the titration flask. Add 0.15 mL of phenolphthalein indicator solution to the flask, and titrate with 0.1 N sodium hydroxide. Continue the titration, as the washing proceeds, until 50 mL of eluate requires no further titration. One milliliter of 0.1 N sodium hydroxide corresponds to 0.004199 g of NaF.

$$\% \text{ NaF} = \frac{(\text{mL} \times \text{N NaOH}) \times 4.199}{\text{Sample wt (g)} / 10}$$

Note: All water used in the assay must be carbon dioxide- and ammonia-free.

Insoluble Matter. (Page 25). Use 3.5 g dissolved in 100 mL of warm water in a platinum dish.

Loss on Drying. Weigh accurately about 1 g in a tared, preconditioned platinum dish or crucible, and dry to constant weight at 150 °C.

Chloride. (Page 35). Dissolve 1.0 g plus 1 g of boric acid in 80 mL of water, filter if necessary through a chloride-free filter, and dilute with water to 100 mL. Use 20 mL of this solution.

Titrable Acid. Dissolve 2.0 g in 50 mL of water (free from carbon dioxide) in a platinum dish, add 10 mL of a saturated solution of potassium nitrate, cool the solution to 0 °C, and add 0.15 mL of phenolphthalein indicator solution. If a pink color is produced, omit the titration for free acid, and reserve the solution for the test for titrable base. If no pink color is produced, titrate with 0.01 N sodium hydroxide until the pink color persists for 15 s while the temperature of the solution is near 0 °C. Not more than 6.0 mL of the 0.01 N sodium hydroxide should be required. Reserve the solution for the test for sodium fluosilicate.

Titrable Base. If a pink color was produced in the solution prepared in the test for titrable acid, add 0.01 N hydrochloric acid, stirring the liquid only gently, until the pink color is discharged. Not more than 2.0 mL of the acid should be required. Reserve the solution for the test for sodium fluosilicate.

Sodium Fluosilicate. Heat the solution reserved from one of the preceding tests to boiling, and titrate while hot with 0.01 N sodium hydroxide until a permanent pink color is obtained. Not more than 4.3 mL of the sodium hydroxide should be required.

Sulfate. (Page 41, Method 2). Evaporate 0.17 g with 2 mL of hydrochloric acid in a platinum or Teflon dish. Repeat the evaporation four more times.

Sulfite. Dissolve 8.0 g in 150 mL of water, and add 2 mL of hydrochloric acid and 0.25 mL of starch indicator solution. Titrate immediately with 0.01 N iodine. Not more than 1.0 mL of the 0.01 N iodine should be required to produce a blue color.

Heavy Metals. (Page 36, Method 1). Dissolve 1.0 g in about 25 mL of water, and dilute with water to 30 mL. Use 25 mL to prepare the sample solution, and use the remaining 5.0 mL to prepare the control solution. Use pH paper to adjust the pH.

Iron. Dissolve 0.50 g of sample in 20 mL of water in a 100-mL beaker. Prepare the standards containing 0.01 and 0.015 mg of iron ion (Fe), respectively, and dilute each to 25 mL with water. To each solution, add in the order specified, 15 mL of 4% boric acid, 3 mL of hydrochloric acid, 30–50 mg of ammonium persulfate, and 3 mL of ammonium thiocyanate reagent solution, mixing after each addition. Any red color in the sample must not exceed that of the 0.015 mg standard solution (0.003%).

Potassium. (By flame AAS, page 63).

Sample Stock Solution. Dissolve 1.0 g of sample with water in a 100-mL volumetric flask, and dilute to the mark with water (1 mL = 0.01 g).

Element	Wavelength (nm)	Sample Wt (g)	Standard Added (mg)	Flame Type*	Background Correction
K	766.5	0.10	0.02; 0.04	A/A	No

*A/A is air/acetylene.

Sodium Formate

HCOONa   Formula Wt 68.01   CAS No. 141-53-7

GENERAL DESCRIPTION

Typical appearance: white solid
Analytical use: precipitant for noble metals
Change in state (approximate): melting point, 253 °C
Aqueous solubility: soluble

SPECIFICATIONS

Assay . . . . ≥99.0% HCOONa

Maximum Allowable

Insoluble matter 0.005%
Chloride (Cl) 0.001%
Sulfate (SO$_4$) 0.001%
Calcium (Ca) 0.005%
Iron (Fe). . . . . 5 ppm
Heavy metals (as Pb) . . . . 5 ppm

TESTS

Assay. (By titration of reducing power). Weigh accurately about 0.88 g, transfer to a 250-mL volumetric flask, dissolve with 50 mL of water, dilute with water to volume, and mix thoroughly. To 25.0 mL of this solution, in a 250-mL glass-stoppered flask, add 3 mL of 10% sodium hydroxide reagent solution and 50.0 mL of 0.1 N potassium permanganate volumetric solution. Heat on a hot plate (≈100 °C) for 20 min, and then cool completely to room temperature. Add 4 g of potassium iodide crystals, mix, and add 5 mL of 25% sulfuric acid reagent solution. Titrate the liberated iodine, representing the excess permanganate, with 0.1 N sodium thiosulfate volumetric solution, using 5 mL of starch indicator solution near the end point. Run a blank on 25 mL of water with the same quantities of permanganate and other reagents, in the same manner as the sample. One milliliter of 0.1 N potassium permanganate consumed corresponds to 0.003401 g of HCOONa.

$$\% \text{ HCOONa} = \frac{[(50.0 \times \text{N KMnO}_4) - (\text{mL} \times \text{N Na}_2\text{S}_2\text{O}_3)] \times 3.401}{\text{Sample wt (g)} / 10}$$

Insoluble Matter. (Page 25). Use 20 g dissolved in 200 mL of water.

Chloride. (Page 35). Use 1.0 g.

Sulfate. (Page 40, Method 1). Use 5.0 g of sample. Allow 30 min for turbidity to form.

Calcium and Iron. (By flame AAS, page 63).

> **Sample Stock Solution.** Ignite 20.0 g of sample in a platinum crucible at 600 °C. Dissolve the residue in 10 mL of 20% hydrochloric acid, dilute with water to 50 mL, and filter. Dilute to the mark with water in a 100-mL volumetric flask (1 mL = 0.20 g).

Element	Wavelength (nm)	Sample Wt (g)	Standard Added (mg)	Flame Type*	Background Correction
Ca	422.7	0.40	0.02; 0.04	N/A	No
Fe	248.3	4.0	0.02; 0.04	A/A	Yes

*A/A is air/acetylene; N/A is nitrous oxide/acetylene.

Heavy Metals. (Page 36, Method 1). Substitute glacial acetic acid for 1 N acetic acid to adjust pH. Dissolve 6.0 g in water, and dilute with water to 30 mL. Use 25 mL to prepare the sample solution, and use the remaining 5.0 mL to prepare the control solution.

Sodium Hydrogen Sulfate, Fused

CAS No. 7681-38-1

Note: This product is usually a mixture of sodium pyrosulfate, $Na_2S_2O_7$, and sodium hydrogen sulfate, $NaHSO_4$.

GENERAL DESCRIPTION

Typical appearance: opaque or white solid

Analytical use: fusion matrix

Change in state (approximate): melting point, 315 °C

Aqueous solubility: 50 g in 100 mL at 0 °C

SPECIFICATIONS

Acidity (as H_2SO_4) 39.0–42.0%

	Maximum Allowable
Insoluble matter	0.01%
Chloride (Cl).	0.001%
Phosphate (PO_4).	0.001%
Heavy metals (as Pb).	5 ppm
Iron (Fe).	0.002%
Calcium (Ca)	0.002%
Magnesium (Mg)	0.001%
Potassium (K)	0.005%

TESTS

Acidity. Weigh accurately about 4.0 g, dissolve in 50 mL of water, and titrate with 1 N sodium hydroxide volumetric solution, using 0.15 mL of methyl orange indicator solution. One milliliter of 1 N sodium hydroxide corresponds to 0.04904 g of H_2SO_4.

$$\% \ H_2SO_4 = \frac{(mL \times N \ NaOH) \times 4.904}{Sample \ wt \ (g)}$$

Insoluble Matter. (Page 25). Dissolve 20.0 g in 200 mL of water.

Chloride. (Page 35). Use 1.0 g.

Phosphate. Dissolve 2.0 g in 15 mL of dilute ammonium hydroxide (1 + 2), and evaporate to dryness. Dissolve the residue in 25 mL of 0.5 N hydrochloric acid, add 1 mL of ammonium molybdate–nitric acid reagent solution and 1 mL of 4-(methylamino)phenol sulfate reagent solution, and allow to stand for 2 h at room temperature. Any blue color should not exceed that produced by 0.02 mg of phosphate ion (PO_4) in an equal volume of solution containing the quantities of reagents used in the test, including the residue from evaporation of 5 mL of ammonium hydroxide.

Heavy Metals. (Page 36, Method 1). Dissolve 6.0 g in about 25 mL of water, and dilute with water to 30 mL. Use 25 mL to prepare the sample solution, and use the remaining 5.0 mL to prepare the control solution.

Iron. (Page 38, Method 1). Dissolve 5.0 g in 80 mL of dilute hydrochloric acid $(1 + 3)$, and boil gently for 10 min. Cool, dilute with water to 100 mL, and use 10 mL.

Calcium, Magnesium, and Potassium. (By flame AAS, page 63).

> **Sample Stock Solution.** Dissolve 10.0 g of sample in a 100-mL volumetric flask, and dilute to the mark with water (1 mL = 0.10 g).

Element	Wavelength (nm)	Sample Wt (g)	Standard Added (mg)	Flame Type*	Background Correction
Ca	422.7	1.0	0.02; 0.04	N/A	No
Mg	285.2	1.0	0.01; 0.02	A/A	Yes
K	766.5	0.40	0.01; 0.02	A/A	No

*A/A is air/acetylene; N/A is nitrous oxide/acetylene.

Sodium Hydroxide (Low Chloride)
NaOH   **Formula Wt 40.00**   **CAS No. 1310-73-2**

GENERAL DESCRIPTION
Typical appearance: white solid
Analytical use: alkalimetric titrant; buffers; pH modifiers
Change in state (approximate): melting point, 318 °C
Aqueous solubility: 110 g in 100 mL at 20 °C

SPECIFICATIONS
Assay . . . . $\geq$97.0% NaOH
Sodium carbonate . . . . $\leq$1.0% Na_2CO_3

Maximum Allowable

Sulfate (SO_4) 0.003%
Chloride (Cl) 0.005%
Nitrogen compounds (as N) 0.001%
Phosphate (PO_4) 0.001%
Heavy metals (as Ag) 0.002%
Iron (Fe). 0.001%
Nickel (Ni). 0.001%
Mercury (Hg). . . . . 0.1 ppm
Calcium (Ca) 0.005%
Magnesium (Mg). 0.002%
Potassium (K) 0.02%

TESTS
> *Note:* Special care must be taken in sampling to obtain a representative sample and to avoid absorption of water and carbon dioxide by the sample taken.

Assay and Sodium Carbonate. (By acid–base titrimetry). Weigh accurately about 38 g, dissolve, and dilute to 1 L, using carbon dioxide-free water. Dilute 50.0 mL of this solution with carbon dioxide-free water to 200 mL. Add 5 mL of 12% barium chloride reagent solution, stopper, and allow to stand for 5 min. Add 0.15 mL of phenolphthalein indicator solution, and titrate with 1 N hydrochloric acid volumetric solution to the end point to determine the hydroxide (A mL). One milliliter of 1 N hydrochloric acid corresponds to 0.04000 g of NaOH. Add 0.15 mL of methyl orange indicator solution, and continue the titration with 1 N hydrochloric acid to the end point (B mL) to determine the carbonate. One milliliter of 1 N hydrochloric acid corresponds to 0.05300 g of Na_2CO_3.

$$\% \text{ NaOH} = \frac{(A \text{ mL} \times \text{N HCl}) \times 4.00}{\text{Sample wt (g)} / 20}$$

$$\% \text{ Na}_2\text{CO}_3 = \frac{[(B \text{ mL} - A \text{ mL}) \times \text{N HCl}] \times 5.300}{\text{Sample wt (g)} / 20}$$

Sulfate. Dissolve 3.0 g of sample in 75 mL of water in a 400-mL beaker. Cautiously neutralize with (1 + 1) hydrochloric acid to litmus paper. Add 1 mL of excess hydrochloric acid. Evaporate to dryness. Add 20 mL of water, and re-evaporate to dryness. Dissolve the residue in 20 mL of water, and filter through a small, washed filter paper. Add two 3-mL portions of water through the filter. For the control, take 0.10 mg of sulfate ion (SO_4) in 25 mL of water. To both sample and control, add 1 mL of (1 + 19) hydrochloric acid, and dilute each to 35 mL. Add 1 mL of 12% barium chloride reagent solution to each. Turbidity in the sample solution after 10 min should not exceed that of the control solution.

For the Determination of Chloride, Nitrogen Compounds, Phosphate, Heavy Metals, Iron, and Nickel

Sample Solution A. Dissolve 50.0 g in carbon dioxide- and ammonia-free water, cool, and dilute with the water to 500 mL (1 mL = 0.10 g).

Chloride. (Page 35). Use 2.0 mL of sample solution A (0.2-g sample), diluted with water to 20 mL.

Nitrogen Compounds. (Page 39). Use 20 mL of sample solution A (2-g sample). For the control, use 10 mL of sample solution A and 0.01 mg of nitrogen diluted to 20 mL.

Phosphate. (Page 40, Method 1). To 20 mL of sample solution A (2-g sample), add 5 mL of hydrochloric acid, and evaporate to dryness on a hot plate ($\approx$100 °C). Dissolve the residue in 25 mL of approximately 0.5 N sulfuric acid, and continue as described.

Heavy Metals. To 30 mL of sample solution A (3.0-g sample), cautiously add 10 mL of nitric acid. For the control, add 0.05 mg of silver to 5.0 mL of sample solution A, and cautiously add 10 mL of nitric acid. Evaporate both solutions to dryness over a low flame or on an electric hot plate. Dissolve each residue in about 20 mL of water, filter if necessary through a chloride-free filter, and dilute with water to 25 mL. Adjust the pH of the control and sample solutions to between 3 and 4 (using a pH meter) with 1 N acetic acid or 10% ammonium hydroxide reagent solution, dilute with water to 40 mL, and mix. Add 5 mL of freshly prepared hydrogen sulfide water to each, and mix. Any yellowish brown color in the solution of the sample should not exceed that in the control.

Iron. (Page 38, Method 1). Neutralize 10 mL of sample solution A (1.0-g sample) with hydrochloric acid, using 0.15 mL of phenolphthalein indicator solution, add 2 mL in excess, dilute with water to 50 mL, and use this solution. In preparing the standard, use the residue from evaporation of the quantity of acid used to neutralize the sample solution.

Nickel. Dilute 20 mL of sample solution A (2.0-g sample) to 50 mL with water, and neutralize with hydrochloric acid. Dilute with water to 85 mL, and adjust the pH to 8 with ammonium hydroxide. Add 5 mL of bromine water, 5 mL of 1% dimethylglyoxime reagent solution, and 5 mL of 10% sodium hydroxide reagent solution. Any red color should not exceed that produced by 0.02 mg of nickel in an equal volume of solution containing the quantities of reagents used in the test.

Mercury. To each of two 250-mL conical flasks, add 20 mL of water and 1 mL of 4% potassium permanganate solution. For the sample, add to one flask 5.5 g of sodium hydroxide. For the control, add to the other flask 0.5 g of sample and 0.5 µg of mercury [1.0 mL of a solution prepared freshly by diluting 1.0 mL of mercury ion (Hg) standard solution with water to 100 mL]. To each flask, add slowly, with constant swirling, 18 mL of hydrochloric acid. Heat to boiling, allow to cool, and dilute with water to 100 mL. Determine mercury in 10-mL aliquots by the CVAAS method (page 65), using 1 mL of 10% hydroxylamine hydrochloride reagent solution and 2 mL of 10% stannous chloride solution for the reduction. The peak obtained for the sample should not exceed that obtained for the control.

Calcium, Magnesium, and Potassium. (By flame AAS, page 63).

> **Sample Stock Solution B.** Cautiously dissolve 10.0 g of sample using an ice-water bath in 50 mL of water. Cool to about 15 °C, and slowly add 40 mL of nitric acid (1 + 1) with stirring. Cool to room temperature, transfer to a 100-mL volumetric flask, dilute to the mark with water, and mix (1 mL = 0.10 g).

> **Sample Stock Solution C.** Take 10.0 mL (1.0 g) of sample stock solution B in a 100-mL volumetric flask, dilute to the mark with water, and mix (1 mL = 0.01 g).

Element	Wavelength (nm)	Sample Wt (g)	Standard Added (mg)	Flame Type*	Background Correction
Ca	422.7	2.0	0.05; 0.10	N/A	No
Mg	285.2	1.0	0.01; 0.02	A/A	Yes
K	766.5	0.10	0.02; 0.04	A/A	No

*A/A is air/acetylene; N/A is nitrous oxide/acetylene.

Sodium Hypochlorite Solution

NaOCl   **Formula Wt 74.44**   CAS No. 7681-52-9

Note: Sodium hypochlorite is available in various dilution strengths ranging from 5% to 10%. The use of this product as a standard allows adjusted dilutions to compensate for assay values not equivalent to 2.5%. Always standardize the sodium hypochlorite solution just before use. This solution and subsequent dilutions are unstable.

GENERAL DESCRIPTION

Typical appearance: clear, yellow liquid
Analytical use: chlorine standard

SPECIFICATIONS

Assay (available Cl) . . . . ≥5.0% Cl

Maximum Allowable

Sodium hydroxide (NaOH) 2.5%

TESTS

Assay. (By indirect iodometric titration of available chlorine). Weigh accurately about 3 mL of sample in a glass-stoppered iodine flask, and dilute to about 50 mL with water. Add 2 g of potassium iodide and 10 mL of glacial acetic acid. Stopper immediately, and allow to stand in the dark for 10 min. Add 150 mL of water, and titrate the liberated iodine with 0.1 N sodium thiosulfate, adding 3 mL of starch indicator near the end point. One milliliter of 0.1 N sodium thiosulfate corresponds to 0.00355 g of available chlorine.

$$\% \text{ Cl} = \frac{(\text{mL} \times \text{N Na}_2\text{S}_2\text{O}_3) \times 3.55}{\text{Sample wt (g)}}$$

Sodium Hydroxide. Place 50 mL of 12% barium chloride reagent solution and 30 mL of 3% hydrogen peroxide reagent solution in a 250-mL beaker. Using a pH meter, neutralize to pH 7.5 with 0.1 N sodium hydroxide. Add 10 g of sample, accurately weighed with the use of a Smith weighing buret, stir vigorously for 1 min, and again titrate to pH 7.5 with 0.1 N hydrochloric acid. Calculate the NaOH content using the following equation:

$$\% \text{ NaOH} = \frac{(\text{mL} \times \text{N HCl}) \times 4.0}{\text{Sample wt (g)}}$$

Sodium Iodide

NaI   **Formula Wt 149.89**   **CAS No. 7681-82-5**

GENERAL DESCRIPTION

Typical appearance: white solid; may become brown on exposure to air
Analytical use: iodometric reagent
Change in state (approximate): melting point, 651 °C
Aqueous solubility: 200 g in 100 mL

SPECIFICATIONS

Assay . . . . ≧99.5% NaI
pH of a 5% solution at 25.0 °C 6.0–9.0

Maximum Allowable

Insoluble matter 0.01%
Chloride and bromide (as Cl) 0.01%
Iodate (IO_3) . . . . 3 ppm
Phosphate (PO_4) 0.001%
Sulfate (SO_4) 0.005%
Barium (Ba) 0.002%
Heavy metals (as Pb) . . . . 5 ppm
Iron (Fe) . . . . 5 ppm
Calcium (Ca) 0.002%
Magnesium (Mg) 0.001%
Potassium (K) 0.01%

TESTS

Assay. (By titration of iodide content). Weigh accurately about 0.5 g, and dissolve in 20 mL of water in a glass-stoppered conical flask. Add 30 mL of hydrochloric acid and 5 mL of chloroform, cool if necessary, and titrate with 0.05 M potassium iodate volumetric solution until the iodine color disappears from the aqueous layer. Shake vigorously for 30 s, and continue the titration, shaking vigorously after each addition, until the iodine color in the chloroform is discharged. One milliliter of 0.05 M potassium iodate corresponds to 0.01499 g of NaI.

$$\% \text{ NaI} = \frac{(\text{mL} \times \text{M KIO}_3) \times 29.98}{\text{Sample wt (g)}}$$

pH of a 5% Solution at 25.0 °C. (Page 49).

Insoluble Matter. (Page 25). Use 20 g dissolved in 200 mL of water.

Chloride and Bromide. Dissolve 1.0 g in 100 mL of water in a distillation flask. Add 1 mL of 30% hydrogen peroxide and 1 mL of phosphoric acid, heat to boiling, and boil gently until all the iodine is expelled and the solution is colorless. Cool, wash down the sides of the flask with water, and add 0.5 mL of hydrogen peroxide. If an iodine color develops, boil until the solution is colorless and for 10 min longer. If no color develops, boil for 10 min, filter if necessary through a chloride-free filter, and dilute with water to 100 mL. Dilute 10 mL with water to 23 mL, and add 1 mL of nitric acid and 1 mL of silver nitrate reagent solution. Any turbidity should not exceed that produced by 0.01 mg of chloride ion (Cl) in an equal volume of solution containing the quantities of nitric acid and silver nitrate used in the test.

Iodate. (Page 54). Use 10.0 g of sample in 25 mL of solution. For the standard, add 0.03 mg of iodate ion (IO_3).

Phosphate. (Page 40, Method 1). Dissolve 2.0 g in water, and add 10 mL of nitric acid and 5 mL of hydrochloric acid. Evaporate to dryness on a hot plate (≈100 °C). Dissolve the residue in 25 mL of approximately 0.5 N sulfuric acid, and continue as described. For the standard, include the residue from evaporation of the quantities of acids used with the sample.

Sulfate. (Page 40, Method 1). Use 1.0 g. Any yellow iodine color that develops can be discharged by the addition of a few milligrams of ascorbic acid.

Barium. Dissolve 3.0 g in 10 mL of 6 M hydrochloric acid reagent solution, and add 5 mL of nitric acid. For the control, dissolve 0.5 g of sample and 0.05 mg of barium in 10 mL of dilute hydrochloric acid (1 + 1), and add 5 mL of nitric acid. Evaporate both solutions to dryness. Dissolve the residues in 10 mL of 6 M hydrochloric acid, add 5 mL of nitric acid, and again evaporate to dryness. Dissolve each residue in water, and dilute with water to 23 mL. To each solution, add 2 mL of potassium dichromate reagent solution and 10% ammonium hydroxide reagent solution until the orange color is just dissipated and the yellow color persists. To each solution, add with constant stirring 25 mL of methanol. Any turbidity in the solution of the sample should not exceed that in the control.

Heavy Metals. (Page 36, Method 1). Dissolve 6.0 g in water, and dilute with water to 30 mL. Use 25 mL to prepare the sample solution, and use the remaining 5.0 mL to prepare the control solution.

Iron. (Page 38, Method 2). Use 2.0 g.

Calcium, Magnesium, and Potassium. (By flame AAS, page 63).

> **Sample Stock Solution.** Dissolve 50 g in about 70 mL of water, and dilute to 100 mL with water (1 mL = 0.50 g).

Element	Wavelength (nm)	Sample Wt (g)	Standard Added (mg)	Flame Type*	Background Correction
Ca	422.7	1.00	0.02; 0.04	N/A	No
Mg	285.2	1.00	0.01; 0.02	A/A	Yes
K	766.5	0.20	0.01; 0.02	A/A	No

*A/A is air/acetylene; N/A is nitrous oxide/acetylene.

Sodium Metabisulfite
Disodium Disulfite

$Na_2S_2O_5$ **Formula Wt 190.11**   **CAS No. 7681-57-4**

GENERAL DESCRIPTION

Typical appearance: white solid
Analytical use: reducing agent
Aqueous solubility: 64 g in 100 mL at 20 °C

SPECIFICATIONS

Assay . . . . $\geq$97.0% $Na_2S_2O_5$

Maximum Allowable

Insoluble matter 0.005%
Chloride (Cl) 0.05%
Thiosulfate (S_2O_3) 0.05%
Heavy metals (as Pb) 0.001%
Iron (Fe). 0.002%

TESTS

Assay. (By titration of reductive capacity). Weigh accurately about 0.45 g, and add to a mixture of 100.0 mL of 0.1 N iodine volumetric solution and 5 mL of 10% hydrochloric acid solution. Swirl gently until the sample is dissolved completely. Titrate the excess of iodine with 0.1 N sodium thiosulfate volumetric solution, adding 3 mL of starch indicator solution near the end of the titration. One milliliter of 0.1 N iodine corresponds to 0.004753 g of $Na_2S_2O_5$.

$$\% \ Na_2S_2O_5 = \frac{[(mL \times N \ I_2) - (mL \times N \ Na_2S_2O_3)] \times 4.753}{Sample \ wt \ (g)}$$

Insoluble Matter. (Page 25). Use 20 g dissolved in 200 mL of water.

Chloride. Dissolve 1.0 g in 10 mL of water, filter if necessary through a small chloride-free filter, and add 6 mL of 30% hydrogen peroxide. Add 1 N sodium hydroxide volumetric solution until the solution is slightly alkaline to phenolphthalein, and dilute with water to 100 mL. Dilute 2.0 mL of this solution with water to 20 mL, and add 1 mL of nitric acid and 1 mL of silver nitrate reagent solution. Any turbidity should not exceed that produced by 0.01 mg of chloride ion (Cl) in an equal volume of solution containing the quantities of reagents used in the test.

Thiosulfate. In a 50-mL beaker, dissolve 2.20 g of the sample in 30 mL of water. Simultaneously, for the control solution, dissolve 1.10 g of the sample in 30 mL of water, add 0.50 mL of 0.01 N sodium thiosulfate solution (freshly made from 0.1 N sodium thiosulfate volumetric solution). To both solutions, add 1.0 mL of 0.1 N silver nitrate and 5.0 mL of sulfuric acid (1 + 1). Mix each solution, and allow to stand for 5 min. The color of the sample solution should not exceed the brownish color of the control solution.

Heavy Metals. (Page 36, Method 1). Dissolve 3.0 g in 25 mL of dilute hydrochloric acid (2 + 3), and evaporate to dryness on a hot plate ($\approx$100 °C). Add 5 mL of hydrochloric acid, and again evaporate to dryness. Dissolve the residue in 20 mL of water, and dilute with water to 25 mL. For the control, add 0.02 mg of lead to 1.0 g of sample, and treat exactly as the 3.0 g of sample.

Iron. (Page 38, Method 1). Dissolve 0.50 g in 14 mL of dilute hydrochloric acid (2 + 5), and evaporate on a hot plate ($\approx$100 °C) to dryness. Dissolve the residue in 7 mL of dilute hydrochloric acid (2 + 5), and again evaporate to dryness. Dissolve the residue in 2 mL of hydrochloric acid, dilute with water to 50 mL, and use the solution without further acidification.

Sodium Methoxide, 0.5 M Methanolic Solution
Sodium Methylate
CH_3ONa Formula Wt 54.02   CAS No. 124-41-4

GENERAL DESCRIPTION
Typical appearance: colorless liquid
Analytical use: nonaqueous titrations

SPECIFICATIONS
Assay . . . . .0.48–0.52 mole of CH_3ONa
per liter of solution (25.9–28.1 g/L)
Clarity of solution . . . . .Passes test

TESTS
Assay. (By acid–base titrimetry). Pipet 5.0 mL of the sample into a conical flask, and slowly add 100 mL of water, with stirring. Add 0.10 mL of phenolphthalein indicator solution, and titrate with 0.1 N hydrochloric acid to a colorless end point. One milliliter of 0.1 N hydrochloric acid corresponds to 0.005402 g of CH_3ONa.

$$\% \ CH_3ONa = \frac{(mL \times N \ HCl) \times 5.402}{Sample \ wt \ (g)}$$

Clarity of Solution. Place 50 mL in a 50-mL color comparison tube. The sample should not have more than a trace of turbidity or insoluble matter.

Sodium Molybdate Dihydrate
$Na_2MoO_4 \cdot 2H_2O$ Formula Wt 241.95   CAS No. 10102-40-6

GENERAL DESCRIPTION
Typical appearance: white solid
Analytical use: alkaloid reagent
Change in state (approximate): loses its water of crystallization at 100 °C
Aqueous solubility: 56 g in 100 mL at 0 °C

SPECIFICATIONS
Assay . . . . .99.5–103.0% $Na_2MoO_4 \cdot 2H_2O$
pH of a 5% solution at 25.0 °C 7.0–10.5

	Maximum Allowable
Insoluble matter	0.005%
Chloride (Cl).	0.005%
Phosphate (PO_4).	5 ppm

Sulfate (SO_4) 0.015%
Ammonium (NH_4) 0.001%
Heavy metals (as Pb) . . . . 5 ppm
Iron (Fe) 0.001%

TESTS

Assay. (By oxidative titration after reduction of Mo^{VI}). Weigh accurately about 0.3 g, and dissolve in 10 mL of water in a 150-mL beaker. Activate the zinc amalgam of a Jones reductor by passing 100 mL of 1 N sulfuric acid through the column. Discard this acid, and place 25 mL of ferric ammonium sulfate solution (described below) in the receiver under the column. To the sample solution, add 100 mL of 1 N sulfuric acid, and pass this mixture through the reductor, followed by 100 mL of 1 N sulfuric acid and then 100 mL of water (using these to rinse the beaker). Add 5 mL of phosphoric acid to the solution in the receiver, and titrate with 0.1 N potassium permanganate volumetric solution. Run a blank, and make any necessary correction. One milliliter of 0.1 N potassium permanganate corresponds to 0.008066 g of $Na_2MoO_4 \cdot 2H_2O$.

$$\% \ Na_2MoO_4 \cdot 2H_2O = \frac{\{[mL \ (sample) - mL \ (blank)] \times N \ KMnO_4\} \times 8.066}{Sample \ wt \ (g)}$$

Ferric Ammonium Sulfate Solution. Dissolve 50 g of ferric ammonium sulfate dodecahydrate in 500 mL of solution containing 25 mL of sulfuric acid.

pH of a 5% Solution at 25.0 °C. (Page 49).

Insoluble Matter. (Page 25). Use 20 g dissolved in 200 mL of water.

Chloride. Dissolve 1.0 g in 50 mL of water, and mix. Pipet 10 mL of the solution into a color-comparison tube, add 2 mL of nitric acid, dilute with water to 25 mL, and add 1 mL of silver nitrate reagent solution. Any turbidity should not exceed that produced by 0.01 mg of chloride ion (Cl) in an equal volume of solution containing the quantities of reagents used in the test.

Phosphate. (Page 40, Method 3).

Sulfate. (Page 41, Method 2). Dissolve in a mixture of 5 mL of water and 5 mL of nitric acid. Evaporate to dryness on a hot plate. Digest the residue with a mixture of 1 mL of hydrochloric acid and 10 mL of water. Dilute to 20 mL, and filter through two prewashed fine-porosity papers. Add 12% barium chloride reagent solution to the clear filtrate, and use Method 1.

Ammonium. Dissolve 1.0 g in 20 mL of water, add 10 mL of 10% sodium hydroxide reagent solution, and dilute with water to 50 mL. For the standard, add 0.01 mg of ammonium ion (NH_4) to 20 mL of water, add 10 mL of 10% sodium hydroxide reagent solution, and dilute with water to 50 mL. To each solution, add 2 mL of Nessler reagent. Any color produced in the sample solution should not exceed that in the standard.

Heavy Metals. Dissolve 6.0 g in 35 mL of 10% sodium hydroxide reagent solution, add 5 mL of ammonium hydroxide, and dilute with water to 42 mL. For the control, add 0.02

mg of lead to 7.0 mL of the solution, and dilute with water to 40 mL. For the sample, dilute the remaining 35-mL portion with water to 40 mL. Add 10 mL of freshly prepared hydrogen sulfide water to each, and mix. Any color in the solution of the sample should not exceed that in the control.

Iron. (By flame AAS, page 63).

> **Sample Stock Solution.** Dissolve 10.0 g in water, and dilute with water to 100 mL (1 mL = 0.10 g).

Element	Wavelength (nm)	Sample Wt (g)	Standard Added (mg)	Flame Type*	Background Correction
Fe	248.3	2.50	0.05; 0.10	A/A	Yes

*A/A is air/acetylene.

Sodium Nitrate

NaNO$_3$ **Formula Wt 84.99**   **CAS No. 7631-99-4**

GENERAL DESCRIPTION

Typical appearance: colorless or white solid
Analytical use: oxidizer
Change in state (approximate): melting point, 308 °C
Aqueous solubility: 88 g in 100 mL at 20 °C

SPECIFICATIONS

Assay . . . . ≥99.0% NaNO$_3$
pH of a 5% solution at 25.0 °C 5.5–8.3

	Maximum Allowable
Insoluble matter	0.005%
Chloride	0.001%
Iodate (IO$_3$)	5 ppm
Nitrite (NO$_2$)	0.001%
Phosphate (PO$_4$)	5 ppm
Sulfate (SO$_4$)	0.003%
Calcium (Ca)	0.005%
Magnesium (Mg)	0.002%
Heavy metals (as Pb)	5 ppm
Iron (Fe)	3 ppm

TESTS

Assay. (By indirect acid–base titrimetry). Weigh, to the nearest 0.1 mg, about 0.3 g of sample, and dissolve in 100 mL of water. Pass the solution through a cation-exchange column (see page 29) at a rate of about 5 mL/min, and collect the eluate in a 500-mL titration flask. Wash the resin in the column with water at a rate of about 10 mL/min, and collect in the same titration flask. Add 0.15 mL of phenolphthalein indicator solution to the flask, and titrate with 0.1 N sodium hydroxide. Continue the titration, as the washing proceeds, until 50 mL of eluate requires no further titration. One milliliter of 0.1 N sodium hydroxide corresponds to 0.008499 g of $NaNO_3$.

$$\% \; NaNO_3 = \frac{(mL \times N \; NaOH) \times 8.499}{Sample \; wt \; (g)}$$

Note: All water used in the assay must be carbon dioxide- and ammonia-free.

pH of a 5% Solution at 25.0 °C. (Page 49).

Insoluble Matter. (Page 25). Use 20 g dissolved in 100 mL of water.

Chloride. (Page 35). Use 1.0 g.

Iodate. (Page 54). Use 10.0 g of sample, and dissolve in 25 mL. For the standard, add 0.05 mg of iodate ion (IO_3).

Nitrite. (Page 55). Use 1.0 g of sample and 0.01 mg of nitrite.

Phosphate. (Page 40, Method 1). Dissolve 4.0 g in 25 mL of approximately 0.5 N sulfuric acid, and continue as described.

Sulfate. (Page 41, Method 2). Evaporate 3.0 g using 7 mL of dilute hydrochloric acid (1 + 1). Repeat the evaporation. Allow 30 min for turbidity to form.

Calcium and Magnesium. (By flame AAS, page 63).

    Sample Stock Solution. Dissolve 5.0 g of sample in water in a 100-mL volumetric flask, and dilute to the mark with water (1 mL = 0.05 g).

Element	Wavelength (nm)	Sample Wt (g)	Standard Added (mg)	Flame Type*	Background Correction
Ca	422.7	0.50	0.025; 0.05	N/A	No
Mg	285.2	0.50	0.01; 0.02	A/A	Yes

*A/A is air/acetylene; N/A is nitrous oxide/acetylene.

Heavy Metals. (Page 36, Method 1). Dissolve 6.0 g in about 20 mL of water, and dilute with water to 30 mL. Use 25 mL to prepare the sample solution, and use the remaining 5.0 mL to prepare the control solution.

Iron. (Page 38, Method 1). Use 3.3 g.

Sodium Nitrite

$NaNO_2$ Formula Wt 69.00   CAS No. 7632-00-0

GENERAL DESCRIPTION

Typical appearance: white or slightly yellow solid

Analytical use: reducing agent

Change in state (approximate): melting point, 271 °C; decomposes above 320 °C

Aqueous solubility: 82 g in 100 mL at 20 °C

SPECIFICATIONS

Assay . . . . $\geq$97.0% $NaNO_2$

Maximum Allowable

Insoluble matter 0.01%

Chloride (Cl). 0.005%

Sulfate (SO_4) 0.01%

Heavy metals (as Pb). 0.001%

Iron (Fe). 0.001%

Calcium (Ca) 0.01%

Potassium (K) 0.005%

TESTS

For the Assay and Determination of Chloride, Sulfate, Heavy Metals, and Iron

Sample Solution A. Weigh accurately about 10.0 g, dissolve in water, and dilute with water to 100 mL in a volumetric flask (1 mL = 0.1 g).

Assay. (By indirect titration of reductive capacity). Dilute a 10.0-mL aliquot of sample solution A to 100 mL in a volumetric flask. Add 5 mL of sulfuric acid to 300 mL of water, and while the solution is still warm, add 0.1 N potassium permanganate volumetric solution until a faint pink color that persists for 2 min is produced. Then add 40.0 mL of 0.1 N potassium permanganate, and mix gently. Add, slowly and with constant agitation, 10.0 mL of the diluted sample solution, holding the tip of the pipet well below the surface of the liquid. Add 15.0 mL of 0.1 N ferrous ammonium sulfate volumetric solution, allow the solution to stand for 5 min, and titrate the excess of ferrous ammonium sulfate with 0.1 N potassium permanganate volumetric solution. Each milliliter of 0.1 N potassium permanganate consumed by the sodium nitrite corresponds to 0.003450 g of $NaNO_2$.

$$\% \ NaNO_2 = \frac{[(40.0 + V) \times N \ KMnO_4] - [15.0 \times N \ Fe(NH_4)_2(SO_4)_2] \times 3.450}{\text{Sample wt (g)} \ / \ 100}$$

where V = mL of potassium permanganate required to titrate the excess of ferrous ammonium sulfate.

Insoluble Matter. (Page 25). Use 10 g dissolved in 100 mL of water.

Chloride. (Page 35). To 2.0 mL of sample solution A (0.2-g sample), add 10 mL of water, slowly add 1 mL of glacial acetic acid, and boil gently for 5 min.

Sulfate. (Page 41, Method 2). Use 5.0 mL of sample solution A plus 1 mL of hydrochloric acid.

Heavy Metals. (Page 36, Method 1). To 30 mL of sample solution A (3-g sample), add 5 mL of hydrochloric acid, and evaporate to dryness on a steam bath. Add 5 mL of hydrochloric acid, and again evaporate to dryness. Dissolve in about 20 mL of water, and dilute with water to 25 mL. For the control, add 0.02 mg of lead ion (Pb) to 10 mL of sample solution A (1-g sample), add 5 mL of hydrochloric acid, and treat exactly as the 30 mL of sample solution A.

Iron. (Page 38, Method 1). To 10 mL of sample solution A (1-g sample), add 5 mL of hydrochloric acid, and evaporate on a hot plate ($\approx$100 °C) to dryness. Dissolve the residue in 2 mL of hydrochloric acid plus about 20 mL of water, filter if necessary, and dilute with water to 50 mL. Use the solution without further acidification.

Calcium and Potassium. (By flame AAS, page 63).

Sample Stock Solution. Dissolve 10.0 g of sample with water in a 100-mL volumetric flask, and dilute to the mark with water (1 mL = 0.10 g).

Element	Wavelength (nm)	Sample Wt (g)	Standard Added (mg)	Flame Type*	Background Correction
Ca	422.7	0.50	0.05; 0.10	N/A	No
K	766.5	0.50	0.025; 0.05	A/A	No

*A/A is air/acetylene; N/A is nitrous oxide/acetylene.

Sodium Nitroferricyanide Dihydrate
**Disodium Pentakis(cyano-C)nitrosylferrate(2–) Dihydrate;
Sodium Pentacyanonitrosoferrate(III) Dihydrate**

$Na_2Fe(CN)_5NO \cdot 2H_2O$ **Formula Wt 297.95**   **CAS No. 13755-38-9**

GENERAL DESCRIPTION

Typical appearance: ruby-red solid

Analytical use: detection of alkali sulfides

Aqueous solubility: 40 g in 100 mL at 20 °C

SPECIFICATIONS

Assay . . . . 99.0–102.0% $Na_2Fe(CN)_5NO \cdot 2H_2O$

Maximum Allowable

Insoluble matter 0.01%

Chloride (Cl) 0.02%

Sulfate (SO$_4$) . . . . Passes test

TESTS

Assay. (By argentimetric titration of ferricyanide content). Weigh accurately about 1.5 g, and transfer to a 100-mL volumetric flask with about 80 mL of water. Dilute to the mark, and mix after complete dissolution. To a 25-mL aliquot of this solution in a 250-mL conical flask, add 50 mL of water and 1 mL of 10% potassium chromate reagent solution. Titrate with 0.1 N silver nitrate volumetric solution to a reddish brown color. One milliliter of 0.1 N silver nitrate corresponds to 0.01490 g of $Na_2Fe(CN)_5NO \cdot 2H_2O$.

$$\% \ Na_2Fe(CN)_5NO \cdot 2H_2O = \frac{(mL \times N \ AgNO_3) \times 14.90}{Sample \ wt \ (g) \ / \ 4}$$

Insoluble Matter. Dissolve 10 g in 50 mL of water at room temperature, filter promptly through a tared, preconditioned filtering crucible, wash thoroughly, and dry at 105 °C.

Chloride. Dissolve 1.0 g in 175 mL of water, add 1.25 g of cupric sulfate crystals dissolved in 25 mL of water, mix thoroughly, and allow to stand until the precipitate has settled. Filter through a chloride-free filter, rejecting the first 50 mL of the filtrate. Dilute 10 mL of the filtrate with water to 20 mL, and add 1 mL of nitric acid and 1 mL of silver nitrate reagent solution. Any turbidity should not exceed that produced by 0.01 mg of chloride ion (Cl) in an equal volume of solution containing the quantities of reagents used in the test, with enough cupric sulfate added to match the color of the test.

Sulfate. Dissolve 5.0 g in 100 mL of water without heating, filter, and to the filtrate add 0.25 mL of glacial acetic acid and 5 mL of 12% barium chloride reagent solution. Stir and pour into a Nessler tube for observation. No turbidity should be produced in 10 min. (Limit about 0.01%)

Sodium Oxalate

Ethanedioic Acid, Disodium Salt

$(COONa)_2$ **Formula Wt 134.00**   **CAS No. 62-76-0**

GENERAL DESCRIPTION

Typical appearance: white solid
Analytical use: standardizing potassium and permanganate solutions; titrimetry
Aqueous solubility: 3.7 g in 100 mL at 20 °C

SPECIFICATIONS

Assay . . . . $\geq$99.5% $(COONa)_2$

Maximum Allowable

Insoluble matter 0.005%
Loss on drying 0.01%

Neutrality. . . . . Passes test
Chloride (Cl) 0.002%
Ammonium (NH$_4$) 0.002%
Sulfate (SO$_4$) 0.002%
Heavy metals (as Pb) 0.002%
Iron (Fe). 0.001%
Potassium (K) 0.005%
Substances darkened by hot sulfuric acid . . . . Passes test

TESTS

Assay. (By indirect titration of reducing power of oxalate against permanganate). Weigh, to the nearest 0.1 mg, about 0.25 g of the dried sample, and dissolve with 100 mL of water in a 250-mL glass-stoppered flask. Add 3 mL of sulfuric acid, and then slowly add 20 mL of 0.1 N potassium permanganate volumetric solution. Heat the solution to 70 °C, and complete the titration with the permanganate until a pale pink color persists for 30 s. One milliliter of 0.1 N potassium permanganate corresponds to 0.0067 grams of Na$_2$C$_2$O$_4$.

$$\% \text{ Na}_2\text{C}_2\text{O}_4 = \frac{(\text{mL} \times \text{N KMnO}_4) \times 6.70}{\text{Sample wt (g)}}$$

Insoluble Matter. (Page 25). Use 20 g dissolved in 500 mL of water.

Loss on Drying. Weigh accurately 10 g in a tared, preconditioned dish or crucible, and dry at 105 °C to constant weight.

Neutrality. Dissolve 2.0 g in 200 mL of water, and add 10.0 mL of 0.01 N oxalic acid and 0.15 mL of phenolphthalein indicator solution. Boil the solution in a flask for 10 min, passing through it a stream of carbon dioxide-free air. Cool the solution rapidly to room temperature while keeping the flow of carbon dioxide-free air passing through it. Titrate with 0.01 N sodium hydroxide. Not less than 9.2 nor more than 10.5 mL of 0.01 N sodium hydroxide should be required to match the pink color of a buffer solution containing 0.15 mL of phenolphthalein indicator solution. For the buffer solution, add 3.1 g of boric acid, 3.8 g of potassium chloride, and 5.90 mL of 1 N sodium hydroxide to a 1-L container, and fill to the mark with water.

Chloride. Dissolve 2.0 g in water plus 10 mL of nitric acid, filter if necessary through a chloride-free filter, and dilute with water to 100 mL. To 25 mL of the solution, add 1 mL of silver nitrate reagent solution. Any turbidity should not exceed that produced by 0.01 mg of chloride ion (Cl) in an equal volume of solution containing the quantities of reagents used in the test.

Ammonium. Dissolve 1.0 g in 50 mL of ammonia-free water, and add 2 mL of Nessler reagent. Any color should not exceed that produced by 0.02 mg of ammonium ion (NH$_4$) in an equal volume of solution containing 2 mL of Nessler reagent.

For the Determination of Sulfate, Heavy Metals, and Iron

Sample Stock Solution A. In a 400-mL beaker, mix 10.0 g of sample with 50 mL of water. Add 5.0 mL of 1% sodium carbonate reagent solution, 15 mL of nitric acid, and 15 mL of 30% hydrogen peroxide. Digest in a covered beaker on a hot plate until the reaction ceases. Remove the cover, and evaporate to dryness. Add 25 mL of dilute hydrochloric acid (1 + 1), and evaporate to dryness. Repeat the evaporation with another 25 mL of dilute hydrochloric acid (1 + 1). Dissolve the residue in 30 mL of water, add 1 mL of dilute hydrochloric acid (1 + 9), filter, and dilute to 60 mL with water (1.0 g = 6.0 mL).

Blank Solution B. In a 400-mL beaker, add 10 mL of 1% sodium carbonate reagent solution to 30 mL of nitric acid and 30 mL of 30% hydrogen peroxide, and evaporate to dryness. Add 100 mL of dilute hydrochloric acid (1 + 1), and evaporate to dryness. Dissolve the residue in 30 mL of water, add 2 mL of dilute hydrochloric acid (1 + 19), filter, and dilute to 120 mL with water.

Sulfate. (Page 40, Method 1). For the sample, use 30 mL (5-g sample) of sample stock solution A. Prepare standards using 30 mL of blank solution B and 0.05 and 0.10 mg of sulfate ion (SO_4), respectively.

Heavy Metals. (Page 36, Method 1). For the sample, use 9 mL (1.5-g sample) of sample stock solution A. Prepare standards by adding 0.01 and 0.02 mg lead ion (Pb) to, respectively, 3 mL (0.5-g sample) of sample stock solution A and 6 mL of blank solution B. Adjust the pH of the standard and sample solutions to between 3 and 4 (using a pH meter) with 1 N acetic acid or 10% ammonium hydroxide reagent solution, dilute with water to 40 mL, and mix. Add 5 mL of freshly prepared hydrogen sulfide water to each, and mix.

Iron. (Page 38, Method 1). For the sample, dilute 12 mL (2-g sample) of sample stock solution A to 40 mL with water. Prepare standards by adding 0.005 and 0.1 mg of iron ion (Fe), respectively, to two 12-mL portions of blank solution B, and dilute both with water to 40 mL. To the sample and standards, add 3 mL of hydrochloric acid, 30–50 mg of ammonium persulfate crystals, and 3 mL of ammonium thiocyanate reagent solution.

Potassium. (By flame AAS, page 63).

Sample Stock Solution. Dissolve 5.0 g of sample in 5 mL of nitric acid in a 200-mL volumetric flask, and dilute to the mark with water (1 mL = 0.025 g).

Element	Wavelength (nm)	Sample Wt (g)	Standard Added (mg)	Flame Type*	Background Correction
K	766.5	0.50	0.025; 0.05	A/A	No

*A/A is air/acetylene.

Substances Darkened by Hot Sulfuric Acid.
Heat 1.0 g in a recently ignited test tube with 10 mL of sulfuric acid until the appearance of dense fumes. The acid, when cooled, should have no more color than a mixture of the following composition: 0.2 mL of cobalt chloride reagent solution, 0.3 mL of ferric chloride reagent solution, 0.3 mL of cupric sulfate reagent solution, and 9.2 mL of water.

Sodium Perchlorate

NaClO₄ (anhydrous)	Formula Wt 122.44	CAS No. 7601-89-0
NaClO₄ · H₂O (monohydrate)	Formula Wt 140.46	CAS No. 7791-07-3

Note: This specification includes the anhydrous form and the monohydrate.

GENERAL DESCRIPTION

Typical appearance: colorless or white solid
Analytical use: electrolyte; oxidimetric standard
Change in state (approximate): melting point, 130 °C
Aqueous solubility: 209 g in 100 mL

SPECIFICATIONS

Assay . . . . 98.0–102.0% NaClO₄ (anhydrous)
85.0–90.0% NaClO₄ · H₂O (monohydrate)

pH of a 5% solution at 25.0 °C 6.0–8.0

Maximum Allowable

Insoluble matter 0.005%
Chloride (Cl) 0.003%
Sulfate (SO₄) 0.002%
Calcium (Ca) 0.02%
Potassium (K) 0.05%
Heavy metals (as Pb) . . . . 5 ppm
Iron (Fe). . . . . 5 ppm

TESTS

Assay. (By argentimetric titration of chloride content after perchlorate reduction to chloride). Weigh accurately about 0.28 g of sample into a preconditioned platinum crucible or dish. Add 3 g of sodium carbonate, and heat, gently at first, then to fusion until a clear melt is obtained. Leach with minimal 10% nitric acid, and boil gently to expel carbon dioxide. Cool to room temperature, add 10 mL of aqueous 30% ammonium acetate solution, and titrate with 0.1 N silver nitrate volumetric solution, using dichlorofluorescein indicator solution to the pink end point. One milliliter of 0.1 N silver nitrate corresponds to 0.01225 g of $NaClO_4$ or 0.01405 g of $NaClO_4 \cdot H_2O$.

$$\% \text{ NaClO}_4 = \frac{(\text{mL} \times \text{N AgNO}_3) \times 12.25}{\text{Sample wt (g)}}$$

$$\% \text{ NaClO}_4 \cdot \text{H}_2\text{O} = \frac{(\text{mL} \times \text{N AgNO}_3) \times 14.05}{\text{Sample wt (g)}}$$

pH of a 5% Solution at 25.0 °C. (Page 49).

Insoluble Matter. (Page 25). Use 5.0 g dissolved in 50 mL of water.

Chloride. (Page 35). Dissolve 0.33 g in 20 mL of water.

Sulfate. (Page 40, Method 1). Dissolve 2.5 g in 5 mL of water.

Calcium and Potassium. (By flame AAS, page 63).

> **Sample Stock Solution.** Dissolve 1.0 g of sample with water in a 100-mL volumetric flask, add 10 mL of hydrochloric acid, and dilute to the mark with water (1 mL = 0.01 g).

Element	Wavelength (nm)	Sample Wt (g)	Standard Added (mg)	Flame Type*	Background Correction
Ca	422.7	0.10	0.02; 0.04	N/A	No
K	766.5	0.10	0.05; 0.10	A/A	No

*A/A is air/acetylene; N/A is nitrous oxide/acetylene.

Heavy Metals. (Page 36, Method 1). Dissolve 3.0 g in water, dilute with water to 60 mL, and use 50 mL to prepare the sample solution. For the control, add 0.01 mg of lead ion (Pb) to the remaining 10 mL, and dilute with water to 50 mL.

Iron. (Page 38, Method 1). Use 2.0 g.

Sodium Periodate
Sodium Metaperiodate; Sodium Tetraoxoiodate(VII)

$NaIO_4$ **Formula Wt 213.89**   **CAS No. 7790-28-5**

GENERAL DESCRIPTION
Typical appearance: white solid
Analytical use: oxidimetric standard; determination of manganese
Change in state (approximate): decomposes at 300 °C
Aqueous solubility: 37 g in 100 mL at 50 °C

SPECIFICATIONS
Assay (dried basis). . . . . 99.8–100.3% $NaIO_4$

Maximum Allowable
Other halogens (as Cl) 0.02%
Manganese (Mn) . . . . 3 ppm

TESTS

Assay. (By indirect titration of reductive capacity). Dry a sample over phosphorus pentoxide or magnesium perchlorate dessicant for 6 h. Weigh accurately about 1 g, dissolve in water, and transfer to a 500-mL volumetric flask. Dilute to volume with water, and mix thoroughly. Place a 50.0-mL aliquot of this solution in a glass-stoppered flask, and add 10 g of potassium iodide and 10 mL of a cooled solution of sulfuric acid (1 + 5). Stopper, swirl, allow to stand for 5 min, and add 100 mL of cold water. Titrate the liberated iodine with

0.1 N sodium thiosulfate volumetric solution, adding 3 mL of starch indicator solution near the end of the titration. Correct for a complete blank. One milliliter of 0.1 N sodium thiosulfate corresponds to 0.002674 g of $NaIO_4$.

$$\% \ NaIO_4 = \frac{\{[mL \ (sample) - mL \ (blank)] \times N \ Na_2S_2O_3\} \times 2.674}{Sample \ wt \ (g) \ / \ 10}$$

Other Halogens. Dissolve 0.50 g in 25 mL of a dilute solution of sulfuric acid (3 + 2). Boil for 3 min, cool, and add 5 mL of ammonium hydroxide and 20 mL of 2.5% silver nitrate solution. Allow the precipitate to coagulate, filter through a chloride-free filter, and dilute the filtrate with water to 200 mL. To 20 mL of this solution, add 1.5 mL of nitric acid. Any turbidity should not exceed that produced by 0.01 mg of chloride ion (Cl) in an equal volume of solution containing 0.5 mL of ammonium hydroxide, 1.5 mL of nitric acid, and 1 mL of silver nitrate reagent solution.

Manganese. Dissolve 2.5 g in 40 mL of 10% sulfuric acid reagent solution. For the control, add 0.5 g of the sample and 0.006 mg of manganese to 40 mL of 10% sulfuric acid reagent solution. Add 5 mL of nitric acid and 5 mL of phosphoric acid to each, boil gently for 10 min, and cool. Any pink color in the solution of the sample should not exceed that in the control.

Sodium Peroxide

Na_2O_2 **Formula Wt 77.98**   **CAS No. 1313-60-6**

Caution: This reagent should be stored in airtight containers in a cool place. Avoid contact with organic materials; fire or explosion may result.

Note: When dissolving sodium peroxide, the material should be added slowly and in small portions to well-cooled water; when neutralizing, the acid must be added cautiously in small portions, and the solution kept cool.

GENERAL DESCRIPTION

Typical appearance: slightly yellow solid
Analytical use: oxidizing agent
Aqueous solubility: soluble in cold water; decomposes in warm water

SPECIFICATIONS

Assay . . . . ≥93.0% Na_2O_2

	Maximum Allowable
Chloride (Cl)	0.002%
Phosphate (PO_4)	5 ppm
Sulfate (SO_4)	0.001%
Heavy metals (as Pb)	0.002%
Iron (Fe)	0.005%

TESTS

Assay. (By titration of reducing power against permanganate). Weigh accurately about 0.7 g, and add slowly to 400 mL of dilute sulfuric acid (1 + 99) that has been cooled to 10 °C. Dilute to volume in a 500-mL volumetric flask with dilute sulfuric acid (1 + 99), and titrate 100.0 mL with 0.1 N potassium permanganate volumetric solution. One milliliter of 0.1 N potassium permanganate corresponds to 0.003899 g of Na_2O_2.

$$\% \ Na_2O_2 = \frac{(mL \times N \ KMnO_4) \times 3.899}{Sample \ wt \ (g) \ / \ 5}$$

Chloride. Add 1.0 g in small portions to 35 mL of water, cool, and slowly add 4 mL of nitric acid. Filter if necessary through a chloride-free filter, and dilute the filtrate with water to 40 mL. To 20 mL of the filtrate, add 1 mL of silver nitrate reagent solution. Any turbidity should not exceed that produced by 0.01 mg of chloride ion (Cl) in an equal volume of solution containing the quantities of reagents used in the test.

Phosphate. (Page 40, Method 1). Dissolve 4.0 g in 50 mL of water, cautiously add 10 mL of nitric acid, and evaporate to dryness on a hot plate ($\approx$100 °C). Dissolve the residue in 25 mL of approximately 0.5 N sulfuric acid, and continue as described.

Sulfate. Completely dissolve 20 g in 300 mL of cold water (10 °C), neutralize with hydrochloric acid, and note the volume used. Add an excess of 2 mL of the acid, and evaporate to a volume of about 200 mL. Filter if necessary, heat the filtrate to boiling, add 10 mL of 12% barium chloride reagent solution, digest in a covered beaker on a hot plate ($\approx$100 °C) for 2 h, and allow to stand for at least 8 h. If a precipitate is formed, filter, wash thoroughly, and ignite. Correct for the weight obtained in a complete blank test. For the blank, add about 10 mg of sodium carbonate and 1 mL of bromine water to the volume of hydrochloric acid used to neutralize the sodium peroxide, and evaporate to dryness. Dissolve in 2 mL of hydrochloric acid, add 200 mL of water, boil, and filter if necessary. Add 10 mL of 12% barium chloride reagent solution, and carry along with the sample.

Heavy Metals. (Page 36, Method 1). Add 2.0 g cautiously to 15 mL of dilute hydrochloric acid (1 + 2) cooled to 10 °C. Evaporate the solution to dryness on a hot plate ($\approx$100 °C), add 15 mL of dilute hydrochloric acid, and repeat the evaporation to dryness. Dissolve the residue in 20 mL of water, and dilute with water to 25 mL. For the control, add 0.02 mg of lead ion (Pb) to 15 mL of dilute hydrochloric acid, cool the acid to 10 °C, and cautiously add 1.0 g of the sample, continuing with the evaporations exactly as for the 2.0 g of sample.

Iron. (Page 38, Method 1). Add 10 g, cautiously, to 75 mL of a cooled solution of dilute hydrochloric acid (1 + 2). Evaporate on a hot plate ($\approx$100 °C) to dryness, add 10 mL of hydrochloric acid, and again evaporate to dryness. Dissolve the residue in 80 mL of dilute hydrochloric acid (1 + 40), dilute with water to 100 mL, and use 2.0 mL of the solution.

Sodium Phosphate, Dibasic
Disodium Hydrogen Phosphate

Na_2HPO_4 Formula Wt 141.96   CAS No. 7558-79-4

GENERAL DESCRIPTION

Typical appearance: colorless or white solid

Analytical use: pH buffer solutions

Aqueous solubility: soluble in 8 parts of water, more soluble in hot water

SPECIFICATIONS

Assay . . . . $\geq$99.0% Na_2HPO_4

pH of a 5% solution at 25.0 °C 8.7–9.3

Maximum Allowable

Insoluble matter 0.01%

Loss on drying 0.2%

Chloride (Cl) 0.002%

Sulfate (SO_4) 0.005%

Heavy metals (as Pb) 0.001%

Iron (Fe). 0.002%

TESTS

Assay. (By indirect acid–base titrimetry). Weigh accurately about 6.0 g, and dissolve in 50 mL of water and exactly 50.0 mL of 1 N hydrochloric acid volumetric solution. Titrate the excess acid with 1 N sodium hydroxide volumetric solution to the first inflection point at about pH 4, as measured with a pH meter. Calculate *A*, the equivalent of 1 N hydrochloric acid consumed by the sample. Continue the titration with 1 N sodium hydroxide to the second inflection point at about pH 8.8. Calculate *B*, the equivalent of 1 N sodium hydroxide required in the titration between the two inflection points.

$$A = (50.0 \times N\ HCl) - (mL \times N\ NaOH)$$

$$B = mL \times N\ NaOH$$

If *A* is equal to or less than *B*, then

$$\% \ Na_2HPO_4 = \frac{A \times 0.1420}{Sample\ wt\ (g)} \times 100$$

If *A* is greater than *B*, then

$$\% \ Na_2HPO_4 = \frac{(2B - A) \times 0.1420 \times 100}{Sample\ wt\ (g)}$$

pH of a 5% Solution at 25.0 °C. (Page 49). Use the sample retained from the test for loss on drying.

Insoluble Matter. (Page 25). Use 10.0 g dissolved in 100 mL of water. Do not use a glass filtering crucible.

Loss on Drying. Weigh accurately 6.0 g, and dry in a preconditioned dish at 105 °C to constant weight. Save the material for the determination of pH.

Chloride. (Page 35). Use 0.50 g of sample and 1.5 mL of nitric acid.

Sulfate. Dissolve 1.0 g in 15 mL of water, and add 4.0 mL of 10% hydrochloric acid (solution pH approximately 2). Filter through washed filter paper, and add 10 mL of water through the same filter. For the control, add 0.05 mg of sulfate ion (SO_4) in 20 mL of water, and add 1.0 mL of (1 + 19) hydrochloric acid. Dilute both solutions to 35 mL with water, add 5.0 mL of 40% barium chloride reagent solution, and mix. Observe turbidity after 10 min. Sample solution turbidity should not exceed that of the control solution.

Heavy Metals. (Page 36, Method 1). Dissolve 4.0 g in about 20 mL of water, add 14.5 mL of 2 N hydrochloric acid, and dilute with water to 40 mL. Use 30 mL to prepare the sample solution, and use the remaining 10 mL to prepare the control solution.

Iron. (Page 38, Method 2). Dissolve 1.0 g in water, dilute with water to 20 mL, and use 10 mL of the solution.

Sodium Phosphate, Dibasic, Heptahydrate
Disodium Hydrogen Phosphate Heptahydrate
$Na_2HPO_4 \cdot 7H_2O$ Formula Wt 268.07   CAS No. 7782-85-6

GENERAL DESCRIPTION
Typical appearance: colorless or white solid
Analytical use: buffering agent
Aqueous solubility: soluble in 4 parts of water

SPECIFICATIONS
Assay . . . . 98.0–102.0% $Na_2HPO_4 \cdot 7H_2O$
pH of a 5% solution at 25.0 °C 8.7–9.3

	Maximum Allowable
Insoluble matter	0.005%
Chloride (Cl)	0.001%
Sulfate (SO_4)	0.005%
Heavy metals (as Pb)	0.001%
Iron (Fe)	0.001%

TESTS

Assay. (By indirect acid–base titrimetry). Weigh accurately about 10.5 g, and dissolve in 50 mL of water and exactly 50.0 mL of 1 N hydrochloric acid volumetric solution. Titrate the excess acid with 1 N sodium hydroxide volumetric solution to the first inflection point at about pH 4, as measured with a pH meter. Calculate A, the equivalent of 1 N hydrochloric acid consumed by the sample. Continue the titration with 1 N sodium hydroxide volumetric solution to the second inflection point at about pH 8.8. Calculate B, the equivalent of 1 N sodium hydroxide required in the titration between the two inflection points.

$$A = (50.0 \times \text{N HCl}) - (\text{mL} \times \text{N NaOH})$$

$$B = \text{mL} \times \text{N NaOH}$$

If A is equal to or less than B, then

$$\% \text{ Na}_2\text{HPO}_4 \cdot 7\text{H}_2\text{O} = \frac{A \times 0.2681}{\text{Sample wt (g)}} \times 100$$

If A is greater than B, then

$$\% \text{ Na}_2\text{HPO}_4 \cdot 7\text{H}_2\text{O} = \frac{(2B - A) \times 0.2681}{\text{Sample wt (g)}} \times 100$$

pH of a 5% Solution at 25.0 °C. (Page 49).

Insoluble Matter. (Page 25). Use 20 g dissolved in 200 mL of water.

Chloride. (Page 35). Use 1.0 g of sample and 3 mL of nitric acid.

Sulfate. Dissolve 1.0 g in 15 mL of water, and add 2.5 mL of 10% hydrochloric acid (solution approximately pH 2). Filter through washed filter paper, and add 10 mL of water through the same filter. For the control, add 0.05 mg of sulfate ion (SO_4) in 20 mL of water, and add 1.0 mL of (1 + 19) hydrochloric acid. Dilute both to 35 mL with water, add 5.0 mL of 40% barium chloride solution, and mix. Observe turbidity after 10 min. Sample solution turbidity should not exceed that of the control solution.

Heavy Metals. (Page 36, Method 1). Dissolve 4.0 g in about 20 mL of water, add 8 mL of 2 N hydrochloric acid, and dilute with water to 32 mL. Use 24 mL to prepare the sample solution, and use the remaining 8.0 mL to prepare the control solution.

Iron. (Page 38, Method 2). Use 1.0 g.

Sodium Phosphate, Monobasic, Monohydrate
Sodium Dihydrogen Phosphate Monohydrate

$NaH_2PO_4 \cdot H_2O$	Formula Wt 137.99	CAS No. 10049-21-5

GENERAL DESCRIPTION

Typical appearance: white or colorless solid
Analytical use: buffering agent
Change in state (approximate): loses all water at 100 °C
Aqueous solubility: freely soluble

SPECIFICATIONS

Assay . . . . 98.0–102.0% $NaH_2PO_4 \cdot H_2O$

pH of a 5% solution at 25.0 °C 4.1–4.5

Maximum Allowable

Insoluble matter 0.01%

Chloride (Cl). . . . . 5 ppm

Sulfate (SO_4) 0.003%

Calcium (Ca) 0.005%

Potassium (K) 0.01%

Heavy metals (as Pb). 0.001%

Iron (Fe). 0.001%

TESTS

Assay. (By acid–base titrimetry). Weigh accurately about 5.0 g, and dissolve in 50 mL of water. Titrate with 1 N sodium hydroxide volumetric solution to the inflection point at about pH 8.8, as measured with a pH meter. One milliliter of 1 N sodium hydroxide corresponds to 0.1380 g of $NaH_2PO_4 \cdot H_2O$.

$$\% \ NaH_2PO_4 \cdot H_2O = \frac{(mL \times N \ NaOH) \times 13.80}{Sample \ wt \ (g)}$$

pH of a 5% Solution at 25.0 °C. (Page 49).

Insoluble Matter. (Page 25). Dissolve 10.0 g in 100 mL of water.

Chloride. (Page 35). Use 2.0 g of sample and 2 mL of nitric acid.

Sulfate. Dissolve 1.0 g in 15 mL of water, and add 2.5 mL of 10% hydrochloric acid (solution pH approximately 2). Filter through washed filter paper, and add 10 mL of water through the same filter. For control, add 0.03 mg of sulfate ion (SO_4) in 20 mL of water, and add 1.0 mL of (1 + 19) hydrochloric acid. Dilute both to 35 mL with water, add 5.0 mL of 40% barium chloride solution, and mix. Observe turbidity after 10 min. Sample solution turbidity should not exceed that of the control solution.

Calcium and Potassium. (By flame AAS, page 63).

    Sample Stock Solution. Dissolve 10.0 g of sample in 70 mL of water, and transfer to a 100 mL volumetric flask. Add 5 mL of nitric acid, and dilute to the mark with water (1 mL = 0.10 g).

Element	Wavelength (nm)	Sample Wt (g)	Standard Added (mg)	Flame Type*	Background Correction
Ca	422.7	0.40	0.02; 0.04	N/A	No
K	766.5	0.40	0.02; 0.04	A/A	No

*A/A is air/acetylene; N/A is nitrous oxide/acetylene.

Heavy Metals. (Page 36, Method 1). Dissolve 4.0 g in about 20 mL of water, and dilute with water to 32 mL. Use 24 mL to prepare the sample solution, and use the remaining 8.0 mL to prepare the control solution.

Iron. (Page 38, Method 2). Use 1.0 g.

Sodium Phosphate, Tribasic, Dodecahydrate

$Na_3PO_4 \cdot 12H_2O$ Formula Wt 380.12   CAS No. 10101-89-0

Note: This reagent, when stored under ordinary conditions, may lose some water of crystallization. Any loss in water will result in an assay of more than 100% $Na_3PO_4 \cdot 12H_2O$ but does not affect the determination of the relative amount of free alkali present.

GENERAL DESCRIPTION

Typical appearance: colorless or white solid

Analytical use: buffer; sequestering agent

Change in state (approximate): melting point, 75 °C

Aqueous solubility: soluble in 3.5 parts water, 1 part boiling water

SPECIFICATIONS

Assay . . . . .98.0–102.0% $Na_3PO_4 \cdot 12H_2O$

Maximum Allowable

Excess alkali (as NaOH) 2.5%

Insoluble matter 0.01%

Chloride (Cl) 0.001%

Sulfate (SO_4) 0.01%

Heavy metals (as Pb) 0.001%

Iron (Fe). 0.001%

TESTS

Assay. (By acid–base titrimetry). Weigh accurately about 7.5 g, and dissolve in 50 mL of water and exactly 50.0 mL of 1 N hydrochloric acid volumetric solution. Heat to boiling to expel carbon dioxide. While protecting the solution from absorbing carbon dioxide from the air, cool, and titrate the solution with 1 N sodium hydroxide volumetric solution to the inflection point at about pH 4, as measured with a pH meter. Calculate A, the equivalent of 1 N hydrochloric acid consumed by the sample. Continue the titration with 1 N sodium hydroxide to the second inflection point at about pH 8.8. Calculate B, the equivalent of 1 N sodium hydroxide required in the titration between the two inflection points.

$$A = (50.0 \times N\ HCl) - (mL \times N\ NaOH)$$

$$B = mL \times N\ NaOH$$

If A is equal to or greater than $2B$, then

$$\% \ Na_3PO_4 \cdot 12H_2O = \frac{B \times 0.3801}{\text{Sample wt (g)}} \times 100$$

If A is less than $2B$, then

$$\% \ Na_3PO_4 \cdot 12H_2O = \frac{(A - B) \times 0.3801}{\text{Sample wt (g)}} \times 100$$

Excess Alkali. Calculate the amount of excess alkali as sodium hydroxide from the titration values obtained in the test for assay. If A is equal to or less than $2B$, there is no excess alkali present when the salt is dissolved. If A is greater than $2B$, then

$$\% \, \text{NaOH} = \frac{(A - 2B) \times 0.040}{\text{Sample wt (g)}} \times 100$$

Insoluble Matter. (Page 25). Dissolve 10.0 g in 100 mL of water, add 0.10 mL of methyl red indicator solution, add hydrochloric acid until the solution is slightly acid to the methyl red indicator, and continue as described.

Chloride. (Page 35). Use 1.0 g of sample and 3 mL of nitric acid.

Sulfate. Dissolve 1.0 g in 15 mL of water, and add 4.0 mL of 10% hydrochloric acid (solution approximately pH 2). Filter through washed filter paper, and add 10 mL of water through the same filter. For the control, add 0.10 mg of sulfate ion (SO_4) in 20 mL of water, and add 1.0 mL of (1 + 19) hydrochloric acid. Dilute both to 35 mL with water, add 5.0 mL of 40% barium chloride solution, and mix. Observe turbidity after 10 min. Sample solution turbidity should not exceed that of the control solution.

Heavy Metals. (Page 36, Method 1). Dissolve 4.0 g in about 20 mL of water, add 6 mL of 2 N hydrochloric acid, and dilute with water to 32 mL. Use 24 mL to prepare the sample solution, and use the remaining 8.0 mL to prepare the control solution.

Iron. (Page 38, Method 2). Dissolve 1.0 g in 10 mL of water, and add 3 mL of 1 N hydrochloric acid. Omit the addition of acid to the standard.

Sodium Pyrophosphate Decahydrate
Diphosphoric Acid, Tetrasodium Salt, Decahydrate
$Na_4P_2O_7 \cdot 10H_2O$ Formula Wt 446.06   CAS No. 13472-36-1

GENERAL DESCRIPTION
Typical appearance: colorless solid
Analytical use: fusion matrix
Change in state (approximate): melting point, 79 °C
Aqueous solubility: 10 g in 100 mL at 20 °C

SPECIFICATIONS
Assay . . . . .99.0–103.0% $Na_4P_2O_7 \cdot 10H_2O$
pH of a 5% solution at 25.0 °C 9.5–10.5

	Maximum Allowable
Insoluble matter	0.01%
Chloride (Cl)	0.002%
Sulfate (SO_4)	0.005%
Nitrogen compounds (as N)	0.001%

Heavy metals (as Pb) 0.001%
Iron (Fe). 0.001%

TESTS

Assay. (By acid–base titrimetry). Weigh accurately about 8.0 g, and dissolve in 100 mL of water. Titrate with 1 N hydrochloric acid volumetric solution to a green color with 0.15 mL of bromphenol blue indicator solution. One milliliter of 1 N hydrochloric acid corresponds to 0.2230 g of $Na_4P_2O_7 \cdot 10H_2O$.

$$\% \ Na_4P_2O_7 \cdot 10H_2O = \frac{(mL \times N \ HCl) \times 22.30}{Sample \ wt \ (g)}$$

pH of a 5% Solution at 25.0 °C. (Page 49).

Insoluble Matter. (Page 25). Use 10.0 g dissolved in 150 mL of water.

Chloride. (Page 35). Use 0.50 g of sample and 3 mL of nitric acid.

Sulfate. Dissolve 1.0 g in 15 mL of water, and add 5.0 mL of 10% hydrochloric acid (solution pH approximately 2). Filter through washed filter paper, and add 10 mL of water through the same filter. For control, add 0.05 mg of sulfate ion (SO_4) in 20 mL of water, and add 1.0 mL of (1 + 19) hydrochloric acid. Dilute both to 35 mL with water, add 5.0 mL of 40% barium chloride solution, and mix. Observe turbidity after 10 min. Sample solution turbidity should not exceed that of the control solution.

Nitrogen Compounds. (Page 39). Use 1.0 g. For the standard, use 0.01 mg of nitrogen ion (N).

Heavy Metals. (Page 36, Method 1). Dissolve 3.0 g in about 20 mL of water, warming if necessary, add 8 mL of 2 N hydrochloric acid, and dilute with water to 42 mL. Use 35 mL to prepare the sample solution, and use the remaining 7.0 mL to prepare the control solution.

Iron. (Page 38, Method 2). Use 1.0 g.

Sodium Sulfate, Anhydrous
Na_2SO_4   **Formula Wt 142.04**   **CAS No. 7757-82-6**

Suitable for general use or in extraction–concentration analysis. Product labeling shall designate the uses for which suitability is represented on the basis of meeting the relevant specifications and tests. The extraction–concentration suitability specifications include all of the specifications for general use.

GENERAL DESCRIPTION

Typical appearance: white solid
Analytical use: Kjeldahl nitrogen determination; drying agent

Change in state (approximate): melting point, 884 °C
Aqueous solubility: 20 g in 100 mL at 20 °C

SPECIFICATIONS
General Use
Assay . . . . ≥99.0% Na_2SO_4
pH of a 5% solution at 25.0 °C 5.2–9.2

	Maximum Allowable

Insoluble matter 0.01%
Loss on ignition 0.5%
Chloride (Cl). 0.001%
Nitrogen compounds (as N). . . . . 5 ppm
Phosphate (PO_4). 0.001%
Heavy metals (as Pb). . . . . 5 ppm
Iron (Fe). 0.001%
Calcium (Ca) 0.01%
Magnesium (Mg) 0.005%
Potassium (K) 0.01%

Specific Use
Extraction–concentration suitability . . . . Passes test

TESTS

Assay. (By indirect acid–base titrimetry after ion exchange). Weigh accurately about 0.3 g of sample, and dissolve in about 50 mL of water. Pass the sample solution through the column at a rate of about 5 mL/min, collecting the eluate in a 500-mL titration flask. Wash the column with water at a rate of about 10 mL/min, collecting the washings in the same flask. Add 0.15 mL of phenolphthalein indicator solution, and titrate the eluate with 0.1 N sodium hydroxide. Continue to wash the column and to titrate the eluate until the passage of 50 mL of water requires no further titrant. One milliliter of 0.1 N sodium hydroxide corresponds to 0.007102 g of Na_2SO_4.

$$\% \ Na_2SO_4 = \frac{(mL \times N \ NaOH) \times 7.102}{Sample \ wt \ (g)}$$

Note: All water used in the assay must be free of carbon dioxide and ammonia.

pH of a 5% Solution at 25.0 °C. (Page 49).

Insoluble Matter. (Page 25). Use 10.0 g dissolved in 100 mL of water.

Loss on Ignition. Weigh accurately about 2 g in a tared, preconditioned dish or crucible, and ignite at 600 ± 25 °C for 15 min.

Chloride. (Page 35). Use 1.0 g.

Nitrogen Compounds. (Page 39). Use 2.0 g. For the standard, use 0.01 mg of nitrogen ion (N).

Phosphate. (Page 40, Method 1). Dissolve 2.0 g in 20 mL of 0.5 N sulfuric acid. For the standard, dilute 0.02 mg of phosphate ion (PO_4) to 20 mL with 0.5 N sulfuric acid.

Heavy Metals. (Page 36, Method 1). Dissolve 6.0 g in water, and dilute with water to 42 mL. Use 35 mL to prepare the sample solution, and use the remaining 7 mL to prepare the control solution.

Iron. (Page 38, Method 1). Use 1.0 g of sample.

Calcium, Magnesium, and Potassium. (By flame AAS, page 63).

> **Sample Stock Solution.** Dissolve 4.0 g in 50 mL of water, and transfer to a 100-mL volumetric flask. Dilute to the mark with water (1 mL = 0.04 g).

Element	Wavelength (nm)	Sample Wt (g)	Standard Added (mg)	Flame Type*	Background Correction
Ca	422.7	0.40	0.02; 0.04	N/A	No
Mg	285.2	0.40	0.01; 0.02	A/A	Yes
K	766.5	0.10	0.01; 0.02	A/A	No

*A/A is air/acetylene; N/A is nitrous oxide/acetylene.

Extraction–Concentration Suitability. (Page 84). Prepare a method blank to check the glassware and dichloromethane before beginning the sample analysis; follow the sample procedure below, except omit the sample.

Dry about 55 g of sample in a shallow tray at 400 °C for 1 h, and allow to cool in a desiccator. Transfer 50 ± 0.1 g of the dried sample, accurately weighed, to a 250-mL conical flask, add 60 mL of extraction–concentration grade dichloromethane, and allow to stand for 2–3 min. Swirl every 15 s to mix. Allow to stand until any fines that may be present have settled, and then decant the dichloromethane into a Kuderna–Danish flask equipped with a 3-ball Snyder column and a 10-mL concentrator tip. Extract the sample with two additional 60-mL portions of dichloromethane, as described above, combine the extracts, and concentrate the contents of the flasks to 1 mL. Perform solvent exchange using 15 mL of extraction–concentration grade hexanes in the Kuderna–Danish flask, and concentrate to 1 mL. Place a micro Snyder column on the 10-mL tip, perform two additional solvent exchanges, using 5 mL of hexanes each time, and concentrate to 1.0 mL. Inject 5.0 μL of the concentrate from the method blank into a gas chromatograph equipped with an electron capture detector (ECD) and a flame ionization detector (FID), as described on page 85. Dilute each standard specified in the ECD and FID suitability tests with an equal volume of hexanes, giving concentrations of 0.5 μg/L heptachlor epoxide and 0.5 mg/L of 2-octanol, respectively, and inject 1.0 μL of each into the chromatograph. Finally, inject 5.0 μL of the 180:1 concentrate from the sample into the chromatograph. Calculate the concentration of contaminants in the sample by comparing the sample peak heights with the height of the heptachlor epoxide and 2-octanol peaks from the injection of the standards, correcting for any contribution of the solvents as determined by the method blank. No peak in the extract of the sample shall exceed the height of the standard when measured as described on page 86.

Sodium Sulfate, Decahydrate

$Na_2SO_4 \cdot 10H_2O$ Formula Wt 322.19   CAS No. 7727-73-3

GENERAL DESCRIPTION

Typical appearance: colorless solid

Analytical use: Kjeldahl analysis

Change in state (approximate): melting point, 32 °C

Aqueous solubility: 36 g in 100 mL at 15 °C

SPECIFICATIONS

General Use

Assay . . . . ≥99.0% $Na_2SO_4 \cdot 10H_2O$

pH of a 5% solution at 25.0 °C 5.2–9.2

	Maximum Allowable
Insoluble matter	0.01%
Chloride (Cl)	5 ppm
Phosphate (PO_4)	5 ppm
Heavy metals (as Pb)	3 ppm
Iron (Fe)	5 ppm
Calcium (Ca)	0.005%
Magnesium (Mg)	0.003%
Potassium (K)	0.005%

TESTS

Assay. (By indirect acid–base titrimetry after ion exchange). Weigh accurately about 0.7 g of sample, and dissolve in about 50 mL of water. Pass the sample solution through the column at a rate of about 5 mL/min, collecting the eluate in a 500-mL titration flask. Wash the column with water at a rate of about 10 mL/min, collecting the washings in the same flask. Add 0.15 mL of phenolphthalein indicator solution, and titrate the eluate with 0.1 N sodium hydroxide. Continue to wash the column and to titrate the eluate until the passage of 50 mL of water requires no further titrant. One milliliter of 0.1 N sodium hydroxide corresponds to 0.01611 g of $Na_2SO_4 \cdot 10H_2O$.

$$\% \ Na_2SO_4 \cdot 10H_2O = \frac{(mL \times N \ NaOH) \times 16.11}{Sample \ wt \ (g)}$$

pH of a 5% Solution at 25.0 °C. (Page 49).

Insoluble Matter. (Page 25). Use 10.0 g dissolved in 100 mL of water.

Chloride. (Page 35). Use 2.0 g.

Phosphate. (Page 40, Method 1). Dissolve 4.0 g in 20 mL of 0.5 N sulfuric acid. For the standard, dilute 0.02 mg of phosphate ion (PO_4) to 20 mL with 0.5 N sulfuric acid.

Heavy Metals. (Page 36, Method 1). Dissolve 8.0 g in water, and dilute with water to 48 mL. Use 42 mL to prepare the sample solution, and use the remaining 6 mL to prepare the control solution.

Iron. (Page 38, Method 1). Use 2.0 g.

Calcium, Magnesium, and Potassium. (By flame AAS, page 63).

> *Sample Stock Solution.* Dissolve 4.0 g in 50 mL of water in a 100-mL volumetric flask. Dilute to the mark with water (1 mL = 0.04 g).

Element	Wavelength (nm)	Sample Wt (g)	Standard Added (mg)	Flame Type*	Background Correction
Ca	422.7	0.40	0.02; 0.04	N/A	No
Mg	285.2	0.40	0.01; 0.02	A/A	Yes
K	766.5	0.40	0.01; 0.02	A/A	No

*A/A is air/acetylene; N/A is nitrous oxide/acetylene.

Sodium Sulfide, Nonahydrate
$Na_2S \cdot 9H_2O$ **Formula Wt 240.18**   **CAS No. 1313-84-4**

Note: To enhance stability, storage below 10 °C is recommended.

GENERAL DESCRIPTION
Typical appearance: colorless or slightly yellow solid
Analytical use: source of hydrogen sulfide for heavy metals tests
Change in state (approximate): melting point, 50 °C
Aqueous solubility: 47 g in 100 mL at 10 °C

SPECIFICATIONS
Assay . . . . $\geq$98.0% $Na_2S \cdot 9H_2O$

Maximum Allowable
Ammonium (NH_4) 0.005%
Sulfite and thiosulfate (as SO_4) 0.1%
Iron . . . . Passes test

TESTS
Assay. (By titration of reductive capacity). Weigh accurately about 2.5 g, and in a 250-mL volumetric flask, dilute with water through which high-purity nitrogen gas has been freshly bubbled to expel oxygen. Dilute to volume with similar water, and displace air from the head-space with nitrogen gas. Stopper, and mix thoroughly. Place a 50.0-mL aliquot of this solution into a mixture of 50.0 mL of 0.1 N iodine volumetric solution and 25 mL of 0.1 N hydrochloric acid in 400 mL of water. Titrate the excess iodine with 0.1 N sodium thiosulfate

volumetric solution, adding 3 mL of starch indicator solution near the end of the titration. One milliliter of 0.1 N iodine consumed corresponds to 0.01201 g of $Na_2S \cdot 9H_2O$.

$$\% \, Na_2S \cdot 9H_2O = \frac{[(mL \times N \, I_2) - (mL \times N \, Na_2S_2O_3)] \times 12.01}{Sample \, wt \, (g) \, / \, 5}$$

Ammonium. (By colorimetry, page 32). Dissolve 1.0 g in 80 mL of water, add 20 mL of lead acetate reagent solution, and allow to stand until the precipitate has settled. Decant 50 mL of the clear supernatant liquid, and use the 50 mL for distillation. For the standard, use 0.025 mg of ammonium ion (NH_4) treated exactly as the sample.

Sulfite and Thiosulfate. (*Note:* The water used in this test must be free from dissolved oxygen.) Dissolve 3.0 g in 200 mL of water and add 100 mL of 5% zinc sulfate solution. Mix thoroughly; allow to stand for 30 min. Filter, and titrate 100 mL of the filtrate with 0.01 N iodine, using starch indicator solution. Not more than 4.0 mL of the iodine solution should be required.

Iron. Dissolve 5.0 g in 100 mL of water. The solution should be clear and colorless.

Sodium Sulfite
Sodium Sulfite, Anhydrous
Na_2SO_3   **Formula Wt 126.04**   **CAS No. 7757-83-7**

GENERAL DESCRIPTION
Typical appearance: white solid
Analytical use: reducing agent
Aqueous solubility: 28 g in 100 mL at 84 °C

SPECIFICATIONS
Assay . . . . ≥98.0% Na_2SO_3

	Maximum Allowable
Insoluble matter	0.005%
Free acid	Passes test
Titrable free base	0.03 meq/g
Chloride (Cl)	0.02%
Heavy metals (as Pb)	0.001%
Iron (Fe)	0.001%

TESTS

Assay. (By titration of reductive capacity). Place about 0.5 g, accurately weighed, into 100 mL of 0.1 N iodine volumetric solution. Mix, allow to stand for 5 min, and add 1 mL of hydrochloric acid. Titrate the excess of iodine with 0.1 N sodium thiosulfate volumetric solution, adding 3 mL of starch indicator solution near the end of the titration. Perform a complete blank titration. One milliliter of 0.1 N iodine consumed corresponds to 0.006302 g of Na_2SO_3.

$$\% \ Na_2SO_3 = \frac{\{[mL \ (blank) - mL \ (sample)] \times N \ Na_2S_2O_3\} \times 6.302}{Sample \ wt \ (g)}$$

Insoluble Matter. (Page 25). Use 20.0 g dissolved in 200 mL of water.

Free Acid. Dissolve 1.0 g in 10 mL of water, and add 0.10 mL of phenolphthalein indicator solution. A pink color should be produced.

Titrable Free Base. Dissolve 1.0 g in 10 mL of water, and add 3 mL of 30% hydrogen peroxide that has been previously neutralized using 0.15 mL of methyl red indicator solution. Shake well, allow to stand for 5 min, and titrate with 0.01 N hydrochloric acid. Not more than 3.0 mL of the acid should be required to neutralize the solution.

Chloride. Dissolve 1.0 g in 10 mL of water, filter if necessary through a small chloride-free filter, add 3 mL of 30% hydrogen peroxide, and dilute with water to 100 mL. Dilute 5 mL of this solution with water to 20 mL, and add 1 mL of nitric acid and 1 mL of silver nitrate reagent solution. Any turbidity should not exceed that produced by 0.01 mg of chloride ion (Cl) in an equal volume of solution containing the quantities of nitric acid and silver nitrate used in the test.

Heavy Metals. (Page 36, Method 1). Dissolve 3.0 g in 20 mL of dilute hydrochloric acid (1 + 1), and evaporate the solution to dryness on a hot plate ($\approx$100 °C). Add 10 mL more of dilute hydrochloric acid (1 + 1), and again evaporate to dryness. Dissolve the residue in about 20 mL of water, and dilute with water to 25 mL. For the control, add 0.02 mg of lead ion (Pb) to 1.0 g of sample, and treat exactly as the 3.0 g of sample.

Iron. (Page 38, Method 1). Dissolve 1.0 g in 20 mL of dilute hydrochloric acid (1 + 9), and evaporate on a hot plate ($\approx$100 °C) to dryness. Dissolve the residue in 5 mL of dilute hydrochloric acid (1 + 1), and again evaporate to dryness. Dissolve the residue in 4 mL of dilute hydrochloric acid (1 + 1), dilute with water to 50 mL, and use the solution without further acidification.

Sodium Tartrate Dihydrate
2,3-Dihydroxybutanedioic Acid, Disodium Salt, Dihydrate

$$NaO-\overset{\overset{\displaystyle O}{\|}}{C}-\overset{\overset{\displaystyle OH}{|}}{CH}-\overset{\overset{\displaystyle OH}{|}}{CH}-\overset{\overset{\displaystyle O}{\|}}{C}-ONa \cdot 2H_2O$$

$(CHOHCOONa)_2 \cdot 2H_2O$	**Formula Wt 230.08**	**CAS No. 6106-24-7**

Suitable for standardization of Karl Fischer reagent as used for the determination of trace amounts of water (less than 1%).

GENERAL DESCRIPTION

Typical appearance: white solid
Analytical use: standardizing Karl Fischer reagent
Change in state (approximate): dehydrates at 120 °C
Aqueous solubility: 69 g in 100 mL at 43 °C

SPECIFICATIONS

Assay . . . . 99.0–101.0% $Na_2C_4H_4O_6 \cdot 2H_2O$
Loss on drying 15.61%–15.71%
pH of a 5% solution at 25.0 °C 7.0–9.0

Maximum Allowable

Insoluble matter 0.005%
Chloride (Cl). . . . . 5 ppm
Phosphate (PO_4). . . . . 5 ppm
Sulfate (SO_4) 0.005%
Ammonium (NH_4) 0.003%
Calcium (Ca) 0.01%
Heavy metals (as Pb). . . . . 5 ppm
Iron (Fe). 0.001%

TESTS

Assay. (Total alkalinity by nonaqueous titration). Weigh accurately about 0.45 g, and dissolve in 100 mL of glacial acetic acid. Titrate with 0.1 N perchloric acid in glacial acetic acid volumetric solution to a green end point using crystal violet indicator solution. One milliliter of 0.1 N perchloric acid corresponds to 0.01151 g of $Na_2C_4H_4O_6 \cdot 2H_2O$.

$$\% \ Na_2C_4H_4O_6 \cdot 2H_2O = \frac{(mL \times N \ HClO_4) \times 11.51}{Sample \ wt \ (g)}$$

Loss on Drying. Weigh accurately 3.0 g into a preconditioned, low-form weighing bottle, and dry at 150 °C to constant weight (minimum 4 h).

pH of a 5% Solution at 25.0 °C. (Page 49).

Insoluble Matter. (Page 25). Use 20 g dissolved in 200 mL of water.

Chloride. (Page 35). Use 2.0 g.

Phosphate. (Page 40, Method 1). Ignite 4.0 g in a platinum dish. Dissolve the residue in 5 mL of water, add 5 mL of nitric acid, and evaporate to dryness. Dissolve the residue in 25 mL of approximately 0.5 N sulfuric acid, and continue as described.

Sulfate. Drive off any moisture on a hot plate, and then carefully ignite 1.0 g in an electric muffle furnace until it is nearly free of carbon. Boil the residue with 10 mL of water and 0.5 mL of 30% hydrogen peroxide for 5 min. Add 5 mg of sodium carbonate and 1 mL of hydrochloric acid, and evaporate on a hot plate ($\approx$100 °C) to dryness. Dissolve the residue in 4 mL of hot water to which has been added 1 mL of dilute hydrochloric acid (1 + 19), filter through a small filter, wash with two 2-mL portions of water, and dilute the filtrate to 10 mL. Add 1 mL of 12% barium chloride reagent solution, and mix well. Any turbidity produced should not be greater than that in a standard prepared as described and to which the 12% barium chloride reagent solution is added at the same time it is added to the sample solution. To 10 mL of water, add 5 mg of sodium carbonate, 1 mL of hydrochloric acid, 0.5 mL of 30% hydrogen peroxide, and 0.05 mg of sulfate ion (SO_4), and evaporate on a hot plate ($\approx$100 °C) to dryness. Dissolve the residue in 9 mL of water, and add 1 mL of dilute hydrochloric acid (1 + 19). If necessary, adjust with water to the same volume as the sample solution, add 1 mL of 12% barium chloride reagent solution, and mix well.

Ammonium. Dissolve 1.0 g in 60 mL of water. To 20 mL, add 1 mL of 10% sodium hydroxide reagent solution, dilute with water to 50 mL, and add 2 mL of Nessler reagent. Any color should not exceed that produced by 0.01 mg of ammonium ion (NH_4) in an equal volume of solution containing the quantities of reagents used in the test.

Calcium. (By flame AAS, page 63).

> ***Sample Stock Solution.*** Dissolve 10.0 g of sample with water in a 100-mL volumetric flask, add 5 mL of hydrochloric acid, and dilute to the mark with water (1 mL = 0.1 g).

Element	Wavelength (nm)	Sample Wt (g)	Standard Added (mg)	Flame Type*	Background Correction
Ca	422.7	0.50	0.05; 0.10	N/A	No

*N/A is nitrous oxide/acetylene.

Heavy Metals. (Page 36, Method 1). Dissolve 6.0 g in about 20 mL of water, and dilute with water to 30 mL. Use 25 mL to prepare the sample solution, and use the remaining 5.0 mL to prepare the control solution.

Iron. (Page 38, Method 1). Use 1.0 g.

Sodium Tetraphenylborate
Sodium Tetraphenylboron

$NaB(C_6H_5)_4$ **Formula Wt 342.22**   **CAS No. 143-66-8**

GENERAL DESCRIPTION
Typical appearance: white solid
Analytical use: determination of potassium, ammonium, rubidium, and cesium
Aqueous solubility: freely soluble

SPECIFICATIONS

Assay . . . . $\geq$99.5% $NaB(C_6H_5)_4$

Maximum Allowable

Loss on drying 0.5%

Clarity of solution . . . . Passes test

TESTS

Assay. (By gravimetry). Weigh accurately about 0.5 g, dissolve in 100 mL of water, and add 1 mL of acetic acid. Add, with constant stirring, 25 mL of 5% potassium hydrogen phthalate solution, and allow to stand for 2 h. Filter through a tared, fine-porosity filtering crucible, wash with three 5-mL portions of saturated potassium tetraphenylborate solution, and dry at 105 °C for 1 h. The weight of the potassium tetraphenylborate multiplied by 0.9551 corresponds to the weight of the sodium tetraphenylborate.

$$\% \, NaB(C_6H_5)_4 = \frac{\text{Residue wt (g)} \times 95.51}{\text{Sample wt (g)}}$$

Loss on Drying. Weigh accurately about 1.5 g, and dry at 105 °C to constant weight.

Clarity of Solution. Weigh 1.5 g, add 250 mL of water and 0.75 g of hydrated aluminum oxide, stir for 5 min, and filter through a filter paper. Filter the first 25-mL portion again. The filtrate should be clear.

Sodium Thiocyanate

NaSCN   **Formula Wt 81.07**   **CAS No. 540-72-7**

GENERAL DESCRIPTION

Typical appearance: colorless or white solid
Analytical use: titrimetry
Change in state (approximate): melting point, 300 °C
Aqueous solubility: soluble in 0.6 parts of water

SPECIFICATIONS

Assay . . . . $\geq$98.0% NaSCN

Maximum Allowable

Insoluble matter 0.005%

Carbonate (as Na_2CO_3). 0.2%

Chloride (Cl). 0.01%

Sulfate (SO_4) 0.01%

Sulfide (S). 0.001%

Ammonium (NH_4) 0.002%

Heavy metals (as Pb). . . . . 5 ppm

Iron (Fe). . . . . 2 ppm

TESTS

Assay. (By indirect argentimetric titration). Weigh, to the nearest 0.1 mg, about 6.0 g of sample, dissolve with 100 mL of water in a 1000-mL volumetric flask, dilute to volume with water, and mix well. To 50.0 mL in a 250-mL glass-stoppered flask, add while agitating exactly 50.0 mL of 0.1 N silver nitrate volumetric solution, then add 3 mL of nitric acid and 2 mL of ferric ammonium sulfate indicator solution, and titrate the excess silver nitrate with 0.1 N ammonium thiocyanate volumetric solution. One milliliter of 0.1 N silver nitrate corresponds to 0.008107 g of NaSCN.

$$\% \text{ NaSCN} = \frac{[(50.0 \times N \text{ AgNO}_3) - (mL \times N \text{ NH}_4\text{SCN})] \times 8.107}{\text{Sample wt (g)} / 20}$$

Insoluble Matter. (Page 25). Use 20.0 g dissolved in 150 mL of water.

Carbonate. Dissolve 10.0 g in 100 mL of carbon dioxide-free water, add 0.10 mL of methyl orange indicator solution, and titrate with 0.1 N sulfuric acid. Not more than 3.8 mL of the acid should be required.

Chloride. Dissolve 0.50 g in 20 mL of water, filter if necessary through a chloride-free filter, and add 10 mL of 25% sulfuric acid reagent solution and 7 mL of 30% hydrogen peroxide. Evaporate to 20 mL by boiling in a well-ventilated hood, add 15–20 mL of water, and evaporate again. Repeat until all of the cyanide has been volatilized, cool, and dilute with water to 100 mL. To 20 mL of this solution, add 1 mL of nitric acid and 1 mL of silver nitrate reagent solution. Any turbidity should not exceed that produced by 0.01 mg of chloride ion (Cl) in an equal volume of solution containing the quantities of reagents used in the test.

Sulfate. (Page 40, Method 1).

Sulfide. Add 5.0 g to 80 mL of solution prepared by mixing 10 mL of ammonium hydroxide and 10 mL of silver nitrate reagent solution in 60 mL of water. Heat on a hot plate ($\approx$100 °C) with occasional shaking for 20 min, and transfer to a color comparison tube. Any color should not exceed that produced by 0.05 mg of sulfide ion (S) in an equal volume of solution containing the quantities of reagents used in the test.

Ammonium. (By colorimetry, page 32). Dilute the distillate from a 1.0-g sample to 100 mL, and use 50 mL of the dilution. For the standard, use 0.01 mg of ammonium ion (NH$_4$).

Heavy Metals. (Page 36, Method 1). Dissolve 6.0 g in about 20 mL of water, and dilute with water to 30 mL. Use 25 mL to prepare the sample solution, and use the remaining 5.0 mL to prepare the control solution.

Iron. Dissolve 10.0 g in 60 mL of water. To 30 mL of the solution, add 0.1 mL of hydrochloric acid, 6 mL of hydroxylamine hydrochloride reagent solution, and 4 mL of 1,10-phenanthroline reagent solution, and then add ammonium hydroxide to bring the pH to approximately 5. Any red color should not exceed that produced by 0.01 mg of iron in an equal volume of solution containing the quantities of reagents used in the test. Compare 1 h after adding the reagents to the sample and standard solutions.

Sodium Thiosulfate Pentahydrate

$Na_2S_2O_3 \cdot 5H_2O$ Formula Wt 248.19   CAS No. 10102-17-7

GENERAL DESCRIPTION

Typical appearance: colorless solid
Analytical use: titrimetry
Change in state (approximate): melting point, 48 °C, dehydrates completely at 100 °C
Aqueous solubility: very soluble

SPECIFICATIONS

Assay . . . . .99.5–101.0% $Na_2S_2O_3 \cdot 5H_2O$
pH of a 5% solution at 25.0 °C 6.0–8.4

Maximum Allowable

Insoluble matter 0.005%
Nitrogen compounds 0.002%
Sulfate and sulfite (as SO_4) 0.1%
Sulfide (S) . . . . .Passes test

TESTS

Assay. (By titration of reductive capacity). Weigh accurately about 1.0 g, and dissolve in 30 mL of water. In the pH of a 5% solution test, if a value greater than 7.5 was obtained, add 0.05 mL of 1 N acetic acid volumetric solution. Titrate with 0.1 N iodine volumetric solution to a blue color with starch indicator solution. One milliliter of 0.1 N iodine corresponds to 0.02482 g of $Na_2S_2O_3 \cdot 5H_2O$.

$$\% \, Na_2S_2O_3 \cdot 5H_2O = \frac{(mL \times N \, I_2) \times 24.82}{\text{Sample wt (g)}}$$

pH of a 5% Solution at 25.0 °C. (Page 49).

Insoluble Matter. (Page 25). Use 10.0 g dissolved in 100 mL of water.

Nitrogen Compounds. Dissolve 1.5 g in 10 mL of water in a flask, and add 10 mL of 25% sulfuric acid reagent solution. Connect the flask to a water-cooled reflux condenser, heat to boiling, and reflux for about 5 min. Cool, and rinse down the condenser with a small amount of water. Filter, using a filter that has been washed free of ammonia, into a flask, and wash the residue with 10–15 mL of water. Add to the flask 30 mL of freshly boiled 10% sodium hydroxide reagent solution and distill. To the distillate, add 2 mL of 10% sodium hydroxide reagent solution, dilute with water to 100 mL, and use 50 mL of the dilution. For the control, use 0.01 mg of nitrogen ion (N) and 0.25 g of sample.

Sulfate and Sulfite. Dissolve 1.0 g in 50 mL of water, and add 0.1 N iodine volumetric solution until the liquid has a faint yellow color. Dilute with water to a volume of 100 mL, and mix thoroughly. To 5.0 mL of the solution, add 5 mL of water, 1 mL of 1 N hydrochloric acid, and 1 mL of 12% barium chloride reagent solution. Any turbidity should not exceed that produced by 0.05 mg of sulfate ion (SO_4) in an equal volume of solution containing the quantities of reagents used in the test. Compare 10 min after adding the barium chloride to the sample and standard solutions.

Sulfide. Dissolve 1.0 g in 10 mL of water, and add 0.5 mL of alkaline lead solution (made by adding sufficient 10% sodium hydroxide reagent solution to 10% lead acetate reagent solution to redissolve the precipitate that is first formed). No dark color should be produced in 1 min. (Limit about 1 ppm)

Sodium Tungstate Dihydrate

$Na_2WO_4 \cdot 2H_2O$ Formula Wt 329.84   CAS No. 10213-10-2

GENERAL DESCRIPTION

Typical appearance: colorless or white solid
Analytical use: precipitant for alkaloids and proteins in biological materials
Change in state (approximate): dehydrates at 100 °C
Aqueous solubility: 124 g in 100 mL at 100 °C

SPECIFICATIONS

Assay . . . . 99.0–101.0% $Na_2WO_4 \cdot 2H_2O$

Maximum Allowable

Insoluble matter 0.01%
Titrable free base. . . . . 0.02 meq/g
Chloride (Cl) 0.005%
Molybdenum (Mo) 0.001%
Sulfate (SO_4) 0.01%
Heavy metals and iron (as Pb). 0.001%

TESTS

Assay. (By gravimetry). Weigh accurately about 1.0 g, and dissolve in 50 mL of water in a 150-mL beaker. Add 2 mL of 30% hydrogen peroxide and 45 mL of nitric acid. Heat on a steam bath for 90 min, cool, and filter through a fine filter paper. Wash with 50 mL of hot 1% nitric acid, dry in a tared, preconditioned crucible, burn off the paper at a low temperature, and ignite at 600 ± 25 °C to constant weight. The weight of the WO_3 multiplied by 1.4228 corresponds to the weight of the of $Na_2WO_4 \cdot 2H_2O$.

$$\% \, Na_2WO_4 \cdot 2H_2O = \frac{\text{Residue wt (g)} \times 142.28}{\text{Sample wt (g)}}$$

Insoluble Matter. (Page 25). Use 10.0 g dissolved in 100 mL of water.

Titrable Free Base. Dissolve 2.0 g in 50 mL of cold water, and add 0.10 mL of thymol blue indicator solution. A blue color should be produced, which is changed to yellow by the addition of not more than 4.0 mL of 0.01 N hydrochloric acid.

Chloride. Dissolve 1.0 g in 20 mL of water, filter if necessary through a chloride-free filter, add 3 mL of phosphoric acid, mix well, and dilute with water to 50 mL. Dilute 10 mL of the solution with water to 25 mL, and add 1 mL of nitric acid and 1 mL of silver nitrate reagent solution. Any turbidity should not exceed that produced by 0.01 mg of chloride ion (Cl) in an equal volume of solution containing the quantities of reagents used in the test.

Molybdenum.

> *Ammonium Citrate Solution.* Dissolve 300 g of dibasic ammonium citrate in 500 mL of water, and filter.
>
> *Sample Solution A.* Dissolve 0.50 g in 15 mL of ammonium citrate solution.
>
> *Standard Solution B.* Add a molybdenum standard solution containing 5 μg of molybdenum ion (Mo) to 15 mL of the ammonium citrate solution.
>
> *Blank Solution C.* Use 15 mL of the ammonium citrate solution.

Dilute the three solutions with water to 18 mL, then to each add the following reagents in order while stirring: 30 mL of dilute hydrochloric acid (1 + 1), 0.05 mL of 0.1 M cupric chloride in dilute hydrochloric acid (1 + 99), 3 mL of ammonium thiocyanate reagent solution (filtered if necessary), 3 mL of potassium iodide solution (50 g in 100 mL of water), and 2 mL of 1% sodium sulfite solution. Set a spectrophotometer at 460 nm, and using 1-cm cells, adjust the instrument to read 0 absorbance with blank solution C in the light path, then determine the absorbance of sample solution A and standard solution B. The absorbance of sample solution A should not exceed that of standard solution B. Compare within 30 min after development of the colors.

Sulfate. Dissolve 2.0 g in 100 mL of water, and add slowly with stirring 5 mL of hydrochloric acid. Evaporate to dryness, and heat for 20 min at 110 °C. Add 30 mL of water, 2.5 mL of hydrochloric acid, and 2 mL of cinchonine solution (described below), and heat just below the boiling point for 30 min. Dilute with water to 30 mL, and allow to stand until cool. Filter, and to 15 mL of the filtrate, add ammonium hydroxide dropwise until a slight permanent precipitate forms. Add just enough hydrochloric acid to redissolve the precipitate, and add 2 mL of 12% barium chloride reagent solution. Any turbidity should not exceed that produced in a standard made as follows: to 1 mL of the cinchonine solution, add 0.1 mg of sulfate ion (SO_4), dilute with water to 15 mL, and add 2 mL of 12% barium chloride reagent solution. Compare 10 min after adding the barium chloride to the sample and standard solutions.

> *Cinchonine Solution.* Dissolve 5 g of cinchonine in 50 mL of dilute hydrochloric acid (1 + 3).

Heavy Metals and Iron. Dissolve 2.0 g in 90 mL of water plus 2 mL of ammonium hydroxide, and dilute with water to 100 mL. To 10 mL of this solution in a separatory funnel, add 10 mL of dithizone extraction solution in chloroform. Shake well for 20 s, and discard the aqueous layer. Wash the chloroform layer with successive 10-mL portions of 0.5% ammonium hydroxide until the ammonium hydroxide solution remains nearly colorless. The pink color remaining in the chloroform should not exceed that produced by 0.002 mg of lead in 10 mL of dilute ammonium hydroxide (1 + 99) treated exactly as the 10-mL portion of the solution of the sample.

SPADNS

4,5-Dihydroxy-3-(4-sulfophenylazo)-2,7-naphthalene Disulfonic Acid, Trisodium Salt

$C_{16}H_9N_2Na_3O_{11}S_3$ **Formula Wt 570.42**   **CAS No. 23647-14-5**

GENERAL DESCRIPTION

Typical appearance: red to brown solid
Analytical use: fluoride indicator
Aqueous solubility: moderate (1–10%)

SPECIFICATIONS

Clarity of solution . . . . Passes test
Sensitivity to fluoride. . . . . Passes test

TESTS

Clarity of Solution. Dissolve 0.30 g in 100 mL of water. The solution is complete.

Sensitivity to Fluoride. To 50 mL of the fluoride working solution (described below), add 10 mL of the acid zirconyl–SPADNS reagent solution (described below). A distinct blue-violet color develops (λ maximum at 570 nm).

> *Acid Zirconyl–SPADNS Reagent Solution.* First, make SPADNS solution by dissolving 0.10 g of sample in 50 mL of water. Then make zirconyl acid reagent solution by dissolving 25 mg of zirconyl chloride octahydrate in 25 mL of water, add 70 mL of hydrochloric acid, and dilute to 100 mL with water. Finally, mix equal portions of SPADNS solution and zirconyl acid reagent solution.

> *Fluoride Working Solution.* Dilute 1 mL of the fluoride ion (F) standard solution to 10.0 mL with water. Dilute 1.0 mL of this solution to 50.0 mL.

Stannous Chloride Dihydrate
Tin(II) Chloride Dihydrate
$SnCl_2 \cdot 2H_2O$ Formula Wt 225.65   CAS No. 10025-69-1

GENERAL DESCRIPTION

Typical appearance: white to off-white solid
Analytical use: reducing agent
Change in state (approximate): melting point, 37–38 °C when heated rapidly
Aqueous solubility: 118 g in 100 mL at 0 °C

SPECIFICATIONS

Assay . . . . .98.0–103.0% $SnCl_2 \cdot 2H_2O$

Maximum Allowable

Solubility in hydrochloric acid. . . . . .Passes test
Sulfate (SO_4) . . . . .Passes test
Calcium (Ca) 0.005%
Iron (Fe). 0.003%
Lead (Pb). 0.01%
Potassium (K) 0.005%
Sodium (Na). 0.01%

TESTS

Assay. (By titration of reductive capacity of Sn^{II}). In a weighing bottle, weigh accurately about 2.0 g. Transfer to a 250-mL volumetric flask, first with three 5-mL portions of hydrochloric acid and then with water through which high-purity nitrogen gas has been freshly bubbled to expel oxygen. Dilute to volume with similar water. Displace air from the headspace with nitrogen gas. Stopper and mix thoroughly. Place a 50.0-mL aliquot of this solution in a 500-mL conical flask filled with nitrogen gas. Add 5 g of potassium sodium tartrate tetrahydrate and 60 mL of a cold saturated solution of sodium bicarbonate. Titrate promptly with 0.1 N iodine volumetric solution to a blue color with starch indicator. One milliliter of 0.1 N iodine corresponds to 0.01128 g of $SnCl_2 \cdot 2H_2O$.

$$\% \ SnCl_2 \cdot 2H_2O = \frac{(mL \times N \ I_2) \times 11.28}{\text{Sample wt (g)} / 5}$$

Solubility in Hydrochloric Acid. Dissolve 5.0 g in 5 mL of hydrochloric acid, heat to 40 °C if necessary, and dilute with 5 mL of water. The salt should dissolve completely.

Sulfate. Dissolve 5.0 g in 5 mL of hydrochloric acid, dilute the solution with water to 50 mL, filter if necessary, and heat to boiling. Add 5 mL of 12% barium chloride reagent solution, digest in a covered beaker on a hot plate ($\approx$100 °C) for 2 h, and allow to stand for at least 8 h. No precipitate should be formed. (Limit about 0.003%)

Calcium, Iron, Lead, Potassium, and Sodium. (By flame AAS, page 63).

Sample Stock Solution. Dissolve 20.0 g of sample in a mixture of 50 mL of water and 20 mL of hydrochloric acid. Heat to 50 °C if necessary. Cool to room temperature, transfer to a 100-mL volumetric flask, and dilute to the mark with water (1 mL = 0.20 g).

Element	Wavelength (nm)	Sample Wt (g)	Standard Added (mg)	Flame Type*	Background Correction
Ca	422.7	1.0	0.05; 0.10	N/A	No
Fe	248.3	4.0	0.12; 0.24	A/A	Yes
Pb	217.0	1.0	0.10; 0.20	A/A	Yes
K	766.5	0.20	0.01; 0.02	A/A	No
Na	589.0	0.20	0.01; 0.02	A/A	No

*A/A is air/acetylene; N/A is nitrous oxide/acetylene.

Starch, Soluble

CAS No. 9005-84-9

GENERAL DESCRIPTION

Typical appearance: white solid
Analytical use: indicator; iodometry
Aqueous solubility: soluble in water

SPECIFICATIONS

Solubility . . . . Passes test
pH of a 2% solution at 25.0 °C 5.0–7.0
Residue after ignition . . . . ≤0.4%
Sensitivity . . . . Passes test

TESTS

Solubility. Prepare a paste of 2.0 g of the sample with a little cold water, and add to it with stirring 100 mL of boiling water. The solution should be no more than opalescent, and on cooling it should remain liquid and not increase in opalescence.

pH of a 2% Solution at 25.0 °C. (Page 49). Use the solution obtained in the test for solubility.

Residue after Ignition. (Page 26). Ignite 1.0 g, and moisten the char with 1 mL of sulfuric acid.

Sensitivity. Prepare a paste of 1.0 g of the sample with a little cold water, and add it with stirring to 200 mL of boiling water. Cool, and add 5 mL of this solution to 100 mL of water containing 50 mg of potassium iodide, and add 0.05 mL of 0.1 N iodine volumetric solution. A deep blue color should be produced, which will be discharged by 0.05 mL of 0.1 N sodium thiosulfate volumetric solution.

Strontium Chloride Hexahydrate

$SrCl_2 \cdot 6H_2O$ Formula Wt 266.62   CAS No. 10025-70-4

GENERAL DESCRIPTION

Typical appearance: white to off-white solid

Analytical use: preparation of strontium standard solutions

Change in state (approximate): melting point, 61 °C when rapidly heated

Aqueous solubility: 198 g in 100 mL at 40 °C

SPECIFICATIONS

Assay . . . . 99.0–103.0% $SrCl_2 \cdot 6H_2O$

pH of a 5% solution at 25.0 °C 5.0–7.0

	Maximum Allowable
Insoluble matter	0.005%
Sulfate (SO_4)	0.001%
Barium (Ba)	0.05%
Calcium (Ca)	0.05%
Magnesium (Mg)	2 ppm
Heavy metals (as Pb)	5 ppm
Iron (Fe)	5 ppm

TESTS

Assay. (By complexometric titration of strontium). Weigh accurately about 1.0 g of sample, transfer to a 250-mL beaker, and dissolve in 50 mL of water. Add 5 mL of diethylamine and 35 mg of methylthymol blue indicator mixture. Titrate immediately with 0.1 M EDTA volumetric solution until the blue color turns to colorless or gray. One milliliter of 0.1 M EDTA corresponds to 0.02666 g of $SrCl_2 \cdot 6H_2O$.

$$\% \ SrCl_2 \cdot 6H_2O = \frac{(mL \times M \ EDTA) \times 26.66}{Sample \ wt \ (g)}$$

pH of a 5% Solution at 25.0 °C. (Page 49).

Insoluble Matter. (Page 25). Use 20.0 g dissolved in 150 mL of water.

Sulfate. (Page 40, Method 1). Use a 5.0-g sample. Allow 30 min for turbidity to form.

Barium, Calcium, and Magnesium. (By flame AAS, page 63).

> **Sample Stock Solution.** Dissolve 10.0 g of sample with water in a 100-mL volumetric flask, add 5 mL of nitric acid, and dilute to the mark with water (1 mL = 0.10 g).

Element	Wavelength (nm)	Sample Wt (g)	Standard Added (mg)	Flame Type*	Background Correction
Ba	553.6	0.20	0.10; 0.20	N/A	No
Ca	422.7	0.10	0.05; 0.10	N/A	No
Mg	285.2	2.50	0.005; 0.01	A/A	No

*A/A is air/acetylene; N/A is nitrous oxide/acetylene.

Heavy Metals. (Page 36, Method 1). Dissolve 6.0 g in about 20 mL of water, and dilute with water to 30 mL. Use 25 mL to prepare the sample solution, and use the remaining 5.0 mL to prepare the control solution.

Iron. Dissolve 2.0 g in 25 mL of water, add 2 mL of hydrochloric acid and 0.10 mL of 0.1 N potassium permanganate volumetric solution, and allow to stand for 5 min. Add 3 mL of ammonium thiocyanate reagent solution. Any red color should not exceed that produced by 0.01 mg of iron in an equal volume of solution containing the quantities of reagents used in the test.

Strontium Nitrate

$Sr(NO_3)_2$ **Formula Wt 211.63**   **CAS No. 10042-76-9**

GENERAL DESCRIPTION

Typical appearance: white solid
Analytical use: preparation of strontium standard solutions
Change in state (approximate): melting point, 570 °C
Aqueous solubility: 66 g in 100 mL at 20 °C

SPECIFICATIONS

Assay . . . . $\geq$99.0% $Sr(NO_3)_2$
pH of a 5% solution at 25.0 °C 5.0–7.0

	Maximum Allowable
Insoluble matter	0.01%
Loss on drying	0.1%
Chloride (Cl)	0.002%
Sulfate (SO_4)	0.005%
Barium (Ba)	0.05%
Calcium (Ca)	0.05%
Magnesium (Mg)	0.10%
Sodium (Na)	0.10%
Heavy metals (as Pb)	5 ppm
Iron (Fe)	5 ppm

TESTS

Assay. (By complexometric titration of strontium). Weigh accurately about 0.8 g, transfer to a 250-mL beaker, and dissolve in 50 mL of water. Add 5 mL of diethylamine and 25 mg of methylthymol blue indicator mixture. Titrate immediately with 0.1 M EDTA volumetric solution until the blue color just turns to colorless or gray. One milliliter of 0.1 M EDTA corresponds to 0.02116 g of $Sr(NO_3)_2$.

$$\% \ Sr(NO_3)_2 = \frac{(mL \times M \ EDTA) \times 21.16}{Sample \ wt \ (g)}$$

pH of a 5% Solution at 25.0 °C. (Page 49).

Insoluble Matter. (Page 25). Use 10.0 g dissolved in 100 mL of water.

Loss on Drying. Weigh accurately 2.0 g, and dry at 105 °C for 4 h.

Chloride. (Page 35). Use 0.50 g.

Sulfate. (Page 41, Method 2). Use 6.0 mL of dilute hydrochloric acid (1 + 1), and do two evaporations. Allow 30 min for turbidity to form.

Barium, Calcium, Magnesium, and Sodium. (By flame AAS, page 63).

> *Sample Stock Solution.* Dissolve 1.0 g of sample in 50 mL of water in a 100-mL volumetric flask, add 5 mL of nitric acid, and dilute to the mark with water (1 mL = 0.01 g).

Element	Wavelength (nm)	Sample Wt (g)	Standard Added (mg)	Flame Type*	Background Correction
Ba	553.6	0.20	0.10; 0.20	N/A	No
Ca	422.7	0.05	0.02; 0.04	N/A	No
Mg	285.2	0.02	0.01; 0.02	A/A	Yes
Na	589.0	0.02	0.01; 0.02	A/A	No

*A/A is air/acetylene; N/A is nitrous oxide/acetylene.

Heavy Metals. (Page 36, Method 1). Dissolve 6.0 g in about 20 mL of water, and dilute with water to 30 mL. Use 25 mL to prepare the sample solution, and use the remaining 5.0 mL to prepare the control solution.

Iron. Dissolve 2.0 g in 10 mL of dilute hydrochloric acid (1 + 1), and evaporate to dryness. Repeat the evaporation. Dissolve the residue in 20 mL of water, and add 2 mL of hydrochloric acid. Add 0.10 mL of 0.1 N potassium permanganate volumetric solution, dilute with water to 50 mL, and allow to stand for 5 min. Add 3 mL of ammonium thiocyanate reagent solution. Any red color should not exceed that produced by 0.01 mg of iron ion (Fe) treated exactly as the sample.

Succinic Acid
Butanedioic Acid

$$HO-\overset{\overset{\displaystyle O}{\|}}{C}-CH_2CH_2-\overset{\overset{\displaystyle O}{\|}}{C}-OH$$

HOOCCH₂CH₂COOH   **Formula Wt 118.09**   **CAS No. 110-15-6**

GENERAL DESCRIPTION

Typical appearance: colorless solid
Analytical use: internal standard in chromatography
Change in state (approximate): melting point, 185–191 °C

Aqueous solubility: 6.8 g in 100 mL at 20 °C; 121 g in 100 mL at 100 °C

pK_a: 4.2

SPECIFICATIONS

Assay . . . . ≥99.0% HOOCCH$_2$CH$_2$COOH
Melting point . . . . 185.0–191.0 °C

Maximum Allowable

Insoluble matter 0.01%
Residue after ignition 0.02%
Chloride (Cl) 0.001%
Phosphate (PO$_4$) 0.001%
Sulfate (SO$_4$) 0.003%
Nitrogen compounds (as N) 0.001%
Heavy metals (as Pb) . . . . 5 ppm
Iron (Fe). . . . . 5 ppm

TESTS

Assay. (By acid–base titrimetry). Weigh accurately about 0.3 g, dissolve in 25 mL of water in a conical flask, add 0.15 mL of phenolphthalein indicator solution, and titrate with 0.1 N sodium hydroxide. One milliliter of 0.1 N sodium hydroxide corresponds to 0.005905 g of HOOCCH$_2$CH$_2$COOH.

$$\% \text{ HOOCCH}_2\text{CH}_2\text{COOH} = \frac{(\text{mL} \times \text{N NaOH}) \times 5.905}{\text{Sample wt (g)}}$$

Melting Point. (Page 45).

Insoluble Matter. (Page 25). Use 10.0 g dissolved in 150 mL of water.

Residue after Ignition. (Page 26). Ignite 10.0 g, omitting the use of sulfuric acid.

Chloride. (Page 35). Use 1.0 g.

Phosphate. (Page 40, Method 1). Dissolve 1.0 g of sample in 40 mL of water. Add 5 mL of 5 N sulfuric acid, and continue as directed, using equal amounts of reagents and volume for the control solution.

Sulfate. Dissolve 1.0 g of sample in 20 mL of water, and filter through small, fine-porosity filter paper. Add two 2-mL portions of water through the filter, and collect the sample. For the control, use 0.03 mg of sulfate ion (SO$_4$) in the same volume of water. To the sample and control solutions, add 5 mL of ethyl alcohol, 1 mL of dilute hydrochloric acid (1 + 19), and 1 mL of 12% barium chloride reagent solution. After 30 min, the turbidity of the sample should not exceed that of the control.

Nitrogen Compounds. (Page 39). Use 1.0 g. For the standard, use 0.01 mg of nitrogen ion (N).

Heavy Metals. To 30 mL of ammonium hydroxide (1 + 9), add 25 mL of water. While stirring, add 6.0 g of sample, dissolve, and dilute to 60 mL with water. For the sample, take 50 mL of sample solution. For the control, take 10 mL of sample solution, add 0.02 mg of lead ion (Pb), and dilute to 50 mL. Add 10 mL of freshly prepared hydrogen sulfide water to each, and mix. Any brown color produced in the test solution within 5 min should not be darker than that produced in the control.

Iron. (Page 38, Method 1). Use 2.0 g.

Sucrose

$C_{12}H_{22}O_{11}$ **Formula Wt 342.30**   **CAS No. 57-50-1**

GENERAL DESCRIPTION

Typical appearance: white solid
Analytical use: optical rotation standard
Change in state (approximate): decomposes at 160–186 °C
Aqueous solubility: 200 g in 100 mL at 20 °C

SPECIFICATIONS

Specific rotation $[\alpha]_D^{25°C}$ . . . . +66.3° to +66.8°

	Maximum Allowable
Insoluble matter	0.005%
Loss on drying	0.03%
Residue after ignition.	0.01%
Titrable acid.	0.0008 meq/g
Chloride (Cl).	0.005%
Sulfate and sulfite (as SO_4)	0.005%
Heavy metals (as Pb).	5 ppm
Iron (Fe).	5 ppm
Invert sugar	0.05%

TESTS

Specific Rotation. (Page 46). Weigh accurately 26.0 g, and dissolve in 90 mL of water in a 100-mL volumetric flask. Dilute to volume with water at 25 °C. Observe the optical rotation in a polarimeter at 25 °C using sodium light, and calculate the specific rotation.

Insoluble Matter. (Page 25). Use 40.0 g dissolved in 150 mL of water. Reserve the filtrate, without the washings, for preparation of sample solution A.

Loss on Drying. Weigh accurately 4.9–5.1 g in a tared, preconditioned dish or crucible, and dry at 105 °C for 2 h.

Residue after Ignition. (Page 26). Ignite 10.0 g, and moisten the char with 1 mL of sulfuric acid.

Titrable Acid. To 100 mL of carbon dioxide-free water, add 0.10 mL of phenolphthalein indicator solution and 0.01 N sodium hydroxide until a pink color is produced. Dissolve 10.0 g of the sample in this solution, and titrate with 0.01 N sodium hydroxide to the same end point. Not more than 0.83 mL should be required.

For the Determination of Chloride, Sulfate and Sulfite, Heavy Metals, and Iron

Sample Solution A. Transfer the filtrate from the test for insoluble matter to a 200-mL volumetric flask, and dilute to volume with water (1 mL = 200 mg).

Chloride. Dilute 10 mL of sample solution A (2.0-g sample), with water to 50 mL. Dilute 5 mL of this solution with water to 25 mL, and add 1 mL of nitric acid and 1 mL of silver nitrate reagent solution. Any turbidity should not exceed that produced by 0.01 mg of chloride ion (Cl) in an equal volume of solution containing the quantities of reagents used in the test.

Sulfate and Sulfite. (Page 41, Method 2). Use 5.0 mL of sample solution A (1.0-g sample), add 1 mL of bromine water, and boil. Omit evaporation to dryness, and follow Method 1 for the remainder of the test, starting with adding 1 mL of dilute hydrochloric acid (1 + 19).

Heavy Metals. (Page 36, Method 1). Use 25.0 mL of sample solution A (5.0-g sample) to prepare the sample solution, and use 5.0 mL of sample solution A to prepare the control solution.

Iron. (Page 38, Method 1). Use 10.0 mL of sample solution A (2.0-g sample).

Invert Sugar. Prepare 1 L of a reagent solution containing 150 g of potassium bicarbonate, 100 g of potassium carbonate, and 6.928 g of cupric sulfate pentahydrate. Transfer 50 mL of this reagent to a 400-mL beaker, cover, heat to boiling, and allow to boil for 1 min. Dissolve 10.0 g of the sample in water, dilute with water to 50 mL, and add this solution to the solution in the 400-mL beaker. Heat to boiling, and boil for 5 min. At the end of this period, stop the reaction by adding 100 mL of cold, recently boiled water. Filter through a tared, preconditioned filtering crucible, wash thoroughly, and dry at 105 °C. The weight of the precipitate should not exceed 0.0277 g.

Sulfamic Acid

NH$_2$SO$_3$H   **Formula Wt 97.09**   **CAS No. 5329-14-6**

GENERAL DESCRIPTION

Typical appearance: white to off-white solid
Analytical use: acidimetric standard
Change in state (approximate): melting point, 205 °C, with decomposition
Aqueous solubility: 40 g in 100 mL at 70 °C

SPECIFICATIONS

Assay (dried basis).. 99.3–100.3% NH$_2$SO$_3$H

Maximum Allowable

Insoluble matter 0.01%
Residue after ignition..0.01%
Chloride (Cl)..0.001%
Sulfate (SO$_4$) 0.05%
Heavy metals (as Pb)...0.001%
Iron (Fe)..5 ppm

TESTS

Assay. (By acid–base titrimetry). Weigh accurately about 0.4 g, previously dried over sulfuric acid for 2 h, and dissolve in about 30 mL of water. Add 0.15 mL of phenolphthalein indicator solution, and titrate with 0.1 N sodium hydroxide. One milliliter of 0.1 N sodium hydroxide corresponds to 0.009709 g of NH$_2$SO$_3$H.

$$\% \; NH_2SO_3H = \frac{(mL \times N \; NaOH) \times 9.709}{Sample \; wt \; (g)}$$

Insoluble Matter. (Page 25). Use 10.0 g dissolved in 200 mL of water.

Residue after Ignition. (Page 26). Ignite 10.0 g without addition of sulfuric acid.

Chloride. (Page 35). Use 1.0 g.

Sulfate. (Page 40, Method 1). Dissolve 1.0 g of sample in 100 mL of water, and use 10 mL (0.1 g) of this solution for the test.

Heavy Metals. (Page 36, Method 1). Dissolve 4.0 g in 30 mL of water, neutralize to litmus paper with ammonium hydroxide, and dilute with water to 40 mL. Use 30 mL to prepare the sample solution, and use the remaining 10 mL to prepare the control solution.

Iron. (Page 38, Method 1). Use 2.0 g.

Sulfanilic Acid

4-Aminobenzenesulfonic Acid

NH$_2$C$_6$H$_4$SO$_3$H (anhydrous) **Formula Wt 173.19** **CAS No. 121-57-3**
NH$_2$C$_6$H$_4$SO$_3$H · H$_2$O (monohydrate) **Formula Wt 191.21**

Note: This reagent is available in both the anhydrous and monohydrate forms. The identity should be indicated on the label.

GENERAL DESCRIPTION

Typical appearance: white solid

Analytical use: nitrite determination

Change in state (approximate): decomposes at 288 °C without melting

Aqueous solubility: 1 g in 100 mL at 20 °C

SPECIFICATIONS

Assay . . . . 98.0–102.0% of the form offered

Maximum Allowable

Residue after ignition 0.01%

Insoluble in sodium carbonate solution 0.02%

Chloride (Cl) 0.002%

Nitrite (NO$_2$) . . . . 0.5 ppm

Sulfate (SO$_4$) 0.01%

TESTS

Assay. (By acid–base titrimetry). Weigh accurately about 0.7 g, and dissolve by warming gently in 50 mL of water. Cool, and titrate with 0.1 N sodium hydroxide, using 0.15 mL of phenolphthalein indicator solution. One milliliter of 0.1 N sodium hydroxide corresponds to 0.01732 g of NH$_2$C$_6$H$_4$SO$_3$H and to 0.01912 g of NH$_2$C$_6$H$_4$SO$_3$H · H$_2$O.

$$\% \ NH_2C_6H_4SO_3H = \frac{(mL \times N \ NaOH) \times 17.32}{Sample \ wt \ (g)}$$

$$\% \ NH_2C_6H_4SO_3H \cdot H_2O = \frac{(mL \times N \ NaOH) \times 19.12}{Sample \ wt \ (g)}$$

Residue after Ignition. Gently ignite 10.0 g in a tared, preconditioned crucible or dish until charred. Slowly raise the temperature until all carbon is removed, and finally heat at 600 ± 25 °C for 15 min.

Insoluble in Sodium Carbonate Solution. Dissolve 5.0 g in 50 mL of a clear 5% sodium carbonate solution, and allow to stand in a covered beaker for 1 h. If an insoluble residue remains, filter through a tared, preconditioned porous porcelain or a platinum filtering crucible, wash with cold water, and dry at 105 °C.

Chloride. Boil 5.0 g with 100 mL of water until dissolved. Cool, dilute with water to 100 mL, mix well, and filter through a chloride-free filter. Dilute 10 mL of the filtrate with water to 20 mL, and add 1 mL of nitric acid and 1 mL of silver nitrate reagent solution. Any turbidity should not exceed that produced by 0.01 mg of chloride ion (Cl) in an equal volume of solution containing the quantities of reagents used in the test. Save the remaining filtrate for the test for sulfate.

Nitrite. Dissolve 0.70 g in 100 mL of water, warming if necessary but keeping the temperature below 30 °C. For the control, dissolve 0.20 g in about 75 mL of water, add 0.00025 mg of nitrite ion (NO_2), and dilute with water to 100 mL. To each solution, add 5 mL of sulfanilic-1-naphthylamine solution (described below), and allow to stand for 10 min. Any pink color produced in the solution of the sample should not exceed that in the control.

> **Sulfanilic-1-Naphthylamine Solution.** Dissolve 0.5 g of sulfanilic acid in 150 mL of 36% acetic acid. Dissolve 0.1 g of N-(1-naphthyl)ethylenediamine dihydrochloride in 150 mL of 36% acetic acid. Mix the two solutions. If a pink color develops on standing, it may be discharged with a little zinc dust.

Sulfate. Cool about 30 mL of the solution reserved from the test for chloride to about 0 °C, and filter. To 10 mL of the filtrate, add 1 mL of dilute hydrochloric acid (1 + 19) and 1 mL of 12% barium chloride reagent solution. Any turbidity should not exceed that produced by 0.05 mg of sulfate ion (SO_4) in an equal volume of solution containing the quantities of reagents used in the test. Compare 10 min after adding the barium chloride to the sample and standard solutions.

5-Sulfosalicylic Acid Dihydrate
2-Hydroxy-5-sulfobenzoic Acid Dihydrate

$HOC_6H_3(COOH)SO_3H \cdot 2H_2O$ **Formula Wt 254.22**   **CAS No. 5965-83-3**

GENERAL DESCRIPTION
Typical appearance: white solid
Analytical use: detection of iron (ferric)
Change in state (approximate): melting point, anhydrous, 120 °C
Aqueous solubility: very soluble

SPECIFICATIONS

Assay . . . . .99.0–101.0% $HOC_6H_3(COOH)SO_3H \cdot 2H_2O$

Maximum Allowable

Insoluble matter 0.02%
Residue after ignition 0.1%
Chloride (Cl) 0.001%
Salicylic acid (HOC_6H_4COOH) 0.04%
Sulfate (SO_4) 0.02%
Heavy metals (as Pb) 0.002%
Iron (Fe). 0.001%

TESTS

Assay. (By acid–base titrimetry). Weigh accurately about 5.0 g, dissolve in 50 mL of water, add 0.15 mL of phenolphthalein indicator solution, and titrate with 1 N sodium hydroxide volumetric solution. One milliliter of 1 N sodium hydroxide corresponds to 0.1271 g of $HOC_6H_3(COOH)SO_3H \cdot 2H_2O$.

$$\% \ HOC_6H_3(COOH)SO_3H \cdot 2H_2O = \frac{(mL \times N \ NaOH) \times 12.71}{Sample \ wt \ (g)}$$

Insoluble Matter. (Page 25). Use 5.0 g dissolved in 50 mL of water.

Residue after Ignition. (Page 26). Ignite 1.0 g in a preconditioned dish, other than platinum, and moisten the char with 1 mL of sulfuric acid.

Chloride. (Page 35). Use 1.0 g.

Salicylic Acid. Dissolve 5.0 g in 15 mL of cold water, add 10 mL of dilute hydrochloric acid (1 + 1), and extract with 50 mL of benzene in a 250-mL separatory funnel. Shake for 1.5 min, allow the layers to separate, and discard the aqueous layer. Add about 2 g of anhydrous sodium sulfate to the benzene layer, shake for 1 min, and allow the sodium sulfate to settle. Pour some of the clear benzene extract into a beaker, transfer 10 mL to a test tube, and add 10 mL of ferric ammonium sulfate solution (described below). Shake the tube vigorously for 15 s, and centrifuge at about 2500 rpm (about 1700 rcf) for 3 min. The color in the clear lower aqueous layer should not exceed that produced by 2.0 mg of salicyclic acid treated in exactly the same manner as the sample. In cases of borderline results or apparent nonconformity, aspirate off the benzene layer, and determine the absorbance of the aqueous layer from the sample and the standard in 1-cm cells at 540 nm, using the ferric ammonium sulfate solution as the reference liquid in the spectrophotometer. Calculate the percent of salicylic acid as follows:

$$\% \ Salicylic \ acid = \frac{C \times (A_u / A_s)}{W \times 10}$$

where C = weight, in mg, of salicylic acid in standard; A_u = absorbance of sample solution; A_s = absorbance of standard solution; and W = weight, in g, of sample.

Ferric Ammonium Sulfate Solution. To 200 mL of water in a beaker, add 10% sulfuric acid until the pH is about 3, dissolve 200 mg of ferric ammonium sulfate dodecahydrate in the solution, and adjust the pH to 2.45 (using a pH meter) with 10% sulfuric acid.

Sulfate. (Page 40, Method 1).

For the Determination of Heavy Metals and Iron

Sample Solution A. Gently ignite 4.0 g in a preconditioned crucible or dish, other than platinum, until charred. Cool, moisten the char with 1 mL of sulfuric acid, and ignite again slowly until all the carbon and excess sulfuric acid have been volatilized. Add 3 mL of hydrochloric acid and 0.5 mL of nitric acid, cover, and digest on a hot plate ($\approx$100 °C) until dissolved. Remove the cover, evaporate to dryness on a hot plate ($\approx$100 °C), dissolve the residue in 4 mL of 1 N acetic acid, and dilute with water to 100 mL (1 mL = 0.04 g).

Heavy Metals. (Page 36, Method 1). Use 25 mL of sample solution A (1.0 g) to prepare the sample solution.

Iron. (Page 38, Method 1). Use 25 mL of sample solution A (1.0 g).

Sulfuric Acid

H_2SO_4 Formula Wt 98.08   CAS No. 7664-93-9

GENERAL DESCRIPTION

Typical appearance: clear, colorless liquid
Analytical use: digestion of organic matter; pH modification; titration of bases
Change in state (approximate): boiling point, 290 °C
Aqueous solubility: miscible; generates heat
Density: 1.84
pK_a: ~ −3

SPECIFICATIONS

Appearance . . . . Free from suspended or insoluble matter
Assay . . . . 95.0–98.0% H_2SO_4

Maximum Allowable

Color (APHA) 10
Residue after ignition. . . . . 5 ppm
Chloride (Cl). . . . . 0.2 ppm
Nitrate (NO_3) . . . . 0.5 ppm
Ammonium (NH_4) . . . . 2 ppm
Substances reducing permanganate (as SO_2) . . . . 2 ppm
Arsenic (As) . . . . 0.01 ppm
Heavy metals (as Pb). . . . . 1 ppm
Iron (Fe). . . . . 0.2 ppm
Mercury (Hg) . . . . 5 ppb

TESTS

Appearance. Mix the material in the original container, pour 10 mL into a test tube (20 × 150 mm), and compare with distilled water in a similar tube. The liquids should be equally clear and free from suspended matter.

Assay. (By acid–base titrimetry). Tare a small glass-stoppered flask, add about 1 mL of the sample, and weigh accurately. Cautiously add 30 mL of water, cool, add 0.15 mL of methyl orange indicator solution, and titrate with 1 N sodium hydroxide volumetric solution. One milliliter of 1 N sodium hydroxide corresponds to 0.04904 g of H_2SO_4.

$$\% \ H_2SO_4 = \frac{(mL \times N \ NaOH) \times 4.904}{\text{Sample wt (g)}}$$

Color (APHA). (Page 43).

Residue after Ignition. (Page 26). Evaporate 200.0 g (110 mL) to dryness in a tared, preconditioned platinum dish, and ignite at 600 ± 25 °C for 15 min.

Chloride. Place 40 mL of water in each of three beakers. To one, carefully add 50.0 g (27.2 mL) of sample, and to the others, carefully add 50.0 g (27.2 mL) of chloride-free sulfuric acid. To the second beaker, add 0.01 mg of chloride ion (Cl). Cool to room temperature, mix, and to each beaker, add 1 mL of nitric acid and 1 mL of silver nitrate reagent solution. Mix well, and if necessary, make the volume of each solution identical by adding water. Transfer equal portions of each solution to Nessler tubes. After 10 min, any turbidity of the sample solution should not exceed that of the standard. The blank should be free of turbidity. The comparison can be aided by use of a nephelometer.

Nitrate. (Page 38, Method 1).

Sample Solution A. Cautiously add 40.0 g (22 mL) to 2.0 mL of water, dilute to 50 mL with brucine sulfate reagent solution, and mix.

Control Solution B. Cautiously add 40.0 g (22 mL) to 2.0 mL of nitrate ion (NO_3) standard solution, dilute to 50 mL with brucine sulfate reagent solution, and mix.

Continue with the procedure, starting with the preparation of blank solution C.

Ammonium. (By differential pulse polarography, page 53). Carefully add 2.0 g (1.1 mL) to 4 mL of water in a 25-mL volumetric flask in an ice bath. Use 9.5 mL of ammonia-free 6 N sodium hydroxide reagent solution for neutralization and 0.004 mg of ammonium ion (NH_4) for the standard.

Substances Reducing Permanganate. Cool 50 mL of water in a 150-mL beaker in an ice bath. Slowly add 25 mL (46 g) of sample while stirring and keeping the solution cool. Remove from the bath, and add 0.30 mL of 0.01 N potassium permanganate. The solution should remain pink for at least 5 min.

Arsenic. (Page 34). To 300.0 g (165 mL), add 3 mL of nitric acid, and evaporate on a hot plate (≈100 °C) to about 10 mL. Cool, cautiously dilute with about 20 mL of water, and evaporate just to dense fumes of sulfur trioxide. Cool, and cautiously wash the solution into a generator flask with the aid of 50 mL of water. Omit the addition of dilute sulfuric acid. For the standard, use 0.003 mg of arsenic ion (As).

Heavy Metals. (Page 36, Method 1). Add 20.0 g (11 mL) to about 10 mg of sodium carbonate dissolved in a small quantity of water. Heat over a low flame until nearly dry, then add 1 mL of nitric acid. Evaporate to dryness, add about 20 mL of water, and dilute with water to 25 mL.

Iron. (Page 38, Method 1). Add 50.0 g (27 mL) to about 10 mg of sodium carbonate dissolved in a small quantity of water. Evaporate to dryness by heating on an electric hot plate. Cool, add 5 mL of dilute hydrochloric acid (1 + 1), cover with a watch glass, and digest on a hot plate ($\approx$100 °C) for 15 min. Cool, and dilute with water to 25 mL.

Mercury. To each of two 125-mL conical flasks, add 20 mL of water and 5 mL of 4% potassium permanganate solution. For the sample, add to one flask, slowly and with cooling, 11.0 g (6.0 mL) of the sulfuric acid. For the control, add to the second flask 1.0 g (0.5 mL) of the sulfuric acid and 0.05 μg of mercury ion (Hg) [0.5 mL of a solution freshly prepared by diluting 1 mL of mercury ion (Hg) standard solution, with water to 500 mL]. Place both flasks on a hot plate ($\approx$100 °C) for 15 min, then cool to room temperature. Determine the mercury in each by CVAAS (page 65), using 1 mL of 10% hydroxylamine hydrochloride reagent solution and 2 mL of 10% stannous chloride reagent solution for the reduction. Any mercury found in the sample should not exceed that in the control.

Sulfuric Acid, Fuming

CAS No. 8014-95-7

Note: This specification applies to fuming sulfuric acid with nominal contents of 15%, 20%, or 30% free SO_3.

GENERAL DESCRIPTION

Typical appearance: colorless to light-brown viscous liquid
Analytical use: dehydrating agent
Aqueous solubility: miscible; generates heat
pK_a: ~ –3

SPECIFICATIONS

Assay (free SO_3) . . . . 12.0–17.0%, 18.0–24.0%, or 26.0–29.5%

	Maximum Allowable
Residue after ignition	0.002%
Nitrate (NO_3)	1 ppm
Ammonium (NH_4)	3 ppm
Arsenic (As)	0.03 ppm
Iron (Fe)	2 ppm

TESTS

Assay. (By acid–base titrimetry). Weigh accurately about 4.0 g in a tared Dely weighing tube. Carefully transfer to a casserole containing 100 mL of carbon dioxide-free water by placing the tip of the tube beneath the surface and flushing the tube with carbon dioxide-free water. Cool, add 0.15 mL of phenolphthalein indicator solution, and titrate with 1 N sodium hydroxide volumetric solution. One milliliter of 1 N sodium hydroxide corresponds to 0.04904 g of H_2SO_4.

$$\% \ H_2SO_4 = \frac{(mL \times N \ NaOH) \times 4.904}{Sample \ wt \ (g)}$$

where % free SO_3 = 4.445 × (% H_2SO_4 − 100).

> *Note:* For accurate results, a weight buret should be used. The Dely weighing tube and its use are described in standard reference books.

Residue after Ignition. (Page 26). Evaporate 50.0 g (27 mL) to dryness in a tared, pre-conditioned platinum dish, and ignite at 600 ± 25 °C for 15 min.

Nitrate. (Page 38, Method 1).

> *Sample Solution A.* Cautiously add 10.0 g (5 mL) to 1.0 mL of water, dilute to 50 mL with brucine sulfate reagent solution, and mix.

> *Control Solution B.* Cautiously add 10.0 g (5 mL) to 1.0 mL of nitrate ion (NO_3) standard solution, dilute to 50 mL with brucine sulfate reagent solution, and mix.

Continue with the procedure, starting with the preparation of blank solution C.

Ammonium. (By differential pulse polarography, page 53). Carefully add 2.0 g (1.1 mL) to 4 mL of water in a 25-mL volumetric flask in an ice bath. Use approximately 10 mL of ammonia-free 6 N sodium hydroxide reagent solution for neutralization and 0.006 mg of ammonium ion (NH_4) for the standard. The volume of buffer solution may be lowered to 4 mL.

Arsenic. (Page 34). To 100.0 g (52 mL), add 3 mL of nitric acid, and evaporate on a hot plate (≈100 °C) to about 10 mL. Cool, cautiously dilute with about 20 mL of water, and evaporate to about 5 mL. Cool, cautiously dilute again with about 20 mL of water, and evaporate just to dense fumes of sulfur trioxide. Cool, and cautiously wash the solution into a generator flask with the aid of 50 mL of water. Omit the addition of 20 mL of dilute sulfuric acid (1 + 4). For the standard, use 0.003 mg of arsenic ion (As).

Iron. (Page 38, Method 1). Cautiously add 5.0 g (2.6 mL) to about 10 mg of sodium carbonate, and evaporate to dryness by heating on a hot plate (≈100 °C). Cool, add 5 mL of dilute hydrochloric acid (1 + 1), cover with a watch glass, and digest on a hot plate (≈100 °C) for 15 min. Cool, and dilute with water to 25 mL.

Sulfuric Acid, Ultratrace

H_2SO_4 Formula Wt 98.08   CAS No. 7664-93-9

Suitable for use in ultratrace elemental analysis.

Note: Reagent must be packaged in a preleached Teflon bottle and used in a clean laboratory environment to maintain purity.

GENERAL DESCRIPTION

Typical appearance: clear, colorless liquid
Analytical use: digestion for trace analysis
Change in state (approximate): boiling point, 290 °C
Aqueous solubility: miscible; generates heat
Density: 1.84
pK_a: ~ –3

SPECIFICATIONS

Assay . . . . .85–98% H_2SO_4

	Maximum Allowable
Chloride (Cl)	0.2 ppm
Nitrate (NO_3)	0.5 ppm
Mercury (Hg)	1 ppb
Selenium (Se)	100 ppb
Aluminum (Al)	1 ppb
Barium (Ba)	1 ppb
Boron (B)	5 ppb
Cadmium (Cd)	1 ppb
Calcium (Ca)	1 ppb
Chromium (Cr)	1 ppb
Cobalt (Co)	1 ppb
Copper (Cu)	1 ppb
Iron (Fe)	5 ppb
Lead (Pb)	1 ppb
Lithium (Li)	1 ppb
Magnesium (Mg)	1 ppb
Manganese (Mn)	1 ppb
Molybdenum (Mo)	1 ppb
Potassium (K)	1 ppb
Silicon (Si)	5 ppb
Sodium (Na)	5 ppb
Strontium (Sr)	1 ppb
Tin (Sn)	1 ppb
Titanium (Ti)	1 ppb
Vanadium (V)	1 ppb
Zinc (Zn)	1 ppb
Zirconium (Zr)	1 ppb

TESTS

Assay. (By acid–base titrimetry). Tare a small, glass-stoppered flask, add about 1 mL of the sample, and weigh accurately. Cautiously add 30 mL of water, cool, add 0.15 mL of methyl orange indicator solution, and titrate with 1 N sodium hydroxide volumetric solution. One milliliter of 1 N sodium hydroxide corresponds to 0.04904 g of H_2SO_4.

$$\% \ H_2SO_4 = \frac{(mL \times N \ NaOH) \times 4.904}{Sample \ wt \ (g)}$$

Chloride. (Page 35).

Nitrate. (Page 38, Method 1).

Mercury. (By CVAAS, page 65). Cool each of three 100-mL volumetric flasks containing about 35 mL of water in an ice-water bath for 15 min, and then slowly and cautiously add 10.0 g of sample to each. To the second and third flasks, add mercury ion (Hg) standards of 10 ng (1.0 ppb) and 20 ng (2.0 ppb), respectively. Add 5 mL of nitric acid to all, and dilute to the mark with water. Mix. Zero the instrument with a blank, and determine the mercury content using a suitable mercury analyzer system. Cool to room temperature (1.0 mL of 0.01 μg/mL Hg = 10 ng).

Selenium. (By HGAAS, page 66). To a set of three 100-mL volumetric flasks, transfer 10.0 g (5.4 mL) of sample. To two flasks, add selenium ion (Se) standards of 1.0 μg (100 ppb) and 2.0 μg (200 ppb), respectively. Dilute each to 100 mL with (1 + 1) hydrochloric acid. Mix. Prepare fresh working standard before use.

Trace Metals. Determine the aluminum, barium, boron, cadmium, calcium, chromium, cobalt, copper, iron, lead, lithium, magnesium, manganese, molybdenum, potassium, silicon, sodium, strontium, tin, titanium, vanadium, zinc, and zirconium by the ICP–OES method described on page 69.

Sulfurous Acid

(a solution of SO₂ in water)   CAS No. 7782-99-2

GENERAL DESCRIPTION

Typical appearance: colorless, clear liquid
Analytical use: reducing agent
Density: 1.03
pK_a: 1.8

SPECIFICATIONS

Assay . . . . ≥6.0% SO₂

	Maximum Allowable
Residue after ignition.	0.005%
Chloride (Cl).	5 ppm
Heavy metals (as Pb).	2 ppm
Iron (Fe).	5 ppm

TESTS

Assay. (By iodometric titration). Tare a glass-stoppered conical flask containing 50.0 mL of 0.1 N iodine. Quickly introduce about 2 mL of the sample, stopper, and weigh again. Titrate the excess iodine with 0.1 N sodium thiosulfate volumetric solution, adding 3 mL of starch indicator solution near the end of the titration. One milliliter of 0.1 N iodine consumed corresponds to 0.003203 g of SO_2.

$$\% \ SO_2 = \frac{[(mL \times N \ I_2) - (mL \times N \ Na_2S_2O_3)] \times 3.203}{Sample \ wt \ (g)}$$

Residue after Ignition. Evaporate 20.0 g (20 mL) to dryness on a hot plate ($\approx$100 °C) in a tared, preconditioned crucible or dish, and ignite at 600 $\pm$ 25 °C for 15 min.

Chloride. Digest 10.0 g (10 mL) with 2 mL of nitric acid on a hot plate ($\approx$100 °C) for 1 h. Cool, and dilute with water to 100 mL. To 20 mL of the solution, add 1 mL of nitric acid and 1 mL of silver nitrate reagent solution. Any turbidity should not exceed that produced by 0.01 mg of chloride ion (Cl) in an equal volume of solution containing the quantities of reagents used in the test.

Heavy Metals. (Page 36, Method 1). To 10.0 g (10 mL) add 10 mL of water, boil to expel the sulfur dioxide, and dilute with water to 25 mL.

Iron. (Page 38, Method 1). To 2.0 g (2.0 mL), add about 10 mg of sodium carbonate, and evaporate to dryness. Dissolve the residue with 0.5 mL of hydrochloric acid, add 0.5 mL of nitric acid, and evaporate again to dryness. Dissolve the residue in 2 mL of hydrochloric acid, dilute with water to 50 mL, and use the solution without further acidification.

Tannic Acid

CAS No. 1401-55-4

Note: For analytical purposes, tannic acid should be the hydrolyzable type, such as that isolated from nutgalls, sumac, or seed pods of Tara.

GENERAL DESCRIPTION

Typical appearance: off-white to light brown solid
Analytical use: clarifying agent; pH control
Aqueous solubility: 1 g dissolves in 0.35 g

SPECIFICATIONS

Identification . . . . Passes test

Maximum Allowable

Loss on drying 12.0%
Residue after ignition 0.5%
Heavy metals (as Pb) 0.003%
Zinc (Zn) 0.005%
Sugars, dextrin . . . . Passes test

TESTS

Identification. To 2 mL of an aqueous 10% tannic acid solution, add 0.1 mL of an aqueous 5% lead nitrate solution. A precipitate forms immediately but completely dissolves on continued swirling to yield a clear solution.

Loss on Drying. Weigh accurately 1.0 g, and dry at 105 °C to constant weight.

Residue after Ignition. (Page 26). Ignite 5.0 g and moisten the char with 1 mL of sulfuric acid. Retain the residue for the test for zinc.

Heavy Metals. (Page 36, Method 1). Transfer 0.67 g into a 150-mL beaker, and cautiously add 15 mL of nitric acid and 5 mL of 70% perchloric acid. Evaporate the mixture to dryness on a hot plate ($\approx$100 °C) in a suitable hood, cool, add 2 mL of hydrochloric acid, and wash down the sides of the beaker with water. Carefully evaporate the solution to dryness on a hot plate ($\approx$100 °C), rotating the beaker to avoid spattering. Repeat the addition of 2 mL of hydrochloric acid, wash down the sides of the beaker with water, and again evaporate to dryness on a hot plate ($\approx$100 °C). Cool the residue, take up in 1 mL of hydrochloric acid and 10 mL of water, and dilute with water to 25 mL.

Zinc. Dissolve the residue from the residue after ignition test in 2 mL of glacial acetic acid, dilute with 8 mL of water, and filter if necessary. Add 0.5 g of sodium acetate and 5 mL of freshly prepared hydrogen sulfide water, and mix. Any white turbidity should not exceed that produced by 0.25 mg of zinc ion (Zn) treated exactly as the sample residue.

Sugars, Dextrin. Dissolve 2.0 g in 10 mL of water, and add 20 mL of ethyl alcohol. The mixture should be clear and should remain clear after standing for 1 h. Add 0.5 mL of ether. No turbidity should be produced.

Tartaric Acid

2,3-Dihydroxybutanedioic Acid

$$\underset{HO-\overset{O}{\underset{\parallel}{C}}-\overset{OH}{\underset{\mid}{C}}H-\overset{OH}{\underset{\mid}{C}}H-\overset{O}{\underset{\parallel}{C}}-OH}{}$$

HOOC(CHOH)₂COOH   **Formula Wt 150.09**   **CAS No. 87-69-4**

GENERAL DESCRIPTION

Typical appearance: colorless solid

Analytical use: complexing agent

Change in state (approximate): melting point, 170 °C

Aqueous solubility: 139 g in 100 mL at 20 °C

SPECIFICATIONS

Assay . . . . ≥99.0% $C_4H_6O_6$

Maximum Allowable

Insoluble matter 0.005%

Residue after ignition. 0.02%

Chloride (Cl). 0.001%

Oxalate (C_2O_4). . . . . .Passes test

Phosphate (PO_4). 0.001%

Sulfur compounds (as SO_4) 0.002%

Heavy metals (as Pb). . . . . .5 ppm

Iron (Fe). . . . . .5 ppm

TESTS

Assay. (By acid–base titrimetry). Weigh accurately about 3.0 g of sample, and transfer to a 250-mL conical flask with about 75 mL of carbon dioxide-free water. Add 0.15 mL of phenolphthalein indicator solution, and titrate with 1 N sodium hydroxide volumetric solution to a faint-pink end point that persists for at least 30 s. One milliliter of 1 N sodium hydroxide corresponds to 0.07504 g of $C_4H_6O_6$.

$$\% \ C_4H_6O_6 = \frac{(mL \times N \ NaOH) \times 7.504}{Sample \ wt \ (g)}$$

Insoluble Matter. (Page 25). Use 20.0 g dissolved in 200 mL of water.

Residue after Ignition. (Page 26). Ignite 5.0 g, and moisten the char with 2 mL of sulfuric acid.

Chloride. (Page 35). Use 1.0 g.

Oxalate. Dissolve 5.0 g in 30 mL of water, and divide into two equal portions. Neutralize one portion with ammonium hydroxide, using litmus paper as the indicator. Add the other portion, and dilute with water to 40 mL. Shake well, cool, and allow to stand at 15 °C for 15 min. Filter, and to 20 mL of the filtrate, add an equal volume of a saturated, filtered solution of calcium sulfate. No turbidity or precipitate should appear in 2 h. (Limit about 0.1%)

For the Determination of Phosphate, Sulfur Compounds, and Heavy Metals

Sample Solution A. To 10.0 g, add about 10 mg of sodium carbonate, 5 mL of nitric acid, and 5 mL of 30% hydrogen peroxide. Digest in a covered beaker on a hot plate ($\approx$100 °C) until reaction ceases. Wash down the cover glass and the sides of the beaker, and evaporate to dryness. Repeat the treatment with nitric acid and peroxide, and again evaporate to dryness. Dissolve the residue in about 30 mL of water, filter if necessary, and dilute with water to 50 mL (1 mL = 0.2 g).

Phosphate. (Page 40, Method 1). Dilute 10 mL of sample solution A (2-g sample) with water to 20 mL, add 25 mL of 0.5 N sulfuric acid, and continue as described.

Sulfur Compounds. (Page 40, Method 1). Use 4.2 mL of sample solution A (0.84-g sample).

Heavy Metals. (Page 36, Method 1). Dilute 20 mL of sample solution A (4-g sample) with water to 25 mL.

Iron. (Page 38, Method 1). Use 2.0 g.

Tetrabromophenolphthalein Ethyl Ester

$C_{22}H_{14}Br_4O_4$ (free acid)	**Formula Wt 661.97**	**CAS No. 1176-74-5**
$C_{22}H_{13}Br_4O_4K$ (potassium salt)	**Formula Wt 700.08**	**CAS No. 62637-91-6**

GENERAL DESCRIPTION

Typical appearance: yellow to red solid
Analytical use: indicator

SPECIFICATIONS

Assay . . . . $\geq$95.0%
Melting point (free acid) . . . . 208.0–213.0 °C (within an interval
of no more than 3.0°)
Solubility . . . . Passes test

TESTS

Assay. (By spectrophotometry). Weigh accurately 50.0 mg, transfer to a 100-mL volumetric flask, dissolve in about 75 mL of a 50/50 mixture of ethyl alcohol and pH 7.00 buffer solution (described below), and dilute to volume with the mixture. Pipet 2 mL of this solution into a 100-mL volumetric flask, and dilute to volume with buffer solution. Determine the absorbance of this solution in 1.00-cm cells at the wavelength of maximum absorption, approximately 593 nm, using the pH 7.00 buffer solution as the blank. Determine the molar absorptivity by dividing the absorbance by 1.5×10^{-5} (that is, the molar concentration of the tetrabromophenolphthalein ethyl ester) and calculate the assay value as follows:

$$\% \ C_{22}H_{14}Br_4O_4 = \frac{\text{Molar absorptivity at the maximum}}{76,922} \times 100$$

$$\% \ C_{22}H_{13}Br_4O_4K = \frac{\text{Molar absorptivity at the maximum}}{76,922} \times 100 \times 1.0575$$

pH 7.00 Buffer Solution. Add 291 mL of 0.1 M sodium hydroxide to 500 mL of 0.1 M monobasic potassium phosphate, and dilute with water to 1 L.

Note: Because the solutions are light-sensitive, protect them from direct light.

Melting Point. (Page 45).

Solubility. For the free acid, dissolve 0.1 g in 100 mL of toluene. For the potassium salt, dissolve 0.1 g in 100 mL of water. For either, the solution should be clear, and dissolution should be complete.

Tetrabutylammonium Bromide
TBAB
(CH₃CH₂CH₂CH₂)₄NBr   **Formula Wt 322.37**   **CAS No. 1643-19-2**

GENERAL DESCRIPTION
Typical appearance: white solid
Analytical use: ion pair reagent

SPECIFICATIONS
General Use
Assay . . . . ≥98.0% (CH₃CH₂CH₂CH₂)₄NBr

Maximum Allowable

Tributylamine (Bu₃N) 0.5%
Tributylamine hydrobromide (Bu₃N · HBr) 0.5%

Specific Use
Polarographic test. . . . . .Passes test

TESTS

Assay. (By argentimetric titration of bromide content). Weigh accurately about 1.0 g, and dissolve in a 250-mL conical flask with 50 mL of water. Add 0.05 g of sodium bicarbonate and 0.2 mL of eosin Y reagent solution. Titrate with 0.1 N silver nitrate volumetric solution to the first color change. One milliliter of 0.1 N silver nitrate volumetric solution corresponds to 0.03224 g of tetrabutylammonium bromide.

$$\% \ (CH_3CH_2CH_2CH_2)_4NBr = \frac{(mL \times N \ AgNO_3) \times 32.24}{Sample \ wt \ (g)}$$

Tributylamine and Tributylamine Hydrobromide. (By potentiometric titration, page 27).

Sample Preparation. Accurately weigh and dissolve 20.0 g of sample in 90 mL of isopropyl alcohol in a 400-mL beaker.

Measure the pH of the sample solution on a standardized pH meter. If the pH is less than 5, then no tributylamine is present. Potentiometrically titrate with 1 N sodium hydroxide volumetric solution to determine the tributylamine hydrobromide content. The first inflection point, A mL, will be approximately pH 5. A second inflection point, B mL, will occur at approximately pH 10.5.

$$\% \ Bu_3N \cdot HBr = \frac{[(B-A) \times N \ NaOH] \times 26.6265}{Sample \ wt \ (g)}$$

If the pH is greater than 5, then potentiometrically titrate with 1 N hydrochloric acid volumetric solution to determine the tributylamine content. The inflection point, C mL, will be approximately pH 5. In the calculations, D mL is the total mL of 1 N hydrochloric acid added. Titrate the same sample with 1.0 N sodium hydroxide volumetric solution to an inflection point approximately pH 10.5 at E mL to determine the tributylamine hydrobromide content.

$$\% \ Bu_3N = \frac{(C \times N \ HCl) \times 18.5353}{Sample \ wt \ (g)}$$

$$\% \ Bu_3N \cdot HBr = \frac{[(E \times N \ NaOH) - (D \times N \ HCl)] \times N \ NaOH \times 26.6265}{Sample \ wt \ (g)}$$

Polarographic Test. (Page 50).

Cathode: Dropping Hg electrode

Anode: Pt wire

Reference: Similar saturated calomel electrode: make up with 1.0 M tetrabutylammonium chloride.

Prepare a 1.0 M solution (32.24 g in 100 mL of water), and add 0.2 mL of 1.0 M tetrabutylammonium hydroxide. Deoxygenate with high-purity nitrogen gas until minimum background current is achieved (approximately 5 min). By using a drop time of 1 s, scan at 0.01 V/s from −1.2 to −2.6 V vs. electrode with a sensitivity of 0.5 μA full scale with modulation amplitude of 50 mV. The material is acceptable if no impurity current greater than 0.05 μA is observed.

Tetrabutylammonium Hydroxide, 1.0 M Aqueous Solution

TBAH

$(CH_3CH_2CH_2CH_2)_4 \cdot NOH$ Formula Wt 259.47   CAS No. 2052-49-5

Suitable for general use or in ultraviolet spectrophotometry. Product labeling shall designate the uses for which suitability is represented on the basis of meeting the relevant specifications and tests. The ultraviolet spectrophotometry specifications include all of the specifications for general use.

GENERAL DESCRIPTION

Typical appearance: colorless liquid
Analytical use: buffer; titrant; ion pair reagent

SPECIFICATIONS

General Use

Assay . . . . 1.0 ± 0.02 M as $(CH_3CH_2CH_2CH_2)_4 \cdot NOH$

	Maximum Allowable
Color (APHA)	30
Carbonate (CO_3)	0.1%
Tributylamine (Bu_3N)	0.2%
Halide (as Cl)	0.08%
Potassium (K)	2 ppm
Sodium (Na)	2 ppm

Specific Use

Ultraviolet Spectrophotometry

Wavelength (nm)	Absorbance (AU)
280	0.04
254	0.07
245	0.10
240	0.15
Polarographic test	Passes test

TESTS

Assay. (By acid–base titrimetry). Pipet 25.0 mL into a 250-mL conical flask. Add 100 mL of water and 0.15 mL of phenolphthalein indicator solution, and titrate with 1 N hydrochloric acid volumetric solution until the solution becomes completely colorless.

$$M\ (CH_3CH_2CH_2CH_2)_4 \cdot NOH = \frac{mL \times N\ HCl}{25.0\ mL}$$

Color (APHA). (Page 43).

Carbonate and Tributylamine. (By a two-step potentiometric titration, page 27).

Part A. Pipet 25.0 mL of sample into a 400-mL beaker, and dilute with water to 200 mL. Standardize a pH meter with a glass electrode and reference electrode, usually a saturated calomel electrode (SCE) or a combination pH electrode. Use a magnetic stir plate and magnetic stir bar or equivalent agitation throughout the titration. Add dilute nitric acid $(1 + 9)$ to the sample just short of the phenolphthalein end point (or pH 9.0). Titrate with 1 N hydrochloric acid volumetric solution potentiometrically to the first end point. Save the sample for the tributylamine determination in Part B.

Part B. Deoxygenate the sample from Part A with high-purity nitrogen gas for at least 5 min. Dropwise, add 1 N sodium hydroxide volumetric solution to a pH of 11–12. Titrate with standardized 0.1 N hydrochloric acid.

$$\text{ppm CO}_3 \text{ (part A)} = \frac{(\text{mL of 1.0 N HCl}) \times \text{N HCl} \times 60{,}000}{25.0 \text{ mL}}$$

$$\text{ppm CO}_3 \text{ (part B)} = \frac{(\text{mL of 0.1 N HCl}) \times \text{N HCl} \times 60{,}000)}{25.0 \text{ mL}}$$

$$\text{ppm Carbonate} = \text{ppm CO}_3 \text{ (Part A)} - \text{ppm CO}_3 \text{ (Part B)}$$

$$\text{ppm Tributylamine} = \frac{\text{ppm CO}_3 \text{ (Part B)} \times 185}{60}$$

Halide. (By potentiometric titration, page 27). Pipet 25.0 mL of sample into a 400-mL beaker, and dilute to 200 mL. Add 0.05 mL of phenolphthalein indicator solution to the solution, and agitate with a magnetic stirrer and stir bar. Add nitric acid dropwise until the sample is acidic. Titrate with 0.1 N silver nitrate volumetric solution using a silver electrode and a suitable millivolt meter.

$$\% \text{ Halide} = \frac{\text{mL of 0.1 N AgNO}_3 \times \text{N AgNO}_3 \times 3.5453}{25.0 \text{ g}}$$

Potassium and Sodium. (By flame AAS, page 63).

Sample Stock Solution. Evaporate 50.0 g to dryness in a platinum crucible on a hot plate ($\approx$100 °C), and take up the residue in 10 mL of 5% nitric acid. Transfer to a 100-mL volumetric flask, and dilute to the mark with 5% nitric acid (1 mL = 0.50 g).

Element	Wavelength (nm)	Sample Wt (g)	Standard Added (mg)	Flame Type*	Background Correction
K	766.5	10.0	0.02; 0.04	A/A	No
Na	589.0	10.0	0.01; 0.02	A/A	No

* A/A is air/acetylene.

Ultraviolet Spectrophotometry. Use the procedure on page 86 to determine absorbance.

Polarographic Test. (Page 50).

Cathode: Dropping Hg electrode

Anode: Pt wire

Reference: Saturated calomel electrode: make up with 1.0 M tetrabutylammonium chloride

Warm 1.0 M TBAH (25.95 g in 100 mL) to 30–35 °C. Deoxygenate with high-purity nitrogen gas until minimum background current is achieved (approximately 5 min). By using a drop time of 1 s, scan at 0.01 V/s from −1.2 to −2.6 V vs. SCE with a sensitivity of 0.5 μA full scale with modulation amplitude of 50 mV. The material is acceptable if no impurity current greater than 0.05 μA is observed.

Tetrahydrofuran

C_4H_8O **Formula Wt 72.11**   **CAS No. 109-99-9**

Note: Generally, a stabilizer is present to retard peroxide formation.

GENERAL DESCRIPTION

Typical appearance: clear, colorless liquid
Analytical use: solvent
Change in state (approximate): boiling point, 67 °C
Aqueous solubility: miscible
Density: 0.89
pK_a: −2.2

SPECIFICATIONS

Assay . . . . ≥99.0% C_4H_8O

Maximum Allowable

Color (APHA) 20
Peroxide (as H_2O_2) 0.015%
Residue after evaporation 0.03%
Water (H_2O) 0.05%

TESTS

Assay. Analyze the sample by gas chromatography using the general parameters cited on page 80. The following specific conditions are also required.

Column: Type I, methyl silicone

Measure the area under all peaks, and calculate the tetrahydrofuran content in area percent. Correct for water content.

Color (APHA). (Page 43).

Peroxide. To 44.0 g (50 mL) in a conical flask, add 10 mL of 10% potassium iodide solution and 2 mL of 10% sulfuric acid reagent solution. Titrate the liberated iodine with 0.1 N sodium thiosulfate volumetric solution. Not more than 3.9 mL should be required.

Caution: If peroxide is present, do not perform the test for residue after evaporation.

Residue after Evaporation. (Page 25). Evaporate 20.0 g (22.5 mL) to dryness in a tared, preconditioned dish on a hot plate ($\approx$100 °C), and dry the residue at 105 °C for 30 min.

Water. (Page 31, Method 2). Use 100 μL (88 mg).

Tetramethylammonium Bromide
TMAB
$(CH_3)_4NBr$ Formula Wt 154.05   CAS No. 64-20-0

GENERAL DESCRIPTION
Typical appearance: white solid
Analytical use: ion pair reagent
Aqueous solubility: soluble

SPECIFICATIONS
General Use
Assay . . . . $\geq$98.0% as $(CH_3)_4NBr$

Maximum Allowable

Trimethylamine (Me_3N) 0.5%
Trimethylamine hydrobromide ($Me_3N \cdot HBr$) 0.5%

Specific Use
Polarographic test . . . . Passes test

TESTS

Assay. (By argentimetric titration of bromide content). Weigh accurately about 0.5 g, and dissolve in a 250-mL conical flask with 50 mL of water. Add 0.05 g of sodium bicarbonate and 0.2 mL of eosin Y reagent solution. Titrate with 0.1 N silver nitrate volumetric solution to the first color change. One milliliter of 0.1 N silver nitrate corresponds to 0.01540 g of tetramethylammonium bromide.

$$\% \ (CH_3)_4NBr = \frac{(mL \times N \ AgNO_3) \times 15.40}{\text{Sample wt (g)}}$$

Trimethylamine and Trimethylamine Hydrobromide. (By potentiometric titration, page 27).

Sample Preparation. Accurately weigh and dissolve about 20 g of sample in 90 mL of isopropyl alcohol in a 400-mL beaker.

Measure the pH of the sample solution on a standardized pH meter. If the pH is less than 5, then no trimethylamine is present. Potentiometrically titrate with 1 N sodium hydroxide volumetric solution to determine the trimethylamine hydrobromide content. The first inflection point, A mL, will be approximately pH 5. A second inflection point, B mL, will occur at approximately pH 10.5.

$$\% \ Me_3N \cdot HBr = \frac{(B-A) \times N \ NaOH \times 14.0023}{Sample \ wt \ (g)}$$

If the pH is greater than 5, then potentiometrically titrate with 1 N hydrochloric acid volumetric solution to determine the trimethylamine content. The inflection point, C mL, will be approximately pH 5. In the calculations, D mL is the total milliliters of 1 N hydrochloric acid added. Titrate the same sample with 1 N sodium hydroxide volumetric solution to an inflection point at approximately pH 10.5 at E mL.

$$\% \ Me_3N = \frac{C \times N \ HCl \times 8.9111}{Sample \ wt \ (g)}$$

$$\% \ Me_3N \cdot HBr = \frac{(E-D) \times N \ NaOH \times 14.0023}{Sample \ wt \ (g)}$$

Polarographic Test. (Page 50).

  Cathode: Dropping Hg electrode

  Anode: Pt wire

  Reference: Similar saturated calomel electrode: make up with 1.0 M tetrabutylammonium chloride

Prepare a 1.0 M solution (15.4 g in 100 mL of water), and add 0.2 mL of 1.0 M tetrabutylammonium hydroxide. Deoxygenate with nitrogen until minimum background current is achieved (approximately 5 min). By using a drop time of 1 s, scan at 0.01 V/s from −1.2 to −2.6 V vs. SCE with a sensitivity of 0.5 μA full scale with modulation amplitude of 50 mV. The material is acceptable if no impurity current greater than 0.05 μA is observed.

Tetramethylammonium Hydroxide, 1.0 M Aqueous Solution
TMAH; *N,N,N*-Trimethylmethanaminium Hydroxide
$(CH_3)_4NOH$ **Formula Wt 91.16**   **CAS No. 75-59-2**

GENERAL DESCRIPTION
Typical appearance: colorless to straw-colored liquid
Analytical use: buffer; titrant; ion pair reagent
Density: 1.0

SPECIFICATIONS

General Use

Assay . . . . 1.0 ± 0.02 M as $(CH_3)_4NOH$

Maximum Allowable

Color (APHA). 10
Chloride (Cl) 0.03%
Residue after ignition . . . . 5 ppm
Potassium (K) . . . . 2 ppm
Sodium (Na) . . . . 2 ppm

Specific Use

Polarographic test. . . . . Passes test

TESTS

Assay. (By acid–base titrimetry). Pipet 25.0 mL into a 250-mL conical flask. Add 100 mL of water and 0.15 mL of phenolphthalein indicator solution, and titrate with 1 N hydrochloric acid volumetric solution until the solution becomes completely colorless.

$$M\ (CH_3)_4NOH = \frac{mL \times N\ HCl}{25.0\ mL}$$

Color (APHA). (Page 43).

Chloride. (Page 35). Dilute 1.0 mL to 10.0 mL with water in a volumetric flask. Mix well. Pipet 1.0 mL into 15 mL of dilute nitric acid (1 + 15). Filter, if necessary, through a small chloride-free filter, and add 1 mL of silver nitrate reagent solution. Any turbidity in the sample solution should not exceed that produced by 0.03 mg of chloride ion (Cl) in an equal volume of solution containing the quantities of reagents used in the test.

Residue after Ignition. (Page 26). Evaporate 200.0 g in a tared, preconditioned platinum dish to dryness in a hood, and ignite.

Potassium and Sodium. (By flame AAS, page 63).

Sample Stock Solution. Evaporate 50.0 g to dryness in a platinum crucible, and take up the residue in 5 mL of 5% nitric acid solution. Transfer to a 100-mL volumetric flask, and dilute to the mark with 5% nitric acid solution (1 mL = 0.50 g).

Element	Wavelength (nm)	Sample Wt (g)	Standard Added (mg)	Flame Type*	Background Correction
K	766.5	10.0	0.02; 0.04	A/A	No
Na	589.0	10.0	0.01; 0.02	A/A	No

*A/A is air/acetylene.

Polarographic Test. (Page 50).

Cathode: Dropping Hg electrode

Anode: Pt wire

Reference: Similar saturated calomel electrode: make up with 1.0 M tetrabutylammonium chloride

Deoxygenate the TMAH with nitrogen until minimum background current is achieved (approximately 5 min). By using a drop time of 1 s, scan at 0.01 V/s from –1.2 to –2.6 V vs. SCE with a sensitivity of 0.5 μA full scale with modulation amplitude of 50 mV. The material is acceptable if no impurity current greater than 0.05 μA is observed.

Tetramethylsilane

$$CH_3 - \underset{\underset{CH_3}{|}}{\overset{\overset{CH_3}{|}}{Si}} - CH_3$$

(CH₃)₄Si   **Formula Wt 88.22**   **CAS No. 75-76-3**

GENERAL DESCRIPTION

Typical appearance: clear, colorless liquid
Analytical use: silylation of sample for gas chromatography analysis
Change in state (approximate): boiling point, 26–28 °C

SPECIFICATIONS

Residue after evaporation . . . . ≤0.05%
Suitability for proton NMR position reference. . . . . Passes test

TESTS

Residue after Evaporation. (Page 25). Evaporate 2.0 g (3.1 mL) to dryness in a tared, preconditioned dish, and dry the residue at 105 °C for 30 min.

Suitability for Proton NMR Position Reference. An undiluted sample should display no impurity peak in its proton NMR reference spectrum more intense than 20% of the ^{13}C satellite line at ±59.1 Hz from tetramethylsilane.

Thioacetamide

$$CH_3 - \overset{\overset{S}{\|}}{C} - NH_2$$

CH₃CSNH₂   **Formula Wt 75.13**   **CAS No. 62-55-5**

Note: This reagent, when stored under ordinary conditions, may decompose slightly and fail the test for clarity of a 2% solution.

GENERAL DESCRIPTION

Typical appearance: white to off-white solid
Analytical use: sulfide generation
Change in state (approximate): melting point, 111–114 °C
Aqueous solubility: 16.3 g in 100 mL at 25 °C
pK_a: 13.4

SPECIFICATIONS

Assay . . . . ≥99.0% CH_3CSNH_2
Melting point . . . . 111–114 °C
Clarity of a 2% solution . . . . Passes test

Maximum Allowable

Residue after ignition 0.05%

TESTS

Assay. (By argentimetric titration). Weigh accurately 1.5 g, dissolve in water, and dilute with water to 500 mL in a volumetric flask. To 50.0 mL of this solution, add 1 mL of ammonium hydroxide and 50.0 mL of 0.1 N silver nitrate volumetric solution. Allow to stand for 20 min, carefully filter through a filtering crucible or a sintered glass funnel that has been cleaned with dilute nitric acid, and wash the funnel and flask well with water. To the clear solution, add 5 mL of nitric acid and 2 mL of ferric ammonium sulfate indicator solution, and titrate with 0.1 N potassium thiocyanate volumetric solution. One milliliter of 0.1 N silver nitrate corresponds to 0.003757 g of CH_3CSNH_2.

$$\% \ CH_3CSNH_2 = \frac{[(50.0 \times N \ AgNO_3) - (mL \times N \ KSCN)] \times 3.757}{Sample \ wt \ (g) \ / \ 10}$$

Melting Point. (Page 45).

Clarity of a 2% Solution. Dissolve 2.0 g in 100 mL of water. The solution should be clear and colorless.

Residue after Ignition. (Page 26). Ignite 2.0 g, and moisten the char with 1 mL of sulfuric acid.

Thiophene
Thiofuran

C_4H_4S   **Formula Wt 84.14**   **CAS No. 110-02-1**

GENERAL DESCRIPTION

Typical appearance: liquid
Analytical use: preparation of standards

Change in state (approximate): boiling point, 84 °C
Aqueous solubility: insoluble
Density: 1.05

SPECIFICATIONS

Assay . . . . ≥99.0% C_4H_4S

Maximum Allowable

Color (APHA) 20
Water (H_2O) 0.05%

TESTS

Assay. Analyze the sample by gas chromatography using the general parameters cited on page 80. The following specific conditions are also required.

 Column: Type I, methyl silicone

Measure the area under all peaks, and calculate the thiophene content in area percent. Correct for water content.

Color (APHA). (Page 43).

Water. (Page 31, Method 2). Use 100 μL (105 mg) of the sample.

Thiourea

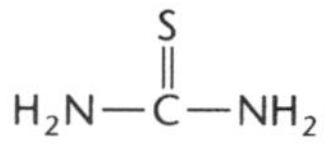

$$H_2N-\overset{\overset{\textstyle S}{\|}}{C}-NH_2$$

NH_2CSNH_2   **Formula Wt 76.12**   **CAS No. 62-56-6**

GENERAL DESCRIPTION

Typical appearance: white solid
Analytical use: determination of bismuth
Change in state (approximate): melting point, 174 –177 °C
Aqueous solubility: 16.5 g in 100 mL at 25 °C

SPECIFICATIONS

Assay (dried basis) . . . . ≥99.0% NH_2CSNH_2
Melting point . . . . 174–177 °C
Solubility in water . . . . Passes test

Maximum Allowable

Residue after ignition. 0.1%
Loss on drying 0.5%

TESTS

Assay. (By argentimetric titration). Weigh accurately about 1.0 g of sample from the loss on drying test, dissolve in water, and dilute to 250 mL in a volumetric flask. To 20.0 mL of this solution in a preconditioned glass-stoppered flask, add 25.0 mL of 0.1 N silver nitrate volumetric solution and 10 mL of 10% ammonium hydroxide reagent solution. Stopper the flask, shake vigorously for 2 min, heat to boiling, and cool. To the cooled solution, add 10 mL of 10% nitric acid, shake vigorously, and filter through a filtering crucible or a sintered glass funnel that has been cleaned with 10% nitric acid, and wash the funnel and flask well with water. To the filtrate plus washings, add 2 mL of ferric ammonium sulfate indicator solution, and titrate with 0.1 N potassium thiocyanate volumetric solution. One milliliter of 0.1 N silver nitrate corresponds to 0.003806 g of NH_2CSNH_2.

$$\% \ NH_2CSNH_2 = \frac{[(mL \times N \ AgNO_3) - (mL \times N \ KSCN)] \times 3.806}{Sample \ wt \ (g) \ / \ 12.5}$$

Melting Point. (Page 45).

Solubility in Water. Dissolve 1 g in 20 mL of water (50 °C). The solution should be clear and colorless.

Residue after Ignition. (Page 26). Ignite 10.0 g.

Loss on Drying. Weigh accurately about 1.5 g, and dry at 105 °C for 2 h.

Thorium Nitrate Tetrahydrate

$Th(NO_3)_4 \cdot 4H_2O$ **Formula Wt 552.12**   **CAS No. 13470-07-0**

GENERAL DESCRIPTION

Typical appearance: white solid
Analytical use: titrant in fluoride determination
Aqueous solubility: very soluble

SPECIFICATIONS

Assay . . . . 98.0–102.0% $Th(NO_3)_4 \cdot 4H_2O$

	Maximum Allowable
Insoluble matter	0.01%
Chloride (Cl)	0.002%
Sulfate (SO₄)	0.01%
Heavy metals (as Pb)	0.002%
Iron (Fe)	0.002%
Rare earth elements (as La)	0.2%

Titanium (Ti) 0.01%
Calcium (Ca) 0.01%
Magnesium (Mg) 0.005%
Potassium (K) 0.005%
Sodium (Na). 0.05%

TESTS

Assay. (By gravimetry). Weigh accurately about 1.0 g, and dissolve in 100 mL of water in a beaker. Add 2 mL of 10% sulfuric acid, heat to boiling, and add 20 mL of a hot 10% oxalic acid solution. Cool, filter, wash with a little water, and ignite the precipitate to constant weight.

$$\% \ Th(NO_3)_4 \cdot 4H_2O = \frac{\text{Precipitate wt (g)} \times 209.0}{\text{Sample wt (g)}}$$

Insoluble Matter. (Page 25). Use 10.0 g dissolved in 100 mL of water. Reserve the filtrate and washings for the preparation of sample solution A.

For the Determination of Chloride, Sulfate, Heavy Metals, Iron, Rare Earth Elements, and Titanium

Sample Solution A. Transfer the filtrate and washings reserved from the test for insoluble matter to a 200-mL volumetric flask, and dilute with water to 200 mL (1 mL = 0.05 g).

Chloride. (Page 35). Dilute 10 mL (0.5-g sample) of sample solution A with water to 20 mL.

Sulfate. Dilute 10 mL (0.5-g sample) of sample solution A with water to 20 mL. Add 5 mL of 2 N ammonium acetate solution (described below), and transfer the solution to a separatory funnel. Extract with successive 10-mL portions of 0.2 M *N*-phenylbenzohydroxamic acid (described below) in chloroform until the color of the chloroform layer remains unchanged. Wash the aqueous layer twice with 5-mL portions of chloroform. Transfer the aqueous solution to a beaker, and evaporate on a hot plate ($\approx$100 °C) to 2–3 mL. Add 2 mL of nitric acid and 2 mL of hydrochloric acid, cover, and digest on the hot plate ($\approx$100 °C) until any reaction ceases. Uncover, wash down the sides of the beaker, and evaporate to dryness. Dissolve the residue in 4 mL of water plus 1 mL of dilute hydrochloric acid (1 + 19). Filter if necessary through a small filter, wash with two 2-mL portions of water, and dilute with water to 10 mL. Add 1 mL of 12% barium chloride reagent solution. Any turbidity should not exceed that produced by 0.05 mg of sulfate ion (SO_4) in an equal volume of solution that has been treated exactly like the sample. Compare 10 min after adding the barium chloride to the sample and standard solutions.

Ammonium Acetate Solution, 2 N. Dissolve 15.4 g of ammonium acetate in water, and dilute to 100 mL.

N-*Phenylbenzohydroxamic Acid Solution, 0.2 M.* Dissolve 4.26 g of *N*-phenylbenzohydroxamic acid in chloroform, and dilute to 100 mL with chloroform.

Heavy Metals. (Page 36, Method 1). Use 30 mL of sample solution A (1.5-g sample) to prepare the sample solution, and use 10 mL of sample solution A to prepare the control solution.

Iron. (Page 38, Method 1). Use 10 mL of sample solution A (0.5-g sample).

Rare Earth Elements. Dilute 5.0 mL (0.25-g sample) of sample solution A to 90 mL with water. Use a pH meter to adjust to pH 4.5 with sodium acetate–acetic acid buffer solution (described below), and dilute with water to 100 mL. To 10 mL in a separatory funnel, add 10 mL of 0.2 M *N*-phenylbenzohydroxamic acid solution (described above), shake vigorously, allow the layers to separate, and draw off and discard the chloroform layer. Repeat the extraction to 5-mL portions of 0.2 M *N*-phenylbenzohydroxamic acid solution two more times, drawing off and discarding the chloroform layer each time. Wash the aqueous layer with 5 mL of chloroform, and draw off and discard the chloroform. Add 2 mL of alizarin red S reagent solution, and dilute with water to 50 mL. Adjust the pH of the solution to 4.68 with 0.5 N sodium acetate. Any red color should not exceed that produced by 0.05 mg of lanthanum that has been treated exactly like the 10 mL of the sample.

Sodium Acetate–Acetic Acid Buffer Solution. Dissolve 54.4 g of sodium acetate in water, add 23 mL of acetic acid, and dilute with water to 200 mL.

Titanium. Dilute 10 mL (0.5-g sample) of sample solution A with water to 20 mL and add 0.5 mL of 30% hydrogen peroxide. Any yellow color should not exceed that produced by 0.05 mg of titanium ion (Ti) in an equal volume of solution containing the quantities of reagents used in the test.

Calcium, Magnesium, Potassium, Sodium. (By flame AAS, page 63).

Sample Stock Solution B. Dissolve 10.0 g of sample in 80 mL of water, add 2 mL of nitric acid (1 + 1), transfer to a 100-mL volumetric flask, dilute to the mark with water, and mix (1 mL = 0.10 g).

Sample Stock Solution C. Transfer 5.0 mL (0.50 g) of sample stock solution B to a 100-mL volumetric flask, dilute to the mark with water, and mix (1 mL = 0.005 g).

Element	Wavelength (nm)	Sample Wt (g)	Standard Added (mg)	Flame Type*	Background Correction
Ca	422.7	1.0	0.05; 0.10	N/A	No
Mg	285.2	0.20	0.005; 0.01	A/A	Yes
K	766.5	1.0	0.025; 0.05	A/A	No
Na	589.0	0.04	0.01; 0.02	A/A	No

*A/A is air/acetylene; N/A is nitrous oxide/acetylene.

Thymol Blue

Thymolsulfonphthalein; 4,4'-(3H-2,1-Benzoxathiol-3-ylidene)bis-[5-methyl-2-(1-methylethyl)phenol] S,S-Dioxide

$C_{27}H_{30}O_5S$	Formula Wt 466.59	CAS No. 76-61-9

Note: This specification applies to both the free acid form and the salt form of this indicator.

GENERAL DESCRIPTION

Typical appearance: brownish-green solid
Analytical use: acid–base indicator
Aqueous solubility: insoluble

SPECIFICATIONS

Clarity of solution . . . . Passes test
Visual transition interval (acid range) . . . . From pH 1.2 (pink) to pH 2.8 (yellow)
Visual transition interval (alkaline range) . . . . From pH 8.0 (yellow) to pH 9.2 (blue)

TESTS

Clarity of Solution. If the indicator is the acid form, dissolve 0.1 g in 100 mL of alcohol. If the indicator is a salt form, dissolve 0.1 g in 100 mL of water. Not more than a faint trace of turbidity or insoluble matter should remain. Reserve the solution for the tests for visual transition interval.

Visual Transition Interval (Acid Range). Dissolve 1 g of potassium chloride in 100 mL of water. Adjust the pH of the solution to 1.20 (using a pH meter) with 1 N hydrochloric acid volumetric solution. Add 0.1–0.3 mL of the 0.1% solution reserved from the test for clarity of solution. The color of the solution should be pink. Titrate the solution with 1 N sodium hydroxide volumetric solution until the pH is 2.2 (using the pH meter). The color of the solution should be orange. Continue the titration until the pH is 2.8. The color of the solution should be yellow.

Visual Transition Interval (Alkaline Range). Dissolve 1 g of potassium chloride in 100 mL of water. Adjust the pH of the solution to 8.0 (using a pH meter) by adding 0.01 N hydrochloric acid or sodium hydroxide. Add 0.1–0.3 mL of the 0.1% solution reserved from the test for clarity of solution. The color of the solution should be yellow. Titrate the solution with 0.01 N sodium hydroxide until the pH is 8.4 (using the pH meter). The color of the solution should be green. Continue the titration until the pH is 9.2. The color of the solution should be blue.

Thymolphthalein

5′,5″-Diisopropyl-2′,2″-dimethylphenolphthalein; 3,3-Bis[4-hydroxy-2-methyl-5-(1-methylethyl)phenyl]-1(3*H*)-isobenzofuranone

$C_{28}H_{30}O_4$ Formula Wt 430.54   CAS No. 125-20-2

GENERAL DESCRIPTION

Typical appearance: white to off-white solid

Analytical use: acid–base indicator

Change in state (approximate): melting point, 253 °C

Aqueous solubility: insoluble

SPECIFICATIONS

Clarity of solution . . . . Passes test

Visual transition interval. . . . . .From pH 8.8 (colorless) to pH 10.5 (blue)

TESTS

Clarity of Solution. Dissolve 0.1 g in 100 mL of alcohol. Not more than a faint trace of turbidity or insoluble matter should remain. Reserve the solution for the test for visual transition interval.

Visual Transition Interval. Dissolve 1 g of potassium chloride in 100 mL of water. Adjust the pH of the solution to 8.80 (using a pH meter) with 0.01 N sodium hydroxide. Add 0.5–1.0 mL of the 0.1% solution reserved from the test for clarity of solution. The solution should be colorless. Titrate the solution with 0.01 N sodium hydroxide to pH 9.4 (using the pH meter). The solution should have a pale grayish-blue color. Continue the titration to pH 9.9. The solution should have a blue color.

Tin

Sn   Atomic Wt 118.71   CAS No. 7440-31-5

GENERAL DESCRIPTION

Typical appearance: silver-white metal

Analytical use: reducing agent

Change in state (approximate): melting point, 232 °C

SPECIFICATIONS

Assay . . . . ≥99.5% Sn

Maximum Allowable

Antimony (Sb). 0.02%
Copper (Cu) 0.005%
Iron (Fe). 0.01%
Lead (Pb). 0.005%
Arsenic (As) . . . . .1 ppm

TESTS

Assay. (By complexometric titration). Weigh, to the nearest 0.1 mg, about 0.25 g of sample, and digest on a hot plate ($\approx$100 °C) in a covered beaker with 5 mL of hydrochloric acid until the sample is dissolved. Add 30.0 mL of 0.1 M EDTA volumetric solution, and heat the solution almost to boiling. Cool, and adjust the pH to 5.5, using a pH meter, with a saturated hexamethylenetetramine reagent solution. Dilute with water to about 150 mL, and add a few milligrams of xylenol orange indicator mixture. Titrate with 0.1 M lead nitrate volumetric solution to a change from yellow to reddish purple. One milliliter of 0.1 M EDTA corresponds to 0.01187 g of Sn.

$$\% \ Sn = \frac{[(30.0 \times M \ EDTA) - (mL \times M \ Pb)] \times 11.87}{Sample \ wt \ (g)}$$

Antimony, Copper, Iron, and Lead. (By flame AAS, page 63).

Sample Stock Solution. Dissolve 10.0 g of sample in 30 mL of hydrochloric acid and 3 mL of nitric acid by allowing to stand in a covered beaker until dissolution is complete. Transfer to a 100-mL volumetric flask, and dilute to the mark with water (1 mL = 0.10 g).

Element	Wavelength (nm)	Sample Wt (g)	Standard Added (mg)	Flame Type*	Background Correction
Sb	217.6	1.0	0.20; 0.40	A/A	Yes
Cu	324.8	0.50	0.025; 0.05	A/A	Yes
Fe	248.3	0.50	0.05; 0.10	A/A	Yes
Pb	217.0	1.0	0.05; 0.10	A/A	Yes

*A/A is air/acetylene.

Arsenic. Dissolve 1.0 g in a mixture of 5 mL of water and 10 mL of nitric acid in a generator flask in a hood. Add 10 mL of dilute sulfuric acid (1 + 1), evaporate on a hot plate ($\approx$100 °C) to about 5 mL, and heat just to fumes of sulfur trioxide. (At this point, the metastannic acid should go into solution.) Cool, cautiously wash down the flask with 5–10 mL of water, and again heat just to fumes of sulfur trioxide. Cool, cautiously wash down the flask with 5–10 mL of water, and repeat the fuming. Cool, dilute with water to 55 mL, and proceed as described in the general method for arsenic on page 34, omitting the addition of the dilute sulfuric acid and using 2 mL of 48% hydrobromic acid instead of the potassium iodide solution specified in the procedure. Swirl the contents of the flask occasionally to break up the zinc mass. When evolution of gas stops, open the generator flask, and add 1-g portions of granulated zinc until the solution is clear and colorless. Any red color in the silver diethyldithiocarbamate solution of the sample should not exceed that in a standard containing 0.001 mg of arsenic ion (As).

Titanium Tetrachloride
Titanium(IV) Chloride

$TiCl_4$	Formula Wt 189.69	CAS No. 7550-45-0

Note: This reagent reacts violently with water, producing hydrogen chloride, titanium oxychloride, and titanium oxides. It fumes strongly when exposed to moist air, and the reaction is very exothermic. The reagent will darken with age.

GENERAL DESCRIPTION

Typical appearance: liquid
Analytical use: peroxide detection; preparation of standards
Change in state (approximate): boiling point, 136 °C
Aqueous solubility: soluble in cold water
Density: 1.73

SPECIFICATIONS

Assay . . . . ≥99.0% $TiCl_4$

Maximum Allowable

Color (APHA) 50

TESTS

Assay. (By oxidative titration after reduction of titanium). Weigh accurately 0.8 g with the use of a Smith weighing buret, and add cautiously to 100 mL of 10% sulfuric acid reagent solution. Activate the zinc amalgam of a Jones reductor by passing 100 mL of 10% sulfuric acid reagent solution through the column. Discard the acid, and place 25 mL of ferric ammonium sulfate indicator solution in the receiving beaker under the column. Pass the sample solution through the reductor, followed by 100 mL of 10% sulfuric acid reagent solution and then 100 mL of water (using these to rinse the sample beaker). Collect the eluent in the receiving beaker, making sure the column is maintained wet. Add 10 mL of phosphoric acid, and titrate with 0.1 N potassium permanganate volumetric solution to a pink end point. Run a blank, and make any necessary correction. One milliliter of 0.1 N potassium permanganate, corrected for the blank, corresponds to 0.01897 g of $TiCl_4$.

$$\% \, TiCl_4 = \frac{\{[mL \, (sample) - mL \, (blank)] \times N \, KMnO_4\} \times 18.97}{Sample \, wt \, (g)}$$

Color (APHA). (Page 43).

o-Tolidine Dihydrochloride

3,3′-Dimethyl-[1,1′-biphenyl]-4,4′-diamine Dihydrochloride

$$H_2N-\!\!\!\bigcirc\!\!\!-\!\!\!\bigcirc\!\!\!-NH_2 \cdot 2HCl$$
$$CH_3 \qquad CH_3$$

[-C₆H₃(CH₃)-4-NH₂]₂ · 2HCl   **Formula Wt 285.21**   **CAS No. 612-82-8**

GENERAL DESCRIPTION

Typical appearance: off-white solid
Analytical use: indicator

SPECIFICATIONS

Sensitivity to chlorine . . . . Passes test
Solubility . . . . Passes test

                                                                     Maximum Allowable
Residue after ignition. 0.1%

TESTS

Sensitivity to Chlorine. Dissolve 0.7 g in 50 mL of water. Add this solution, with constant stirring, to 50 mL of dilute hydrochloric acid (3 + 7). Add 1 mL of this solution to 20 mL of water containing 5 mL of chlorine ion (Cl) standard solution. A distinct yellow color is produced that, when measured in a 1-cm cell against the chlorine standard at 430 nm, has an absorbance of not less than 0.42. (Absorbance readings should be made within 10 min.) (Limit 0.1 ppm free chlorine)

Solubility. Dissolve 0.50 g in 50 mL of water. The resulting solution should be complete and colorless.

Residue after Ignition. Place 1.0 g in a tared, preconditioned porcelain crucible, add 0.5 mL of sulfuric acid, and slowly ignite in a hood. Heat in a furnace at 600 ± 25 °C to constant weight.

Toluene
Methylbenzene

C$_6$H$_5$CH$_3$ **Formula Wt 92.14**   **CAS No. 108-88-3**

Suitable for general use or in ultraviolet spectrophotometry. Product labeling shall designate the uses for which suitability is represented on the basis of meeting the relevant specifications and tests. The ultraviolet spectrophotometry specifications include all of the specifications for general use.

GENERAL DESCRIPTION

Typical appearance: clear, colorless liquid

Analytical use: solvent

Change in state (approximate): boiling point, 111 °C

Aqueous solubility: 0.045 g in 100 mL at 25 °C

Density: 0.87

pK_a: 37

SPECIFICATIONS

General Use

Assay . . . . ≥99.5% C$_6$H$_5$CH$_3$

Maximum Allowable

Color (APHA) 10
Residue after evaporation 0.001%
Substances darkened by sulfuric acid . . . . Passes test
Sulfur compounds (as S) 0.003%
Water (H$_2$O) 0.03%

Specific Use

Ultraviolet Spectrophotometry

Wavelength (nm)	Absorbance (AU)
350–400	0.01
335	0.02
310	0.05
300	0.10
293	0.20
288	0.50
286	1.00

TESTS

Assay. Analyze the sample by gas chromatography using the general parameters cited on page 80. The following specific conditions are also required.

Column: Type I, methyl silicone

Measure the area under all peaks, and calculate the toluene content in area percent. Correct for water content.

Color (APHA). (Page 43).

Residue after Evaporation. (Page 25). Evaporate 100.0 g (115 mL) to dryness in a tared, preconditioned dish on a hot plate ($\approx$100 °C), and dry the residue at 105 °C for 30 min.

Substances Darkened by Sulfuric Acid. Shake 15 mL with 5 mL of sulfuric acid for 15–20 s, and allow to stand for 15 min. The toluene layer should be colorless, and the color of the acid should not exceed that of a color standard composed of 2 volumes of water plus 1 volume of a color standard solution (described below).

> **Color Standard Solution.** In a 1-L volumetric flask, add 5 g of cobalt chloride hexahydrate, 40 g of ferric chloride hexahydrate, and 20 mL of hydrochloric acid. Dilute to the mark with water.

Sulfur Compounds. Place 30 mL of approximately 0.5 N potassium hydroxide in methanol in a conical flask, add 5.0 g (6.0 mL) of the sample, and boil the mixture gently for 30 min under a reflux condenser, avoiding the use of a rubber stopper or connection. Detach the condenser, dilute with 50 mL of water, and heat on a hot plate ($\approx$100 °C) until the toluene and methanol are evaporated. Add 50 mL of bromine water, and heat for 15 min longer. Transfer the solution to a beaker, neutralize with dilute hydrochloric acid (1 + 3), add an excess of 1 mL of the acid, and evaporate to about 50 mL. Filter if necessary, heat the filtrate to boiling, add 5 mL of 12% barium chloride reagent solution, digest in a covered beaker on a hot plate ($\approx$100 °C) for 2 h, and allow to stand for at least 8 h. If a precipitate is formed, filter, wash thoroughly, and ignite. Correct for the weight obtained in a complete blank test.

Water. (Page 31, Method 2). Use 100 μL (87 mg) of the sample.

Ultraviolet Spectrophotometry. Use the procedure on page 86 to determine the absorbance.

para-Toluenesulfonic Acid Monohydrate
4-Methylbenzenesulfonic Acid Monohydrate

$CH_3C_6H_4SO_3H \cdot H_2O$ Formula Wt 190.22   CAS No. 6192-52-5

GENERAL DESCRIPTION

Typical appearance: solid
Analytical use: derivatizing agent
Aqueous solubility: freely soluble

SPECIFICATIONS

Assay . . . . ≥98.5% $CH_3C_6H_4SO_3H \cdot H_2O$
Water (H_2O) 9.5–11.5%

Maximum Allowable

Clarity of solution . . . . Passes test
Residue after ignition 0.1%
Sulfate (SO_4) 0.3%
Heavy metals (as Pb) 0.001%
Iron (Fe). 0.01%
Sodium (Na) 0.002%

TESTS

Assay. (By acid–base titrimetry). Weigh accurately about 0.8 g, dissolve in 100 mL of water, add 0.10 mL of phenolphthalein indicator solution, and titrate with 0.1 N sodium hydroxide volumetric solution to a pink end point. One milliliter of 0.1 N sodium hydroxide corresponds to 0.01902 g of $CH_3C_6H_4SO_3H \cdot H_2O$.

$$\% \ CH_3C_6H_4SO_3H \cdot H_2O = \frac{(mL \times N \ NaOH) \times 19.02}{Sample \ wt \ (g)}$$

Water. (Page 31, Method 1). Use 0.5 g.

Clarity of Solution. Dissolve 10 g in 50 mL of water. The solution should be clear and complete.

Residue after Ignition. (Page 26). Use 5.0 g. Retain the residue for the test for heavy metals.

Sulfate. Dissolve 0.70 g in 180 mL of water, add 10 mL of hydrochloric acid, and heat to boiling. Add 10 mL of 12% barium chloride reagent solution, digest in a covered beaker on a hot plate (≈100 °C) for 2 h, and allow to stand for at least 8 h. If a precipitate is formed, filter, wash thoroughly, and ignite. The weight of the precipitate should not be more than 0.005 g greater than the weight obtained in a complete blank test.

Heavy Metals. (Page 36, Method 1). Dissolve the residue from the test for residue after ignition with a mixture of 2 mL of nitric acid and 4 mL of hydrochloric acid, and evaporate to dryness on a hot plate (≈100 °C). Take up the residue with 1 mL of dilute hydrochloric acid (1 + 1) and 8 mL of hot water. Cool, and dilute to 50 mL with water. Use 20 mL of this solution.

Iron and Sodium. (By flame AAS, page 63).

Sample Stock Solution. Dissolve 10.0 g in water, and dilute with water to 100 mL (1 mL = 0.10 g).

Element	Wavelength (nm)	Sample Wt (g)	Standard Added (mg)	Flame Type*	Background Correction
Fe	248.3	1.0	0.05; 0.10	A/A	Yes
Na	589.0	0.50	0.005; 0.01	A/A	No

*A/A is air/acetylene.

Trichloroacetic Acid

$$Cl_3C-\overset{\overset{\textstyle O}{\|}}{C}-OH$$

CCl₃COOH   **Formula Wt 163.39**   **CAS No. 76-03-9**

GENERAL DESCRIPTION

Typical appearance: colorless to white solid

Analytical use: detection of albumin

Change in state (approximate): melting point, 57–58 °C; boiling point, 197–198 °C

Aqueous solubility: very soluble

pK_a: 0.7

SPECIFICATIONS

Assay . . . . ≥99.0% CCl₃COOH
Clarity of solution . . . . Passes test

	Maximum Allowable

Insoluble matter 0.01%
Residue after ignition. 0.03%
Chloride (Cl). 0.002%
Nitrate (NO₃) 0.002%
Phosphate (PO₄). . . . . 5 ppm
Sulfate (SO₄) 0.02%
Heavy metals (as Pb). 0.002%
Iron (Fe). 0.001%
Substances darkened by sulfuric acid . . . . Passes test

TESTS

Assay. (By acid–base titrimetry). Weigh accurately about 5.0 g, previously dried for 24 h over an efficient desiccant, and dissolve in 20 mL of water in a 500-mL reflux flask. Add 0.15 mL of methyl red indicator solution, and titrate with 1 N sodium hydroxide volumetric solution to a distinct yellow color, then add 50.0 mL of 1 N sulfuric acid volumetric solution. Add sufficient dioxane to make the final solution at least 55% dioxane. Add silicon carbide boiling chips, and boil gently under complete reflux for 1 h. Cool, add 0.15 mL of methyl red indicator solution, and titrate the excess of sulfuric acid with 1 N sodium hydroxide, the end point being the distinct transition from a light orange color to a definite yellow color. Correct for the acid present in the dioxane by running a complete blank, containing 2.5 mL of chloroform. One milliliter of 1 N sodium hydroxide corresponds to 0.1634 g of CCl₃COOH.

$$\% \ CCl_3COOH = \frac{\{[mL \ (blank) - mL \ (sample)] \times N \ NaOH\} \times 16.34}{Sample \ wt \ (g)}$$

Clarity of Solution. Dissolve 10.0 g in 100 mL of water in a 250-mL conical flask. The solution should be clear and colorless and contain no significant amount of insoluble matter, such as minute fibers and/or other particles causing haze or reduced transparency. Reserve the solution for the test for insoluble matter.

Insoluble Matter. (Page 25). Use the solution reserved from the clarity of solution test.

Residue after Ignition. (Page 26). Ignite 5.0 g.

Chloride. (Page 35). Use 0.5 g.

Nitrate. (Page 38, Method 1). For sample solution A, use 0.25 g. For control solution B, use 0.25 g with 0.5 mL of nitrate ion (NO_3) standard solution.

For the Determination of Phosphate, Sulfate, Heavy Metals, and Iron

Sample Solution A. Evaporate 5.0 g to dryness in a beaker on a hot plate ($\approx$100 °C). Dissolve in about 10 mL of water, filter if necessary, and dilute with water to 100 mL (1 mL = 0.05 g).

Phosphate. (Page 40, Method 2). Dilute 40 mL of sample solution A (2.0-g sample) with water to about 80 mL. Add 0.5 g of ammonium molybdate, and adjust the pH to 1.8 (using a pH meter) with dilute hydrochloric acid (1 + 9). Heat to boiling, cool to room temperature, dilute with water to 90 mL, and add 10 mL of hydrochloric acid. Continue as described, and carry along a standard containing 0.01 mg of phosphate ion (PO_4) treated in the same manner as the 40 mL of sample solution A.

Sulfate. (Page 40, Method 1). Use 5.0 mL of sample solution A (0.25-g sample).

Heavy Metals. (Page 36, Method 1). Use 20 mL of sample solution A (1-g sample) with water to 25 mL.

Iron. (Page 38, Method 1). Use 20 mL of sample solution A (1-g sample).

Substances Darkened by Sulfuric Acid. Dissolve 1.0 g in 10 mL of sulfuric acid, and digest in a covered beaker on a hot plate ($\approx$100 °C) for 15 min. Any color should not exceed that produced when 10 mL of sulfuric acid is added to 1 mL of water containing 0.1 mg of sucrose and heated in the same way as the sample.

1,2,4-Trichlorobenzene

$C_6H_3Cl_3$	Formula Wt 181.46	CAS No. 120-82-1

Suitable for general use or in ultraviolet spectrophotometry. Product labeling shall designate the uses for which suitability is represented on the basis of meeting the relevant speci-

fications and tests. The ultraviolet spectrophotometry specifications include all the specifications for general use.

GENERAL DESCRIPTION

Typical appearance: liquid
Analytical use: HPLC solvent; general solvent; preparation of standards
Change in state (approximate): boiling point, 213 °C
Aqueous solubility: insoluble
Density: 1.46

SPECIFICATIONS

General Use
Assay . . . . ≥99.0% $C_6H_3Cl_3$

	Maximum Allowable
Color (APHA)	10
Titrable acid.	0.0003 meq/g
Water (H_2O)	0.02%

Specific Use

Ultraviolet Spectrophotometry

Wavelength (nm)	Absorbance (AU)
385	0.01
325	0.15
310	0.60
308	1.00

TESTS

Assay. Analyze the sample by gas chromatography using the general parameters cited on page 80. The following specific conditions are also required.

   Column: Type I, methyl silicone

Measure the area under all the peaks, and calculate the 1,2,4-trichlorobenzene content in area percent. Correct for water content.

Color (APHA). (Page 43).

Titrable Acid. Place 50 mL of isopropyl alcohol in a flask, and add 0.10 mL of phenolphthalein indicator solution. Titrate with 0.01 N sodium hydroxide to a faint pink color. Add 23 mL (33 g) of sample, and titrate with 0.01 N sodium hydroxide to the same pink end point. Not more than 1.0 mL of the 0.01 N sodium hydroxide should be required.

Water. (Page 31, Method 2). Use 0.2 mL (0.29 g) of the sample.

Ultraviolet Spectrophotometry. Use the procedure on page 86 to determine the absorbance.

1,1,1-Trichloroethane
Methyl Chloroform
CH₃CCl₃   **Formula Wt 133.40**   **CAS No. 71-55-6**

Note: This solvent usually contains additives to retard decomposition or to inhibit the corrosion of metals, particularly aluminum, with which it might come into contact. The nature and concentrations of such compounds should be stated on the label. Typical stabilizers include dioxane and alcohols such as 2-butanol.

GENERAL DESCRIPTION

Typical appearance: clear liquid

Analytical use: solvent

Change in state (approximate): boiling point, 74 °C

Aqueous solubility: insoluble

Density: 1.34

SPECIFICATIONS

Assay . . . . 98.5–101.0% (CH_3CCl_3 + active additives)

Maximum Allowable

Color (APHA) 10

Water (H_2O) 0.10%

Residue after evaporation 0.001%

Titrable acid . . . . 0.00030 meq/g

TESTS

Assay. Analyze the sample by gas chromatography using the general parameters cited on page 80. The following specific conditions are also required.

> *Column:* 6 m × 3.2 mm stainless steel, packed with 15% SP 1000 on Supelcoport
>
> *Column Temperature:* 70 °C for 8 min, then programmed at 8 °C/min to 200 °C
>
> *Injection Port Temperature:* 200 °C
>
> *Detector Temperature:* 250 °C
>
> *Carrier Gas:* Nitrogen at 18 mL/min
>
> *Sample Size:* 1 μL
>
> *Detector:* Flame ionization
>
> *Range:* 100
>
> *Approximate Retention Times (min):* Vinylidene chloride, 4.8; butylene oxide, 7.9; 1,1-dichloroethane, 8.9; carbon tetrachloride, 9.0; 1,1,1-trichloroethane, 9.8; 2-propanol, 10.5; dichloromethane, 10.8; 1,1,2-trichloroethylene, 13.5; chloroform, 14.2; *n*-propanol, 14.7; perchloroethylene, 15.2; 1,2-dichloroethane, 16.0; dioxane, 16.2; 2-

methyl-1-propanol, 17.0; *n*-butanol, 17.8; 1,1,2-trichloroethane, 21.3. Carbon tetra-chloride and 1,1-dichloroethane are not resolved from 1,1,1-trichloroethane.

Measure the area under all peaks, and calculate the 1,1,1-trichloroethane content in area percent plus intentional additives. Correct for water content.

Color (APHA). (Page 43).

Water. (Page 31, Method 1). Use 13.3 g (10 mL) of the sample.

Residue after Evaporation. (Page 25). Evaporate 100.0 g (75 mL) to dryness in a tared, preconditioned dish on a hot plate ($\approx$100 °C), and dry the residue at 105 °C for 30 min.

Titrable Acid. Measure 100.0 g (75 mL) into a dry conical 250-mL flask. Add 0.2 mL of bromthymol blue indicator solution, and titrate with 0.02 N sodium hydroxide in metha-nol to a greenish-blue color. Not more than 1.50 mL should be required.

Trichloroethylene
Trichloroethene

$$Cl_2C=CHCl$$

CHCl:CCl$_2$ **Formula Wt 131.39**   **CAS No. 79-01-6**

Note: This material usually contains a stabilizer. If a stabilizer is present, the amount and type should be stated on the label.

GENERAL DESCRIPTION

Typical appearance: clear, colorless liquid
Analytical use: solvent
Change in state (approximate): boiling point, 87 °C
Aqueous solubility: 0.1 g in 100 mL at 25 °C
Density: 1.46

SPECIFICATIONS

Assay .... ≥99.5% CHCl:CCl$_2$

	Maximum Allowable
Color (APHA)	10
Residue after evaporation	0.001%
Titrable acid	0.0001 meq/g
Titrable base	0.0003 meq/g
Water (H$_2$O)	0.02%
Heavy metals (as Pb)	1 ppm
Free halogens	Passes test

TESTS

Assay. Analyze the sample by gas chromatography using the general parameters cited on page 80. The following specific conditions are also required.

Column: Type I, methyl silicone

Measure the area under all peaks, and calculate the trichloroethylene content in area percent. Correct for water content and added stabilizer.

Color (APHA). (Page 43).

Residue after Evaporation. (Page 25). Evaporate 100.0 g (69 mL) to dryness in a tared, preconditioned dish on a hot plate ($\approx$100 °C) in a well-ventilated hood, and dry the residue at 105 °C for 30 min.

Titrable Acid and Titrable Base. To 25 mL of water and 0.10 mL of phenolphthalein indicator solution in a 250-mL glass-stoppered flask, add 0.01 N sodium hydroxide until a slight pink color appears. Add 36.0 g (25 mL) of sample, and shake for 30 s. If the pink color persists, titrate with 0.01 N hydrochloric acid, shaking repeatedly, until the pink color just disappears. Not more than 0.90 mL of 0.01 N hydrochloric acid should be required (titrable base). If the pink color is discharged when the sample is added, titrate with 0.01 N sodium hydroxide until the pink color is restored. Not more than 0.50 mL of 0.01 N sodium hydroxide should be required (titrable acid).

Water. (Page 31, Method 2). Use 100 μL (140 mg) of the sample.

Heavy Metals. Evaporate 10.0 g (7 mL) to dryness in a glass evaporating dish on a hot plate ($\approx$100 °C) inside a well-ventilated hood. For the standard, evaporate a solution containing 0.01 mg of lead to dryness. Cool, add 2 mL of hydrochloric acid to each, and slowly evaporate to dryness on a hot plate ($\approx$100 °C). Moisten the residues with 0.05 mL of hydrochloric acid, add 10 mL of hot water, and digest for 2 min. Filter if necessary through a small filter, wash the evaporating dish and the filter with about 10 mL of water, and dilute each with water to 25 mL. Adjust the pH of the standard and sample solutions to between 3 and 4 (using a pH meter) with 1 N acetic acid or 10% ammonium hydroxide reagent solution, dilute with water to 40 mL, and mix. Add 10 mL of freshly prepared hydrogen sulfide water to each, and mix. Any color in the solution of the sample should not exceed that in the standard.

Free Halogens. Shake 10 mL for 2 min with 10 mL of water to which 0.10 mL of 10% potassium iodide reagent solution has been added, and allow to separate. The lower layer should not show a violet tint.

2,2,4-Trimethylpentane

Isooctane

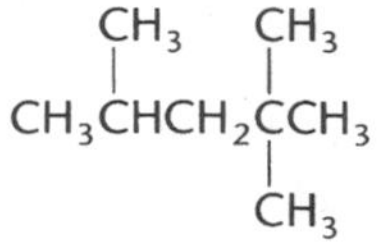

$$CH_3CHCH_2CCH_3$$

with CH_3 groups on the 2nd and 4th positions and a CH_3 branch below.

(CH₃)₃CCH₂CH(CH₃)₂   **Formula Wt 114.23**   **CAS No. 540-84-1**

Suitable for general use or in extraction–concentration analysis or ultraviolet spectrophotometry. Product labeling shall designate the uses for which suitability is represented based on meeting the relevant specifications and tests. The ultraviolet spectrophotometry specifications include all of the specifications for general use. The extraction–concentration suitability specifications include only the general use specification for color.

GENERAL DESCRIPTION

Typical appearance: clear liquid
Analytical use: solvent
Change in state (approximate): boiling point, 99 °C
Aqueous solubility: practically insoluble
Density: 0.69

SPECIFICATIONS

General Use

Assay . . . . ≥99.0% (CH₃)₃CCH₂CH(CH₃)₂

Maximum Allowable

Color (APHA) 10
Residue after evaporation 0.001%
Water-soluble titrable acid . . . . 0.0003 meq/g
Sulfur compounds (as S) 0.005%

Specific Use

Ultraviolet Spectrophotometry

Wavelength (nm)	Absorbance (AU)
250–400	0.01
240	0.04
230	0.10
220	0.20
210	1.0

Extraction–Concentration Suitability

Absorbance . . . . Passes test
GC–FID . . . . Passes test
GC–ECD . . . . Passes test

TESTS

Assay. Analyze the sample by gas chromatography using the general parameters cited on page 80. The following specific conditions are also required.

   Column: Type I, methyl silicone

Measure the area under all peaks, and calculate the 2,2,4-trimethylpentane content in area percent. Correct for water content.

Color (APHA). (Page 43).

Residue after Evaporation. (Page 25). Evaporate 100.0 g (145 mL) to dryness in a tared, preconditioned dish on a hot plate ($\approx$100 °C), and dry the residue at 105 °C for 30 min.

Water-Soluble Titrable Acid. To 30.0 g (44 mL) in a separatory funnel, add 50 mL of water, and shake vigorously for 2 min. Allow the layers to separate, draw off the aqueous layer, and add 0.15 mL of phenolphthalein indicator solution to the aqueous layer. Not more than 1.0 mL of 0.01 N sodium hydroxide should be required to produce a pink color.

Sulfur Compounds. To 30 mL of 0.5 N potassium hydroxide in methanol in a conical flask, add 5.5 g (8 mL) of the sample, and boil the mixture gently for 30 min under a reflux condenser, avoiding the use of a rubber stopper or connection. Detach the condenser, dilute with 50 mL of water, and heat on a hot plate ($\approx$100 °C) until the 2,2,4-trimethylpentane and methanol are evaporated. Add 50 mL of bromine water, heat for 15 min longer, and transfer to a beaker. Neutralize with dilute hydrochloric acid (1 + 3), add an excess of 1 mL of the acid, evaporate to about 50 mL, and filter if necessary. Heat the filtrate to boiling, add 5 mL of 12% barium chloride reagent solution, digest in a preconditioned covered beaker on a hot plate ($\approx$100 °C) for 2 h, and allow to stand for at least 8 h. If a precipitate is formed, filter, wash thoroughly, ignite, cool, and weigh. Correct for the weight obtained in a complete blank test.

$$\% \, S = \frac{\text{Residue wt (g)} \times 13.7}{5.5 \, \text{g}}$$

Ultraviolet Spectrophotometry. Use the procedure on page 86 to determine the absorbance.

Extraction–Concentration Suitability. Analyze the sample using the general procedure cited on page 85. Use the procedure on page 86 to determine the absorbance.

Tris(hydroxymethyl)aminomethane

2-Amino-2-(hydroxymethyl)-1,3-propanediol; "Tris"

$$CH_2OH$$
$$HOCH_2\overset{|}{C}CH_2OH$$
$$\underset{|}{NH_2}$$

$NH_2C(CH_2OH)_3$	Formula Wt 121.14	CAS No. 77-86-1

GENERAL DESCRIPTION

Typical appearance: white solid
Analytical use: buffering agent; acidimetric standard
Change in state (approximate): melting point, 171 °C
Aqueous solubility: 55 g in 100 mL at 25 °C
pK_a: 8.3

SPECIFICATIONS

Assay (dry basis). . . . . 99.8–100.1% $C_4H_{11}NO_3$

Maximum Allowable

Absorbance. . . . . .Passes test
Water (H_2O) 2%
Insoluble matter 0.005%
Heavy metals (as Pb). . . . . .5 ppm
Iron (Fe). . . . . .5 ppm

TESTS

Assay. (By acid–base titrimetry). Weigh accurately about 0.5 g of sample previously dried at 105 °C for 3 h. Transfer to a 200-mL beaker, dissolve in 50 mL of carbon dioxide- and ammonia-free water, and titrate the solution with 0.1 N hydrochloric acid to a pH of 4.7, using a suitable pH meter. One milliliter of 0.1 N hydrochloric acid corresponds to 0.012114 g of $C_4H_{11}NO_3$.

$$\% \ C_4H_{11}NO_3 = \frac{(mL \times N \ HCl) \times 12.114}{Sample \ wt \ (g)}$$

Absorbance. Determine the absorbance of a 40% solution of the sample in water in a 1.00-cm cell at 290 nm against water in a similar matched cell set at zero absorbance as the reference liquid. The absorbance should not exceed 0.2 absorbance units.

Water. (Page 31, Method 1). Use 1.0 g of the sample.

Insoluble Matter. Dissolve 20.0 g in 200 mL of water. Filter through a tared, preconditioned filtering crucible, wash thoroughly, and dry at 105 °C.

Heavy Metals. (Page 36, Method 1). Substitute glacial acetic acid for 1 N acetic acid to adjust pH in Method 1. Dissolve 6.0 g in about 20 mL of water, and dilute with water to 30 mL. Use 25 mL to prepare the sample solution, and use the remaining 5.0 mL to prepare the control solution.

Iron. (Page 38, Method 1). Use a 2.0-g sample.

Uranyl Acetate Dihydrate
$UO_2(CH_3COO)_2 \cdot 2H_2O$ **Formula Wt 424.15**   **CAS No. 6159-44-0**

Note: The formula weight of this reagent is likely to deviate from the value cited because the natural distribution of uranium isotopes is often altered in current sources of uranium compounds.

GENERAL DESCRIPTION

Typical appearance: greenish-yellow to bright yellow solid

Analytical use: precipitant for sodium

Change in state (approximate): dehydrates, 110 °C

Aqueous solubility: 7.7 g in 100 mL at 15 °C; but solution slowly becomes cloudy due to formation of insoluble basic salt

SPECIFICATIONS

Assay . . . . .98.0–102.0% $UO_2(CH_3COO)_2 \cdot 2H_2O$

Maximum Allowable

Insoluble matter 0.01%
Chloride (Cl) 0.003%
Sulfate (SO$_4$) 0.01%
Alkalis and alkaline earths (as sulfates) 0.05%
Heavy metals (as Pb) 0.002%
Iron (Fe). 0.001%
Substances reducing permanganate (as U^{IV}) 0.06%

TESTS

Assay. (Total uranium by oxidimetry). Weigh accurately about 0.6 g, dissolve in 100 mL of 1 N sulfuric acid volumetric solution, and add enough 0.1 N potassium permanganate volumetric solution to give a distinct pink color. Pass 100 mL of 1 N sulfuric acid through a Jones reductor (page 102), followed by 100 mL of water, and discard. Pass the sample through the reductor, followed in succession by 100 mL each of 1 N sulfuric acid and water, and collect the effluent in a titration flask. Bubble air through the solution for 5–10 min, then titrate with 0.1 N potassium permanganate volumetric solution to a pink color. Run a reagent blank, and subtract it from the sample titration. One milliliter of 0.1 N potassium permanganate corresponds to 0.02121 g of $UO_2(CH_3COO)_2 \cdot 2H_2O$.

$$\% \ UO_2(CH_3COO)_2 \cdot 2H_2O = \frac{\{[mL \ (sample) - mL \ (blank)] \times N \ KMnO_4\} \times 21.21}{Sample \ wt \ (g)}$$

Insoluble Matter. Dissolve 10.0 g in 185 mL of water plus 5 mL of glacial acetic acid at room temperature. Filter through a tared, preconditioned filtering crucible, wash thoroughly with water, and dry at 105 °C.

For the Determination of Chloride, Sulfate, Alkalis and Alkaline Earths, Heavy Metals, and Iron

Sample Solution A. Dissolve 15 g in 275 mL of water plus 5 mL of glacial acetic acid, add 10 mL of 30% hydrogen peroxide, and heat to coagulate the precipitate. Decant through a fritted-glass filter, without washing, and dilute with water to 300 mL (1 mL = 0.05 g).

Chloride. (Page 35). Dilute 6.7 mL of sample solution A (0.33-g sample) with water to 20 mL.

Sulfate. (Page 40, Method 1). Use 10 mL of sample solution A (0.5-g sample).

Alkalis and Alkaline Earths. To 40 mL of sample solution A (2-g sample), add 0.10 mL of sulfuric acid, evaporate to dryness, and gently ignite at about 450 °C. Digest the residue with 25 mL of hot water, filter, evaporate the filtrate to dryness in a tared, preconditioned dish, and ignite at 600 ± 25 °C for 15 min.

Heavy Metals. (Page 36, Method 1). Evaporate 20 mL of sample solution A (1-g sample) to dryness, dissolve the residue in about 20 mL of water, and dilute with water to 25 mL.

Iron. (Page 38, Method 1). Evaporate 20 mL of sample solution A (1-g sample) to dryness, dissolve the residue in 2 mL of hydrochloric acid, and dilute with water to 50 mL. Use the solution without further acidification.

Substances Reducing Permanganate. Prepare two solutions, each containing 3.0 g in 200 mL of dilute sulfuric acid (1 + 99). Titrate one solution with 0.1 N potassium permanganate volumetric solution. Not more than 0.20 mL of permanganate should be required to cause a color change that can be observed by comparison with the solution not titrated. The 0.20 mL includes the 0.05 mL allowed to produce the color change in the absence of uranous compounds.

Uranyl Nitrate Hexahydrate

$UO_2(NO_3)_2 \cdot 6H_2O$ Formula Wt 502.13   CAS No. 13520-83-7

Note: The formula weight of this reagent is likely to deviate from the value cited, since the natural distribution of uranium isotopes is often altered in current sources of uranium compounds.

GENERAL DESCRIPTION

Typical appearance: yellow solid
Analytical use: precipitation of sodium and magnesium

Change in state (approximate): melting point, 60 °C

Aqueous solubility: 122 g in 100 mL

SPECIFICATIONS

Assay . . . . 98.0–102.0% $UO_2(NO_3)_2 \cdot 6H_2O$

Maximum Allowable

Insoluble matter	0.005%
Chloride (Cl)	0.002%
Sulfate (SO_4)	0.005%
Alkalis and alkaline earths (as sulfates)	0.1%
Heavy metals (as Pb)	0.002%
Iron (Fe).	0.002%
Substances reducing permanganate (as U^{IV})	0.06%

TESTS

Assay. (By gravimetry for uranium). Weigh accurately about 0.5 g, and dissolve in 100 mL of water in a 250-mL beaker. Heat the solution to boiling, and add ammonium hydroxide, free from carbonate, until precipitation is complete. Filter through a tared Gooch crucible, wash with 1% ammonium nitrate solution, and ignite gently with free access to air, to constant weight.

$$\% \ UO_2(NO_3)_2 \cdot 6H_2O = \frac{\text{Residue wt (g)} \times 178.9}{\text{Sample wt (g)}}$$

Insoluble Matter. (Page 25). Use 20 g dissolved in 200 mL of water.

For the Determination of Chloride, Sulfate, Alkalis and Alkaline Earths, Heavy Metals, and Iron

Sample Solution A. Dissolve 10.0 g in 200 mL of water, add 7 g of ammonium acetate and 10 mL of 30% hydrogen peroxide, and heat to coagulate the precipitate. Decant through a fritted-glass filter, without washing, and dilute with water to 250 mL in a volumetric flask (1 mL = 0.025 g).

Chloride. (Page 35). Dilute 10 mL of sample solution A (0.5-g sample) with water to 20 mL.

Sulfate. (Page 40, Method 1). Evaporate 20 mL of sample solution A (1-g sample) to 10 mL.

Alkalis and Alkaline Earths. To 20 mL of sample solution A (1-g sample), add 0.10 mL of sulfuric acid, evaporate to dryness in a tared, preconditioned evaporating dish, and ignite at 600 ± 25 °C for 15 min.

Heavy Metals. (Page 36, Method 1). Evaporate 20 mL of sample solution A (1-g sample) to dryness, dissolve the residue in about 20 mL of water, and dilute with water to 25 mL.

Iron. (Page 38, Method 1). Use 10 mL of sample solution A (0.5-g sample).

Substances Reducing Permanganate. Prepare two solutions, each containing 3.0 g in 200 mL of dilute sulfuric acid (1 + 99). Titrate one solution with 0.1 N potassium permanganate volumetric solution. Not more than 0.20 mL of permanganate should be required to cause a color change that can be observed by comparison with the solution not titrated. The 0.20 mL includes 0.05 mL allowed to produce the color change in the absence of uranous compounds.

Urea

$$H_2N-\overset{\overset{\displaystyle O}{\displaystyle \|}}{C}-NH_2$$

NH$_2$CONH$_2$ **Formula Wt 60.06**   **CAS No. 57-13-6**

GENERAL DESCRIPTION

Typical appearance: white solid
Analytical use: buffer solutions
Change in state (approximate): melting point, 132–135 °C
Aqueous solubility: 109 g in 100 mL at 100 °C
pK_a: 1.0

SPECIFICATIONS

Assay . . . . .99.0–100.5% NH$_2$CONH$_2$
Melting point . . . . .132–135 °C

Maximum Allowable

Insoluble matter 0.01%
Residue after ignition. 0.01%
Chloride (Cl). . . . . .5 ppm
Sulfate (SO$_4$) 0.001%
Heavy metals (as Pb). 0.001%
Iron (Fe). 0.001%

TESTS

Assay. (By nonaqueous titration). Accurately weigh about 0.06 g of sample into a 150-mL beaker. Add 75 mL of acetonitrile and about 2 mL of 0.1 N perchloric acid in glacial acetic acid volumetric solution from a buret. When the sample is completely dissolved, finish the titration potentiometrically using a glass electrode set, with 10% lithium perchlorate in acetonitrile (described below) as the filling solution in a fast-flow reference electrode. One milliliter of the 0.1 N perchloric acid titer is equivalent to 0.006006 g of urea.

$$\% \text{ NH}_2\text{CONH}_2 = \frac{(\text{mL} \times \text{N HClO}_4) \times 6.006}{\text{Sample wt (g)}}$$

Lithium Perchlorate in Acetonitrile, 10% Solution. Dissolve 5.0 g of lithium perchlorate in 400 mL of acetonitrile, and dilute to 500 mL with acetonitrile.

Melting Point. (Page 45).

Insoluble Matter. (Page 25). Use 10.0 g dissolved in 100 mL of water.

Residue after Ignition. (Page 26). Ignite 10.0 g.

Chloride. (Page 35). Use 2.0 g.

Sulfate. Dissolve 4.0 g and about 10 mg of sodium carbonate in 10 mL of hydrochloric acid. Add 10 mL of nitric acid, and digest in a covered beaker on a hot plate ($\approx$100 °C) until reaction ceases. Add 5 mL of hydrochloric acid and 10 mL of nitric acid, and again digest in a covered beaker on a hot plate ($\approx$100 °C) until reaction ceases. Remove the cover, and evaporate to dryness. Dissolve the residue in 4 mL of water plus 1 mL of dilute hydrochloric acid (1 + 19), and filter through a small filter. Wash with two 2-mL portions of water, dilute with water to 10 mL, and add 1 mL of 12% barium chloride reagent solution. Any turbidity should not exceed that produced when a solution containing 0.04 mg of sulfate ion (SO_4) is treated exactly like the 4.0-g sample. Compare 10 min after adding the barium chloride to the sample and standard solutions.

Heavy Metals. (Page 36, Method 1). Dissolve 2.0 g in about 20 mL of water, and dilute with water to 25 mL.

Iron. (Page 38, Method 1). Use 1.0 g.

Variamine Blue B
N-[*p*-Methoxyphenyl]-*p*-phenylenediamine Hydrochloride

$C_{13}H_{14}N_2O \cdot HCl$ Formula Wt 250.72   CAS No. 3566-44-7

GENERAL DESCRIPTION
Typical appearance: gray-blue solid
Analytical use: indicator

SPECIFICATIONS
Suitability as a complexometric ion indicator . . . . Passes test

TESTS

Suitability as a Complexometric Ion Indicator. Weigh 0.2 g of sample, and suspend in 20 mL of water. Stir for 5 min. Mix 200 mL of water, 10 mL of 0.1 M EDTA volumetric solution, and 2 mL of 20% hydrochloric acid. Heat the solution to 40–50 °C. Adjust the pH to 2.5 $\pm$ 0.2 with 10% sodium acetate reagent solution, using a pH meter. Add 0.2

mL of the sample solution, and titrate with ferric chloride hexahydrate solution (described below). The color change is from bright yellow to burgundy. Add back gradually 0.10 ± 0.05 mL of 0.1 M EDTA in 0.05-mL increments. The color should return to bright yellow.

> **Ferric Chloride Hexahydrate Solution.** Dissolve 2.5 g of ferric chloride hexahydrate in 100 mL of dilute hydrochloric acid (1 + 99).

Water, Reagent
Distilled Water, Deionized Water

H_2O	Formula Wt 18.02	CAS No. 7732-18-5

Note: The specifications for water are for freshly prepared material and not intended for bottled or stored product.

GENERAL DESCRIPTION

Typical appearance: clear, colorless liquid
Analytical use: solvent; diluent
Change in state (approximate): boiling point, 100 °C
Density: 0.99
pK_a: 15.7

SPECIFICATIONS
General Use

	Maximum Allowable
Specific conductance at 25 °C	2.0×10^{-6} ohm^{-1} cm^{-1}
Silicate (as SiO_2)	0.01 ppm
Heavy metals (as Pb)	0.01 ppm
Substances reducing permanganate	Passes test
Chloride (Cl)	0.4 ppm
Nitrate (NO_4)	0.4 ppm
Phosphate (PO_4)	1.0 ppm
Sulfate (SO_4)	1.0 ppm

Specific Use

Ultraviolet Spectrophotometry

Wavelength (nm)	Absorbance (AU)
300	0.005
254	0.005
210	0.010
200	0.010

Liquid Chromatography Suitability

Absorbance	Passes test
Gradient elution	Passes test

TESTS

Specific Conductance. Fill a conductance cell that has been properly standardized. Take precautions to prevent the water from absorbing carbon dioxide, ammonia, hydrogen chloride, or any other gases commonly present in a laboratory. Immerse in a constant-temperature bath at 25 °C, equilibrate, and measure the resistance with a suitable conductance bridge. The calculated specific conductance should not exceed 2.0×10^{-6} ohm^{-1} cm^{-1}.

Silicate. Evaporate 1500 mL to 80 mL in a platinum dish on a hot plate (≈ 100 °C) (in portions, if necessary). Add 10 mL of silica-free ammonium hydroxide (described below), cool, and add 5 mL of a 10% ammonium molybdate reagent solution. Adjust the pH to between 1.7 and 1.9 (using a pH meter) with hydrochloric acid or silica-free ammonium hydroxide. Heat the solution just to boiling, cool to room temperature, and dilute with water to 90 mL. Transfer the solution to a separatory funnel, add 10 mL of hydrochloric acid and 35 mL of ethyl ether, and shake vigorously for a few minutes. Allow the two layers to separate, draw off the aqueous layer into another separatory funnel, and add 10 mL of hydrochloric acid and 50 mL of butyl alcohol. Shake vigorously, allow to separate, draw off the aqueous phase, and discard. Wash the butyl alcohol phase three times with 20 mL of dilute hydrochloric acid (1 + 99), discarding the aqueous solution each time. Add 0.5 mL of freshly prepared 2% stannous chloride reagent solution to the butyl alcohol. Any blue color should not exceed that produced by 0.015 mg of silica ion (SiO_2) (use the silicate solution in ammonium hydroxide, described below) when treated exactly like the sample. Because the blue color fades on standing, compare immediately after the reduction with stannous chloride. Treatment again with stannous chloride will restore a fading blue color to its original intensity.

> **Ammonium Hydroxide, Silica-Free.** Aqueous ammonia in contact with glass, even momentarily, will dissolve sufficient silica to make the reagent useless for this test. Because ammonium hydroxide is ordinarily supplied in glass containers by commercial sources, silica-free ammonium hydroxide must be prepared in the laboratory. The most convenient method is the saturation of water with ammonia gas from a cylinder of compressed anhydrous ammonia. Plastic tubing and bottles—for example, polyethylene—must be used throughout.

> **Silicate Solution in Ammonium Hydroxide.** (0.01 mg of SiO_2 in 1 mL). To prepare a stock solution of silica, dissolve 0.946 g of assayed sodium silicate in 9 mL of silica-free ammonium hydroxide, and dilute to 100 mL (99.7 g at 25 °C) with water. Dilute 5.0 mL of this solution plus 5 mL of silica-free ammonium hydroxide to 1 L (997.1 g at 25 °C) with water. If volumetric polyolefin ware is unavailable, dilute the silica solution by weight.

Heavy Metals. (Page 36, Method 1). Add 10 mg of sodium chloride and 1 mL of 1 N hydrochloric acid to 2 L of sample, and evaporate to dryness on a hot plate (≈ 100 °C). Dissolve the residue in about 20 mL of water, and dilute with water to 25 mL. For the control, evaporate 1 mL of 1 N hydrochloric acid to dryness on a hot plate (≈ 100 °C), cool, add 0.02 mg of lead ion (Pb), and dilute with water to 25 mL.

Substances Reducing Permanganate. To 500 mL, add 1 mL of sulfuric acid and 0.30 mL of 0.01 N potassium permanganate, and allow to stand for 1 h at room temperature. The pink color should not be entirely discharged.

Chloride, Nitrate, Phosphate, and Sulfate. To each of three 100-mL beakers, add 2.0 mg of sodium carbonate. To two beakers, add 50 mL of the sample. To one of these beakers (the standard), add 0.02 mg each of chloride and nitrate ions and 0.05 mg each of phosphate and sulfate ions. To the third beaker (blank), add 20 mL of 0.06 micromho cm^{-1} water. Evaporate each beaker on a steam bath or low-temperature hot plate to about 20 mL. Cool. Dilute to 25 mL. Analyze 100-μL aliquots by ion chromatography, and measure the various peaks. The differences between the sample and the blank should not be greater than the corresponding differences between the standard and the sample.

Ultraviolet Spectrophotometry. Use the procedure on page 86 to determine the absorbance.

Liquid Chromatography Suitability. Analyze the sample by using the general procedure on page 84. Use the procedure on page 86 to determine the absorbance.

Water, Ultratrace

H_2O Formula Wt 18.02   CAS No. 7732-18-5

Suitable for use in ultratrace elemental analysis.

Note: This reagent must be packaged in a precleaned Teflon or polyethylene bottle and used in a clean laboratory environment to maintain purity.

GENERAL DESCRIPTION

Typical appearance: clear, colorless liquid
Analytical use: solvent in trace analysis
Change in state (approximate): boiling point, 100 °C
Density: 0.99
pK_a: 15.7

SPECIFICATIONS   *Maximum Allowable*

Mercury (Hg) . . . . 0.5 ppb
Selenium . . . . 5 ppb
Aluminum (Al) . . . . 0.5 ppb
Barium (Ba) . . . . 0.5 ppb
Boron (B). . . . . 0.5 ppb
Cadmium (Cd) . . . . 0.5 ppb
Calcium (Ca) . . . . 0.5 ppb
Chromium (Cr) . . . . 0.5 ppb
Cobalt (Co). . . . . 0.5 ppb
Copper (Cu) . . . . 0.5 ppb

Iron (Fe)... 0.5 ppb
Lead (Pb) .... 0.5 ppb
Lithium (Li) .... 0.5 ppb
Magnesium (Mg).. 0.5 ppb
Manganese (Mn).. 0.5 ppb
Molybdenum (Mo) .... 0.5 ppb
Potassium (K) .... 0.5 ppb
Silicon (Si) .... 0.5 ppb
Sodium (Na) .... 0.5 ppb
Strontium (Sr).. 0.5 ppb
Tin (Sn) .... 0.5 ppb
Titanium (Ti)... 0.5 ppb
Vanadium (V) .... 0.5 ppb
Zinc (Zn)... 0.5 ppb
Zirconium (Zr) .... 0.5 ppb

TESTS

Mercury. (By CVAAS, page 65). To each of three 100-mL volumetric flasks, add 80.0 g of sample. To the second and third flasks, add mercury ion (Hg) standards of 20 ng (0.25 ppb) and 40 ng (0.5 ppb), respectively. Add 5 mL of nitric acid to all three flasks, and dilute to the mark with water. Mix. Zero the instrument with a blank, and determine the mercury content using a suitable mercury analyzer system (1.0 mL of 0.01 μg/mL Hg = 10 ng).

Selenium. (By HGAAS, page 66). To a set of three 100-mL volumetric flasks, transfer 50.0 g (50 mL) of sample. To two flasks, add selenium ion (Se) standards of 0.25 μg (5 ppb) and 0.50 μg (10 ppb), respectively. Dilute each to 100 mL with (1 + 1) hydrochloric acid. Mix. Prepare fresh working standard before use.

Trace Metals. Determine the aluminum, barium, boron, cadmium, calcium, chromium, cobalt, copper, iron, lead, lithium, magnesium, manganese, molybdenum, potassium, silicon, sodium, strontium, tin, titanium, vanadium, zinc, and zirconium by the ICP–OES method described on page 69.

Xylenes
Dimethylbenzenes

$C_6H_4(CH_3)_2$ **Formula Wt 106.17**   **CAS No. 1330-20-7**

Note: This reagent is generally a mixture of the *ortho*, *meta*, and *para* isomers and may contain some ethylbenzene.

GENERAL DESCRIPTION

Typical appearance: clear liquid
Analytical use: solvent
Change in state (approximate): boiling point, 140 °C
Aqueous solubility: insoluble
Density: 0.87

SPECIFICATIONS

Assay . . . . ≥98.5% xylene isomers plus ethylbenzene
(ethylbenzene not to exceed 25%)

Maximum Allowable

Color (APHA) 10
Residue after evaporation 0.002%
Substances darkened by sulfuric acid . . . . Passes test
Sulfur compounds (as S) 0.003%
Water (H$_2$O) 0.05%

TESTS

Assay. Analyze the sample by gas chromatography using the general parameters cited on page 80. The following specific conditions are also required.

Column: Type I, methyl silicone

Measure the area under all peaks, and calculate the area percent of each peak. The total percent of xylenes is the sum of the percents of *ortho*, *meta*, and *para* isomers and of ethylbenzene. Correct for water content.

Color (APHA). (Page 43).

Residue after Evaporation. (Page 25). Evaporate 100 g (115 mL) to dryness in a tared, preconditioned dish on a hot plate (≈100 °C), and dry the residue at 105 °C for 30 min.

Substances Darkened by Sulfuric Acid. Shake 15 mL with 5 mL of sulfuric acid for 15–20 s, and allow to stand for 15 min. The xylene layer should be colorless, and the color of the acid should not exceed that of a color standard composed of 1 volume of water and 3 volumes of a color standard solution (described below).

> **Color Standard Solution.** In a 1-L volumetric flask, add 5 g of cobalt chloride hexahydrate, 40 g of ferric chloride hexahydrate, and 20 mL of hydrochloric acid. Dilute to the mark with water.

Sulfur Compounds. Place 30 mL of approximately 0.5 N potassium hydroxide in methanol in a conical flask, add 6.0 mL of the sample, and boil the mixture gently for 30 min under a reflux condenser, avoiding the use of a rubber stopper or connection. Detach the condenser, dilute with 50 mL of water, and heat on a hot plate (≈100 °C) until the xylenes and methanol have evaporated. Add 50 mL of bromine water, and heat for 15 min longer. Transfer the solution to a beaker, neutralize with dilute hydrochloric acid (1 + 3), add an excess of 1 mL of the acid, and concentrate to about 50 mL. Filter if necessary, heat the filtrate to boiling, add 5 mL of 12% barium chloride reagent solution, digest in a covered beaker on a hot plate ($\approx$100 °C) for 2 h, and allow to stand for at least 8 h. If a precipitate is formed, filter, wash thoroughly, and ignite. Correct for the weight obtained in a complete blank test.

Water. (Page 31, Method 2). Use 100 µL (87 mg) of the sample.

Xylenol Orange

$R = H$ or Na

$C_{31}H_{30}N_2O_{13}S$ (free acid)   **Formula Wt 672.66**   **CAS No. 1611-35-4**
$C_{31}H_{28}N_2Na_4O_{13}S$ (salt)   **Formula Wt 760.59**   **CAS No. 3618-43-7**

Note: This standard applies to both the free acid form and the salt form of this reagent.

GENERAL DESCRIPTION

Typical appearance: dark solid
Analytical use: indicator

SPECIFICATIONS

Clarity of solution . . . . Passes test
Suitability for zinc titration . . . . Passes test

TESTS

Clarity of Solution. If the indicator is the acid form, dissolve 0.1 g in 100 mL of alcohol. If the indicator is a salt form, dissolve 0.1 g in 100 mL of water. Not more than a faint trace of turbidity or insoluble matter should remain. Reserve the solution for the test for suitability for zinc titration.

Suitability for Zinc Titration. Dissolve 2.5 g of hexamethylenetetramine in 75 mL of water. Add 3 mL of 10% nitric acid and 1 mL of 0.1% sample solution reserved from the test for clarity of solution. The color should be yellow. Add 0.03 mL of 0.1 M zinc chloride volumetric solution, which should change the color to a magenta purple. Addition of 0.05 mL of 0.1 M volumetric EDTA solution should restore the yellow color.

Zinc

Zn   **Atomic Wt 65.41**   **CAS No. 7440-66-6**

GENERAL DESCRIPTION

Typical appearance: silver-gray metal
Analytical use: reducing agent; determination of arsenic
Change in state (approximate): melting point, 419 °C

SPECIFICATIONS

Assay . . . . ≥99.8% Zn
Suitability for determination of arsenic (As) . . . . Passes test

Maximum Allowable

Iron (Fe) 0.01%
Lead (Pb) 0.01%

TESTS

Assay. (By complexometric titration for zinc). Weigh, to the nearest 0.1 mg, about 0.2 g of sample, and dissolve with 10 mL of (1 + 1) nitric acid. When completely dissolved, dilute with water to about 300 mL, and with the aid of magnetic stirring, neutralize with 10% sodium hydroxide to methyl red. Add a few milligrams of ascorbic acid, 15 mL of pH 10 ammoniacal buffer solution, and about 50 mg of Eriochrome Black T indicator mixture. Titrate immediately with 0.1 M EDTA volumetric solution to a blue end point. One milliliter of 0.1 M EDTA corresponds to 0.006539 g of Zn.

$$\% \; Zn = \frac{(mL \times M \; EDTA) \times 6.539}{Sample \; wt \; (g)}$$

Suitability for Determination of Arsenic. The apparatus used for this test is described in the general method for arsenic on page 34. Add 13 g of the sample to a generator flask containing a mixture of 35 mL of water, 10 mL of hydrochloric acid, 2 mL of 16.5% potassium iodide reagent solution, and 0.5 mL of 40% stannous chloride reagent solution. Immediately connect the scrubber–absorber assembly to the generator flask. When the gas evolution stops, disconnect the flask, and add successive 5-mL portions of hydrochloric acid. Immediately connect the scrubber–absorber assembly to the flask after each addition. After four or five portions of acid have been added, decant the spent acid, and add a mixture of 35 mL of water, 10 mL of hydrochloric acid, 2 mL of 16.5% potassium iodide reagent solution, and 0.5 mL of 40% stannous chloride reagent solution. Immediately reconnect the scrubber–absorber assembly to the flask. Continue the addition of 5-mL portions of hydrochloric acid and the decantations, if necessary, until all of the zinc sample is dissolved. Disconnect the tubing from the generator flask, and transfer the silver diethyldithiocarbamate solution to a suitable color-comparison tube. Any red color in the silver diethyldithiocarbamate solution of the sample should not exceed that in a control prepared with 3 g of the zinc sample and containing 0.001 mg of arsenic ion (As).

Iron and Lead. (By flame AAS, page 63).

    Sample Stock Solution. Dissolve 10.0 g of sample in 75 mL of (1 + 1) nitric acid. When dissolution is nearly complete, add 5 mL of nitric acid, and heat to boiling, or until any residue from the zinc is dissolved. Cool, transfer with water to a 100-mL volumetric flask, and dilute to the mark with water (1 mL = 0.1 g).

Element	Wavelength (nm)	Sample Wt (g)	Standard Added (mg)	Flame Type*	Background Correction
Fe	248.3	1.0	0.05; 0.1	A/A	Yes
Pb	217.0	1.0	0.05; 0.1	A/A	Yes

*A/A is air/acetylene.

Zinc Acetate Dihydrate

(CH₃COO)₂Zn · 2H₂O　　　　**Formula Wt 219.53**　　　　**CAS No. 5970-45-6**

GENERAL DESCRIPTION

Typical appearance: colorless or white solid

Analytical use: determination of albumin, tannin, urobilin, phosphate, and blood

Change in state (approximate): melting point, 327 °C

Aqueous solubility: 43 g in 100 mL at 20 °C

SPECIFICATIONS

Assay . . . . 98.0–101.0% $(CH_3COO)_2Zn \cdot 2H_2O$

pH of a 5% solution at 25.0 °C 6.0–7.0

	Maximum Allowable
Insoluble matter	0.005%
Chloride (Cl)	5 ppm
Sulfate (SO_4)	0.005%
Calcium (Ca)	0.005%
Magnesium (Mg)	0.005%
Potassium (K)	0.01%
Sodium (Na)	0.05%
Iron (Fe)	5 ppm
Lead (Pb)	0.002%

TESTS

Assay. (By complexometric titration for zinc). Weigh accurately about 0.8 g, transfer to a 500-mL beaker, and dilute to about 300 mL with water. Add 15 mL of saturated hexamethylenetetramine reagent solution and 50 mg of xylenol orange indicator mixture. Titrate with standard 0.1 M EDTA volumetric solution to a color change of purple-red to lemon yellow. One milliliter of 0.1 M EDTA corresponds to 0.02195 g of $(CH_3COO)_2Zn \cdot 2H_2O$.

$$\% \, (CH_3COO)_2Zn \cdot 2H_2O = \frac{(mL \times M \, EDTA) \times 21.95}{Sample \, wt \, (g)}$$

pH of a 5% Solution at 25.0 °C. (Page 49).

Insoluble Matter. (Page 25). Use 20 g dissolved in a mixture of 2 mL of glacial acetic acid and 200 mL of water.

Chloride. (Page 35). Use 2.0 g.

Sulfate. Dissolve 20 g in 200 mL of water, add 1 mL of hydrochloric acid, filter, heat to boiling, and add 10 mL of 12% barium chloride reagent solution. Digest in a covered beaker on a hot plate ($\approx$100 °C) for 2 h, and allow to stand for at least 8 h. If a precipitate forms, filter, wash with water containing about 0.2% hydrochloric acid, wash finally with plain water, and ignite. Correct for a complete blank determination. One gram of $BaSO_4$ so obtained corresponds to 0.41 g of sulfate ion (SO_4).

Calcium, Magnesium, Potassium, and Sodium. (By flame AAS, page 63).

Sample Stock Solution. Dissolve 10.0 g in water in a 100-mL volumetric flask, add 0.5 mL of hydrochloric acid, and dilute with water to the mark (1 mL = 0.10 g).

Element	Wavelength (nm)	Sample Wt (g)	Standard Added (mg)	Flame Type*	Background Correction
Ca	422.7	0.80	0.02; 0.04	N/A	No
Mg	285.2	0.20	0.005; 0.01	A/A	Yes
K	766.5	0.20	0.01; 0.02	A/A	No
Na	589.0	0.04	0.01; 0.02	A/A	No

*A/A is air/acetylene; N/A is nitrous oxide/acetylene.

Iron. (Page 38, Method 1). Use 2.0 g.

Lead. (Page 54). Use 5.0 g of sample and 0.25 mL of hydrochloric acid. For the standard, add 0.10 mg of lead ion (Pb).

Zinc Chloride

ZnCl$_2$ **Formula Wt 136.30**   **CAS No. 7646-85-7**

GENERAL DESCRIPTION

Typical appearance: white solid
Analytical use: dehydrating agent
Change in state (approximate): melting point, 318 °C
Aqueous solubility: 368 g in 100 mL

SPECIFICATIONS

Assay . . . . ≥97.0% ZnCl$_2$

Maximum Allowable

Oxychloride. . . . . Passes test
Insoluble matter 0.005%
Nitrate (NO$_3$) 0.003%
Sulfate (SO$_4$) 0.01%
Ammonium (NH$_4$) 0.005%
Calcium (Ca) 0.06%
Iron (Fe). 0.001%
Lead (Pb) 0.005%
Magnesium (Mg). 0.01%
Potassium (K) 0.02%
Sodium (Na) 0.05%

TESTS

Assay. (By argentimetric titration). Using precautions to avoid absorption of moisture, weigh accurately about 0.3 g, dissolve in about 100 mL of water in a 200-mL volumetric flask, and add 5 mL of nitric acid and 50.0 mL of 0.1 N silver nitrate volumetric solution. Shake vigorously, dilute to volume, mix well, and filter through a dry paper into a dry flask or beaker, rejecting the first 20 mL of the filtrate. To 100 mL of the filtrate subsequently collected, add 2 mL of ferric ammonium sulfate indicator solution, and titrate the excess of silver nitrate with 0.1 N ammonium thiocyanate volumetric solution. One milliliter of 0.1 N silver nitrate consumed corresponds to 0.006816 g of ZnCl$_2$.

$$\% \ ZnCl_2 = \frac{[(mL \times N \ AgNO_3) - (2 \times mL \times N \ NH_4SCN)] \times 6.816}{Sample \ wt \ (g)}$$

Oxychloride. Dissolve 20 g in 200 mL of water. On the addition of 6 mL of 1 N hydrochloric acid, any flocculent precipitate should entirely dissolve. Retain the solution for the test for insoluble matter.

Insoluble Matter. To the solution obtained in the test for oxychloride, add 6 mL of 1 N hydrochloric acid. If insoluble matter is present, filter through a tared, preconditioned filtering crucible, wash thoroughly with water containing about 0.2% hydrochloric acid, and dry at 105 °C.

Nitrate. (Page 38, Method 1). For sample solution A, use 0.50 g. For control solution B, use 0.50 g and 1.5 mL of nitrate ion (NO$_3$) standard solution.

Sulfate. (Page 40, Method 1). Allow 30 min for turbidity to form.

Ammonium. Dissolve 1.0 g in water and dilute with water to 50 mL. Pour 10 mL of the solution into 10 mL of freshly boiled 10% sodium hydroxide reagent solution. Dilute with water to 50 mL, and add 2 mL of Nessler reagent. Any color should not exceed that produced by 0.01 mg of ammonium ion (NH$_4$) in an equal volume of solution containing the quantities of reagents used in the test.

Calcium, Iron, Lead, Magnesium, Potassium, and Sodium. (By flame AAS, page 63).

Sample Stock Solution. Dissolve 10.0 g in water in a 100-mL volumetric flask, add 0.5 mL of hydrochloric acid, and dilute with water to the mark (1 mL = 0.10 g).

Element	Wavelength (nm)	Sample Wt (g)	Standard Added (mg)	Flame Type*	Background Correction
Ca	422.7	0.10	0.03; 0.06	N/A	No
Fe	248.3	2.0	0.02; 0.04	A/A	Yes
Pb	217.0	2.0	0.05; 0.10	A/A	Yes
Mg	285.2	0.10	0.005; 0.01	A/A	Yes
K	766.5	0.10	0.01; 0.02	A/A	No
Na	589.0	0.02	0.005; 0.01	A/A	No

*A/A is air/acetylene; N/A is nitrous oxide/acetylene.

Zinc Oxide

ZnO   **Formula Wt 81.41**   **CAS No. 1314-13-2**

GENERAL DESCRIPTION

Typical appearance: white solid
Analytical use: preparation of zinc standard solutions
Aqueous solubility: practically insoluble

SPECIFICATIONS

Assay . . . . ≥99.0% ZnO

Maximum Allowable
Insoluble in dilute sulfuric acid 0.01%
Alkalinity . . . . .Passes test
Chloride (Cl) 0.001%
Nitrate (NO_3) 0.003%
Sulfur compounds (as SO_4) 0.01%
Calcium (Ca) 0.005%
Iron (Fe) 0.001%
Lead (Pb) 0.005%
Magnesium (Mg) 0.005%
Manganese (Mn) . . . . .5 ppm
Potassium (K) 0.01%
Sodium (Na) 0.05%

TESTS

Assay. (By complexometric titration of zinc). Weigh accurately 0.25 g, and transfer to a 500-mL beaker. Dissolve in 15 mL of hydrochloric acid (1 + 1), and dilute to about 300 mL with water. Add 0.15 mL of methyl red indicator solution, and neutralize (to a yellow color) by dropwise addition of 10% sodium hydroxide reagent solution. Add 15 mL of saturated hexamethylenetetramine reagent solution and 50 mg of xylenol orange indicator mixture. Titrate with standard 0.1 M EDTA volumetric solution to a color change of purple-red to lemon yellow. One milliliter of 0.1 M EDTA corresponds to 0.008141 g of ZnO.

$$\% \text{ZnO} = \frac{(\text{mL} \times \text{M EDTA}) \times 8.141}{\text{Sample wt (g)}}$$

Insoluble in Dilute Sulfuric Acid. Dissolve 10 g by heating in a covered beaker on a steam bath with 160 mL of dilute sulfuric acid (1 + 15) for 1 h. Filter through a tared filtering crucible, wash thoroughly, and dry at 105 °C.

Alkalinity. Suspend 2 g in 20 mL of water, boil for 1 min, and filter. Add 0.10 mL of phenolphthalein indicator solution to 10 mL of the filtrate. No red color should be produced.

Chloride. (Page 35). Suspend 1.0 g in 20 mL of water, and dissolve by adding 3 mL of nitric acid.

Nitrate. (Page 38, Method 1). For sample solution A, dissolve 0.5 g of sample in 5 mL of hydrochloric acid (1 + 1). For control solution B, use 5 mL of hydrochloric acid (1 + 1) and 1.5 mL of nitrate ion (NO_3) standard solution.

Sulfur Compounds. Suspend 5.0 g in 50 mL of water, and add about 50 mg of sodium carbonate and 1 mL of bromine water. Boil for 5 min, cautiously add hydrochloric acid in small portions until the zinc oxide is dissolved, then add 1 mL more of the acid. Filter, wash, and dilute the filtrate with water to about 150 mL. Heat the filtrate to boiling, add 5 mL of 12% barium chloride reagent solution, digest in a covered beaker on a hot plate (≈ 100 °C) for 2 h, and allow to stand for at least 8 h. If a precipitate is formed, filter, wash two or three times with dilute hydrochloric acid (1 + 99), complete the washing with water alone, and ignite. Correct for the weight obtained in a complete blank test.

$$\% \text{SO}_4 = \frac{\text{Residue wt (g)} \times 13.7}{5.0 \text{ g}}$$

Calcium, Iron, Lead, Magnesium, Manganese, Potassium, and Sodium. (By flame AAS, page 63).

 Sample Stock Solution. Dissolve 10.0 g in 10 mL of hydrochloric acid in a 100-mL volumetric flask, and dilute with water to volume (1 mL = 0.10 g).

Element	Wavelength (nm)	Sample Wt (g)	Standard Added (mg)	Flame Type*	Background Correction
Ca	422.7	0.80	0.02; 0.04	N/A	No
Fe	248.3	2.0	0.02; 0.04	A/A	Yes
Pb	217.0	2.0	0.05; 0.10	A/A	Yes
Mg	285.2	0.20	0.005; 0.01	A/A	Yes
Mn	279.5	2.0	0.01; 0.02	A/A	Yes
K	766.5	0.20	0.01; 0.02	A/A	No
Na	589.0	0.04	0.01; 0.02	A/A	No

*A/A is air/acetylene; N/A is nitrous oxide/acetylene.

Zinc Sulfate Heptahydrate

$ZnSO_4 \cdot 7H_2O$ **Formula Wt 287.58**   **CAS No. 7446-20-0**

GENERAL DESCRIPTION

Typical appearance: colorless or white solid
Analytical use: preparation of complexometric titrant
Change in state (approximate): melting point, 40 °C; dehydrates completely at 200 °C
Aqueous solubility: 96 g in 100 mL at 20 °C

SPECIFICATIONS

Assay . . . . .99.0–103.0% $ZnSO_4 \cdot 7H_2O$
pH of a 5% solution at 25.0 °C 4.4–6.0

	Maximum Allowable
Insoluble matter	0.01%
Chloride (Cl).	5 ppm
Nitrate (NO_3)	0.002%
Ammonium (NH_4)	0.001%
Calcium (Ca)	0.005%
Iron (Fe).	0.001%
Lead (Pb).	0.003%
Magnesium (Mg)	0.005%
Manganese (Mn)	3 ppm
Potassium (K)	0.01%
Sodium (Na).	0.05%

TESTS

Assay. (By complexometric determination of zinc). Weigh accurately about 1.0 g, transfer to a 500-mL beaker, and dissolve in about 300 mL of water. Add 15 mL of saturated hexamethylenetetramine reagent solution and 50 mg of xylenol orange indicator mixture. Titrate with standard 0.1 M EDTA volumetric solution to a color change of purple-red to lemon yellow. One milliliter of 0.1 M EDTA corresponds to 0.02876 g of $ZnSO_4 \cdot 7H_2O$.

$$\% \ ZnSO_4 \cdot 7H_2O = \frac{(mL \times M \ EDTA) \times 28.76}{Sample \ wt \ (g)}$$

pH of a 5% Solution at 25.0 °C. (Page 49).

Insoluble Matter. (Page 25). Use 10.0 g dissolved in 100 mL of water.

Chloride. (Page 35). Use 2.0 g.

Nitrate. (Page 38, Method 1). For sample solution A, use 1.0 g. After sample dissolution, add 2.0 mL of hydrochloric acid. For control solution B, after sample dissolution, add 2.0 mL of hydrochloric acid and 2.0 mL of nitrate ion (NO_3) standard solution.

Ammonium. Dissolve 1.0 g in 25 mL of water, pour into 25 mL of freshly boiled and cooled 10% sodium hydroxide reagent solution, and add 2 mL of Nessler reagent. Any color should not exceed that produced by 0.01 mg of ammonium ion (NH_4) in an equal volume of solution containing the quantities of reagents used in the test.

Calcium, Iron, Lead, Magnesium, Manganese, Potassium, and Sodium.   (By flame AAS, page 63).

Sample Stock Solution. Dissolve 10.0 g of sample in water in a 100-mL volumetric flask, add 0.5 mL of hydrochloric acid, and dilute with water to the mark (1 mL = 0.10 g).

Element	Wavelength (nm)	Sample Wt (g)	Standard Added (mg)	Flame Type*	Background Correction
Ca	422.7	0.80	0.02; 0.04	N/A	No
Fe	248.3	2.0	0.02; 0.04	A/A	Yes
Pb	217.0	2.0	0.06; 0.12	A/A	Yes
Mg	285.2	0.20	0.005; 0.01	A/A	Yes
Mn	279.5	2.0	0.01; 0.02	A/A	Yes
K	766.5	0.20	0.01; 0.02	A/A	No
Na	589.0	0.04	0.01; 0.02	A/A	No

*A/A is air/acetylene; N/A is nitrous oxide/acetylene.

Zincon
2-{[α-(2-Hydroxy-5-sulfophenylazo)-benzylidene]-hydrazino}-benzoic Acid Monosodium Salt

$C_{20}H_{15}N_4NaO_6S$ (anhydrous)   **Formula Wt 462.42**   **CAS No. 62625-22-3**
$C_{20}H_{15}N_4NaO_6S \cdot H_2O$ (monohydrate)   **Formula Wt 480.43**

Note: This standard applies to both the anhydrous and monohydrate form of the indicator.

GENERAL DESCRIPTION

Typical appearance: purple solid
Analytical use: reagent for zinc, mercury, and copper

SPECIFICATIONS

Clarity of solution . . . . Passes test
Suitability for complexometric titration (Zn) . . . . Passes test

TESTS

Clarity of Solution. Weigh 0.10 g, and dissolve in 100 mL of water. Not more than a faint trace of turbidity or insoluble matter should remain. Reserve the solution for the suitability for complexometric titration test.

Suitability for Complexometric Titration. To 25 mL of water, add 5 mL of pH 10 ammoniacal buffer solution and 0.10 mg of zinc ion. Add 0.5 mL of solution from the test for clarity of solution. A blue color develops. Add 0.10 mL of 0.1 M EDTA volumetric solution. The solution should become red.

Zirconyl Chloride Octahydrate

Zirconium Oxide Chloride, Octahydrate

$ZrOCl_2 \cdot 8H_2O$ **Formula Wt 322.25**   **CAS No. 13520-92-8**

GENERAL DESCRIPTION

Typical appearance: white solid
Analytical use: reagent for fluoride determination in water analyses; reagent for reduction of nitro compounds to primary amines

SPECIFICATIONS

Assay . . . . $\geq$99.0% $ZrOCl_2 \cdot 8H_2O$

Maximum Allowable

Sulfate (SO_4) 0.005%
Heavy metals (as Pb). 0.001%
Iron (Fe). 0.001%
Titanium (Ti) 0.005%

TESTS

Assay. (By indirect complexometric titration of zirconium). Accurately weigh 1.0 g of sample, and transfer to a 250-mL Erlenmeyer flask. Add 5 mL of hydrochloric acid and then 50.0 mL of 0.1 M EDTA volumetric solution. Heat to boiling, cool, and add ammonium acetate buffer solution to a pH of 4.5 (using a pH meter). Add 75 mL of alcohol and 2 mL of dithizone indicator solution, and titrate with 0.1 M zinc chloride volumetric solution to a pink end point. One milliliter of 0.1 M EDTA corresponds to 0.03223 g of $ZrOCl_2 \cdot 8H_2O$.

$$\% \ ZrOCl_2 \cdot 8H_2O = \frac{[(50.0 \times M \ EDTA) - (mL \times M \ ZnCl_2)] \times 32.23}{\text{Sample wt (g)}}$$

Sulfate. (Page 40, Method 1). Dissolve 1 g in 30 mL of hot water, and pour slowly, with stirring, into a mixture of 3 mL of ammonium hydroxide and 40 mL of water. Boil for 10 min, and filter through a sintered glass crucible. Wash the filtrate with small portions of water. Boil one half of the filtrate down to about 10 mL, and neutralize with hydrochloric acid.

Heavy Metals. (Page 36, Method 1). Dissolve 2 g in 30 mL of water, and add 2 g of sodium acetate, trihydrate. Adjust the pH of the solution to between 3 and 4 with acetic acid, and dilute to 40 mL. Any brown color should not exceed that produced by a 0.02-mg lead ion (Pb) standard solution treated exactly as the sample.

Iron. (Page 38, Method 1). Use 1.0 g.

Titanium. Dissolve 1.0 g in 5 mL of water, and add 0.2 mL of 25% hydrochloric acid and 1 mL of 3% hydrogen peroxide reagent solution. Compare the immediate resulting color with that of an equally treated control solution containing 0.05 mg of titanium.

Part 5:
Monographs for Standard-Grade Reference Materials
Specifications and Tests

Specifications and analytical methods for standard-grade reference materials have been developed and were first introduced in the Ninth Edition of *Reagent Chemicals*. The development of this new grade of chemicals was a result of needs from the analytical testing community.

Materials designated as standard grade are suitable for preparation of analytical standards used for a variety of applications, including instrument calibration, quality control, analyte identification, and measuring method performance. These compounds are also suitable for other general applications requiring high-purity materials.

Part 5 of this book is formatted differently than Part 4. Rather than having an individual monograph for each standard-grade material, materials are grouped by compound class. Each class has general specifications that are applicable to the entire group. In addition to assay requirements, standard-grade materials also have an identity requirement. The tests section includes analytical techniques and methods to support the specifications. Following the specifications and tests are tables of compound-specific information, including physical properties, analytical data, and references to analytical methods.

Chlorinated Phenols

SPECIFICATIONS

Identity by

Infrared spectroscopy . . . . Passes test

Gas chromatography/mass spectrometry . . . . Passes test

Assay by

Gas chromatography. . . . . ≥98.0%

Acid–base titrimetry 98–102%

Thin-layer chromatography. . . . . Passes test

TESTS

Identity by Infrared Spectroscopy. Analyze the sample using the general procedure cited on page 92. Identifying absorbances for the compound listed in Table 5-1 must be present. The spectrum is compared to the NIST standardized library of compounds, if available. Absorbances found in the NIST spectrum should be present in the test spectrum; however, absorbance frequency and intensity may be different due to different analysis techniques. The spectrum may also be compared to other published libraries of infrared spectra, if available.

Identity by Gas Chromatography/Mass Spectrometry. Analyze the sample by gas chromatography/mass spectrometry using method GCMS-CP1 described below.

Method GCMS-CP1

Ionization Mode: Electron ionization/70 eV

Column: 100% Dimethylpolysiloxane, 30 M × 0.25 mm i.d., 0.25-μm film thickness

Temperature Program: 80 °C for 12 min, then 5 °C/min to 250 °C

Carrier Gas: Helium at 0.8 mL/min

Injector: Split/splitless (20:1 split)

Injector Temperature: 250 °C

Scan Range: 45–500 amu

Sample Size: 1 μL of a 400 μg/mL solution

Identifying ions for the compound listed in Table 5-1 must be present. The spectrum is compared to the NIST standardized library of compounds, if available. Ions in the test spectrum must match those found in the NIST spectrum.

Table 5-1. Chlorinated Phenols Compound Data

Name	CAS No.	Chemical Formula	Formula Weight	IR-1	IR-2	IR-3	MS-1	MS-2	MS-3
2-Chlorophenol	95-57-8	C_6H_5ClO	128.56	3584	1486	740	128	64	130
3-Chlorophenol[a]	108-43-0	C_6H_5ClO	128.56	3649	1600	885	128	65	130
4-Chlorophenol	106-48-9	C_6H_5ClO	128.56	3650	1495	825	128	65	130
2,3-Dichlorophenol	576-24-9	$C_6H_4Cl_2O$	163.00	3575	1450	770	162	164	126
2,4-Dichlorophenol	120-83-2	$C_6H_4Cl_2O$	163.00	3584	1480	720	162	164	63
2,5-Dichlorophenol	583-78-8	$C_6H_4Cl_2O$	163.00	3579	1478	906	162	164	63
2,6-Dichlorophenol[a]	87-65-0	$C_6H_4Cl_2O$	163.00	3575	1458	770	162	164	63
3,4-Dichlorophenol	95-77-2	$C_6H_4Cl_2O$	163.00	3450	1470	650	162	164	99
3,5-Dichlorophenol	591-35-5	$C_6H_4Cl_2O$	163.00	3640	1590	828	162	164	63
2,3,4,5,6-Pentachlorophenol	87-86-5	C_6HCl_5O	266.34	3557	1383	770	266	264	268
2,3,4,5-Tetrachlorophenol	4901-51-3	$C_6H_2Cl_4O$	231.89	3520	1410	725	232	230	234
2,3,4,6-Tetrachlorophenol	58-90-2	$C_6H_2Cl_4O$	231.89	3570	1445	750	131	232	230
2,3,5,6-Tetrachlorophenol	935-95-5	$C_6H_2Cl_4O$	231.89	3563	1404	700	232	230	234
2,3,4-Trichlorophenol	15950-66-0	$C_6H_3Cl_3O$	197.45	3575	1453	783	196	198	97
2,3,5-Trichlorophenol	933-78-8	$C_6H_3Cl_3O$	197.45	3571	1585	840	196	198	97
2,3,6-Trichlorophenol	933-75-5	$C_6H_3Cl_3O$	197.45	3567	1449	800	196	198	200
2,4,5-Trichlorophenol	95-95-4	$C_6H_3Cl_3O$	197.45	3584	1458	870	196	198	97
2,4,6-Trichlorophenol	88-06-2	$C_6H_3Cl_3O$	197.45	3575	1470	730	196	198	97
3,4,5-Trichlorophenol	609-19-8	$C_6H_3Cl_3O$	197.45	3653	1420	816	196	198	133

[a] Use GC–CP2.

Assay by Gas Chromatography. Analyze the sample by gas chromatography using method GC-CP1 described below. Alternate method GC-CP2 should be used for specific compounds as noted in Table 5-1.

Method GC-CP1

Column: 5% Diphenyl–95% dimethylpolysiloxane, 30 M × 0.53 mm i.d., 1.5-μm film thickness

Detector: Thermal conductivity

Detector Temperature: 300 °C

Injector: Split/splitless (10:1 split)

Injector Temperature: 250 °C

Sample Size: 2 μL of a 1000 μg/mL solution

Carrier Gas: Helium at 3.0 mL/min

Temperature Program: 80 °C for 12 min, then 5 °C/min to 240 °C

Method GC-CP2

Column: 100% Dimethylpolysiloxane, 30 M × 0.25 mm i.d., 0.25 μm film thickness

Detector: Thermal conductivity

Detector Temperature: 300 °C

Injector: Split/splitless (10:1 split)

Injector Temperature: 250 °C

Sample Size: 2 μL of a 1000 μg/mL solution

Carrier Gas: Helium at 1.0 mL/min

Temperature Program: 80 °C for 12 min, then 5 °C/min to 240 °C

Measure the area under each peak (excluding the solvent peak), and calculate the analyte content in area percent.

Assay by Acid–Base Titrimetry. Follow the general procedure cited on page 27. The following specific conditions are also required.

Weigh enough sample so that a minimum of 25 mL of titrant will be used. Dissolve the sample in 2-propanol. Titrate the sample using a standardized solution of potassium hydroxide in methanol.

Assay by Thin-Layer Chromatography. Analyze the sample by thin-layer chromatography using the general procedure described on page 86. The following specific conditions are also required.

Stationary Phase: Silica

Mobile Phase: 1:1 (v/v) Ethyl acetate:hexane

Detection Methods: UV at 254 nm and iodine

Compound must exhibit a single spot.

Explosives
by Gas Chromatography

SPECIFICATIONS
Identity by
Gas chromatography/mass spectrometry .... Passes test
Boiling point (liquids) .... Passes test
Melting point (solids).. Passes test
Assay by
Gas chromatography... ≥98%
Thin-layer chromatography.. Passes test

TESTS

Identity by Gas Chromatography/Mass Spectrometry. Analyze the sample by gas chromatography/mass spectrometry using GCMS-EXP1 described below.

Method GCMS-EXP1

Ionization Mode: Electron ionization/70 eV

Column: 5% Phenyl–95% dimethylpolysiloxane, 30 M × 0.25 mm i.d., 0.25-μm film thickness

Temperature Program: 75 °C for 1 min, then 10 °C/min to 275 °C, hold 5 min

Carrier Gas: Helium at 1 mL/min

Injector: Split/splitless (50:1 split)

Injector Temperature: 250 °C

Scan Range: 40–650 amu

Sample Size: 1.0 μL of a 1 mg/mL solution

Identifying ions listed in Table 5-2 must be present. The spectrum is compared to the NIST standardized library of compounds, if available. Ions in the test spectrum must match those found in the reference spectrum.

Identity by Boiling Point. Analyze the sample using the general procedure described on page 42. Assigned values must be within the ranges listed in Table 5-2.

Identity by Melting Point. Analyze the sample using the general procedure described on page 45. Assigned values must be within the ranges listed in Table 5-2.

Assay by Gas Chromatography. Analyze the sample by gas chromatography using method GCFID-EXP1 described below.

Method GCFID-EXP1

Column: 5% Phenyl–95% dimethylpolysiloxane, 30 M × 0.53 mm i.d., 1.0-μm film thickness

Temperature Program: 75 °C for 1 min, then 10 °C/min to 275 °C, hold 5 min

Carrier Gas: Helium at 5 mL/min

Injector: Splitless

Injector Temperature: 250 °C

Detector Temperature: 280 °C

Sample Size: 1.0 μL of a 1 mg/mL solution

Measure the area under each peak (excluding the solvent peak) and calculate the analyte content in area percent.

Assay by Thin-Layer Chromatography. Analyze the sample by thin-layer chromatography using the general procedure described on page 86. Compound must exhibit a single spot.

Table 5-2. Explosives by Gas Chromatography Compound Data

Name	CAS No.	Chemical Formula	Formula Weight	MS (m/z)			MP(°C)	BP(°C)
				MS-1	MS-2	MS-3		
2-Amino-4,6-dinitrotoluene	35572-78-2	$C_7H_7N_3O_4$	197.15	197	180	78	182–187	
4-Amino-2,6-dinitrotoluene	19406-51-0	$C_7H_7N_3O_4$	197.15	197	180	163	183–188	
2,4-Diamino-6-nitrotoluene	6629-29-4	$C_7H_9N_3O_2$	167.17	167	150	121	131–136	
2,6-Diamino-4-nitrotoluene	59229-75-3	$C_7H_9N_3O_2$	167.17	167	121	94	216–221	
2,3-Dimethyl-2,3-dinitrobutane	3964-18-9	$C_6H_{12}N_2O_4$	176.17	100	69	57	211–216	
1,2-Dinitrobenzene	528-29-0	$C_6H_4N_2O_4$	168.11	168	76	63	115–120	
1,3-Dinitrobenzene	99-65-0	$C_6H_4N_2O_4$	168.11	168	122	75	294–299	
3,5-Dinitroaniline	618-87-1	$C_6H_5N_3O_4$	183.12	183	137	91	162–167	
2,4-Dinitrotoluene	121-14-2	$C_7H_6N_2O_4$	182.13	165	119	89	66–71	
2,6-Dinitrotoluene	606-20-2	$C_7H_6N_2O_4$	182.13	165	89	63	62–67	
3,4-Dinitrotoluene	610-39-9	$C_7H_6N_2O_4$	182.13	182	89	63	53–58	
Nitrobenzene	98-95-3	$C_6H_5NO_2$	123.11	123	77	51		208–213
2-Nitrotoluene	88-72-2	$C_7H_7NO_2$	137.14	137	120	65		119–124
3-Nitrotoluene	99-08-1	$C_7H_7NO_2$	137.14	137	91	65		235–240
4-Nitrotoluene	99-99-0	$C_7H_7NO_2$	137.14	137	91	65	52–57	
Picramic acid	831-52-7	$C_6H_5N_3O_5$	199.12	199	153	106	166–171	

Note: MP = melting point; BP = boiling point.

Halogenated Hydrocarbons

SPECIFICATIONS

Identity by
Gas chromatography/mass spectrometry . . . . Passes test

Assay by
Gas chromatography (using two methods) . . . . ≥98.0%

TESTS

Identity by Gas Chromatography/Mass Spectrometry. Analyze the sample by gas chromatography/mass spectrometry using method GCMS-HHC1 described below.

Table 5-3. Halogenated Hydrocarbons Compound Data

Name	CAS No.	Chemical Formula	Formula Weight	MS (m/z)		
				MS-1	MS-2	MS-3
Allyl chloride	107-05-1	C_3H_5Cl	76.53	41	76	78
1-Bromo-2-chloroethane	107-04-0	C_2H_4BrCl	143.42	63	142	65
2-Bromo-1-chloropropane	3017-95-6	C_3H_6BrCl	157.44	77	156	79
Bromochloromethane	74-97-5	CH_2BrCl	129.39	49	128	130
Bromodichloromethane	75-27-4	$CHBrCl_2$	163.83	83	127	85
Bromoethane	74-96-4	C_2H_5Br	108.97	108	110	79
Bromoform	75-25-2	$CHBr_3$	252.75	173	250	171
Carbon tetrachloride	56-23-5	CCl_4	153.82	117	119	82
3-Chloro-2-methylpropene	563-47-3	C_3H_7Cl	90.55	55	90	75
1-Chlorobutane	109-69-3	C_4H_8Cl	92.57	56	63	41
Chlorodibromomethane	124-48-1	$CHBr_2Cl$	208.29	127	206	129
Chloroform	67-66-3	$CHCl_3$	119.38	83	117	85
1-Chlorohexane	544-10-5	$C_6H_{13}Cl$	120.62	91	93	69
1,2-Dibromo-3-chloropropane	96-12-8	$C_3H_5ClBr_2$	236.34	157	155	75
1,2-Dibromoethane	106-93-4	$C_2H_4Br_2$	187.87	107	186	109
Dibromofluoromethane	1868-53-7	$CHBr_2F$	191.84	111	190	113
Dibromomethane	74-95-3	CH_2Br_2	173.85	174	172	93
cis-1,4-Dichloro-2-butene	1476-11-5	$C_4H_6Cl_2$	125.00	53	88	75
trans-1,4-Dichloro-2-butene	110-57-6	$C_4H_6Cl_2$	125.00	75	124	53
1,3-Dichlorobutane	1190-22-3	$C_4H_8Cl_2$	127.01	55	90	63
1,4-Dichlorobutane	110-56-5	$C_4H_8Cl_2$	127.01	55	90	62
1,1-Dichloroethane	75-34-3	$C_2H_4Cl_2$	98.96	63	98	83
1,2-Dichloroethane	107-06-2	$C_2H_4Cl_2$	98.96	62	98	64
1,1-Dichloroethene	75-35-4	$C_2H_2Cl_2$	96.94	61	96	63

Continued on next page

Method GCMS-HHC1

Ionization Mode: Electron ionization/70 eV

Column: 5% Diphenyl–95% dimethylpolysiloxane, 30 M × 0.25 mm i.d., 0.25-μm film thickness

Temperature Program: 35 °C for 10 min, then 4 °C/min to 200 °C, hold 10 min

Carrier Gas: Helium at 0.8 mL/min

Injector: Split/splitless (100:1 split)

Injector Temperature: 200 °C

Scan Range: 35–400 amu

Sample Size: 1 μL of a 100 μg/mL solution

Table 5-3. *Continued*

Name	CAS No.	Chemical Formula	Formula Weight	MS (m/z)		
				MS-1	MS-2	MS-3
cis-1,2-Dichloroethene	156-59-2	C₂H₂Cl₂	96.94	61	96	63
trans-1,2-Dichloroethene	156-60-5	C₂H₂Cl₂	96.94	61	96	98
1,2-Dichloropropane	78-87-5	C₃H₆Cl₂	112.99	63	112	76
1,3-Dichloropropane	142-28-9	C₃H₆Cl₂	112.99	76	112	78
2,2-Dichloropropane	594-20-7	C₃H₆Cl₂	112.99	77	97	41
1,1-Dichloropropene	563-58-6	C₃H₄Cl₂	110.97	75	109	77
Diiodomethane	75-11-6	CH₂I₂	267.84	268	141	127
Hexachlorobutadiene	87-68-3	C₄Cl₆	260.76	225	258	190
Hexachlorocyclopentadiene	77-47-4	C₅Cl₆	272.77	237	270	235
Hexachloroethane	67-72-1	C₂Cl₆	236.74	117	201	119
Hexachloropropene	1888-71-7	C₃Cl₆	248.75	213	246	141
Iodomethane	74-88-4	CH₃I	141.94	142	127	141
Methylene chloride	75-09-2	CH₂Cl₂	84.93	49	84	86
Pentachloroethane	76-01-7	C₂HCl₅	202.29	117	167	119
Tetrachloroethene	127-18-4	C₂Cl₄	165.83	166	164	131
1,1,1,2-Tetrachloroethane	630-20-6	C₂H₂Cl₄	167.85	131	117	95
1,1,2,2-Tetrachloroethane	79-34-5	C₂H₂Cl₄	167.85	83	166	95
1,1,1-Trichloroethane	71-55-6	C₂H₃Cl₃	133.41	97	99	61
1,1,2-Trichloroethane	79-00-5	C₂H₃Cl₃	133.41	97	132	83
Trichloroethene	79-01-6	C₂HCl₃	131.39	95	130	132
1,2,3-Trichloropropane	96-18-4	C₃H₅Cl₃	147.43	75	110	61

Identifying ions for the compound listed in Table 5-3 must be present. The spectrum is compared to the NIST standardized library of compounds, if available. Ions in the test spectrum must match those found in the NIST spectrum.

Assay by Gas Chromatography. Analyze the sample by gas chromatography using method GC-HHC1 described below.

Method GC-HHC1

Column: 3% Diphenyl–3% cyanopropyl–94% dimethylpolysiloxane, 30 M × 0.53 mm i.d., 3.0-μm film thickness

Detector: Thermal conductivity

Detector Temperature: 300 °C

Injector: Split/splitless (10:1 split)

Injector Temperature: 200 °C

Sample Size: 2 μL of a 1000 μg/mL solution

Carrier Gas: Helium at 3.0 mL/min

Temperature Program: 35 °C for 10 min, then 4 °C/min to 200 °C, hold 9 min

Measure the area under all peaks (excluding the solvent peak), and calculate the analyte content in area percent for each analysis.

Methyl Haloacetates

SPECIFICATIONS

Identity by

Gas chromatography/mass spectrometry . . . . Passes test

Assay by

Gas chromatography . . . . ≥98.0%

Liquid chromatography . . . . ≥98.0%

TESTS

Identity by Gas Chromatography/Mass Spectrometry. Analyze the sample by gas chromatography/mass spectrometry using method GCMS-MHA1 described below.

Method GCMS-MHA1

Ionization Mode: Electron ionization/70 eV

Column: 5% Diphenyl–95% dimethylpolysiloxane, 30 M × 0.25 mm ID, 0.25-μm film thickness

Temperature Program: 50 °C, hold 1 min, then 10 °C/min to 270 °C

Carrier Gas: Helium at 0.8 mL/min

Injector: Split/splitless (100:1 split)

Injector Temperature: 250 °C

Scan Range: 50–550 amu

Sample Size: 0.5 μL of a 50 μg/mL solution

Identifying ions for the compounds listed in Table 5-4 must be present. The spectrum is compared to the NIST standardized library of compounds, if available. Ions in the test spectrum must match those found in the NIST spectrum.

Assay by Gas Chromatography. Analyze the sample by gas chromatography using method GC-MHA1 described below.

Method GC-MHA1

Column: 5% Diphenyl–95% dimethylpolysiloxane, 30 M × 0.53 mm i.d., 1.5-μm film thickness

Table 5-4. Methyl Haloacetates Compound Data

Name	CAS Number	Chemical Formula	Formula Weight	MS (m/Z)		
				MS-1	MS-2	MS-3
Methyl bromoacetate	96-32-2	$C_3H_5BrO_2$	152.98	152	93	59
Methyl bromochloroacetate	20428-74-4	$C_3H_4BrClO_2$	187.42	129	127	59
Methyl bromodichloroacetate	20428-76-6	$C_3H_3BrCl_2O_2$	221.87	163	141	59
Methyl chloroacetate	96-34-4	$C_3H_5ClO_2$	108.52	108	77	59
Methyl chlorodibromoacetate	20428-75-5	$C_3H_3Br_2ClO_2$	266.32	207	187	59
Methyl dibromoacetate	6482-26-4	$C_3H_4Br_2O_2$	231.88	232	173	59
Methyl dichloroacetate	116-54-1	$C_3H_4Cl_2O_2$	142.97	85	83	59
Methyl tribromoacetate	3222-05-7	$C_3H_3Br_3O_2$	310.78	310	251	59
Methyl trichloroacetate	598-99-2	$C_3H_3Cl_3O_2$	177.42	141	117	59

Detector: Flame ionization

Detector Temperature: 300 °C

Injector: Splitless

Injector Temperature: 250 °C

Sample Size: 1 μL of a 1000 μg/mL solution

Carrier Gas: Hydrogen at 3.5 mL/min

Temperature Program: 40 °C, then 5 °C/min to 165 °C, then 15 °C/min to 280 °C, hold 5 min

Measure the area under all peaks (excluding the solvent peak), and calculate the analyte content in area percent.

Assay by Liquid Chromatography. Analyze the sample by liquid chromatography using method HPLC-MHA1 described below.

Method HPLC-MHA1

Column: Octadecyl on high-purity silica, 250 × 4.6 mm i.d., 5 μm

Mobile Phase: Acetonitrile: 0.01 M phosphate buffer (KH_2PO_4) 80:20

Flow Rate: 2.0 mL/min

Detector: UV, 254 nm

Sensitivity: 0.05 AUFS

Sample Size: 1 μL of a 100 μg/mL solution in acetonitrile

Measure the area under all peaks (excluding the solvent peak), and calculate the analyte content in area percent.

Monocyclic Aromatic Hydrocarbons

SPECIFICATIONS

Identity by
Gas chromatography/mass spectrometry . . . . Passes test
Boiling point (liquids). . . . . Passes test
Melting point (solids) . . . . Passes test
Relative retention index (for *m*-, *o*-, and *p*-xylene only) . . . . Passes test

Assay by
Gas chromatography (using two methods) . . . . ≥98.0%

TESTS

Identity by Gas Chromatography/Mass Spectrometry. Analyze the sample by gas chromatography/mass spectrometry using method GCMS-MCA1 described below.

Method GCMS-MCA1

Ionization Mode: Electron ionization/70 eV

Column: 5% Diphenyl–95% dimethylpolysiloxane, 30 M × 0.25 mm i.d., 0.25-μm film thickness

Temperature Program: 35 °C for 5 min, then 5 °C/min to 300 °C, hold 10 min

Carrier Gas: Helium 1.0 mL/min

Injector: Split/splitless (100:1 split)

Injector Temperature: 240 °C

Scan Range: 35–550 amu

Sample Size: 2 μL of a 1000 μg/mL solution

Identifying ions for the compound listed in Table 5-5 must be present. The spectrum is compared to the NIST standardized library of compounds, if available. Ions in the test spectrum must match those found in the NIST spectrum.

Identity by Boiling Point. Analyze the liquid sample using the procedure cited on page 42. Boiling point values must be within the ranges listed in Table 5-5.

Identity by Melting Point. Analyze the solid sample using the procedure cited on page 45. Melting point values must be within the ranges listed in Table 5-5.

Identity by Relative Retention Index. Prepare a solution by adding equal amounts of the sample and toluene in a suitable solvent for a final concentration of 500 μg/mL. Analyze the sample by gas chromatography using method GC-MCA1 described below. Calculate the relative retention time of the sample versus toluene by dividing the retention time of the sample by the retention time of toluene. Assayed values must be within 1% of 1.63 for *m*-xylene, 1.90 for *o*-xylene, and 1.59 for *p*-xylene.

Table 5-5. Monocyclic Aromatic Hydrocarbons Compound Data

Name	CAS No.	Chemical Formula	Formula Weight	MS (m/z)			MP (°C)	BP (°C)
				MS-1	MS-2	MS-3		
Benzal chloride	98-87-3	$C_7H_6Cl_2$	161.03	125	160	89		202–207
Benzene	71-43-2	C_6H_6	78.11	78	52			77–82
Benzyl chloride	100-44-7	C_7H_7Cl	126.59	91	126	65		176–181
Bromobenzene	108-86-1	C_6H_5Br	157.02	77	156	51		153–158
n-Butylbenzene	104-51-8	$C_{10}H_{14}$	134.22	91	134	65		180–185
sec-Butylbenzene	135-98-8	$C_{10}H_{14}$	134.22	105	134	91		170–175
tert-Butylbenzene	98-06-6	$C_{10}H_{14}$	134.22	119	91	134		166–171
Chlorobenzene	108-90-7	C_6H_5Cl	112.56	112	77	51		129–134
2-Chlorotoluene	95-49-8	C_7H_7Cl	126.59	91	126	63		156–161
4-Chlorotoluene	106-43-4	C_7H_7Cl	126.59	91	126	63		159–164
1,2-Dichlorobenzene	95-50-1	$C_6H_4Cl_2$	147.00	146	111	75		178–183
1,3-Dichlorobenzene	541-73-1	$C_6H_4Cl_2$	147.00	146	111	75		170–175
1,4-Dichlorobenzene	106-46-7	$C_6H_4Cl_2$	147.00	146	111	75	50–55	
Ethylbenzene	100-41-4	C_8H_{10}	106.17	91	106	51		134–139
Hexachlorobenzene	118-74-1	C_6Cl_6	284.78	284	142	249	227–232	
Isopropylbenzene	98-82-8	C_9H_{12}	120.20	105	120	77		150–155
4-Isopropyltoluene	99-87-6	$C_{10}H_{14}$	134.22	119	134	91		174–179
Pentachlorobenzene	608-93-5	C_6HCl_5	250.34	250	215	108	83–88	
n-Propylbenzene	103-65-1	C_9H_{12}	120.20	91	120	38		156–161
Styrene	100-42-5	C_8H_8	104.15	104	91	78		142–147
1,2,3,4-Tetrachlorobenzene	634-66-2	$C_6H_2Cl_4$	215.89	216	179	109		251–256
1,2,3,5-Tetrachlorobenzene	634-90-2	$C_6H_2Cl_4$	215.89	216	179	108	52–57	
1,2,4,5-Tetrachlorobenzene	95-94-3	$C_6H_2Cl_4$	215.89	216	179	108	137–142	
Toluene	108-88-3	C_7H_8	92.14	91	65	51		108–113
1,2,3-Trichlorobenzene	87-61-6	$C_6H_3Cl_3$	181.45	180	145	109	51–56	
1,2,4-Trichlorobenzene	120-82-1	$C_6H_3Cl_3$	181.45	180	145	109		211–216
1,3,5-Trichlorobenzene	108-70-3	$C_6H_3Cl_3$	181.45	180	145	109	61–66	
a,a,a-Trichlorotoluene	98-07-7	$C_7H_5Cl_3$	195.48	159	194	123		218–223
1,2,3-Trimethylbenzene	526-73-8	C_9H_{12}	120.20	105	120	77		173–178
1,2,4-Trimethylbenzene	95-63-6	C_9H_{12}	120.20	105	120	77		166–171
1,3,5-Trimethylbenzene	108-67-8	C_9H_{12}	120.20	105	120	77		162–167
m-Xylene	108-38-3	C_8H_{10}	106.17	91	106	51		136–141
o-Xylene	95-47-6	C_8H_{10}	106.17	91	106	51		141–146
p-Xylene	106-42-3	C_8H_{10}	106.17	91	106	51		135–140

Note: MP = melting point; BP = boiling point.

Assay by Gas Chromatography. Analyze the sample by gas chromatography using method GC-MCA1 and method GC-MCA2 described below.

Method GC-MCA1

Column: Polyethylene glycol, 30 M × 0.53 mm i.d., 1.0-μm film thickness

Detector: Flame ionization

Detector Temperature: 200 °C

Injector: Splitless

Injector Temperature: 200 °C

Sample Size: 2 μL of a 1000 μg/mL solution

Carrier Gas: Helium at 5 mL/min

Temperature Program: 35 °C for 10 min, then 2.5 °C/min to 200 °C

Method GC-MCA2

Column: 6% Cyanopropylphenyl–94% dimethylpolysiloxane, 75 M × 0.53 mm i.d., 3.0-μm film thickness

Detector: Flame ionization

Detector Temperature: 200 °C

Injector: Splitless

Injector Temperature: 200 °C

Sample Size: 2 μL of a 1000 μg/mL solution

Carrier Gas: Helium at 10 mL/min

Temperature Program: 70 °C, then 5 °C/min to 210 °C, hold 10 min

Measure the area under all peaks (excluding the solvent peak), and calculate the analyte content in area percent for each analysis.

Organochlorine Pesticides

SPECIFICATIONS

Identity by
Infrared spectroscopy . . . . Passes test
Gas chromatography/mass spectrometry . . . . Passes test

Assay by
Gas chromatography . . . . ≥98.0%
Thin-layer chromatography . . . . Passes test

Specific Use
GC-ECD suitability . . . . Passes test

TESTS

Identity by Infrared Spectroscopy. Analyze the sample using the general procedure cited on page 92. Identifying absorbances for the compound listed in Table 5-6 must be present. The spectrum is compared to the NIST standardized library of compounds, if available. Absorbances found in the NIST spectrum should be present in the test spectrum; however, absorbance frequency and intensity may be different due to different analysis techniques. The spectrum may also be compared to other published libraries of infrared spectra, if available.

Identity by Gas Chromatography/Mass Spectrometry. Analyze the sample by gas chromatography/mass spectrometry using method GCMS-OCP1 described below.

Method GCMS-OCP1

Ionization Mode: Electron ionization/70 eV

Column: 5% Diphenyl–95% dimethylpolysiloxane, 30 M × 0.25 mm i.d., 0.25-µm film thickness

Temperature Program: 45 °C for 1 min, then 10 °C/min to 300 °C, hold 10 min

Carrier Gas: Helium at 1.0 mL/min

Injector: Split/splitless (100:1 split)

Injector Temperature: 250 °C

Scan Range: 33–525 amu

Sample Size: 1 µL of a 500 µg/mL solution

Identifying ions for the compound listed in Table 5-6 must be present. The spectrum is compared to the NIST standardized library of compounds, if available. Ions in the test spectrum must match those found in the NIST spectrum.

Assay by Gas Chromatography. Analyze the sample by gas chromatography using method GC-OCP1 described below.

Method GC-OCP1

Column: 5% Diphenyl–95% dimethylpolysiloxane, 30 M × 0.53 mm i.d., 1.0-µm film thickness

Detector: Flame ionization

Detector Temperature: 200 °C

Injector: Splitless

Injector Temperature: 200 °C

Sample Size: 2 µL of a 1000 µg/mL solution

Carrier Gas: Helium at 3.0 mL/min

Temperature Program: 70 °C, then 5 °C/min to 210 °C, hold 10 min

Measure the area under all peaks (excluding the solvent peak), and calculate the analyte content in area percent.

Table 5-6. Organochloride Pesticides Compound Data

Name	CAS No.	Chemical Formula	Formula Weight	IR (cm⁻¹)			MS (m/z)		
				IR-1	IR-2	IR-3	MS-1	MS-2	MS-3
Aldrin	309-00-2	$C_{12}H_8Cl_6$	364.91	2987	1597	693	66	362	263
a-BHC	319-84-6	$C_6H_6Cl_6$	290.83	2963	1340	793	183	219	181
b-BHC	319-85-7	$C_6H_6Cl_6$	290.83	2952	1309	754	109	253	181
d-BHC	319-86-8	$C_6H_6Cl_6$	290.83	2932	1236	774	109	219	183
e-BHC	6108-10-7	$C_6H_6Cl_6$	290.83	930	788	714	219	183	181
g-BHC	58-89-9	$C_6H_6Cl_6$	290.83	2948	1344	689	181	288	219
Butachlor	23184-66-9	$C_{17}H_{26}ClNO_2$	311.85	2971	1703	1074	176	311	57
cis-Chlordane	5103-71-9	$C_{10}H_6Cl_8$	409.78	2963	1604	534	373	406	272
trans-Chlordane	5103-74-2	$C_{10}H_6Cl_8$	409.78	2964	1603	560	373	406	272
Chlordecone	143-50-0	$C_{10}Cl_{10}O$	490.63	1814	1043	646	272	274	237
Chlorobenzilate	510-15-6	$C_{16}H_{14}Cl_2O_3$	325.20	3495	1719	1485	251	139	111
Chloroneb	2675-77-6	$C_8H_8Cl_2O_2$	207.05	2950	1490	775	191	206	141
Chlorthal	1861-32-1	$C_{10}H_6Cl_4O_4$	331.96	2963	1762	1240	301	330	299
2,4'-DDD	53-19-0	$C_{14}H_{10}Cl_4$	320.04	3075	1490	770	235	318	165
4,4'-DDD	72-54-8	$C_{14}H_{10}Cl_4$	320.04	3090	1494	765	235	318	165
2,4'-DDE	3424-82-6	$C_{14}H_8Cl_4$	318.02	3075	1590	862	246	316	248
4,4'-DDE	72-55-9	$C_{14}H_8Cl_4$	318.02	3090	1490	858	246	316	248
2,4'-DDT	789-02-6	$C_{14}H_9Cl_5$	354.48	3075	1490	777	335	352	165
4,4'-DDT	50-29-3	$C_{14}H_9Cl_5$	354.48	3094	1494	774	235	352	165
Dibutylchlorendate	1770-80-5	$C_{17}H_{20}Cl_6O_4$	501.06	2960	1745	845	501	499	427
Dicamba	1918-00-9	$C_8H_6Cl_2O_3$	221.04	3565	1776	1167	173	220	175
Diclofop methyl	51338-27-3	$C_{16}H_{14}Cl_2O_4$	341.20	3000	1750	1219	253	340	281
Dicofol	115-32-2	$C_{14}H_9Cl_5O$	370.50	3532	805	770	139	251	111
Dieldrin	60-57-1	$C_{12}H_8Cl_6O$	380.91	2986	1599	850	79	378	263
Endosulfan I	959-98-8	$C_9H_6Cl_6O_3S$	406.93	2937	1605	752	195	404	339
Endosulfan II	33213-65-9	$C_9H_6Cl_6O_3S$	406.92	2959	1205	673	195	404	339
Endosulfan sulfate	1031-07-8	$C_9H_6Cl_6O_4S$	422.93	1437	1201	808	272	420	387
Endrin	72-20-8	$C_{12}H_8Cl_6O$	380.91	2979	1600	855	81	378	263
Endrin aldehyde	7421-93-4	$C_{12}H_8Cl_6O$	380.91	2940	1741	835	67	345	250
Endrin ketone	53494-70-5	$C_{12}H_7Cl_6O$	380.91	2963	1748	754	317	378	250
Heptachlor	76-44-8	$C_{10}H_5Cl_7$	373.32	2959	1602	789	100	370	272
Heptachlor epoxide	1024-57-3	$C_{10}H_5Cl_7$	389.32	3044	1602	858	81	386	353
Hexachlorophene	70-30-4	$C_{13}H_6Cl_6O_2$	406.90	3502	1438	1281	196	404	209
Isodrin	465-73-6	$C_{12}H_8Cl_6$	364.91	2963	1603	1468	193	362	263
Mecoprop	7085-19-0	$C_{10}H_{11}ClO_3$	214.65	3575	1803	1490	142	214	107
4,4'-Methoxychlor	72-43-5	$C_{16}H_{15}Cl_3O_2$	345.65	2938	1511	1252	227	344	228
Metolachlor	51218-45-2	$C_{15}H_{22}ClNO_2$	283.79	2982	1691	1113	162	238	146
Mirex	2385-85-5	$C_{10}Cl_{12}$	545.54	1148	1059	650	272	274	237
cis-Nonachlor	5103-73-1	$C_{10}H_5Cl_9$	444.23	2980	1599	1267	409	440	237
trans-Nonachlor	39765-80-5	$C_{10}H_5Cl_9$	444.23	2954	1599	1257	409	440	272
Pentachloroanisole	1825-21-4	$C_7H_3Cl_5O$	280.36	2948	1375	1032	280	278	265
Propachlor	1918-16-7	$C_{11}H_{14}ClNO_4$	211.69	2982	1691	704	120	211	176

Assay by Thin-Layer Chromatography. Analyze the sample by thin-layer chromatography using the general procedure described on page 86. The following specific conditions are also required.

Stationary Phase: Silica

Mobile Phase: Hexane

Detection Method: UV at 254 nm and iodine

Compound must exhibit a single spot.

GC-ECD Suitability. Prepare a 200 ng/mL solution of the sample in an ECD nonresponsive solvent. Analyze the sample using method GC-OCP2 described below.

Method GC-OCP2

Column: 5% Diphenyl–95% dimethylpolysiloxane, 30 M × 0.53 mm i.d., 1.5-μm film thickness

Detector: Electron capture

Detector Temperature: 325 °C

Injector: Splitless

Injector Temperature: 250 °C

Sample Size: 1 μL

Carrier Gas: Nitrogen at 3.5 mL/min

Temperature Program: 45 °C for 1 min, then 10 °C/min to 250 °C, hold 10 min

Measure the area under all peaks (excluding the solvent peak), and calculate the analyte and total impurities content in area percent. The sum of all impurity peaks should not exceed 5%, with no single peak greater than 3%.

Organophosphorous Pesticides

SPECIFICATIONS

Identity by
Gas chromatography/mass spectrometry . . . . Passes test
Infrared spectroscopy . . . . Passes test

Assay by
Gas chromatography . . . . ≥98.0%
Thin-layer chromatography . . . . Passes test

TESTS

Identity by Gas Chromatography/Mass Spectrometry. Analyze the sample by gas chromatography/mass spectrometry using method GCMS-OP1 described below.

Method GCMS-OP1

Ionization Mode: Electron ionization/70 eV

Column: 5% Diphenyl–95% dimethylpolysiloxane, 30 M × 0.25 mm i.d., 0.25-μm film thickness

Temperature Program: 50 °C, hold 1 min, then 10 °C/min to 300 °C

Carrier Gas: Helium at 1 mL/min

Injector: Split/splitless (20:1 split)

Injector Temperature: 250 °C

Scan Range: 45–500 amu

Sample Size: 1 μL of a 5000 μg/mL solution

Identifying ions for the compounds listed in Table 5-7 must be present. The spectrum is compared to the NIST standardized library of compounds, if available. Ions in the test spectrum must match those found in the NIST spectrum.

Identity by Infrared Spectroscopy. Analyze the sample using the general procedure cited on page 92 or by method FTIR-OP1, if a Fourier transform (FT-IR) and an attenuated total reflectance (ATR) accessory are available.

Method FTIR-OP1

Accessory: Single reflectance ATR

Scan range: 4000–600 cm^{-1}

Number of scans: 16

Resolution: 4 cm^{-1}

Identifying absorbances for the compounds listed in Table 5-7 must be present. The spectrum is compared to the NIST standardized library of compounds, if available. Absorbances found in the NIST spectrum should be present in the test spectrum; however, absorbance frequency and intensity may be different due to different analysis techniques. The spectrum may also be compared to other published libraries of IR spectra, if available.

Assay by Gas Chromatography. Analyze the sample by gas chromatography using method GC-OP1 described below.

Method GC-OP1

Column: 5% Diphenyl–95% dimethylpolysiloxane, 30 M × 0.53 mm i.d., 1.5-μm film thickness

Detector: Flame ionization

Detector Temperature: 300 °C

Table 5-7. Organophosphorous Pesticides Compound Data

Name	CAS No.	Chemical Formula	Formula Weight	MS (m/z)			IR (cm⁻¹)		
				MS-1	MS-2	MS-3	IR-1	IR-2	IR-3
Acephate	30560-19-1	$C_4H_{10}NO_3PS$	183.17	136	94	95	1221	999	951
Chlorpyrifos	2921-88-2	$C_9H_{11}Cl_3NO_3PS$	350.59	197	199	314	829	965	1016
Diazinon	333-41-5	$C_{12}H_{21}N_2O_3PS$	304.35	137	179	304	1017	815	977
Dichlorvos	62-73-7	$C_4H_7Cl_2O_4P$	220.98	109	79	185	1033	847	975
Dimethoate	60-51-5	$C_5H_{12}NO_3PS_2$	229.26	87	93	125	638	1001	825
Fenthion	55-38-9	$C_{10}H_{15}O_3PS_2$	278.33	278	125	109	818	1027	960
Malathion	121-75-5	$C_{10}H_{19}O_6PS_2$	330.36	125	127	173	652	1007	1732
Methamidophos	10265-92-6	$C_2H_8NO_2PS$	141.13	94	95	141	773	1038	1211
Paraoxon	311-45-5	$C_{10}H_{14}NO_6P$	275.20	109	81	149	1022	922	1346
Parathion-ethyl	56-38-2	$C_{10}H_{14}NO_5PS$	291.26	97	109	291	913	1015	1345
Parathion-methyl	298-00-0	$C_8H_{10}NO_5PS$	263.21	263	125	109	1025	765	831
Phenthoate	2597-13-7	$C_{12}H_{17}O_4PS_2$	320.36	274	125	121	651	1007	819
Phorate	298-02-2	$C_7H_{17}O_2PS_3$	260.38	75	97	260	650	1008	954
Pirimiphos-methyl	29232-93-7	$C_{11}H_{20}N_3O_3PS$	305.33	290	276	305	816	1014	1033
Terbufos	13071-79-9	$C_9H_{21}O_2PS_3$	288.43	231	57	97	1009	652	954

Injector: Split/splitless (10:1 split)

Injector Temperature: 250 °C

Sample Size: 1 μL of a 5000 μg/mL solution

Carrier Gas: Helium at 3.5 mL/min

Temperature Program: 50 °C, hold 1 min, then 10 °C/min to 300 °C, hold 5 min

Measure the area under all peaks (excluding the solvent peak), and calculate the analyte content in area percent.

Assay by Thin-Layer Chromatography. Analyze the sample by thin-layer chromatography using the general procedure described on page 86. The following specific conditions are also required.

Stationary Phase: Silica

Mobile Phase: Ethyl acetate hexane 1:6 (the ratio of ethyl acetate may be varied to achieve desired R_f as needed for specific compounds).

Detection Method: UV at 254 nm and iodine

Compound must exhibit a single spot.

Polychlorinated Biphenyl Congeners

SPECIFICATIONS

Identity by
Infrared spectroscopy . . . . Passes test
Gas chromatography/mass spectrometry . . . . Passes test

Assay by
Gas chromatography (using two methods) . . . . ≥98.0%
Melting point (for solids). . . . . Passes test

TESTS

Identity by Infrared Spectroscopy. Analyze the sample using the general procedure cited on page 92. Identifying absorbances for the compound listed in Table 5-8 must be present. The spectrum is compared to the NIST standardized library of compounds, if available. Absorbances found in the NIST spectrum should be present in the test spectrum; however, absorbance frequency and intensity may be different due to different analysis techniques. The spectrum may also be compared to other published libraries of infrared spectra, if available.

Identity by Gas Chromatography/Mass Spectrometry. Analyze the sample by gas chromatography/mass spectrometry using method GCMS-PCB1 described below.

Method GCMS-PCB1

Ionization Mode: Electron ionization/70 eV

Column: 5% Diphenyl–95% dimethylpolysiloxane, 30 M × 0.25 mm i.d., 0.25-μm film thickness

Temperature Program: 140 °C for 1 min, then 10 °C/min to 320 °C, then hold 5 min

Carrier Gas: Helium at 1 mL/min

Injector: Split/splitless (50:1 split)

Injector Temperature: 280 °C

Scan Range: 40–600 amu

Sample Size: 0.5 μL of a 2000 μg/mL solution

Identifying ions for the compound listed in Table 5-8 must be present. The spectrum is compared to the NIST standardized library of compounds, if available. Ions in the test spectrum must match those found in the NIST spectrum.

Assay by Gas Chromatography. Analyze the sample by gas chromatography using method GCMS-PCB1 described above and method GC-PCB1 described below.

Method GC-PCB1

Column: 50% Diphenyl–50% dimethylpolysiloxane, 25 M × 0.53 mm i.d., 1.0-μm film thickness

Detector: Flame ionization

Detector Temperature: 300 °C

Injector: Splitless

Injector Temperature: 280 °C

Sample Size: 1 μL of a 1000 μg/mL solution

Carrier Gas: Helium at 3 mL/min

Temperature Program: 100 °C for 1 min, then 10 °C/min to 300 °C, then hold 5 min

Measure the area under all peaks (excluding the solvent peak), and calculate the analyte content in area percent for each analysis.

Assay by Melting Point. Analyze the sample using the general procedure listed on page 45. Assayed values must be within the temperature ranges stated in Table 5-8. The melting point assay should have a melting range of no more than 3 °C

Table 5-8. Polychlorinated Biphenyl Cogeners Compound Data

BZ No.	Name	CAS No.	Chemical Formula	Formula Weight	IR (cm⁻¹)					MS (m/z)			MP
					IR-1	IR-2	IR-3	IR-4	IR-5	MS-1	MS-2	MS-3	
1	2-Chlorobiphenyl	2051-60-7	$C_{12}H_9Cl$	188.65	750	699	1471	1039	771	188	152	190	28–33
2	3-Chlorobiphenyl	2051-61-8	$C_{12}H_9Cl$	188.65	756	698	1477	1597	1570	188	152	190	—
3	4-Chlorobiphenyl	2051-62-9	$C_{12}H_9Cl$	188.65	1482	760	1097	834	1009	188	152	190	74–79
4	2,2'-Dichlorobiphenyl	13029-08-8	$C_{12}H_8Cl_2$	223.10	752	762	1467	1462	1064	222	152	224	58–63
5	2,3-Dichlorobiphenyl	16605-91-7	$C_{12}H_8Cl_2$	223.10	759	699	1453	1410	1044	222	224	152	—
6	2,3'-Dichlorobiphenyl	25569-80-6	$C_{12}H_8Cl_2$	223.10	756	1463	1467	1040	825	222	224	152	—
7	2,4-Dichlorobiphenyl	33284-50-3	$C_{12}H_8Cl_2$	223.10	1469	817	700	1107	825	222	224	152	—
8	2,4'-Dichlorobiphenyl	34883-43-7	$C_{12}H_8Cl_2$	223.10	757	1471	830	1098	1008	222	224	152	42–47
9	2,5-Dichlorobiphenyl	34883-39-1	$C_{12}H_8Cl_2$	223.10	1463	1101	700	766	815	222	224	152	—
10	2,6-Dichlorobiphenyl	33146-45-1	$C_{12}H_8Cl_2$	223.10	1428	698	781	761	1441	222	224	152	33–38
11	3,3'-Dichlorobiphenyl	2050-67-1	$C_{12}H_8Cl_2$	223.10	722	1597	780	1468	1566	222	224	152	27–32
12	3,4-Dichlorobiphenyl	2974-92-7	$C_{12}H_8Cl_2$	223.10	1467	761	1033	1138	1139	222	224	152	46–51
13	3,4'-Dichlorobiphenyl	2974-90-5	$C_{12}H_8Cl_2$	223.10	1098	803	785	1475	832	222	224	152	—
14	3,5-Dichlorobiphenyl	34883-41-5	$C_{12}H_8Cl_2$	223.10	1561	761	806	1596	697	222	224	152	30–35
15	4,4'-Dichlorobiphenyl	2050-68-2	$C_{12}H_8Cl_2$	223.10	1097	815	1477	1488	1007	222	224	152	148–153
16	2,2',3-Trichlorobiphenyl	38444-78-9	$C_{12}H_7Cl_3$	257.55	756	1411	787	810	1035	256	186	258	—
17	2,2',4-Trichlorobiphenyl	37680-66-3	$C_{12}H_7Cl_3$	257.55	1463	821	758	1590	1105	256	258	186	—
18	2,2',5-Trichlorobiphenyl	37680-65-2	$C_{12}H_7Cl_3$	257.55	758	1463	1100	1019	816	256	258	186	42–47
19	2,2',6-Trichlorobiphenyl	38444-73-4	$C_{12}H_7Cl_3$	257.55	754	1431	779	794	1444	256	258	186	88–93
20	2,3,3'-Trichlorobiphenyl	38444-84-7	$C_{12}H_7Cl_3$	257.55	781	1451	1398	1561	702	256	258	186	42–47
21	2,3,4-Trichlorobiphenyl	55702-46-0	$C_{12}H_7Cl_3$	257.55	1444	1450	700	1364	837	256	258	186	100–105
22	2,3,4'-Trichlorobiphenyl	38444-85-8	$C_{12}H_7Cl_3$	257.55	1094	1452	804	785	1494	256	258	186	70–75
23	2,3,5-Trichlorobiphenyl	55720-44-0	$C_{12}H_7Cl_3$	257.55	1414	699	766	1554	1121	256	258	186	37–42
24	2,3,6-Trichlorobiphenyl	55702-45-9	$C_{12}H_7Cl_3$	257.55	1433	1181	698	1385	763	256	258	186	54–59
25	2,3',4-Trichlorobiphenyl	55712-37-3	$C_{12}H_7Cl_3$	257.55	1463	1109	836	1604	787	256	258	186	34–39

No.	Name	CAS	Formula	MW									
26	2,3',5-Trichlorobiphenyl	38444-81-4	$C_{12}H_7Cl_3$	257.55	1458	1103	1032	1464	696	256	258	186	40-45
27	2,3',6-Trichlorobiphenyl	38444-76-7	$C_{12}H_7Cl_3$	257.55	786	1434	1444	695	1561	256	258	186	—
28	2,4,4'-Trichlorobiphenyl	7012-37-5	$C_{12}H_7Cl_3$	257.55	1469	1098	816	738	1006	256	258	186	56-61
29	2,4,5-Trichlorobiphenyl	15862-07-4	$C_{12}H_7Cl_3$	257.55	1458	699	1092	1048	1447	256	258	186	75-80
30	2,4,6-Trichlorobiphenyl	35693-92-6	$C_{12}H_4Cl_3$	257.55	833	1544	1581	698	1371	222	258	186	50-65
31	2,4',5-Trichlorobiphenyl	16606-02-3	$C_{12}F_7Cl_3$	257.55	1460	1098	1499	1027	833	256	258	186	63-68
32	2,4',6-Trichlorobiphenyl	38444-77-4	$C_{12}H_7Cl_3$	257.55	787	1433	780	1440	1114	256	258	186	53-58
33	2',3,4-Trichlorobiphenyl	38444-86-9	$C_{12}H_7Cl_3$	257.55	1461	758	1136	1035	1041	256	258	186	59-64
34	2',3,5-Trichlorobiphenyl	37680-68-5	$C_{12}H_7Cl_3$	257.55	757	1562	1590	809	1045	256	258	186	55-60
35	3,3',4-Trichlorobiphenyl	37680-69-6	$C_{12}H_7Cl_3$	257.55	1465	737	786	1138	1553	256	258	186	64-69
36	3,3',5-Trichlorobiphenyl	38444-87-0	$C_{12}H_7Cl_3$	257.55	1561	1592	807	786	712	256	258	186	76-81
37	3,4,4'-Trichlorobiphenyl	38444-90-5	$C_{12}H_7Cl_3$	257.55	1466	1098	815	1015	1376	256	258	186	85-91
38	3,4,5-Trichlorobiphenyl	53555-66-1	$C_{12}H_7Cl_3$	257.55	1434	1546	761	1377	813	256	258	186	69-74
39	3,4',5-Trichlorobiphenyl	38444-88-1	$C_{12}H_7Cl_3$	257.55	1097	1498	1559	804	827	256	258	186	85-90
40	2,2',3,3'-Tetrachlorobiphenyl	38444-93-8	$C_{12}H_6Cl_4$	291.99	1441	782	1407	1420	1041	290	292	220	121-126
41	2,2',3,4-Tetrachlorobiphenyl	52663-59-9	$C_{12}H_6Cl_4$	291.99	1444	1365	757	1436	1177	290	292	220	46-51
42	2,2',3,4'-Tetrachlorobiphenyl	36559-22-5	$C_{12}H_6Cl_4$	291.99	837	1444	787	1452	1417	290	292	220	66-71
43	2,2',3,5-Tetrachlorobiphenyl	70362-46-8	$C_{12}H_6Cl_4$	291.99	758	1412	1388	1123	1557	290	292	220	43-48
44	2,2',3,5'-Tetrachlorobiphenyl	41464-39-5	$C_{12}H_6Cl_4$	291.99	1450	1033	1102	788	755	290	292	220	46-51
45	2,2',3,6-Tetrachlorobiphenyl	70362-45-7	$C_{12}H_6Cl_4$	291.99	1434	757	1386	1177	1440	290	292	220	78-83
46	2,2',3,6'-Tetrachlorobiphenyl	41464-47-5	$C_{12}H_6Cl_4$	291.99	779	1435	1412	1563	811	290	292	220	123-128
47	2,2',4,4'-Tetrachlorobiphenyl	2437-79-8	$C_{12}H_6Cl_4$	291.99	1465	791	1107	1588	817	294	292	294	44-49
48	2,2',4,5-Tetrachlorobiphenyl	70362-47-9	$C_{12}H_6Cl_4$	291.99	1456	1075	760	1347	735	290	292	220	84-89
49	2,2',4,5'-Tetrachlorobiphenyl	41464-40-8	$C_{12}H_6Cl_4$	291.99	1459	1100	1106	843	1018	290	292	220	64-59
50	2,2',4,6-Tetrachlorobiphenyl	62796-65-0	$C_{12}H_6Cl_4$	291.99	757	1546	1429	1582	1373	290	292	220	45-50
51	2,2',4,6'-Tetrachlorobiphenyl	68194-04-7	$C_{12}H_6Cl_4$	291.99	1435	790	820	1107	780	290	292	220	—
52	2,2',5,5'-Tetrachlorobiphenyl	35693-99-3	$C_{12}H_6Cl_4$	291.99	1463	1102	817	1456	1026	290	292	220	84-89
53	2,2',5,6'-Tetrachlorobiphenyl	41464-41-9	$C_{12}H_6Cl_4$	291.99	1434	1444	793	1099	780	290	292	220	99-104

Continued on next page

Table 5-8. Continued

BZ No.	Name	Chemical Formula	Formula Weight	IR-1	IR-2	IR-3	IR-4	IR-5	MS-1	MS-2	MS-3	MP
54	2,2',6,6'-Tetrachlorobiphenyl	$C_{12}H_6Cl_4$	291.99	798	1434	777	1564	1084	290	292	294	195–200
55	2,3,3',4-Tetrachlorobiphenyl	$C_{12}H_6Cl_4$	291.99	1446	1359	789	743	1178	290	292	294	84–89
56	2,3,3',4'-Tetrachlorobiphenyl	$C_{12}H_6Cl_4$	291.99	1449	1036	787	1136	752	290	292	294	96–101
57	2,3,3',5-Tetrachlorobiphenyl	$C_{12}H_6Cl_4$	291.99	1555	1414	1576	1125	697	290	292	294	85–90
58	2,3,3',5'-Tetrachlorobiphenyl	$C_{12}H_6Cl_4$	291.99	785	1560	1387	1570	1585	290	292	294	123–128
59	2,3,3',6-Tetrachlorobiphenyl	$C_{12}H_6Cl_4$	291.99	1434	1442	786	709	812	290	292	294	—
60	2,3,4,4'-Tetrachlorobiphenyl	$C_{12}H_6Cl_4$	291.99	1447	1093	816	836	1363	290	292	294	143–148
61	2,3,4,5-Tetrachlorobiphenyl	$C_{12}H_6Cl_4$	291.99	1414	1350	708	766	824	290	292	294	91–96
62	2,3,4,6-Tetrachlorobiphenyl	$C_{12}H_6Cl_4$	291.99	1422	1345	697	819	770	290	292	294	76–81
63	2,3,4',5-Tetrachlorobiphenyl	$C_{12}H_6Cl_4$	291.99	1433	1098	1543	808	1373	290	292	294	159–164
64	2,3,4',6-Tetrachlorobiphenyl	$C_{12}H_6Cl_4$	291.99	1437	787	1498	1094	1181	290	292	294	85–90
65	2,3,5,6-Tetrachlorobiphenyl	$C_{12}H_6Cl_4$	291.99	700	1392	1062	680	1380	290	292	294	75–80
66	2,3',4,4'-Tetrachlorobiphenyl	$C_{12}H_6Cl_4$	291.99	1463	817	1107	1036	777	290	292	294	122–127
67	2,3',4,5-Tetrachlorobiphenyl	$C_{12}H_6Cl_4$	291.99	1459	1453	886	697	789	290	292	294	60–65
68	2,3',4,5'-Tetrachlorobiphenyl	$C_{12}H_6Cl_4$	291.99	847	805	1482	1569	1597	290	292	294	90–95
69	2,3',4,6-Tetrachlorobiphenyl	$C_{12}H_6Cl_4$	291.99	845	1446	1580	1372	787	290	292	294	50–55
70	2,3',4',5-Tetrachlorobiphenyl	$C_{12}H_6Cl_4$	291.99	1460	1033	1139	1105	824	290	292	294	102–107
71	2,3',4',6-Tetrachlorobiphenyl	$C_{12}H_6Cl_4$	291.99	7 89	1439	1435	781	1136	290	292	294	34–39
72	2,3',5,5'-Tetrachlorobiphenyl	$C_{12}H_6Cl_4$	291.99	1558	808	1487	1104	1569	290	292	294	104–109
73	2,3',5',6-Tetrachlorobiphenyl	$C_{12}H_6Cl_4$	291.99	789	1413	1568	1600	811	290	292	294	68–73
74	2,4,4',5-Tetrachlorobiphenyl	$C_{12}H_6Cl_4$	291.99	1458	1099	1088	1016	830	290	292	294	125–130
75	2,4,4',6-Tetrachlorobiphenyl	$C_{12}H_6Cl_4$	291.99	1099	1583	1447	1544	1430	290	292	294	60–65
76	2',3,4,5-Tetrachlorobiphenyl	$C_{12}H_6Cl_4$	291.99	757	1430	1544	1378	812	290	292	294	132–137
77	3,3',4,4'-Tetrachlorobiphenyl	$C_{12}H_6Cl_4$	291.99	1464	1137	816	1033	1363	290	292	294	178–183
78	3,3',4,5-Tetrachlorobiphenyl	$C_{12}H_6Cl_4$	291.99	1546	1436	811	1369	786	290	292	294	116–121

No.	Compound	CAS	Formula	MW									mp (°C)
79	3,3',4,5'-Tetrachlorobiphenyl	41464-48-6	$C_{12}H_6Cl_4$	291.99	806	1590	1552	1481	1434	290	292	294	119–124
80	3,3',5,5'-Tetrachlorobiphenyl	33284-52-5	$C_{12}H_6Cl_4$	291.99	1559	1589	805	1129	1378	290	292	294	169–174
81	3,4,4',5-Tetrachlorobiphenyl	70362-50-4	$C_{12}H_6Cl_4$	291.99	1495	1121	830	1416	1040	290	292	294	86–91
82	2,2',3,3',4-Pentachlorobiphenyl	52663-62-4	$C_{12}H_5Cl_5$	326.43	1439	1415	791	753	1179	324	326	328	116–121
83	2,2',3,3',5-Pentachlorobiphenyl	60145-20-2	$C_{12}H_5Cl_5$	326.43	1413	787	1386	1041	1558	324	326	328	81–86
84	2,2',3,3',6-Pentachlorobiphenyl	52663-60-2	$C_{12}H_5Cl_5$	326.43	1434	1044	1178	812	785	324	326	328	107–112
85	2,2',3,4,4'-Pentachlorobiphenyl	65510-45-4	$C_{12}H_5Cl_5$	326.43	1440	794	1108	1179	817	324	326	328	46–51
86	2,2',3,4,5-Pentachlorobiphenyl	55312-69-1	$C_{12}H_5Cl_5$	326.43	1409	1350	759	736	1178	324	326	328	83–88
87	2,2',3,4,5'-Pentachlorobiphenyl	38380-02-8	$C_{12}H_5Cl_5$	326.43	1444	819	1101	1363	1179	324	326	328	108–113
88	2,2',3,4,6-Pentachlorobiphenyl	55215-17-3	$C_{12}H_5Cl_5$	326.43	1346	1424	743	761	1570	324	326	328	64–69
89	2,2',3,4,6'-Pentachlorobiphenyl	73575-57-2	$C_{12}H_5Cl_5$	326.43	1430	791	1435	1368	778	324	326	328	83–88
90	2,2',3,4',5-Pentachlorobiphenyl	68194-07-1	$C_{12}H_5Cl_5$	326.43	850	1416	819	1109	1389	324	326	328	49–54
91	2,2',3,4',6-Pentachlorobiphenyl	68194-05-8	$C_{12}H_5Cl_5$	326.43	1438	846	809	1107	1178	324	326	328	60–65
92	2,2',3,5,5'-Pentachlorobiphenyl	52663-61-3	$C_{12}H_5Cl_5$	326.43	1475	1101	1036	832	1390	324	326	328	58–63
93	2,2',3,5,6-Pentachlorobiphenyl	73575-56-1	$C_{12}H_5Cl_5$	326.43	1395	741	678	1044	1166	324	326	328	94–99
94	2,2',3,5,6'-Pentachlorobiphenyl	73575-55-0	$C_{12}H_5Cl_5$	326.43	1408	794	1436	1385	780	324	326	328	80–85
95	2,2',3,5',6-Pentachlorobiphenyl	38379-99-6	$C_{12}H_5Cl_5$	326.43	816	1033	1180	1436	1098	324	326	328	91–96
96	2,2',3,6,6'-Pentachlorobiphenyl	73575-54-9	$C_{12}H_5Cl_5$	326.43	1429	796	1437	1179	812	324	326	328	117–122
97	2,2',3',4,5-Pentachlorobiphenyl	41464-51-1	$C_{12}H_5Cl_5$	326.43	1445	1417	789	1060	889	324	326	328	75–81
98	2,2',3',4,6-Pentachlorobiphenyl	60233-25-2	$C_{12}H_5Cl_5$	326.43	848	1431	1413	1547	785	324	326	328	92–97
99	2,2',4,4',5-Pentachlorobiphenyl	38380-01-1	$C_{12}H_5Cl_5$	326.43	1456	1450	801	1098	1107	324	326	328	57–62
100	2,2',4,4',6-Pentachlorobiphenyl	39485-83-1	$C_{12}H_5Cl_5$	326.43	800	1432	1580	1098	1547	324	326	328	—
101	2,2',4,5,5'-Pentachlorobiphenyl	37680-73-2	$C_{12}H_5Cl_5$	326.43	1457	1101	1075	1144	1480	324	326	328	76–81
102	2,2',4,5,6'-Pentachlorobiphenyl	68194-06-9	$C_{12}H_5Cl_5$	326.43	1438	793	781	1096	1477	324	326	328	69–74
103	2,2',4,5',6-Pentachlorobiphenyl	60145-21-3	$C_{12}H_5Cl_5$	326.43	1439	851	1097	818	1371	324	326	328	67–72
104	2,2',4,6,6'-Pentachlorobiphenyl	56558-16-8	$C_{12}H_5Cl_5$	326.43	1434	1416	836	1579	795	324	326	328	85–90
105	2,3,3',4,4'-Pentachlorobiphenyl	32598-14-4	$C_{12}H_5Cl_5$	326.43	1443	1136	1355	787	1036	324	326	328	115–120
106	2,3,3',4,5-Pentachlorobiphenyl	70424-69-0	$C_{12}H_5Cl_5$	326.43	1412	1348	1400	698	789	324	326	328	83–38

Continued on next page

Table 5-8. Continued

BZ No. Name	CAS No.	Chemical Formula	Formula Weight	IR (cm⁻¹)					MS (m/z)			MP
				IR-1	IR-2	IR-3	IR-4	IR-5	MS-1	MS-2	MS-3	
107 2,3,3',4',5-Pentachlorobiphenyl	70424-68-9	$C_{12}H_5Cl_5$	326.43	1480	1138	821	1415	1036	324	326	328	94–99
108 2,3,3',4,5'-Pentachlorobiphenyl	70362-41-3	$C_{12}H_5Cl_5$	326.43	806	1567	1427	1357	1594	324	326	328	118–123
109 2,3,3',4,6-Pentachlorobiphenyl	74472-35-8	$C_{12}H_5Cl_5$	326.43	704	1343	1430	1415	824	324	326	328	68–73
110 2,3,3',4',6-Pentachlorobiphenyl	38380-03-9	$C_{12}H_5Cl_5$	326.43	1435	1177	1037	813	1135	324	326	328	81–86
111 2,3,3',5,5'-Pentachlorobiphenyl	39635-32-0	$C_{12}H_5Cl_5$	326.43	1555	1414	1377	1568	809	324	326	328	105–110
112 2,3,3',5,6-Pentachlorobiphenyl	74472-36-9	$C_{12}H_5Cl_5$	326.43	719	1390	1063	1165	708	324	326	328	89–94
113 2,3,3',5',6-Pentachlorobiphenyl	68194-10-5	$C_{12}H_5Cl_5$	326.43	1568	814	1178	1377	701	324	326	328	54–59
114 2,3,4,4',5-Pentachlorobiphenyl	74472-37-0	$C_{12}H_5Cl_5$	326.43	1413	1496	1346	738	1018	324	326	328	96–101
115 2,3,4,4',6-Pentachlorobiphenyl	74472-38-1	$C_{12}H_5Cl_5$	326.43	1342	1419	747	1097	821	324	326	328	63–68
116 2,3,4,5,6-Pentachlorobiphenyl	18259-05-7	$C_{12}H_5Cl_5$	326.43	701	1351	1329	1386	1372	324	326	328	121–126
117 2,3,4',5,6-Pentachlorobiphenyl	68194-11-6	$C_{12}H_5Cl_5$	326.43	1495	1059	1385	677	1095	324	326	328	166–171
118 2,3',4,4',5-Pentachlorobiphenyl	31508-00-6	$C_{12}H_5Cl_5$	326.43	1455	1094	1142	1051	824	324	326	328	109–114
119 2,3',4,4',6-Pentachlorobiphenyl	56558-17-9	$C_{12}H_5Cl_5$	326.43	1431	853	1447	1579	1544	324	326	328	73–78
120 2,3',4,5,5'-Pentachlorobiphenyl	68194-12-7	$C_{12}H_5Cl_5$	326.43	1470	1565	808	1430	1569	324	326	328	129–134
121 2,3',4,5',6-Pentachlorobiphenyl	56558-18-0	$C_{12}H_5Cl_5$	326.43	854	1596	1564	801	1372	324	326	328	91–96
122 2',3,3',4,5-Pentachlorobiphenyl	76842-07-4	$C_{12}H_5Cl_5$	326.43	1430	785	1544	1164	817	324	326	328	115–120
123 2',3,4,4',5-Pentachlorobiphenyl	65510-44-3	$C_{12}H_5Cl_5$	326.43	1431	810	1365	1483	1110	324	326	328	132–137
124 2',3,4,5,5'-Pentachlorobiphenyl	70424-70-3	$C_{12}H_5Cl_5$	326.43	1433	1544	817	814	1102	324	326	328	114–119
125 2',3,4,5,6'-Pentachlorobiphenyl	74472-39-2	$C_{12}H_5Cl_5$	326.43	1419	790	1434	817	1547	324	326	328	124–129
126 3,3',4,4',5-Pentachlorobiphenyl	57465-28-8	$C_{12}H_5Cl_5$	326.43	1432	811	1033	1144	1359	324	326	328	158–163
127 3,3',4,5,5'-Pentachlorobiphenyl	39635-33-1	$C_{12}H_5Cl_5$	326.43	1545	808	1420	1586	1416	324	326	328	150–155
128 2,2',3,3',4,4'-Hexachlorobiphenyl	38380-07-3	$C_{12}H_4Cl_6$	360.88	1434	1184	796	1356	817	358	360	362	148–153
129 2,2',3,3',4,5-Hexachlorobiphenyl	52215-18-4	$C_{12}H_4Cl_6$	360.88	1410	1348	1420	788	735	358	360	362	100–105
130 2,2',3,3',4,5'-Hexachlorobiphenyl	52663-66-8	$C_{12}H_4Cl_6$	360.88	1413	818	1390	1460	1361	358	360	362	112–117
131 2,2',3,3',4,6-Hexachlorobiphenyl	61798-70-7	$C_{12}H_4Cl_6$	360.88	1412	1345	1425	740	788	358	360	362	133–138

No.	Name	CAS	Formula	MW									
132	2,2',3,3',4,6'-Hexachlorobiphenyl	38380-05-1	$C_{12}H_4Cl_6$	360.88	1431	1181	811	1365	875	358	360	362	115–120
133	2,2',3,3',5,6'-Hexachlorobiphenyl	35694-04-3	$C_{12}H_4Cl_6$	360.88	1557	1125	835	1378	869	358	360	362	125–130
134	2,2',3,3',5,6-Hexachlorobiphenyl	52704-70-8	$C_{12}H_4Cl_6$	360.88	735	1397	1071	1049	1167	358	360	362	130–135
135	2,2',3,3',5,6'-Hexachlorobiphenyl	52744-13-5	$C_{12}H_4Cl_6$	360.88	1043	1417	1181	812	1369	358	360	362	100–105
136	2,2',3,3',6,6'-Hexachlorobiphenyl	38411-22-2	$C_{12}H_4Cl_6$	360.88	1179	1430	813	1049	1407	358	360	362	111–116
137	2,2',3,4,4',5-Hexachlorobiphenyl	35694-06-5	$C_{12}H_4Cl_6$	360.88	1413	1350	789	1106	821	358	360	362	79–84
138	2,2',3,4,4',5'-Hexachlorobiphenyl	35065-28-2	$C_{12}H_4Cl_6$	360.88	1440	805	1368	1054	1180	358	360	362	77–82
139	2,2',3,4,4',6-Hexachlorobiphenyl	56030-56-9	$C_{12}H_4Cl_6$	360.88	1424	801	1346	1108	819	358	360	362	76–81
140	2,2',3,4,4',6'-Hexachlorobiphenyl	59291-64-4	$C_{12}H_4Cl_6$	360.88	1426	804	859	1366	1576	358	360	362	66–71
141	2,2',3,4,5,5'-Hexachlorobiphenyl	52712-04-6	$C_{12}H_4Cl_6$	360.88	1421	1348	1099	1474	819	358	360	362	86–91
142	2,2',3,4,5,6-Hexachlorobiphenyl	41411-61-4	$C_{12}H_4Cl_6$	360.88	1353	742	1329	1387	690	358	360	362	132–137
143	2,2',3,4,5,6'-Hexachlorobiphenyl	68194-15-0	$C_{12}H_4Cl_6$	360.88	1404	1349	798	1436	1565	358	360	362	87–92
144	2,2',3,4,5',6-Hexachlorobiphenyl	68194-14-9	$C_{12}H_4Cl_6$	360.88	1344	1433	1428	1099	818	358	360	362	70–75
145	2,2',3,4,6,6'-Hexachlorobiphenyl	74472-40-5	$C_{12}H_4Cl_6$	360.88	1409	1346	780	796	1572	358	360	362	134–139
146	2,2',3,4',5,5'-Hexachlorobiphenyl	51908-16-8	$C_{12}H_4Cl_6$	360.88	1470	1416	1079	1146	1389	358	360	362	86–91
147	2,2',3,4',5,6-Hexachlorobiphenyl	68194-13-8	$C_{12}H_4Cl_6$	360.88	1400	848	1482	1386	1167	358	360	362	134–139
148	2,2',3,4',5,6'-Hexachlorobiphenyl	74472-41-6	$C_{12}H_4Cl_6$	360.88	1409	857	1588	1126	1371	358	360	362	79–84
149	2,2',3,4',5',6-Hexachlorobiphenyl	38380-04-0	$C_{12}H_4Cl_6$	360.88	1436	1475	1181	1049	1392	358	360	362	76–81
150	2,2',3,4',6,6'-Hexachlorobiphenyl	68194-08-1	$C_{12}H_4Cl_6$	360.88	854	1430	819	1549	1374	358	360	362	74–79
151	2,2',3,5,5',6-Hexachlorobiphenyl	52663-63-5	$C_{12}H_4Cl_6$	360.88	1393	1099	1409	1045	1087	358	360	362	96–101
152	2,2',3,5,6,6'-Hexachlorobiphenyl	68194-09-2	$C_{12}H_4Cl_6$	360.88	1397	782	675	1436	797	358	360	362	124–129
153	2,2',4,4',5,5'-Hexachlorobiphenyl	35065-27-1	$C_{12}H_4Cl_6$	360.88	1449	1455	1087	1148	1048	358	360	362	101–106
154	2,2',4,4',5,6'-Hexachlorobiphenyl	60145-22-4	$C_{12}H_4Cl_6$	360.88	1434	818	1105	1578	1083	358	360	362	66–71
155	2,2',4,4',6,6'-Hexachlorobiphenyl	33979-03-2	$C_{12}H_4Cl_6$	360.88	1577	817	1419	1550	859	358	360	362	109–114
156	2,3,3',4,4',5-Hexachlorobiphenyl	38380-08-4	$C_{12}H_4Cl_6$	360.88	1413	1136	1343	1036	773	358	360	362	128–133
157	2,3,3',4,4',5'-Hexachlorobiphenyl	69782-90-7	$C_{12}H_4Cl_6$	360.88	1426	810	1353	1167	789	358	360	362	159–164
158	2,3,3',4,4',6-Hexachlorobiphenyl	74472-42-7	$C_{12}H_4Cl_6$	360.88	1414	1343	820	1132	1037	358	360	362	108–113
159	2,3,3',4,5,5'-Hexachlorobiphenyl	39635-35-3	$C_{12}H_4Cl_6$	360.88	1404	1342	809	1569	1595	358	360	362	147–152

Continued on next page

Table 5-8. Continued

BZ No. Name	CAS No.	Chemical Formula	Formula Weight	IR (cm⁻¹)					MS (m/z)			MP
				IR-1	IR-2	IR-3	IR-4	IR-5	MS-1	MS-2	MS-3	
160 2,3,3',4,5,6-Hexachlorobiphenyl	41411-62-5	$C_{12}H_4Cl_6$	360.88	722	1328	1349	1381	1366	358	360	362	90–95
161 2,3,3',4,5',6-Hexachlorobiphenyl	74474-43-8	$C_{12}H_4Cl_6$	360.88	1409	1567	1341	810	1595	358	360	362	103–108
162 2,3,3',4',5,5'-Hexachlorobiphenyl	39635-34-2	$C_{12}H_4Cl_6$	360.88	816	1546	1413	1391	1367	358	360	362	141–146
163 2,3,3',4',5,6-Hexachlorobiphenyl	74472-44-9	$C_{12}H_4Cl_6$	360.88	1400	1383	729	1066	1479	358	360	362	119–124
164 2,3,3',4',5',6-Hexachlorobiphenyl	74472-45-0	$C_{12}H_4Cl_6$	360.88	1425	818	1180	808	1371	358	360	362	90–95
165 2,3,3',5,5',6-Hexachlorobiphenyl	74472-46-1	$C_{12}H_4Cl_6$	360.88	1078	724	1385	1569	812	358	360	362	149–154
166 2,3,4,4',5,6-Hexachlorobiphenyl	41411-63-6	$C_{12}H_4Cl_6$	360.88	1329	1379	1353	751	737	358	360	362	162–167
167 2,3',4,4',5,5'-Hexachlorobiphenyl	52663-72-6	$C_{12}H_4Cl_6$	360.88	1431	1066	1471	813	1553	358	360	362	123–128
168 2,3',4,4',5',6-Hexachlorobiphenyl	59291-65-5	$C_{12}H_4Cl_6$	360.88	1420	1552	1367	862	822	358	360	362	107–112
169 3,3',4,4',5,5'-Hexachlorobiphenyl	32774-16-6	$C_{12}H_4Cl_6$	360.88	1423	1538	808	1357	1529	358	360	362	206–211
170 2,2',3,3',4,4',5-Heptachlorobiphenyl	35065-30-6	$C_{12}H_3Cl_7$	395.32	1409	1181	794	1348	666	392	394	396	135–140
171 2,2',3,3',4,4',6-Heptachlorobiphenyl	52663-71-5	$C_{12}H_3Cl_7$	395.32	1419	1347	806	1177	818	392	394	396	115–120
172 2,2',3,3',4,5,5'-Heptachlorobiphenyl	52663-74-8	$C_{12}H_3Cl_7$	395.32	1346	1409	1382	841	1558	392	394	396	133–138
173 2,2',3,3',4,5,6-Heptachlorobiphenyl	68194-16-1	$C_{12}H_3Cl_7$	395.32	1352	737	1328	1384	1372	392	394	396	202–207
174 2,2',3,3',4,5,6'-Heptachlorobiphenyl	38411-25-5	$C_{12}H_3Cl_7$	395.32	1418	1182	814	1345	1380	392	394	396	122–127
175 2,2',3,3',4,5',6-Heptachlorobiphenyl	40186-70-7	$C_{12}H_3Cl_7$	395.32	1407	1343	1389	823	1124	392	394	396	119–124
176 2,2',3,3',4,6,6'-Heptachlorobiphenyl	52663-65-7	$C_{12}H_3Cl_7$	395.32	1423	1343	1180	898	813	392	394	396	100–105
177 2,2',3,3',4',5,6-Heptachlorobiphenyl	52663-70-4	$C_{12}H_3Cl_7$	395.32	1401	1378	1367	686	1166	392	394	396	150–155
178 2,2',3,3',5,5',6-Heptachlorobiphenyl	52663-67-9	$C_{12}H_3Cl_7$	395.32	1393	1358	1095	1048	1401	392	394	396	108–113
179 2,2',3,3',5,6,6'-Heptachlorobiphenyl	52663-64-6	$C_{12}H_3Cl_7$	395.32	1062	1399	813	1410	1348	392	394	396	127–132
180 2,2',3,4,4',5,5'-Heptachlorobiphenyl	35065-29-3	$C_{12}H_3Cl_7$	395.32	1413	1469	1353	847	1066	392	394	396	111–116
181 2,2',3,4,4',5,6-Heptachlorobiphenyl	74472-47-2	$C_{12}H_3Cl_7$	395.32	1330	1390	803	1352	1368	392	394	396	123–128
182 2,2',3,4,4',5,6'-Heptachlorobiphenyl	60145-23-5	$C_{12}H_3Cl_7$	395.32	1406	1349	800	1548	1587	392	394	396	106–111
183 2,2',3,4,4',5',6-Heptachlorobiphenyl	52663-69-1	$C_{12}H_3Cl_7$	395.32	1424	1133	1095	1350	1357	392	394	396	91–96
184 2,2',3,4,4',6,6'-Heptachlorobiphenyl	74472-48-3	$C_{12}H_3Cl_7$	395.32	1413	813	1346	1569	1373	392	394	396	113–118

No.	Congener	CAS No.	Formula	MW									MP
185	2,2',3,4,5,5',6-Heptachlorobiphenyl	52712-05-7	$C_{12}H_3Cl_7$	395.32	1353	1330	1361	1408	1098	392	394	396	146–151
186	2,2',3,4,5,6,6'-Heptachlorobiphenyl	74472-49-4	$C_{12}H_3Cl_7$	395.32	1353	1363	782	1328	1437	392	394	396	193–198
187	2,2',3,4',5,5',6-Heptachlorobiphenyl	52663-68-0	$C_{12}H_3Cl_7$	395.32	1398	1468	1389	1167	908	392	394	396	102–107
188	2,2',3,4',5,6,6'-Heptachlorobiphenyl	74487-85-7	$C_{12}H_3Cl_7$	395.32	1400	859	1587	1369	672	392	394	396	132–137
189	2,3,3',4,4',5,5'-Heptachlorobiphenyl	39635-31-9	$C_{12}H_3Cl_7$	395.32	1406	817	1339	771	840	392	394	396	160–165
190	2,3,3',4,4',5,6-Heptachlorobiphenyl	41411-64-7	$C_{12}H_3Cl_7$	395.32	1329	731	1368	1350	1387	392	394	396	120–125
191	2,3,3',4,4',5',6-Heptachlorobiphenyl	74472-50-7	$C_{12}H_3Cl_7$	395.32	1415	1341	1405	817	1547	392	394	396	111–116
192	2,3,3',4,5,5',6-Heptachlorobiphenyl	74472-51-8	$C_{12}H_3Cl_7$	395.32	1365	1327	1350	728	1569	392	394	396	169–174
193	2,3,3',4',5,5',6-Heptachlorobiphenyl	69782-91-8	$C_{12}H_3Cl_7$	395.32	1373	1167	724	1401	819	392	394	396	137–142
194	2,2',3,3',4,4',5,5'-Octachlorobiphenyl	35694-08-7	$C_{12}H_2Cl_8$	429.77	1403	1349	1181	1342	852	426	430	428	153–158
195	2,2',3,3',4,4',5,6-Octachlorobiphenyl	52663-78-2	$C_{12}H_2Cl_8$	429.77	1372	1328	1350	806	1368	426	430	428	158–173
196	2,2',3,3',4,4',5',6-Octachlorobiphenyl	42740-50-1	$C_{12}H_2Cl_8$	429.77	1402	1341	1350	802	1335	426	430	428	125–130
197	2,2',3,3',4,4',6,6'-Octachlorobiphenyl	33091-17-7	$C_{12}H_2Cl_8$	429.77	1403	814	1564	1356	1336	426	430	428	135–140
198	2,2',3,3',4,5,5',6-Octachlorobiphenyl	68194-17-2	$C_{12}H_2Cl_8$	429.77	1360	1353	1331	1401	749	426	430	428	153–198
199	2,2',3,3',4,5,6,6'-Octachlorobiphenyl	52663-73-7	$C_{12}H_2Cl_8$	429.77	1353	1329	1411	1181	1073	426	430	428	173–178
200	2,2',3,3',4,5',6,6'-Octachlorobiphenyl	40186-71-8	$C_{12}H_2Cl_8$	429.77	1403	1338	1368	685	1169	426	430	428	139–144
201	2,2',3,3',4',5,5',6-Octachlorobiphenyl	52663-75-9	$C_{12}H_2Cl_8$	429.77	1403	1338	1377	1169	745	426	430	428	155–160
202	2,2',3,3',5,5',6,6'-Octachlorobiphenyl	2136-99-4	$C_{12}H_2Cl_8$	429.77	1406	1331	1073	1169	760	426	430	428	155–160
203	2,2',3,4,4',5,5',6-Octachlorobiphenyl	52663-76-0	$C_{12}H_2Cl_8$	429.77	1356	1329	1333	1383	1393	426	430	428	109–114
204	2,2',3,4,4',5,6,6'-Octachlorobiphenyl	74472-52-9	$C_{12}H_2Cl_8$	429.77	1365	817	1353	1331	1550	426	430	428	174–179
205	2,3,3',4,4',5,5',6-Octachlorobiphenyl	74472-53-0	$C_{12}H_2Cl_8$	429.77	1370	1327	1383	1327	727	426	430	428	195–200
206	2,2',3,3',4,4',5,5',6-Nonachlorobiphenyl	40186-72-9	$C_{12}HCl_9$	464.23	1378	1354	1331	1341	805	460	464	462	200–205
207	2,2',3,3',4,4',5,6,6'-Nonachlorobiphenyl	52663-79-3	$C_{12}HCl_9$	464.23	1374	1338	1357	818	693	460	464	462	211–216
208	2,2',3,3',4,5,5',6,6'-Nonachlorobiphenyl	52663-77-1	$C_{12}HCl_9$	464.23	1343	1338	1409	1082	691	460	464	462	178–183
209	2,2',3,3',4,4',5,5',6,6'-Decachlorobiphenyl	2051-24-3	$C_{12}Cl_{10}$	498.66	1345	1328	829	696	760	494	214	213	316–321

Note: MP = melting point.

Polycyclic Aromatic Hydrocarbons

SPECIFICATIONS

Identity by
Gas chromatography/mass spectrometry . . . . Passes test
Melting point . . . . Passes test

Assay by
Gas chromatography . . . . ≥98.0%
Thin-layer chromatography . . . . Passes test

Specific Use
Liquid chromatography suitability . . . . Passes test

TESTS

Identity by Gas Chromatography/Mass Spectrometry. Analyze the sample by gas chromatography/mass spectrometry using method GCMS-PAH1 described below.

Method GCMS-PAH1

Ionization Mode: Electron ionization/70 eV

Column: 5% Diphenyl–95% dimethylpolysiloxane, 30 M × 0.25 mm i.d., 0.25-μm film thickness

Temperature Program: 50 °C, then 10 °C/min to 310 °C, hold 10 min

Carrier Gas: Helium at 0.8 mL/min

Injector: Split/splitless (100:1 split)

Injector Temperature: 250 °C

Scan Range: 50–550 amu

Sample Size: 0.5 μL of a 1000 μg/mL solution

Identifying ions for the compound listed in Table 5-9 must be present. The spectrum is compared to the NIST standardized library of compounds, if available. Ions in the test spectrum must match those found in the NIST spectrum.

Identity by Melting Point. Analyze the sample by using the general procedure cited on page 45. Assayed values must be within the ranges stated in Table 5-9.

Assay by Gas Chromatography. Analyze the sample by gas chromatography using method GC-PAH1 described below.

Method GC-PAH1

Column: 5% Diphenyl–95% dimethylpolysiloxane, 30 M × 0.53 mm i.d., 1.5-μm film thickness

Detector: Flame ionization

Table 5-9. Polycyclic Aromatic Hydrocarbons Compound Data

Name	CAS No.	Chemical Formula	Formula Weight	MS (m/z)			MP (°C)
				MS-1	MS-2	MS-3	
Acenaphthene	83-32-9	$C_{12}H_{10}$	154.21	154	153	76	91–96
Acenaphthylene	208-96-8	$C_{12}H_8$	152.20	152	151	76	90–95
Anthracene	120-12-7	$C_{14}H_{10}$	178.23	178	176	89	214–219
Benz[a]anthracene	56-55-3	$C_{18}H_{12}$	228.29	228	226	114	158–163
Benzo[a]pyrene	50-32-8	$C_{20}H_{12}$	252.31	252	126	113	173–178
Benzo[b]fluoranthene	205-99-2	$C_{20}H_{12}$	252.31	252	126	113	166–171
Benzo[g,h,i]perylene	191-24-2	$C_{22}H_{12}$	276.34	276	277	138	277–282
Benzo[j]fluoranthene	205-82-3	$C_{20}H_{12}$	252.31	252	250	126	161–166
Benzo[k]fluoranthene	207-08-9	$C_{20}H_{12}$	252.31	252	126	133	215–220
Chrysene	218-01-9	$C_{18}H_{12}$	228.29	228	114	101	254–259
Dibenz[a,h]acridine	226-36-8	$C_{21}H_{13}N$	279.34	279	280	139	226–231
Dibenz[a,h]anthracene	53-70-3	$C_{22}H_{14}$	278.35	278	279	139	266–271
Dibenz[a,j]acridine	224-42-0	$C_{21}H_{13}N$	279.34	279	280	139	217–222
Dibenzo[a,e]pyrene	192-65-4	$C_{24}H_{14}$	302.37	302	303	151	242–247
Dibenzo[a,i]pyrene	189-55-9	$C_{24}H_{14}$	302.37	302	303	152	280–285
Fluoranthene	206-44-0	$C_{16}H_{10}$	202.26	202	101	88	108–113
Fluorene	86-73-7	$C_{13}H_{10}$	166.22	166	165	83	114–119
Indeno[1,2,3-c,d]pyrene	193-39-5	$C_{22}H_{12}$	276.34	276	274	138	158–163
3-Methylcholanthrene	56-49-5	$C_{20}H_{16}$	268.36	268	252	126	177–182
Naphthalene	91-20-3	$C_{10}H_8$	128.17	128	127	102	78–83
Phenanthrene	85-01-8	$C_{14}H_{10}$	178.23	178	176	152	97–102
Pyrene	129-00-0	$C_{16}H_{10}$	202.26	202	203	101	150–155

Note: MP = melting point.

Detector Temperature: 325 °C

Injector: Splitless

Injector Temperature: 250 °C

Sample Size: 1 μL of a 1000 μg/mL solution

Carrier Gas: Hydrogen at 3.5 mL/min

Temperature Program: 70 °C, then 10 °C/min to 310 °C, hold 10 min

Measure the area under all peaks (excluding the solvent peak), and calculate the analyte content in area percent.

Assay by Thin-Layer Chromatography. Analyze the sample by thin-layer chromatography using the general procedure described on page 86. The following specific conditions are also required.

Stationary Phase: Silica

Mobile Phase: 9:1 (v/v) Hexane:dichloromethane

Detection Methods: UV at 254 nm and iodine

Compound must exhibit a single spot.

Liquid Chromatography Suitability. Analyze the sample by liquid chromatography using method HPLC-PAH1 described below.

Method HPLC-PAH1

Column: Octadecyl, 250 × 4.6 mm i.d., 5 μm

Mobile Phase: 80:20 Acetonitrile:water

Flow Rate: 2.0 mL/min

Detector: UV, 254 nm

Sensitivity: 0.05 AUFS

Sample Size: 1 μL of a 100 μg/mL solution in acetonitrile

Measure the area under all peaks (excluding the solvent peak), and calculate the analyte and total impurities content in area percent. The sum of all impurity peaks should not exceed 5%, with no single peak greater than 3%.

Straight-Chain Hydrocarbons

SPECIFICATIONS

Identity by
Gas chromatography/mass spectrometry . . . . Passes test
Relative retention index . . . . Passes test

Assay by
Gas chromatography (using two methods) . . . . ≥98.0%

TESTS

Identity by Gas Chromatography/Mass Spectrometry. Analyze the sample by gas chromatography/mass spectrometry using method GCMS-SCHC1 described below.

Method GCMS-SCHC1

Ionization Mode: Electron ionization/70 eV

Column: 5% Diphenyl–95% dimethylpolysiloxane, 30 M × 0.25 mm i.d., 0.25-μm film thickness

Temperature Program: 35 °C for 5 min, then 30 °C/min to 320 °C, then hold 5 min

Carrier Gas: Helium at 1 mL/min

Injector: Split/splitless (100:1 split)

Injector Temperature: 250 °C

Scan Range: 32–600 amu

Sample Size: 1 μL of a 1000 μg/mL solution

Identifying ions for the compound listed in Table 5-10 must be present. The spectrum is compared to the NIST standardized library of compounds, if available. Ions in the test spectrum must match those found in the NIST spectrum.

Identity by Relative Retention Index. Prepare a solution by adding equal amounts of the sample and *n*-decane in a suitable solvent for a final concentration of 500 μg/mL. Analyze the sample by gas chromatography using method GC-SCHC1 described below.

Method GC-SCHC1

Column: 5% Diphenyl–95% dimethylpolysiloxane, 30 M × 0.53 mm i.d., 1.5-μm film thickness

Detector: Flame ionization

Detector Temperature: 300 °C

Injector: Splitless

Injector Temperature: 250 °C

Sample Size: 1 μL

Carrier Gas: Helium at 5 mL/min

Temperature Program: 35 °C for 5 min, then 30 °C/min to 320 °C, then hold 10 min

Calculate the relative retention time of the sample versus *n*-decane by dividing the retention time of the sample by the retention time of *n*-decane. Assayed values must be within 1% of the stated values in Table 5-10.

Assay by Gas Chromatography. Analyze the sample by gas chromatography using method GC-SCHC1 described above and method GC-SCHC2 described below.

Method GC-SCHC2

Column: 100% Dimethylpolysiloxane, 25 M × 0.25 mm i.d., 0.25-μm film thickness

Detector: Flame ionization

Detector Temperature: 300 °C

Injector: Split/splitless (100:1 split)

Injector Temperature: 250 °C

Sample Size: 1 μL of a 1000 μg/mL solution

Carrier Gas: Helium at 5 mL/min

Temperature Program: 35 °C for 5 min, then 30 °C/min to 320 °C, then hold 10 min

Measure the area under all peaks (excluding the solvent peak), and calculate the analyte content in area percent.

Table 5-10. Straight-Chain Hydrocarbons Compound Data

Name	CAS No.	Chemical Formula	Formula Weight	MS (m/z)			Relative Retention
				MS-1	MS-2	MS-3	
Decane	124-18-5	$C_{10}H_{22}$	142.28	43	142	57	1.000
Docosane	629-97-0	$C_{22}H_{46}$	310.61	43	57	71	1.578
Dodecane	112-40-3	$C_{12}H_{26}$	170.34	43	170	57	1.132
Dotriacontane	544-85-4	$C_{32}H_{66}$	450.88	43	57	71	1.989
Eicosane	112-95-8	$C_{20}H_{42}$	282.55	57	282	71	1.503
Heneicosane	629-94-7	$C_{21}H_{44}$	296.58	43	57	71	1.541
Hentriacontane	630-04-6	$C_{31}H_{64}$	436.85	43	57	71	1.927
Heptacosane	593-49-7	$C_{27}H_{56}$	380.74	43	57	71	1.752
Heptadecane	629-78-7	$C_{17}H_{36}$	240.47	57	240	71	1.382
Heptane	142-82-5	C_7H_{16}	100.20	43	100	71	0.528
Heptatriacontane	NA	$C_{37}H_{76}$	521.01	43	57	71	2.514
Hexacosane	630-01-3	$C_{26}H_{54}$	366.71	43	57	71	1.718
Hexadecane	544-76-3	$C_{16}H_{34}$	226.45	57	226	71	1.337
Hexane	110-54-3	C_6H_{14}	86.18	57	86	43	0.268
Hexatriacontane	630-06-8	$C_{36}H_{74}$	506.98	43	57	71	2.362
Nonacosane	630-03-5	$C_{29}H_{60}$	408.80	43	57	71	1.828
Nonadecane	629-92-5	$C_{19}H_{40}$	268.53	57	268	71	1.465
Nonane	111-84-2	C_9H_{20}	128.26	43	128	57	0.913
Octacosane	630-02-4	$C_{28}H_{58}$	394.77	43	57	71	1.788
Octadecane	593-45-3	$C_{18}H_{38}$	254.50	57	254	43	1.424
Octane	111-65-9	C_8H_{18}	114.23	43	114	85	0.791
Octatriacontane	7194-85-6	$C_{38}H_{78}$	535.04	43	57	71	2.666
Pentacosane	629-99-2	$C_{25}H_{52}$	352.69	43	57	71	1.684
Pentadecane	629-62-9	$C_{15}H_{32}$	212.42	57	212	71	1.291
Pentatriacontane	630-07-9	$C_{35}H_{72}$	492.96	43	57	71	2.243
Tetracontane	4181-95-7	$C_{40}H_{82}$	563.09	43	57	71	3.097
Tetracosane	646-31-1	$C_{24}H_{50}$	338.66	43	57	71	1.649
Tetradecane	629-59-4	$C_{14}H_{30}$	198.39	43	198	57	1.241
Tetratriacontane	14167-59-0	$C_{34}H_{70}$	478.93	43	57	71	9.88
Triacontane	638-68-6	$C_{30}H_{62}$	422.82	43	57	71	1.874
Tricosane	638-67-5	$C_{23}H_{48}$	324.63	43	57	71	1.614
Tridecane	629-50-5	$C_{13}H_{28}$	184.37	43	184	57	1.189
Tritriacontane	630-05-7	$C_{33}H_{68}$	464.90	43	57	71	2.060
Undecane	1120-21-4	$C_{11}H_{24}$	156.31	43	156	57	1.070

Part 6:
Lists and Indexes

List of Registered Trademarks

The tests in *Reagent Chemicals* employ some products having registered trademarks. These are listed below with the corporate owners.

Amberlite (Rohm and Haas Corp.)
Dowex (Dow Chemical Co.)
Eriochrome (Ciba-Geigy Corp.)
Porapak (Waters Corporation)
Supelcoport (Sigma-Aldrich Co.)
Teflon (E.I. du Pont de Nemours & Co., Inc.)
Triton (Dow Chemical Co.)
Whatman (Whatman International Ltd.)

List of Works Cited

The references listed below are specifically cited in the text of this book. For bibliographies on specific subjects, see List of Bibliographies on page 772.

APHA (American Public Health Association). 1998. *Standard Methods for the Examination of Water and Wastewater*, 20th ed.; APHA, American Water Works Association, and Water Pollution Control Federation: Washington, DC.

ASTM (American Society for Testing and Materials). 1999. *Standard Specification for Reagent Water*, D1193-99e1; ASTM: West Conshohocken, PA.

ASTM. 2000. *Standard Test Method for Color of Clear Liquids (Platinum–Cobalt Scale)*, D1209; ASTM: West Conshohocken, PA.

ASTM. 2003. *Standard Test Method for Distillation Range of Volatile Organic Liquids*, D1078; ASTM: West Conshohocken, PA.

ASTM. 2004. *Standard Specification for Wire Cloth and Sieves for Testing Purposes*, E11; ASTM: West Conshohocken, PA.

IUPAC (International Union of Pure and Applied Chemistry). 2003. Atomic Weights of the Elementsf 2001. *Pure Appl. Chem.* 75(8):1107–1122.

Kenkel, J. 2003. *Analytical Chemistry for Technicians*, 3rd ed.; Lewis Publishers: Boca Raton.

Lind, J.E., Zwolenik, J.J., and Fuoss, R.M. 1959. Calibration of Conductance Cells at 25 °C with Aqueous Solutions of Potassium Chloride. *J. Am. Chem. Soc.*, 81:1557.

Mitchell, J., and Smith, D.M. 1984. *Aquametry: A Treatise on Methods for the Determination of Water*, 2nd ed.; Wiley-Interscience: New York.

Moody, J.R. 1982. NBS Clean Laboratories for Trace Element Analysis. *Anal. Chem.* 54(13):1358A–1376A.

National Research Council. 1981. *Food Chemicals Codex*, 3rd ed; National Academies Press: Washington, DC.

NIST (National Institute of Standards and Technology). 1974. *The Calibration of Small Volumetric Laboratory Glassware*, NBSIR 74–461; NIST: Gaithersburg, MD.

PerkinElmer, Inc. 2004. *Guide to Inorganic Analysis*; PerkinElmer LAS: Shelton, CT.

Rosin, J. 1967. *Reagent Chemicals and Standards*, 5th ed.; Van Nostrand: Princeton, NJ.

Schilt, A.A. 1991. *Moisture Measurement by Karl Fischer Titrimetry*. GFS Chemicals: Columbus, OH.

Scholz, E. 1984. *Karl Fischer Titration*. Springer-Verlag: New York.

Schwedt, Georg. 1997. *The Essential Guide to Analytical Chemistry*. B. Haderlie, trans. John Wiley: New York.

SEMI (Semiconductor Equipment and Materials International). 2005. *Guide for Determination of Method Detection Limits*, Standard C10-0305 (supercedes C10-0299); SEMI: San Jose, CA.

Smith, H.M., and others. 1950. Measurement of Density of Hydrocarbon Liquids by the Pycnometer. *Anal. Chem.* 22:1452.

Zief, M., and Mitchell, J.W. 1976. Contamination Control in Trace Element Analysis. *Chemical Analyses;* Elving, J.P., Ed.; Wiley: New York.

List of Bibliographies

The works listed below direct the reader to additional information on specific subjects. For the works cited specifically in the text of this book, see List of Works Cited on page 770.

Analytical Chemistry

APHA (American Public Health Association). 1998. *Standard Methods for the Examination of Water and Wastewater*, 20th ed.; APHA, American Water Works Association, and Water Pollution Control Federation: Washington, DC.

APHA. 2001. *Supplement to Standard Methods for the Examination of Water and Wastewater*, 20th ed.; APHA, American Water Works Association, and Water Pollution Control Federation: Washington, DC.

ASTM (American Society for Testing and Materials). 2001. *Standard Practice for Preparation, Standardization, and Storage of Standard and Reagent Solutions for Chemical Analysis*, E200-97(2001)e1; ASTM: West Conshohocken, PA.

Dux, J.P. 1990. *Handbook of Quality Assurance for the Analytical Chemistry Laboratory*, 2nd ed.; Van Nostrand Reinhold: New York.

Friebolin, H. 2005. *Basic One- and Two-Dimensional NMR Spectroscopy*, 4th ed.; Wiley-VCH: New York.

Funk, W., Dammann, V., and Donnevert, G. 1996. *Quality Assurance in Analytical Chemistry;* Wiley-VCH: New York.

Günzler, H., and Gremlich, H.-U. 2002. *IR Spectroscopy: An Introduction.* Wiley-VCH: New York.

Harris, D.C. 2002. *Quantitative Chemical Analysis*, 6th ed.; W.H. Freeman: New York.

Huber, L. 1998. *Validation and Qualification in Analytical Laboratories;* CRC Press: Boca Raton, FL.

Kenkel, J. 2003. *Analytical Chemistry for Technicians*, 3rd ed.; Lewis Publishers: Boca Raton.

Kingston, H.M., and Haswell, S.J., Eds. 1997. *Microwave-Enhanced Chemistry: Fundamentals, Sample Preparation, and Applications;* American Chemical Society: Washington, DC.

Latimer, W.M., and Hildebrand, J.H. 1951. *Reference Book of Inorganic Chemistry*, 3rd ed.; Macmillan: New York.

Potts, P.J. 1987. *A Handbook of Silicate Rock Analysis;* Chapman & Hall: New York.

Schomburg, G. 1990. *Gas Chromatography: A Practical Course;* Wiley-VCH: New York.

Schwedt, Georg. 1997. *The Essential Guide to Analytical Chemistry.* B. Haderlie, trans. John Wiley: New York.

Settle, F.A. 1997. *Handbook of Instrumental Techniques for Analytical Chemistry;* Prentice-Hall: Upper Saddle River, NJ.

Skoog, D.A., Holler, F.J., and Nieman, T.A. 1998. *Principles of Instrumental Analysis,* 5th ed.; Saunders College Publishing: Philadelphia.

Vogel, A.I., and BAssett, J., eds. 1980. *Vogel's Testbook of Quantitative Inorganic Analysis, Including Elementary Instrumental Analysis,* 4th ed.; Longmans: New York.

Willard, H.H., Dean, J.A., Settle, F.A., and Merritt, L.L., Eds. 1988. *Instrumental Methods of Analysis,* 7th ed.; Wadsworth: Belmont, CA.

ICP Mass Spectroscopy

Houk, R.S. 1986. The Mass Spectrometry of Inductively Coupled Plasmas. *Anal. Chem.* 6:97A.

Hutton, R., Walsh, A., Milton, D., and Cantle, J. 1991. Ultratrace Elemental Analysis by Plasma Source High Resolution Mass Spectrometry. *ChemSA,* 17:213–215.

Jiang, S.J., Houk, R.S., and Stevens, M.A. 1988. The Determination of K Isotope Ratios by ICP–MS. *Anal. Chem.* 60:1217.

Montaser, A. 1998. *Inductively Coupled Plasma Mass Spectrometry;* Wiley-VCH: New York.

Tanner, S.D. 1995. Characterization of Ionization and Matrix Suppression in Inductively Coupled 'Cold' Plasma Mass Spectrometry. *J. Anal. Atomic Spectrom.* 10:905.

Tanner, S.D., and Baranov, V.I. 1999. Theory, Design and Operation of a Dynamic Reaction Cell ICP–MS. *Atomic Spectrosc.* 20(2):45–52.

Turner, P., Merren, T., Speakman, J., and Haines, C. 1996. *Plasma Source Mass Spectrometry: Developments and Applications;* Royal Society of Chemistry: Cambridge, U.K., pp. 28–34.

Limits of Detection

PerkinElmer, Inc. 2004. *Guide to Inorganic Analysis;* PerkinElmer: Shelton, CT.

Semiconductor Equipment and Materials International (SEMI). 2005. *Guide for Determination of Method Detection Limits,* Standard C10-0305 (supercedes C10-0299); SEMI: San Jose, CA.

Measurement Techniques

NIST (National Institute of Standards and Technology). 1974. *NBSIR 74-461: The Calibration of Small Volumetric Laboratory Glassware;* NIST: Gaithersburg, MD.

NIST. 1986. *NIST 145: Handbook for the Quality Assurance of Metrological Measurements;* Oppermann, H.V., Taylor, J.K.; National Bureau of Standards: Gaithersburg, MD, 1986. (superceded by NIST IR 6969)

NIST. *NIST 105 Series Handbooks: Specifications and Tolerances for Reference Standards and Field Standard Weights and Measures*, 105-1 through 105-8; NIST: Gaithersburg, MD.

NIST. *NIST IR 6969: Selected Laboratory and Measurement Practices and Procedures To Support Basic Mass Calibrations*; NIST: Gaithersburg, MD. (supercedes NIST 145)

Mercury Determination

McIntosh, S. 1993. The Determination of Mercury at Ultratrace Levels Using an Automated Amalgamation Technique. *Atomic Spectrosc.* 14:47.

Physical Properties

BHD Chemicals, Ltd. 1984. *"AnalaR" Standards for Laboratory Chemicals*, 8th ed.; Whitefriars Press: London and Tonbridge, U.K.

CambridgeSoft Corp. *ChemFinder.com: Database and Internet Searching*. http://www.chemfinder.com (accessed March 2005).

Food Chemicals Codex, 5th ed. 2003. Committee on Food Chemicals Codex, Food and Nutrition Board, Institute of Medicine; National Academies Press, Washington, DC.

Merck Index: An Encyclopedia of Chemicals, Drugs, and Biologicals, 13th ed. 2001. O'Neil, M.J., Smith, A., Heckelman, P.E., Budavari, S., Eds.; Merck: Whitehouse Station, NJ.

Societe Prolabo. 1981. *Prolabo Analytical Standards for Reagents*. Rhone-Poulenc: Paris, France.

United States Pharmacopeia and National Formulary. 2005. USP 28–NF 23; U.S. Pharmacopeial Convention: Rockville, MD.

Plasma Emission Spectroscopy

Barnard, T., Crockett, M.J., Ivaldi, M., and Lundberg, P. 1993. Design of an Echelle Optical System for ICP–OES. *Anal. Chem.* 60:9.

Barnard, T., Crockett, M., Ivaldi, J., Lundberg, P., Yates, D., Levine, P., and Sauer, D. 1993. Solid State Detectors for ICP–OES. *Anal. Chem.* 60:9.

Ivaldi, J., and Barnard, T. 1992. Advantages of Coupling Multivariate Data Techniques for Simultaneous ICP-OES Spectra. *Spectrochim. Acta* 48B:12.

Trace Analysis

Butcher, D.J., and Sneddon, J. 1998. *A Practical Guide to Graphite Furnace Atomic Absorption Spectrometry*; Wiley-Interscience: New York.

Montasser, A., and Golightly, D.W. 1992. *Inductively Coupled Plasmas in Analytical Atomic Spectrometry*; Wiley-VCH: New York.

Slavin, W., Manning, D.C., and Carnrick, G.R. 1981. The Stabilized Temperature Platform Furnace. *Atomic Spectrosc.* 2:137.

Thomas, R. 2003. *Practical Guide to ICP-MS*; Marcel Dekker: New York.

Welz, B., and Sperling, M. 1999. *Atomic Absorption Spectrometry*; Wiley-VCH: New York.

Ultratrace Environment

Boutron, C.F. 1990. A Clean Laboratory for Ultra-low Concentration Metal Analysis. *Fresenius J. Anal. Chem,* 337:482–491.

Hanley, Q.S., Earle, C.W., Pennebaker, F.M., Madden, S.P., and Denton, M.B. 1996. Charge-Transfer Devices in Analytical Instrumentation. *Anal. Chem.* 68(21): 661A–667A.

ISO (International Organization for Standardization). 1999. *ISO 14644-1: Cleanrooms and Associated Controlled Environments. Part 1. Classification of air cleanliness.* ISO: Geneva, Switzerland.

ISO. 2000. *ISO 14644-2: Cleanrooms and Associated Controlled Environments. Part 2. Specifications for testing and monitoring to prove continued compliance with ISO 14644-1.* ISO: Geneva, Switzerland.

Moody, J., 1982. The NBS Clean Laboratories for Trace Element Analysis. *Anal. Chem.* 54:13, 1358A–1376A.

General Index

This index includes entries on general subjects, test methods, solutions and mixtures used in tests, the reagent monographs in Part 4, and the classes of standard-grade reference materials that are included in Part 5. The index of standard-grade reference materials (page 794) lists each compound that appears in the tables in Part 5. The index by CAS number (page 801) lists the CAS number for every compound that has specifications in this book.

Index of Standard-Grade Reference Materials

This index lists each compound that appears in the tables in Part 5, as well as some common alternate names. Each compounds is also indexed by its class in the general index and by its CAS number in the index by CAS number.

Acenaphthene, 763
Acenaphthylene, 763
Acephate, 751
Aldrin, 748
Allyl chloride, 740
2-Amino-4,6-dinitrotoluene, 739
4-Amino-2,6-dinitrotoluene, 739
Anthracene, 763

Benz[a]anthracene, 763
Benzal chloride, 745
1,2-Benzanthracene. See Benz[a]anthracene
1,2:5,6-Benzanthracene. See Dibenz[a,h]anthracene
3,4-Benz[a]pyrene. See Benzo[a]pyrene
Benzene, 745
α-Benzenehexachloride. See a-BHC
β-Benzenehexachloride. See b-BHC
δ-Benzenehexachloride. See d-BHC
γ-Benzenehexachloride. See g-BHC
2,3-Benzindene. See Fluorene
Benzo[a]phenanthrene. See Chrysene
Benzo[a]pyrene, 763
Benzo[b]fluoranthene, 763
Benzo[b]phenanthrene. See Benz[a]anthracene
Benzo[d,e,f]chrysene. See Benzo[a]pyrene
Benzo[d,e,f]phenanthrene. See Pyrene
Benzo[g,h,i]perylene, 763
Benzo[j]fluoranthene, 763

Benzo[j,k]fluorene. See Fluoranthene
Benzo[k]fluoranthene, 763
Benzo[r,s,t]pentaphene. See Dibenzo[a,i]pyrene
2,3-Benzofluoranthene. See Benzo[b]fluoranthene
10,11-Benzofluoranthene. See Benzo[j]fluoranthene
11,12-Benzofluoranthene. See Benzo[k]fluoranthene
1,12-Benzoperylene. See Benzo[g,h,i]perylene
Benzyl chloride, 745
a-BHC, 748
b-BHC, 748
d-BHC, 748
e-BHC, 748
g-BHC, 748
1,1-Bis(p-chlorophenyl)-2,2-dichloroethane. See 4,4'-DDD
1,1-Bis(p-chlorophenyl)-2,2,2-trichloroethanol. See Dicofol
1-Bromo-2-chloroethane, 740
2-Bromo-1-chloropropane, 740
Bromobenzene, 745
Bromochloromethane, 740
Bromodichloromethane, 740
Bromoethane, 740
Bromoform, 740
Butachlor, 748

Index by CAS Number

This index by CAS number lists the numbers in order digit by digit to the first hyphen, then in order digit by digit to the second hyphen.

Useful Equations and Conversions

Henderson–Hasselbalch equation	$$pH = pK_a + \log \frac{\text{fraction neutralized}}{\text{fraction unneutralized}}$$
	$$pH = pK_w - pK_b + \log \frac{\text{fraction unneutralized}}{\text{fraction neutralized}}$$
Beer's law	$$A = \epsilon bc = \log \frac{1}{\%\,T/100} = 2 - \log \%\,T$$
Faraday's law	$$w = \frac{i \times t \times \text{equiv wt}}{F}$$
Molarity	$$M = \frac{\text{moles of solute (mole)}}{\text{liter (L)}} = \frac{\text{millimoles of solute (mmole)}}{\text{milliliter (mL)}}$$
Normality	$$N = \frac{\text{equivalents (eq)}}{\text{liter (L)}} = \frac{\text{milliequivalents (meq)}}{\text{milliliter (mL)}}$$
Assay	$$\%\,A = \frac{mL_B \times F_B \times \text{reaction ratio} \times \text{formula wt of A (mg/mmole)} \times 100}{\text{sample wt (mg)}}$$

Parts per Million		Parts per Billion		Percent
10,000	ppm			1.0%
1,000	ppm	1,000,000	ppb	0.1%
100	ppm	100,000	ppb	0.01%
10	ppm	10,000	ppb	0.001%
1	ppm	1,000	ppb	0.0001%
0.1	ppm	100	ppb	0.00001%
0.01	ppm	10	ppb	0.000001%

grams/milliliter (g/mL)	= milligrams/microliter (mg/µL)
micrograms/milliliter (µg/mL)	= nanograms/microliter (ng/µL)
parts per million (ppm)	= micrograms/gram (µg/g)
	= micrograms/milliliter (µg/mL)
	= nanograms/milligram (ng/mg)
	= picograms/microgram (pg/µg)
	= 10^{-6}
parts per billion (ppb)	= nanograms/gram (ng/g)
	= nanograms/milliliter (ng/mL)
	= picograms/milligram (pg/mg)
	= 10^{-9}